Geology and Paleontology of the Ellsworth Mountains, West Antarctica

Frontispiece. Mount Gardner (78°23′S, 86°W; 4,685 m), Sentinel Range of the Ellsworth Mountains, Antarctica, looking east. The mountain is composed of folded rocks of the 3,200-m-thick Crashsite Quartzite (Upper Cambrian to Devonian), a rock highly resistant to erosion that forms spectacular horns and aretes. Photo by John Splettstoesser, January 1962. Published with permission from Blackwell Scientific Publications, Ltd. (cover photo, *Geology Today,* March/April 1985).

Geological Society of America
Memoir 170

Geology and Paleontology of the Ellsworth Mountains, West Antarctica

Edited by

Gerald F. Webers
Department of Geology
Macalester College
St. Paul, Minnesota 55105

Campbell Craddock
Department of Geology and Geophysics
University of Wisconsin
Madison, Wisconsin 53706

and

John F. Splettstoesser
One Jameson Point Road
Rockland, Maine 04841

THE GEOLOGICAL SOCIETY OF AMERICA
1888

1992

Published by The Geological Society of America, Inc.
3300 Penrose Place, P.O. Box 9140, Boulder, Colorado 80301

GSA Books Science Editor Richard A. Hoppin
Accepting Officer, GSA Science Editor, Campbell Craddock

Printed in U.S.A.

Library of Congress Cataloging-in-Publication Data

Geology and paleontology of the Ellsworth Mountains, West Antarctica / edited by Gerald F. Webers, Campbell Craddock, and John F. Splettstoesser.
p. cm. — (Memoir / Geological Society of America ; 170)
Includes bibliographical references and index.
ISBN 0-8137-1170-3
1. Geology, Stratigraphic—Paleozoic. 2. Geology—Ellsworth Mountains (Antarctic regions) 3. Paleontology—Paleozoic. 4. Paleontology—Ellsworth Mountains (Antarctic regions) I. Webers, Gerald F., 1932– . II. Craddock, Campbell, 1930– III. Splettstoesser, John F. IV. Series: Memoir (Geological Society of America) ; 170.
QE654.G42 1992
559.8'9—dc20 92-16871
CIP

10 9 8 7 6 5 4 3 2 1

Contents

Plates
(in pockets)

Dedication

This volume is dedicated to Mortimer D. Turner—geologist, administrator, philosopher, and friend—for his long and effective service to the geology, and the geologists, of Antarctica.

Mort was born October 24, 1920, in Greeley, Colorado. He attended the University of California-Berkeley and graduated with a B.S. in geological engineering in 1943. He then entered the U.S. Army and served as a technician in the Ordnance Department until his discharge in 1946; early in his service he was sent to Virginia Polytechnic Institute for a semester of graduate courses in metallurgical engineering. After three months as a civilian engineer at the Aberdeen Proving Grounds in 1946, he reentered the University of California-Berkeley. He was a mining geologist at the California Division of Mines from 1948 to 1954, and he received his M.S. in geology in 1954. Mort's interests are in regional geology, tectonics, economic geology, and engineering geology, and he has many publications on these topics.

In 1954 Mort went to Puerto Rico where he founded a geological survey and served as state geologist. In addition to a substantial administrative responsibility, he also was active in geological mapping, assessment of mineral resources, and studies of beach erosion.

The U.S. Antarctic Research Program was founded within the National Science Foundation in 1959, and Mort became an assistant to the program director that year; he was responsible for a large fraction of the science program, including geology. He took leave of NSF from 1962 to 1965 for graduate study at the University of Kansas, receiving his Ph.D. later in geology and metallurgical engineering. On his return to NSF in 1965, Mort became program director, Antarctic Earth Sciences, responsible for geology, geophysics, glaciology, and oceanography. In 1970 he was made program manager, Polar Earth Sciences—with additional responsibility for the Arctic.

Mortimer D. Turner

Mort has been active in several organizations and has found a little time for some personal interests (besides his family). He is a Fellow of the Geological Society of America, was a vice-president of the International Geological Congress in 1956, and has served as president of the Antarctican Society. He was a part-time lecturer in geology at George Washington University on occasion after 1972. He is an avid collector of highway maps, paperback science fiction books, and philatelic cachets; a visit to his basement will testify to his success in acquisitions!

Mort retired from NSF in December 1984. He soon moved from Maryland back to his home state of Colorado where he lives happily as a mountain man. He now has more time to spend with his wife Joanne and their children, to travel, and to work on his hobbies. He has also revived another interest—the geology of early man in North America, and Mort and Joanne made a recent trip to China to pursue this problem in the field.

Mort Turner spent 22 years at NSF in the administration of Antarctic research. While he was able to carry out some personal research at the beginning, he really surrendered himself to administration so that other scientists would have the chance to work in Antarctica. It is likely that no other administrator anywhere has had as much direct impact on advancing our understanding of Antarctic geology; the Turner Hills, Antarctica, were named by the U.S. Board on Geographic Names.

It was not unusual to find large stacks of documents on Mort's desk at NSF. This logjam was the result of 1) Mort's limited interest in paperwork, 2) his periodic visits to Antarctic field camps, and 3) in particular, his tendency to delay some grant decisions to the latest possible date. This latter trait drove his superiors to despair and his proposing scientists to anxiety, or sometimes worse. But geologists in his program gradually realized that his real motive for delay was his deep desire to support as many of the worthy as possible, and he waited in the hope that his pleas would produce more funds for his intended grants. He listened with interest to all who contacted him, and he gave all the moral and financial support that he could.

Mort Turner came to NSF to serve his community, and not to rule. While he was over us, at the same time he was one of us. It is because of his earnest, compassionate service that Mort is regarded so highly by us and by many other Antarctic earth scientists.

Preface

The Ellsworth Mountains lie in the interior of West Antarctica, and this long mountain chain is centered at 79°S, 85°W. The landscape is dominated by rugged, angular peaks including the Vinson Massif, the highest mountain in Antarctica. These remote mountains are of special interest because of their strategic location between the East Antarctic craton and the tectonically active zone of coastal West Antarctica. They are an important key to the Phanerozoic history of West Antarctica.

The Ellsworth Mountains were discovered by Lincoln Ellsworth on November 23, 1935, during his flight from Dundee Island near the tip of the Antarctic Peninsula to a point just short of Admiral Richard E. Byrd's Little America Base on the Ross Ice Shelf (Ellsworth, 1936, 1937; Joerg, 1936a, 1936b, 1937). Ellsworth probably saw only the northern part of these mountains, the Sentinel Range. The mountains were next seen on January 12, 1958, during an IGY oversnow traverse from Byrd Station (80°S, 120°W) led by Charles R. Bentley; rocks were collected from two outlying nunataks west of the northern Sentinel Range—Fisher Nunatak and Helfert Nunatak—and positions of major peaks were established by ground surveying (Anderson, 1960, 1961). The first geologic observations in the Ellsworth Mountains (folded quartzites) were obtained during this brief visit. The southern Ellsworth Mountains (Heritage Range) were first seen on December 14, 1959, during a triangular reconnaissance flight from Byrd Station to the Thiel Mountains (85°S) to the Ellsworth Mountains to Byrd Station; C. Craddock and the late Edward C. Thiel were on board. A snow landing was made on December 26, 1959, at Pipe Peak in the northwestern Heritage Range (Thiel, 1961); folded quartzite beds were observed, and rock specimens were collected by Craddock, Thiel, and Edwin S. Robinson.

On December 14, 1959, U.S. Navy Squadron VX-6 photographed the Sentinel Range in trimetrogon aerial photographic coverage. These pictures and subsequent ground surveys were the basis for a series of 1:250,000 topographic maps that were published by the U.S. Geological Survey (Newcomer Glacier, Nimitz Glacier, and Vinson Massif sheets, 1962). On November 4, 1961, the U.S. Navy photographed the Heritage Range in trimetrogon coverage. Topographic maps of the Heritage Range (Liberty Hills and Union Glacier Sheets) were compiled and printed in 1967 from these photographs and from additional photographic missions in 1962, 1964, and 1966. These maps were vital during the field mapping and in later production of geologic maps, including the colored geologic map in this volume (Craddock and others, 1986).

In November 1961, the first of three successive University of Minnesota expeditions in the Ellsworth Mountains began under the direction of C. Craddock. In the 1961–62 season, under the field leadership of geologist John J. Anderson, most of the western Sentinel Range and some of the eastern Heritage Range were traversed by motor toboggans (snowmobiles), and the basic structure and the upper stratigraphic units of the Ellsworth Mountains were determined (Anderson and others, 1962; Anderson, 1965). Other members of the 1961–62 group were geologists Thomas W. Bastien, Paul G. Schmidt, and John F. Splettstoesser.

Motor toboggans were again used in the 1962–63 season for further reconnaissance from a base camp near the Minnesota Glacier to the ends of the ranges and for some site-specific studies (Soholt and Craddock, 1964). The 1962–63 group included C. Craddock, geologist and leader; geologists Peter J. Barrett, Jerry Dolence, Ben Drake, Harvey J. Meyer, Donald E. Soholt, and Gerald F. Webers; geophysicist Glenn Bowie; and U.S. Geological Survey topographic engineers Rob Collier and Dean Edson. That season Craddock, Bowie, and Dolence also traveled south by snowmobile to the Pirrit Hills, the Martin Hills, and the Nash Hills for geologic reconnaissance work.

In 1963–64 the expedition had the support of U.S. Army UH-1B turbine helicopters at a base camp on the Minnesota Glacier (Rutford and Smith, 1966), allowing visits to outcrops throughout the mountains and great advances in the geologic work. The scientific participants in the 1963–64 season were Robert H. Rutford, geologist and leader; geologists Thomas W. Bastien, John P. Evans, W. D. Michael Hall, and K. Bernhard Spörli; geophysicist Barton D. Gross; biologist Kelly Rennell; and U.S.G.S. topographic engineer Al Zavis.

Significant chronostratigraphic discoveries were made in both the second and third seasons—an important fossil fauna in 1962–63 (Craddock and Webers, 1964; Webers, 1966, 1972b; and others), and the first occurrence of *Glossopteris* in West Antarctica in 1963–64 (Craddock and others, 1965). Geologic results of these expeditions were reported by Craddock in several papers (1962, 1964, 1966, 1971, 1983) and a geologic map (1969), and by Craddock and others (1964).

In the 1964–65 season motor toboggans were used to work in some mountains and nunataks farther south and southwest of the Ellsworth Mountains (Hart Hills, Pagano Nunatak, Whitmore Mountains) as part of a continuing geologic reconnaissance of this part of West Antarctica by University of Minnesota expeditions (Webers and others, 1982, 1983). This was the last of such field investigations in the Ellsworth Mountains region under the direction of C. Craddock.

Geologic field parties from other U.S. institutions conducted studies during the 1966–67 season in the Sentinel Range (Matthews and others, 1967; Schopf, 1967; Tasch, 1967; Frakes and others, 1971), and an expedition from the Norsk Polarinstitutt carried out geologic studies in the Heritage Range in the 1974–75 season (Hjelle and others, 1978, 1982). In 1975 the British Antarctic Survey conducted airborne ice-sounding surveys of the Ellsworth Mountains (Swithinbank, 1977), and they performed other field work there in the 1978–80 and 1984–86 seasons using twin-engined Otter aircraft. In 1976 the U.S. Geological Survey published a satellite image map of the Ellsworth Mountains at a scale of 1:500,000 based on imagery from the NASA Earth Resources Technology Satellite (ERTS-1). The Soviet Antarctic Expedition conducted an aerial photography mission of the area in the 1978–79 season.

A U.S. mountaineering expedition made the first ascents of the highest peaks in the 1966–67 season (Clinch, 1967; Silversteen, 1967), including the Vinson Massif (4,897 m), the highest peak in Antarctica. Cursory geologic studies were made by geologists on the expedition (Evans, 1973). Private mountaineering expeditions also visited the Ellsworth Mountains in the 1980s using air support from the Antarctic Peninsula. The Vinson Massif was climbed by three different groups in the 1985–86 season and many times since. Beginning in the 1987–88 season, a Douglas DC-4 (and later, a DC-6) aircraft had been used to transport mountaineering and tourist groups from Chile to the Ellsworth Mountains, using a blue-ice area adjacent to the Patriot Hills in the southern Heritage Range for wheeled landings (Swithinbank, 1988).

The largest geologic field program in the Ellsworth Mountains was conducted in the 1979–80 season, when 42 geologists and other investigators were supported by U.S. Navy UH-1N helicopters operating from a base camp (Camp Macalester) in the northwestern Heritage Range (Splettstoesser and Webers, 1980; Splettstoeser and others, 1982). The largest contingent of scientific personnel was under the direction of G. F. Webers, and it included numerous senior investigators who studied various topics related to completing the first geological survey of the Ellsworth Mountains. A team from the U.S. Geological Survey resurveyed parts of the mountains and established better position and elevation control. One of the results of the resurvey was a new elevation for Vinson Massif (4,897 m, compared with the earlier figure of 5,140 m). Subsequently, investigators from other institutions studied the area in the 1980–81, 1983–84, and 1986–87 seasons as part of a regional study of West Antarctica and the Antarctic Peninsula (Dalziel, 1981; Doake and others, 1981; Dalziel and Pankhurst, 1984, 1987).

Results of the 1979–80 field season were reported at a special symposium on the Ellsworth Mountains, organized by C. Craddock, G. F. Webers, J. Splettstoesser, and M. D. Turner for the Geological Society of America annual meeting, New Orleans, October 1982. The 13 papers presented at the symposium provided a stimulus to compile a volume of papers on the geology and paleontology of the Ellsworth Mountains. An annotated bibliography of the Ellsworth Mountains was begun as an aid to investigators preparing manuscripts for the volume (Webers and Splettstoeser, 1982). The bibliography of citations (without annotations) is included here as an appendix to this volume (it contains all the references cited in this Preface).

The work on this volume eventually came to fruition, and a total of 23 papers are included here. This memoir is the culmination of geologic field studies begun in 1959 by C. Craddock, and continued in the 1960s and 1979–80 field seasons by C. Craddock, G. F. Webers, and others. Further field studies are needed to fill in some of the remaining gaps in the geology, and additional sampling of the fossil faunas is required to establish a more complete understanding of the paleoenvironment and the life that existed in the Ellsworth Mountains area in Paleozoic time.

We gratefully acknowledge the financial support of the National Science Foundation Division of Polar Programs and the logistic support of the U.S. Navy. We also thank all the scientists and support personnel in the Ellsworth Mountains expeditions; without these individuals this volume would not have been possible. Finally, we thank the numerous reviewers of various drafts of the manuscripts included here as chapters.

G. F. Webers
C. Craddock
J. F. Splettstoesser

Geological Society of America
Memoir 170
1992

Chapter 1

Geologic history of the Ellsworth Mountains, West Antarctica

Gerald F. Webers
Department of Geology, Macalester College, St. Paul, Minnesota 55105
Campbell Craddock
Department of Geology and Geophysics, University of Wisconsin, Madison, Wisconsin 53706
John F. Splettstoesser*
Minnesota Geological Survey, University of Minnesota, St. Paul, Minnesota 55114

ABSTRACT

The stratigraphic succession in the Ellsworth Mountains includes strata from Cambrian to Permian in age. No definite evidence of major unconformities in the Ellsworth succession is known, and it is possible that continuous deposition took place from Cambrian to Permian time.

The oldest stratigraphic unit, the Heritage Group, was deposited in Middle to Late Cambrian time. More than half of the 13,000+-m-thick stratigraphic succession of the Ellsworth Mountains was deposited during this time interval. Basic igneous volcanism and tectonic activity occurred in both the source and accumulation areas throughout the deposition of this group. Shallow-marine conditions prevailed during the deposition of the overlying 3,000-m-thick Upper Cambrian to Devonian Crashsite Group. This group indicates a period of tectonic stability that continued through the remainder of Ellsworth Mountains sedimentation. Deposition of the glaciomarine Permo-Carboniferous Whiteout Conglomerate and the Permian Polarstar Formation completed the sedimentary sequence.

Major deformation of the Ellsworth Mountains sedimentary succession, the Ellsworth (Gondwanide) Orogeny, took place in Late Permian or early Mesozoic time. The original location of these rocks is unclear, but they probably accumulated near the margin of East Antarctica. With the breakup of Gondwanaland, the Ellsworth Mountains, and their southern neighbors likely comprised a microplate that translated and rotated to its present position sometime in late Mesozoic or early Cenozoic time. The uplift of the mountains may have accompanied these postulated movements.

The geomorphic evolution of the Ellsworth Mountains in Cretaceous and Cenozoic time includes the development of an integrated stream valley pattern and, later, valley and continental glaciation, followed by moderate deglaciation.

INTRODUCTION

The Ellsworth Mountains are a rugged group of mountains 350 km long and 80 km wide in central West Antarctica. They are divided by the east-flowing Minnesota Glacier into the Sentinel Range to the north and the Heritage Range to the south (Fig. 1). The stratigraphic succession (Fig. 2) is somewhat greater than 13,000 m in thickness and ranges in age from Cambrian to Permian. The present position of the Ellsworth Mountains is structurally anomalous, lying between, and nearly at right angles to, the Transantarctic Mountains and coastal West Antarctica to the west of the Antarctic Peninsula.

*Present address: 1 Jameson Point Road, Rockland, Maine 04841

Webers, G. F., Craddock, C., and Splettstoesser, J. F., 1992, Geologic history of the Ellsworth Mountains, West Antarctica, *in* Webers, G. F., Craddock, C., and Splettstoesser, J. F., Geology and Paleontology of the Ellsworth Mountains, West Antarctica: Boulder, Colorado, Geological Society of America Memoir 170.

Only a few minor economic deposits of any kind are known in the Ellsworth Mountains. Minor occurrences of coal exist in the Polarstar Formation in the northern Sentinel Range, and a variety of sulfide minerals and their weathering products occur in the southern Heritage Range. There is no indication of any hydrocarbon potential in this part of Antarctica. The high rank of coals and the low-grade metamorphism of the rocks indicate that these strata are quite unlikely to contain petroleum.

GEOLOGIC HISTORY

It is not known when sedimentation began at the site of the future Ellsworth Mountains. Although Yoshida (1982) reports a K-Ar age of 935 ± 47 Ma on some metasedimentary rocks at the southeastern end of the northeastern side of the Heritage Range, definite basement rocks are not exposed. The earliest exposed sedimentary rocks in the Ellsworth Mountains succession were

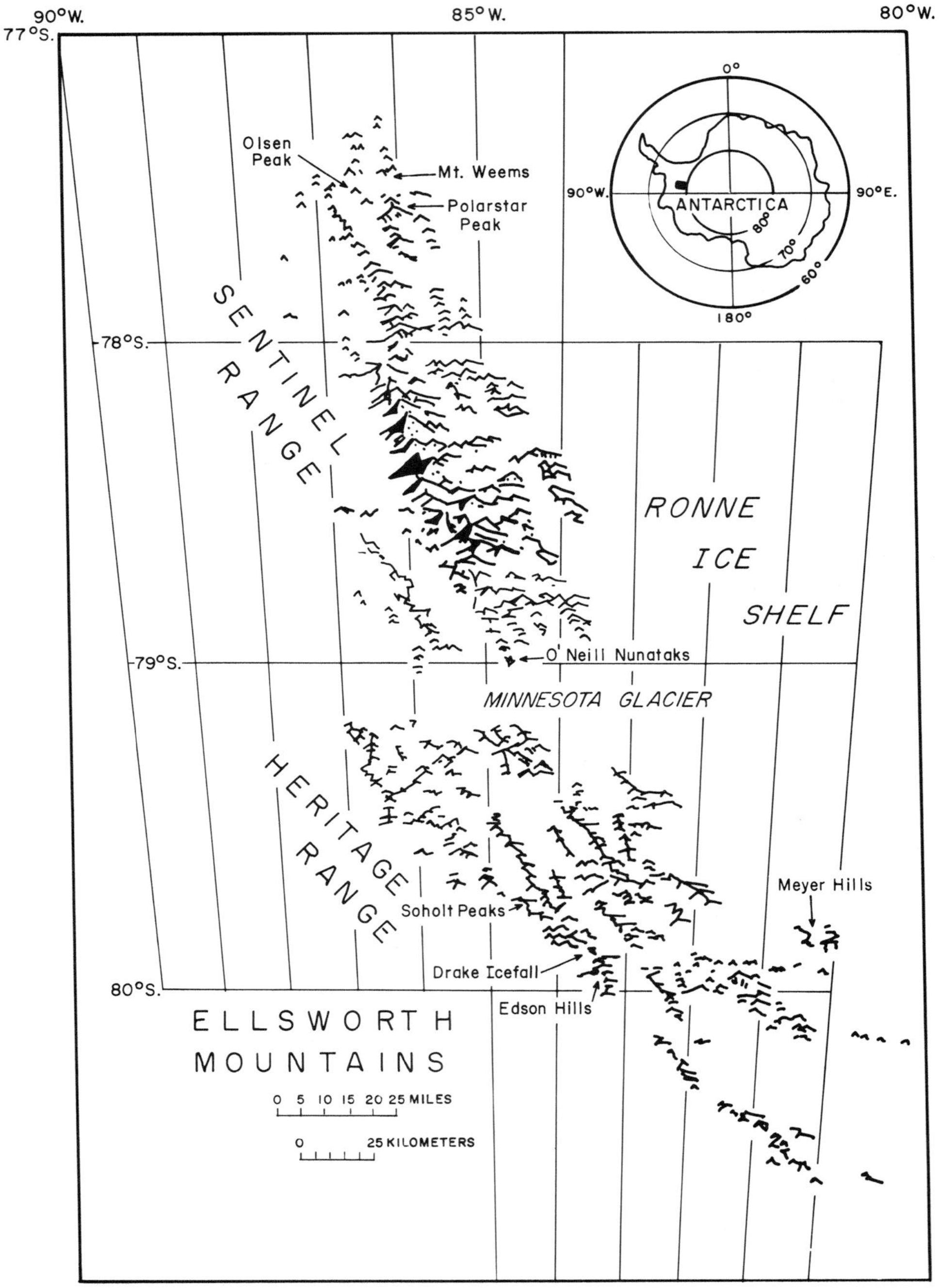

Figure 1. General location map of the Ellsworth Mountains.

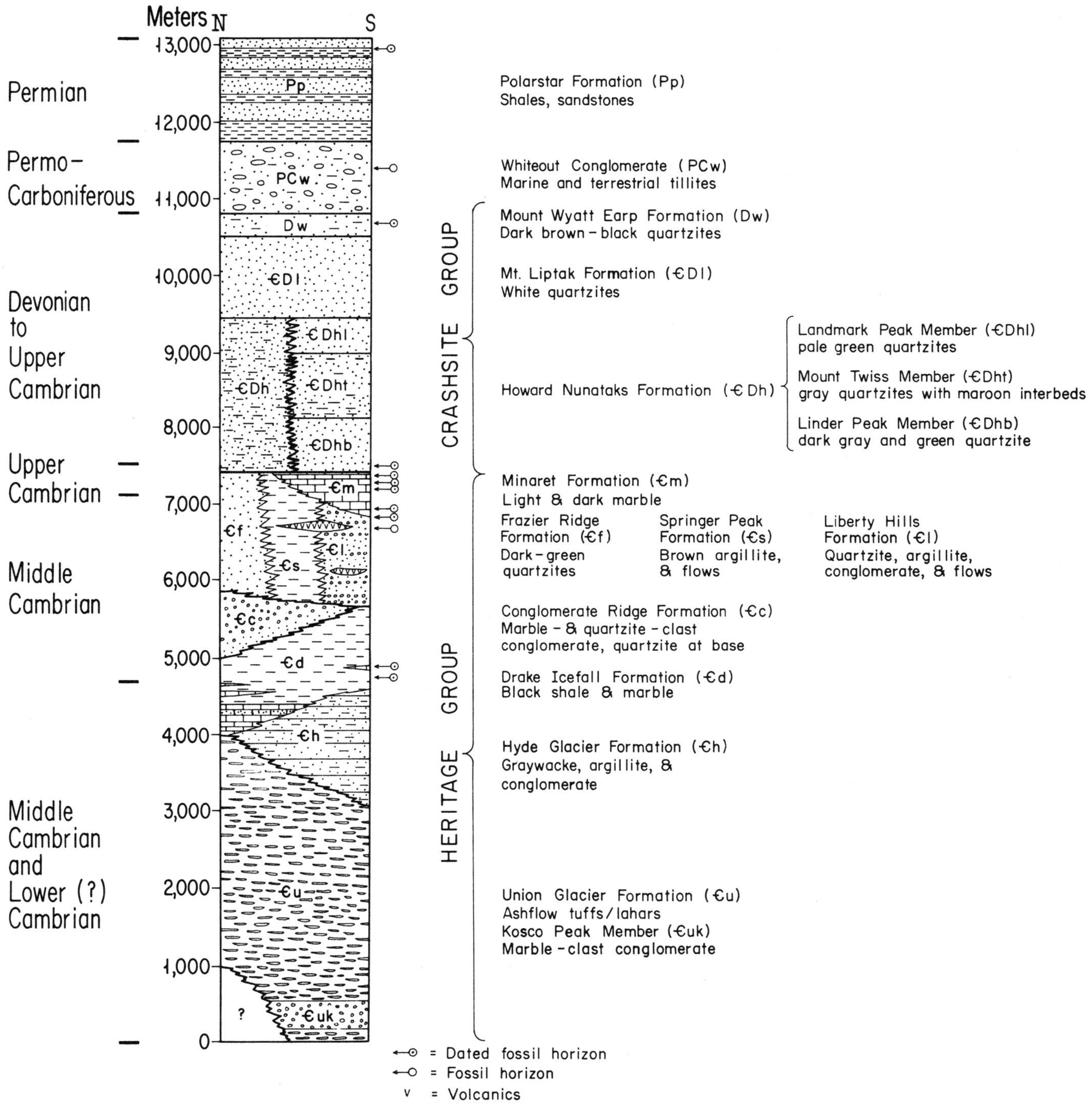

Figure 2. Columnar section of the Paleozoic rocks exposed in the Ellsworth Mountains, West Antarctica.

deposited in late Early or early Middle Cambrian time. These strata indicate marked volcanic activity and crustal instability, and these conditions continued for the remainder of Cambrian time. During this time the 7,500 m of the Heritage Group were deposited (Webers, Bauer, and others, this volume). Active volcanic centers rapidly built up thick accumulations of ashflow tuff and/or lahar deposits, forming most of the Union Glacier Formation (more than 3,000 m) of the Heritage Group (Fig. 2). Abundant disk-shaped clasts as large as 30 cm in diameter probably represent flattened pumice. Volcanic bombs are common in some strata.

The Kosco Peak Member, a thick (390-m) coarse, clast-supported, polymictic conglomerate considered to be of fluvial origin, is near the base of the Union Glacier Formation. The composition and coarseness of the clasts (as long as 1 m) provide a picture of the surrounding region; these source areas were of moderate relief and consisted predominantly of Lower Cambrian dolostone and lesser amounts of buff orthoquartzite. Dolostone clasts from these areas contain Lower Cambrian fossils (Buggisch and Webers, 1982, and this volume; Debrenne, this volume). Small exposures of granitic and metamorphic rocks, presumably of Precambrian age, were also present. These source areas provided the coarse clasts found not only in the Kosco Peak Member of the Union Glacier Formation but also in the Conglomerate Ridge and Liberty Hills formations and in the Whiteout Conglomerate.

Deposition of the overlying 1,750-m-thick Hyde Glacier Formation marks a transition from fluvial-deltaic to shallow-marine, deltaic conditions. The formation contains a wide variety of rock types including green and maroon argillite and phyllite, quartzite, metagraywacke, and polymict conglomerate. Nearby source areas of moderate relief are indicated by (1) a 30-m-thick channel structure filled with coarse polymict conglomerate (clasts as long as 15 cm) and (2) bedded polymict conglomerates as thick as 50 m. The formation also contains as much as 300 m of deltaic graywacke. The Hyde Glacier Formation is known only from the Edson Hills area of the Heritage Range and is not present some 10 km to the north in the Drake Icefall area. Its absence to the north could be due to faulting, but it is more likely a result of rapid facies change along a tectonically active coastline.

Continued subsidence under restricted marine conditions is indicated by the overlying Drake Icefall Formation. Black shales with a few limestone beds make up this unit, which is estimated to be as thick as 800 m. A 200-m-thick gray limestone unit is present at the base of the formation in the type section in the Drake Icefall area. This is followed by thick black shales with interbedded lenses of black limestone and finally by continuous black shale. As with the Hyde Glacier Formation, rapid facies changes are found in the Edson Hills area some 10 km to the south. The 200-m-thick limestone at the base is here represented by a 5-m-thick silicified limestone in black shale, and the lenses of black limestone are represented by a zone of carbonate nodules in black shale. A Middle Cambrian trilobite fauna was collected from this formation in both the Edson Hills and the Drake Icefall areas (Jago and Webers, this volume).

The overlying Conglomerate Ridge Formation is about 600 m thick and represents shallow-marine and/or fluvial conditions. In general appearance the buff polymict conglomerate of the Conglomerate Ridge Formation is similar to the Kosco Peak Member of the Union Glacier Formation, and it also reflects nearby source areas of moderate relief exposing bedrock that is dominantly composed of carbonate rocks and quartzites. In spite of its thickness, the Conglomerate Ridge Formation is found only in the Drake Icefall area. Rapid facies changes are thought to account for the lack of this formation in strata of equivalent age some 10 km to the south in the Edson Hills.

The Liberty Hills, Springer Peak, and Frazier Ridge formations are laterally equivalent and make up a sizable fraction of the exposures in the Heritage Range. Trilobite faunas (Jago and Webers, this volume) indicate that they are Middle Cambrian in age and probably extend up to the base of the Upper Cambrian. Sedimentation took place during active volcanism. Basalt flows estimated to be as thick as 500 m are found in the Liberty Hills Formation, and basalt flows at least 200 m thick are present in the Springer Peak Formation. Active volcanic centers were located in at least three and perhaps as many as six localities in the Heritage and southernmost Sentinel areas. Rapid lateral and vertical variations in sedimentary rock types suggest fluvial, deltaic, and shallow-marine environments and extrusive and possibly intrusive igneous activity. The Liberty Hills and Springer Park formations are estimated to be about 1,000 m thick, and they represent fluvial and deltaic sedimentation from a source area to the southwest. The Frazier Ridge Formation is estimated to be about 800 m thick and is a shallow-marine orthoquartzite derived from a western source area.

The Upper Cambrian Minaret Formation reflects a gentle subsidence of the region with relatively little influx of sediment. Fossiliferous limestones (Webers, 1972, 1982; Buggisch and Webers, this volume; Shergold and Webers, this volume; Henderson and others, this volume; Webers, Pojeta, and Yochelson, this volume) were deposited in agitated marine environments. Less common dolomites reflect more restricted conditions. The formation is as thick as 700 m in the southern Heritage Range and pinches out in the northwestern Heritage and southern Sentinel ranges. A volcanic center was probably active during this time in the O'Neill Nunataks area of the southernmost Sentinel Range.

With the end of deposition of the Heritage Group, major tectonic activity ceased in the Ellsworth Mountains area for the remainder of Paleozoic time. More than half of the 13,000+-m total stratigraphic column had been deposited in less than a geologic period. From Late Cambrian through part of Permian time the region underwent gentle subsidence and deposition under predominantly shallow marine conditions.

The Heritage Group is overlain by the 3,000-m Crashsite Group (Spörli, this volume). The Crashsite Group has the greatest area of surface exposure of any rock unit in the Ellsworth Mountains. In ascending order the group is subdivided into the Howard

Nunataks, the Mount Liptak, and the Mount Wyatt Earp formations. A Late Cambrian trilobite fauna is present in the basal transition beds of the Howard Nunataks Formation (Shergold and Webers, this volume) and a Devonian fauna is present in the Mount Wyatt Earp Formation (Webers and others, this volume).

The Howard Nunataks Formation (previously known as the lower dark member of the Crashsite Quartzite) is as thick as 2,250 m and consists dominantly of pale green, greenish gray, and maroon orthoquartzites with some interbeds of argillite. These strata have an eastern source (Spörli, this volume, Fig. 5) and are relatively high in garnet compared to the overlying Mt. Liptak and Mt. Wyatt Earp formations. This source area probably consisted of metaquartzites and other metamorphic rocks and was low in relief. An unnamed member in the northwestern Heritage Range, consisting of wedges of impure gray quartzite, marks the influence of a westerly source area that is apparently the same as that of the Frazier Ridge Formation of the Heritage Group. This quartzite from the western source is significantly higher in matrix and rock fragments than other members of the formation and probably reflects higher relief in the source area. The Howard Nunataks Formation shows an increase in textural and compositional maturity from bottom to top. The entire unit is more mature than the underlying clastic strata of the Heritage Group. Abundant cross-bedding, current and oscillation ripple marks, and a few mud cracks reflect a stable, shallow, marine environment.

The Mount Liptak Formation (previously known as the middle light member of the Crashsite Quartzite) is a very mature, well-sorted, light gray to white orthoquartzite estimated to be as much as 900 m thick. Cross-bedding is common at some localities. Tracks and trails found in the northern Heritage Range are the only fossils. A local unconformity may be present in the northern Heritage Range. There is no direct evidence for the age of the formation. If deposition of the Crashsite Group was continuous, the age is probably Ordovician and Silurian. The Mount Liptak Formation represents shallow-marine sedimentation in a tectonically stable area, with detritus from surrounding source areas of low relief.

The youngest formation of the Crashsite Group is the Mount Wyatt Earp Formation (previously known as the upper dark member of the Crashsite Quartzite). The formation is as thick as 300 m in the Sentinel Range. The formation is made up dominantly of dark gray, red, and dark brown lithic quartzites, with interbeds of brown argillite. The quartzites are poorly sorted and high in matrix and rock fragments. The formation is quite variable laterally. In the southeastern Heritage Range the top of the formation exposes interbedded quartzite and diamictite identical to that of the overlying Whiteout Conglomerate. In the northern Sentinel Range isolated pebbles present in the uppermost beds of the formation presumably represent ice-rafted dropstones similar to those of the overlying Whiteout Conglomerate. An Early Devonian orbiculoid brachiopod fauna was recovered from this formation in the northern Heritage Range (Boucot and others, 1967). A second collection of this site has yielded a diverse but sparse fauna of orbiculoid brachiopods. cephalopods, bivalves, gastropods, a conularid, a trilobite, and a fish spine (Webers, Glenister, Pojeta, and Young, this volume). The Mount Wyatt Earp Formation was deposited under shallow marine conditions. Marked lateral variations in rock type are apparently due to variations in source directions, although no detailed work on this subject has yet been done. The marked increase in argillite and the marked increase in matrix and rock fragment percentages in the quartzites compared to the underlying Mount Liptak Formation could indicate an uplift of the source area and/or the onset of glacial conditions associated with the Permo-Carboniferous glaciations. The presumed dropstones and the interbedded diamictite in the upper part of the Mount Wyatt Earp Formation would support the latter conclusion. The presence of a dacitic intrusive stock of Devonian age in the west-central Heritage Range (Vennum and others, this volume) may indicate limited tectonic activity in the region at that time.

The Whiteout Conglomerate is a gray to black diamictite that represents the Permo-Carboniferous Gondwanaland glaciation (Ojakangas and Matsch, 1981; Matsch and Ojakangas, this volume) in the Ellsworth Mountains. A 1,000-m-thick complete section in the Sentinel Range exposes as many as six massive diamictite units separated by thin beds of shale, mudstone, and sandstone. In the east-central Heritage Range a 250-m-thick partial section exposes thinner diamictites in a succession punctuated by striated boulder pavements and thin laminated beds of shale, sandstone, and lag pebbles. Clasts in the Sentinel Range diamictites are dominantly quartzite (66%), with granite (17%), shale-argillite (8%), and carbonate rock (5%) making up the majority of the remainder. In the Heritage Range the diamictite clasts are also predominantly quartzite (47 to 86%) with granite and carbonate rock each about 20%. Carbonate rock clasts in the Heritage Range have yielded Lower Cambrian archaeocyathids (Debrenne, this volume). The Heritage Range section represents both direct deposition of lodgement till by glacial ice and deposition from floating ice and by currents in a zone of fluctuating grounding lines. The Sentinel Range section reflects deeper glacial-marine and sub–ice shelf conditions with periodic recessions of the ice. Source areas for the Whiteout Conglomerate were probably low-relief regions exposing metamorphic rocks primarily of quartzite and carbonate rock composition and igneous rocks primarily of granitic composition.

The 1,000-m-thick Polarstar Formation is the youngest sedimentary unit in the Ellsworth Mountains section. It consists of dark gray to black argillite, siltstone, sandstone, and coal (Collinson and others, this volume). The lower part is mostly argillite, and the middle part consists of sedimentary cycles of argillite to sandstone that coarsen upward. The upper part of the Polarstar Formation fines upward from medium-grained sandstone to fine-grained sandstone, siltstone, argillite, and coal. Argillite in the upper part contains a fossil flora dominated by *Glossopteris* foliage. The argillite also includes sphenophyte stems and leaf sheaths, *Gangamopteris* leaves, and *Vertebraria*-type axes (Taylor and Taylor, this volume). We suggest a Middle to Late Permian age for the flora. Tasch (1968) collected trace fossils from the Polarstar Formation at Mount Weems and also a homopterous insect wing (Tasch and Riek, 1969) from Polarstar Peak.

Castle and Craddock (1975) suggested that Polarstar strata were deposited in a prograding delta. Further work by Collinson and others (this volume) suggested that the sedimentary environment changed upward from prodelta, to delta, to coastal plain under conditions ranging from anaerobic to dysaerobic in a stratified inland sea. Source areas for the Polarstar Formation are markedly different from those of the underlying formations. Rock fragments in Polarstar sandstones suggest a source area dominated by silicic and andesitic volcanic rocks. If this volcanic activity reflects the development of a back-arc basin, it represents the beginning of tectonic activity culminating in the formation of the Ellsworth Mountains fold belt by Jurassic time. The Polarstar Formation includes the youngest known strata in the Ellsworth Mountains. If younger strata were deposited, they have been removed by erosion or they remain undiscovered.

The Ellsworth Mountains in Gondwanaland reconstructions

The original position of the Ellsworth Mountains within Gondwanaland has been discussed by a number of authors. Craddock (1970) included the Ellsworth Mountains, the nunataks and small mountain groups south of the Ellsworth Mountains (Pirrit, Nash, and Martin Hills; Whitmore Mountains; Pagano Nunatak), and the Pensacola Mountains in the Ellsworth Orogen and linked it with the Cape Orogen of South Africa and the Sierra Orogen of South America. Schopf (1969) was the first to suggest that the Ellsworth Mountains were a crustal fragment that had been translated to its present position. Noting that the middle and upper Paleozoic sedimentary rocks of the Ellsworth Mountains were similar to those of the Pensacola, Horlick, and other parts of the Transantarctic Mountains, he suggested an original position of the Ellsworth Mountains north of the Pensacola Mountains near the East Antarctic Shield.

Clarkson and Brook (1977), on the basis of the Precambrian age of the Haag Nunataks rocks and the structure and stratigraphy of the Ellsworth Mountains, suggested an original position of the Ellsworth Mountains west of the Littlewood Nunataks (north of the Pensacola Mountains). Watts and Bramall (1980, 1981) suggested a 90-degree counterclockwise rotation of the Ellsworth Mountains fragment based on paleomagnetic work on early Paleozoic strata in the Ellsworth Mountains. Restoration of this postulated rotation would bring the Ellsworth Mountains parallel to the Transantarctic Mountains; such a restoration is consistent with the Schopf model. Dalziel and Elliot (1982) postulated that the Ellsworth Mountains and the nunataks to the south represent a microplate that had originally been adjacent to the Falkland Plateau in an orientation parallel to the Sierra de la Ventana of South America and the Cape Fold Belt of South Africa. However, Schmidt and Rowley (1986) argued that if the Ellsworth Mountains have undergone rotation, it has been in a clockwise direction. On the basis of paleomagnetic work in the nunataks south of the Ellsworth Mountains, Dalziel and others (1987) suggested close association of the Ellsworth Mountains–Whitmore Mountains block (EWM) with other microplates of West Antarctica and an original orientation of the EWM at a high angle to the Sierra de la Ventana and the Cape Fold Belt.

In our opinion the original location of the Ellsworth Mountains remains to be established.

Deformation of the Ellsworth Mountains

Craddock (1969, 1972) suggested that the Ellsworth Mountains were deformed in a single orogenic event of early Mesozoic age. Hjelle and others (1978, 1982) suggested an additional late Paleozoic deformation based on differences in fold style and intensity between the Crashsite Quartzite and the Whiteout Conglomerate, and they suggested that an unconformity existed between the two formations. Yoshida (1982, 1983) argued for a polyorogenic sequence of deformations as follows: (1) a Precambrian event exemplified by a K-Ar date of 935 Ma, (2) low-grade recrystallization after the deposition of the Heritage Group during either the early Borchgrevink Orogeny (Devonian) or more probably the late Ross Orogeny (Ordovician), (3) low-grade recrystallization of existing rocks after the deposition of the Crashsite Quartzite and before or during the deposition of the Whiteout Conglomerate at about 300 Ma (Carboniferous) during a late phase of the Borchgrevink Orogeny, and (4) very low grade metamorphic recrystallization at about 235 Ma after the deposition of the Polarstar Formation during the early Mesozoic orogeny.

We hold that the sedimentary and igneous rocks of the Ellsworth Mountains and their southern extensions in the Pirrit, Nash, and Martin Hills, Pagano Nunatak, and the Whitmore Mountains were deformed mainly in a single orogenic event in early Mesozoic time (Craddock, 1972; Craddock and others, 1982; Webers and others, 1982, 1983). Earlier deformational events, if present, had a relatively minor influence on the development of the mountain belt.

The emplacement of a Middle Devonian dacite stock (Vennum and others, this volume) in the Soholt Peaks area of the Heritage Range seems to represent a minor igneous event, and likely does not imply regional metamorphic activity. The time of emplacement of the dacite stock is about coeval with the end of deposition of the Crashsite Group, and thus it appears to support the mid-Paleozoic deformation suggested by Hjelle and others (1982). The latter authors predicated the existence of this deformation on a suspected unconformity between the Crashsite Group and the Whiteout Conglomerate. Field work by Webers in 1963 and in 1980, however, showed the contact to be conformable. The contact at Olsen Peak in the northern Sentinel Range is gradational, and in the Meyer Hills of the Heritage Range interbedded quartzite and diamictite layers occur in the gradational contact zone.

Eleven isotopic apparent ages (K-Ar) from granite batholiths emplaced in the southern extensions of the Ellsworth Mountains (Pagano Nunatak, the Whitmore Mountains, and the Nash,

Martin, and Pirrit Hills) fall in a range of 163 to 190 Ma and cluster around 175 Ma (Craddock and others, 1982). Thus the Ellsworth Mountains were deformed prior to Middle Jurassic time, as these granites appear to be posttectonic.

A series of cavelike features occurs in the southwestern Heritage Range in areas exposing limestone of the Minaret Formation (Craddock and Webers, 1981). These deposits or breccia bodies (Craddock and others, 1964) are distributed over a length of 80 km and occur as cylindrical, podlike, lenslike, or irregular masses. They are as large as 200 m high and 50 m in diameter. It is probable that more than 100 such breccia bodies of various sizes occur in the Heritage Range.

The breccia bodies were formed by a combination of cavelike and low-temperature hydrothermal processes during the latest stage of the regional deformation of the Ellsworth Mountains. Field crosscutting relations suggest that at least three phases of breccia body formation occurred. Some breccia bodies are slightly offset by thrust faults but are not folded, suggesting that their formation was contemporaneous with the last stage of the regional deformation. Fluid inclusion analyses of some vein calcite crystals indicate that the crystallization temperature was 160 ± 5° C (Spörli, Craddock, and others, this volume).

Post-Jurassic history

The time of uplift of the Ellsworth Mountains is unclear from regional geology, but it was probably not before late Triassic nor after middle Tertiary time. It may have been related to translation/rotation of the Ellsworth Mountains, likely in middle Cretaceous to early Tertiary time. Recently Fitzgerald and Stump (1991) have carried out apatite fission-track analysis of quartzite specimens from the western escarpment of the highest peaks (Vinson Massif); they conclude that uplift of at least 4 km occurred during a 20-m.y. interval in Early Cretaceous time and no more than 3 km of additional uplift since then.

Rutford (1972a, 1972b) and Rutford and others (1980, 1982) have traced the posttectonic geomorphic history of the Ellsworth Mountains and outlined the following sequence of events: (1) fluvial erosion and the development of the major valley systems present today, (2) onset of valley glaciation that has continued until the present as the dominant process of landform development in the higher parts of the range, (3) continental glaciation resulting in the inundation of the lower part of the range by continental ice, and (4) partial deglaciation with a lowering of the level of the continental ice by as much as 300 m. This sequence of events was confirmed by field work conducted during the 1979–1980 field season (Denton and others, this volume).

As a small archipelago surrounded by ocean, the Ellsworth Mountains and their southward extensions must have had a maritime climate in their present position. However, the timing of events following the uplift of the Ellsworth Mountains is difficult to pinpoint. Fluvial erosion probably occurred during Late Cretaceous and early Tertiary times. Valley glaciation and subsequent continental glaciation, probably followed in middle Tertiary time.

REFERENCES CITED

Boucot, A. J., Doumani, G. A., Johnson, J. G., and Webers, G. F., 1967, Devonian of Antarctica, *in* Oswald, D. H., ed., International Symposium on the Devonian System, Calgary, Alberta, 1967, Papers, Vol. 1: Calgary, Alberta Society of Petroleum Geologists, p. 639–648.

Buggisch, W., and Webers, G. F., 1982, Zur Facies der Karbonatgesteine in den Ellsworth Mountains (Paläozoikum, Westantarktis): Facies (Erlangen), v. 7, p. 199–228.

Castle, J. W., and Craddock, C., 1975, Deposition and metamorphism of the Polarstar Formation (Permian), Ellsworth Mountains: Antarctic Journal of the United States, v. 10, p. 239–241.

Clarkson, P. D., and Brook, M., 1977, Age and position of the Ellsworth Mountains crustal fragment, Antarctica: Nature, v. 265, p. 615–616.

Craddock, C., 1969, Geology of the Ellsworth Mountains, *in* Bushnell, V. C., and Craddock, C., eds., Geologic maps of Antarctica: New York, American Geographical Society, Antarctic Map Folio Series, Folio 12, Plate 4.

——, 1970, Map of Gondwanaland, *in* Bushnell, V. C., and Craddock, C., eds., Geologic maps of Antarctica: New York, American Geographical Society, Antarctic Map Folio Series, Folio 12, Plate 23.

——, 1972, Antarctic tectonics, *in* Adie, R. J., ed., Antarctic geology and geophysics: Oslo, Universitetsforlaget, p. 449–455.

Craddock, C., Anderson, J. J., and Webers, G. F., 1964, Geologic outline of the Ellsworth Mountains, *in* Adie, R. J., ed., Antarctic geology, Proceedings of the First International Symposium on Antarctic Geology, Cape Town, 16–21 September 1963: Amsterdam, North-Holland, p. 155–170.

Craddock, C., Webers, G. F., and Anderson, J. J., 1982, Geology of the Ellsworth Mountains–Thiel Mountains Ridge [abs.], *in* Craddock, C., ed., Antarctic geoscience: Madison, University of Wisconsin Press, p. 849.

Craddock, J., and Webers, G. F., 1981, Probable cave deposits in the Ellsworth Mountains of West Antarctica, *in* Beck, B. F., ed., Proceedings of the Eighth International Congress of Speleology, Bowling Green, Kentucky, July 1981, p. 395–397.

Dalziel, I.W.D., and Elliot, D. H., 1982, West Antarctica, problem child of Gondwanaland: Tectonics, v. 1, p. 3–19.

Dalziel, I.W.D., Garrett, S. W., Grunow, A. M., Pankhurst, R. J., Storey, B. C., and Vennum, W. R., 1987, The Ellsworth–Whitmore Mountains crustal block: Its role in the tectonic evolution of West Antarctica, *in* McKenzie, G. D., ed., Gondwana six: Structure, tectonics, and geophysics: American Geophysical Union Geophysical Monograph 40, p. 173–182.

Fitzgerald, P. G., and Stump, E., 1991, Early Cretaceous uplift in the Ellsworth Mountains of West Antarctica: Science, v. 254, p. 92–94.

Hjelle, A., Ohta, Y., and Winsnes, T. S., 1978, Stratigraphy and igneous petrology of southern Heritage Range, Ellsworth Mountains, Antarctica: Norsk Polarinstitutt Skrifter, no. 169, p. 5–43.

——, 1982, Geology and petrology of the southern Heritage Range, Ellsworth Mountains, *in* Craddock, C., ed., Antarctic geoscience: Madison, University of Wisconsin Press, p. 599–608.

Ojakangas, R. W., and Matsch, C. L., 1981, The Late Palaeozoic Whiteout Conglomerate: A glacial and glaciomarine sequence in the Ellsworth Mountains, West Antarctica, *in* Hambrey, M. J., and Harland, W. B., eds., Earth's pre-Pleistocene glacial record: Cambridge, Cambridge University Press, p. 241–244.

Rutford, R. H., 1972a, Drainage systems of the Ellsworth Mountains area [abs.], *in* Adie, R. J., ed., Antarctic geology and geophysics: Oslo, Universitetsforlaget, p. 233.

——, 1972b, Glacial geomorphology of the Ellsworth Mountains, *in* Adie, R. J., ed., Antarctic geology and geophysics: Oslo, Universitetsforlaget, p. 225–232.

Rutford, R. H., Denton, G. H., and Andersen, B. G., 1980, Glacial history of the Ellsworth Mountains: Antarctic Journal of the United States, v. 15, p. 56–57.

Rutford, R. H., Bockheim, J. G., Andersen, B. G., and Denton, G. H., 1982, Glacial history of the Ellsworth Mountains, West Antarctica: Geological

Society of America Abstracts With Programs, v. 14, p. 605–606.
Schmidt, D. L., and Rowley, P. D., 1986, Continental rifting and transform faulting along the Jurassic Transantarctic rift: Tectonics, v. 5, p. 279–291.
Schopf, J. M., 1969, Ellsworth Mountains: Position in West Antarctica due to sea-floor spreading: Science, v. 164, p. 63–66.
Tasch, P., 1968, Trace fossils from the Permian Polarstar Formation, Sentinel Mountains, Antarctica: Kansas Academy of Science Transactions, v. 71, p. 184–194.
Tasch, P., and Riek, E. F., 1969, Permian insect wing from Antarctic Sentinel Mountains: Science, v. 164, p. 1529–1530.
Watts, D. R., and Bramall, A. M., 1980, Paleomagnetic investigation in the Ellsworth Mountains: Antarctic Journal of the United States, v. 15, p. 34–36.
——, 1981, Palaeomagnetic evidence for a displaced terrain in Western Antarctica: Nature, v. 293, p. 638–641.
Webers, G. F., 1972, Unusual Upper Cambrian fauna from West Antarctica, *in* Adie, R. J., ed., Antarctic geology and geophysics: Oslo, Universitetsforlaget, p. 235–237.
——, 1982, Upper Cambrian mollusks from the Ellsworth Mountains, *in* Craddock, C., ed., Antarctic geoscience: Madison, University of Wisconsin Press, p. 635–638.
Webers, G. F., Craddock, C., Rogers, M. A., and Anderson, J. J., 1982, Geology of the Whitmore Mountains, *in* Craddock, C., ed., Antarctic geoscience: Madison, University of Wisconsin Press, p. 841–847.
——, 1983, Geology of Pagano Nunatak and the Hart Hills, *in* Oliver, R. L., James, P. R., and Jago, J. B., eds., Antarctic earth science: Canberra, Australian Academy of Science, p. 251–255.
Yoshida, M., 1982, Superposed deformation and its implication to the geologic history of the Ellsworth Mountains, West Antarctica: Tokyo, National Institute of Polar Research Memoirs, Special issue no. 21, p. 120–171.
——, 1983, Structural and metamorphic history of the Ellsworth Mountains, West Antarctica, *in* Oliver, R. L., James, P. R. and Jago, J. B., eds., Antarctic earth science: Canberra, Australian Academy of Science, p. 266–269.

MANUSCRIPT ACCEPTED BY THE SOCIETY FEBRUARY 4, 1992

Printed in U.S.A.

Geological Society of America
Memoir 170
1992

Chapter 2

The Heritage Group of the Ellsworth Mountains, West Antarctica

Gerald F. Webers
Department of Geology, Macalester College, St. Paul Minnesota 55105
Robert L. Bauer
Department of Geology, University of Missouri at Columbia, Columbia, Missouri 65211
John M. Anderson
Department of Geology, Victoria University, Wellington, New Zealand
Werner Buggisch
Institut für Geologie, Universität Erlangen, D-8500 Erlangen, Federal Republic of Germany
Richard W. Ojakangas
Department of Geology, University of Minnesota, Duluth, Minnesota 55812
K. Bernhard Spörli
Department of Geology, University of Auckland, Auckland, New Zealand

ABSTRACT

The Heritage Group is composed of about 7,500 m of sedimentary strata exposed in the Heritage Range of the Ellsworth Mountains, West Antarctica. The Heritage Group is here redefined to include the Minaret Formation as the uppermost unit. New formations within the Heritage Group are formally described; they are, from the bottom upward, the Union Glacier, Hyde Glacier, Drake Icefall, Conglomerate Ridge, Springer Peak, Liberty Hills, and Frazier Ridge Formations. The Kosco Peak Member of the Union Glacier Formation is also formally described.

Deposition of the Heritage Group took place in Middle and early Late Cambrian time in a rapidly subsiding basin bordered by carbonate rock and quartzite source areas of moderate relief. Sediment transport directions were dominantly from the present south and west. Thick, volcaniclastic terrestrial strata lie at the exposed base of the group, and these rocks grade upward into deltaic black shale and normal marine sediments. A number of active volcanic centers were present in the Heritage Range during the deposition of upper Heritage Group strata.

INTRODUCTION

The Heritage Group is exposed only in the Heritage Range of the Ellsworth Mountains (Fig. 1), except for a small area in the southern Sentinel Range (see geologic map in pocket). It is second only to the Crashsite Group in area of exposure within the Heritage Range.

The Minaret and Heritage Groups were first described by Craddock (1969). The Minaret Group was considered to be possibly of Precambrian age, to be the oldest unit in the Ellsworth Mountains stratigraphic section, and to be overlain by the Heritage Group. Hjelle and others (1978) revised the description of the Minaret and the Heritage Groups, placing the upper part of the Minaret Group in the Heritage Group and describing it as the Middle Horseshoe Formation. The Heritage Group of Hjelle and others (1978) then contained the Middle Horseshoe Formation and a new formation called the Edson Hills Formation. A new formation, the Dunbar Ridge Formation, was erected by Hjelle and others (1978) between the Heritage Group and the Crashsite Quartzite.

Webers, G. F., Bauer, R. L., Anderson, J. M., Buggisch, W., Ojakangas, R. W., and Spörli, K. B., 1992, The Heritage Group of the Ellsworth Mountains, West Antarctica, *in* Webers, G. F., Craddock, C., and Splettstoesser, J. F., Geology and Paleontology of the Ellsworth Mountains, West Antarctica: Boulder, Colorado, Geological Society of America Memoir 170.

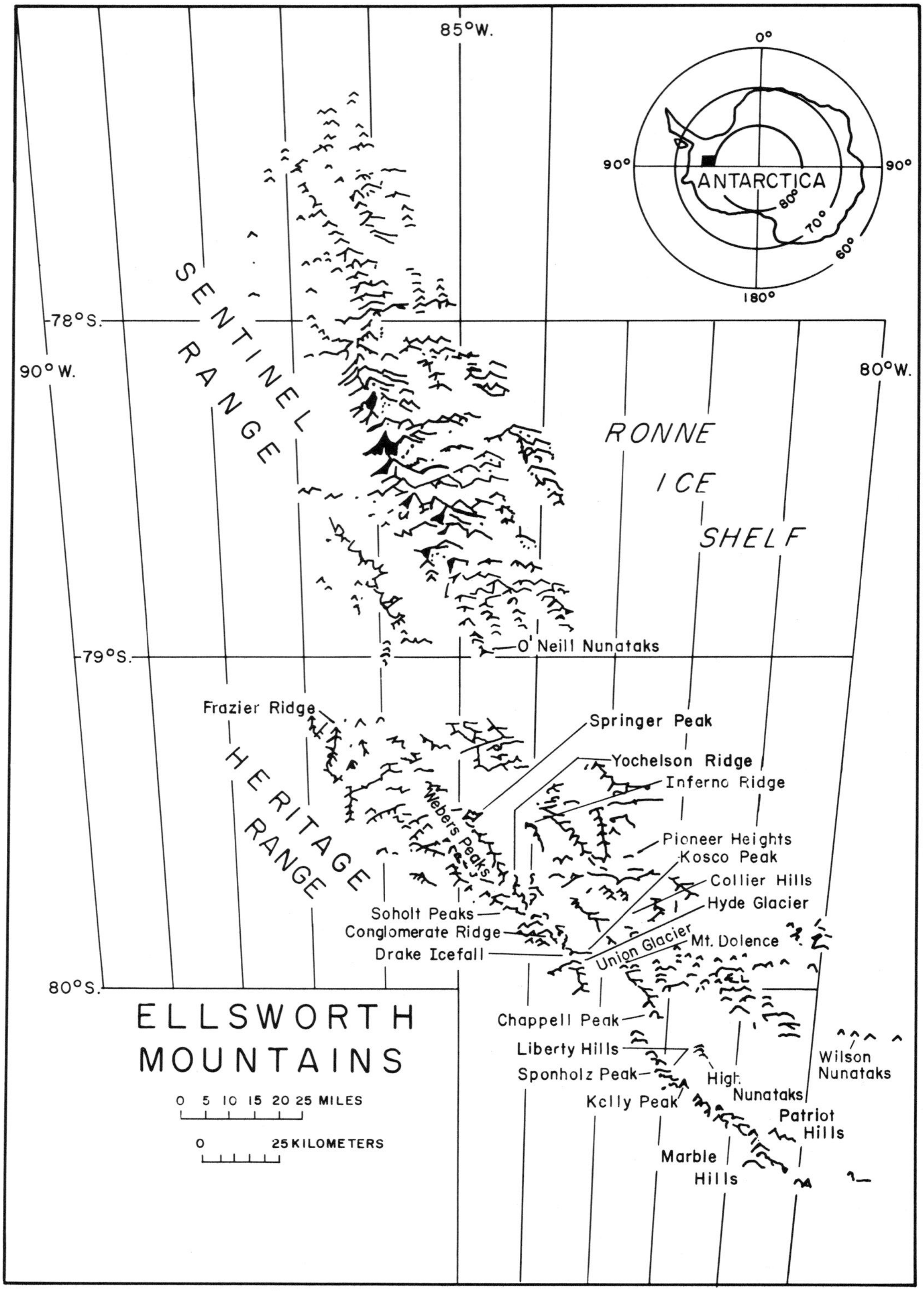

Figure 1. General location map of the Ellsworth Mountains, West Antarctica.

During the 1979–1980 United States Ellsworth Mountains Expedition, paleontological finds, sedimentary structures, and structural studies indicated that the Minaret and Heritage Groups were in reverse stratigraphic order from that previously described (Splettstoesser and Webers, 1980). This and other detailed stratigraphic investigations conducted during the 1979–1980 field season raised major problems with the stratigraphy of Hjelle and others (1978). It thus was considered necessary to return to the stratigraphic units of Craddock (1969). Webers and Spörli (1983), in a preliminary version of this chapter, revised the Minaret Group to the Minaret Formation and placed it above the Heritage Group. The Heritage Group was informally divided into formations. These formations are formally described below following the stratigraphy outlined in Webers and Spörli (1983). Generalized geologic sections of Craddock (1969), Hjelle and others (1978), Webers and Spörli (1983), and this chapter are shown in Figure 2.

The Heritage Group is primarily Middle Cambrian in age but ranges into the early Late Cambrian; it may extend into the Lower Cambrian at the base of the unit. Figure 3 shows the relation of the formations of the Heritage Group.

The detail of the information available on the formations of the Heritage Group varies considerably from unit to unit, and thus the description of each in this chapter varies correspondingly. For example, the Union Glacier and the Hyde Glacier Formations were studied during a three-week period by four geologists who measured the 5,000-m stratigraphic section at 1.5-m intervals. By contrast, the abrupt end of the field season resulted in only a three-hour examination of a 1,500-m stratigraphic section of the Conglomerate Ridge Formation and part of the Drake Icefall Formation. Detailed sampling and stratigraphic study of the section along the Hyde Glacier permits detailed petrographic descriptions for the Union Glacier and the Hyde Glacier Formations. This chapter may be considered a first step in understanding the Heritage Group.

STRATIGRAPHY OF THE HERITAGE GROUP

The Union Glacier Formation

The type section for the Union Glacier Formation, as formally described here, is along the valley of the Hyde Glacier, and the type area is in the Edson Hills in the Heritage Range. It is exposed only in the Edson Hills and the area of the Drake Icefall in the Heritage Range.

The Union Glacier Formation is the lowermost unit of the

Craddock, 1969	Hjelle and others, 1978	Webers and Spörli, 1983		This Chapter		
Heritage Group Cambrian and Precambrian 6,710 m	Dunbar Ridge Formation Middle and Upper Cambrian 700 to 1,600 m	Minaret Formation Upper Cambrian 0 to 600 m		Minaret Formation Middle and Upper Cambrian 0 to 700 m		
	Heritage Group		Heritage Group	Frazier Ridge Formation Middle Cambrian 800 m	Springer Peak Formation Middle Cambrian 500 m	Liberty Hills Formation Middle Cambrian 1,000 m
	Edson Hills Formation Middle Cambrian 3,100 to 3,900 m	Heritage Group		Conglomerate Ridge Formation Middle Cambrian 650 m		
Minaret Group Precambrian 975 m	Middle Horshoe Formation Precambrian 530 m	Middle and Upper Cambrian 7,200 m		Drake Icefall Formation Middle Cambrian 800 m		
	Minaret Group			Hyde Glacier Formation Middle Cambrian 1,750 m		
				Union Glacier Formation Lower Cambrian(?) and Middle Cambrian 3,000+ m		

Figure 2. Comparison of stratigraphic classifications in the southern Heritage Range.

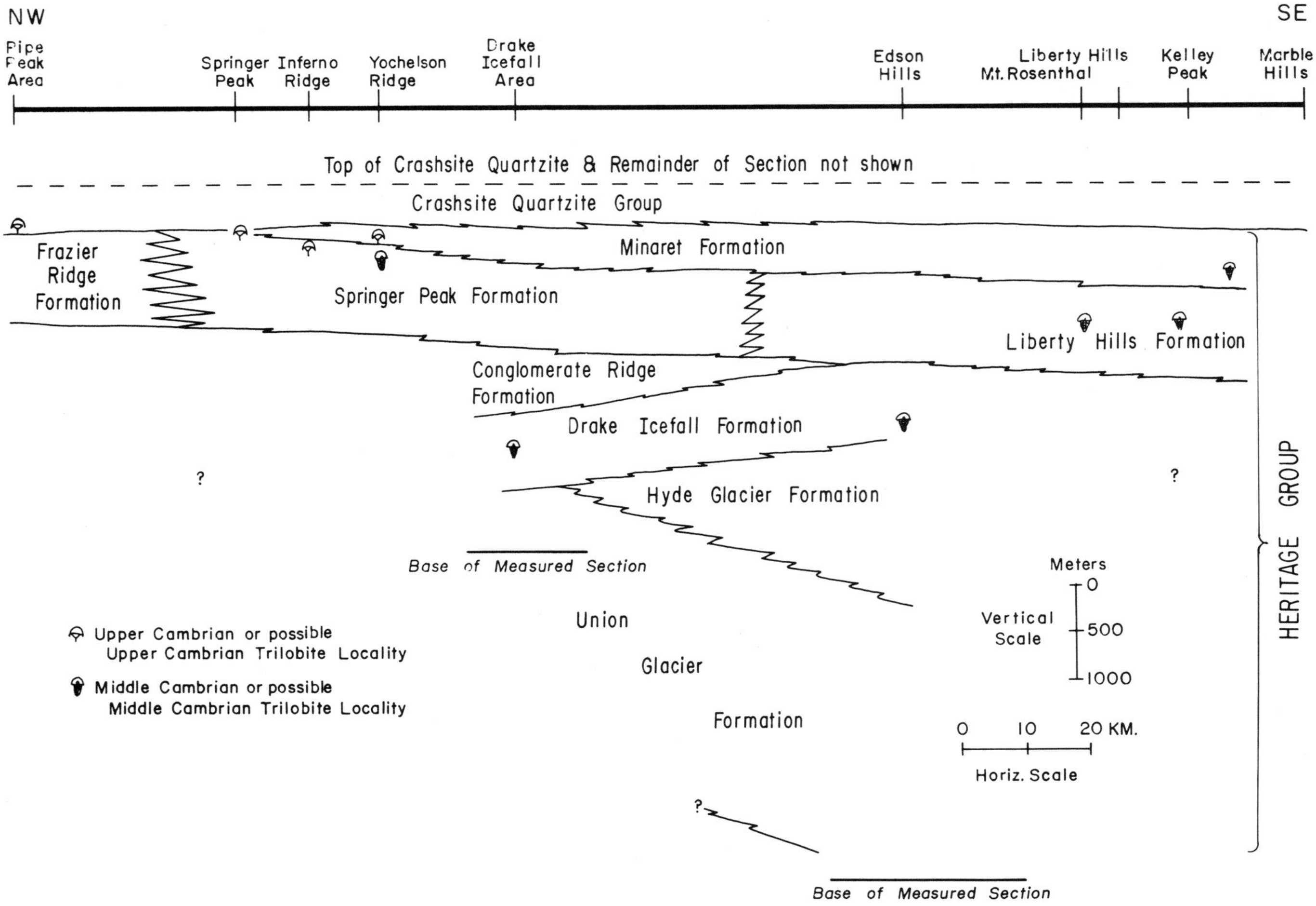

Figure 3. Diagrammatic sketch of the relations among formations of the Heritage Group. Note: The section between the Drake Icefall Area and the Edson Hills is enlarged to show detail; it represents a distance of 8 km.

Heritage Group. The known thickness is about 3,000 m, but its base is not exposed. The dominant rock type (perhaps as much as 90 percent) is dark green tuffaceous diamictite, with the remainder consisting of tuffaceous rocks, metamorphosed buff calcareous sandstone, and buff calcareous conglomerate. The tuffaceous diamictite contains slightly deformed basalt clasts as large as 60 cm in diameter, which compose 5 to 30 percent of the volume. Some of these clasts have been identified as volcanic bombs. Surrounding the basalt clasts is tuffaceous matrix, which contains abundant "stretched pebbles"—highly deformed maroon and green elongate clasts of scoriaceous, slaty and crystal tuff fragments as large as 30 cm long, 15 cm wide, and 2 cm thick. The abundance of these deformed clasts ranges from none to 70 percent of the material. Matrix surrounding clasts is similar in composition to the finer-grained tuffaceous rocks described below. The tuffaceous diamictite is considered to be a terrestrial deposit of lahar and/or ash-flow tuff.

In addition to tuffaceous diamictites, finer-grained tuffaceous rocks include crystal tuff, lithic tuff, and lapilli tuff. The crystal tuffs (Fig. 4) generally contain sand-size clasts of subhedral to anhedral plagioclase crystals and opaque mineral clasts as well as grains or aggregates of anhedral epidote; quartz clasts occur as a minor constituent of some of the tuffs. All of the clasts occur in recrystallized matrix consisting of chlorite, white mica, fine-grained dusky aggregates of epidote and sphene, and fine-grained quartz.

Deformation and metamorphic recrystallization largely obscure the original nature of the tuff matrix. Relative proportions of the matrix phases vary as a function of tuff chemistry, with high proportions of white mica and quartz in the more felsic tuffs and greater proportions of chlorite and epidote in the more mafic tuffs. Quartz in the matrix and some of the quartz clasts display undulatory extinction and/or undulatory polycrystalline forms. Such grains contain indistinct, curved, or less commonly serrated boundaries, resulting in an undulatory polygonal extinction pattern. Large epidote aggregates consist of yellow to brownish

Figure 4. Crystal tuff from the Union Glacier Formation. Clasts of plagioclase, opaques, and sparse quartz occur in a matrix of fine-grained quartz, white mica, chlorite, and ribbons of calcite. Strain shadows containing various combinations of the matrix phases occur around some of the clasts. (Specimen W-79-97Z, x-polars.)

epidote grains that are commonly rimmed by, or intergrown with, fine-grained, dusky aggregates of sphene. Less commonly the aggregates have opaque rims of magnetite or hematite. Some of this metamorphic epidote is probably an alteration product after mafic clasts such as clinopyroxene. Although no remnant clinopyroxene was observed in the samples, similar alteration products (± actinolite) replace clinopyroxene in basaltic flows within the Heritage Group (see Bauer, this volume). Actinolite (± chlorite, ± epidote) occurs in some samples as an alteration product, probably after clinopyroxene, and also as fine acicular prisms throughout the groundmass. Plagioclase clasts, now entirely albite, are partially altered to white mica, epidote, and/or calcite.

The lithic tuffs, lapilli tuffs, and tuffaceous diamictites are heterolithologic, containing lithoclasts of crystal tuff, basalt, shale, and probable scoria, all in crystal tuff matrix. The crystal tuff clasts are generally less mafic than the surrounding matrix and contain more quartz and secondary white mica and less chlorite and epidote than the matrix. Clasts of probable scoriaceous material contain small, flattened vesicles with dusky opaque outlines filled with chlorite and subhedral to anhedral epidote.

The tuffs are invariably foliated and contain both lithic and crystal clasts flattened in the foliation plane. Oriented clusters of chlorite, white mica, and less commonly actinolite define the foliation in the tuff matrix. Strain shadows containing chlorite, white mica and/or quartz are especially common around crystal clasts, some of which display microboudinage structure with chlorite or white mica between the boudins.

About 180 m above the lowest exposed beds of the Union Glacier Formation is a 390-m-thick unit composed of metamorphosed calcareous sandstone and conglomerate with minor amounts of orthoquartzite and arkose. This unit is here given formal member status as the Kosco Peak Conglomerate. It would have been given formation status except for the fact that the volcanics above and below the conglomerates are indistinguishable from each other.

The conglomerates of the Kosco Peak Conglomerate member are typically clast-supported, with clasts making up as much as 70 percent of the material. Clasts average about 5 cm in diameter and range in size from granules to boulders as large as 1 m in diameter. Carbonate pebbles dominate and make up from 70 to 95 percent of the clasts. Other pebbles include quartzite and argillite. Carbonate pebbles are dominantly gray and consist of dolomicrite, dolodismicrite, and dolomitized calcitic biomicrite, sparite, and microsparite in a matrix of dolomite and quartz sand. The matrix dolomite is recrystallized and commonly occurs as euhedral to subhedral rhombohedrons partially replacing quartz. Much of the dolomite is cloudy; some grains have cloudy euhedral cores and clear rims, and still others display rhombohedral zoning with marginal concentrations of reddish-brown iron oxide overgrown by clear dolomite. Local dedolomitization is evidenced by staining that shows partial replacement of dolomite rhombohedrons by calcite. This matrix dolomite was clearly detrital and deposited as silt- and sand-sized grains of terrestrially derived dolomite.

Sandstones as thick as 20 m occur within the conglomerate sequences and are dominantly foliated, medium-crystalline doloarenites. Some are similar in appearance to the matrix in the conglomerates. Quartzite and arkose make up the remaining sandstones. The quartzite contains sand-sized quartz clasts surrounded by fine-grained (silty) polygonal quartz, intergrown medium-crystalline dolomite rhombs, and sparse white mica. Some quartz clasts contain well-rounded nuclei, outlined by thin layers of dark inclusions, which are surrounded by epitaxial quartz overgrowths. The monocrystalline clasts invariably display undulose extinction; some occur as composite grains with curved boundaries. Most of the grains contain few vacuoles, and some contain abundant rutile needles or microlites. Clast margins are generally irregular as a result of marginal recrystallization to fine-grained polycrystalline quartz or partial replacement by dolomite. Staining indicates that the dolomite has undergone local dedolomitization; this also is made evident by the replacement or partial replacement of dolomite rhombohedrons by calcite.

The arkose is a tightly interlocked mosaic of plagioclase clasts with lesser amounts of quartz, opaque fragments, and rare micaceous rock fragments. Sparry calcite commonly partially replaces both plagioclase and quartz clasts. The quartz clasts are similar in most respects to those found in the quartzite; however, none was observed to contain older clastic nuclei. Plagioclase clasts are commonly twinned and partially replaced by epidote, calcite, and white mica. Chlorite and white mica occur locally along clast boundaries or as the principal constituents of micaceous rock fragments.

The Union Glacier Formation is considered to be a terrestrial accumulation of predominantly volcaniclastic material. Bordering source areas of moderate relief were dominantly composed of dolostone, with smaller amounts of quartzite and limestone. Active volcanism produced thick accumulations of terrestrial tuff and lahar. The Kosco Peak Conglomerate member and rare quartzites record the periodic influx of sediments from adjacent source areas. The age of the Union Glacier Formation is considered to be Middle Cambrian, although it could be as old as Early Cambrian. This age is based on fossiliferous carbonate clasts derived from adjacent source areas that periodically contributed sediments to the Ellsworth Mountains stratigraphic section. Limestone clasts in the Kosco Peak Member contain stromatolites of the type LLS, fenestral fabrics of the *Stromatactis* type, and calcareous algae of the *Renalcis* type (Buggisch and Webers, this volume). The best dated fossils in carbonate clasts are found higher in the stratigraphic column in the Permo-Carboniferous Whiteout Conglomerate; these clasts contain Lower Cambrian archaeocyathids (Debrenne, this volume).

Figure 5. Graywacke from the Hyde Glacier Formation containing clasts of quartz, plagioclase, and rock fragments in a matrix of fine-grained quartz, white mica, and chlorite.

The Hyde Glacier Formation

The Hyde Glacier Formation is here formally described. The type section is along the Hyde Glacier in the Edson Hills, and the type area (with the only known exposures) is the Edson Hills.

The Hyde Glacier Formation is a clastic unit about 1,750 m thick that conformably overlies the Union Glacier Formation. A wide variety of volcanigenic sedimentary rock types occur in this formation, including argillite, phyllite, quartzite, metagraywacke, and polymict conglomerates. In the type section, the sedimentary rocks were locally intruded by dikes and sills of porphyritic, greenish-gray basalt as thick as 35 m and sparse dikes of porphyritic dacite. The formation could be subdivided into three formal members, but they will be described informally because the stratigraphy of the Hyde Glacier Formation is dominantly based on a study of a single stratigraphic section and the formation is found only in the Edson Hills area.

The lowermost member is about 1,170 m thick and consists dominantly of green and maroon phyllite, green and maroon argillite, metagraywacke, and dark polymict conglomerates. Near the base of the member is a 70-m-thick, green, clast-supported, polymict conglomerate sequence. Clasts ranging in size from 0.5 to 60 cm make up about 75 percent of the conglomerate and include granite, basalt, buff and red quartzite, and "stretched pebbles" similar to those of the Union Glacier Formation. This conglomerate is unusual in that it contains coarse-grained granitic clasts and appears to represent sediment derived from a localized source area. Overlying this conglomerate are 760 m of green or locally maroon phyllite and 340 m of green, locally conglomerate graywacke, with lesser amounts of green phyllite and green polymict conglomerate. Five paleocurrent measurements suggest that currents moved toward the southwest.

The metagraywacke contains clasts of rock fragments, quartz, plagioclase, and widely dispersed opaque grains in a matrix of white mica, chlorite, and fine-grained quartz, plagioclase, and carbonate (Fig. 5). The quartz clasts are generally angular to subangular single crystals. Rare grains have subhedral boundaries indicative of a volcanic origin. Plagioclase clasts are both twinned and untwinned and commonly display irregular replacement boundaries adjacent to white mica or carbonate in the matrix. Sedimentary lithic clasts of quartz siltstone and micaceous quartz siltstone are locally abundant. Numerous fine-grained, foliated and unfoliated, micaceous fragments of uncertain origin occur in all the graywacke samples. These fragments could be metamorphic slate or phyllite clasts or sedimentary shale fragments that were recrystallized during the metamorphism affecting the graywacke. Lithic clasts of definite igneous origin are sparse but include granitic fragments or, less commonly, basalt and granophyre clasts. Rare mafic mineral fragments, including greenish brown to blue-green hornblende and brown or green biotite altering to chlorite, occur in some samples. Both calcite and dolomite are present in variable amounts in most of the samples. They occur most commonly replacing quartz or plagioclase clasts, which are also commonly altered to sericite. Tourmaline and zircon are commonly present in trace amounts.

The middle member of the formation is a 280-m-thick sequence of red and buff polymict conglomerate with a graywacke matrix and metagraywackes similar to those in the lower member. Conglomerate clasts include phyllite, quartzite sandstone and siltstone, and shale. Coarsely crystalline dolomite and/or calcite is probably due to dedolomitization as is evident from the numerous dolomite rhombohedrons partially or completely replaced by calcite. A 10-m-thick cut-and-fill structure occurs at the base of this member in the measured section and is cut into alternating bands of red polymict conglomerate and green conglomeratic sandstone. The entire sequence is comformable. A total of 47 cross-bed measurements show variable paleocurrent directions, but paleocurrent movement toward the southeast dominates. Two-thirds of the cross-beds are thicker than 30 cm.

The upper member of the formation is about 300 m thick and is composed primarily of green graywacke similar to that described from the lower member of the formation. Sand-size material makes up about 80 percent of this unit, but some of the coarser beds are conglomeratic, and slate interbeds are common.

The entire Hyde Glacier Formation appears to represent a transitional sequence from a fluvial-deltaic environment at the beginning of sedimentation to a shallow-marine-deltaic environment in the upper member. Nearby source areas of moderate relief are indicated by the amount and coarseness of the conglomerates. Conglomerate clasts indicate that the source areas continued to expose bedrock primarily of carbonate and quartzite composition with localized coarse-grained igneous exposures. Small amounts of "stretched pebble" clasts similar to those in the underlying Union Glacier Formation indicate probable erosion of volcanic deposits equivalent in age to the latter formation.

The Hyde Valley Formation is known only from the Edson Hills area. Its absence from the Drake Icefall area only 8 km to the north could be due to rapid facies changes and/or to faulting. We consider rapid facies changes to be the more probable explanation in a tectonically active area with sedimentation in a transitional environment, but the possibility of omission due to faulting cannot be ruled out.

The Drake Icefall Formation

The Drake Icefall Formation is here formally described. The type section is just north of the Drake Icefall, and the type area includes the Drake Icefall and the Edson Hills. It is a unit of black shale and interbedded limestone that conformably overlies the Hyde Glacier Formation. Total thickness is difficult to determine because of the intense shearing and parasitic folds within the formation, but it is estimated to be 500 to 800 m.

In the type section, the base of the unit is composed of a 200-m-thick gray limestone sequence. The limestone is massive and sheared and has not yielded fossils to date. Above the limestone is a zone estimated to be about 250 m thick that is composed of black calcareous shale. The shale is extremely sheared and weathers to thin black flakes. Above this is a strongly sheared and parasitically folded unit composed of black calcareous shale similar to the underlying unit, with sparse, pod-like, light-gray to black limestone lenses as thick as 10 m and as long as 90 m. They contain beds of relatively pure gray limestone, calcareous black shale, and black shaly limestone but are quite variable laterally and stratigraphically. One black limestone lens has yielded a Middle Cambrian trilobite fauna (Jago and Webers, this volume). The upper portion of the Drake Icefall Formation is composed almost entirely of calcareous black shale. In the upper part of this unit, about 75 m below the contact with the overlying Conglomerate Ridge Formation, is a 6-m-thick gray limestone, overlain by brown argillite with occasional beds of brown impure quartzite.

In the Hyde Glacier area of the Edson Hills the base of the Drake Icefall Formation is marked by a 5-m-thick siliceous unfossiliferous limestone. This bed is overlain by about 20 m of brown impure quartzite, siltstone, and argillite. The remainder of the formation is composed almost entirely of black calcareous shale that weathers to thin black flakes. About 50 m above the base of the formation is a 20-m-thick zone in the black calcareous shale that contains abundant elongate carbonate concretions averaging about 20 cm in length. These concretions contain a well-preserved Middle Cambrian trilobite fauna that correlates with the fauna from the Drake Icefall area (Jago and Webers, this volume). About 170 m above the base of the formation is a gabbroic sill of about 180 m in thickness (see Vennum and others, this volume). The thickness of the Drake Icefall Formation is considered to be about 500 m in the Hyde Glacier area.

The two exposures of the Drake Icefall Formation in the Drake Icefall and the Hyde Glacier areas correlate well paleontologically and in overall rock type but vary considerably in thickness and in the proportion of carbonate beds present. The sections are separated by only about 8 km. The carbonates of the Drake Icefall area are represented in the Hyde Glacier area by only a 5-m-thick siliceous limestone and the presence of carbonate nodules in black shale. The thickness of the formation in the Hyde Glacier area is estimated to be at least 300 m less than in the Drake Icefall area. These differences are considered to be facies variations, although the omission of a portion of the formation in the Hyde Glacier area through faulting cannot be ruled out.

The Drake Icefall Formation is considered to represent shallow marine sedimentation in a restricted euxinic environment and a deepening of the basin following the deposition of the Hyde Glacier Formation.

The Conglomerate Ridge Formation

The Conglomerate Ridge Formation, here formally described, is a thick clastic unit that conformably overlies the Drake Icefall Formation. Thickness is estimated to be about 600 m. The type section is just north of the Drake Icefall, and the formation is known only from the Soholt Peaks–Drake Icefall area. The lower 150 m is composed dominantly of sheared gray, green, and buff argillaceous quartzite. Some quartzite beds are conglomeratic, and clast-supported buff polymict conglomerates as thick as 6 m are present in the quartzite. Clasts are dominantly red and buff quartzite, and buff and gray limestone and dolostone.

The lower unit is overlain by a 450-m-thick buff polymict clast-supported conglomerate with interbeds of fine to coarse quartzite. Clasts make up about 70 percent of the material and are angular to rounded. They are composed dominantly of buff and gray limestone and minor buff quartzite. The matrix is a poorly sorted, fine-to-coarse, buff sand composed of rather variable percentages of carbonate minerals, quartz, and a variety of rock fragments. Interbeds of maroon quartzite and green polymict conglomerate are present near the top of the formation. In general appearance the buff polymict conglomerate of the Conglomerate Ridge Formation is similar to the buff polymict conglomerate near the base of the Union Glacier Formation and similarly reflects nearby source areas of moderate relief, which exposed bedrock composed dominantly of carbonates and quartzites. The deposition of the Conglomerate Ridge Formation represents a marked change from the lower energy marine environment of the Drake Icefall Formation, and could be fluvial and/or shallow marine in origin.

The absence of the Conglomerate Ridge Formation in the Hyde Glacier area, 8 km to the south, is considered to be the result of rapid facies changes, although omission due to faulting cannot be discounted.

Springer Peak, Liberty Hills, and Frazier Ridge Formations

These three formations are considered to be laterally equivalent and are here described formally. The three formations make up a sizable portion of the exposures in the Heritage Range (see geologic map in rear pocket) and most of the exposures of the Heritage Group. The formations are almost entirely clastic in composition and include a wide variety of rock types, including argillite, graywacke, conglomerate, and quartzite. Thick lava flows are present in the Liberty Hills and Springer Peak Formations. Most deposition occurred during the Middle Cambrian with deposition of the upper beds during the beginning of the Late Cambrian time.

Sedimentation took place during active volcanism as is evidenced by thick basalt lava flows in the Liberty Hills and the Springer Peak formations. Active volcanic centers were located in or near the Liberty Hills, Anderson Massif, and Soholt Peaks and possibly High Nunatak, O'Neill Nunataks, and Gross Hills. Rapid lateral and vertical variations in sedimentary rock types suggesting fluvial, deltaic, and shallow marine environments and extrusive and possibly intrusive basic igneous activity would suggest a tectonically active coastal region. The sequence of deltaic facies in the Liberty Hills Formation (discussed below) suggests source areas for the sediments from the south and west. Paleocurrent studies of the cross-bedding and other sedimentary structures are needed to confirm this hypothesis.

The Springer Peak Formation. The Springer Peak Formation is estimated to be about 1,000 m thick and conformably overlies the Conglomerate Ridge Formation in the area north of the Drake Icefall. The type section is east of the base of Springer Peak at the northern end of Webers Peaks. Actual thickness is very difficult to estimate because of shearing and abundant parasitic folds within the formation. A 60-m-thick limestone with black shale interbeds is present at the base of the unit in the Drake Icefall area. It is overlain by an estimated 500 m of brown argillite with interbeds of black shale, buff quartzite, and graywacke. The top of the unit was not seen.

At Springer Peak (northern end of Webers Peaks) an 800-m-thick exposure, composed dominantly of brown argillite with lesser amounts of buff graywacke, is in conformable contact with the Minaret Formation, but the base of the Springer Peak Formation was not exposed. Thin (2 cm thick) interbeds of quartzite and occasionally argillaceous limestone are present near the upper contact. Graded bedding, load casts, Bouma intervals, and soft-sediment deformation of bedding are common in the graywacke beds. Calcareous nodules in argillites from this formation in the Inferno Ridge exposures contain strongly deformed trilobites (Shergold and Webers, this volume) and conodonts (Buggisch and others, this volume). Late Middle Cambrian trilobites have been recovered near the top of this formation, and Late Cambrian faunas have been recovered just above the base of the overlying Minaret Formation at Yochelson Ridge (Shergold and Webers, this volume).

The Springer Peak Formation appears to have been deposited under marine-deltaic conditions.

The Liberty Hills Formation. The Liberty Hills Formation is best exposed in the Liberty Hills and is laterally equivalent to the Springer Peak Formation in the area north of the Drake Icefall. It is at least 1,000 m thick and is composed of basaltic lava flows, volcanic breccias, polymict conglomerate, conglomeratic quartzite, quartzite, argillite, and minor amounts of limestone. Facies changes are extremely rapid and complex. Figure 6 gives a somewhat simplified view of the detailed relations among these rock types. The clastic stratigraphic sequence fines upward and to the northeast, suggesting a southwest source area. The massive conglomerate at the base of the sequence grades upward into more uniformly bedded sandy conglomerate and eventually into buff and green conglomeratic quartzite. The conglomeratic quartzites, in turn, grade into quartzite, sandy argillite, and finally to argillite.

A clast-supported polymict conglomerate of the formation is as thick as 800 m at Kelley and Sponholz Peaks, and 500 m thick in the southeastern Liberty Hills. The bases of these sections are not exposed. The conglomerate is the lowest stratigraphic unit in the Liberty Hills, but it is not exposed in the northern Liberty Hills. Clasts are variable in size, average about 5 cm in length, and are as large as 90 cm in length. They are composed of granite, vein quartz, quartzite, siltstone, limestone, and volcanics. In the lower exposures of the conglomerate, bedding is massive, clasts are moderately rounded and relatively large (up to 90 cm), and sorting is poor. These conglomeratic sediments show rapid lateral and vertical changes and are considered to be of fluvial origin.

The massive conglomerates grade upward into well-bedded buff sandy conglomerate and buff conglomerate quartzite in which the clasts tend to be smaller and more rounded than in the massive unit. Lithologies are interbedded and lensing with rapid variations in stratigraphic thickness. Sand-size particles are composed predominantly of quartz and are coarse to very coarse grained. Clasts are as much as 50 cm in diameter and are similar to the lithologies in the massive conglomerate. Trough cross-bedding is common, and numerous cut-and-fill structures indicate active channel activity. The channel deposits fine upward. The percentage of pebble clasts in the conglomeratic quartzites decreases upward and northward, and these strata finally interbed with finer grained sediment. The sandy conglomerates and con-

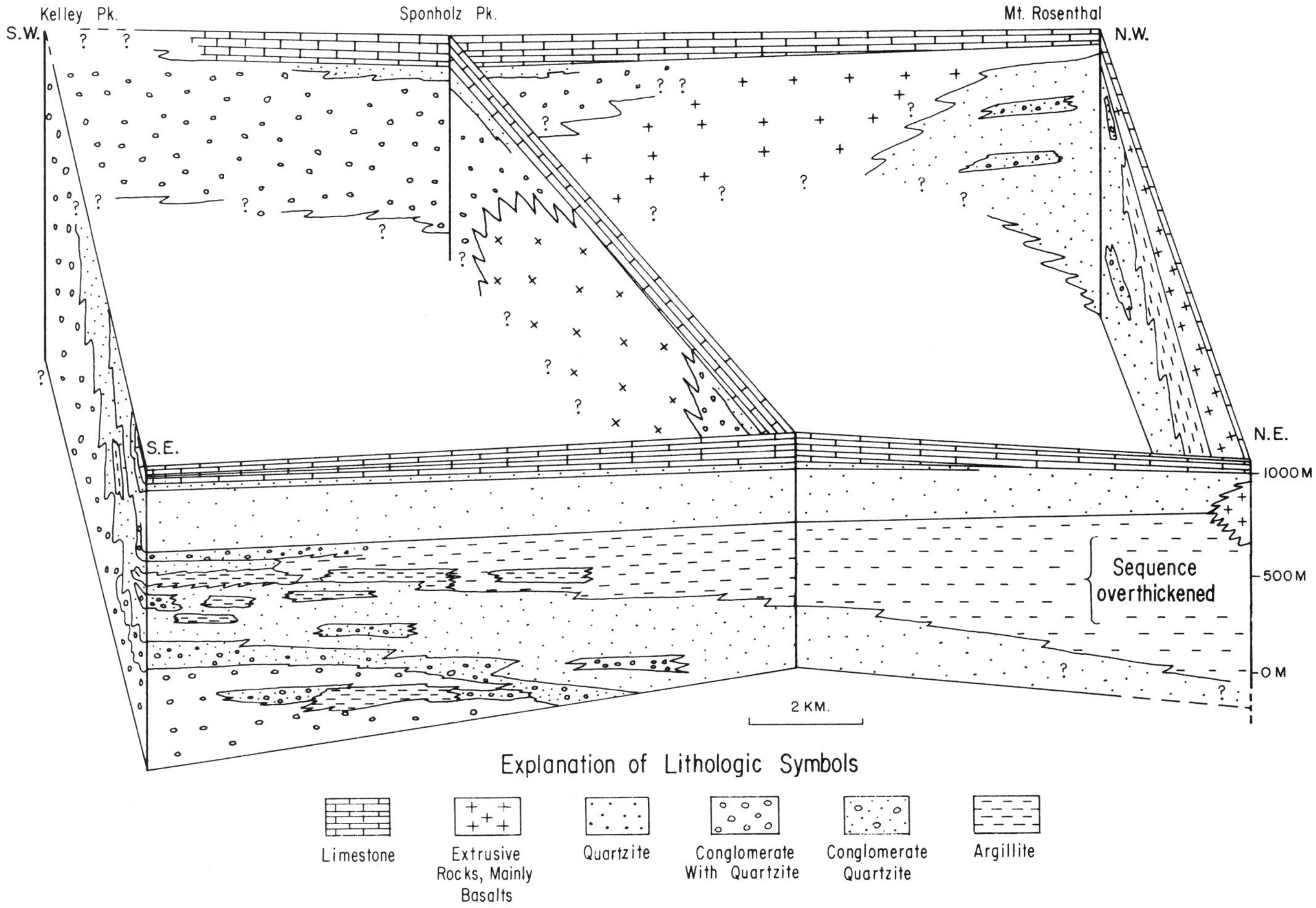

Figure 6. Fence diagram showing facies relations among sedimentary rocks of the Liberty Hills Formation in the Heritage Range of the Ellsworth Mountains.

glomeratic quartzites represent fluvial and possibly shallow marine environments.

Overlying the conglomerates are interbedded quartzite, siltstone, and green argillite. Micritic limestone beds occur in the upper levels. The units are well bedded and only rarely exhibit cross-bedding. One of the more noteworthy features of this sequence is the gradational lateral change from predominantly buff argillite and quartzite in the southeastern and northwestern Liberty Hills to a green argillite in the northeastern Liberty Hills. These rocks are considered to be of shallow marine origin.

An active volcanic center in the central Liberty Hills is evidenced by mafic volcanic breccias as much as 800 m thick. Following the deposition of the green argillite, mafic extrusions and intrusions from this center, and possibly also from a center in the High Nunatak area, affected most of the Liberty Hills area. Mafic lava flows as much as 300 m thick are exposed near the top of the formation on the eastern and northern portions of the Liberty Hills. The Liberty Hills Formation is in conformable contact with the Minaret Formation in the Liberty Hills.

The Liberty Hills Formation is also exposed in the Patriot Hills, High Nunatak, Wilson Nunataks, and the Chappell Peak–Mount Dolence area. In the Patriot Hills and the Chappell Peak–Mount Dolence area, the upper beds of the Liberty Hills Formation are conformably overlain by the Minaret Formation. High Nunatak and the Wilson Nunataks exhibit the variety of sedimentary lithologies found in the Liberty Hills, as well as intrusive and extrusive mafic igneous rocks.

Sediments in the Liberty Hills Formation were probably deposited in both fluvial and shallow marine deltaic environments. Source areas were to the south and southwest. Nonmarine coarse-grained fluvial deposits graded seaward into deltaic transitional sediments and finally to fine-grained shallow marine accumulations. The sedimentary sequence onlaps gently to the southwest (Fig. 6).

The Frazier Ridge Formation. The Frazier Ridge Formation is exposed primarily in the northwestern Heritage Range, with a tongue of the formation present in the Pioneer Hills of the northeastern Heritage Range. The type area of this formation is in the Frazier Ridge area of the northwestern Heritage Range, but a type section is not named because detailed stratigraphic sections have not been measured. The formation is primarily a fine- to medium-grained green quartzite with occasional beds of green argillite and black shale. It is estimated to be at least 500 m thick. The base of the formation is not exposed in the northwestern Heritage Range. Where the Frazier Ridge Formation is exposed, the Minaret Formation that usually overlies the formation is absent, and the top of the formation is in conformable contact with the overlying Crashsite Group. The contact with the Crashsite Group is marked by a series of interbedded black shales. The "quartzites," on the basis of limited petrographic observations, are arkoses with an abundance of large detrital chlorite, muscovite, and muscovite-chlorite intergrowths. An incipient foliation with alignment of fine-grained sericite and chlorite at a moderate angle to the bedding is also visible in thin section. Sedimentary structures are abundant. Limited cross-bedding data (n = 23) from gently dipping rocks at Frazier Ridge and Pipe Peak indicate that the paleocurrents moved toward the southern half of the compass, and 20 other paleocurrent measurements of ripples and parting lineation are compatible. One example of herringbone cross-bedding, with the two directions 215° apart, was observed. At Pipe Peak, cross-bed sets were generally less than 10 cm thick; at Frazier Ridge, they were 20 to 45 cm thick. The environment of deposition of the Frazier Ridge Formation appears to have been shallow marine—perhaps tidally influenced.

The age of the Frazier Ridge Formation probably ranges from late Middle Cambrian to Late Cambrian. A Late Cambrian trilobite fauna (possibly of Franconian age) was recovered just above the Frazier Ridge Formation in the transition beds of the overlying Howard Nunataks Formation of the Crashsite Group in the northwestern Heritage Range (Shergold and Webers, this volume).

The Minaret Formation. The Minaret Group was first described by Craddock (1969) and considered to be of probable Precambrian age. Splettstoesser and Webers (1980) reported the age to be Late Cambrian on the basis of trilobite faunas, and Webers and Spörli (1983) downgraded the group to formational status.

The Minaret Formation is here included as the highest formation in the Heritage Group. It is composed almost entirely of white to gray marble. It includes a variety of carbonate rocks including biosparite, oncosparite, pelsparite, oolite, biomicrite, and dolomicrite, representing a variety of environments from high-energy normal marine to low-energy restricted marine (Buggisch and Webers, this volume).

The type area is the Marble Hills in the southern Heritage Range. Here, the formation reaches its maximum thickness of at least 600 m and could be as thick as 900 m. Shearing and intense folding are common in many exposures (Spörli, this volume). Areas of light and dark marble can be distinguished readily here, but they do not appear to have stratigraphic significance. Sedimentary structures include oolites, oncolites, flat-pebble conglomerates, and occasional ripples and cross-beds. A poorly preserved trilobite-mollusc fauna of possible Middle Cambrian age has been recovered.

The Minaret thins almost uniformly to the north. In the Mount Dolence–Chappell Peak area it is at least 120 m thick and may be as thick as 400 m. Shearing and a steep-angle reverse fault traversing the Minaret along its length preclude easy measurements of thickness. The Minaret Formation is here in contact with the underlying Liberty Hills Formation and the overlying Crashsite Group. Oncolites and oolites are common in the formation in this area. Fossil faunas were not seen in the field, but trilobite and inarticulate brachiopod fragments have been observed in thin section.

At Yochelson Ridge the formation is considerably thinner. An 80-m section is exposed, and the total thickness is only slightly larger. Trilobite faunas (Shergold and Webers, this volume) indicate a Late Cambrian age. Oolites and oncolites are common sedimentary structures. The Minaret Formation is in conformable contact here with the underlying Springer Peak Formation. Black shales in the upper Springer Peak Formation just below the contact contain a Middle Cambrian trilobite fauna (Jago and Webers, this volume).

At Springer Peak in the northern Webers Peaks the Minaret Formation thins to a feather edge of only 8 m thick. Here biosparites have produced a well-preserved Late Cambrian fauna of trilobites, monoplacophorans, rostroconchs, gastropods, hyoliths, articulate and inarticulate brachiopods, archeocyathids, and pelmatozoans (Webers, 1972; Henderson and others, this volume). Oolites and oncolites are common sedimentary structures.

There are also exposures of the Minaret Formation in the Liberty Hills, Collier Hills, Anderson Massif, Soholt Peaks, O'Neill Nunataks, and possibly the Wilson Nunataks. In the Anderson Massif, highly sheared exposures of the Minaret are in contact and possibly interbedded with thick basalt flows. Deformed fossils including articulate brachiopods, a trilobite, and an archaeocyathid were collected near Huggler Peak but could not be identified in detail. Their presence is consistent with correlation with the fauna at Springer Peak. The Minaret Formation is not present in the northeastern Heritage Range. A small exposure in the O'Neill Nunataks in the southern Sentinel Range marks the only occurrence of the formation in that Range.

Sedimentary structures indicate that the deposition of Minaret sediments was for the most part in open, shallow marine, agitated environments. Sedimentation of the Minaret Formation probably began in the late Middle Cambrian in the Marble Hills area and continued into the Late Cambrian. To the north, sedimentation did not begin until the beginning of the Late Cambrian.

SUMMARY

The Heritage Group is a succession of predominantly clastic sedimentary rocks. Volcanic and tectonic activity was present during most of the time of deposition of the Heritage Group.

Thick terrestrial volcaniclastic sediments were deposited as tuffs and possibly lahars during Middle Cambrian time to form the Union Glacier Volcanics. The Kosco Peak Member of the Union Glacier Formation represents a fluvial deposit derived from nearby carbonate source areas of moderate relief. Deposition of the Hyde Valley Formation followed subsidence in the area of the Edson Hills and represents a sequence from nonmarine fluvial deposits to shallow marine deltaic sediments. Continued subsidence resulted in the deposition of the carbonate rocks and black shales of the Drake Icefall Formation in a restricted marine environment during Middle Cambrian time.

The Conglomerate Ridge Formation in the Drake Icefall area represents fluvial conditions similar to those of the Kosco Peak Member of the Union Glacier Formation.

The correlative Liberty Hills, Springer Peak, and Frazier Ridge Formations, and the overlying Minaret Formation, were deposited in shallow marine environments in Middle to Late Cambrian time during a period of volcanic activity. A number of volcanic centers were than active in the Heritage Range. Basaltic and gabbroic intrusions were presumably also emplaced during this time.

The Liberty Hills Formation represents a thick fluvial and shallow marine to deltaic sequence deposited in a subsiding basin under gentle onlap conditions. Source areas of moderate relief were to the south and west. The Liberty Hills Formation grades into the Springer Peak Formation. The black shale, graywacke, and argillite of the latter formation represent a basinward restricted marine deltaic environment. The green quartzite of the Frazier Ridge Formation represents a shallow marine deposit presumably associated with the above deltaic deposits, but it requires further study before its relations can be assigned.

The Minaret Formation reflects a gentle subsidence of the region with relatively little influx of sediment. Late Cambrian fossiliferous limestones were deposited in agitated shallow marine environments. Less common dolomites reflect more restricted environments. A volcanic center was active in the Anderson Massif area during this time.

Following the deposition of the Heritage Group the region became relatively stable, with gentle subsidence resulting in the deposition of the overlying Crashsite Group. The volcanic activity that was active throughout the deposition of the Group was not present during subsequent deposition in the Ellsworth Mountains until volcanigenic sediments again became abundant in Polarstar time.

REFERENCES CITED

Craddock, C., 1969, Geology of the Ellsworth Mountains, *in* Bushnell, V. C., and Craddock, C., eds., Geologic maps of Antarctica: New York, American Geographical Society, Antarctic Map Folio Series, Folio 12, Plate 4.

Hjelle, A., Ohta, Y., and Winsnes, T. S., 1978, Stratigraphy and igneous petrology of southern Heritage Range, Ellsworth Mountains, Antarctica: Norsk Polarinstitutt Skrifter, no. 169, p. 5–43.

Splettstoesser, J. F., and Webers, G. F., 1980, Geological investigations and logistics in the Ellsworth Mountains, 1979–80: Antarctic Journal of the United States, v. 15, no. 5, p. 36–39.

Webers, G. F., 1972, Unusual Upper Cambrian fauna from West Antarctica, *in* Adie, R. J., ed., Antarctic geology and geophysics: Oslo, Universitetsforlaget, p. 235–237.

Webers, G. F., and Spörli, K B., 1983, Palaeontological and stratigraphic investigations in the Ellsworth Mountains, West Antarctica, *in* Oliver, R. L., James, P. R., and Jago, J. B. eds., Antarctic earth science: Canberra. Australian Academy of Science p. 261–264.

Manuscript Accepted by the Society March 1, 1989

Printed in U.S.A.

Geological Society of America
Memoir 170
1992

Chapter 3

Stratigraphy of the Crashsite Group, Ellsworth Mountains, West Antarctica

K. B. Spörli
Department of Geology, University of Auckland, Auckland, New Zealand

ABSTRACT

The Crashsite Group of the Ellsworth Mountains, a 3,000-m-thick sequence of shallow-water, mostly marine, tan, green, and red quartzose sandstones (quartzites) and argillites, is here subdivided, in ascending order, into the Howard Nunataks Formation (1,630 m), the Mount Liptak Formation 1,070 m), and the Mount Wyatt Earp Formation (300 m). The Mount Wyatt Earp Formation has yielded Devonian fossils. Late Cambrian trilobites occur in the transition beds at the base of the Crashsite Group, and it is probable that the lower parts of the Group include Ordovician and Silurian strata.

INTRODUCTION

The Crashsite Group of the Ellsworth Mountains is underlain by the Cambrian Heritage Group, consisting of quartzite, argillite, conglomerate, limestone, and basic igneous rocks, and is overlain by the glacigene Whiteout Conglomerate, which in turn is overlain by the Permian *Glossopteris*-bearing Polarstar Formation. The Ellsworth Mountains sequence was deformed and metamorphosed in Early Jurassic time (Craddock, 1969). The stratigraphic sequence is typical of Du Toit's SAMFRAU geosyncline; the deformation fits into the concept of a Gondwanide Orogeny and has been used as one of the keys in reassemblies of Gondwanaland (Du Toit, 1927, 1937; Dalziel and Elliot, 1982).

Rocks of the Crashsite Group crop out extensively throughout the length of the Ellsworth Mountains, mostly in rugged peaks and ranges. The type area lies at the northern end of the Sentinel Range, in the vicinity of a wrecked aircraft near the Newcomer Glacier. The total thickness is about 3,000 m in the Sentinel Range and ranges from about 2,700 m to 3,000 m in the Heritage Range. Interlayered quartz-rich sandstone and argillite make up the entire section. Lithologic variations both in vertical and lateral directions are much less pronounced than in the underlying Heritage Group. Limestones and igneous rocks are absent. The sandstones of the Crashsite Group generally contain fewer lithic fragments than those of the Heritage Group. Subdivision is entirely based on the field aspect of the rocks, including coloration of the sandstones and argillites and the maturity of the sandstones. In the field the units described below are distinctive enough to be mapped. The term *quartzite* is here used in a sedimentological sense, to emphasize the quartz-rich nature of the sandstones.

SUBDIVISION

In this paper the Crashsite Quartzite of previous authors (e.g., Craddock, 1969) has been elevated to group status. The three-fold subdivision, originally established in the Sentinel Range, is also applicable in the Heritage Range. However, the units formerly recognized only as informal members are here named as formations (Table 1).

Only some of the new names are supported by complete type sections (see Appendix). For the rest it was not possible to find a complete section. Instead, type areas with relatively easy access have been chosen.

The lower boundary of the Crashsite Group is exposed in the main anticlines of the Heritage Range and can be traced into the Nimitz Glacier area of the southern Sentinel Range (Fig. 1). No angular discordance has been recognized between the Heritage Group and the Crashsite Group. However, the variability of the rocks immediately below and above the contact and the presence of a gradational contact between the two groups in one locality in contrast to erosional contacts in other places may indicate local gaps in the sedimentary record.

At Webers Peaks (Fig. 1) the boundary between the Heritage and Crashsite groups is marked by a 60-m section of "transition beds," consisting of dark gray argillite and sandstone, many with

Spörli, K. B., 1992, Stratigraphy of the Crashsite Group, Ellsworth Mountains, West Antarctica, *in* Webers, G. F., Craddock, C., and Splettstoesser, J. F., Geology and Paleontology of the Ellsworth Mountains, West Antarctica: Boulder, Colorado, Geological Society of America Memoir 170.

TABLE 1. SUBDIVISION OF THE CRASHSITE GROUP

Craddock, 1969		This Paper	
Whiteout Conglomerate		Whiteout Conglomerate	
Crashsite Quartzite	upper dark member	Mount Wyatt Earp Formation	Crashsite Group
	light member	Mount Liptak Formation	
	lower dark member	Howard Nunataks Formation	
		Landmark Peak Member	
		Mount Twiss Member	
		Linder Peak Member	
Heritage Group		Heritage Group	

slump structures. A similar argillite unit occurs in the northwestern Heritage Range at Pipe Peak and the area to the south (Fig. 1); however, it is mostly green and, in the south, tends to be subdivided into three units by intercalation of two units of more massive quartzite. The argillite unit of the northwestern Heritage Range can reasonably be correlated with the "transition beds" of Webers Peaks.

The contact of the Crashsite Group and the overlying Whiteout Conglomerate is exposed in the Meyer Hills of the Heritage Range and in the northern and eastern Sentinel Range (Fig. 1). The contact is locally conformable. The uppermost units of the Crashsite Group in the Heritage Range appear to be highly varied.

The age of the Crashsite Group is not well known. The Heritage Group limestone directly underlying "transition beds" between the Crashsite Group and Heritage Group in the Webers Peaks has a Late Cambrian age. Late Cambrian trilobites (Shergold and Webers, this volume) have been recovered from the transition beds at the base of the Crashsite Group in the northwestern Heritage Range. A poorly preserved fossil from the Mount Twiss Member in the hills west of the Nimitz Glacier may be an orthoceratid cephalopod. An Early Devonian fauna was recovered from a nunatak northwest of Planck Point in strata of the Mount Wyatt Earp Formation (Boucot and others, 1967; Webers and others, this volume).

HOWARD NUNATAKS FORMATION

The Howard Nunataks Formation (named after Howard Nunataks) has a thickness of 1,630 m in the northern Sentinel Range (near 77°30′S, 87°W). In the northeastern Heritage Range, the total thickness increases to more than 2,250 m. However, its equivalents in the northwestern Heritage Range are together only 1,350 m thick. The majority of the Crashsite exposures in the Heritage Range belong to the Howard Nunataks Formation (Fig. 1).

The Howard Nunataks Formation of the Sentinel Range can be subdivided, from top to bottom, into three members of about equal thickness:

Landmark Peak Member: pale green quartzite with intercalated green argillite
Mount Twiss Member: characterized by interlayered beds of red and maroon quartzite and argillite
Linder Peak Member: impure greenish gray quartzite

The same subdivision can be carried southward into the Heritage Range, although the thickness of the individual members is more variable.

Northwestern Heritage Range

In the northwestern part of the Heritage Range, a large part of the Howard Nunataks Formation of the Crashsite Group appears to be replaced by a rather uniform sequence of dark grayish-green, rather impure clastic rocks that in part can be regarded as graywacke because of their high proportion of matrix and lithic fragments. These rocks occur in the form of wedges that all terminate toward the southeast and that appear to merge in the most northwesterly areas. The Frazier Ridge Formation of the underlying Heritage Group (Webers and Spörli, 1983) may be yet another wedge of the same sedimentary complex.

One wedge (A, Fig. 2), directly above the transition beds, can be considered to be a lateral equivalent of the Linder Peak Member; it consists of greenish-gray quartzite and graywacke with graded bedding, both interbedded with dark-green argillite. It is approximately 300 m thick at Skelly Peak (Fig. 1) but may become thicker toward the west.

The next higher wedge (B, Fig. 2) is intercalated between two subunits of the Mount Twiss Member and ranges from 300 m thick in the Windy Peak syncline (Fig. 1) to 200 m in the Landmark Peak syncline to 30 m at Webers Peaks. The highest wedge (C, Fig. 2) overlies and laterally replaces the Mount Twiss Member. Its thickness in the Windy Peak syncline is 180 m, but it is reduced to zero on the west limb of the Landmark syncline. However, a 300-m-thick unit of similar strata reappears at the top of the section south of Webers Peaks.

Lithologies in the two highest wedges are dark to medium greenish-gray, finely laminated, medium- to fine-grained, cross-bedded, graywackelike quartzite with argillite flakes, interlayered regularly with dark green argillite. On the western limb of Landmark Peak syncline (Fig. 1), the rocks of the lower wedge appear to grade eastward into thinly laminated, grayish-green quartzite.

These units have not been given formal names because their relation to the Crashsite Group and to similar rocks in the Heritage Group has not yet been sufficiently investigated.

Linder Peak Member

Named after Linder Peak in the Heritage Range (79°53′S, 83°9′W), the Linder Peak Member is present only in sections outside the northwestern part of the range. It lies above the "transition beds" or on limestones at the top of the Heritage Group and below quartzites with conspicuous intercalations of maroon argillites that belong to the Mount Twiss Member. In some cases it is

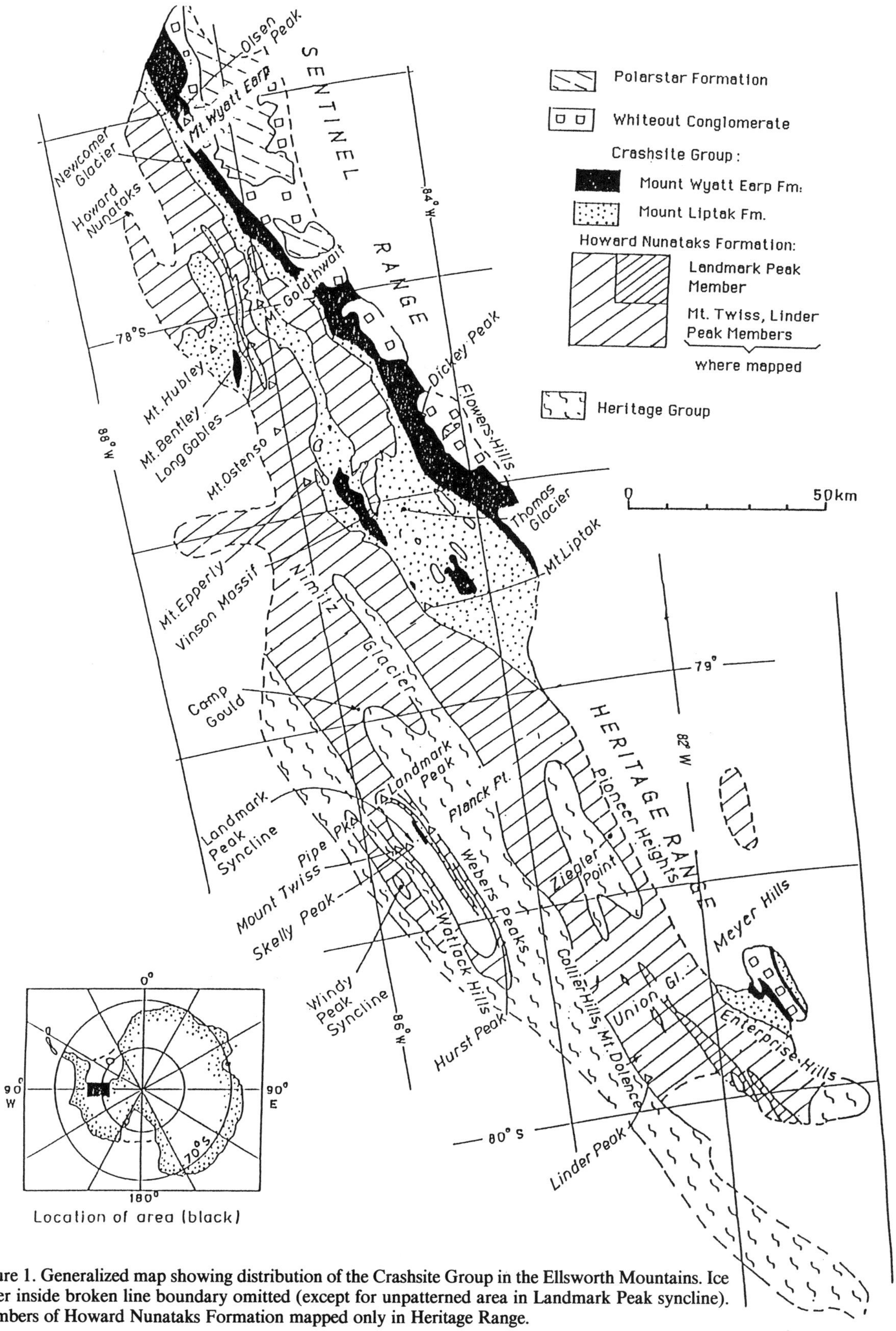

Figure 1. Generalized map showing distribution of the Crashsite Group in the Ellsworth Mountains. Ice cover inside broken line boundary omitted (except for unpatterned area in Landmark Peak syncline). Members of Howard Nunataks Formation mapped only in Heritage Range.

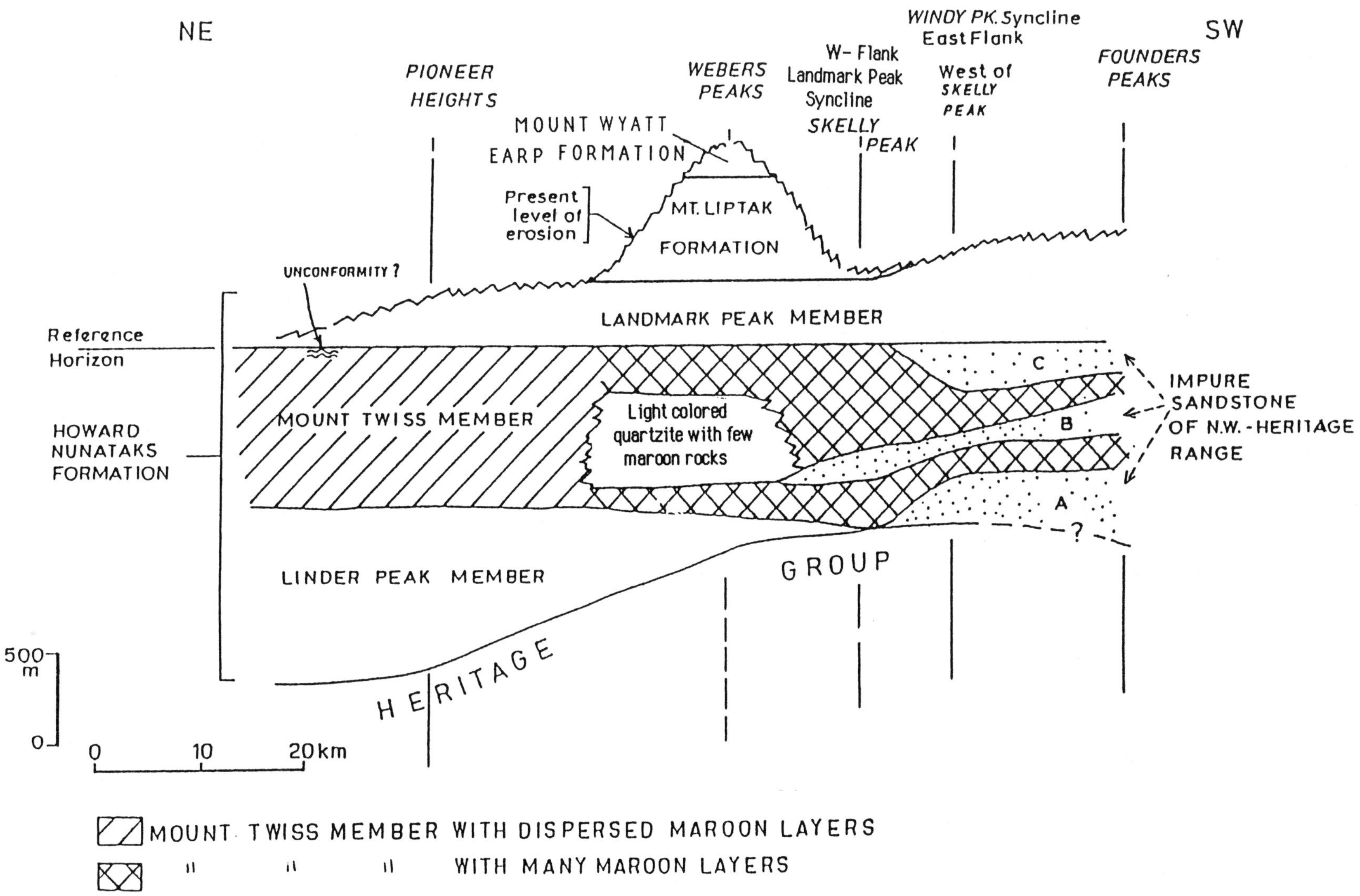

Figure 2. Variation of Crashsite Group across the Heritage Range. A, B, C denote three wedges of impure sandstones recognizable in the northwestern Heritage Range.

debatable whether some quartzites included in the "transition beds" (e.g., Webers Peaks section) should not also be included in the basal Crashsite Group. In most areas the lower contact is conformable, either a sharp change from transition beds into quartzite or an upward gradation into quartzite (Webers Peaks?). In the Mount Dolence area (Fig. 1), on the other hand, the contact is erosional, with the Linder Peak Member overlying marble cut by erosional pockets.

Because of the way the upper contact has been defined, it is not clear whether this contact represents an identical stratigraphic level throughout the Heritage Range. The somewhat differing rock types of the Linder Peak Member in various areas of the Heritage Range may indicate that this unit, rather than being a uniform stratum, may consist of interfingering lenses of different clastic rocks.

The thickness of the Linder Peak Member is 900 m in the southeastern Heritage Range and decreases toward the northwest. Lithologies are mainly dark green, locally light gray, coarse- to medium-grained quartzite in beds 20 cm to several meters thick. The regularly intercalated green argillites are from 5 cm to more than 1 m thick. Quartzites at the southern end of the Mount Dolence area (Fig. 1) are mottled green and red, and argillite partings are very thin.

Conglomerates occur in the ranges of Pioneer Heights and in the Webers Peaks area. Those in the Pioneer Heights area weather greenish brown, and matrix material is scarce to absent. Components include quartz pebbles, green slate, and feldspars. At Webers Peaks these horizons appear to be replaced by greenish-gray, medium-grained, thinly laminated quartzite containing argillite, flat pebble conglomerate, and layers of concentrated quartz fragments. Conglomerates are confined to the eastern part of the northern Heritage Range and appear to be better sorted toward the west.

The Linder Peak Member of the Sentinel Range has not been studied in detail, but it can be mapped above the Heritage Group contact and below the Mount Twiss Member on both sides of the Nimitz Glacier. North of Camp Gould (Fig. 1), the Mount Twiss Member with typical maroon interbeds is underlain by dark green argillite and sandstone, with a few beds of light gray quartzite. Bedding surfaces display interference ripples and fucoid trace fossils. These rock types resemble more the wedges of impure sandstone of the northwestern Heritage Range than the Linder Peak Member. The location of the boundary between impure sandstone and the Howard Nunataks Formation in the southern Sentinel Range is one of the outstanding problems of Crashsite stratigraphy. A unit of the Crashsite Group older than the Mount Twiss Member may occupy the core of an anticline in the area between Mounts Ostenso and Epperly (Fig. 1).

Mount Twiss Member

Named after Mount Twiss, in the northwestern Heritage Range (79°23′S, 85°37′W), this unit typically has abundant maroon and red argillite, intercalated with quartzite.

In the Heritage Range, rocks of this unit occur in the core of the Windy Peak syncline, on the flanks of the Landmark Peak syncline, in the highest summits of Pioneer Heights, in the area north and south of Union Glacier, and in the Enterprise Hills.

Except in the northwestern Heritage Range, the boundaries of the Mount Twiss Member are defined by the upper and lower limit of the occurrence of abundant maroon argillite and of maroon and red mottling within the quartzite beds. The two contacts are gradational in most areas; nevertheless, there is no difficulty in distinguishing the Mount Twiss Member from the other stratigraphic units.

In the northwestern Heritage Range two prominent horizons with maroon material can be distinguished. They are separated and overlain by units without red horizons (Fig. 2). In this area the contacts between the Mount Twiss Member and the wedges of impure sandstone are sharp and well defined.

In the eastern and southeastern part of the Heritage Range, the Mount Twiss Member consists of regular couplets of (1) 30-cm- to 2-m-thick green and red mottled quartzite, and (2) dominantly green but locally maroon argillite. In the area of Pioneer Heights two conspicuous horizons can be recognized in this otherwise uniform sequence:

- At the base of the Mount Twiss Member there are 15 to 30 m of reddish brown–weathering rocks.
- Approximately in the middle of the sequence there is a 15-m band of buff-weathering massive quartzite.

Both horizons have only been observed from a distance, and no samples are available.

The Mount Twiss Member at Ziegler Point and at Webers Peaks (Fig. 1) is traversed by a number of clastic dikes, whose rock types resemble those observed in the Whiteout Conglomerate and in sandstone in the Mount Wyatt Earp Formation at Meyer Hills (Fig. 1).

From Webers Peaks to the west, the Mount Twiss Member consists of two subunits approximately corresponding to the two wedges in Figure 2. In the lower of these, the thickness increases from 150 m at Webers Peaks (east flank of Landmark Peak syncline) to 200 m on the west flank of Landmark Peak syncline, only to decrease to 150 m again in the Windy Peak syncline farther to the northwest.

At Webers Peaks, lithologies in this subunit are generally fine-grained, sometimes red and green mottled quartzite. Also included are a few beds of pinkish-gray quartzite and a few layers of quartz-pebble conglomerate, some well-developed mudcracks and a 60-cm-thick, northwest-trending clastic dike perpendicular to bedding and filled with a dark, greenish-gray sandstone containing conglomerate components. These components include granitic clasts as much as 8 cm in diameter. A slight nonparallelism of bedding is visible in the southeast-facing cliff of Hurst Peak (Fig. 1). This discordance may be due to large-scale cross-bedding in a deltaic environment rather than to tectonic causes.

On the west limb of Landmark Peak syncline, the subunit is less distinctively developed; it mainly consists of pale green to light brown, finely laminated quartzite intercalated with 3-m-thick maroon argillite. However, parts of the section are composed of finely laminated red and green quartzite with intercalations of thin, dark green argillite, spaced 15 to 30 m apart.

The upper subunit of the Mount Twiss Member in the northwestern Heritage Range consists of massive pale greenish-gray quartzite with some red and green mottling, intercalated with maroon argillite. Slumps and some 3-cm spherical concretions occur in some of the sandstone beds. In the Windy Peak syncline, finely laminated red and pale green quartzite in 30-cm beds, containing scattered mudflakes, predominate.

At the northern end of Webers Peaks the section is somewhat different: it contains light gray quartzite and maroon, mottled quartzite, both intercalated with maroon argillite. Slump features and cross-bedding are abundant. A thin pebble-conglomerate band that occurs at the top of the sequence contains components of limestone, quartzite, and quartz. Part of the section contains very few maroon beds (Fig. 2).

In the Sentinel Range the Mount Twiss Member consists mainly of pale green, fine- to medium-grained quartzite with maroon mottling, interbedded with maroon or green argillite including some layers with argillaceous flat pebbles. Other rock types are thin-bedded, light to dark gray quartzite and interlayered light green and gray micaceous quartzite. Cross-bedding and ripple marks are common. Polygonal, distorted mud fragments found in exposures west of the Nimitz Glacier are most likely the product of subaerial desiccation. Some of the maroon rocks have conspicuous fine heavy mineral laminations enriched in detrital garnets.

The contact with the overlying Landmark Peak Member is rather sharp. The transition between the two units, if any, takes place within a few meters.

Landmark Peak Member

The Landmark Peak Member—named after Landmark Peak in the Heritage Range (79°10′S, 85°39′W)—forms a uniform rock unit both in the Sentinel and the Heritage ranges. In the Heritage Range only minimal thicknesses can be given because the Landmark Peak Member usually forms the highest part of the section. At Webers Peaks there are at least 420 m of this unit. On the western limb of the Landmark Peak syncline the minimal thickness is 300 m, and in the northwestern Heritage Range it is 600 m.

The Landmark Peak Member occurs in the core of the Windy Peak syncline, on the flanks of the Landmark Peak syncline, in the area of Collier Hills and the massifs east of it, and in the Enterprise Hills and the nunataks between them and Meyer Hills. Two slightly different rock types can be distinguished, one occurring in the northwestern Heritage Range, the other in the remainder of the Heritage Range.

In the main part of the Heritage Range the Landmark Peak Member consists of fine- to coarse-grained, pale grayish-green quartzite interbedded with thin layers of green argillite. In contrast to the two lower members of the Howard Nunataks Formation, where the argillite/quartzite ratios are commonly near 1, quartzites make up more than 60 or 70 percent of the section. Thicknesses of individual quartzite beds range from 1 to 3 m. Within this sequence are some thicker beds consisting of very pure, pale green to white quartzite with layers and lenses of conglomerate, including pebble-sized clasts of quartz and green argillite flakes.

In the northwestern Heritage Range the sequence is somewhat less regular. The following rock types are interbedded with each other:

1. massive to thinly laminated, cross-bedded, white to pale greenish-gray quartzite with rusty brown iron nodules.
2. medium-grained, pale grayish-green quartzite, weathering yellowish gray, and containing slump structures and 3-cm-diameter spheroidal calcareous(?) concretions.
3. dark brown, laminated, brown- to red-stained quartzite with layers of light- to dark-gray quartzite, including quartz-pebble conglomerate and conglomerates with components of dark limestone, quartzite, and quartz. Lenses of coarser quartzite are associated with the pebble conglomerates.
4. fine- to medium-grained, thick-bedded, massive, buff quartzite that weathers yellowish brown.

The two facies of the Landmark Peak Member described above grade into each other. The sequence in the northwestern Heritage Range displays the following variations as one proceeds from east to west: (1) conglomerates disappear; (2) number of white quartzite beds decrease; and (3) number of layers with ripple marks appears to increase.

Rusty staining, locally along complicated joint patterns, is common in the Landmark Peak Member. Similar phenomena have been observed in the uppermost layers of the Crashsite Group in the Sentinel Range.

In the Sentinel Range the Landmark Peak Member consists of thick-bedded, olive greenish-gray, cross-bedded quartzite with some grayish-green medium-grained, rather clean quartzite, interbedded with thin argillite. Weathering colors are reddish. In the vicinity of Long Gables a dark red weathering unit occurs about 15 m below the contact with the overlying Mount Liptak Formation. It occurs in a sequence of light grayish-brown, medium-grained quartzite that weathers light gray with mustard yellow stains. The dark red color appears to decrease in intensity toward the east.

Petrology

The outstanding petrographic feature of the Howard Nunataks Formation is the high percentage of heavy minerals. The assemblage is dominated by colorless garnet and a less common pink garnet. Other heavy minerals are epidote, clinozoisite, sphene, and rounded opaques, probably mostly magnetite (Fig. 3).

The feldspar content ranges from less than 1 percent to more than 10 percent. Strongly vacuolized orthoclase is the most diagnostic constituent, along with fresh microcline, fresh plagioclase, and some sericitized plagioclase (Fig. 4).

Matrix is present in most of the rocks with arenaceous texture, and it probably is of both primary and authigenic origin.

Provenance

Variation in sorting of Linder Peak Member conglomerate in the Heritage Range may indicate derivation from a northeasterly direction. The geometry of the wedges of impure sandstone in the northwestern Heritage Range (Fig. 2) suggests that their source area lay to the northwest (Fig. 5). Variations in sedimentology and the westward disappearance of conglomerate in the Landmark Peak Member of the Heritage Range may favor an easterly source for this unit. Lithologic variations in the Mount Twiss Member of the Heritage Range are too complex to yield a unique direction to the source area. Too few detailed observations of variations within the Howard Nunataks Formation of the Sentinel Range are available for useful speculation about provenance.

Environment of deposition

A trend toward greater textural and mineralogical maturity of the sediments from the Linder Peak Member upward to the Landmark Peak Member is easily discernible. However, even in the lowest members of the Howard Nunataks Formation, sands are better sorted and more mature than in those of the Heritage Group below. Oscillation ripplemarks and mudcracks indicate very shallow conditions of deposition, especially in the middle part of the sequence (Mount Twiss Member) where emergence may have taken on an especially great importance. The maroon mottling and staining along jointlike surfaces are diagenetic effects that may have been influenced by repeated emergence of strata. The presence of layers of garnetiferous sands may indicate marine redeposition of nearby beach sands in a shallow coastal environment. Currents are indicated by cross-bedding, which ranges from centimeter to meter scale. The presence of clastic dikes and a few slump features may indicate local rapid sedimentation and instability.

Sediments of the wedges in the northwestern Heritage Range (A, B, C in Fig. 2) are more impure and graywackelike, with some turbidite features, indicating a more rugged area of provenance than that of the Howard Nunataks Formation. Thus a hypothetical paleogeography for the lower part of the Crashsite Group would include well-eroded, mature (high-grade metamorphic?) borderlands to the east and a more rugged area to the west. The latter appears to have disappeared by the time of deposition of the Landmark Peak Member. Its existence from the upper part of the Heritage Group to the lower Crashsite Group probably makes it a Cambrian (and Ordovician?) feature.

MOUNT LIPTAK FORMATION

The type section of this unit lies in the Sentinel Range at Mount Liptak (78°45′S, 83°53′W; Fig. 1). In the Heritage Range a typical section is exposed in the nunataks between the Enterprise Hills and the Meyer Hills. The thickness in the Sentinel Range is about 750 m. In the Heritage Range the thickness can only be estimated because of structural complications. If the bottom and top contacts between Enterprise Hills and Meyer Hills are dipping 10°, then the maximum possible thickness for this area is 1,350 m. If 450 m are taken off this value to take account of the folding, an estimated thickness of 900 m remains. In the vicinity of Landmark Peak (Fig. 1) approximately 750 m of the section belong to the Mount Liptak Formation.

In the Heritage Range, the lower contact of the formation is gradational in the Enterprise Hills, but it is probably marked by a slight unconformity on Landmark Peak. The contact against the Mount Wyatt Earp Formation is only exposed at Planck Point (Fig. 1) and is also a sharp boundary.

Lithologically the formation consists of thick-bedded pale gray to whitish, fine- to coarse-grained, mostly featureless quartzite with only a few laminated or cross-bedded units. Argillite units, where present, are very thin. Argillite pebble layers are present in an exposure on the west side of Webers Peaks. Wine red and black stains following joint patterns are a characteristic feature of the rocks, and in part they consist of specular hematite. They may be due to a groundwater effect predating deposition of the Mount Wyatt Earp Formation. In the southeastern Heritage Range the unit can be recognized by its orange-red weathering colors.

On Landmark Peak abundant trace fossils mark the bedding planes.

In the southern Sentinel Range (Mount Liptak), the Mount Liptak Formation consists of buff-colored quartzite with intercalations of grayish-green quartzite that give way toward the top to pinkish-gray quartzite. Bedding thickness ranges from 30 cm to 6 cm. Near Mount Goldthwait the sequence can be subdivided into the following three subunits (from bottom [1] to top):

1. Thin- to thick-bedded interlayers of salmon-colored medium-grained, cross-bedded quartzite.
2. Strongly weathering, light-gray, medium- to coarse-grained, cross-bedded quartzite with reddish interbeds; weathers to orange-red.
3. Light-gray, medium-grained, clean, cross-bedded quartzite in thick, massive units.

In the Mount Hubley area the lowest unit of the Mount Liptak Formation is a pink, medium-grained, clean, cross-bedded quartzite.

Both the upper and lower boundaries of the formation are well defined. At Mount Liptak a clastic dike of dark gray, fine-grained quartzite cuts across the boundary between the Howard Nunataks Formation and the Mount Liptak Formation. In the Mount Goldthwait area the lower contact is marked by a 30-m-thick transition zone of interlayered beds. On the east flank of Mount Wyatt Earp a light-gray, fine- to medium-grained, cross-bedded massive quartzite is overlain with possible unconformity by the Mount Wyatt Earp Formation.

Petrology

Petrologically this unit is characterized by the extreme maturity of the detrital suite. Feldspathic grains and crystal fragments are rare. Quartz commonly is rutilated. Muscovite is a

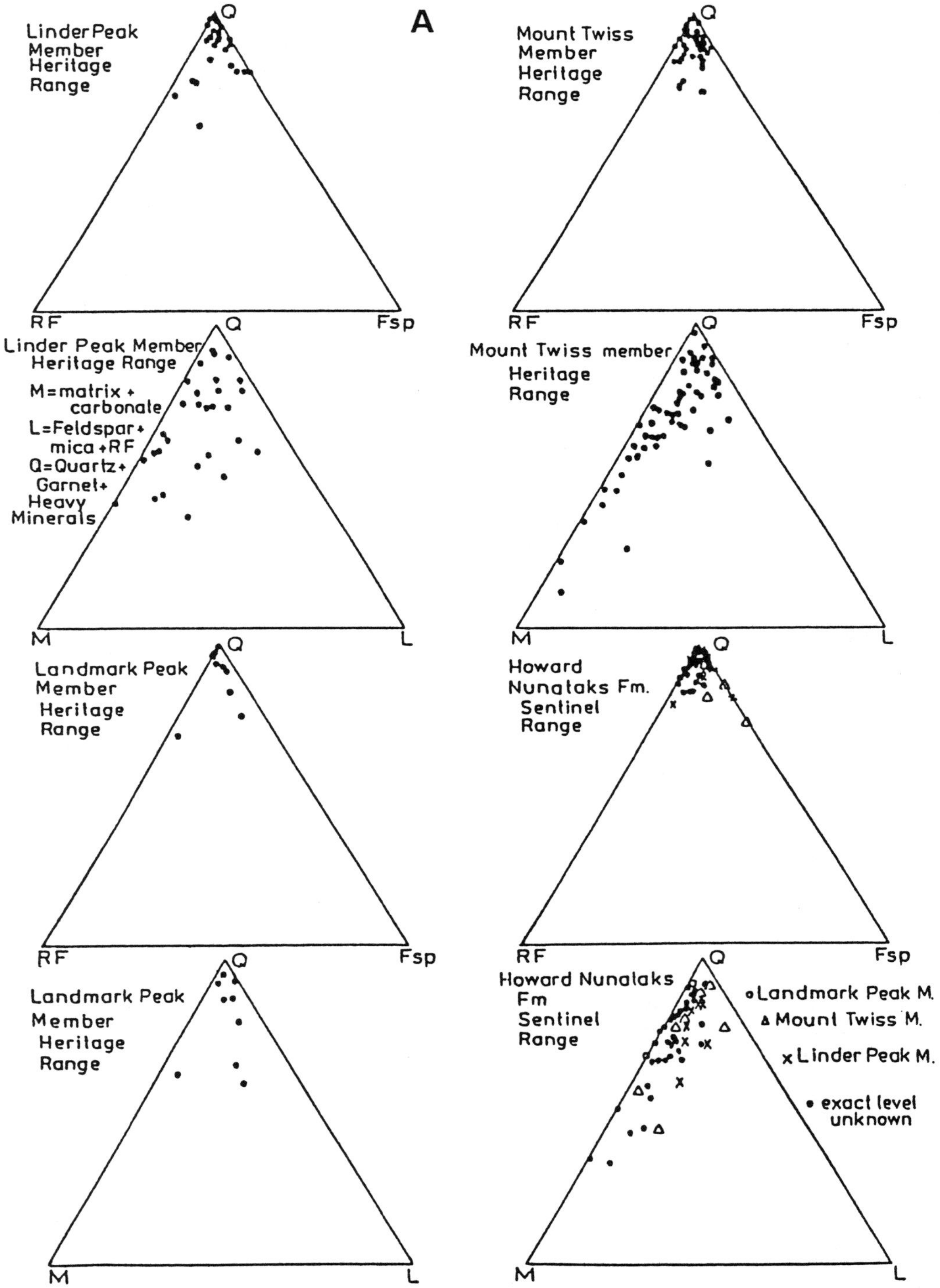

Figure 3A, B. Modal compositions of sandstones in the Crashsite Group. RF = rock fragments, Fsp = feldspar, Q = quartz + heavy minerals, L = feldspar + mica + rock fragments, M = matrix + carbonate. Data are from a number of investigators (P. Barrett, Sonja Hills, H. Meyer, K. B. Spörli). Number of points counted: 200 to 1,000 per slide. Based on 93 Howard Nunataks Formation specimens from the Heritage Range and 49 Howard Nunataks Formation specimens from the Sentinel Range, 23 Mount Liptak Formation specimens from the Heritage Range and 25 Mount Liptak Formation specimens from the Sentinel Range, 18 Mount Wyatt Earp Formation specimens from the Heritage Range and 19 Mount Wyatt Earp Formation specimens from the Sentinel Range.

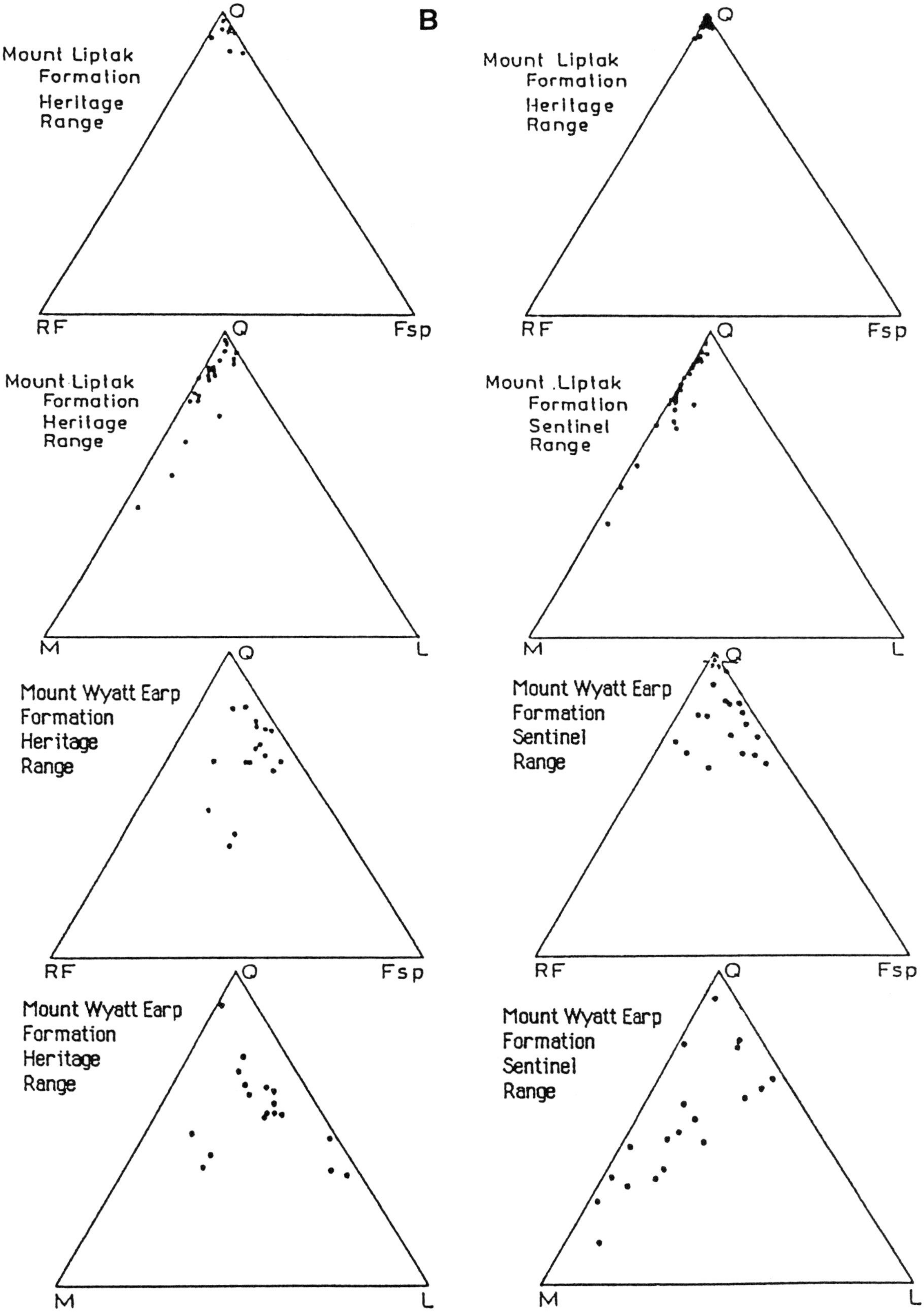
B
Q
Mount Liptak Formation Heritage Range
RF
Fsp
Q
Mount Liptak Formation Heritage Range
RF
Fsp
Q
Mount Liptak Formation Heritage Range
M
L
Q
Mount Liptak Formation Sentinel Range
M
L
Q
Mount Wyatt Earp Formation Heritage Range
RF
Fsp
Q
Mount Wyatt Earp Formation Sentinel Range
RF
Fsp
Q
Mount Wyatt Earp Formation Heritage Range
M
L
Q
Mount Wyatt Earp Formation Sentinel Range
M
L

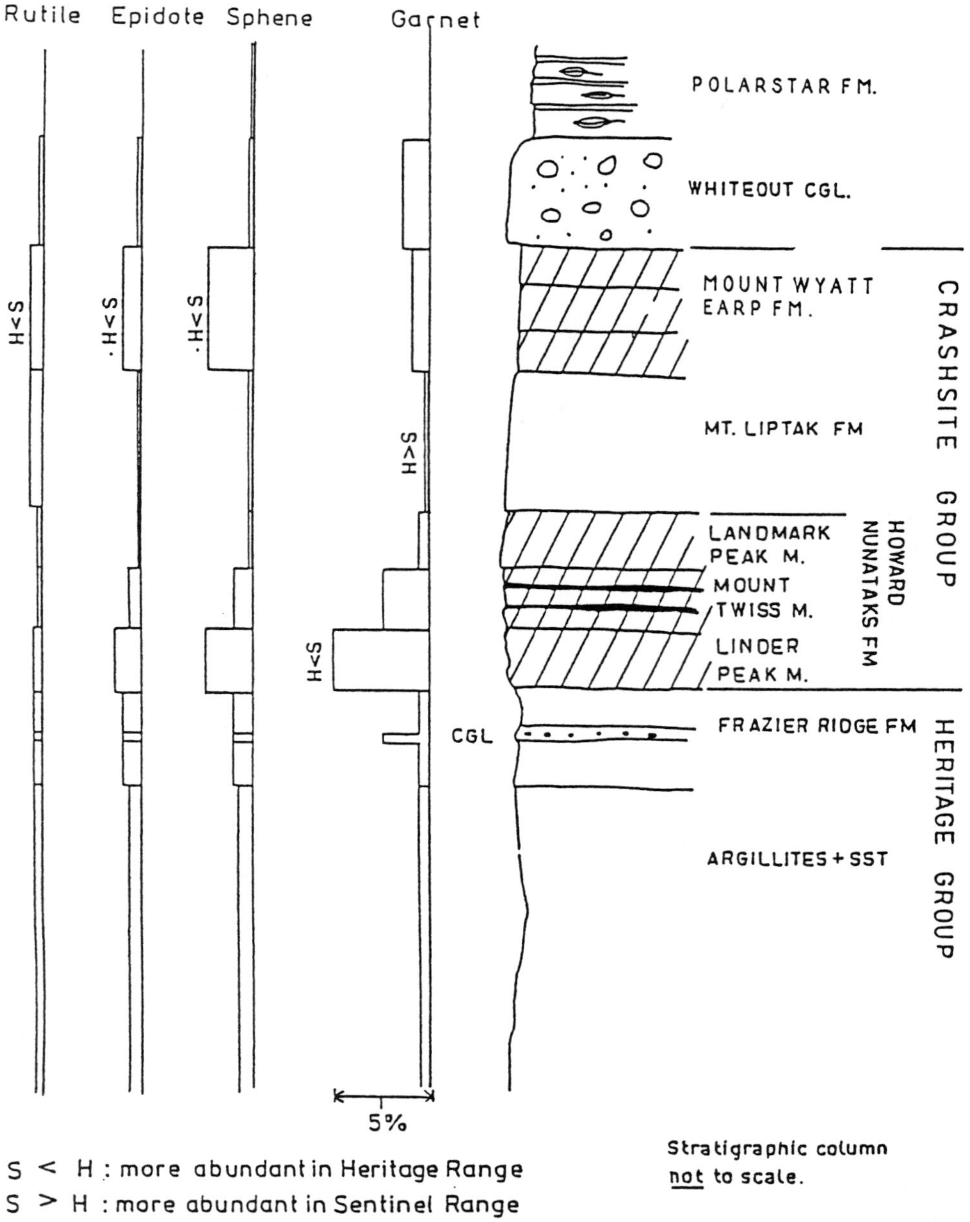

Figure 4. Modal abundance of heavy minerals in the Crashsite Group, compared with other stratigraphic units in the Ellsworth Mountains. Crashsite Group data based on 142 Howard Nunataks Formation specimens, 49 Mount Liptak Formation specimens, and 37 Mount Wyatt Earp Formation specimens.

common constituent. Among the accessory detrital minerals, garnet and epidote are rare compared to rutile (Fig. 4), a result either of scarcity in the area of provenance or of secondary alteration.

Provenance

No clear evidence for provenance has been collected. Conditions over the whole area appear to have been very uniform, as is indicated by the rather constant thickness and the lithological uniformity. Rutilated quartz in the detrital suite suggests plutonic material in the source area.

Environment of deposition

The high degree of sorting of the detrital material and the trace fossils probably indicate a rather shallow depositional environment. Because of the advanced stage of erosion in the source area and because of current and wave action in the area of

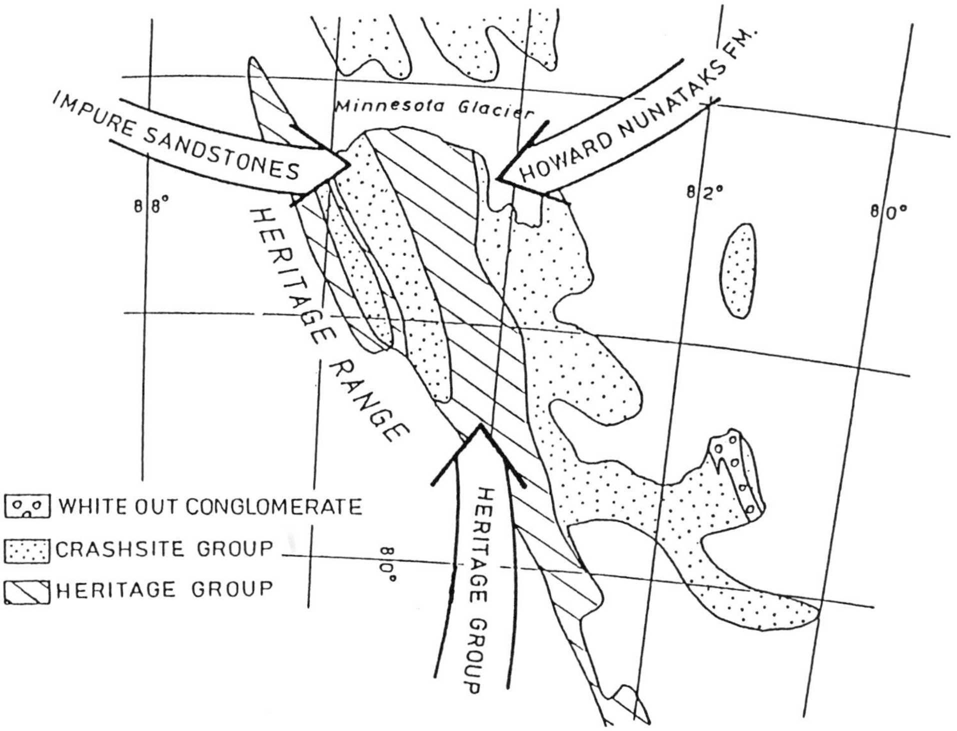

Figure 5. Possible provenance of clastics in the Heritage Range, based on thickness and grain size variations. "Impure sandstones" of northwest Heritage Range interfinger with Howard Nunataks Formation (see A, B, C, Fig. 2). Hjelle and others (1982) determined current flow from SSE to NNW for Crashsite Group of the southern Heritage Range.

deposition, most of the erosional products except quartz were eliminated from the sediment.

MOUNT WYATT EARP FORMATION

The formation is named after Mount Wyatt Earp (formerly Mount Earp), in the northern Sentinel Range (77°34′S; 86°25′W). Thicknesses range from 150 m to 300 m in the Sentinel Range (Mount Wyatt Earp–Flowers Hills area). Although no continuous section is exposed in the Heritage Range and lithological variations are extreme, a maximum of 150 m, or possibly 270 m, of this formation has been recorded in one section (Meyer Hills).

In the Heritage Range, the Mount Wyatt Earp Formation is exposed at Planck Point and surrounding the Whiteout Conglomerate of the Meyer Hills.

The small outcrop at Planck Point consists of black micaceous argillite and quartzite with burrow trails and possibly some shredded plant matter. The unit contains black micaceous sandstone nodules as much as 10 cm in diameter and has yielded a Devonian fauna (Boucot and others, 1967; Webers and others, this volume).

In the Meyer Hills the rocks below the Whiteout Conglomerate are variable. Conglomerate beds within the Formation are similar to the diamictites of the Whiteout Conglomerate and suggest that the two units are interbedded at this locality. Argillite layers immediately below the base of the Whiteout Conglomerate are strongly contorted. Reddish, gray, buff, and light gray medium- to fine-grained quartzite with discontinuous argillite layers and some 6-m-thick continuous argillite layers are the main rock type. Some of the units are stained red. While dark quartzites underlie the Whiteout Conglomerate in most sections, at one locality light buff quartzites form the substratum. It is not yet clear whether these light-colored rocks belong to the Mount Wyatt Earp Formation or to the underlying Mount Liptak Formation. If the latter is the case, an unconformable relationship between the Crashsite Group and the Whiteout Conglomerate would be indicated there. The variability of lithologies within the Mount Wyatt Earp Formation may be due either to complex interdigitation of lensoid sand bodies or to tectonic unrest prior to deposition of the Whiteout Conglomerate, or perhaps both.

In the southern Sentinel Range, the Mount Wyatt Earp Formation is exposed in a continuous belt along the eastern side of the range. Toward the north this belt traverses the range obliquely (Fig. 1). Isolated synclinal outliers are present south of Thomas Glacier, east of Vinson Massif, and north of Long Gables.

Dark gray to brown weathering colors and red staining are

characteristic of the Mount Wyatt Earp Formation in the Sentinel Range. On field evidence, a division into two equally thick subunits may be possible, the upper unit being darker and displaying stronger red weathering. It has not been possible to recognize this subdivision petrologically. The twofold subdivision is well recognizable in the exposures along Newcomer Glacier and on Mount Wyatt Earp, where light-gray, cross-bedded quartzite is overlain by approximately 60 m of dark-gray, medium-grained impure massive quartzite with indistinct bedding and reddish weathering colors.

At Dickey Peak 225 to 300 m of the Mount Wyatt Earp Formation is exposed. The following section covers the accessible upper part:

7. Whiteout Conglomerate

Mount Wyatt Earp Formation

6. 6 m Dark-gray quartzite, may grade upward into Whiteout Conglomerate
5. 3 m Flat-shale-pebble conglomerate in a light-gray quartzite matrix
4. 63 m Thin- to thick-bedded dark gray quartzite
3. 150 m Mottled light- and dark-gray quartzite with mottles as much as 6 inches
2. 30 m Dark quartzite with circular mottles
1. Dark gray medium- to fine-grained quartzite.

The light and dark gray mottling is characteristic of the Mount Wyatt Earp Formation elsewhere.

On the arete west of Mount Bentley, the Mount Wyatt Earp Formation consists of dark-gray, reddish-brown weathering, fine-grained quartzite interbedded with pale-gray quartzite. In the Flowers Hills area dark-gray argillite, argillite conglomerate, and quartzite, probably as much as 300 m thick, belong to the Mount Wyatt Earp Formation.

An isolated, probably ice-rafted pebble occurs 8 m below the Crashsite/Whiteout contact in the Olsen Peak area (Fig. 1).

Petrology

Prominent components of fresh angular plagioclase and of basaltic volcanic rock fragments characterize this formation. A large percentage of the quartz grains contain rutile. Despite the dark colors of the rocks, sulfides and carbonaceous matter are rare. The heavy mineral suite is dominated by garnet and epidote (Fig. 4). The sandstones of the Mount Wyatt Earp Formation can be classified as arkoses and quartzites. Petrologic similarities exist both with the Whiteout Conglomerate and with the Polarstar Formation; however, the affinity to the Whiteout Conglomerate is greater, in that there is a greater abundance of rock fragments, and the heavy mineral suite is more diverse.

Provenance

Not enough data are available to deduce paleocurrent directions of the Mount Wyatt Earp Formation. Judging from the detrital suite, the source terrane consisted of basic volcanic rocks associated with plutonic and high-grade metamorphic rocks.

Environment of deposition

The rocks of the Mount Wyatt Earp Formation are more impure than those of the Mount Liptak Formation and much of the Howard Nunataks Formation. This suggests an increase in rate of sedimentation and/or a decrease in winnowing by currents.

Not much evidence on depth of deposition is available, but like the rest of the Crashsite Group these rocks were probably deposited in rather shallow water. The dark color of the sediments results from finely disseminated sulfides, indicating somewhat stagnant conditions. Gradual cooling of the climate toward the top of the Mount Wyatt Earp Formation is suggested by the first influxes of ice-rafted glacial material.

DISCUSSION

Correlation with other areas

There appear to be no large hiatuses in the deposition of the Crashsite Group, and only weak evidence exists for a sedimentary break between the Crashsite and Heritage Groups. The immediately underlying parts of the Heritage Group are Upper Cambrian, and the top of the Crashsite Group may span Ordovician, Silurian, and at least part of Devonian time.

In the Pensacola Mountains, the sequence covering the same time span shows considerable lithological similarity to the Crashsite Group. Although the Pensacola Mountain sequence is probably even more quartzose (Williams, 1969), it lies unconformably on Cambrian formations and underlies the glacially derived Gale Mudstone (Schmidt and Ford, 1969). A lower unit (Neptune Group) is separated by a disconformity from the overlying, probably Devonian, Dover Sandstone, which has a phosphate pebble-bearing basal conglomerate. The two sequences are dominated by quartz sandstone and quartzite, displaying buff and tan colors. Conglomerates are more common there than in the Crashsite Group. The Elliott Sandstone and the Elbow Formation of the Neptune Group in the Pensacola Mountains may be correlated with the Mount Twiss Member of the Howard Nunataks Formation in the Heritage Range because of their high content of red argillaceous siltstone and argillite. The Dover Sandstone, and especially an unnamed equivalent at the southwestern end of the Patuxent Range, consisting of thick-bedded quartz sandstone and quartzite with minor thin black carbonaceous beds, may be an equivalent of the Mount Wyatt Earp Formation in the Ellsworth Mountains. Perhaps the uppermost unit of the Neptune Group, the Heiser Sandstone, bearing *Scolithus*-like tubes, may be correlated with the Mount Liptak Formation.

The combined thickness of the Neptune Group plus the Dover Sandstone (2,100 m) is somewhat less than that of the Crashsite Group (approximately 3,000 m). The depositional area of the Neptune Group and Dover Sandstone in the Pensacola Mountains probably was closer to the source area of the sediments and also closer to a tectonically active area than that of the Crashsite Group, as is indicated by the greater abundance of conglomerates and the presence of unconformities.

Along the Transantarctic Mountains, Devonian sedimentary rocks are too thin and discontinuous to be lithologically compared with the Crashsite Group. Boucot and others (1967) show possible correlations of Devonian rocks throughout Antarctica. In the Horlick Mountains (Mirsky, 1969), the Horlick Formation consists of as much as 50 m of interbedded feldspathic and quartzite sandstone with dark carbonaceous siltstone and silty shale that has yielded Lower Devonian brachiopods and spores. Along the Shackleton Coast, the Alexandra Formation consists of 270 m of unfossiliferous pale-yellow, pebbly, quartz arenites (Boucot and others, 1967). In the Darwin Mountains 405 m of Hatherton Sandstone overlies the Brown Hills Conglomerate (30 m thick). Farther north (Terra Nova Bay–McMurdo Sound area, Warren, 1969) pre-Permian rock types are very variable. Coarse roundstone conglomerate, breccia, and interbedded arkosic sandstone and siltstone may be of Silurian or Devonian age. In most sections of the Beacon Supergroup the lowest units are of Middle to Late Devonian age and are massive and cross-bedded quartz arenite with minor siltstone, attaining thicknesses of as much as 900 m (Boucot and others, 1967).

The Bowers Group or Supergroup (Sturm and Carryer, 1970; Bradshaw and Laird, 1983; Weaver and others, 1984) of northern Victoria Land is a sequence with great similarity to the Heritage Group and the Crashsite Group of the Ellsworth Mountains. This is the only other area in Antarctica where there is no unconformity between Cambrian and Ordovician strata. The upper 1,650 m consist of the Camp Ridge Quartzite, a white to red orthoquartzite and quartz conglomerate, with minor cleaved red argillite and grit; these rocks may be correlated with parts of the Crashsite Group. Lithologically the Camp Ridge quartzites are most similar to the Mount Twiss Member of the Howard Nunataks Formation in the Ellsworth Mountains, but because of the large distances separating the two areas, such a detailed correlation would be unreasonable. The Camp Ridge Quartzite is considered to be of Ordovician age (Bradshaw and Laird, 1983). It is interesting to note that the Bowers Group is localized along the boundary between the Ross and the Borchgrevink orogenic belts (Craddock, 1970).

The Cape Fold Belt in South Africa has long been regarded as a stratigraphic and tectonic equivalent of the Ellsworth Mountains (Craddock, 1970). Rocks of the Cape System (Silurian–lower Carboniferous) (Du Toit, 1937) would approximately correspond to the Crashsite Group and are about 3,000 m thick (Theron, 1962). The Silurian part of the section (Table Mountain Series) consists of white sandstone with quartz pebbles, and the Devonian/Carboniferous part (Bokkeveld Series, Witteberg Series) of interlayered ripplemarked sandstones and dark to red shales. The presences of thick, traceable units of shale and the relatively higher fossil content of the Cape System are the main points of difference from the Crashsite Group.

Some remarks on the environment of deposition

Sedimentary features indicate that the Crashsite Group was deposited in shallow to very shallow water. The question of whether the environment was totally marine or partially freshwater has not yet been resolved. Grindley and Warren's (1964) interpretation of the Camp Ridge Quartzite as the product of littoral marine and deltaic sedimentation also fits the Crashsite Group rather well. Williams (1969) suggested that the Elbow, Heiser, and Dover sandstones in the Pensacola Mountains were probably deposited at shallow depths in the epineritic environment, on a slowly subsiding epicontinental shelf. The scarcity of fossils in both the Crashsite Group and its equivalents in the Pensacola Mountains is puzzling in view of the fossiliferous nature of deposits of approximately equal age and roughly similar lithology in both Australia (Brown and others, 1968) and South Africa (e.g., Theron, 1962). One of the reasons for this may be that a considerable fraction of the rocks in the Ellsworth Mountains and in the Pensacola Range may be of freshwater origin.

Interdigitation of the Howard Nunataks Formation of the Crashsite Group with less mature graywacke formations, probably deposited from the west (Fig. 5), is another puzzling feature. It could indicate the presence of land areas on both sides of the area of Crashsite deposition. It is interesting to note that the probable extension of the SAMFRAU Geosyncline into South America is bordered to the south and north by metamorphic Precambrian basement. The upper Paleozoic graywacke-type rocks of the Antarctic Peninsula (Trinity Peninsula Series) apparently overlie high-grade metamorphic basement (Smellie, 1981) and could perhaps be regarded as part of a tectonic/stratigraphic belt whose equivalent would be to the west of the Ellsworth Mountains.

ACKNOWLEDGMENTS

The petrologic information is in part based on reports by H. Meyer, P. Barrett, J. J. Anderson, and Sonja Hills. Field notes from all the University of Minnesota expeditions formed the basis for the stratigraphic scheme. M. Hall and T. Bastien were part of the team that unraveled the Crashsite Group stratigraphy in the Heritage Range. I am grateful to Rosemary Bunker and Robyn Curham for typing and retyping the manuscript and to Roy Harris for drafting the figures.

APPENDIX

Representative stratigraphic sections

Stratigraphic units are numbered in order of sequence, with 1) the lowest. Specimen numbers refer to collections presently held at the Department of Geology and Geophysics, University of Wisconsin–Madison.

I. Section north of Bingham Peak, Webers Peaks, Heritage Range (East flank of Landmark Peak Syncline)

Mount Liptak Formation

12) 60 m + Massive, white to buff, coarse-grained quartzite with mud pebbles. Bedding planes stained black. Specimen 64-MH-61.

Landmark Peak Member (Howard Nunataks Formation)

11) 120 m Dark-gray quartzite alternating with light-gray to white massive quartzite containing quartz pebbles. Lowest parts of unit show slump folding. Specimens 64-MH-62, 63.

10) 84 m Light-colored coarse-grained quartzite with lenses of sandstone with carbonate matrix. Some layers of pebble conglomerate.

9) 150 m + Light-green quartzite alternating with dark-gray quartzite. The quartzites contain brown spherical concretionary bodies as much as 3 cm in diameter, giving the rocks a spotted appearance. The upper half of this unit contains a layer of maroon argillite with pebbles of dark limestone, quartz, and quartzite.

Mount Twiss Member (Howard Nunataks Formation)

8) 300 m Buff quartzite with argillite pebbles and concretionary bodies as in 9, intercalated with maroon argillite. Some of the quartzite display red mottling. Large-scale cross-bedding occurs at the base.

7) 150 m Green quartzite with little maroon material both in the quartzite and in the argillite intercalations. Specimen 64-MH-64. Some layers of this unit are slumped.

6) 330 m Red, pale green, and pink quartzite alternating with conspicuous layers of maroon argillite.

5) 30 m Massive unit with quartz-pebble conglomerate.

4) 9 m Massive fine-grained quartzite.

3) 120 m Quartzite, mottled red and green. Specimen 64-MH-65. Uppermost layers display distinctive ripple marks. One bed near the middle of the section has well-developed mud cracks; the polygons have average diameters of about 1 m.

Folded interval approx. 60 m.

Linder Peak Member (Howard Nunataks Formation)

2) 150 m Well-bedded pale quartzite with thin argillite beds.

1) 60 m Black argillite.

Marble of the Heritage Group.

II. Section on the western part of Skelly Peak (east limb of Windy Peak syncline, Heritage Range)

Landmark Peak Member (Howard Nunataks Formation)

12) 126 m Pale grayish-green quartzite with some intercalations of white to pale brown quartzite. Specimens 64-BS-11, 12

11) 405 m Thin-bedded, light greenish-gray quartzite with rusty iron-rich nodules. Green argillite partings. Some ripple marks present. Specimen 64-BS-13.

Impure sandstones (C, Fig. 2)

10) 300 m Medium to dark greenish-gray, finely laminated quartzite with cross-bedding but no well developed ripple marks. Some layers with argillite pebbles. Argillite interbeds are thin. Specimen 64-BS-14.

Mount Twiss Member (Howard Nunataks Formation)

9) 270 m Finely laminated red and green quartzite with scattered argillite pebbles.

Impure sandstones (B, Fig. 2)

8) 30 m Medium to dark greenish-gray, finely laminated quartzite.

Mount Twiss Member (Howard Nunataks Formation)

7) 60 m Green quartzite with only a few maroon argillite partings.

6) 60 m Pale green quartzite with 3-m bands of maroon argillite.
Some of the quartzites are mottled red and green. Samples 64-BS-15, 16.

5) 30 m Green quartzite with dark laminae and some intercalations of brown quartzite. Thin maroon argillites.

Impure sandstones (A, Fig. 2)

4) 120 m Impure quartzite, graywackelike, interbedded with argillite. Sandstone layers grade upward into argillite.

3) 24 m Finely bedded quartzite with ripple marks and thin argillite partings.

2) 120 m Impure quartzite, graywackelike, forming a regular intercalation with argillite. Sandstone layers grade upward into argillite. Some maroon bands present.

1) 18 m *Transition beds* Specimens 64-BS-17, 19.

Heritage Group, equivalent of Frazier Ridge Formation.

III. Mount Liptak, Sentinel Range (Type section)—Synclinal axis

10) 90 m Mount Wyatt Earp Formation

Mount Liptak Formation

9) 159 m Similar to 8 with a stronger reddish tinge; bedding is thinner (30 cm to 1.5 m).

8) 107 m Pinkish buff, medium-grained, pure quartzite, slightly more uniform and lighter in color than lower units. In massive beds 60 cm to 3 m thick.

7) 161 m Pinkish gray medium-grained quartzite in massive beds from 60 cm to 3 m thick.

6) 108 m Gray-buff to pale olive, medium-grained quartzite. Beds massive with some cross-bedding.

5) 108 m Buff medium-grained quartzite with grayish-green tinge in massive beds 60 cm to 3 m thick.

4) 54 m Medium-grained light grayish-green quartzite in beds 60 cm to 3 m thick.

3) 54 m Buff quartzite with grayish-green tinge. Some massive beds of pure medium-grained quartzite as much as 6 m thick.

2) 30 m Medium grayish-green quartzite, medium to coarse-grained thick-bedded with 15 cm (or less) thick beds of buff quartzite. Weathering produces red patches. The whole unit can be subdivided into subunits. From bottom to top they are:
a) medium grayish-dark olive quartzite
b) slightly finer-grained quartzite
c) buff-medium quartzite

1) *Howard Nunataks Formation*

IV. Mount Wyatt Earp Formation, Heritage Range

a) *West flank of syncline, Meyer Hills, 1962–63 expedition, stop 19*

6)		Whiteout Conglomerate.

Mount Wyatt Earp Formation

5)	75 m	Light buff, massive bedded, limonite stained quartzite weathering light tan. (63-W-218A)
4)	203 m	Light buff, massive, medium-grained to coarse-grained quartzite weathering light tan with 15-cm stringers of black argillite.
3)	5 m	Dark-gray argillite weathering grayish-brown.
2)	23 m	Light buff, massive quartzite.
1)	5 m	Layer with predominant argillite and some stringers of massive quartzite.

b) *West flank of syncline, Meyer Hills, 1962–63 expedition, stop 16*

6)		Whiteout Conglomerate.

Mount Wyatt Earp Formation

5)	18 m	Medium-grained, massive bedded, limonite-stained quartzite weathering reddish (63-W-216m).
4)	9 m	Buff, thinly laminated quartzite (63-W-216n).
3)	8 m	Black conglomerate unit. Matrix medium to coarse grained. (63-W-216o). Resembles Whiteout Conglomerate.
2)	8 m	Buff, massive-bedded, thinly laminated quartzite with stringers of black argillite.
1)	~150 m	Reddish-gray, massive bedded, thinly laminated quartzite.

c) *East flank of syncline, Meyer Hills, 1962–63 expedition, stop 20*

4)		Whiteout Conglomerate with dark gray matrix (63-W-220e).

Mount Wyatt Earp Formation

3)	15 m	Medium-gray, fine-grained quartzite, weathering dark gray (63-W-220d).
2)	18 m	Dark-gray, fine-grained quartzite, weathering reddish brown (64-W-220c).
1)		Light to medium-gray, fine-grained quartzite, weathering light tan (63-W-220a + b).

d) *East flank of syncline, south end of Meyer Hills*

6)		Whiteout Conglomerate with light green, sandy matrix. Abundant limestone pebbles. Sharp contact with underlying Crashsite Group.

Mount Wyatt Earp Formation

5)	6 m	Gray to pale brown, rusty weathering quartzite with contorted argillite laminae (64-BS-102).
4)	6 m	Conglomerate with gray to green sandy matrix, weathering rusty red (64-BS-103). Whiteout Conglomerate.
3)	15 m	Massive dark gray, fine-grained quartzite.
2)	60-90 m	Light gray, finely laminated, massive, buff weathering, quartzite (64-BS-104), intercalated with occasional beds of dark gray, brown weathering quartzite (64-BS-105).
1)		Same as 2) with two 6-m-thick layers of dark gray to black argillite. The two layers are spaced approximately 16 m apart (64-BS-106).

REFERENCES CITED

Boucot, A. J., Doumani, G.A., Johnson, J. A., and Webers, G. F., 1967, Devonian of Antarctica, *in* Oswald, D. H., ed., International Symposium on the Devonian System, Calgary, Alberta, 1967, v. 1: Calgary, Alberta Society of Petroleum Geologists, p. 639–648.

Bradshaw, J. D., and Laird, M. G., 1983, The pre-Beacon geology of northern Victoria Land; A review, *in* Oliver, R. L., James, P. R., and Jago, J. B., eds., Antarctic earth science: Canberra, Australian Academy of Science, p. 98–101.

Brown, D. A., Campbell, K.S.W., and Crook, K.A.W., 1968, The geological evolution of Australia and New Zealand: Oxford, Pergamon Press, 409 p.

Craddock, C., 1969, Geology of the Ellsworth Mountains, *in* Bushnell, V. C.,and Craddock, C., eds., Geologic maps of Antarctica: New York, American Geographical Society Antarctic Map Folio Series, Folio 12, Plate 4.

——, 1970, Antarctic geology and Gondwanaland: Antarctic Journal of the United States, v. 5, no. 3, p. 53.

Dalziel, I.W.D., and Elliot, D.H., 1982, West Antarctica; Problem child of Gondwanaland: Tectonics, v. 1, no. 1, p. 3–119.

Du Toit, A. L., 1927, A geological comparison of South America with South Africa, with paleontological contribution by F.R.C. Reed: Carnegie Institution Publication 381, 158 p.

——, 1937, Our wandering continents: Edinburgh, Oliver and Boyd.

Grindley, G. W., and Warren, G., 1964, Stratigraphic nomenclature and correlation in the western Ross Sea region, *in* Adie, R. J., ed., Antarctic geology: Amsterdam, North-Holland, p. 314–333.

Hjelle, A., Ohta, Y., and Winsnes, T. S., 1982, Geology and petrology of the southern Heritage Range, Ellsworth Mountains, *in* Craddock, C., ed., Antarctic geoscience: Madison, University of Wisconsin Press, p. 599–608.

Mirsky, A., 1969, Geology of the Ohio Range–Liv Glacier area, *in* Bushnell, V. C., and Craddock, C., eds., Geologic maps of Antarctica: New York, American Geographical Society Antarctic Map Folio Series, Folio 12, Plate 17.

Schmidt, D. L., and Ford, A. B., 1969, Geology of the Pensacola and Thiel Mountains, *in* Bushnell, V. C., and Craddock, C., eds., Geologic maps of Antarctica: New York, American Geographical Society Antarctic Map Folio Series, Folio 12, Plate 5.

Smellie, J. L., 1981, A complete arc-trench system recognized in Gondwana sequences of the Antarctic Peninsula region: Geological Magazine, v. 118, p. 139–159.

Sturm, A., and Carryer, S. J., 1970, Geology of the region between the Matusevich and Tucker glaciers, North Victoria Land, Antarctica: New Zealand Journal of Geology and Geophysics, v. 13, no. 2, p. 408–435.

Theron, J. N., 1962, An analysis of the Cape folding in the District of Willowmore, Cape Province: Annals of the University of Stellenbosch, series A, v. 37, no. 5.

Warren, G., 1969, Geology of the Terra Nova Bay, McMurdo Sound area, Victoria Land, *in* Bushnell, V. C., and Craddock, C., eds., Geologic maps of Antarctica: New York, American Geographical Society Antarctic Map Folio Series, Folio 12, Plate 14.

Weaver, S. D., Bradshaw, J. D., and Laird, M. G., 1984, Geochemistry of Cambrian volcanics of the Bowers Supergroup and implications for the Early Palaeozoic tectonic evolution of northern Victoria Land, Antarctica: Earth and Planetary Science Letters, v. 6B, p. 128–140.

Webers, G., and Spörli, K. B., 1983, Palaeontological and stratigraphic investigations in the Ellsworth Mountains, West Antarctica, *in* Oliver, R. L., James, P. R., and Jago, J. B., eds., Antarctic earth science: Canberra, Australian Academy of Science, p. 261–264.

Williams, P. L., 1969, Petrology of Upper Pre-Cambrian and Paleozoic sandstones in the Pensacola Mountains, Antarctica: Journal of Sedimentary Petrology, v. 39, no. 4, p. 1455–1465.

MANUSCRIPT ACCEPTED BY THE SOCIETY MARCH 1, 1989

Printed in U.S.A.

Geological Society of America
Memoir 170
1992

Chapter 4

Stratigraphy and sedimentology of the Whiteout Conglomerate; An upper Paleozoic glacigenic unit, Ellsworth Mountains, West Antarctica

Charles L. Matsch and Richard W. Ojakangas
Department of Geology, University of Minnesota, Duluth, Minnesota 55812

ABSTRACT

The upper Paleozoic Whiteout Conglomerate, situated conformably upon the Crashsite Group and overlain conformably by the Polarstar Formation, may represent the entire period of Gondwanaland glaciation in West Antarctica. The formation, mainly massive diamictites, makes up about 1,000 m of a 13,000-m-thick total stratigraphic column in the Ellsworth Mountains.

In the northern Sentinel Range, the Whiteout Conglomerate contains as many as six dark gray to black, thick, massive-bedded diamictites that are separated by recessive layers of laminated to thin-bedded shale, mudstone, and sandstone as thick as 15 m. Clasts greater than 5 mm in diameter compose 3 to 10 percent of most diamictites. Quartzite (66 percent) is the dominant clast type, with granite (17 percent), shale-argillite (8 percent), and carbonate (5 percent) composing most of the remainder. Some clasts are faceted and striated.

A truncated Whiteout section of about 250 m in the Meyer Hills to the south contains thinner diamictites with more variable rock types and a stratigraphic sequence punctuated by striated boulder pavements and thin laminated beds of shale, sandstone, and pebble lags. Of the pebbles in the diamictites, 47 to 86 percent are quartzites. Limestone and granite each compose about 20 percent. Conspicuous changes in pebble compositions occur across contacts marked by boulder pavements.

Petrographically, the diamictites are 62 percent silt and clay matrix with 38 percent sand grains and granules. The sand-sized grains have an average quartz/feldspar/lithic (Q:F:L:) ratio of 75:9:16; the studied specimens displayed no significant petrographic variations between the northern Sentinel Range and the Meyer Hills. However, there is a difference in megaclast compositions; in the northern Sentinel Range, shale-argillite and vein quartz are abundant, but they are absent in the Meyer Hills where carbonate is abundant. Till pellets, products of ablation on glaciers and icebergs, are common in bedded units and rare in the diamictites.

The Crashsite Group quartzite and Whiteout Conglomerate are dominated by quartz and quartz-rich lithic fragments, whereas the overlying Polarstar Formation contains abundant volcanogenic detritus, reflecting a dramatic change in source area. Contact zones with both enclosing formations contain scattered oversized clasts.

Matsch, C. L., and Ojakangas, R. W., 1992, Stratigraphy and sedimentology of the Whiteout Conglomerate; An upper Paleozoic glacigenic unit, Ellsworth Mountains, West Antarctica, *in* Webers, G. F., Craddock, C., and Splettstoesser, J. F., Geology and Paleontology of the Ellsworth Mountains, West Antarctica: Boulder, Colorado, Geological Society of America Memoir 170.

Striated boulder pavements in the Meyer Hills generally indicate paleo-ice flow to the present northwest. With the Ellsworth block restored to its probable late Paleozoic position via a 90° clockwise rotation, the glacier source area was the East Antarctica landmass now situated south of the Ellsworth Mountains. Sedimentologic similarities with formations in the Pensacola Mountains to the south indicate that the Ellsworth Mountains have moved northward at least 500 km from a hypothetical former position just east of the Pensacolas.

Our glacial model suggests that ice sheets generated the glacial sediments of the Whiteout Conglomerate under a polar climatic regime. The Meyer Hills sequence represents deposition directly by glaciers and also by floating ice and currents in a zone of fluctuating grounding line. Deeper water and sub-ice shelf conditions, i.e., glacial-marine, are postulated for the thick sequence of the northern Sentinel Range. There, periodic retreat of the ice shelf margin was accompanied by iceberg rafting and bottom current activity, probably in response to climatic warming.

INTRODUCTION

The Whiteout Conglomerate of the Ellsworth Mountains (Fig. 1) is one of several upper Paleozoic glacial deposits in Gondwanaland (Craddock and others, 1964; Frakes and others, 1971). In Antarctica, the Whiteout Conglomerate has been correlated with the Gale Mudstone of the Pensacola Mountains 500 km to the south (Frakes and others, 1971). It has also been correlated with glacigenic units of the Transantarctic Mountains to the southwest, including the Buckeye Tillite of the Wisconsin and Ohio Ranges (Frakes and others, 1971), with the Pagoda Formation of the southern Ross Ice Shelf area (Frakes, 1981a), and with unnamed units in southern and northern Victoria Land (Barrett and McKelvey, 1981; Laird and Bradshaw, 1981). Glacigenic rocks of this age have been noted at 14 different localities in Antarctica (Hambrey and Harland, 1981, p. 219).

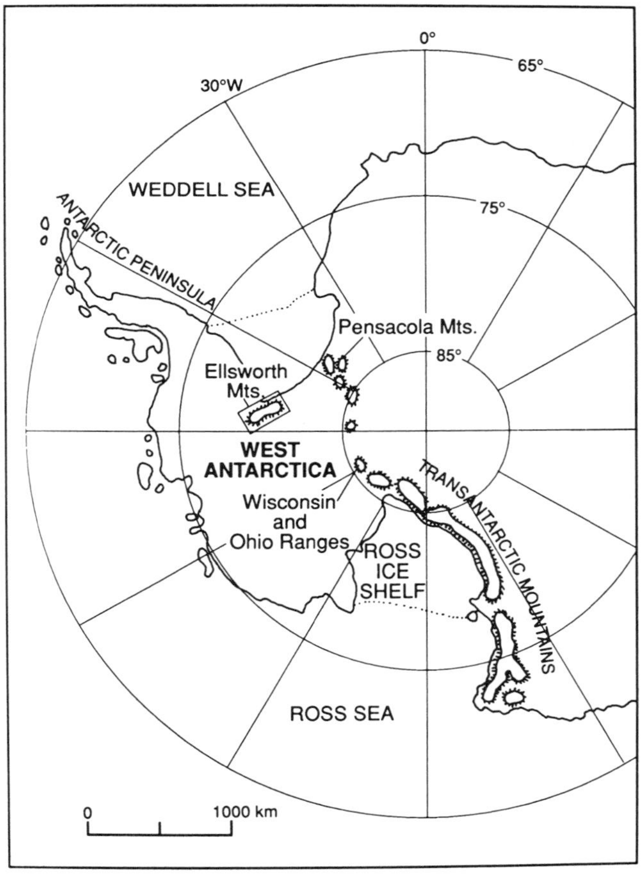

Figure 1. Generalized location map of Antarctica (from Nelson, 1981).

The Whiteout Conglomerate of the Ellsworth Mountains is exposed in the Sentinel Range and in the Meyer Hills of the Heritage Range (Fig. 2). We did detailed fieldwork in the northern part of the Sentinel Range and in the Meyer Hills but were unable to visit outcrops in the Flowers Hills and on Barnes Ridge in the southeastern part of the Sentinel Range, between the two study areas. This is unfortunate, for the Whiteout Conglomerate is quite different in the two study areas, and knowledge of the intervening area would have been useful in erecting the depositional model.

The Whiteout Conglomerate is Permo-Carboniferous in age, conformably situated between the overlying Permian Polarstar Formation with its Permian *Glossopteris* flora and the Crashsite Group quartzite (Spörli, this volume) that contains Devonian fossils at one locality (Webers and Spörli, 1983). The thickness of the Whiteout Conglomerate, therefore probably represents the entire record of the late Paleozoic glaciation of this part of Gondwanaland.

Paleomagnetic data do not yet allow direct determination of a paleolatitude for the rocks of the Ellsworth Mountains. Data from other continents, however, indicate that the pole was located in Antarctica during Late Carboniferous and Early Permian time (e.g., Crowell and Frakes, 1972; Caputo and Crowell, 1985).

REGIONAL STRATIGRAPHY AND STRUCTURE

The Whiteout Conglomerate is situated high in the 13,000-m-thick stratigraphic column of the Ellsworth Mountains, as shown in Figure 3 (Craddock and others, 1964; Webers and

Spörli, 1983). Clastic rocks dominate the sequence. Carbonates are present, however, in the 600-m-thick Upper Cambrian Minaret Formation in the southern Heritage Range. The Middle and Upper Cambrian Heritage Group forms the lower half of the column, consisting of five formations of volcanic and clastic rocks (Webers and Spörli, 1983). The upper part of the Heritage Group is the Minaret Formation. This is overlain by the 3,200-m-thick Upper Cambrian to Devonian Crashsite Group quartzite. The Permo-Carboniferous Whiteout Conglomerate, as thick as 1,000 m, overlies the Crashsite Group quartzite and is overlain by the 1,370-m-thick Permian Polarstar Formation. The column consists of a multitude of rock types that appear to have been deposited in shallow marine to terrestrial environments. No unconformities have been noted.

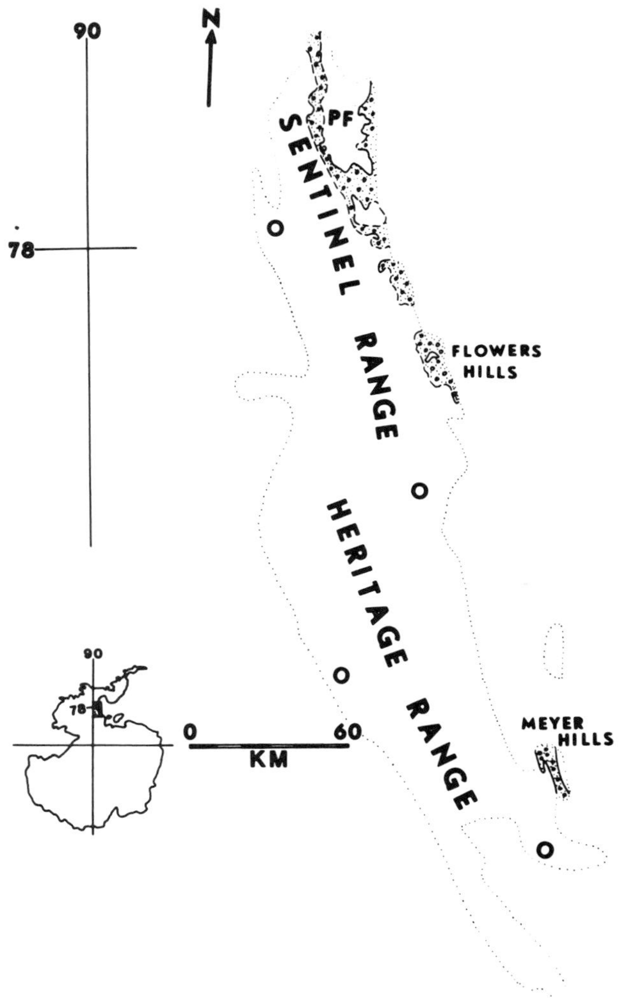

Figure 2. Generalized geologic map of the Ellsworth Mountains. The Whiteout Conglomerate (upper Paleozoic) is shown by the pattern. The younger Polarstar Formation (Permian) is designated "PF" on the largest outcrop area. Older units (Cambrian to Devonian) compose the rest of the two ranges and are designated by "O." The dashed lines are Whiteout contacts; the dotted line shows the boundary between rock exposures and glacial ice. (After Craddock and others, 1964.)

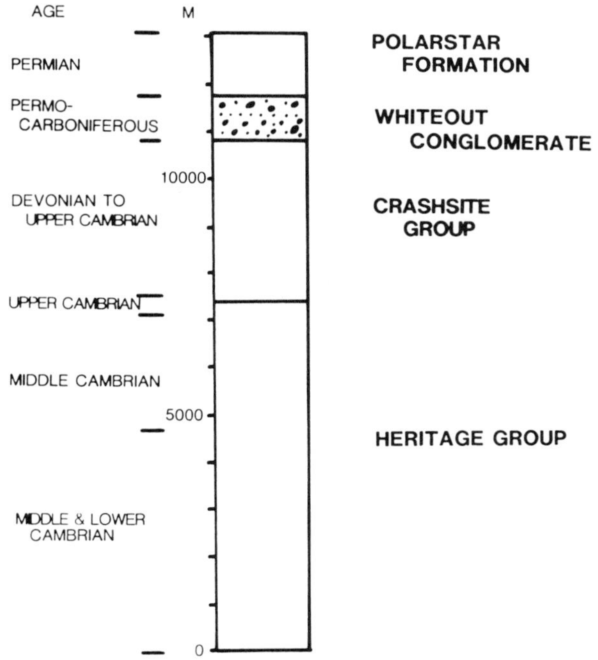

Figure 3. Generalized stratigraphic column, Ellsworth Mountains.

Frakes and others (1971) stated that the Paleozoic deposits of West Antarctica are located in three distinct basins, including the Ellsworth-Pensacola basin. Kamenev and Ivanov (1983), on the basis of various geophysical and geological surveys, portrayed the Ellsworth Mountains (Fig. 1) as the western portion of the sedimentary fill of the southern Weddell Sea Basin. They depicted the total sedimentary thickness at approximately 13.5 km, with the Early to Middle Paleozoic and the less dense Late Paleozoic to Early Mesozoic sequences composing 6 to 7 km of the section. Kamenev and Ivanov (1983) also stated that major faults are present within the basin, including one along the east side of the Ellsworth Mountains.

The Ellsworth Mountains may be on the site of an aulacogen that originated in the Late Proterozoic, underwent extensive downwarping in Early–Middle Paleozoic time, and then stabilized as a platform region in the Late Paleozoic (Kamenev and Ivanov, 1983). It was during this relatively stable time that the Whiteout Conglomerate and the overlying Polarstar Formation were deposited.

The Paleozoic sedimentary rocks of the region now make up the post-Permian Ellsworth Mountains fold belt, which probably formed during the Triassic Period (Craddock, 1983). Other workers have presented evidence for more than one deformation (Hjelle and others, 1982; Yoshida, 1982, 1983). The Whiteout Conglomerate was deformed into broad open folds with a pervasive cleavage, and continuous homoclinal ridges are well exposed. In the Sentinel Range, the dips of the two most intensively studied sections range from 10° to vertical; in the Meyer Hills, studied sections have dips of 20 to 50°.

There are general similarities in the stratigraphy and structure of the Ellsworth Mountains and the Pensacola Mountains (Craddock, 1983). The Ellsworth Mountains are separated from the Pensacola Mountains, which trend at right angles to the Ellsworth Mountains (Fig. 1), by a deep subglacial depression that may be a major fault whose presence has been interpreted on the basis of geophysical data to the west of this area (Jankowski and others, 1983). Displacement and rotation of the Ellsworth Mountains have been suggested by several workers, as summarized by Craddock (1983). Watts and Bramall (1981), on the basis of paleomagnetic data, suggested a 90-degree counterclockwise rotation of the Ellsworth block with respect to the East Antarctic craton. Doake and others (1983) reported that a major rift system separates the Ellsworth block from the Antarctic Peninsula to the north and suggested that this is a boundary between microcontinents.

FIELD DESCRIPTION: NORTHERN SENTINEL RANGE

Thickness and general lithology

Several field studies indicate that the Whiteout Conglomerate in the northern Sentinel Range (Fig. 4) is about 1,000 m thick. Craddock and others (1964) measured an incomplete section of 910 m on Mt. Lymburner, and a complete section of 1,122 m was reported to be exposed a few kilometers south on Mt. Warren (Frakes and others, 1971) (Fig. 4). We observed a composite section of about 1,000 m on Olsen Peak and Mt. Washburn (Ojakangas and Matsch, 1981). Detailed studies were accomplished on Whiteout Nunatak, which is just west of Matsch Ridge, and on the northwestern edge of Mt. Ojakangas, which lies between Mt. Ulmer and Mt. Washburn; the column was measured with a Jacob staff and a compass. There, as elsewhere in the Sentinels, the formation is made up largely of dark gray to black massive-bedded diamictite that displays well-developed penetrative rock cleavage. Minor but prominent layers of laminated to thin-bedded shale, mudstone, and sandstone containing scattered megaclasts interrupt the monotonous sequence of diamictites (Fig. 5). This heterolithic facies is concentrated in discrete stratigraphic intervals as thick as 15 m, but composes less than 5 percent of the total sections that we have observed. Backwasting of these thin units on steep exposures has produced narrow topographic benches. Two such units are displayed on the cirque-wall truncating Mt. Ojakangas (Fig. 6), but more are prominent on an inaccessible face of Mt. Warren (Fig. 7).

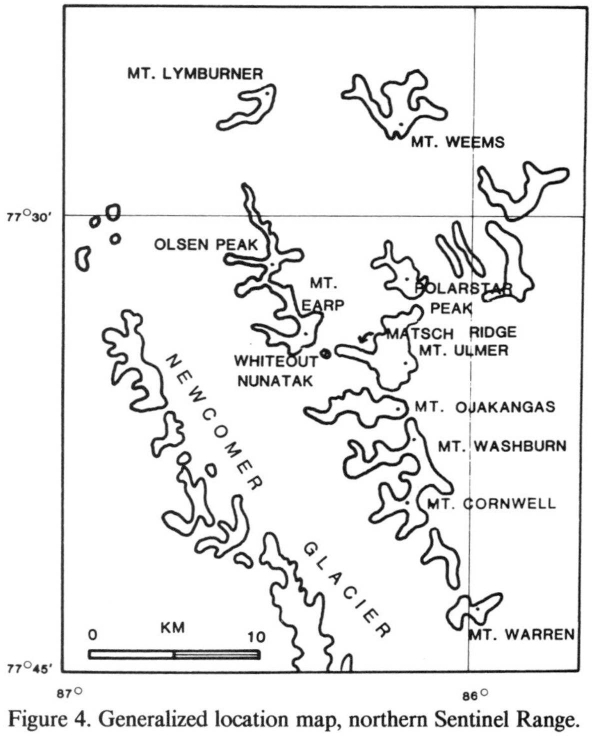

Figure 4. Generalized location map, northern Sentinel Range.

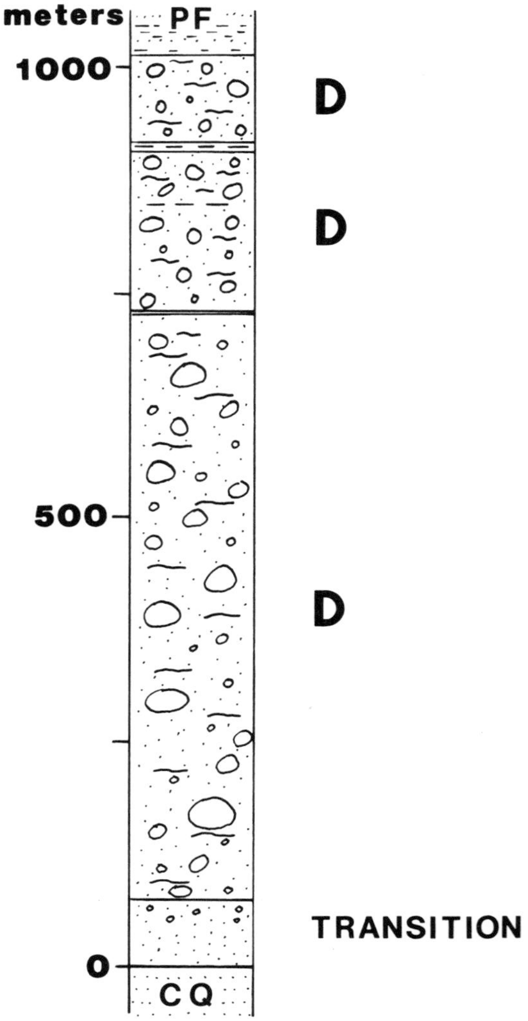

Figure 5. Generalized measured section of Whiteout Conglomerate, Sentinel Range. Units labeled "D" consist of diamictite. Heterolithic (laminated) units are shown by continuous horizontal lines. Note the transitional contact zone between the Crashsite Group quartzite (CQ) and the Whiteout Conglomerate.

Figure 6. Exposure of Whiteout Conglomerate at cirque on west side of Mt. Ojakangas in northern Sentinel Range (Fig. 4). Note three major diamictite members separated by benches. The upper laminated unit of Figure 5 is visible as the prominent snow-covered bench. The lower bench is barely visible just above the big snow patches at the base of the slope. The diamictite member between the two benches is 175 m thick.

Figure 8. Lonestone (upper left) in poorly bedded, fine- to medium-grained sandstone of Polarstar Formation. Photo taken about 15 m above contact with Whiteout Conglomerate, on Whiteout Nunatak (Fig. 4). Note prominent foliation parallel to hammer.

Contact relations

The contact between the Whiteout Conglomerate and the underlying Crashsite Group quartzite appears to be gradational (Fig. 5). It is best exposed on the ridge that extends westward from Olsen Peak. There the quartzite is typical of the formation, a light brown to gray bedded metasandstone, but the uppermost 75 m just beneath the Whiteout is a more massive, darker gray, finer-grained, sandy and clayey rock. In the upper third of this transitional unit, rare oversized quartzite clasts as much as 15 cm in diameter are present; they are most abundant in the upper 2 m. This transitional unit passes abruptly, yet apparently conformably, into a massive diamictite. The beds are nearly vertical, and no bedding surfaces were available to inspect for striations at the actual contact.

Figure 7. Aerial view of Mt. Warren (Fig. 4). The Crashsite Group quartzite is the thinner bedded unit exposed at the base, partially covered by ice. This is overlain by the Whiteout Conglomerate, which here consists of six thick diamictite members separated by thin units of laminated rocks. The contact with the overlying Polarstar Formation is at the upper right, at the bottom edge of the prominent snow-covered zone.

On a western spur of Mt. Washburn, the contact is slightly overturned and is knife sharp, but beds above and below it are parallel. We also interpret this contact to be a conformable one. Gradational contacts are also mentioned by Spörli (this volume).

The upper contact zone of the Whiteout with the Polarstar Formation is exposed on Whiteout Nunatak, where the beds dip at an angle of 50°. Some poorly developed thin black shales are present about 3 m below the top of the otherwise massive diamictite. The diamictite of the uppermost Whiteout Conglomerate is overlain abruptly by several meters of black shale of the basal Polarstar Formation. About 3 m above the contact, a 3-cm-thick graded sandstone bed contains oversized pebbles. A sandstone layer 10 m above the diamictite displays scattered oversized clasts as large as 10 cm in diameter (Fig. 8). We interpret the Whiteout-Polarstar contact as gradational.

Massive diamictite facies

In outcrop, the diamictite in the northern Sentinel Range appears to be remarkably homogeneous in texture and in clast composition (Fig. 9). Homogeneous diamictite layers are up to hundreds of meters thick. Clasts ranging from 2 cm to 2 m (rare) in diameter make up an estimated 3 to 10 percent by volume of the units (Fig. 10). Field studies of the composition of clasts in this size range indicate some variation in and between units (Table 1). In all units, quartzite is the dominant clast rock type, averaging 66 percent of the fragments. Clasts of granitic composition, including gneiss, are second in abundance, averaging 17 percent; lesser amounts of black shale-argillite (8 percent), carbonate (mainly dark gray limestone) (5 percent), and other minor

Figure 9. Massive diamictite of Whiteout Conglomerate. Location is upper diamictite member (Fig. 5) on Whiteout Nunatak, northern Sentinel Range (Fig. 4).

Figure 11. Faceted and striated graywacke boulder in diamictite of Whiteout Conglomerate. Location is Holt Peak, Meyer Hills; similar clasts are also present in the Sentinel Range.

rock types, including white vein quartz, phyllite, and chert, are present. See Frakes and others (1971) for comparison.

Clasts are generally subangular to subrounded, with angular and rounded shapes also well represented. Field time did not permit a systematic quantitative inventory of shape. Quartzite and limestone clasts tend to be better rounded than the other rock types, and large clasts more rounded than small ones. Some faceted and/or striated clasts (Fig. 11) are present. Some stones with crescentic gouges were also noted. Patches or clusters of pebbles are present but not common.

The black matrix, originally a mixture of mud and sand, apparently owes its color to the formation of chlorite and sericite during low-grade metamorphism. Some faint and wispy relicts of bedding in the form of clay-rich horizons (Fig. 12) suggest that the massive appearance of the diamictites is partially the result of metamorphic homogenization. A well-developed, steeply dipping cleavage is present (Fig. 13). Flat clasts tend to be oriented parallel to the cleavage planes. On Whiteout Nunatak and Mt. Ojakangas, calcareous concretions as large as 1 m by 30 cm were observed, with their long dimensions parallel to the cleavage.

Heterolithic facies

In the 450-m-thick section in the northern cirque on Mt. Ojakangas, two prominent topographic benches mark the locations of 10- to 15-m-thick sections of thinly bedded shale, siltstone, and sandstone (Fig. 14). These units (Figs. 5, 6, and 7) provide a means of subdividing the diamictite. The contacts with the diamictites are generally sharp.

The thin-bedded unit that is marked by the uppermost bench offers the best exposures of the heterolithic facies. It displays (1) graded beds of sand- to silt-sized grains, 2 to 3 cm thick; (2) pebble layers, commonly one stone thick; (3) black mud

Figure 10. Massive diamictite of Whiteout Conglomerate at Mt. Ojakangas, northern Sentinel Range (Fig. 4). The faint linear structures are the result of metamorphic foliation.

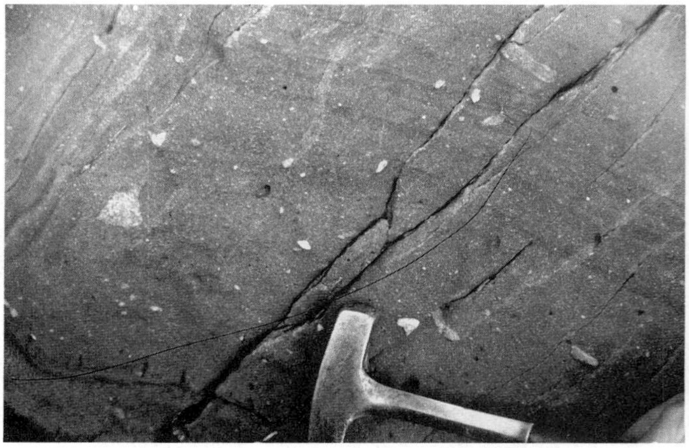

Figure 12. Indistinct and discontinuous "wispy bedding" in lower diamictite member at cirque on Mt. Ojakangas, northern Sentinel Range (Fig. 4).

layers, generally 0.25 to 2 cm thick; (4) dropstones, commonly 5 to 8 cm but as large as 20 cm (Fig. 15), with underlying laminae bowed downward and the overlying laminae draped over the stones, some of which pierce the underlying laminae; (5) thin diamictite layers as thick as 30 cm; and (6) soft-sediment deformational features including faults, load casts, and disturbed zones as thick as 2.5 m.

About 128 m above the base of the middle diamictite unit is a heterolithic sub-unit 2.5 m thick, with a great variety of thin beds and lonestones, some of which are clearly dropstones. Some beds are graded, some are cross-bedded, and others are horizontally laminated (Fig. 16). A few shallow ENE–trending channels with a few centimeters of relief, smaller-scale scour features, and convolute beds are also present. Some 17 m higher are two thin-bedded silty units about 10 cm thick, separated by 1 m of diamictite.

TABLE 1. TWENTY COUNTS OF MEGACLASTS (>2 cm) IN DIAMICTITES OF THE WHITEOUT CONGLOMERATE*

	Quartzite	Granitic	Shale-argillite	Carbonate	Vein quartz	Other
Upper Member						
Whiteout Nunatak	71	8	8	10	3	—
Whiteout Nunatak	75	12	7	2	3	1
Whiteout Nunatak	73	14	10	—	3	—
(N = 3) Average	73	11	8	6	3	X
Middle Member						
Mt. Ojakangas	58	18	6	7	9	2
Mt. Ojakangas	66	20	4	3	7	—
Mt. Ojakangas	58	17	13	2	4	6
(N = 3) Average	61	18	8	4	6	3
Lower Member						
Mt. Ojakangas	69	22	3	—	6	—
Mt. Ojakangas	53	20	18	5	8	—
Mt. Ojakangas	54	20	11	7	7	1
Olsen Peak	58	26	1	2	12	1
Olsen Peak	55	24	7	7	3	4
Olsen Peak	66	18	5	5	6	—
(N = 6) Average	59	22	8	4	7	1
Northern Sentinels Average	63	17	8	4	6	1
Meyer Hills—East Side						
Uppermost	48	25	1	23	3	—
Above Boulder Pavement 3	86	7	—	3	—	4
Above Boulder Pavement 2	47	21	—	25	7	1
Below Boulder Pavement 2	72	21	—	4	2	1
Above Boulder Pavement 1	60	21	—	14	7	2
Below Boulder Pavement 1	47	15	—	34	—	3
(N = 6) East Average	60	18	X	17	3	1
Meyer Hills—West Side						
Above Boulder Pavement 2	63	18	—	18	1	—
Above Boulder Pavement 1	58	14	—	27	1	—
(N = 2) West Average	60	16	—	23	1	—
Meyer Hills Average	60	18	X	18	2	1
Overall Average	62	18	5	10	5	1

*Each count represents 100 or more clasts.
X denotes trace amounts.

Figure 13. Subvertical foliation (cleavage) in massive diamictite of Whiteout Conglomerate, Mt. Ojakangas. Note geologist at lower left for scale, faint bedding at upper left, and sparse megaclasts.

Figure 15. Boulder-sized dropstone in laminated unit of Whiteout Conglomerate. Note pencil for scale. Location is in upper laminated units (Fig. 5) on Mt. Ojakangas, northern Sentinel Range (Fig. 4).

Figure 14. Well-bedded pebbly sandstone, siltstone, and black shale of Whiteout Conglomerate. Note few small lonestones. Coin is 18 mm in diameter. Location is in upper laminated unit (Fig. 5) on Mt. Ojakangas, northern Sentinel Range (Fig. 4).

Figure 16. Bedding in thin laminated unit of Whiteout Conglomerate. Note small-scale cross-beds at lower center, coarser graded sandstone bed at center, thinner graded beds above and below coin, and dropstone. Coin is 20 mm in diameter. Location is within middle diamictite member (Fig. 5) on Mt. Ojakangas (Fig. 4).

FIELD DESCRIPTION: MEYER HILLS, SOUTHERN HERITAGE RANGE

Thickness and general lithology

A partial section encompassing about 250 m of the basal part of the Whiteout Conglomerate was measured on the east side of the Meyer Hills (Fig. 17). The Polarstar Formation is not present.

The Whiteout Conglomerate is markedly different from that noted in the northern Sentinel Range: The stratigraphic sequence is more variable; individual diamictite units are thinner and display greater diversity in megaclast composition, volume, and original matrix texture; the sequence is punctuated by striated and faceted one-stone-thick boulder pavements; and it contains at least one channel-shaped conglomerate.

Two laminated sequences and four boulder pavements were noted within the diamictites. These stratigraphic breaks are continuous for more than 1 km (Fig. 18).

Contact relations

The contact of the Whiteout Conglomerate with the underlying Crashsite Group quartzite is well exposed on the eastern edge of the Meyer Hills (Figs. 18 and 19). An exposed thickness of 60 m of light gray quartzite is overlain by 6 m of sparsely pebbly diamictite with a clayey sand matrix, which is in turn overlain by 40 m of dark gray to black fine-grained quartzite beneath the main diamictite (Figs. 17 and 20). This interbedding of diamictitelike rock and quartzite was also noted in the 1962–1963 University of Minnesota expedition (Spörli, this volume, appendix).

A poorly exposed wavy contact with up to 15 m of relief was observed between the quartzite and the basal diamictite. This may be an erosional surface, but there is no evidence that the sand was lithified prior to such an erosional event. Therefore, we consider the contact as essentially a transitional one.

Massive diamictite units

Six massive diamictite units make up over 95 percent of the section measured in the Meyer Hills (Fig. 17). The dominant lithology is a dark gray to black diamictite with clasts greater than 0.5 cm in diameter composing between 3 and 25 percent of the total volume. Clasts greater than 25 cm in diameter are common. The largest clast observed has a long dimension of about 5 m. Most clasts are subangular to subrounded.

Quartzite clasts, predominantly white, light gray, and tan, make up an average of 60 percent of the fragments larger than 0.5 cm and range between 47 and 86 percent (Table 1). Next in importance are limestone and granite, averaging about 20 percent each. Minor constituents include white vein quartz, argillite, and phyllite. Variations in both quantity and lithology of megaclasts are noticeable in the field, with conspicuous changes occurring across contacts marked by boulder pavements. The diamictite below the second boulder pavement, for example, contains 72 percent quartzite and 4 percent carbonate, whereas that above has 47 percent quartzite and 25 percent carbonate. Lateral variations beyond a few hundred meters were not determined. Although significant differences in relative quantities of quartzite and carbonate exist between units, the percentage of granite clasts is fairly constant throughout the section. Our investigations did not detect the appearance across contacts of any new lithologies worthy of consideration as provenance indicators.

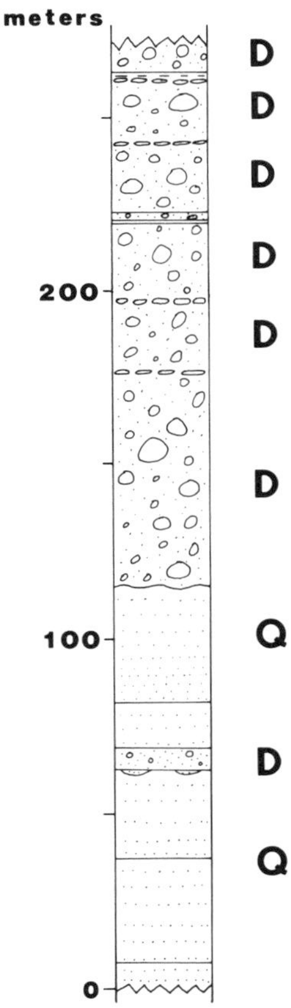

Figure 17. Generalized measured section of Whiteout Conglomerate in Meyer Hills. Units labeled "D" are diamictites, and those labeled "Q" are quartzites. Note also the four boulder pavements and the two laminated units. More rock is exposed above and below the measured section. At least the lowest quartzite unit is part of the Crashsite Group.

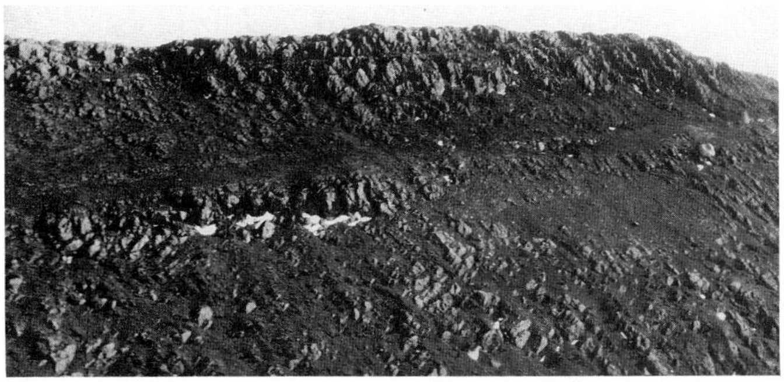

Figure 18. Whiteout Conglomerate conformably(?) overlying Crashsite Group quartzite on east side of Holt Peak (Fig. 19) in Meyer Hills. The prominent bench crossing center of view is the contact. Upper subhorizontal line (near horizon) is the second boulder pavement of Figure 17. The intervening diamictite, actually two separate units, is about 80 m thick.

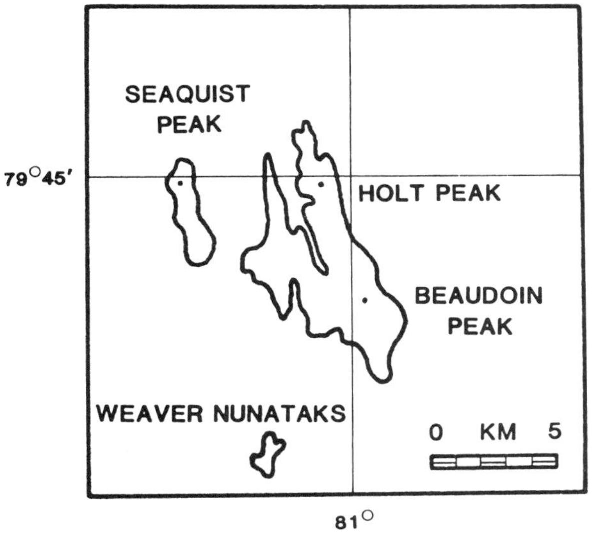

Figure 19. Generalized location map, Meyer Hills.

Figure 20. Crashsite Group quartzite on east side of Holt Peak (Fig. 19) in Meyer Hills. Note quartzite beds at upper left. "Diamictite" unit with lonestones occupies most of view. This is pebbly quartzite shown between quartzites in Figure 17 and indicates a transitional contact.

Figure 21. Laminated unit of well-bedded pebbly sandstone and siltstone between massive diamictite units of Whiteout Conglomerate in Meyer Hills. Note hammer for scale. Backwasting has produced a topographic bench. This laminated unit is shown on Figure 16 between the second and third boulder pavements. Location is east side of Holt Peak (Fig. 19).

Some striated and faceted clasts are present. A fabric study in one of the diamictites revealed a bimodal distribution of the long axes of the megaclasts. The black matrix of the diamictites was a poorly sorted mixture of mud and sand, now somewhat recrystallized.

Noticeable differential weathering and erosion of the diamictite units have resulted in prominent breaks in the slope profile. Small-scale relief up to a meter or so is expressed as miniature tors where uneven spacing of cleavage planes has resulted in alternating zones of weaker and stronger rock.

Heterolithic facies

Laminated to thin-bedded shale, sandstone, diamictite, and pebble bands occur at a few intervals interbedded with the massive diamictites. In the Meyer Hills the highest of these intervals is about 4 m thick (Fig. 21), with a boulder concentration one stone thick at its base. Sorting ranges from good to poor across beds a few centimeters thick. Thin, sandy diamictites are interspersed among poorly sorted conglomerate, conglomeratic sandstone, sandstone, siltstone, and laminated sandstone and siltstone with dropstones.

The pebble bands, made up of generally angular fragments of quartzite and granite (Fig. 22), are expressed on the outcrop as discontinuous lines parallel to more well developed bedding planes. They are a common feature of the thin-bedded shales, siltstones, and sandstones and occur in association with dropstone-pierced laminites. A few pebble bands were noted in thin diamictite beds that otherwise would be without discernible internal bedding.

Boulder pavements

Three well-exposed, extensive boulder pavements, and one less distinct zone of concentrated boulders, are present in the measured section on the eastern ridge north of Beaudoin Peak (Fig. 19). Two pavements also are exposed in the lower part of the Whiteout section on the westernmost ridge south of Seaquist Peak (Fig. 19). The latter are tentatively correlated with the two lowest boulder pavements on the eastern ridge.

The pavements, visible in three dimensions at some spots, are one stone thick (Fig. 23). Individual boulders are generally on the order of 25 cm in diameter, but rare clasts are up to several meters in diameter. The boulders are predominantly quartzite, but some are granite and limestone. In one pavement, 86 percent of the boulders are quartzite, 7 percent granite, and 3 percent carbonate. The upper surfaces of the boulders are characteristically polished and striated facets parallel to the contact with the overlying diamictites. Crescentic-shaped friction marks are displayed on a few of the striated facets (Fig. 24).

Figure 22. Close-up of laminated unit of Whiteout Conglomerate, showing distinctive pebble and cobble concentrations, probably lag deposits produced by current winnowing between episodes of deposition. Some of the clasts are clearly in-place dropstones. Same unit as in Figure 21.

Figure 23. Second boulder pavement in Whiteout Conglomerate of Figure 17 composed largely of white, light gray, and tan quartzite boulders. Black spots in middle distance are two geologists, and large boulder in foreground is about 50 cm in diameter. Location is east side of Holt Peak, Meyer Hills (Fig. 19).

Figure 24. Striated quartzite boulder with crescentic gouges, on second boulder pavement of Whiteout Conglomerate. Glacier ice is interpreted to have moved toward top of photo (Azimuth 325°). This boulder is the same one as in the foreground in Figure 23.

Channel deposit

A small channel-shaped conglomerate and conglomeratic sandstone unit approximately 1 m wide at its top and 0.5 m thick was noted at the top of the diamictite beneath the second boulder pavement (Fig. 25). It grades from a boulder lag at the base of the deposit to coarse sandstone with trough cross-bedding. The quartzite boulders and cobbles are subrounded to rounded. The channel is elongate in a northwest-southeast (310° azimuth) direction.

Figure 25. Sandstone-filled channel cut into diamictite beneath second boulder pavement of Whiteout Conglomerate on east side of Holt Peak, Meyer Hills (Fig. 19).

PETROGRAPHY

Crashsite Group quartzite

Only the uppermost part of the Crashsite Group quartzite was studied. It is composed of sandstones of variable mineralogical maturity, with the Q:F:L (quartz: feldspar: lithic fragment) ratio ranging from 100:0:0 to 63:28:9 (Table 2). The counted samples are orthoquartzite, feldspathic sandstone, and arkose (Figs. 26 and 27).

Most of the quartz is monocrystalline, but small amounts of polycrystalline varieties are present. Orthoclase dominates over plagioclase, and felsic volcanic rock fragments are the dominant type of rock fragment. Quartz is the main cement, and clay matrix (sericitic) is common.

No obvious differences were noted between Crashsite samples of the northern Sentinel Mountains and the Meyer Hills. The muddy sandstone unit near the top of the Crashsite in the Meyer Hills looks like a massive, pebble-poor diamictite in the field, but petrographic study shows it to be a sandstone with 30 percent clayey matrix. Small clots of diaspore(?) were noted in some thin sections; their scattered but uniform distribution suggests a texture developed during low-grade metamorphism.

Whiteout Conglomerate

The matrix of the diamictite is here broadly defined in the field as all grains smaller than 2 cm in diameter. This less-than-2-cm fraction was also studied in thin section to determine the relative abundances of granules of sand, silt, and clay; for this purpose, matrix is defined as the fraction finer than 0.03 mm in diameter. In 10 counted thin sections of diamictite (Table 2), the dark matrix averages 62 percent (56 percent clay—chlorite and sericite; 6 percent silt); the remainder consists of sand grains and coarse silt plus a few granules and small pebbles (Fig. 28). The

TABLE 2. MODAL ANALYSES OF THIN SECTIONS OF DIAMICTITES AND QUARTZITES*

Stratigraphic Unit	Whiteout Conglomerate												Crashsite Quartzite			
Rock Type	Diamictite											SS CH*	Quartzite			
Location	Sentinels			Meyer Hills									Sentinels	Meyer Hills		
Sample Number	OM-79-212	OM-79-211	OM-79-230	OM-79-5B	OM-79-8	OM-79-7	OM-79-83	OM-79-53	OM-79-69	OM-79-70	Aver.	OM-79-232	OM-79-238A	OM-79-1	OM-79-2C	W-79-66
Quartz																
Monocrystalline	25	23	21	23	19	33	26	20	21	21	23	33	56	70	53	82
Polycrystalline	4	3	1	2	X	3	4	3	–	1	2	4	1	5	1	7
Total	29	26	22	24	20	36	30	23	21	22	25	37	57	75	54	89
Feldspar																
Plagioclase (altered)	X	2	23	X	X	X	1	2	1	1	1	1	6	1	4	–
Orthoclase (altered)	1	2	3	1	3	2	2	3	4	3	2	3	10	6	19	–
Perthite (altered)	X	X	–	X	1	1	–	X	X	–	X	1	–	–	2	–
Total	2	4	5	2	4	3	3	5	5	4	4	5	16	7	24	–
Rock Fragments																
Volcanic, felsic	X	1	–	3	1	–	–	X	X	X	1	4	3	–	6	–
Plutonic (Orthoclase-quartz)	–	X	2	–	–	2	2	5	1	X	1	4	1	–	1	–
Metamorphic (schist)	X	X	2	X	1	–	1	3	8	1	2	4	1	–	1	–
Sedimentary-mudst-silt	X	–	–	–	2	–	X	–	–	–	X		1	–	–	–
sandstone	X	1	–	X	4	–	–	–	1	3	1	8	–	–	X	–
carbonate	X	1	3	6	3	1	–	X	X	1	2	1	–	–	–	–
Total	2	3	6	10	11	3	3	9	11	5	6	20	5	–	8	–
Miscellaneous																
Garnet	2	X	–	2	X	2	1	X	X	X	1	–	–	–	–	–
Other heavies, micas	1	X	1	1	–	X	1	X	1	–	1	–	1	–	1	–
Matrix-Cement																
Carbonate	–	1	1	2	X	1	X	2	2	1	1	38	–	X	–	–
Quartz	–	–	–	–	–	–	–	–	–	–	–	–	2	13	7	4
Clay	58	60	59	53	62	53	56	52	50	62	56	–	20	5	6	6
Silt	6	6	5	6	4	3	6	8	9	6	6	–	–	–	–	X

*600 points counted on each thin section. The column headed Aver. gives the averages for the 10 diamictites. The heading SS CH delineates a channel sandstone. X denotes trace amounts.

Figure 26. Photomicrograph of quartz-rich sandstone in Crashsite Group quartzite, Meyer Hills. Large vein quartz pebble at left. Field of view 2.8 mm wide. Crossed nicols. Specimen was collected from east side of Holt Peak.

TABLE 3. THIN-SECTION ANALYSIS OF THE FRAMEWORK GRAINS* IN MATRIX OF DIAMICTITES†

	Grains Q: F: L	Lithic Fragments S: V: P: M
Sentinel Mountains		
OM-79-230 Diamictite (Upper)	67:15:18	46:00:24:30
OM-79-211 Diamictite (Middle)	80:11:09	58:26:05:10
OM-79-212 Diamictite (Lower)	88:06:06	66:16:00:16
Meyer Hills		
OM-79-70 Diamictite	67:06:27	64:34:00:03
OM-79-8 Diamictite	58:11:31	82:08:00:10
OM-79-7 Diamictite	87:06:07	35:00:65:00
Average	71:12:17	44:12:22:22

*Sand and gravel.
†i.e., matrix between megaclasts.
Q:F:L is quartz:feldspar:lithic fragment ratio.
S:V:P:M is ratio of sedimentary:volcanic:plutonic:metamorphic rock fragments, from Table 2.

sand-, granule-, and pebble-sized grains, all floating in the finer-grained matrix, consist largely of quartz; the Q:F:L ratios (with plutonic rock fragments divided equally between feldspar and quartz) average 71:12:17. In general, the diamictites have a mineralogical maturity comparable to parts of the underlying Crashsite Group quartzite (Table 2), which has an average Q:F:L ratio of 82:14:4.

Monocrystalline quartz, subangular to rounded, is the most abundant grain component. Polycrystalline quartz (minor) includes sutured and recrystallized types, plus minor chert. Plagioclase and orthoclase are present in subequal amounts, usually totaling less than 5 percent; although feldspar is indicated on Table 2 as altered, some grains are fresh. Sedimentary fragments, mostly metasandstone, are the dominant type of lithic fragment. The average ratio of sedimentary:volcanic:plutonic:metamorphic rock fragments is 44:12:22:22, although each type has a considerable range (Table 3). The original shapes of sand grains are commonly somewhat modified by recrystallization and tectonic foliation (Fig. 29).

The nature of the sandstone rock fragments varies considerably. Of special interest are some that are quite similar to the sandstones of the Upper Cambrian Heritage Group at Frazier Ridge (Figs. 30 and 31). Other fragments resemble some varieties of the Crashsite Group quartzite (see Figs. 27 and 32). One calcareous sandstone fragment contains a smaller sandstone fragment, thus showing evidence of three cycles of sedimentation (Fig. 33). Calcareous sandstone fragments and clayey sandstone fragments are also present. All of these types of sandstones were included in the field term "quartzite" in the megaclast counts of Table 1. Some of the sand grains classified as polycrystalline quartz grains were probably derived from metaquartzites.

Oolitic limestone fragments resembling those described from the Minaret Formation at the top of the Cambrian Heritage

Figure 27. Photomicrograph of feldspathic sandstone in Crashsite Group quartzite, northern Sentinels. Field of view is 2.8 mm wide. Crossed nicols. Specimen was collected from western ridge of Olsen Peak (Fig. 4).

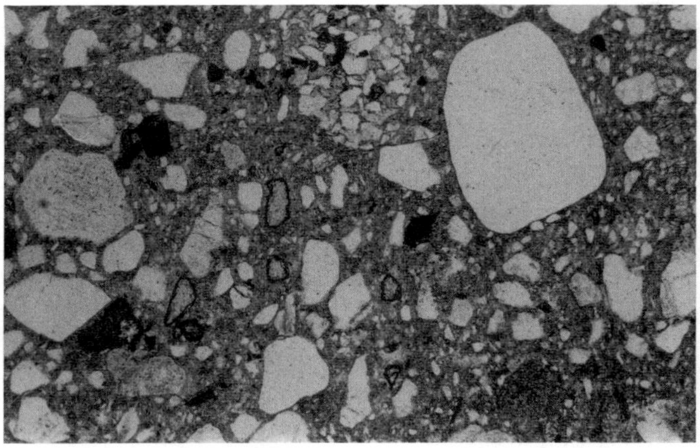

Figure 28. Photomicrograph of representative diamictite of the Whiteout Conglomerate. Note rounded quartz grains, sandstone rock fragment, and grains of garnet. Field of view 2.8 mm wide. Plane polarized light. Specimen was collected from lowest major diamictite unit of Figure 17, on east side of Holt Peak (Fig. 19).

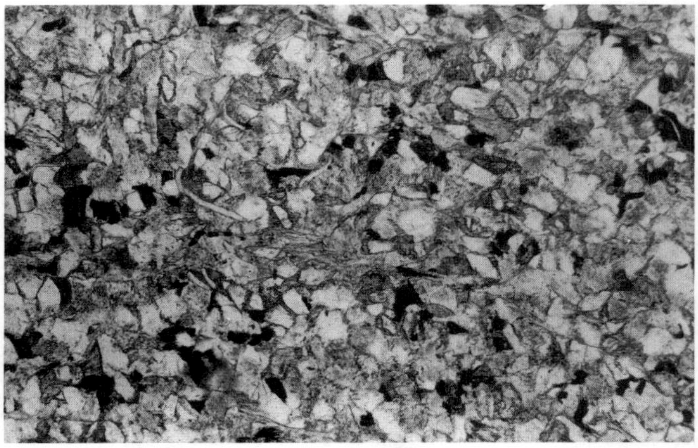

Figure 30. Photomicrograph of sandstone from Cambrian Heritage Group on Frazier Ridge at about center of Ellsworth Mountains. Field of view is 2.8 mm wide. Plane polarized light.

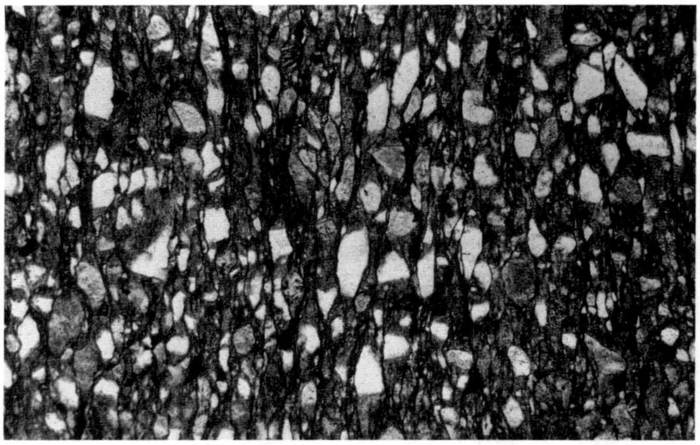

Figure 29. Photomicrograph of foliated diamictite of the Whiteout Conglomerate. Note orientation of grains parallel to foliation. Field of view is 2.8 mm wide. Plane polarized light. Specimen was collected on western ridge of Olsen Peak (Fig. 4), 2 m below lowest diamictite member of Figure 5.

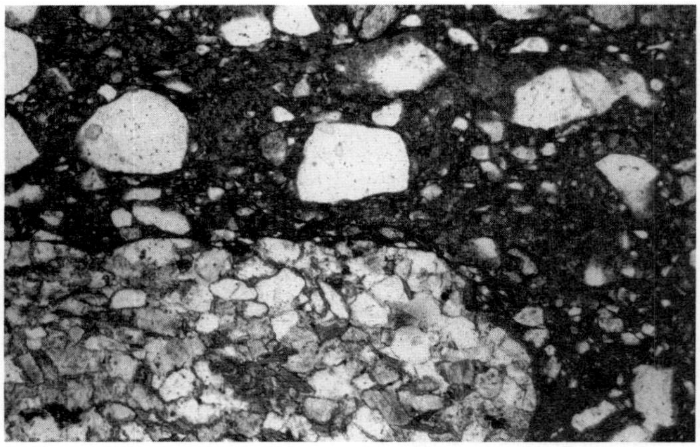

Figure 31. Photomicrograph of sandstone clast in diamictite of Whiteout Conglomerate. Note general resemblance to Cambrian sandstone of Figure 30. Field of view is 2.8 mm wide. Plane polarized light. Specimen was collected from lower part of the lowest diamictite member (Fig. 5) on the western ridge of Olsen Peak (Fig. 4).

Group of the southern Heritage Range by Buggisch and Webers (1982) are present (Fig. 34); some show appreciable silicification. Some carbonate fragments show a micritic texture, but most are recrystallized.

Schistose metamorphic rock fragments (including some slate and argillite, on Table 1) are minor. Garnet is also present in most diamictite samples; in amounts to 2 percent (see Table 2 and Fig. 28) some of the grains were probably derived directly from metamorphic rocks (Fig. 35), whereas others may have been recycled from older garnet-bearing sandstones such as those shown in Figure 36. Some Crashsite sandstones are garnet rich and are possible sources. Zircon, tourmaline, and rutile are also present, generally as subrounded to rounded grains, suggesting derivation in large part from older sandstones. Plutonic rock fragments generally consist of quartz and K-feldspar, but a few contain plagioclase.

Till pellets similar to those described and interpreted by Ovenshine (1970) as products of ablation on glaciers or icebergs were noted in several thin sections from the heterogeneous bedded units of both the Sentinel Range and the Meyer Hills (Fig. 37), but they did not appear in the point counts. Recrystallization of the homogeneous clayey matrix may have obliterated till pellets in the diamictite, but some are still recognizable (Fig. 38). Development of prominent foliation in the diamictites may also have eliminated some pellets and perhaps produced some "pseudo" till pellets.

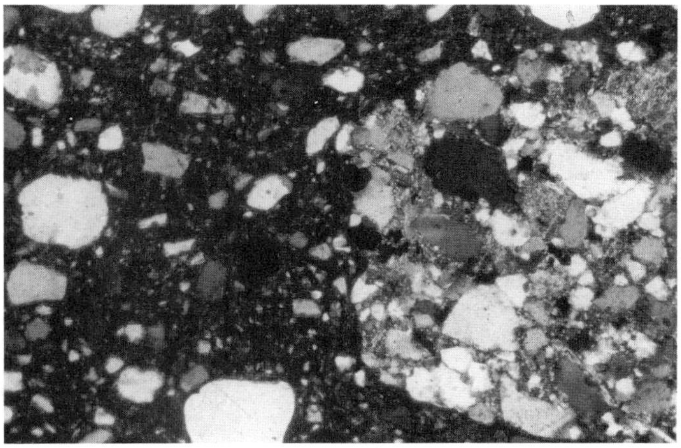

Figure 32. Photomicrograph of sandstone clast in diamictite of Whiteout Conglomerate. Note general resemblance to Crashsite Group quartzite of Figure 27. Field of view 2.8 mm wide. Crossed nicols. Specimen was collected from third major diamictite member (Fig. 17) on east side of Holt Peak, Meyer Hills (Fig. 19).

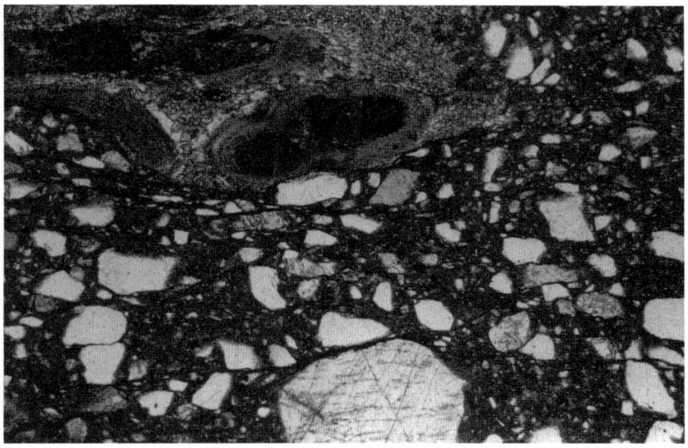

Figure 34. Photomicrograph of oolitic limestone clast (top) and perthite (bottom) in diamictite of Whiteout Conglomerate. Field of view 2.8 mm wide. Plane polarized light. Specimen was collected from upper diamictite member (Fig. 5) on Mt. Ojakangas, northern Sentinel Range (Fig. 4).

Polarstar Formation

The Polarstar Formation has been studied by Collinson and others (1980; this volume) and by Vavra and Collinson (1981). They determined that the formation had a dominant felsic to intermediate calc-alkaline volcanogenic provenance that was probably a volcanic arc along the Pacific margin of Gondwanaland, as modeled, for example, by Elliot (1975). Another possible but probably less likely model is that there was a realignment of the continental drainage on East Antarctica after the Gondwanaland glaciers receded, with volcanic-rich source areas becoming prominent.

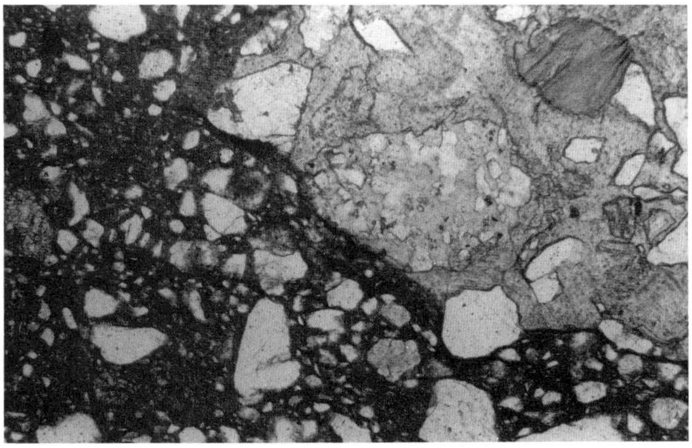

Figure 33. Photomicrograph of calcareous sandstone clast in diamictite of Whiteout Conglomerate. Note clast of finer sandstone in the sandstone clast. Field of view 2.8 mm wide. Plane polarized light. Specimen was collected from the second major diamictite member (Fig. 17) on east side of Holt Peak, Meyer Hills (Fig. 19).

During the present investigation, the lower part of the formation was studied in the northern Sentinel Range where it conformably overlies the Whiteout Conglomerate. Abundant volcanogenic detritus appears in the lowermost sandstones on Whiteout Nunatak within 15 m of the contact (Fig. 39), showing an abrupt injection of volcanogenic detritus into the basin of deposition. The change in source area is also shown by the very low abundance of garnet, a minor but obvious component in the Whiteout Conglomerate, in the Polarstar. Rare megaclasts, together with possible till pellets, occur in the lower 15 m of the Polarstar Formation (Figs. 8 and 40).

PALEOCURRENTS

Paleocurrent data from the Ellsworth Mountains are few, but they indicate broadly consistent paleocurrent trends toward the north in the Crashsite Group quartzite, the Whiteout Conglomerate, and the Polarstar Formation (Fig. 41). Readings were corrected with a one-tilt rotation.

Cross-beds in the upper 100 m of the Crashsite Group quartzite in the Meyer Hills (N = 18) indicate paleocurrents were toward the north. Most measurements were from trough cross-beds; only two trough axes were measurable. In the overlying Whiteout Conglomerate, glacial paleocurrent trends are based on striations measured on three successive boulder pavements (Fig. 42). The lowest pavement shows a strong east-west trend (N = 6), and the next two have a strong northwest-southeast trend (N = 34 and N = 9, respectively). Rare crescentic gouges on the second pavement indicate glacial movement toward an azimuth of 315° to 325° (Fig. 24), with stones aligned generally parallel to the striations scribed onto them.

However, stone orientations *within* the diamictites may be

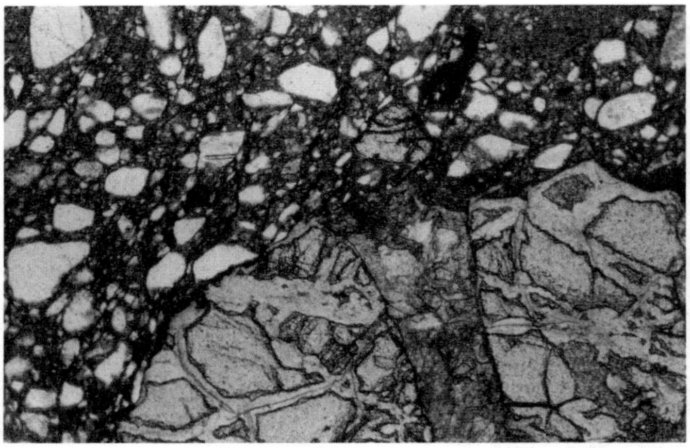

Figure 35. Photomicrograph of garnet-chlorite-carbonate metamorphic rock fragment in diamictite of Whiteout Conglomerate. Also note the free garnet grain just above the rock fragment. Field of view 2.8 mm wide. Plane polarized light. Specimen was collected from fifth major diamictite unit (Fig. 17) on east side of Holt Peak, Meyer Hills (Fig. 19).

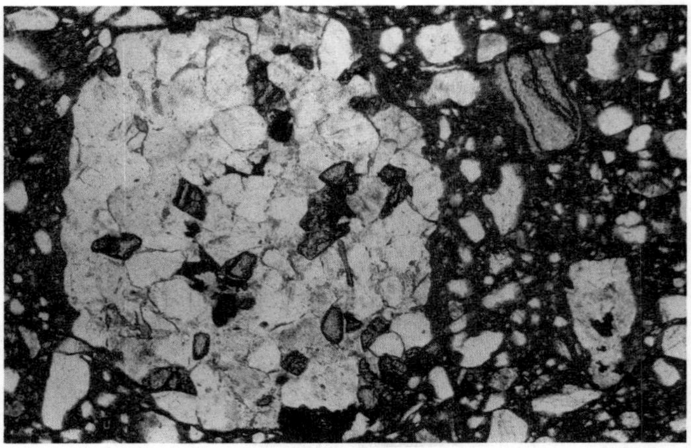

Figure 36. Photomicrograph of garnet-bearing sandstone rock fragment in diamictite of Whiteout Conglomerate. Also note the free garnet grain in upper right. Field of view 2.8 mm wide. Plane polarized light. Specimen was collected from top of lower diamictite member (Fig. 5) on Mt. Ojakangas, northern Sentinel Range (Fig. 4).

unreliable because of tectonic reorientation into the foliation (cleavage) planes, which are subvertical and oriented northwest-southeast. We were concerned that the striations on the boulder pavements might be tectonic in origin and parallel to cleavage, but there is a slight angular difference (Fig. 42), and we interpret the striations and the accompanying crescentic gouges as having been formed by glaciers moving toward the present northeast.

No cross-beds were measured in the Crashsite Group quartzite in the northern Sentinel Range. Three very small-scale cross-bed sets in fine sandstones of the heterolithic facies of the Whiteout gave northwest to southeast transport directions, but such sparse data are not significant. The overlying Polarstar Formation contains abundant small- to medium-scale cross-beds (Collinson and others, 1980) that indicate paleocurrents moved toward the north (Fig. 41).

Most of the scanty data suggest a general paleocurrent trend to the north during deposition of the upper Crashsite, the Whiteout, and the Polarstar. These limited data, especially for the Crashsite Group quartzite, may not be representative of the formation in the Ellsworth Mountains.

PROVENANCE OF THE WHITEOUT CONGLOMERATE

The glacial movements documented in the striated boulder pavements, from southeast to northwest, indicate that the sources of the glaciers and the sediment lay to the southeast. The numerous pre-drift reconstructions all have southern Africa adjacent to West Antarctica, and the striations on the boulder pavements would be compatible with glacier movement onto Antarctica from southern Africa. It has been proposed, however, that the Ellsworth block has been rotated 90° counterclockwise relative to the Transantarctic Mountains and East Antarctica (Watts and Bramall, 1981; Clarkson, 1983), so that the sources of the glaciers might have been in the eastern Antarctica landmass that is now situated south of the Ellsworth Mountains (Fig. 43). This fits the larger picture of glacigenic deposits located at several places in the Transantarctic Mountains, including the Pensacola Mountains and the Wisconsin and Ohio Ranges, all of which parallel the boundary between East and West Antarctica. Studies at these localities (e.g., Frakes and others, 1971; Crowell and Frakes, 1975), indicate that there were some local ice centers and that not all the glaciers moved outward from the East Antarctica landmass.

Because of a thick cover of glacier ice, most of East Antarc-

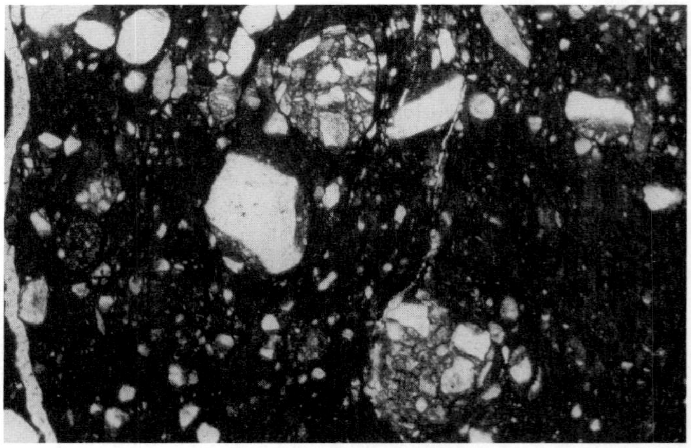

Figure 37. Photomicrograph of till pellets in laminated unit. Field of view 2.8 mm wide. Plane polarized light. Specimen was collected from the upper laminated unit (Fig. 5) on Mt. Ojakangas, northern Sentinel Range (Fig. 4).

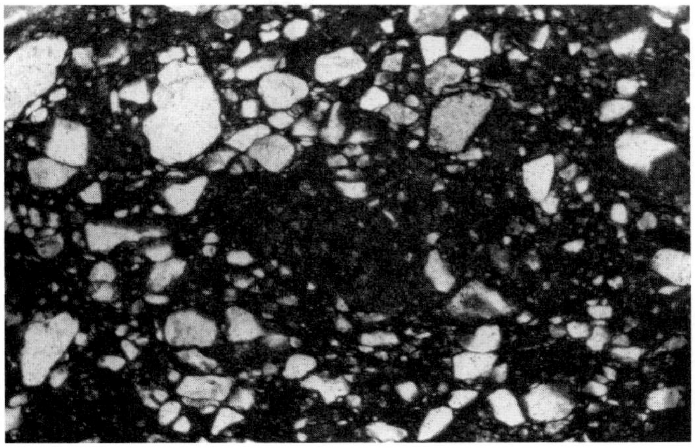

Figure 38. Photomicrograph of till pellet (center) in diamictite. Field of view 2.8 mm wide. Plane polarized light. Specimen was collected from lower diamictite member (Fig. 5) on western ridge of Olsen Peak, northern Sentinel Range (Fig. 4).

Figure 39. Photomicrograph of volcanogenic sandstone of lower Polarstar Formation. Note abundance of fine-grained volcanic rock fragments as well as quartz and feldspar. Field of view 2.8 mm wide. Crossed nicols. Specimen was collected from Whiteout Nunatak, northern Sentinel Range (Fig. 4).

tica's bedrock is not exposed, and knowledge of the lithologies of the proposed East Antarctica source area during Whiteout time must be based on study of the composition of the Whiteout Conglomerate. Three lines of evidence, as follows, suggest that supracrustal rocks were more important than infracrustal rocks in the source area.

1. Megaclasts larger than 2 cm studied in the field are dominantly quartzites and metasandstones (Table 1); they dominate over granite plus vein quartz by a ratio of more than 3:1.

2. In thin sections of the diamictite matrix, sedimentary, metamorphic, and volcanic clasts dominate over plutonic clasts by a 4:1 ratio (Table 2).

3. Although only rare individuals of the abundant monocrystalline quartz grains have silica overgrowths inherited from an older sedimentary cycle, it is estimated that one-third to one-half of the grains are subrounded to rounded (Fig. 28); however, this determination is difficult to make, in part because of replacement of grain boundaries by recrystallized matrix. These rounded grains may have been derived from older sand or sandstones, for they show a longer history of abrasion than do the other monocrystalline quartz grains. Thus, there are indications of direct first-cycle detritus as well as recycled detritus.

The correlative Buckeye Tillite in the Ohio Range 750 km to the southwest contains similar megaclasts; 72 percent are sedimentary, with weakly metamorphosed sandstone as the most common clast type (Long, 1965). In the Gale Mudstone of the Pensacola Mountains, 500 km to the southeast, one-third of the clasts are quartzite and sandstone (Nelson, 1981). Quartzite clasts dominate in the thick diamictites in the southern Karroo Basin (von Brunn and Stratten, 1981), in Argentina (Frakes and Crowell, 1969; Amos and Lopez-Gamundi, 1981), and in the Falkland Islands (Frakes, 1981b).

The presence in the Whiteout of sandstone-quartzite clasts

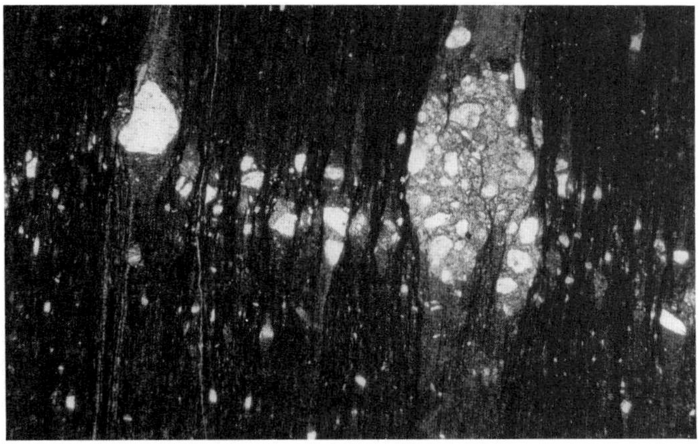

Figure 40. Photomicrograph of possible till pellet in lowest few meters of black shale of Polarstar Formation. Prominent foliation is present and may have been a factor in production of the pellet, thereby possibly making it a pseudopellet. Field of view 2.8 mm wide. Plane polarized light. Specimen was collected from Whiteout Nunatak, northern Sentinel Range (Fig. 4).

that resemble the Crashsite rock types suggests that the Crashsite may have been a source. The observed contact between the Crashsite and the Whiteout is interpreted as gradational, but the same contact, nearer the source area, could have been an erosional one, as it is in the Ohio Range (Bradshaw and others, 1984). Elliot (1975) reviewed the relations of the glacigenic units of the Transantarctic Mountains to the underlying quartzites; disconformities or unconformities exist at all localities except the Pensacola Mountains, where the nature of the contact is unclear. Whereas the stratigraphic sequence in the Ellsworth Mountains is apparently a continuous one without unconformities, unconform-

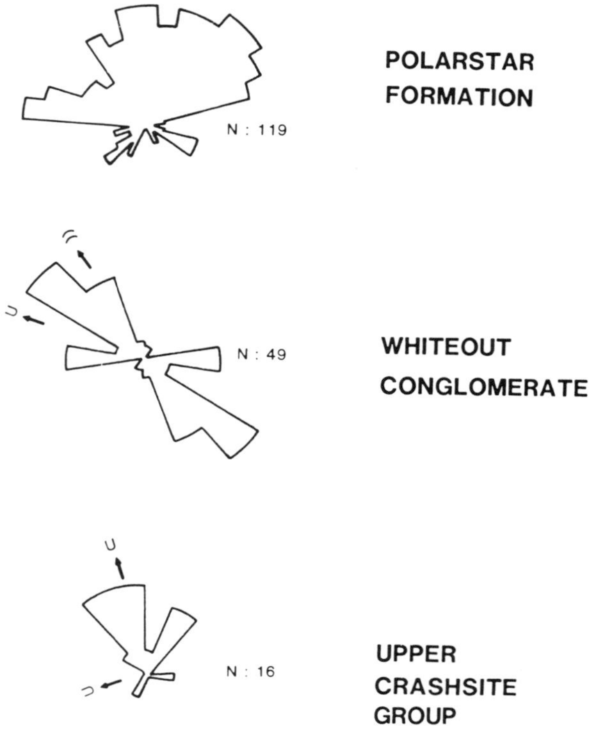

Figure 41. Paleocurrent plots, Ellsworth Mountains. Polarstar Formation measurements are cross-beds (Collinson and others, 1980) from the northern Sentinel Range. Whiteout measurements are striations and 2 crescentic gouges on 3 boulder pavements and 1 channel axis (see Fig. 25), all in the Meyer Hills. The Crashsite measurements are cross-beds and 2 trough fills, all from the Meyer Hills.

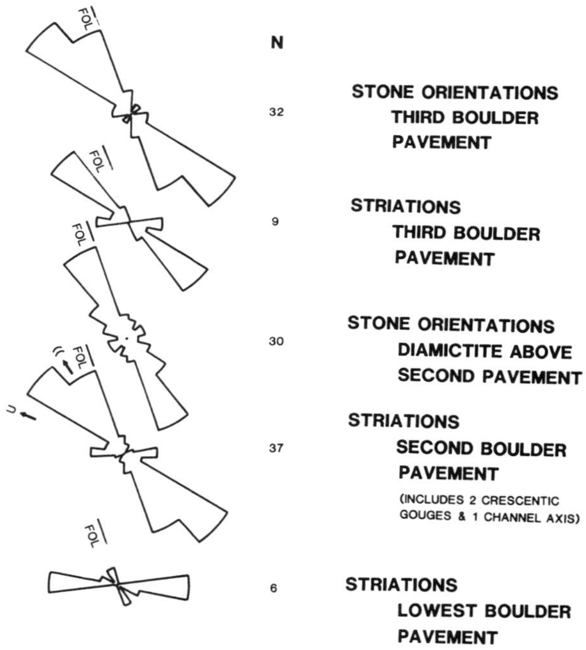

Figure 42. Ice paleocurrent plots of striations on boulder pavements and of oriented stones in and above pavements. The regional foliation is also shown on each rose. Whiteout Conglomerate, Meyer Hills.

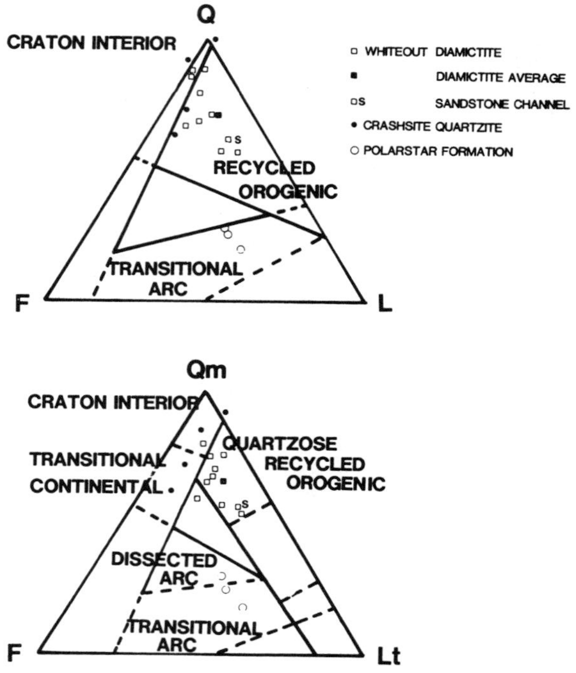

Figure 43. Generalized diagram to show orientations of Ellsworth Mountains before and after tectonic rotation. Arrows in Meyer Hills depict direction of ice movement based on data of Figure 42.

ities could be present to the south nearer East Antarctica, perhaps the result of uplift nearer the basin margin. The 15 m of apparent relief at the Crashsite-Whiteout contact in the Meyer Hills could indicate pre-lithification erosion of Crashsite sand by glaciers near the basin edge. Such an erosion of unlithified sand at the same contact in the Pensacola Mountains has been suggested by Williams (1969). Other clasts resemble sandstones of the Heritage Group, which is stratigraphically beneath the Crashsite. Heavy mineral studies of these possible source rocks and of megaclasts in the Whiteout would be useful in substantiating such derivations.

The oolitic carbonate fragments may have been derived from the Heritage Group. Oolitic carbonates, probably of Early to Middle Cambrian age, must have been present in the vicinity of the southern Heritage Range, for 100-m-thick conglomerates, rich in carbonate clasts, occur near the base of the Heritage Group. Some differences exist in the composition of megaclasts of the northern Sentinel Range and the Meyer Hills; in the former, shale-argillite and vein quartz are relatively abundant, whereas in the Meyer Hills, carbonate is abundant, shale-argillite is absent, and vein quartz is very minor. This suggests somewhat different provenances for the diamictites of the Meyer Hills and the northern Sentinel Range.

The dominance of sedimentary-metasedimentary detritus in the Whiteout Conglomerate, including the most abundant component—clay—could be interpreted to mean derivation from either a deeply weathered, low-lying granitic stable craton, an uplifted platformal sequence, or both. The provenance triangles of Dickinson and others (1985) were utilized to help interpret the sources of the Crashsite sandstones and the sand-pebble fraction of the diamictites of the Whiteout Conglomerate (Fig. 44). The Whiteout and Crashsite modes on the QFL triangle (Qm = total quartz, F = feldspar, and L = lithics) plot in the craton interior portion of the continental block and in the recycled orogen. On the QmFLt triangle (Qm = monocrystalline quartz, F = feldspar, and Lt = total lithics, including polycrystalline quartz and chert), the Crashsite modes fall in the craton interior and transitional continental subfields of the continental block, and the Whiteout modes plot in the quartzose recycled subfield of the recycled orogen. There is a separation of the two formations, most likely reflecting greater abrasion of the detritus in the Crashsite, with the elimination of soft rock fragments. That is, the two formations probably had quite similar source terranes. Provenance based on the triangles fits well with the sources as determined by petrographic observation.

The average modes of the three major facies in the Polarstar Formation (Collinson and others, this volume) were also plotted for comparison; they clearly reflect a different provenance, plotting in the transitional subfield of the magmatic arc field.

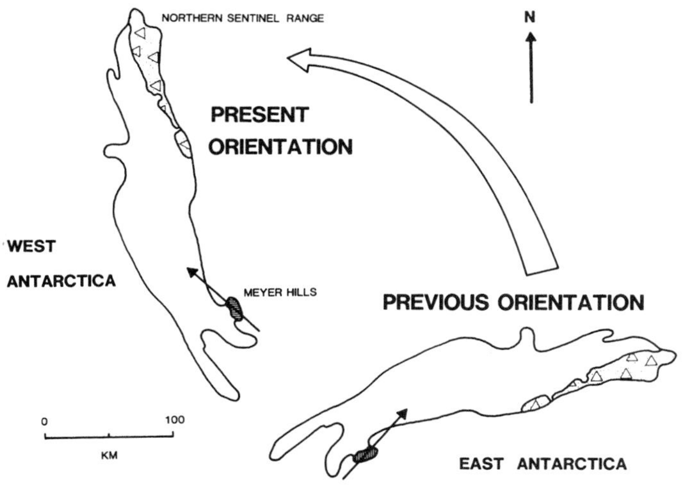

Figure 44. Provenance triangles after Dickinson and others (1985). Diamictites of the Whiteout Conglomerate, quartzites of the Crashsite Group, and sandstones of the Polarstar Formation are plotted. Polarstar data are from Collinson and others (this volume).

PALEOGEOGRAPHY

The extent of deposits of late Paleozoic age that are interpreted to be glacigenic in the Transantarctic Mountains and in the nearby Ellsworth Mountains, as well as the positions of these units in similar stratigraphic sequences, suggests a common origin, even in the absence of precise age determinations. Similarly, the presence of such deposits in South America, the Falkland Islands, and South Africa suggests a common glacial origin. For example, Crowell and Frakes (1972), Crowell (1983), and Caputo and Crowell (1985) concluded that a center of glaciation migrated across South America, South Africa, and Antarctica during 90 m.y. of late Paleozoic time, from 350 to 240 Ma.

At most localities in these Gondwanaland landmasses, the glacigenic deposits are relatively thin and consist of a variety of rock types and/or striated surfaces interpreted to indicate terrestrial or grounded marine ice sheets, as in the Meyer Hills (this chapter), the Ohio Range (Bradshaw and others, 1984), and the northern Karroo Basin (Visser and others, 1978; von Brunn and Stratten, 1981). Thick glacial-marine successions are inferred at a few places including Argentina (1,100 to 1,400 m; Amos and Lopez-Gamundi, 1981), the Falkland Islands (850 m; Frakes 1981b), the southern Karroo Basin of South Africa (1,000 m; Crowell and Frakes, 1972; von Brunn and Stratten, 1981), the Pensacola Mountains at the eastern end of the Transantarctic Mountains (1,200 m; Nelson, 1981), and the Sentinel Range (1,000 m; Ojakangas and Matsch, 1981; this chapter).

Virtually all Gondwanaland reconstructions would place the locations of these thick glacial-marine sequences in proximity to each other (e.g., Crowell and Frakes, 1972; Elliot, 1975), and deposition in a single marine or restricted marine(?) basin is possible. The model proposed by Crowell and Frakes (1972) shows a basin 3,000 km long between the Sentinel Range and Argentina. To our knowledge, radiometric and paleontological data do not negate this possibility, and more data are needed to verify or refute this model. Furthermore, each of these sequences overlies a sandstone sequence, such as the dominantly marine Horlick Formation of the Ohio Range (Bradshaw and others, 1984), the Dover Sandstone of the Pensacola Range (Williams, 1969; Nelson, 1981), the Cape Supergroup sandstone of the Karroo Basin (von Brunn and Stratten, 1981), and the Port Philomel Beds of the Falkland Islands (Frakes, 1981b). Although complete sedimentological studies are lacking, all of these sandstones may be marine, deposited on a subsiding continental shelf or in an intracratonic basin. The Crashsite Group quartzite of the Ellsworth Mountains is also at least in part marine, based on trilobites, cephalopods, and brachiopods (Craddock, 1969; Webers and Spörli, 1983). Furthermore, the transition from the Crashsite to the Whiteout with its abundance of diamictites, herein interpreted to be glacial-marine, necessitates a marine rather than a fluvial origin for at least the upper Crashsite, in order to allow for the presence of floating glaciers and/or icebergs.

The glacigenic formations are conformably overlain by shale-siltstone-sandstone-coal sequences correlative with the Polarstar Formation, which has been determined by Collinson and others (1980; this volume) to be of mixed prodelta-basinal, delta front, and delta plain–fluvial origin. These generally correlative units include the Pecora Formation of the Pensacola Mountains (Nelson, 1981) and the Discovery Ridge and Mt. Glossopteris Formations of the Ohio Range (Bradshaw and others, 1984). In South Africa (von Brunn and Stratten, 1981) and Namibia (Martin, 1981) a similar conformable transitional contact is present between the Dwyka Tillite and the overlying Prince Albert Shale. The same is true in Tasmania (Frakes, 1979) and in the Falkland Islands (Frakes and Crowell, 1967). Elliot (1975) emphasized the similarities between the Polarstar and other postglacial units of the Transantarctic Mountains. Collinson and others (this volume) summarized the paleogeographical interpretations that have been made for the Polarstar Formation and emphasized a possible back-arc depositional setting.

It is generally accepted (e.g., Craddock, 1983) that the abnormal trend of the Ellsworth Mountains (Fig. 1) indicates that the Ellsworth block has been rotated. Striking similarities of the total sedimentary sequence of the Ellsworth Mountains to the total sequence in the Pensacola Mountains, as described below, suggest a close relation of sedimentation in time and space. Frakes and others (1971) and Crowell and Frakes (1975) used the term "Ellsworth-Pensacola basin," thereby emphasizing the general stratigraphic similarities. The present study makes more detailed comparisons possible and strengthens the case for a common history.

The thickness, bedding, and petrographic characteristics of the Crashsite Group quartzite of the Ellsworth Mountains are similar to those of the correlative Dover Sandstone of the Pensacola Mountains (Williams, 1969). These are good stratigraphic and sedimentological reasons for assuming that the depositional strike of the formations of the Transantarctic Mountains was parallel to the trend of the present mountain range. If the two formations were accumulating in the same basin at the same time, and the Crashsite was 500 km or more seaward of the Dover Sandstone, some major facies changes should be present. Likewise, the Whiteout Conglomerate of the Sentinels and the Gale Mudstone of the Pensacolas (Williams, 1969; Nelson, 1981) appear to be very similar. The Gale Mudstone consists of 1,200 m of massive diamictite separated by five thin shale and siltstone sequences (i.e., the heterolithic facies of this study), and striated boulder pavements are present. The latter are concentrated in the northern Pensacolas (Frakes and others, 1971); the column there may be similar to that in the Meyer Hills, but published information is not adequate to confirm this. Furthermore, the Meyer Hills section indicates deposition near the grounding line, and the grounding line would not likely be located 500 km seaward of the thick diamictites of the Pensacola Range.

A necessary corollary of these sedimentological similarities is that not only have the Ellsworth Mountains been rotated about 90°, but they must have moved northward at least 500 km from a hypothetical former position near the Pensacola Mountains. This former position may have been just east of the Pensacola Mountains, making the Ellsworths the eastern extremity of the Transantarctic Mountains. Schopf (1969) suggested that the Ellsworth "continental fragment" migrated from a position in line with the Pensacola and Transantarctic mountains, and Dalziel (1980), on structural grounds, placed the Ellsworths on the reconstructed Gondwana craton margin between the Cape Fold Belt of South Africa and the Transantarctic Mountains.

PALEOCLIMATE

No climatological evidence has yet been determined for the Crashsite Group quartzite as a whole. However, within the upper 200 m or so of the Crashsite Group quartzite in both the Sentinel Range and the Meyer Hills, scattered lonestones suggest the onset of a cool climate that produced ice-rafted detritus. The mechanisms could have been river ice, shore ice, or even organisms rather than glaciers or icebergs (e.g., Crowell, 1964; Ojakangas, 1985). A similar gradation of pebbly sediment into diamictite exists in South Africa (Crowell and Frakes, 1972).

The thick Whiteout Conglomerate, composed of both a heterogeneous facies with dropstones and a massive diamictite facies, contains strong evidence of deposition during a glacial period. Possible minor fluctuations in climate, with accompanying glacial advances and retreats, may be interpreted if the six distinct diamictite units present in the Meyer Hills are indeed lodgement tills. Lindsay (1970) noted 13 advances and retreats based on the stratigraphy of the Pagoda Formation in the Central Transantarctic Mountains. The heterolithic (laminated) facies in the Sentinels could also be related to climatic fluctuations, with deposition occurring during minor glacial retreats.

The lower 15 m of the overlying Polarstar Formation in the northern Sentinels contains rare dropstones and till pellets, indicative of the continued presence of an ice-rafting mechanism. In South Africa, dispersed pebbles are also present in the lower 15 m of shale immediately overlying the Dwyka Tillite (Crowell and Frakes, 1972). Such dropstones occur throughout the 2,000-m-thick Polarstar equivalent in Tasmania (Frakes, 1979). The abundant *Glossopteris* flora and coal preserved in the upper part of the Polarstar Formation (Collinson and others, this volume) suggest a considerably warmer climate.

ORIGIN OF THE WHITEOUT CONGLOMERATE

Glacial-marine sediments and environments of deposition

The climatic setting, basin morphology and bathymetry, glaciology, and type of glacier interface with the marine environment are the primary factors controlling glacial-marine sedimentation and facies patterns (Anderson, 1983; Powell, 1984; Eyles and others, 1985). Fundamental differences in sedimentary facies relate to the climatic setting. Subpolar and temperate glaciers produce meltwater, a major mechanism for delivering sediment to the marine environment. In contrast, polar glaciers do not generate meltwater streams. Consequently, glaciofluvial sediment is not an important component of the sedimentary record of such glacier systems. Depositional models have been developed for tidewater glaciers (Powell, 1981), ice shelves (Drewry and Cooper, 1981; Orheim and Elverhøi, 1981), polar ice sheets (Anderson and others, 1983), and subarctic mountain glacier systems (Molnia, 1983).

Powell (1984) presented a lithofacies model containing

eight different regimes based on controls in the production of glacial-marine sediment, including (1) type of ice source (ice sheet or valley glacier), (2) condition of grounded basal ice (cold or warm and melting/freezing or frozen), and (3) type of glacier-front (ice shelf or ice cliff). In addition to those controls, Eyles and others (1985) cited the importance of relief of the basin margin, ice-flow dynamics, and discharge rates.

An important physical boundary in all of these models is the grounding line. For a tidewater glacier it is where the glacier ends as an ice cliff in the sea. For ice shelves, the grounding line or buoyancy line marks the place where a glacier begins to float (Powell, 1984). Shoreward of the grounding line, subglacial processes prevail for both terrestrial and marine-based glaciers. Seaward, a floating ice shelf may extend the ice margin hundreds of kilometers from the grounding line.

In the subglacial environment, both subglacial erosion and deposition are possible. Sequences of basal tills and subglacial meltwater sediments associated with plucked and abraded bedrock or striated boulder pavements are representative facies. Sequences of sediments associated with direct deposition by grounded glaciers are noted for their uneven distribution, lithologic heterogeneity, limited volume, and association with eroded contacts including abraded bedrock pavements. Many of these characteristics are dependent on the thermal regime of the glaciers. Ice at or near the pressure melting point has a higher capacity for eroding and entraining sediments; meltwater can be an important sediment-transporting agent. Such a glacier regime would produce a complex sedimentary pile. Glacier ice much colder than its melting point does comparatively little erosion, entrains a relatively small volume of sediment at its base, and deposits only thin sediment sequences by lodgement.

The ice-proximal (or grounding-line) environment can be a highly energetic zone, depending on the stability of the ice margin, the style of ablation, and the nature of the ice front. Where steep ice cliffs meet the sea, the interplay among depositional processes, including meltwater, gravity flows, iceberg calving, and marine currents, makes a complex sedimentary facies predictable. If significant melting occurs, accompanied by the concentration of meltwater into streams, glacial outwash is a dominant sediment facies.

Complex sedimentary facies are associated with the grounding-line zone. There the interaction between glacial and marine processes is most direct. Depositional processes include direct release of glacial debris from the ice, meltwater, mass movements, iceberg calving, and reworking by marine currents. Significant morainal banks can accumulate (Lavrushin, 1968; Powell, 1981), and ice push can result in deformation structures. Mass movements in the form of sediment flows and slides seaward of the grounding line, or at local submarine topographic highs, can be a major depositional process (Kurtz and Anderson, 1979).

The extension of ice shelves beyond the grounding line not only provides a mechanism for seaward transport of glacial sediment but significantly modifies the sub–ice shelf marine environment as well. Depending on the thermal conditions of ice and seawater, undermelting may release debris close to the buoyancy line, or seawater will freeze on, resulting in sediment transport to the ice-shelf margin.

The width of iceberg-zone sedimentation ranges considerably, from 1 km to a few hundred kilometers for tidewater-calved bergs to several thousand kilometers beyond ice-shelf margins (Powell, 1984). Sedimentation rates are highly variable, depending mainly on the original debris content of the icebergs as well as on the size, melting rates, and flux of the icebergs. Higher rates of accumulation are generally associated with proximity to the grounding line. The wider the ice shelf, the cleaner are the icebergs produced from it. The highest volume of debris-rich icebergs is produced by outlet glaciers and ice streams because they drain interior ice at high rates, in some cases greater than 1 km/yr (Anderson and others, 1983); the latter source accounts for the major portion of ice-rafted detritus in today's ocean (Drewry and Cooper, 1981). Till pellets, dropstones, dump structures, and grounding structures are criteria for recognizing iceberg rafting in ancient deposits (Ovenshine, 1970; Thomas and Connell, 1985).

Sentinel Range. We have seen no unequivocal indications that the diamictites of the Sentinel Range were deposited directly from grounded glacier ice. Rather, the great thickness and areal extent of the diamictites, together with their remarkable uniformity in lithology, both vertically and laterally, suggest a glacial-marine origin (Craddock and others, 1964; Matthews and others, 1967; Frakes and others, 1971; Ojakangas and Matsch, 1981). A polar ice-shelf model (Fig. 45) for the glacial setting is chosen to explain the general environment of deposition in the northern Sentinel Range for the following reasons: (1) more than 95 percent of the total thickness consists of massive diamictites showing little effect of bottom currents, suggesting a restricted oceanic circulation; (2) the remarkable homogeneity and simplicity of clast composition through a 1,000-m thickness indicates a constant, probably local, sediment source, diminishing the likelihood of distant or changing iceberg sources; (3) regionally, a grounding-line zone has been identified (Meyer Hills); (4) fossils, including burrows and tracks, are absent even in the thin-bedded facies; (5) absence of a meltwater facies suggests a polar climate, and ice shelves occur only under cold-ice regimes because the rate of fracture of a warm-ice glacier, enhanced by intercrystalline water, is too rapid to allow deformation into a floating shelf (Powell, 1984); (6) no iceberg dump and grounding structures are present; (7) no shaped clasts ("bullet boulders") characteristic of lodgement tills were identified; (8) sedimentary structures in the diamictites, characteristic of meltwater-generated sediment plumes and submarine mass-wasting, are absent.

The thin but widespread sorted and well-stratified units indicate that current activity sporadically but rarely dominated the depositional environment. With its abundant dropstones, this ensemble fits a glacial-marine or glacial-lacustrine model, with icebergs or a floating ice shelf transporting the larger-sized clasts to

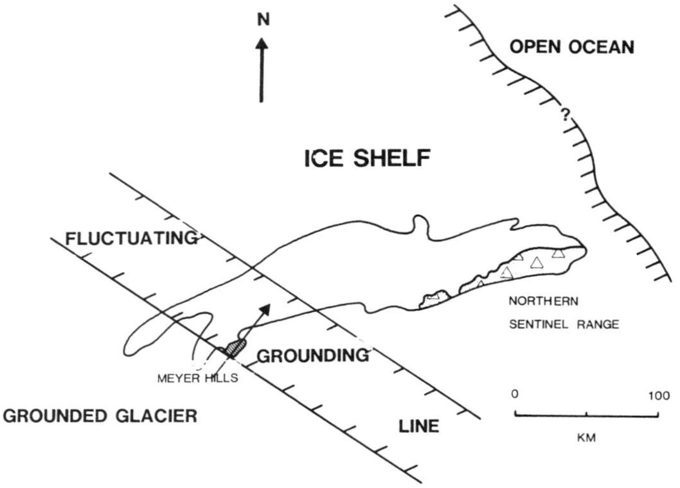

Figure 45. Generalized diagram showing approximate trends and locations of proposed depositional zones of glacigenic sediment in the Ellsworth Mountains during late Paleozoic time. Orientation of the grounding line is drawn perpendicular to paleocurrent data from striated boulder pavements at one locality, so trend and widths of zones could vary appreciably from the map. Glacigenic sediment of the northern Sentinel Range is interpreted to have been deposited offshore from the zone of the grounding line in which the glacigenic sediments of the Meyer Hills were deposited.

the depositional site. Bottom currents produced graded bedding and other small-scale sedimentary structures of deposition and lag concentrations by winnowing. Such sequences may represent interludes of more energetic oceanic circulation accompanying the periodic retreat of the ice-shelf edge during the onset of interglacial climatic conditions.

Massive diamictites in pre-Pleistocene glacigenic sequences have been interpreted in a variety of ways. Late Precambrian tillites in Finmark, North Norway, characterized by their lack of internal sedimentary structures, were considered to be indurated ground moraine or lodgement till (Edwards and Føyn, 1981), as were tillites of similar age in South Norway (Bjørlykke and Nystuen, 1981). Gravenor and others (1984) proposed a subglacial subaquatic depositional environment for massive diamictites of late Paleozoic age in the Paraná Basin in Brazil and the Karroo Basin of South Africa. They introduced the term undermelt diamicton(ite) for thick, massive, poorly sorted glacial sediment "deposited from the sole of the glacier, possibly near the grounding line as a slurry by continuous basal melting" of active ice (Gravenor and others, 1984, p. 144). Deposits of similar origin have been called subaqueous basal till (Link and Gostin, 1981), submarine moraine (Lavrushin, 1968), and waterlain till (Dreimanis, 1979). Thick massive diamictites in the Dwyka of the southern Karroo Basin are attributed to subaqueous downslope movement of debris released from basal ice (von Brunn and Stratten, 1981).

A direct comparison of the stratigraphic sequence in the northern Sentinel Range with modern glacial-marine sediments is difficult because no Quaternary sedimentary sequences of comparable thickness are exposed. From thickness estimates based on geophysical observations and direct observations from short cores, the analysis of one group of glacial-marine sediments from the Antarctic shelf may serve as an appropriate analog from a modern or near-modern glacial-marine environment. Anderson and others (1980) have described massive diamictites from the Antarctic shelf that show only subtle stratification and contain a sparse benthic foraminiferal assemblage. Most recently these diamictites, containing only a few types of rock and mineral clasts, were interpreted as being derived from basal debris zones and "deposited beneath an ice shelf, where circulation is sluggish, and where complete melt-out of basal debris would likely occur . . ." (Anderson and others, 1983). Such an environment is predicted in a narrow zone close to the ice-shelf grounding line (Drewry and Cooper, 1981; Orheim and Elverhøi, 1981). Such glacial-marine sequences are also predicted for open continental shelves peripheral to ice sheets and large glaciers, where a slow flux of glacially derived suspended sediment and ice-rafted detritus produces massive diamictons (Boulton and Deynoux, 1981). We conclude that within the framework of a polar glacial-marine model as proposed by Anderson and others (1983), the diamictites of the northern Sentinel Range represent an ice-shelf glacial-marine facies.

Meyer Hills. We envision a fluctuating grounding line in the vicinity of the Meyer Hills, with an ice shelf extending seaward across the northern Sentinels. The maximum seaward position of the grounding line is not known because bedrock exposures in the intervening terrain have not been studied, notably, the Flowers Hills 140 km to the north. The orientation of the grounding line is therefore somewhat speculative. The sequence in the Meyer Hills is interpreted to consist largely of lodgement tills and glacial-marine sediments deposited by (1) a marine-based glacier near a fluctuating regional grounding line or (2) periodic anchoring and detachment of a large ice shelf on a submarine topographic high.

Well-developed facets and striations on several boulder pavements point to abrasion by sliding, wet-based glacier ice. The occurrence of striated boulder pavements within sequences of glacigenic sediments is not rare. They have been interpreted as lag deposits subsequently overridden and abraded by glacial ice (Ojakangas and Matsch, 1981; Eyles and others, 1985) or possibly as the product of lodgement at the sole of wet-based, sliding glaciers, followed by abrasion after their emplacement during the same glacial advance (Ojakangas and Matsch, 1981). In either case, a grounded glacier is indicated.

The sorted and bedded sediments indicate current activity. Infusions of current-transported sediment resulted in the accumulation of well-sorted silt and sand beds. At the same time, coarser clasts were raining down from floating ice to produce dropstone-bearing strata. Current winnowing resulted in the concentration of coarse lag deposits expressed on the outcrops of the heterolithic facies as pebble bands.

Conspicuous megascopic variations in diamictite lithologies, especially across contacts marked by boulder pavements, are attributable to episodic deposition of tills composed of sediment derived from slightly different source areas or from different stratigraphic levels within a single source area. Similar sequences in areas of Pleistocene glaciation are generally attributed to multiple episodes of glacial advance and retreat.

On the continental shelf of Antarctica, recent studies of sediments interpreted to be lodgement tills resulting from contact with advancing glacial ice indicate that lithologies entrained nearest the site of deposition are most abundantly represented (Domack, 1982). In contrast, glacial-marine sediments from the same sample area display a greater diversity of heavy mineral and megaclast composition, reflecting greater dispersal and mixing from diverse source areas. The variability in clast composition in the Meyer Hills diamictites may therefore reflect different origins for the different diamictites: both glacial and glacial-marine. The close association of diamictites and laminated dropstone deposits supports this multiple-origin hypothesis.

The one channel-filling conglomerate and sandstone deposit set into a diamictite bed and covered by a boulder pavement is a cut-and-fill event probably related to the discharge of a small meltwater stream, either subglacially or subaerially. It is the only clear evidence of fluvial activity in the entire section we examined. Therefore, we conclude that meltwater was not an important mechanism in the depositional environment.

Depositional model for the Whiteout Conglomerate

Several lines of evidence point to glaciers of continental proportion rather than mountain glacier systems as the major sediment source for the Whiteout Conglomerate. We have seen no evidence of significant relief on the underlying Crashsite Group quartzite to indicate glacially eroded troughs (fjords) characteristic of coastal mountain glaciers. The uniform thickness, the stratigraphic continuity with other glacigenic deposits in a 4,000-km-long belt throughout the Transantarctic Mountains, and the quartz-rich nature of the sediments all indicate widespread glaciation of a low-lying, well-weathered, cratonic area dominated by supracrustal rocks (Ojakangas and Matsch, 1984). Such a continental scale was mentioned by Lindsay (1970), although there is evidence of several ice centers (e.g., Crowell and Frakes, 1975).

The almost total absence of meltwater-generated fluvial deposits and meltwater plume–deposited clayey silt facies suggests an arid polar climatic regime. Therefore, we suggest that the subarctic model that typifies the sedimentary environment along the Gulf of Alaska today as described by Molnia (1983) is not applicable here. There, the major unit accumulating today is a clayey silt facies, resulting from the deposition of rock flour originally introduced by rivers into the sea. The abraded boulder pavements of the Meyer Hills indicate that wet-based subglacial conditions existed during their formation.

The impressive volume of glacially derived sediments in the glacial-marine sequence of the northern Sentinels indicates a substantial and constant sediment supply with high rates of deposition. Alternatively, a steady delivery of a high volume of ice with a small load over a long period of time might account for the sediment thickness, as in an ice shelf nourished by rapidly flowing ice streams draining ice sheets.

According to Veevers and Powell (1987), the Gondwanaland glaciation in Antarctica may have existed for as long as 47 m.y. (333 to 286 Ma) or for as short a time as 10 m.y. (296 to 286 Ma). Generalized sedimentation rates can therefore be calculated, based on the 1,000-m thickness of the Whiteout in the northern Sentinels, as 2 to 10 cm per 1,000 years. These are relatively slow rates of sedimentation, comparable to rates on the outer Antarctic shelf today (Elverhøi and Roaldset, 1983).

Whereas our interpretation of the glacial-marine sequences in the Meyer Hills sector of the Heritage Range requires grounded glacier ice, at least periodically, this says nothing specific about the relation of the glacier base to sea level. Marine ice sheets may be grounded many hundreds of meters below sea level. Although no direct indicators of water depths were observed in the Sentinel Range, the association of massive diamictites with thin-bedded sorted units containing dropstones is interpreted to represent an ice-shelf environment requiring a depth sufficient to keep the ice afloat. Icebergs that calve from Antarctic ice shelves today have a draft of between 200 m and 250 m (Anderson and others, 1983). The basin eventually accommodated over 1,000 m of glacial-marine sediment while maintaining a deep-water imprint on most of the sequence. Basin deepening is a predictable accompaniment to glaciation by glaciers of ice-sheet size. A body of ice comparable to the East Antarctic Ice Sheet would create a proglacial depression up to 180 km wide and almost 300 m deep near the ice margin (Anderson and others, 1983; Walcott, 1970). An alternative explanation of the deepening of the basin in Whiteout time, after the shallow-water stable-platform setting of the Devonian Crashsite Group quartzite (Spörli, this volume), is foundering related to the initial extensional movements of the breakup of this portion of Gondwanaland.

CONCLUSIONS

1. The Whiteout Conglomerate is a glacigenic unit more than 1,000 m thick that represents the entire Gondwanaland glacial epoch in West Antarctica. Both the basal contact with the Crashsite Group quartzite and the upper contact with the Polarstar Formation appear to be conformable and transitional, with both contact zones containing ice-rafted detritus.

2. Two distinctly different sedimentary assemblages are present. In the Meyer Hills, diamictites of contrasting lithology interbedded with striated boulder pavements and laminated to thin-bedded sorted sediments containing dropstones represent deposition directly by glaciers and also by floating ice in a fluctuating grounding-line zone. In contrast, the sequence in the northern Sentinel Range is one dominated by massive diamictites of uniform lithology with minor thin-bedded sorted units of diverse type. Deep water and sub–ice shelf conditions are postulated for the deposition of the diamictites. This interpretation

emphasizes the past importance of ice shelves as sediment sources, in contrast to conclusions drawn from studies of modern ice shelves. The sorted units with dropstones could indicate minor iceberg activity along an ice-shelf edge.

3. Field and laboratory studies indicate that supracrustal rocks (quartzites and sandstones) contributed the major volume of framework grains, including megaclasts. A part of the quartz sand fraction may be first cycle detritus.

4. Limited paleocurrent data from glacial abrasion marks on boulder pavements in the Meyer Hills indicate an ice current direction from present southeast to northwest.

5. Synthesis of published paleomagnetic data that indicate both a 90° counterclockwise rotation of the Ellsworth Mountains relative to the Transantarctic Mountains, and stratigraphic-sedimentologic similarities between the Ellsworth and Pensacola Mountains, lead to the conclusion that the two areas had a common sedimentation history. A 500-km northward translation of the Ellsworths is indicated.

6. Ice sheets, rather than valley glaciers, generated the glacigenic component of the Whiteout sediments; the lack of evidence of meltwater deposition suggests a polar climatic regime. The impressive thickness of sediments indicates that under such conditions, glacial erosion can produce significant volumes of sediments.

7. If the 1,000-m section in the northern Sentinel Range represents slow rates of deposition over a long period of time, the environment of deposition remained remarkably constant, indicating a stable glacial regime during the glaciation of this part of Gondwanaland.

ACKNOWLEDGMENTS

We are indebted to Gerald F. Webers for the opportunity to participate in the 1979–1980 Ellsworth Mountains Expedition (NSF Grant No. DPP-7821720) and to John P. Craddock and Patricia Gould Smith for heroic field assistance. Werner Buggisch and Peter von Gizicki furnished useful data from their stratigraphic observations in the northern Sentinels. Lawrence Rosen assisted in the laboratory. Joan Hendershot and Mary Nash oversaw the preparation of the manuscript with great skill and patience, assisted by Elizabeth Metzen. Grant M. Young and Bruce F. Molnia contributed to this report through perceptive critical reviews. To all, thanks!

REFERENCES CITED

Amos, A. J., and Lopez-Gamundi, O., 1981, Late Palaeozoic Sauce Grande Formation of eastern Argentina, *in* Hambrey, M. J., and Harland, W. B., eds., Earth's pre-Pleistocene glacial record: Cambridge, Cambridge University Press, p. 872–877.

Anderson, J. B., 1983, Ancient glacial-marine deposits; Their spatial and temporal distribution, *in* Molnia, B. F., ed., Glacial marine sedimentation: New York, Plenum, p. 3–92.

Anderson, J. B., Kurtz, D. D., Domack, E. W., and Balshaw, K. M., 1980, Glacial and glacial marine sediments of the Antarctic continental shelf: Journal of Geology, v. 88, p. 399–414.

Anderson, J. B., Brake, C., Domack, E., Myers, N., and Wright, R., 1983, Development of a polar glacial-marine sedimentation model from Antarctic Quaternary deposits and glaciological information, *in* Molina, B. F., ed., Glacial-marine sedimentation: New York, Plenum, p. 233–264.

Barrett, P. J., and McKelvey, B. C., 1981, Permian tillites of South Victoria Land, Antarctica, *in* Hambrey, M. J., and Harland, W. B., eds., Earth's pre-Pleistocene glacial record: Cambridge, Cambridge University Press, p. 233–236.

Bjørlykke, K., and Nystuen, J. P., 1981, Late Precambrian tillites of South Norway, *in* Hambrey, M. J., and Harland, W. B., eds., Earth's pre-Pleistocene glacial record: Cambridge, Cambridge University Press, p. 624–627.

Boulton, G. S., and Deynoux, M., 1981, Sedimentation in glacial environments and the identification of tills and tillites in ancient sedimentary sequences: Precambrian Research, v. 15, p. 397–422.

Bradshaw, M. A., Newman, J., and Aitchison, J. C., 1984, Preliminary geological results of the 1983–84 Ohio Range Expedition: New Zealand Antarctic Record, v. 5, p. 1–17.

Buggisch, W., and Webers, G. F., 1982, Zur Facies der Karbonatgesteine in den Ellsworth Mountains (Paläozoikum, Westantarktis): Facies (Erlangen), v. 7, p. 199–228 (in German with English summary).

Caputo, M. V., and Crowell, J. C., 1985, Migration of glacial centers across Gondwana during Paleozoic era: Geological Society of America Bulletin, v. 96, p. 1020–1036.

Clarkson, P. D., 1983, The reconstruction of Lesser Antarctica within Gondwana [abs.], *in* Oliver, R. L., James, P. R., and Jago, J. B., eds., Antarctic earth science: Cambridge, Cambridge University Press, p. 599.

Collinson, J. W., Vavra, C. L., and Zawiskie, J. M., 1980, Sedimentology of the Polarstar Formation (Permian), Ellsworth Mountains: Antarctic Journal of the United States, v. 15, p. 30–32.

Craddock, C., 1969, Geology of the Ellsworth Mountains, *in* Bushnell, V. C., and Craddock, C., Geologic maps of Antarctica: New York, American Geographical Society, Antarctic Map Folio Series, Folio 12, Plate 4.

—— , 1983, The East Antarctica–West Antarctica boundary between the ice shelves; A review, *in* Oliver, R. L., James, P. R., and Jago, J. B., eds., Antarctic earth science: Cambridge, Cambridge University Press, p. 94–97.

Craddock, C., Anderson, J. J., and Webers, G. F., 1964, Geologic outline of the Ellsworth Mountains, *in* Adie, R. J., ed., Antarctic geology proceedings of the first international symposium on Antarctic geology, Cape Town, 16–21 September 1963: Amsterdam, North-Holland, p. 155–170.

Crowell, J. C., 1964, Climatic significance of sedimentary deposits with megaclasts, *in* Nairn, A.E.F., ed., Problems in paleoclimatology: New York, Wiley Interscience, p. 86–99.

—— , 1983, The recognition of ancient glaciations, *in* Medaris, L. G., Jr., Byers, C. W., Mickelson, D. M., and Shanks, W. C., eds., Proterozoic geology; Selected papers from an international Proterozoic symposium: Geological Society of America Memoir 161, p. 289–297.

Crowell, J. C., and Frakes, L. A., 1972, Late Paleozoic glaciation; Part 5, Karroo Basin, South Africa: Geological Society of America Bulletin, v. 83, p. 2887–2912.

—— , 1975, The late Paleozoic glaciation, *in* Campbell, K.S.W., ed., Gondwana geology, third Gondwana symposium: Canberra, Australian National University Press, p. 313–331.

Dalziel, I.W.D., 1980, The Ellsworth Mountains as a displaced fragment of the Transantarctic Mountains; Present status of the Schopf hypothesis [abs.]: EOS American Geophysical Union Transactions, v. 61, p. 1196.

Dickinson, W. R., and 8 others, 1985, Provenance of North American Phanerozoic sandstones in relation to tectonic setting: Geological Society of America Bulletin, v. 94, p. 222–235.

Doake, C.S.M., Crabtree, R. D., and Dalziel, I.W.D., 1983, Subglacial morphology between Ellsworth Mountains and Antarctic Peninsula; New data and tectonic significance, *in* Oliver, R. L., James, P. R., and Jago, J. B., eds., Antarctic earth science: Cambridge, Cambridge University Press, p. 270–273.

Domack, E. W., 1982, Sedimentology of glacial and glacial-marine deposits on the George V–Adelie continental shelf, East Antarctica: Boreas, v. 11, p. 79–97.

Dreimanis, A., 1979, The problems of waterlain tills, *in* Schlüchter, Ch., ed., Moraines and varves: Rotterdam, A. A. Balkema, p. 167–177.

Drewry, D. J., and Cooper, A.P.R., 1981, Processes and models of Antarctic glaciomarine sedimentation: Annals of Glaciology, v. 2, p. 117–122.

Edwards, M. B., and Føyn, S., 1981, Late Precambrian tillites in Finmark, North Norway, *in* Hambrey, M. J., and Harland, W. B., eds., Earth's pre-Pleistocene glacial record: Cambridge, Cambridge University Press, p. 606–610.

Elliot, D. H., 1975, Gondwana basins of Antarctica, *in* Campbell, K.S.W., ed., Gondwana geology, third Gondwana symposium: Canberra, Australian National University Press, p. 493–536.

Elverhøi, A., and Roaldset, E., 1983, Glaciomarine sediments and suspended particulate matter, Weddell Sea Shelf, Antarctica: Polar Research, v. 1, p. 1–21.

Eyles, C. H., Eyles, N., and Miall, A. D., 1985, Models of glaciomarine sedimentation and their application to the interpretation of ancient glacial sequences: Palaeography, Palaeoclimatology, Palaeoecology, v. 51, p. 15–84.

Frakes, L. A., 1979, Climates throughout geologic time: Amsterdam, Elsevier Science Publishing Co., 310 p.

—— , 1981a, Late Palaeozoic tillites near the southern Ross Ice Shelf, Antarctica, *in* Hambrey, M. J., and Harland, W. B., eds., Earth's pre-Pleistocene glacial record: Cambridge, Cambridge University Press, p. 230–232.

—— , 1981b, The Late Palaeozoic Lafonian diamictite, Falkland Islands, *in* Hambrey, M. J., and Harland, W. B., eds., Earth's pre-Pleistocene glacial record: Cambridge University Press, p. 883–885.

Frakes, L. A., and Crowell, J. C., 1967, Facies and paleogeography of Late Paleozoic diamictite, Falkland Islands: Geological Society of America Bulletin, v. 78, p. 37–58.

—— , 1969, Late Paleozoic glaciation; 1, South America: Geological Society of America Bulletin, v. 80, p. 1007–1042.

Frakes, L. A., Matthews, J. L., and Crowell, J. C., 1971, Late Paleozoic glaciation; 3, Antarctica: Geological Society of America Bulletin, v. 82, p. 1581–1603.

Gravenor, C. P., von Brunn, V., and Dreimanis, A., 1984, Nature and classification of waterlain glaciogenic sediments, exemplified by Pleistocene, Late Paleozoic, and Late Precambrian deposits: Earth-Science Reviews, v. 20, p. 105–166.

Hambrey, M. J., and Harland, W. B., 1981, Earth's pre-Pleistocene glacial record: Cambridge, Cambridge University Press, 1004 p.

Hjelle, A., Ohta, Y., and Winsnes, T. S., 1982, Geology and petrology of the southern Heritage Range, Ellsworth Mountains, *in* Craddock, C., ed., Antarctic geoscience: Madison, University of Wisconsin Press, p. 599–608.

Jankowski, E. J., Drewry, D. J., and Behrendt, J. C., 1983, Magnetic studies of upper crustal structure in West Antarctica and the boundary with East Antarctica, *in* Oliver, R. L., James, P. R., and Jago, J. B., eds., Antarctic earth science: Cambridge, Cambridge University Press, p. 197–203.

Kamenev, E. N., and Ivanov, V. L., 1983, Structure and outline of geologic history of the southern Weddell Sea basin, *in* Oliver, R. L., James, P. R., and Jago, J. B., eds., Antarctic earth science: Cambridge, Cambridge University Press, p. 194–196.

Kurtz, D. D., and Anderson, J. B., 1979, Recognition and sedimentologic description of recent debris flow deposits from the Ross and Weddell Seas, Antarctica: Journal of Sedimentary Petrology, v. 49, p. 1159–1170.

Laird, M. G., and Bradshaw, J. D., 1981, Permian tillites of North Victoria Land, Antarctica, *in* Hambrey, M. J., and Harland, W. B., eds., Earth's pre-Pleistocene glacial record: Cambridge, Cambridge University Press, p. 237–240.

Lavrushin, Y. A., 1968, Features of deposition and structures of the glacial-marine deposits under conditions of a fjord coast (based on the example of Spitsbergen): Litologiya i Poleznye Iskopaemye, v. 3, p. 63–79.

Lindsay, J. F., 1970, Depositional environment of Paleozoic glacial rocks in the central Transantarctic Mountains: Geological Society of America Bulletin, v. 81, p. 1149–1172.

Link, P. K., and Gostin, J. A., 1981, Facies and paleogeography of Sturtian glacial strata (Late Precambrian), South Australia: American Journal of Science, v. 281, p. 353–374.

Long, W. E., 1965, Stratigraphy of the Ohio Range, *in* Hadley, J. B., ed., Geology and paleontology of the Antarctic: American Geophysical Union Antarctic Research Series, v. 6, p. 71–116.

Martin, H., 1981, The Late Palaeozoic Dwyka Group of the Karasburg Basin, Namibia, *in* Hambrey, M. J., and Harland, W. B., eds., Earth's pre-Pleistocene glacial record: Cambridge, Cambridge University Press, p. 67–70.

Matthews, J. L., Crowell, J. C., Coates, D. A., Neder, I. R., and Frakes, L. A., 1967, Late Paleozoic glacial rocks in the Sentinel and Queen Alexandra Ranges: Antarctic Journal, v. 2, p. 108.

Molnia, B. F., 1983, Sub-arctic glacial-marine sedimentation; A model, *in* Molnia, B. F., ed., Glacial-marine sedimentation: New York, Plenum, p. 95–144.

Nelson, W. H., 1981, The Gale Mudstone; A Permian(?) tillite in the Pensacola Mountains and neighboring parts of the Transantarctic Mountains, *in* Hambrey, M. J., and Harland, W. B., eds., Earth's pre-Pleistocene glacial record: Cambridge, Cambridge University Press, p. 227–232.

Ojakangas, R. W., 1985, Evidence for Early Proterozoic glaciation; The dropstone unit-diamictite association: Geological Survey of Finland Bulletin, v. 331, p. 51–72.

Ojakangas, R. W., and Matsch, C. L., 1981, The Late Palaeozoic Whiteout Conglomerate; A glacial and glaciomarine sequence in the Ellsworth Mountains, West Antarctica, *in* Hambrey, M. J., and Harland, W. B., eds., Earth's pre-Pleistocene glacial record: Cambridge, Cambridge University Press, p. 241–244.

—— , 1984, Criteria for distinguishing between continental and mountain glaciations of ancient glacial sequences: Geological Society of America Abstracts with Programs, v. 16, p. 612–613.

Orheim, O., and Elverhøi, A., 1981, Model for submarine glacial deposition: Annals of Glaciology, v. 2, p. 123–127.

Ovenshine, A. T., 1970, Observations of iceberg rafting in Glacier Bay, Alaska, and the identification of ancient ice-rafted deposits: Geological Society of America Bulletin, v. 81, p. 891–894.

Powell, R. D., 1981, A model for sedimentation by tidewater glaciers: Annals of Glaciology, v. 2, p. 129–134.

—— , 1984, Glaciomarine processes and inductive lithofacies modelling of ice shelf and tidewater glacier sediments based on Quaternary examples: Marine Geology, v. 57, p. 1–52.

Schopf, J. M., 1969, Ellsworth Mountains; Position in West Antarctica due to sea floor spreading: Science, v. 164, p. 63–66.

Thomas, G.S.P., and Connell, R. J., 1985, Iceberg drop, dump, and grounding structures from Pleistocene glacio-lacustrine sediments, Scotland: Journal of Sedimentary Petrology, v. 55, p. 243–249.

Vavra, C. L., and Collinson, J. W., 1981, Sandstone petrology of the Polarstar Formation (Permian), Ellsworth Mountains: Antarctic Journal of the United States, v. 16, p. 15–16.

Veevers, J. J., and Powell, C. McA., 1987, Late Paleozoic glacial episodes in Gondwanaland reflected in transgressive-regressive depositional sequences in Euramerica: Geological Society of America Bulletin, v. 98, p. 475–487.

Visser, J.N.J., and 9 others, 1978, The Dwyka Formation and Ecca Group, Karoo Sequence, in the northern Karoo Basin, Kimberley–Britstown area: South Africa Geological Survey Annals, v. 12, p. 143–176.

von Brunn, V., and Stratten, T., 1981, Late Palaeozoic tillites of the Karoo Basin of South Africa, *in* Hambrey, M. J., and Harland, W. B., eds., Earth's pre-Pleistocene glacial record: Cambridge, Cambridge University Press, p. 71–79.

Walcott, R. I., 1970, Isostatic response to loading of the crust in Canada: Canadian Journal of Earth Sciences, v. 7, p. 716–727.

Watts, D. R., and Bramall, A. M., 1981, Palaeomagnetic evidence for a displaced terrain in Western Antarctica: Nature, v. 293, p. 638–641.

Webers, G. F., and Spörli, K. B., 1983, Palaeontological and stratigraphic investigations in the Ellsworth Mountains, West Antarctica, *in* Oliver, R. L., James, P. R., and Jago, J. B., eds., Antarctic earth science: Canberra, Australian Academy of Science, p. 261–264.

Williams, P. L., 1969, Petrology of Upper Precambrian and Paleozoic sandstones in the Pensacola Mountains, Antarctica: Journal of Sedimentary Petrology, v. 39, p. 1455–1465.

Yoshida, M., 1982, Tectonic and metamorphic history of the southern part of the Ellsworth Mountains, West Antarctica: Fourth International Symposium on Antarctic Earth Sciences, University of Adelaide, South Australia, 16-20 August 1982, Volume of Abstracts, p. 184.

—— , 1983, Structural and metamorphic history of the Ellsworth Mountains, West Antarctica, *in* Oliver, R. L., James, P. R., and Jago, J. B., eds., Antarctic earth science: Canberra, Australia Academy of Science, p. 266–269.

Manuscript Accepted by the Society March 1, 1989

Printed in U.S.A.

Geological Society of America
Memoir 170
1992

Chapter 5

Sedimentology of the Polarstar Formation (Permian), Ellsworth Mountains, West Antarctica

James W. Collinson
Byrd Polar Research Center and Department of Geology and Mineralogy, Ohio State University, 155 South Oval Mall, Columbus, Ohio 43210
Charles L. Vavra
ARCO Resources Technology, P.O. Box 2819, Dallas, Texas 75221
John M. Zawiskie
Department of Geology, Wayne State University, Detroit, Michigan 48202

ABSTRACT

The Polarstar Formation, a 1-km-thick argillite and sandstone unit, is the uppermost part of a thick Cambrian to Permian sedimentary sequence in the Ellsworth Mountains. The formation gradationally overlies the Whiteout Conglomerate, an Upper Carboniferous–Lower Permian glacial diamictite. The lower part of the Polarstar is mostly argillite, and the middle part consists of coarsening-upward cycles of argillite to sandstone. These cycles begin with lenticular bedding overlain by wavy and flaser bedding and end with ripple-laminated, fine-grained sandstone to cross-bedded, medium-grained sandstone. The upper part of the formation consists of fining-upward cycles of channel-form, cross-bedded, medium-grained sandstone overlain by fine-grained sandstone and of *Glossopteris*-bearing siltstone, argillite, and coal. The sequence of facies suggests that the depositional environment changed temporally in this area from prodelta to delta and coastal plain. The occurrence of a marginal-marine trace fossil fauna in the middle of the formation and the complete absence of a marine shelly fauna suggest depositional conditions ranging from anaerobic to dysaerobic in a stratified inland sea. Detrital grains in Polarstar sandstone indicate a source terrane dominated by silicic to andesitic volcanic rocks, including tuffs, with minor mafic volcanic and low-grade metamorphic and granitic rocks. The Polarstar Formation was probably deposited in a back-arc basin between the Pacific margin of Gondwanaland and the East Antarctic craton.

INTRODUCTION

The Polarstar Formation in the Ellsworth Mountains is similar to postglacial black shale sequences of Early Permian age that occur throughout a large area of the Gondwanaland supercontinent. The stratigraphic sequence of glacial diamictite overlain by black shale and grading upward into deltaic sandstone is familiar to geologists who have worked in the Transantarctic Mountains, the Karoo basin and related basins in southern Africa, or the Paraná basin centered in Brazil. These black shales may have accumulated in a vast inland sea comparable to the present Baltic, Black, or Caspian seas (Du Toit, 1937; Lindsay, 1969; Elliot, 1975; Cooper and Kensley, 1984; Zawiskie, 1984).

If the postglacial black shales in Antarctica, Africa, and South America are related, several questions should be addressed. Under what conditions did such a widespread deposit form? If the black shales represent widespread anoxia, what were the paleogeographic, paleoclimatic, and paleo-oceanographic condi-

Collinson, J. W., Vavra, C. L., and Zawiskie, J. M., 1992, Sedimentology of the Polarstar Formation (Permian), Ellsworth Mountains, West Antarctica, *in* Webers, G. F., Craddock, C., and Splettstoesser, J. F., Geology and Paleontology of the Ellsworth Mountains, West Antarctica: Boulder, Colorado, Geological Society of America Memoir 170.

tions? What were the circulation patterns and salinity conditions? The answers to these questions are not to be found in the Polarstar Formation alone, but the sedimentologic data presented here will contribute to a better understanding of the regional picture. The major objective of this chapter is to document the only known Permian formation in West Antarctica and to compare it to similar correlative sequences in East Antarctica and southern Africa.

Much of the information presented here substantiates and enlarges on findings in previous studies of the Polarstar Formation. Reports that include sedimentologic information are Craddock and others (1964), Castle (1974), and Castle and Craddock (1975). Paleontologic reports include Craddock and others (1965), Schopf (1967), Tasch (1967, 1968), Tasch and Riek (1969), and Rigby and Schopf (1969).

TECTONIC SETTING

The plate relation of West Antarctica to East Antarctica remains one of the major problems in Antarctic geology (Dalziel and Elliot, 1982). The geology of the Ellsworth Mountains, unlike most of the known geology in West Antarctica, appears to have a close relation to that of the Transantarctic Mountains on the adjacent margin of East Antarctica. In the Ellsworth Mountains, however, the Paleozoic sequence is much thicker and possibly more complete than in the Transantarctic Mountains. The greatest similarity is in the Upper Carboniferous to Permian glacigene to fluviodeltaic sequence that is the subject of this chapter.

The Ellsworth Mountains appear to be part of a microplate that also includes the Whitmore Mountains (Jankowski and Drewry, 1981). Schopf (1969), who was impressed by the similarities of the upper Paleozoic sequence in the Ellsworth Mountains to correlative units in the Pensacola Mountains and the Horlick Mountains, suggested that the Ellsworth Mountains plate had once occupied a position between Antarctica and southern Africa in line with the Transantarctic Mountains. The trend of the Ellsworth Mountains block and its structural features—approximately perpendicular to the trend of the Transantarctic Mountains—has led most workers to accept a rotation of the plate, but its original position is a matter of conjecture. Longshaw and Griffiths (1983), because of spatial considerations, favored a simpler explanation that would place the Ellsworth Mountains block close to its current position but rotated clockwise about 90 degrees to agree with the paleomagnetic data reported by Watts and Bramall (1981). This would place the Jurassic plutons of the Whitmore Mountains at the end of the microplate in the vicinity of the hypothetical calc-alkaline arc along the Pacific margin. Dalziel and Elliot (1982) favored translation as well as rotation, partly because rocks in the Ellsworth Mountains appear to have undergone only one episode of major deformation, which would favor a more cratonic position than the present location. Yoshida (1982, 1983), however, argued for multiple episodes of deformation on the basis of detailed studies of folds and cleavage.

More recent paleomagnetic data reported by Grunow and others (1987) are compatible with two different reconstructions: one considers the Antarctic Peninsula and the Ellsworth Mountains blocks as part of the same plate, and the other as separate plates. Neither hypothesis requires as much rotation as that suggested by Watts and Bramall (1981). It is evident that more data are needed to clear up the problems of the relation of the structural trend and thick stratigraphic sequence in the Ellsworth Mountains to that of the Transantarctic Mountains.

STRATIGRAPHY

The Polarstar Formation caps the 13-km-thick, predominantly marine Paleozoic sequence. Exposures are confined to the northeastern Sentinel Range (Craddock, 1969) (Fig. 1). The thickest and most complete sections are in the vicinity of Polarstar Peak (Fig. 2), for which Craddock and others (1964) named the formation. Craddock and others (1964) estimated a thickness of 915 m but later increased it to 1,370 m (Craddock, 1969).

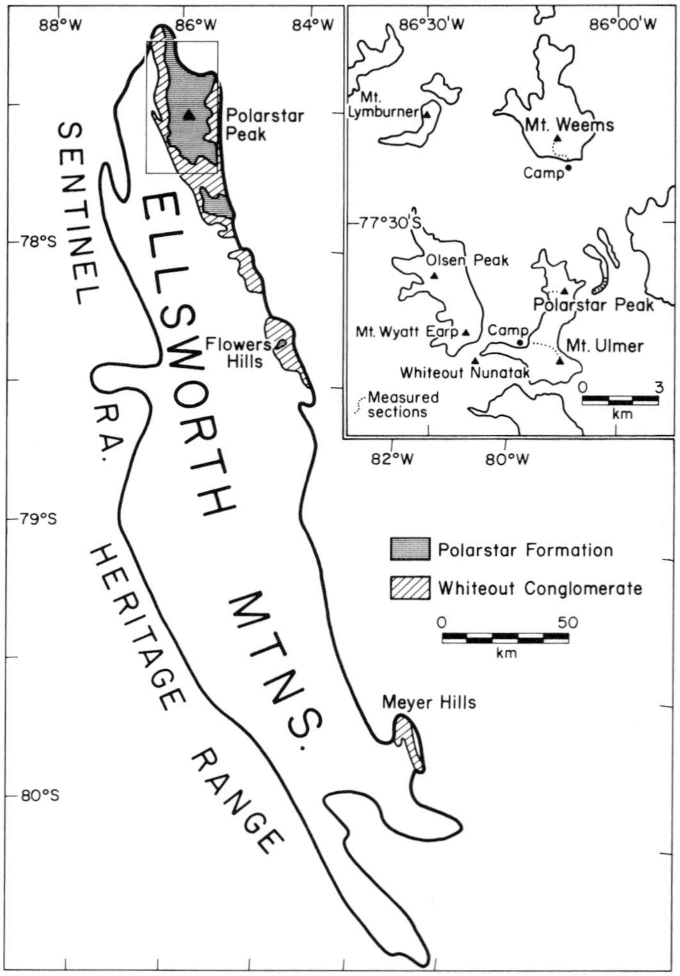

Figure 1. Map of the Ellsworth Mountains showing the distribution of Upper Carboniferous(?)–Permian rocks. Inset shows locations of measured sections.

Figure 2. Folded Polarstar Formation on the north-facing side of Polarstar Peak. Immediate relief is approximately 500 m.

Castle (1974) and Castle and Craddock (1975) revised these figures upward to 1,700 m. Unfortunately, intense deformation of the incompetent shales has made it difficult, if not impossible, to reconstruct the actual thickness. Slaty cleavage obscures bedding in all but axial areas of folds, and thrust faults have repeated the section. The measured sections shown in Figure 3 are all incomplete, but addition of maximum thicknesses from the three informal subdivisions gives a minimum thickness of 800 to 1,000 m. A closely spaced pair of tuff beds occurs in both the Mount Weems and Polarstar Peak sections, permitting correlation between these sections. The section on Polarstar Peak is considerably more sandy than the equivalent section on Mount Weems. The paleocurrents between these two sections are also somewhat divergent.

The Polarstar Formation consists mostly of argillite and sandstone. The formation can be subdivided on the basis of dominant lithology into three parts: (1) argillite at the base, (2) argillite and fine-grained sandstone in the middle, and (3) coal measures at the top. These units are referred to here as the (1) argillite facies, (2) sandstone-argillite facies, and (3) coal measures facies.

The base of the formation is well exposed on the north flank of Whiteout Nunatak, where a 50-m-thick sequence of black argillite conformably overlies the Whiteout Conglomerate (Fig. 4), an Upper Carboniferous–Lower Permian diamictite (Ojakangas and Matsch, 1981). The lithologic change from diamictite in the underlying Whiteout Conglomerate to argillite and interbedded fine-grained sandstone in the basal Polarstar suggests a conformable relationship. Evidence for an abrupt but gradational contact is the occurrence of rare quartz-pebble dropstones and scattered floating sand grains in the lower 10 m of the Polarstar. Sandstone is rare in the lower argillite facies, but a 2.5-m-thick, poorly sorted, fine-grained sandstone occurs 7.5 m above the base. Argillites in the lower part of the formation are massive and rarely show evidence of bioturbation. In addition to the exposure on Whiteout Nuntak, the lower argillite part of the formation is exposed on the lower slopes of Mount Ulmer and along a ridge extending eastward from Mount Ulmer toward Mount Wyatt Earp. It is also found in the Flowers Hills 100 km to the south-southeast. The thickness of the lower argillite member is estimated to be 300 to 500 m.

The middle part of the Polarstar Formation is well exposed on Polarstar Peak, Mount Ulmer, and Mount Weems. The thickest section of the lower to middle Polarstar Formation, 460 m, was measured on the south face of Mount Weems. Three sections were measured on the lower, middle, and upper part of the west face of Mount Ulmer (sections MU1, MU2, and MU3 on Fig. 3).

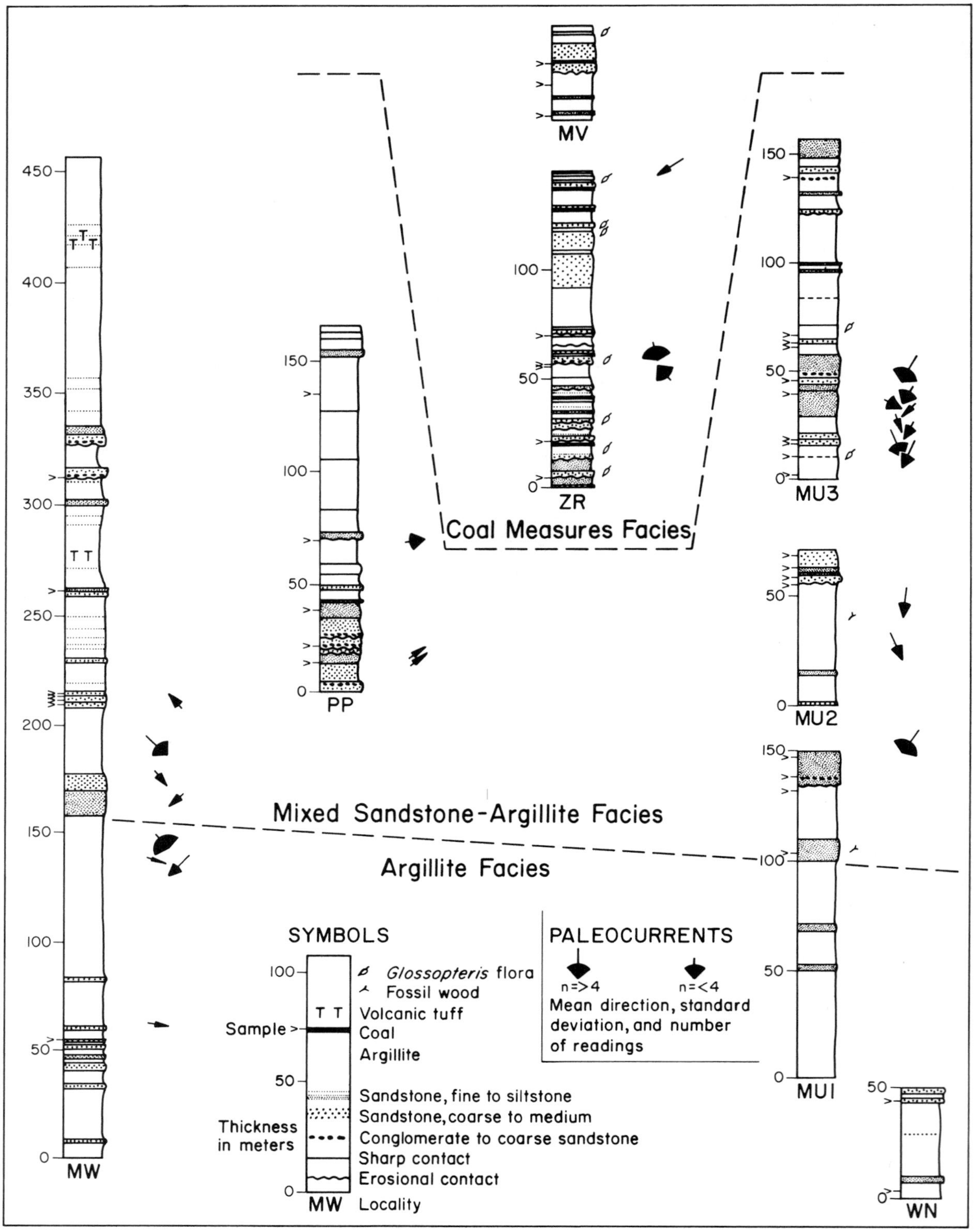

Figure 3. Stratigraphic sections of the Polarstar Formation and their approximate correlation based on comparisons of facies. Section locality designations are: MW = Mount Weems; PP = Polarstar Peak; ZR = lower section on ridge immediately east of Polarstar Peak; MV = upper section on ridge immediately east of Polarstar Peak; MU1 = section near base of Mount Ulmer; MU2 = section midway up Mount Ulmer; MU3 = section near top of Mount Ulmer; WN = Whiteout Nunatak.

These sections are separated by highly cleaved and folded slaty units representing structural discontinuities. Evidence that each section is stratigraphically higher is an upward increase in percentage of sandstone and amount of coalified plant material. Much of the middle Polarstar consists of silty to sandy argillite, but six to 10 major coarsening-upward sequences, 5 to 20 m thick, form the more resistant exposures. The transitions from argillite to sandstone are characterized by lenticular bedding to wavy bedding to flaser bedding. Sandstone sequences grade upward from ripple-laminated, fine-grained sandstone into large-scale, trough–cross-bedded, medium-grained sandstone. Coalified plant fragments occur sparsely in the lower part but are increasingly abundant in the upper part.

The upper part of the Polarstar Formation, which is well exposed on high ridges directly east of Polarstar Peak, is composed of fining-upward sequences of medium-grained sandstone to carbonaceous argillite and coal. Sandstone units commonly have channeled bases and are characterized by large-scale trough cross beds. Coalified plant fossils, including *Glossopteris* leaves and calamitid stems, are abundant on bedding planes. Specimens of *Glossopteris* from the Ellsworth Mountains were first reported by Craddock and others (1965), and the flora was listed by Rigby and Schopf (1969).

A Permian age has been assigned to the Polarstar Formation on the basis of stratigraphic position and the *Glossopteris* flora. The Polarstar Formation has been correlated with similar postglacial sequences in the Pensacola Mountains, Ohio Range (Discovery Ridge and Mt. Glossopteris formations), Wisconsin Range (Weaver Formation), and the Central Transantarctic Mountains (Mackellar, Fairchild, and Buckley formations) (Schopf, 1969; Elliot, 1975). The postglacial shale part of this sequence is much thicker in the Ellsworth Mountains than in the Transantarctic Mountains, suggesting that the Ellsworth Mountains occupied a more basinward position or possibly that Polarstar shale is the facies equivalent of the thick coal measures in East Antarctica. Because the top of the formation has been removed by erosion, it is not possible to ascertain how much of the Permian is represented. No evidence of Triassic rock has been found in this region, and it is likely that this sequence was deformed by Gondwanide folding not long after deposition (Elliot, 1975).

SEDIMENTOLOGY

Many of the sedimentologic features in the Polarstar Formation discussed here were also reported by Castle (1974) from the field notes, measured sections, and collections of J. J. Anderson, T. W. Bastien, and C. Craddock. One of the authors, J. M. Zawiskie, was kindly permitted by C. Craddock to examine the collections at the University of Wisconsin Geology Department to augment trace fossil data.

The sedimentologic features are described in terms of the three broad facies categories that occur in vertical sequence. The general relations of these facies and major characteristics are shown diagrammatically in vertical sequence in Figure 5.

Figure 4. Argillite facies in the lower 50 m of the Polarstar Formation on Whiteout Nunatak. Contact with underlying Whiteout Conglomerate in saddle (arrow) at bottom of photo. View is to the east.

Argillite facies

The lower part of the Polarstar Formation is dominated by the argillite facies. Sedimentary structures are preserved only locally within the fine-grained sandstones where slaty cleavage is parallel to bedding. An overall trend of increasing sandstone upward in the section is evident.

The most common sedimentary structures in the argillite facies are parallel laminae of silt and fine sand and ripple-bedded fine sand. Most of the parallel laminae exhibit sharp bases and irregular tops (Fig. 6). Ripple bedding is in the form of lenticular, wavy, and flaser bedding (classification of Reineck and Singh, 1980).

Undisturbed laminae and bedding characterize most of the argillite facies. In rare cases this regular pattern is interrupted by restricted zones where biogenic sedimentary structures are common. In the basal part of this facies some of the silty to sandy interbeds are densely packed with horizontal unlined burrows

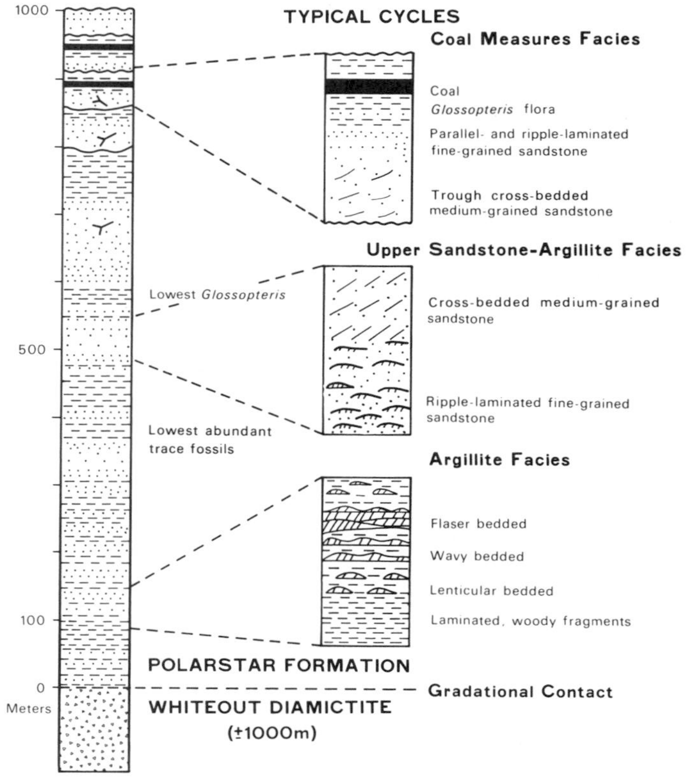

Figure 5. Diagrammatic composite stratigraphic section emphasizing sedimentologic features in the Polarstar Formation.

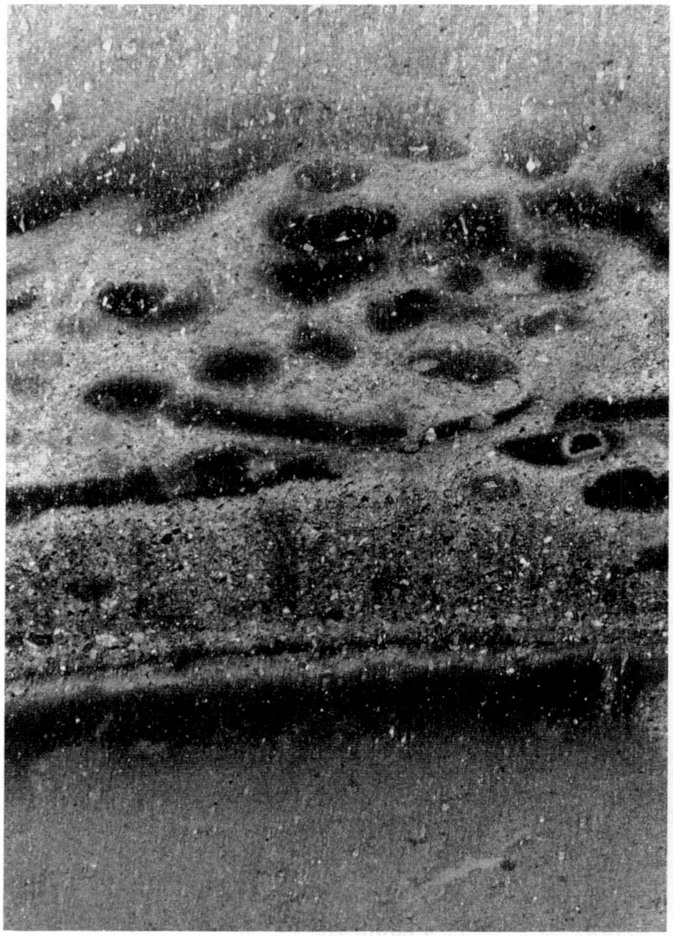

Figure 7. Photomicrograph of silty lamina in lower argillite facies at Mount Weems. Oval to elongate structures are compressed, lined burrows (1 to 3 mm diameter) assignable to *Paleophycus tubularis.* Burrow-linings are composed of framboidal pyrite. Plain light.

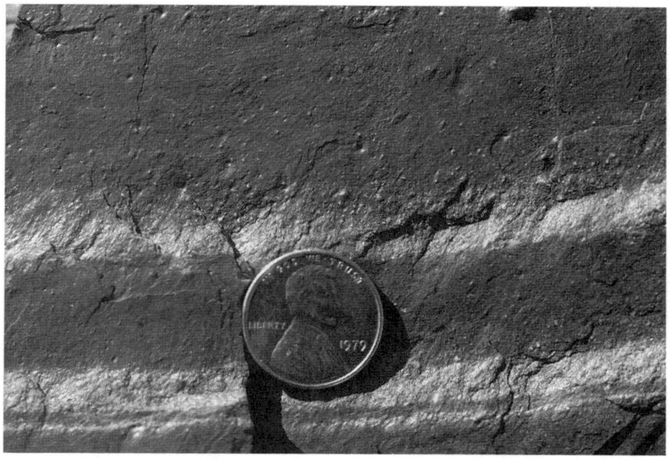

Figure 6. Parallel laminae near the base of the Polarstar Formation on Whiteout Nunatak. Sand grains floating in mud matrix suggest transport by ice-rafting.

filled with sediment contrasting the surrounding matrix; these burrows are assignable to *Planolites annularis* and *P. montanus* (Pemberton and Frey, 1982). At least two types of diagenetically modified horizontal burrows are abundant in restricted zones within the black argillite. One variety resembles the reduction burrows described by Ekdale (1977) from DSDP cores in which portions of the burrows are whitened, ranging from whitening of the burrow wall to whitening of the entire burrow. The continuum of whitening reflects the extent of reduction, with reduction proceeding from the rind to the center. In the second type, the darkened burrow linings composed of framboidal pyrite resemble those described by Byers and Stasko (1978). Taxonomically these burrows from the Polarstar Formation are assigned to *Palaeophycus tubularis* (Fig. 7).

At the base of the section, burrows are associated with distinct, small-scale, normally graded cycles that are composed of a basal, relatively coarse, angular siltstone about 2 mm thick (Fig.

Figure 8. Sandstone with small-scale cross-bedding and current ripples overlain by argillite with lenticular bedding from mixed sandstone-argillite facies in upper section on Mount Ulmer (MU3 in Figure 3). Current direction from left to right.

7) with convolutions of flame structures developed along the contact with the underlying shale. The bases of these cycles are not burrowed, but bioturbation steadily increases upward into the overlying argillite, where discrete biogenic or physical structures have been replaced by a bioturbate texture.

Sandstone-argillite facies

The sandstone-argillite facies, which forms the middle part of the formation, is similar to the argillite facies except that sandstone is more abundant and coarsening-upward cycles are more complete. As in the argillite facies, cycles begin with silty laminae that grade upward into lenticular, wavy, and flaser bedding as the fine-grained sandstone percentage increases. Ripples occur in the lower parts of sand cycles as isolated lenses of sand in the shape of flat-bottomed asymmetrical ripples with pronounced foreset laminae. Upward, isolated sand ripples tend to connect and may form wavy beds of sandstone separated by shale (Fig. 8). In the lower to middle part of the formation, where the sequence is mostly argillite, these cycles seldom develop beyond wavy bedding. Toward the upper part of the argillite facies, some cycles are capped by flaser-bedded fine-grained sandstone.

On rare bedding plane exposures, ripple crest geometries range from straight crested or slightly undulose (Fig. 9) to linguoid (Fig. 10). Most paleocurrent readings were measured in current ripple–bedded sequences. Directions of current ripples were consistently unimodal at all localities. These data are summarized in a rose diagram (Fig. 11) that depicts a consistent pattern of northward transport.

Examples of soft sediment deformation are rare but may have gone undetected because of structural deformation. Convolute bedding and slump structures occur in the lower part of the Mount Weems section in fine-grained ripple-bedded sandstone. Slump folding was also reported by Craddock and others (1964), apparently from the same stratigraphic section (Castle, 1974).

In the upper part of the sandstone-argillite facies these cycles are somewhat thicker, ranging from 2 to 20 m, and are capped by medium-grained trough cross-bedded sandstone in sets as much as 1 m thick. Scour surfaces overlain by intraformational conglomerate channel into the finer sandstone below. Fossil plant material, including coalified logs, becomes increasingly abundant upward.

Trace fossils in the sandstone-argillite facies on Mount Ulmer are concentrated in the upper levels of the coarsening-upward cycles. Some sandy beds are locally packed with horizontal to subhorizontal *Phycodes* (= *Buthrotrephis*) *palmatum,* with some segments showing vertical spreite indicating upward movement (Fig. 12). A large type of horizontal *Chondrites* sp. and a horizontal to slightly oblique form of *Rhizocorallium* sp. are also present.

Tasch (1968) presented a thorough summary of traces in the fine-grained component of this facies at Mount Weems. Two important traces were described: (1) sinusoidal-shaped grooves on upper bedding surfaces assignable to the trace fossil *Cochlichnus* sp. and (2) several unnamed shallow horizontal to subhori-

Figure 9. Weakly undulatory small current-ripples on bedding surface of fine-grained sandstone 65 m above the base of the middle Mount Ulmer section (MU2 in Figure 3).

Figure 10. Linguoid ripples on bedding surface of fine-grained sandstone 136 m above the base of the lower Mount Ulmer section (MU1 in Figure 3).

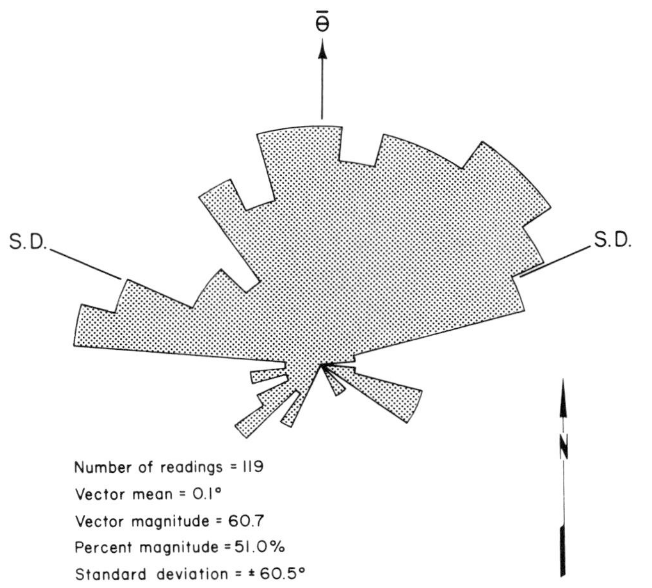

Figure 11. Rose diagram showing distribution of paleocurrent measurements from the Polarstar Formation.

zontal traces that were produced shallowly within the sediment and that display rhythmic annulations. In some cases a cone-in-cone backfilling is discernible in the latter.

Coal measures facies

This facies, which characterizes the uppermost part of the Polarstar Formation, is composed of fining-upward sequences of sandstone to argillite with abundant coalified plant material. Cycles in the lower part, which are on the order of 5 to 10 m thick, begin with an erosional surface overlain by medium-grained, trough cross-bedded sandstone. Intraformational conglomerate typically occurs at the base and along scours. Sandstone units fine upward into parallel and ripple-laminated fine-grained sandstone and siltstone and carbonaceous argillite with abundant plant fossils. Rare mudcracks indicate subaerial exposure. Examples of *Palaeophycus sulcatus* and simple epichnial grooves occur on some bedding planes.

DEPOSITIONAL SETTING

A fluviodeltaic model of deposition for the Polarstar Formation, first proposed by Castle (1974), is well supported by the sequence of lithofacies and sedimentary structures. The great thickness of the sequence suggests a rapidly subsiding basin, such as along a continental margin. The lower part of the sequence was mud dominated, probably because a rising sea level trapped sand higher on the shelf toward East Antarctica. The lower argillite facies represents a mud-dominated shelf, the middle sandstone-argillite facies records the transition from prodelta to delta front, and the upper coal measures accumulated in a deltaic plain to coastal plain environment.

Episodes of delta progradation were recorded as coarsening-upward sequences in the middle sandstone-argillite facies. Laminated shelf mudstone and siltstone occur at the base of cycles and grade upward into siltstone and sandstone, which are interpreted as recording the approach of a distal delta lobe. Transitions from wavy to lenticular to flaser bedding, which occur commonly as cycles, indicate an environment in which both sand and mud were available, but one in which energy conditions varied. During episodes of current activity, mud was carried in suspension and sand was transported and deposited as ripples. Lenticular bedding has been reported from the delta front of the Mississippi River (Coleman and Gagliano, 1965), an environment similar to that envisioned here. Reineck and Singh (1980) noted that such bedding is produced by alternating current and slack water on tidal flats. The generally unimodal paleocurrent directions in the Polarstar Formation argue against the tidal model but do not completely rule it out. Current ripple directions probably reflect downslope transport at the delta front. The cross-bedded medium-grained sandstone at the tops of more complete cycles probably represents stream mouth bar deposits. Paleocurrent directions in these are widely dispersed but indicate a unimodal direction, which we interpret as away from shore.

The coal measures facies was deposited in a delta-plain to alluvial-plain environment. Fining-upward cycles are typically terminated by coal. Channel-form sandstone units were deposited in distributaries and streams. The associated argillite, siltstone, and fine-grained sandstone were deposited in embayments, shallow lakes, and swamps.

Trace fossils in the argillite facies (*Planolites-Palaeophycus* association) were produced by a relatively low-diversity assemblage of small, soft-bodied worm-like deposit feeders. Such an infauna is typical of reduced oxygen or dysaerobic conditions (Theede and others, 1969; Rhoads and Morse, 1971). The interbedding of burrowed and undisturbed layers suggests an environment that was alternately anaerobic and mildly oxygenated (Schäfer, 1972). Similar alternations that occur in the Devonian-Mississippian New Albany and Chattanooga shales of the Appalachian Basin have been attributed to fluctuations in oxygen availability (Ettensohn and Barron, 1981; Miller, 1982).

The *Phycodes-Chondrites* association occurs at the climax of progradational cycles in the sandstone-argillite facies. Ichnogenera in this facies are all feeding burrows, and their horizontal to subhorizontal orientation suggests relatively quiet-water conditions with slow sedimentation rates. This association is interpreted as representing the colonization of an abandoned delta-front environment by deposit feeding invertebrates. The size (diameter) of these organisms (worms and crustaceans) shows an order of magnitude increase over those in the argillite facies, probably signaling an increase in the oxygenation of the water column. Traces in the upper argillaceous units in the sandstone-argillite facies represent both locomotion and feeding activity by worm-like animals in a quiet water regime, possibly lower deltaic

Figure 12. Example of *Phycodes* (= *Buthrotrephis*) *palmatum* with teichnichnid segments from the sandstone-argillite facies in the middle section on Mount Ulmer (MU2 in Figure 3).

plain interdistributary bays. Interdistributary bays and/or lakes higher on the delta to alluvial plain within the coal measures facies also contain horizontal traces, indicating surficial locomotion and shallow tunneling by invertebrates.

The nature of the body of water into which this delta system prograded is less certain, because of the lack of body fossils. This absence could be explained by inadequate search, poor preservation, or scarcity due to rapid sedimentation or by conditions unsuitable to most marine organisms. In southern Africa, Permian marine shelly fauna are restricted to the shales of the southwestern part of the Karoo basin and to the related Kalahari and Warmbad basins (McLachlan and Anderson, 1973). Otherwise, little evidence supports widespread marine conditions in the Karoo basin except perhaps for the trace fossils (Stanistreet and others, 1980). The same relationship holds for the Polarstar Formation. The only evidence of a marine connection to the Ellsworth Mountains basin is the existence of trace fossils that are elsewhere associated with marine sediments.

Using trace fossils as the sole evidence for marine versus nonmarine conditions is fraught with difficulty, because similar patterns of invertebrate behavior occur in marine, brackish, and fresh-water environments (Kamola, 1984). Ichnogenera and patterns of bioturbation in the argillite facies are similar to those observed in DSDP piston cores (Ekdale, 1977); however, similar traces might just as easily be produced in large modern fresh-water lakes (see McCall and Tevesz, 1982). In the sandstone-argillite facies the components of the *Phycodes-Chondrites* association are more suggestive of a marine influence. *Phycodes palmatum,* a typical element of Seilacher's (1967) Cruziana ichnofacies, has been reported from several Paleozoic marine shelf sequences (e.g., Seilacher, 1955; Turner and Benton, 1983). This is the first report of *Phycodes palmatum* from the Gondwana rock succession, and we are not aware of modern or ancient examples from nonmarine settings. Furthermore, specimens referable to *P. palmatum* are known from lower Paleozoic marine strata in the Ellsworth Mountains (University of Wisconsin collection). The morphology of the Polarstar examples of *P. palmatum* and *Chondrites* sp. closely conforms to *Buthrotrephis* (= partim *Phycodes* and *Chondrites*), originally described by Hall (1852) from similar deltaic environments in the Clinton Group (Silurian, New York). Horizontal *Rhizocorallium* sp. is also an element of the Cruziana ichnofacies and occurs in marine units (Fürsich, 1974; Basan and Scott, 1979). However, Fürsich and Mayr (1981) described a possible nonmarine occurrence in the Miocene of Bavaria. *Cochlichnus* sp. from the upper fine-grained sandstone-argillite facies is a notorious facies-breaking form (e.g., Crimes, 1970; Moussa, 1970). However, as noted in the review by Hakes (1976), *Cochlichnus* is a facies indicator in Upper Carboniferous cyclothems of Europe, where it occupies the transition between fresh-water and well-developed brackish conditions. In the southwestern to central Karoo basin, *Cochlichnus* sp. occurs in a marine setting above the tillites (Prince Albert Shale) (Anderson, 1981).

The lack of shelly fauna, along with the occurrence of trace fossils suggesting a marine influence in a thick black shale sequence, suggests that the lower and middle parts of the Polarstar Formation were deposited in a stagnant basin. The deposition of

TABLE 1. MODEL FOR BLACK SHALE DEPOSITION*

SHELF SLOPE	
	aerobic >1.0 ml/l dissolved oxygen
	dysaerobic 0.01 to 1 ml/l dissolved oxygen
	anaerobic <0.01 ml/1 dissolved oxygen

	CHARACTERISTICS	
aerobic	**dysaerobic**	**anaerobic**
calcareous epifauna	no calcareous epifauna	no benthos
highly bioturbated sediments	sediment reworked "due to activities of resistant infauna"—polychaetes, nematodes, poorly calcified crustaceans	laminated sediments—or those deposited from gravity flows

*From Rhoads and Morse, 1971; Byers, 1977.

the dark shales (argillites) characteristic of the lower and middle Polarstar Formation required a source of organic matter, a mechanism for inhibiting oxidation of organic matter, and a lack of dilution by coarser clastic sediment. The abundance of terrestrial plant debris throughout the formation indicates that much of the organic material was land derived. The common occurrence of silty laminations within the argillite, with only sporadic evidence of bioturbation in the lower and middle parts of the formation, suggests low oxygen conditions. Coarse clastic sediment was probably trapped higher on the shelf by the postglacial rise in sea level.

Anoxic environments may develop in a variety of geologic settings, including large lakes, silled basins, cratonic shelves, areas of upwelling, and open oceans where the oxygen-minimum zone impinges on the continental slope or shelf (Tourtelot, 1979; Arthur and Schlanger, 1979; Demaison and Moore, 1980). In the case of the Polarstar Formation, a large lake seems improbable but cannot be ruled out because of the equivocal nature of trace fossils in predicting salinity. The complete lack of a shelly fauna eliminates areas associated with an open ocean setting. Many of the characteristics of the Polarstar sequence can be explained by the depositional model discussed by Rhoads and Morse (1971) and Byers (1977) (see Table 1), which is based on the Black Sea, a silled basin in which stagnation results from density stratification. Anoxic water lies beneath a surficial oxygenated layer. Three biofacies, which are related to stratification of the water column, occur along the gently sloping margin of a stagnant basin in this model. Stagnation occurs when surface and deeper waters do not mix, oxidation of organic material uses up available oxygen, and levels of oxygen therefore decrease with depth. Mixing occurs in surface waters to approximately 50 m depth, and oxygen levels are adequate to support a diverse assemblage of organisms if other conditions are right. Below the mixed surface waters, down to approximately 150 m, is a transitional layer, the pycnocline, which is dysaerobic. Enough oxygen is available to support a variety of soft-bodied burrowing animals but not a shelly fauna. Below approximately 150 m the water is anaerobic, and no fauna is present. These depth figures are approximate and vary according to the setting.

The arc complex along the Pacific margin may have acted as a barrier or sill to restrict the interchange of water with the Pacific Ocean. A high influx of fresh water into the enclosed or semienclosed basin may have produced a pronounced halocline with anoxic conditions below. Large amounts of fresh water were probably available for a limited time from the melting of glaciers. The predominance of coal measures in Permian rocks of Antarctica indicates a generally humid climate, suggesting that rivers could have also provided much fresh-water plant debris to the basin. The lack of shelly fauna in the Polarstar Formation and its correlative shale in the Transantarctic Mountains could also be explained by brackish surface waters. Evidence of bioturbation and trace fossils at some horizons, but a complete lack of body fossils, is supportive of dysaerobic conditions. If normal marine salinities had existed and if oxygen levels were sufficient, a shelly fauna should occur. A diverse cold-water fauna of brachiopods, bryozoans, and bivalves, the *Eurydesma* fauna, does occur in correlative glaciomarine rocks in Tasmania and eastern Australia (Clarke and Banks, 1975).

The upper part of the mixed argillite and sandstone facies, which is interpreted as representing a delta-front environment, was deposited in shallow water or near sea level. The coal measures facies was deposited at or slightly above sea level. Bioturbation is rare in the argillite facies, occurring only near the base of the formation. Evidence of bioturbation and trace fossils is common in the mixed argillite and sandstone facies higher in the sequence, but laminations still occur. These patterns suggest that initial Polarstar deposition began under dysaerobic conditions. Conditions became anoxic as water depth increased during deposition of the argillite facies and returned to dysaerobic to aerobic conditions with the mixed argillite and sandstone facies. The pycnocline may have fluctuated, producing an alternation of anoxic and dysaerobic conditions. Direct biologic evidence for shallow aerobic conditions upward in the sequence is not present, but a shelly fauna might not have occurred in this region under brackish conditions.

PETROGRAPHY

Of the 150 samples collected from the Polarstar Formation, 37 were selected for modal analysis. One of the authors (Vavra) counted a minimum of 300 points per sample; these results are presented in Table 2 and shown graphically in Figure 13. Castle (1974), who also analyzed Polarstar samples, came up with similar results.

Detrital mineralogy

Detrital constituents include quartz, feldspar, lithic fragments, biotite, muscovite and heavy minerals. Detrital grains typically make up 60 to 80 percent of sandstones except where replacement of grains by calcite is extensive.

Unstrained to slightly undulose monocrystalline quartz grains dominate the quartz population. Most grains are subangular to subrounded; however, unstrained grains exhibiting bipyramidal terminations and corrosion embayments occur sparsely in some samples. Polycrystalline quartz grains make up less than 10 percent of the total quartz population.

Plagioclase (andesine-oligoclase) dominates the feldspar population in all samples (P/F = 0.92). Plagioclase grains are subangular to subrounded and most grains are slightly altered to sericite. Replacement by calcite and/or albite is common. Orthoclase typically makes up less than 5 percent of the sandstone, whereas microcline is only present in trace amounts.

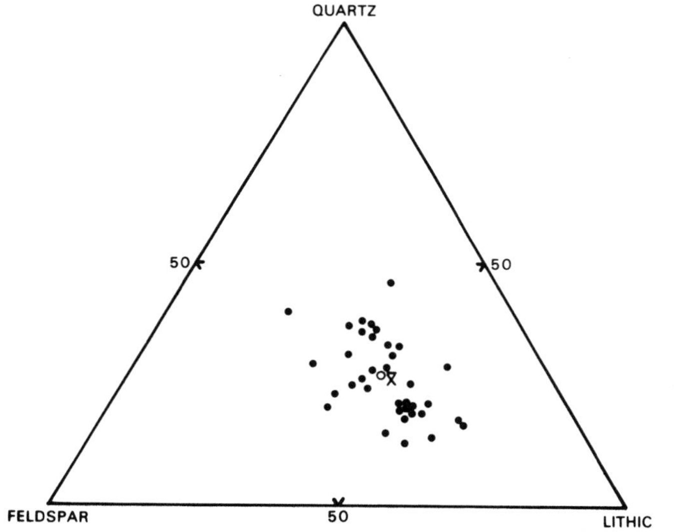

Figure 13. Triangular composition diagram for Polarstar sandstone samples. Solid circles are based on analysis of 300 points per sample. Open circle is mean ($\bar{x}$) sandstone composition.

Biotite and muscovite typically constitute less than 5 percent of sandstones, but in some samples, mica may be as much as 11 percent. Biotite grains are commonly partially to completely altered to chlorite. Muscovite and biotite grains are typically deformed around stable framework grains. Zircon, tourmaline, apatite, garnet, magnetite, and sphene occur in trace amounts.

Lithic fragments, including igneous, metamorphic, and sedimentary types, are the most abundant detrital constituents. The lithic fragment population is typically dominated by felsic volcanic rock fragments, which consist of a microcrystalline mosaic of anhedral quartz and feldspar. Other common volcanic fragments consist of plagioclase microlites, with pilotaxitic or felted textures set in a microcrystalline matrix of quartz and feldspar, and mafic fragments, which consist of plagioclase microlites set in an opaque matrix consisting of magnetite or hematite. Pyroclastic material, including tuffaceous fragments and altered shards, occurs in trace amounts.

Granitic rock fragments constitute a small fraction of the lithic fragment population. Low-grade metamorphic rock fragments, including quartz-mica schist, metaquartzite, and phyllite fragments, make up less than 5 percent of the sandstone. Argillaceous sedimentary rock fragments, which are locally common above scour surfaces in channel-form sandstone in the uppermost part of the formation, are of intraformational origin.

Matrix in Polarstar sandstone consists of orthomatrix, a homogeneous clay matrix formed by recrystallization of detrital matrix, and pseudomatrix, a discontinuous interstitial paste derived from compaction and dispersed on weak lithic fragments into pore space (terminology of Dickinson, 1970). Orthomatrix, which is most common in lenticular and wavy-bedded sandstone in the lower part of the formation, probably represents recrystallization of clay that infiltrated into relatively clay-free sandstone from overlying mud. Pseudomatrix is most common in sandstone that has undergone penetrative deformation. Matrix typically makes up less than 10 percent of the sandstone; however, in lenticular sandstone that has undergone penetrative deformation, orthomatrix, together with pseudomatrix, may make up as much as 27 percent.

Sandstone composition and provenance

Polarstar sandstone is dominantly moderately well sorted, is fine- to medium-grained, and ranges in composition from lithic arkose to feldspathic litharenite (classification of Folk, 1974). Sandstone in the lower part of the formation is typically finer grained and texturally less mature than that in the middle and upper parts. Detrital modes are dominated by silicic to intermediate volcanic rock fragments, plagioclase, and monocrystalline quartz grains. Potassium feldspar, biotite, muscovite, and mafic volcanic, low-grade metasedimentary, granitic, and intraformational sedimentary lithic fragments occur in minor amounts. Although no statistically significant vertical or spatial trends in sandstone composition have been identified, sandstone in the

TABLE 2. MODAL ANALYSES OF SANDSTONE SAMPLES FROM POLARSTAR FORMATION

Specimen	Q	P	K	P/F	V	R	V/L	M	H	Mx	C	A
					COAL MEASURES FACIES							
MV-27.5	10.7	24.7	1.0	0.96	32.0	11.7	0.7	0.7	0.0	6.3	9.0	4.0
MV-3	19.7	17.0	2.7	0.86	23.0	5.7	0.8	0.7	0.3	2.7	1.7	26.7
ZR-70	13.3	18.3	0.3	0.98	19.7	14.3	0.6	0.7	0.0	7.3	2.3	23.7
ZR-58.5	15.7	20.3	0.7	0.97	28.0	10.0	0.7	0.3	0.0	8.3	3.7	13.0
ZR-58	14.0	14.3	2.0	0.88	26.3	23.3	0.5	0.3	0.0	5.3	1.7	12.7
ZR-22	12.7	18.7	2.0	0.90	23.7	12.7	0.7	0.7	0.3	12.7	5.3	11.3
ZR-6	17.3	16.7	2.3	0.88	18.3	13.3	0.6	1.3	0.7	12.0	6.3	11.7
Average	14.8	18.6	1.6	0.92	24.4	13.0	0.7	0.7	0.2	7.8	4.3	14.7
					SANDSTONE-ARGILLITE FACIES							
MU3-139	26.0	18.0	2.3	0.89	16.7	7.7	0.7	4.3	0.3	3.0	14.7	7.0
PP-65	13.0	14.3	1.0	0.93	28.0	16.3	0.6	1.3	0.3	7.7	9.0	9.0
MU3-67.7	17.5	16.3	1.9	0.90	22.8	12.2	0.7	0.3	0.3	14.1	6.3	8.4
MU3-64A	29.7	28.3	0.7	0.98	11.0	2.7	0.8	1.0	0.0	2.0	20.0	4.7
MU3-61	23.3	27.3	5.0	0.85	14.3	10.0	0.6	0.7	1.0	6.0	4.3	8.0
MU3-47.3	17.0	20.7	2.0	0.91	27.0	2.3	0.9	0.3	1.0	5.3	2.3	22.0
MU3-40	30.3	17.7	4.3	0.80	23.0	3.0	0.9	2.3	0.3	13.0	0.7	5.3
MU3-18.7	16.3	16.0	2.3	0.87	34.3	2.7	0.9	3.0	0.3	4.7	5.0	15.3
MU3-17	24.0	15.0	2.0	0.88	25.0	2.0	0.9	4.0	0.3	10.3	6.3	11.0
MU3-12.9	25.0	18.5	2.0	0.90	13.5	4.0	0.8	4.5	0.5	14.5	0.0	17.5
MU3	29.5	19.0	2.0	0.90	19.5	6.0	0.8	4.0	0.5	10.0	1.0	8.5
MW-315	15.7	26.0	1.0	0.96	20.0	4.7	0.8	2.0	1.3	6.7	19.7	3.0
MU2-64	19.0	19.0	5.7	0.77	23.3	4.7	0.8	2.3	0.0	8.7	3.3	14.0
MU2-58.5	26.0	20.3	0.7	0.97	31.3	4.3	0.9	1.3	0.7	4.7	6.7	4.0
MU2-56	22.0	23.3	0.7	0.97	27.7	4.7	0.9	1.0	0.7	5.0	5.3	9.7
MW-260.9	13.7	19.3	0.0	1.0	28.3	7.3	0.8	0.7	0.7	9.3	3.3	17.3
MU2-54.4	27.3	19.3	1.0	0.95	27.7	4.7	0.9	1.7	0.0	1.0	4.7	12.7
PP-34.5	22.3	20.3	3.3	0.86	18.0	5.3	0.8	1.3	0.0	5.3	16.3	7.7
PP-22	23.3	16.7	1.7	0.91	14.3	4.7	0.8	3.7	0.7	0.7	27.3	7.0
PP-17	24.0	9.5	0.0	1.00	9.5	5.0	0.7	3.5	0.5	18.5	1.5	28.0
MW-214.5	9.7	20.3	1.3	0.94	22.0	7.0	0.8	4.0	0.0	13.0	16.7	6.0
MW-213.5	18.7	23.7	3.0	0.89	25.3	4.3	0.9	1.0	0.0	4.0	14.3	5.7
MW-210.5	15.0	28.0	4.0	0.88	19.0	9.0	0.7	1.3	0.0	0.7	12.3	10.7
MW-208.4	14.7	20.3	1.3	0.94	29.3	6.3	0.8	1.7	0.3	8.7	9.7	7.7
MU1-148.8	10.3	20.3	0.0	1.00	34.7	8.3	0.8	1.0	0.0	2.7	17.7	5.0
MU1-139	15.7	21.7	0.3	0.99	33.0	5.7	0.9	0.7	0.0	3.0	16.7	3.3
MU1-134.5	14.7	22.0	0.0	1.00	33.0	8.0	0.8	0.7	0.0	3.0	2.7	16.0
MU1-107.5	22.7	13.7	0.0	1.00	35.7	4.0	0.9	1.7	0.0	9.3	3.3	9.7
Average	20.2	19.8	1.8	0.92	23.8	6.0	0.8	2.0	0.4	7.0	9.0	10.1
					ARGILLITE FACIES							
MW-54.3	11.7	15.0	1.0	0.94	25.3	8.3	0.8	1.0	0.0	8.0	1.0	28.7
WN-44	21.6	15.7	0.0	1.00	8.6	2.2	0.8	10.8	1.1	27.0	0.5	12.4
Average	16.6	15.3	0.5	0.97	17.0	5.2	0.8	5.9	0.5	17.5	0.7	20.6

Sample designations refer to sections in Figure 3 (e.g., WN-44 indicates that this sample was collected 44 m above the base of the measured section on Whiteout Nunatak).

Abbreviations: Q, quartz; P, plagioclase; K, K-feldspar; F, feldspar; V, volcanic lithoclast; R, nonvolcanic lithoclast; L, lithoclast; M, mica; H, heavy minerals; Mx, matrix; C, cement; A, alteration product.

Figure 14. Photomicrograph of glass shards (arrows) in vitric tuff from Mount Weems. Shards are replaced by calcite. Shard in center of photo is 0.24 mm long. Plain light.

lower and upper parts of the formation tends to be more lithic than that in the middle part.

Detrital grains suggest that the source terrane for Polarstar sandstone was dominated by silicic to andesitic volcanic rocks with lesser amounts of mafic volcanic, plutonic, and low-grade metasedimentary rocks. An active volcanic source, possibly a calc-alkaline magmatic arc along the Pacific margin of Gondwanaland, is indicated by the presence of vitric and crystal tuff in the upper part of the formation (Fig. 14).

Postdepositional alterations

Postdepositional alterations to Polarstar mineralogy and fabric reflect diagenesis at near-surface pressure and temperature, and later, very low-grade metamorphism during deep burial and folding associated with the Gondwanide orogenic event. Postdepositional mechanical and chemical modifications have greatly altered the fabric and composition of sandstones. Mechanical modifications include compaction and deformation of unstable rock fragments, fracturing of quartz and feldspar grains, and development of foliation during deep burial and folding. Chemical alterations include formation of clay pore-lining and pore-filling cements; precipitation of quartz, chert, albite, calcite, and dolomite cements; dissolution and replacement of cements and detrital grains by calcite; chloritization of volcanic rock fragments; albitization of detrital plagioclase grains; and formation of dolomite, ferroan dolomite, siderite, pyrite, sphene, and epidote grain replacements and development of veins.

Postdepositional modifications, especially compaction and formation of cements, greatly reduced porosity and permeability so that during deep burial the only effective porosity was fracture porosity.

The fabric of Polarstar sandstones varies vertically through the formation. The fabric of the sandstone in the lower part is characterized by concavo-convex and sutured grain-to-grain contacts and abundant pseudomatrix, a discontinuous paste formed by deformation of weak lithic fragments into intergranular voids. Sandstone in the lower part of the formation commonly displays a foliation that is not parallel to bedding, which probably resulted from ductile deformation and recrystallization during deep burial and folding. Sandstone fabric in the middle and upper parts of the formation is characterized by concavo-convex grain-to-grain contacts and lesser amounts of sutured and long contacts. Bent and wrinkled mica flakes and local pseudomatrix reflect some compaction and ductile deformation, but the main style of deformation is brittle, as illustrated by common fractured feldspar and quartz grains. Throughout the formation, where early calcite cement is present, pseudomatrix is absent, and sandstone is characterized by long and tangential grain-to-grain contacts reflecting depositional packing and fabric.

PALEOGEOGRAPHY

The great thickness of lower Paleozoic rocks suggests that the Ellsworth Mountains area was part of a subsiding continental margin prior to the development of a calc-alkaline arc along the West Antarctic sector of the Pacific plate margin during late Paleozoic time. The upper Paleozoic sequences of the Ellsworth and Transantarctic mountains appear to have accumulated in a major back-arc basin (Barrett and others, 1972; Elliot, 1975; Collinson and others, 1981; Dalziel and Elliot, 1982). Sediments were derived from two major sources: volcanoes in the calc-alkaline arc complex that bordered the Pacific margin, and sedimentary, plutonic, and low-grade metamorphic rocks in the East Antarctic craton (Vavra and others, 1981) (Fig. 15). Areas that were closer to the arc, such as the Ellsworth Mountains, are dominated by volcaniclastic sediments.

Permian drainage along the margin of East Antarctica generally paralleled the trend of the back-arc basin, and streams flowed toward southern Africa in the Transantarctic Mountains south of a drainage divide at the southern end of Victoria Land (Barrett and Kohn, 1975). If the Ellsworth Mountains block is rotated approximately 90 degrees to agree with the paleomagnetic data of Watts and Bramall (1981), paleocurrent directions in the Polarstar Formation line up with those on the adjacent East Antarctic craton. Because Polarstar paleocurrent data represent the migration of deltas rather than stream directions, this possible agreement in directions may be fortuitous. The present directions of paleocurrents in the Polarstar Formation without rotation would agree with a paleogeographic model in which deltas prograded off the East Antarctic craton into the back-arc basin. In either case, it is necessary to account for the dominance of volcaniclastic sandstone, which suggests dispersal from the Pacific margin. The back-arc basin may have filled longitudinally with sediment contributions from both the arc and the craton. An alluvial-deltaic plain may have prograded toward southern Africa, gradually filling the elongate back-arc basin. Paleocurrent data from prograding deltaic deposits in the eastern Karoo basin

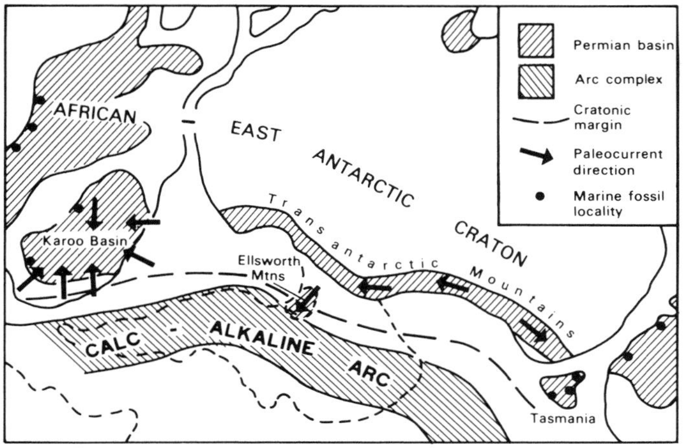

Figure 15. Paleogeographic setting of the Pacific margin of the Antarctic sector of Gondwanaland during Permian time. The Ellsworth Mountains have not been rotated.

suggest dispersal from the direction of Antarctica (Whateley, 1980; Kingsley, 1981).

The histories of Gondwana basins in Antarctica, southern Africa, and South America are very similar. All are characterized by similar upper Paleozoic to lower Mesozoic stratigraphic sequences. The East Antarctic basins, which are seen only in a very narrow cross section along the Transantarctic Mountains, appear to be similar to the intracratonic basins of southern Africa, particularly the Karoo basin. If a cross section could be constructed from the Ellsworth Mountains to the East Antarctic craton, it would probably appear similar to north-south cross sections of the Karoo basin. A thick, upper Paleozoic clastic wedge would thin cratonward, onlapping cratonic basement. In both cases, early Mesozoic folding affected the outer margin of the basins. In both the Ellsworth Mountains and southern Karoo basin, thick glacial diamictites concordantly overlie thick Devonian sandstones and are gradationally overlain by the postglacial black shales that pass upward into fluviodeltaic sandstones. A volcanic source area in the direction of the Pacific margin has also been suggested for volcanic ash beds and rock fragments in Permian rocks of the southern Cape Province (Martini, 1974). Elliot and Watts (1974) discussed the similarities of volcaniclastic material in Karoo and Antarctic rocks.

The Permian postglacial black shale sequence that occurs in Antarctica, in the Karoo basin, and in the Paraná basin of Brazil has been attributed to a large inland sea (Du Toit, 1937; Crowell and Frakes, 1968; Lindsay, 1969; Elliot, 1975). A notable difference between Antarctic Permian deposits and those on other Gondwana continents is the absence of the *Eurydesma* fauna. Marine body fossils are scarce in the Karoo basin but do occur at the base of the shales in the western Karoo basin and subsidiary basins to the west, suggesting a marine incursion into the basin from the west (McLachlan and Anderson, 1973; Tankard and others, 1982). Cooper and Kensley (1984) suggested an oceanic link along the eastern coast of South Africa because of the common occurrence of a Permian bivalve fauna in the southern Cape Province and the Paraná basin. Unless there was a passage across the Pacific margin, this interpretation seems unlikely in light of the evidence presented here. Most of the black shale sequence is regarded as having accumulated in a stratified lacustrine or marginal marine environment (Anderson and others, 1977; Visser and Loock, 1978; Anderson, 1981). However, in addition to the limited occurrence of the *Eurydesma* fauna, trace fossil assemblages similar to those described here in the Polarstar Formation and by Bradshaw and others (1984) in the Discovery Ridge Formation in the Ohio Range have been reported in the Karoo basin (Visser and Loock, 1978; Stanistreet and others, 1980). Because of the marginal-marine aspect of the black shales in Antarctica, it is possible that marine connections of the inland sea to the Pacific were quite distant, perhaps no closer in Gondwanaland reconstructions than along southwestern Africa.

ACKNOWLEDGMENTS

We wish to thank the scientific and support personnel of the 1979–1980 Ellsworth Mountains Expedition for their cooperation. We especially appreciate the advice given in the field by Richard W. Ojakangas. Peter J. Barrett, Robert H. Blodgett, and Molly F. Miller contributed many helpful suggestions pertaining to the manuscript. This research was supported by National Science Foundation Grant DDP 78-21129.

REFERENCES CITED

Anderson, A. M., 1981, The *Umfolozia* arthropod trackways in the Permian Dwyka and Ecca Series of South Africa: Journal of Paleontology, v. 55, p. 84–108.

Anderson, A. M., McLachlan, I. R., and Oelofsen, B. W., 1977, The Lower Permian White Band and the Irati Formation [abs.]: Geological Society of South Africa, 17th Congress, p. 4–7.

Arthur, M. A., and Schlanger, S. A., 1979, Cretaceous "oceanic anoxic events" as causal factors in development of reef-reservoired giant oil fields: American Association of Petroleum Geologists Bulletin, v. 63, p. 870–885.

Barrett, P. J., and Kohn, B. P., 1975, Changing sediment transport directions from Devonian to Triassic in the Beacon Supergroup of south Victoria Land, Antarctica, *in* Campbell, K.S.W., ed., Gondwana geology: Canberra, Australian National University Press, p. 329–332.

Barrett, P. J., Grindley, G. W., and Webb, P. N., 1972, The Beacon Supergroup of East Antarctica, *in* Adie, R. J., ed., Antarctic geology and geophysics: Oslo, Universitetsforlaget, p. 319–332.

Basan, P. B., and Scott, R. W., 1979, Morphology of *Rhizocorallium* and associated traces from the Lower Cretaceous Purgatoire Formation, Colorado: Palaeogeography, Palaeoclimatology, Palaeoecology, v. 28, p. 5–23.

Bradshaw, M. A., Newman, J., and Aitchison, J. C., 1984, Preliminary geological results of the 1983–84 Ohio Range expedition: New Zealand Antarctic Record, v. 5, no. 3, p. 1–17.

Byers, C. W., 1977, Biofacies patterns in euxinic basins; A general model, *in* Cook, H. E., and Enos, P., eds., Deep-water carbonate environments: Society of Economic Paleontologists and Mineralogists Special Publication 25, p. 5–17.

Byers, C. W., and Stasko, L. E., 1978, Trace fossils and sedimentologic interpretation; McGregor Member of Platteville Formation (Ordovician) of Wisconsin: Journal of Sedimentary Petrology, v. 48, p. 1303–1310.

Castle, J. W., 1974, The deposition and metamorphism of the Polarstar Formation (Permian), Ellsworth Mountains, Antarctica [M.Sc. thesis]: Madison, University of Wisconsin, 101 p.

Castle, J. W., and Craddock, C., 1975, Deposition and metamorphism of the Polarstar Formation (Permian), Ellsworth Mountains: Antarctic Journal of the U.S., v. 11, no. 5, p. 239–241.

Clarke, M. J., and Banks, M. R., 1975, The stratigraphy of the lower (Permo–Carboniferous) parts of the Parmeener Super-Group, Tasmania, *in* Campbell, K.S.W., ed., Gondwana geology: Canberra, Australian National University Press, p. 453–467.

Coleman, J. M., and Gagliano, S. M., 1965, Sedimentary structures; Mississippi River deltaic plain, *in* Middleton, G. V., ed., Primary sedimentary structures and their hydrodynamic interpretation; A symposium: Society of Economic Paleontologists and Mineralogists Special Paper 12, p. 133–148.

Collinson, J. W., Stanley, K. O., and Vavra, C. L., 1981, Triassic fluvial depositional systems in the Fremouw Formation, Cumulus Hills, Antarctica, *in* Cresswell, M. M., and Vella, P., eds., Gondwana five; Selected papers and abstracts of papers presented at the Fifth International Gondwana Symposium, Wellington, 1980: Rotterdam, A. A. Balkema, p. 141–148.

Cooper, M. R., and Kensley, B., 1984, Endemic South American Permian bivalve molluscs from the Ecca of South Africa: Journal of Paleontology, v. 58, p. 1360–1363.

Craddock, C., 1969, Geology of the Ellsworth Mountains, *in* Bushnell, V. C., and Craddock, C., eds., Geologic maps of Antarctica: New York, American Geographical Society, Antarctic Map Folio Series, Folio 12, Plate 4.

Craddock, C., Anderson, J. J., and Webers, G. F., 1964, Geologic outline of the Ellsworth Mountains, *in* Adie, R. J., ed., Antarctic geology: Amsterdam, North-Holland, p. 155–170.

Craddock, C., Bastien, T. W., Rutford, R. H., and Anderson, J. J., 1965, *Glossopteris* discovered in West Antarctica: Science, v. 148, no. 3670, p. 634–637.

Crimes, T. P., 1970, The significance of trace fossils in sedimentology, stratigraphy, and palaeoecology with examples from lower Palaeozoic strata, *in* Crimes, T. P., and Harper, J. C., eds., Trace fossils: Liverpool, Seel House Press, Geological Journal Special Issue 3, p. 101–126.

Crowell, J. C., and Frakes, L. A., 1968, Late Paleozoic glacial facies and the origin of the South Atlantic Ocean: International Geological Congress, 23rd, Prague 1968, v. 13, p. 291–302.

Dalziel, I.W.D., and Elliot, D. H., 1982, West Antarctica; Problem child of Gondwanaland: Tectonics, v. 1, no. 1, p. 3–19.

Demaison, G. J., and Moore, G. T., 1980, Anoxic environments and oil source bed genesis: American Association of Petroleum Geologists Bulletin, v. 64, p. 1179–1209.

Dickinson, W. R., 1970, Interpreting detrital modes of graywacke and arkose: Journal of Sedimentary Petrology, v. 40, p. 695–707.

Du Toit, A. L., 1937, Our wandering continents: Edinburgh, Oliver and Boyd, 366 p.

Ekdale, A. A., 1977, Abyssal trace fossils in worldwide Deep Sea Drilling Project cores, *in* Crimes, T. P., and Harper, J. C., eds., Trace fossils 2: Liverpool, Seel House Press, Geological Journal Special Issue 9, p. 163–182.

Elliot, D. H., 1975, Gondwana basins in Antarctica, *in* Campbell, K.S.W., ed., Gondwana geology: Canberra, Australian National University Press, p. 493–536.

Elliot, D. H., and Watts, D. R., 1974, The nature and origin of volcaniclastic material in some Karroo and Beacon rocks: Transactions of the Geological Society of South Africa, v. 77, p. 109–111.

Ettensohn, F. R., and Barron, L. S., 1981, Depositional model for the Devonian–Mississippian black shales of North America; A paleoclimatic-paleogeographic approach, *in* Roberts, T. G., ed., GSA Cincinnati '81 Field Trip Guidebooks, v. 2, Economic geology, structure: p. 344–361.

Folk, R. L., 1974, Petrology of sedimentary rocks: Austin, Hemphill, 182 p.

Fürsich, F. T., 1974, Ichnogenus *Rhizocorallium:* Paläontologische Zeitschrift, v. 48, p. 16–28.

Fürsich, F. T., and Mayr, H., 1981, Non-marine *Rhizocorallium* (trace fossil) from the Upper Freshwater Molasse (Upper Miocene) of southern Germany: Neues Jahrbuch für Geologie und Paläontologie, Monatshefte 1981, p. 321–333.

Grunow, A. M., Dalziel, I.W.D., and Kent, D. V., 1987, Ellsworth–Whitmore Mountains crustal block, western Antarctica: New paleomagnetic results and their tectonic significance, *in* McKenzie, G. D., ed., Gondwana six: Structure, tectonics, and geophysics: American Geophysical Union Geophysical Monograph 40, p. 161–171.

Hakes, W. G., 1976, Trace fossils and depositional environment of four clastic units, Upper Pennsylvanian megacyclothems, northeast Kansas: University of Kansas Paleontological Contributions 63, 46 p.

Hall, J., 1852, Palaeontology of New York, v. 2: Albany, New York, C. Van Benthuysen, 358 p.

Jankowski, E. J., and Drewry, D. J., 1981, The structure of West Antarctica from geophysical studies: Nature, v. 291, p. 17–21.

Kamola, D. L., 1984, Trace fossils from marginal-marine facies of the Spring Canyon Member, Blackhawk Formation (Upper Cretaceous), east-central Utah, *in* Miller, M. F., Ekdale, A. A., and Picard, M. D., eds., Trace fossils and paleoenvironments; Marine carbonate, marginal marine terrigenous, and continental terrigenous settings: Journal of Paleontology, v. 58, p. 529–541.

Kingsley, C. S., 1981, A composite submarine fan delta-fluvial model for the Ecca and Lower Beaufort groups of Permian age in the eastern Cape Province, South Africa: Transactions of the Geological Society of South Africa, v. 84, p. 27–40.

Lindsay, J. F., 1969, Stratigraphy and sedimentation of lower Beacon rocks in the central Transantarctic Mountains: Columbus, Ohio State University Research Foundation, Institute of Polar Studies Report 33, 58 p.

Longshaw, S. K., and Griffiths, D. H., 1983, A palaeomagnetic study of Jurassic rocks from the Antarctic Peninsula and its implications: Journal of the Geological Society of London, v. 140, pt. 6, p. 945–954.

Martini, J.E.J., 1974, On the presence of ash beds and volcanic fragments in the graywackes of the Karroo System in the southern Cape Province (South Africa): Transactions of the Geological Society of South Africa, v. 77, p. 113–116.

McCall, P. L., and Tevesz, M.J.S., 1982, The effects of benthos on physical properties of freshwater sediments, *in* McCall, P. L., and Tevesz, M.J.S., eds., Animal-sediment relations; The biogenic alteration of sediments: New York, Plenum Press, p. 105–175.

McLachlan, I. R., and Anderson, A., 1973, A review of marine conditions in southern Africa during Dwyka times: Palaeontologica Africana, v. 15, p. 37–64.

Miller, M. F., 1982, Biogenic structures as indicators of oxygen availability during deposition of the Chattanooga and Maury Shales (Devonian–Mississippian), central Tennessee: Geological Society of America Abstracts with Programs, v. 15, p. 214.

Moussa, M. T., 1970, Nematode fossil trails from the Green River Formation (Eocene) in the Uinta basin, Utah: Journal of Paleontology, v. 44, p. 304–307.

Ojakangas, R. W., and Matsch, C. L., 1981, The Late Palaeozoic Whiteout Conglomerate; A glacial and glaciomarine sequence in the Ellsworth Mountains, West Antarctica, *in* Hambrey, M. J., and Harland, W. B., eds., Earth's pre-Pleistocene glacial record: Cambridge University Press, p. 241–244.

Pemberton, S. G., and Frey, R. W., 1982, Trace fossil nomenclature and the *Planolites–Palaeophycus* dilemma: Journal of Paleontology, v. 56, no. 4, p. 843–881.

Reineck, H. E., and Singh, I. B., 1980, Depositional sedimentary environments, with reference to terrigenous clastics, second, revised and updated edition: New York, Springer-Verlag, 542 p.

Rhoads, D. C., and Morse, J. W., 1971, Evolutionary and ecologic significance of oxygen-deficient marine basins: Lethaia, v. 4, p. 413–428.
Rigby, J. F., and Schopf, J. M., 1969, Stratigraphic implications of Antarctic paleobotanical studies, *in* Gondwana stratigraphy; IUGS Symposium, Buenos Aires, 1–15 October 1967: Paris, UNESCO, p. 91–106.
Schäfer, W., 1972, Ecology and palaeoecology of marine environments (translated from the German by Irmgard Oertel; edited by G. Y. Craig): Chicago, University of Chicago Press, 568 p.
Schopf, J. M., 1967, Antarctic fossil plant collecting during the 1966–1967 season: Antarctic Journal of the United States, v. 2, no. 4, p. 114–116.
—— , 1969, Ellsworth Mountains; Position in West Antarctica due to sea-floor spreading: Science, v. 164, p. 63–66.
Seilacher, A., 1955, Spuren und fazies in Unterkambrium, *in* Schindewolf, O. H., and Seilacher, A., eds., Beiträge zur kenntnis des Kambriums in der Salt Range (Pakistan): Akademie der Wissenschaften und der Literature Mainz, Mathematisch-naturwissenschaftliche Klasse, Abhandlungen 10, 1955, p. 11–143.
—— , 1967, Bathymetry of trace fossils: Marine Geology, v. 5, p. 413–428.
Stanistreet, I. G., Smith, G. L., and Cadle, A. B., 1980, Trace fossils as sedimentological and palaeoenvironmental indices in the Ecca Group (Lower Permian) of the Transvaal: Transactions of the Geological Society of South Africa, v. 83, p. 333–344.
Tankard, A. J., Jackson, M.P.A., Eriksson, K. A., Hobday, D. K., Hunter, D. R., and Minter, W.E.L., 1982, Crustal evolution of southern Africa; 3.8 billion years of earth history: New York, Springer-Verlag, 523 p.
Tasch, P., 1967, Antarctic fossil conchostracans and the continental drift theory: Antarctic Journal of the United States, v. 2, no. 4, p. 112–113.
—— , 1968, Trace fossils from the Permian Polarstar Formation, Sentinel Mountains, Antarctica: Kansas Academy of Science Transactions, v. 71, no. 2, p. 184–194.
Tasch, P., and Riek, E. F., 1969, Permian insect wing from Antarctic Sentinel Mountains: Science, v. 164, p. 1529–1530.
Theede, H., Ponat, A., Hiroki, K., and Schlieper, C., 1969, Studies on the resistance of marine bottom invertebrates to oxygen deficiency and hydrogen sulfide: Marine Biology, v. 2, p. 325–337.
Tourtelot, H. A., 1979, Black shale; Its deposition and diagenesis: Clays and Clay Minerals, v. 27, p. 313–321.
Turner, B. R., and Benton, M. J., 1983, Paleozoic trace fossils from the Kufra basin, Libya: Journal of Paleontology, v. 57, p. 447–460.
Vavra, C. L., Stanley, K. O., and Collinson, J. W., 1981, Provenance and alteration of Triassic Fremouw Formation, central Transantarctic Mountains, *in* Cresswell, M. M., and Vella, P., eds., Gondwana five; Selected papers and abstracts of papers presented at the Fifth International Gondwana Symposium, Wellington, 1980: Rotterdam, A. A. Balkema, p. 149–153.
Visser, J.N.J., and Loock, J. C., 1978, Water depth in the main Karoo basin, South Africa, during Ecca (Permian) sedimentation: Transactions of the Geological Society of South Africa, v. 81, p. 185–191.
Watts, D. R., and Bramall, A. M., 1981, Palaeomagnetic evidence for a displaced terrain in Western Antarctica: Nature, v. 293, p. 638–641.
Whateley, M.K.G., 1980, Deltaic and fluvial deposits of the Ecca Group, Nongoma graben, northern Zululand: Transactions of the Geological Society of South Africa, v. 83, p. 345–351.
Yoshida, M., 1982, Superposed deformation and its implication to the geologic history of the Ellsworth Mountains, West Antarctica: Tokyo, National Institute of Polar Research Memoirs, Special Issue 21, p. 120–171.
—— , 1983, Structural and metamorphic history of the Ellsworth Mountains, West Antarctica, *in* Oliver, R. L., James, P. R., and Jago, J. B., eds., Antarctic earth science: Canberra, Australian Academy of Science, p. 266–269.
Zawiskie, J. M., 1984, Permian ichnofossils from East Antarctica (Nimrod–Ohio basin) and the Ellsworth Mountains: Geological Society of America Abstracts with Programs, v. 16, p. 704.

Manuscript Submitted March 1985
Manuscript Accepted by the Society March 1, 1989

Printed in U.S.A.

Geological Society of America
Memoir 170
1992

Chapter 6

Facies of Cambrian carbonate rocks, Ellsworth Mountains, West Antarctica

Werner Buggisch
Institut für Geologie, Universität Erlangen, D-8500 Erlangen, Federal Republic of Germany
Gerald F. Webers
Department of Geology, Macalester College, St. Paul, Minnesota 55105

ABSTRACT

Carbonate rocks were studied from the Middle Cambrian Drake Icefall Formation and the Middle to Upper Cambrian Minaret Formation. Additionally, carbonate clasts from the Middle Cambrian Union Glacier Formation and the Permo-Carboniferous Whiteout Conglomerate were examined. No *in situ* Lower Cambrian carbonate rocks are known to crop out in the Ellsworth Mountains; only reworked clasts of such rocks are found in the Heritage Group and the Whiteout Conglomerate. These clasts suggest that during Early (and Middle?) Cambrian time, a wide carbonate platform developed in or close to the Ellsworth Mountains. This was the site of skeletal algae and archaeocyathid boundstones with sparry calcite or mud as matrix—typical sediments of an open marine environment with low to medium hydrodynamic energy. Oolites are common, and a high-energy environment was required to form these oosparites. These grains, with a diameter between 4 and 7 mm, are made up of concentric ooids with simple and complex structures. Nuclei consist of abraded ooids and oolitic intraclasts. Half-moon ooids with collapsed internal structure are attributed to partial solution, and they indicate a hypersaline depositional environment (aragonite or calcium sulfate?) for some parts of the oolitic bank facies.

Laminites with fenestral structures are also present, and they represent low-energy lagoonal deposits. These occur along with dome-shaped and LLH stromatolites that indicate an intertidal environment. Some areas of the carbonate platform were elevated and partly dissolved by fresh water. During the following period of subsidence, voids were filled with fibrous carbonates by rhythmic cementation. Renewed uplift resulted in erosion and destruction of the carbonate platform.

The lowest in situ marly and oolitic carbonates are found in the Middle Cambrian strata of the Heritage Group north of Drake Icefall, but these were not extensively studied. The Upper Cambrian Minaret Formation is also autochthonous, and its thickness increases southward from 8 m in the northern Webers Peaks to several hundred meters in the Marble Hills area. Medium to high hydrodynamic energy conditions prevailed during deposition of the Springer Peak section of the Minaret Formation (biosparite, oncosparite, pelsparite). The high diversity of the fauna indicates an open marine environment. The section at Yochelson Ridge starts with a few meters of still-water carbonate rocks that are overlain by high-energy oolitic carbonate rocks and calcarenites. Farther to the south (south of Mount Dolence), facies fluctuate from medium-energy (oncolite) to high-energy (oolite) environments. Fossils (brachiopods

Buggisch, W., and Webers, G. F., 1992, Facies of Cambrian carbonate rocks, Ellsworth Mountains, West Antarctica, *in* Webers, G. F., Craddock, C., and Splettstoesser, J. F., Geology and Paleontology of the Ellsworth Mountains, West Antarctica: Boulder, Colorado, Geological Society of America Memoir 170.

and trilobites) are rare. Near the top of the sequence a hypersaline milieu is indicated by layers of early diagenetic dolomicrite. Similar facies are exposed in the Liberty and Marble Hills areas. Fast subsidence in this area was compensated by rapid sedimentation of shallow-water carbonates.

INTRODUCTION

The Ellsworth Mountains are divided into a northern range, the Sentinel, and a southern range, the Heritage (Fig. 1). Carbonate rocks are of only minor importance in the Paleozoic stratigraphic succession of the Ellsworth Mountains (Fig. 2). In the Sentinel Range, the only calcareous components are clasts in a glacigenic Permo-Carboniferous diamictite (Whiteout Conglomerate).

In the Heritage Range, however, carbonate rocks are more important. Reworked carbonate clasts are found in the Middle Cambrian Union Glacier and Conglomerate Ridge Formations and in the Permo-Carboniferous Whiteout Conglomerate. Autochthonous carbonates are found in the Middle Cambrian Drake Icefall Formation and the Minaret Formation. The latter formation increases in thickness from north to south, attaining a thickness of as much as 700 m in the Marble Hills. The Cambrian Minaret Formation was originally described by Craddock (1969) as the Minaret Group and was considered upper Proterozoic in age.

One of us (W.B.) had the opportunity to study and sample the carbonate rocks in conjunction with the American Ellsworth Mountains expedition during the Antarctic summer 1979–80. The material from the Marble Hills and the Liberty Hills was collected by J. M. Anderson in the same season. Specimens collected by T. W. Bastien, C. Craddock, K. B. Spörli, and G. F. Webers from previous expeditions were studied at the Department of Geology, University of Wisconsin–Madison.

Carbonate rocks of the Heritage Group

Lower Heritage Group (Lower and Middle Cambrian). The Heritage Group is characterized by distinct and rapid changes in sedimentary facies. The 3,000-m-thick Union Glacier Formation, exposed in the Edson Hills, forms the base of the Heritage Group. It is composed predominantly of mass-flow and mudflow sediments, but it also includes lapilli-tuffs, volcanic bombs, and effusive and intrusive rocks (Webers and others, this volume). The most complete section of the lowest known sedimentary sequence is exposed at the southern margin of the Hyde Glacier in the Edson Hills (Fig. 3). About 180 m above the lowest exposed beds, massive polymict conglomerates and breccias of the 390-m-thick Kosco Peak Member are exposed. The matrix of this conglomerate is composed of quartz and dolomite sand. Individual sand grains are mantled by a micritic envelope ("superficial ooids"). The components range from a few millimeters to almost 1 m in size; clasts from 2 to 8 cm are the most common. The grain-size frequency distribution and the modal composition were measured along a reference line. Gray carbonate rocks are the most abundant clasts; red carbonate rocks are rare. Oolitic and algal limestone clasts occur in the lower portion of the conglomerate. Red pelites as well as red and gray arenites represent the remaining components. In contrast to the tectonically strained volcaniclastic sediments above and below, the carbonate conglomerates of the Kosco Peak Member are only weakly deformed.

The following types were identified among the carbonate clasts:

1. Stromatolites of the type LLS (in the sense of Logan and others, 1964). A further classification was not possible because of the small size of the clasts.

2. Pelmicrites to pelsparites, some with fenestral fabrics (type *Stromatactis*), partly laminated (Plate 1, Fig. 1).

3. Oolites (Plate 2, Figs. 1, 2) with single ooids, complex ooids, and rounded ooid fragments that are mostly recoated. Many details can be observed in the oolite components of the lower Heritage Group. The cores are commonly recrystallized or consist of reworked oolite components like those described from the upper Precambrian oolites of Spitsbergen by Radwanski and Birkenmajer (1977). These authors emphasized—in comparison to recent ooids—the unusual size of the ooids and explained this feature by a change in the physico-chemical conditions. The ooids of the lower Heritage Group, with a diameter of 4 to 7 mm, display a mainly tangential structure. Radial internal fabric is evidenced by cracks caused by the selective solution of an ooid layer during compaction. Frequent reworking is documented by intraclasts and abraded ooids. Half-moon ooids (Carozzi, 1963) were observed in the material collected by Craddock from the adjacent Patriot Hills (Plate 2, Figs. 4, 5). The cavities of the oosparites are filled with drusy cement and block cement.

4. Algal limestones (Plate 1, Figs. 3–5). Micritic tube-like structures are ascribed to calcareous algae of the type *Renalcis.* The interior is composed of sparry carbonate, and the irregular tubes are overgrown by fibrous cement. Rhythmic growth is evidenced by light and dark layers. The remaining open spaces are filled by mud or cement.

5. Cavity fillings (Plate 1, Figs. 2, 3; Plate 2, Fig. 3). Many cavities developed, partly as a result of primary porosity and partly as a result of solution, which cuts into older structures. The cavities were rhythmically filled with marginal fibrous cement and finally with block cement. A late diagenetic dolomitization affected only the original sediment. The texture of the fibrous cement and block cement is preserved. Rare late diagenetic dolomitic rhomboids protrude into the cement.

The grain-size, sorting, and abundance of the carbonate components led to the inference that the constituent particles of

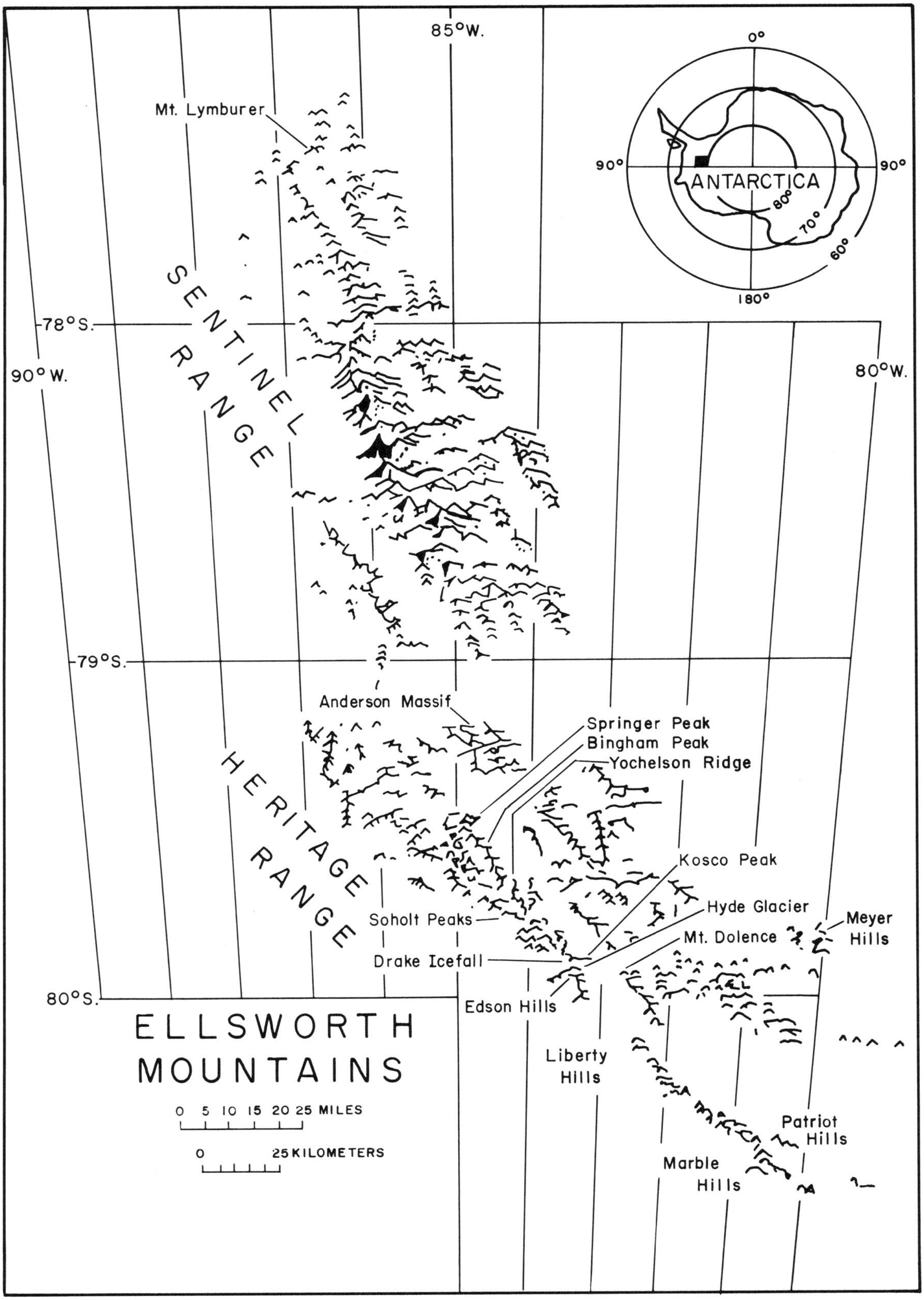

Figure 1. Map showing location and geographic features of the Ellsworth Mountains.

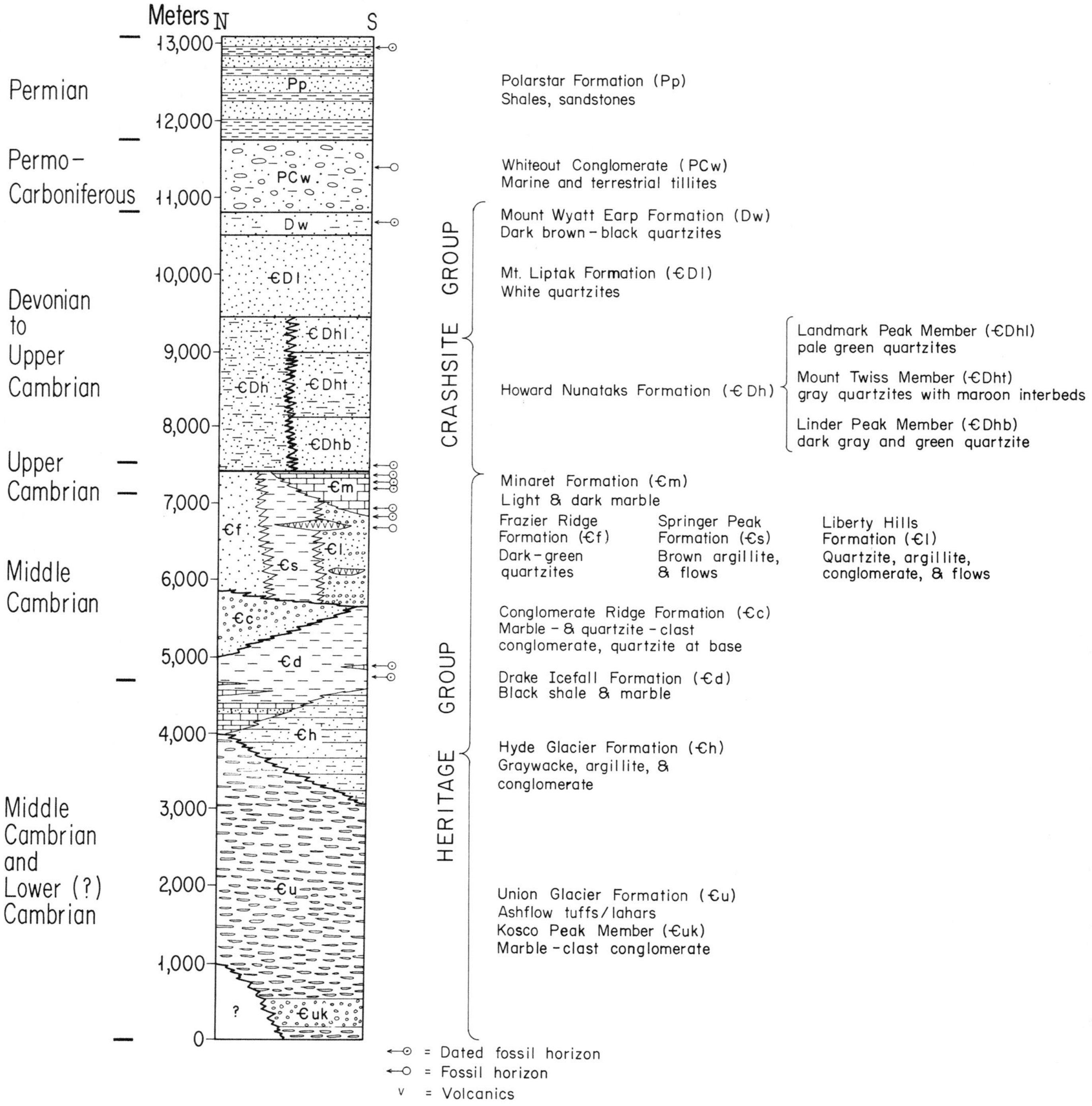

Figure 2. Geologic column of the Ellsworth Mountains.

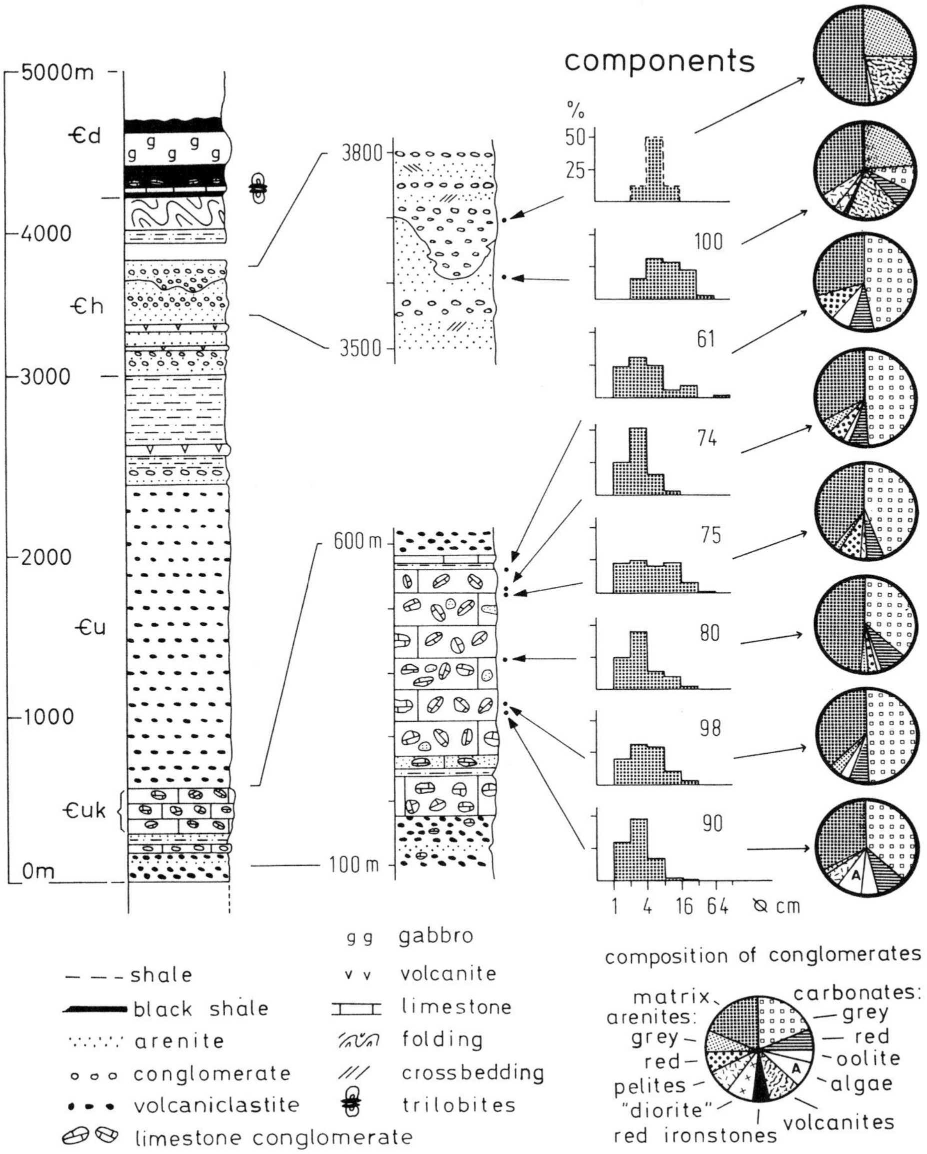

Figure 3. Columnar section of the lower Heritage Group, Hyde Glacier, Edson Hills.

the conglomerate originated in a neighboring source area. A possible source area in or near the Ellsworth Mountains is unknown. Nevertheless, facies analysis of the clasts allows some conclusions about the original area in which the carbonate rocks were formed.

The source area must have been an extended carbonate platform with areas of high and low hydrodynamic energy. The ooids developed in highly agitated water and were subsequently partly destroyed. Final sedimentation and lithification apparently took place in moderately agitated water because almost all particles were coated with carbonates in the last stage without destruction of the envelopes by later abrasion. Half-moon ooids (from the Patriot Hills) indicate a hypersaline environment in some parts of the carbonate platform.

Pelmicrites and pelsparites with fenestral structures may have formed in a lagoonal still water or supratidal area. Stromatolites are attributed to intertidal environments, while the tube-shaped calcareous algae developed in moderately agitated subtidal water. This is substantiated by the good preservation of the tubes and winnowing of the mud. The early cementation documented in thin sections (Plate 1, Figs. 3–5) suggests the possibility of algal-cemented buildups.

The area of deposition ranged from subtidal to supratidal regions. Temporarily, sections of the carbonate platform were elevated, resulting in the development of solution porosity in the vadose zone. After the subsequent phase of submergence, the pores were infilled with radial fibrous cement and finally closed with blocky cement. A renewed uplift resulted in erosion and formation of carbonate rock conglomerate in the lowermost part of the Heritage Group.

The occurrence of algae in the clasts provides information on the age of the Kosco Peak Member of the Union Glacier Formation. Skeletal algae are generally unknown before the Cambrian. Kolosov (1979) and Voronova (1979) recently recovered calcareous algae from the Eo-Cambrian of the Siberian and Russian platforms. Nevertheless, the primary carbonate platform that was the source area of this conglomerate is probably not older than Cambrian in age. In the upper portion of the section at the Hyde Glacier, Middle Cambrian beds are evidenced by agnostids and other trilobites (Jago and Webers, this volume). Consequently the oldest strata (Union Glacier Formation) of the Ellsworth Mountains can be dated as Lower(?) to Middle Cambrian.

Middle Heritage Group. In the southern Soholt Peaks area at the northern rim of the Drake Icefall, dark oolitic limestones occur interbedded with black shales in the Drake Icefall Formation. A Middle Cambrian trilobite fauna has been recovered from these limestones (Jago and Webers, this volume). To a large extent, the ooids of these limestones are recrystallized to radial aggregates, and they have almost or completely lost their concentric structure.

A formation with carbonate rock conglomerates occurs in the southern Soholt Peaks area overlying the Drake Icefall Formation. This unit, the Conglomerate Ridge Formation, is lithologically similar to the Kosco Peak Member of the Union Glacier Formation described from the lower Heritage Group (Webers and others, this volume). The Conglomerate Ridge Formation is estimated to be about 600 m thick. The upper part of this unit is a 450-m-thick buff, polymict, clast-supported conglomerate with interbeds of quartzose sandstone. Clasts make up about 70 percent by volume and are composed dominantly of buff and gray limestone with a lesser amount of buff quartzite. These conglomerates reflect nearby source areas of moderate relief and represent fluvial and/or shallow marine deposition. Only preliminary studies were made of the Conglomerate Ridge Formation because of insufficient time at the end of the season.

Upper Heritage Group. The Minaret Formation of the upper Heritage Group is the most prominent carbonate unit in the Ellsworth Mountains (Figs. 4, 5). An 8- to 10-m-thick limestone bed extends from the northern Webers Peaks toward the south. Its thickness increases considerably to a maximum of as much as 700 m in the Marble Hills of ther southernmost Heritage Range. Because of the rich fossil content (trilobites, mollusks, pelmatozoa, archaeocyathids, and brachiopods) in the area of the Webers Peaks, the earlier studies concentrated on the fauna of these particular limestones (Webers, 1972, 1981; Yochelson and others, 1973). Conodonts discovered in this formation (Buggisch, 1982; Buggisch and others, this volume) represent a first report of conodonts for Antarctica. Additional recent faunal studies of the Minaret Formation include Shergold and Webers (this volume), Webers and others (this volume), and Henderson and others (this volume).

The Minaret Formation was studied at Springer Peak (Fig. 4, Section Ia–e; Fig. 5, Section 1), Bingham Peak (Fig. 4, Section II; Fig. 5, Section 2), South of Mount Dolence (Fig. 5, Section 4), and in the Marble Hills. At Springer Peak, High-energy biosparites with coquinas of trilobites, mollusks, and pelmatozoan sparites (Plate 3, Fig. 1) are interlayered with sediments of low hydrodynamic energy. Locally, fragments of pelmatozoans are micritized at the margins or exhibit syntaxial rim cement (Plate 3, Fig. 3). Other components are coated with algae and developed gradually into oncoids (Plate 3, Fig. 2). The oncoloites commonly contain peloids and are only slightly winnowed. The biopelsparites to biomicrites, with sheltered areas below shells, also point to moderately agitated shallow water (Plate 3, Fig. 4). Locally, the fenestral fabric of the type *Stromatactis* indicates a "lagoonal" environment.

The thickness of the carbonate formation increases considerably from Springer Peak to Bingham Peak, and this thickening is associated with intensified tectonic strain of the rock formations (Plate 4). Accordingly, oncoids and the other components are strongly stretched and recrystallized. Nevertheless, well-preserved trilobites were recovered from the limestones and the overlying black shale and siltstones. In the Yochelson Ridge section, the sedimentation starts above black shales with restrictive carbonates. They are intercalated with fossiliferous limestones and are followed upward by oolites. Silicoclastic grains are commonly coated with carbonate. Only a minimum thickness of 80 m can

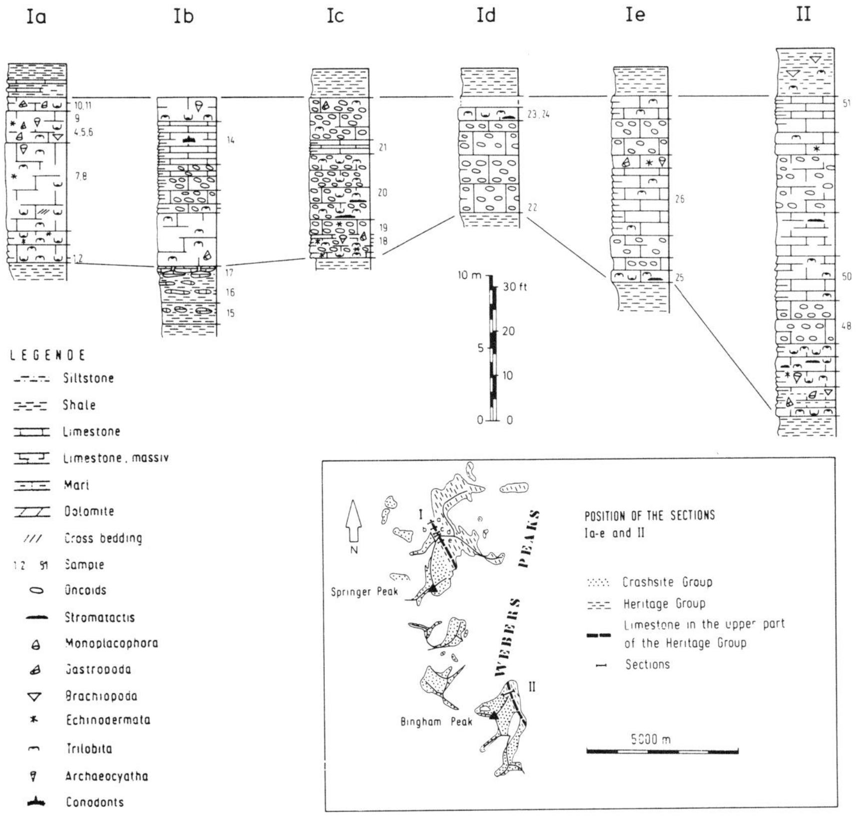

Figure 4. Columnar sections of the Minaret Formation, Webers Peaks.

be recorded because the section is not completely exposed. South of Mount Dolence, the thickness continues to increase, but the complicated structure makes it difficult to measure; a thickness of 120 to 400 m is estimated.

The lithology of the section south of Mount Dolence is characterized by oolites and oncolites. Cross sections of brachiopods and trilobites in thin sections prove—in contrast to earlier assumptions—a Paleozoic age, confirming the result of additional mapping: that these carbonates belong to the same formation that is exposed at Springer Peak. Thin dolomicrite layers withut fossils suggest intervals of increased salinity in the higher portion of the section. At the top of the sequence south of Mount Dolence the contact appears eroded and is overlain by sandstones of the lowermost Crashsite Group.

The strongly deformed carbonate rocks of the Liberty Hills/Marble Hills region, which were dated as Precambrian(?) by Craddock (1969), display similar facies types in spite of their alteration. A well-preserved specimen reveals fossil shells as well as peloids and intraclasts. Furthermore, trilobites and mollusks were found in the Marble Hills (Shergold and Webers, this volume; Webers and others, this volume). This suggests that the Minaret Formation not only thickens to the south but also represents a greater time span of Middle to Late Cambrian (Plate 5, Figs. 2, 3). Additional small outcrops of the Minaret Formation exist in the Soholt Peaks and the Anderson Massif.

A paleogeographic interpretation is difficult because of the chiefly linear distribution of the outcrops. The oolites and oncolites of the Marble Hills, with a thickness of as much as 700 m, were deposited in agitated shallow water with contemporaneous subsidence of the sedimentary platform. Terrestrial detritus was unimportant. South of Mount Dolence, the carbonate sediments were deposited in a similar environment, but the total thickness is less there. Dolomicrites point to a temporary separation from the open sea and to increased salinity. A probable unconformity at the top of this section suggests some emergence of the carbonate platform in this area.

The oolites and silt content of the Minaret Formation at Yochelson Ridge indicate a high-energy environment with significant terrestrial influx. In contrast, the limestones of Webers Peaks are almost lacking in silt. The latter were formed in somewhat agitated water, representing an open marine environment, in contrast to the rather restricted surroundings of the oolitic limestones in the southern Heritage Range.

Carbonate rocks in the Whiteout Conglomerate

The clasts of the Permo-Carboniferous diamictites consist chiefly of quartzites, metamorphic rocks, and magmatic rocks, with only minor carbonate rocks. Plutonic and medium- to high-grade metamorphic rocks are not exposed in the Ellsworth

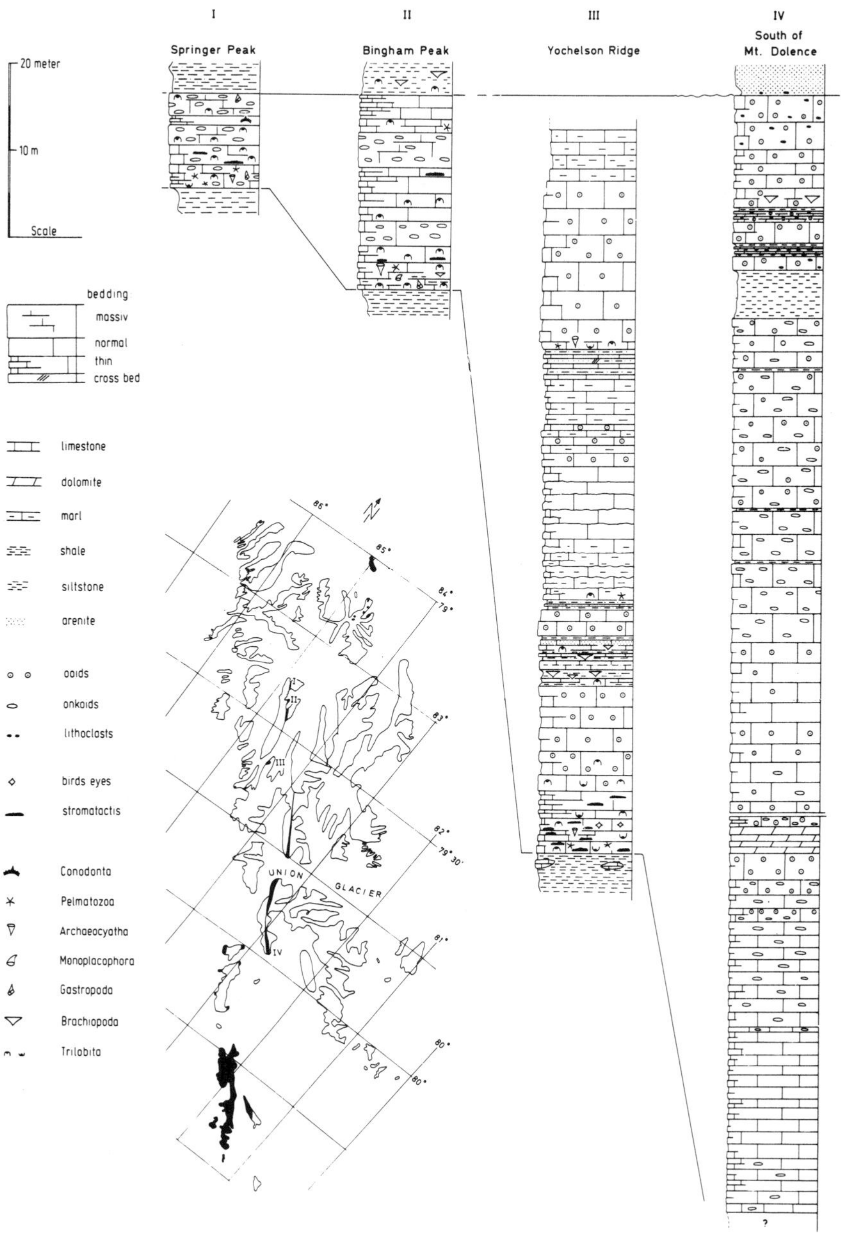

Figure 5. Columnar sections of the Minaret Formation of the Heritage Range.

Mountains, so their source region remains unknown. The following types of carbonate rocks have been identified:

1. Fine-grained limestones without fossils.
2. Dolointrasparites with more or less recrystallized and reworked intraclasts.
3. Archaeocyathid and algal boundstones with *Epiphyton* sp.
5. Stromatolite-archaeocyathid boundstones.
6. Pelsparites with fenestral fabric and trilobites.
7. Oolites with mean grain diameter as much as several millimeters.

Debrenne (this volume) has identified Lower Cambrian archaeocyathids in clasts collected by C. Craddock and G. F. Webers from the Whiteout Conglomerate in the northern Sentinel and southeastern Heritage Ranges, respectively. This is consistent with age estimates based on thin sections in this study. Facies types that are characteristic of the Minaret Formation could not be found. Accordingly, the source area is not known in the Ellsworth Mountains. Similar carbonate materials were recovered during the Transantarctic Expedition (1955–58) in the moraine of the Recovery Glacier in the vicinity of Whichaway Nunataks (about 81°30′S, 28°W), and they were described by Hill (1965). Algal-archaeocyathid limestones are also exposed in the Nelson Limestone of the Pensacola Mountains (Schmidt and Ford, 1969).

The carbonate rock clasts studied give some evidence of the primary sedimentation area. However, the question remains open whether the clasts originated in one or several source areas of the same or of different ages. High hydrodynamic energy is documented by large ooids, and low energy by pelsparites with fenestral fabrics. Archaeocyathid-*Epiphyton* and archaeocyathid-trilobite wackestones are comparable with the concept given by Zhuravleva (in Hill, 1972, Fig. 25) for the Siberian platform. Optimal living conditions may have prevailed for stromatolite-archaeocyathid bioherms in a water depth of 20 to 30 m. In thin section (Plate 5, Fig. 4), the central cavity is filled with dogtooth cement, and the remaining void with a micrite or dismicrite containing crystal silt. This feature indicates the influx of meteoric water or temporary subaerial exposure (Flügel, 1978). Trilobite wackestone with small archaeocyathids is indicative of open marine environments with little current activity.

CONCLUSIONS

The Middle to Upper Cambrian carbonate rocks of the Minaret Formation represent the only autochthonous limestones of the Ellsworth Mountains aside from the rare oolites and marls of the middle Heritage Group. The other Cambrian carbonates are only known as clasts in deposits of Middle Cambrian and Permo-Carboniferous ages.

The facies of the carbonate rocks can be attributed to different areas on large platforms similar to recent shelf areas in warm climates. The isolated position of the Ellsworth Mountains raises the question of the location of the source area from which the Cambrian carbonate rock clasts as well as the plutonic and medium- to high-grade metamorphic rock components originated.

The Lower Cambrian archaeocyathid limestones of the Whichaway Nunataks and the Whiteout Conglomerate suggest a close relation between the Ellsworth Mountains and the Pensacola Mountains in Cambrian time. The other Cambrian carbonate rock clasts may have been deposited at the shelf margin of the East Antarctic continental shield. The preceding observations concerning the carbonate facies are in accordance with the concept of de Wit (1977), Watts and Bramall (1980; 1981), and other authors who postulate a rotation of the Ellsworth microplate from a position close to the margin of East Antarctica into its present geographic location.

PLATE 1
Photomicrographs of limestone conglomerate of the lower Heritage Group: Hyde Glacier Section, Edson Hills
(All photomicrographs taken in plain light.)

Figure 1. Polymict conglomerate with clasts of various facies. Thin section Ant 79-90b × 7.

Figure 2. Oolitic clast with solution vugs. Vugs are closed by rhythmic precipitation of cement. Thin section Ant 79-90c × 7.

Figure 3. Clast with solution vugs. The original sediment is completely replaced by coarse dolomite crystals. The vugs are closed by rhythmic marginal fibrous cements. Thin section Ant 79-90 × 15.

Figure 4. Algal limestone (bindstone). Tubes with varying diameters, but not branched. The wall consists of micrite, the interior of sparite. Thin section Ant 79-171 × 7.

Figure 5. Same as Fig. 4. Thin section Ant 79-109 × 15.

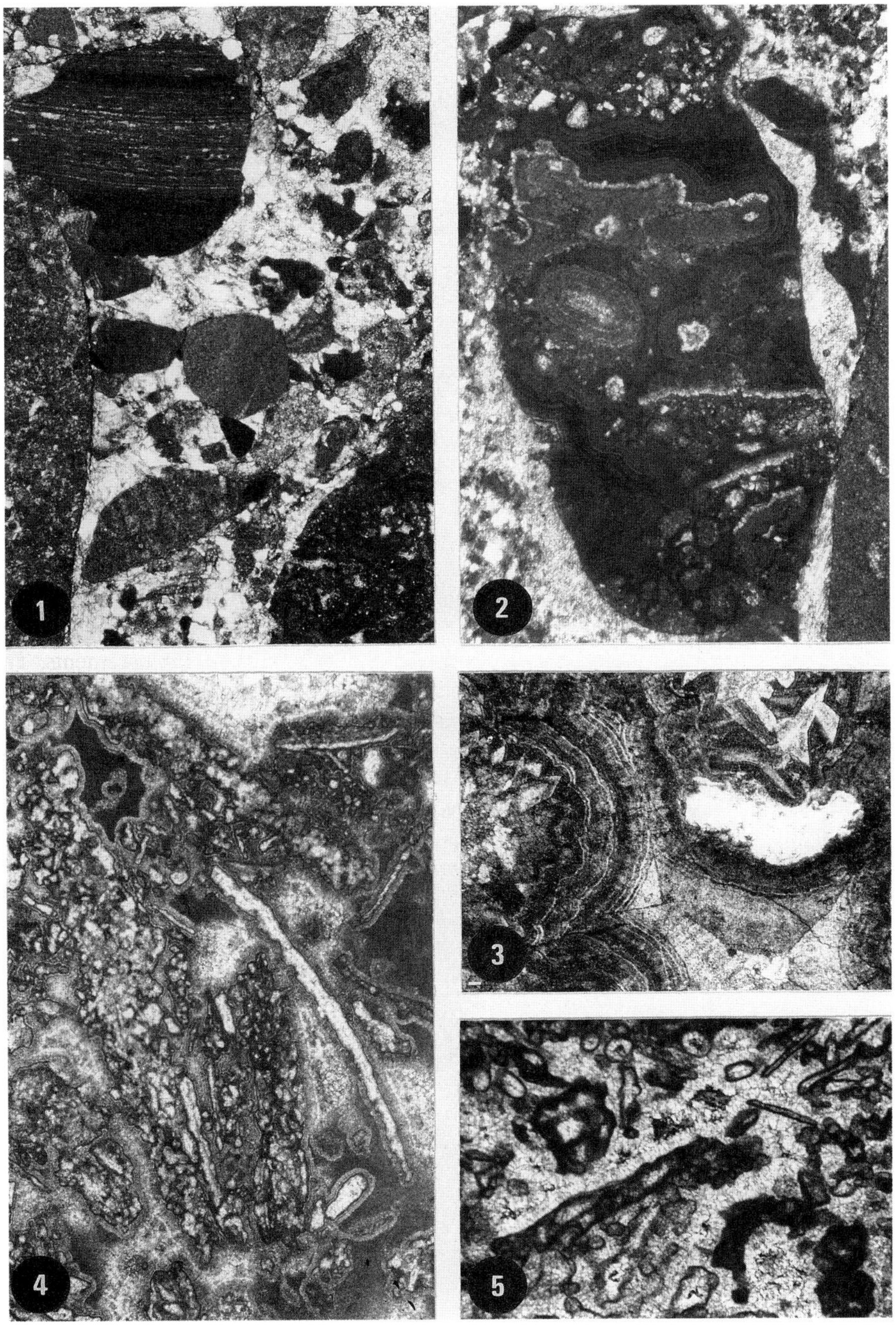
1
2
3
4
5

PLATE 2
Photomicrographs of clasts from the polygenetic conglomerate of the lower Heritage Group (Fig. 1, 2: Hyde Glacier, Edson Hills; Figs. 3–5: Patriot Hills)
(All photomicrographs taken in plain light.)

Figure 1. Oosparite with normal and complex ooids. Particular layers have been dissolved; radial fissures formed during compaction. Thin section Ant 79-168 × 15.

Figure 2. Oolitic intraclasts. Thin section Ant 79-147 × 7.

Figure 3. Dolosparite with vuggy fabric. Solution vugs of the original sediment are cemented rhythmically by marginal fibrous cement; remaining voids are closed by mosaic cement. Between the two stages of cementation vadose silt developed sporadically. During late diagenesis the original sediment was replaced by sparry dolomite, and non-carbonate compounds were displaced at the margins. Thin section Ant 79-223 × 15.

Figures 4, 5. Half-moon ooids with solutions of some parts of the original ooids, which probably consisted of a different mineralogical composition (primary aragonite or calcium sulfate?). Thin section Ant 79-226 × 15.

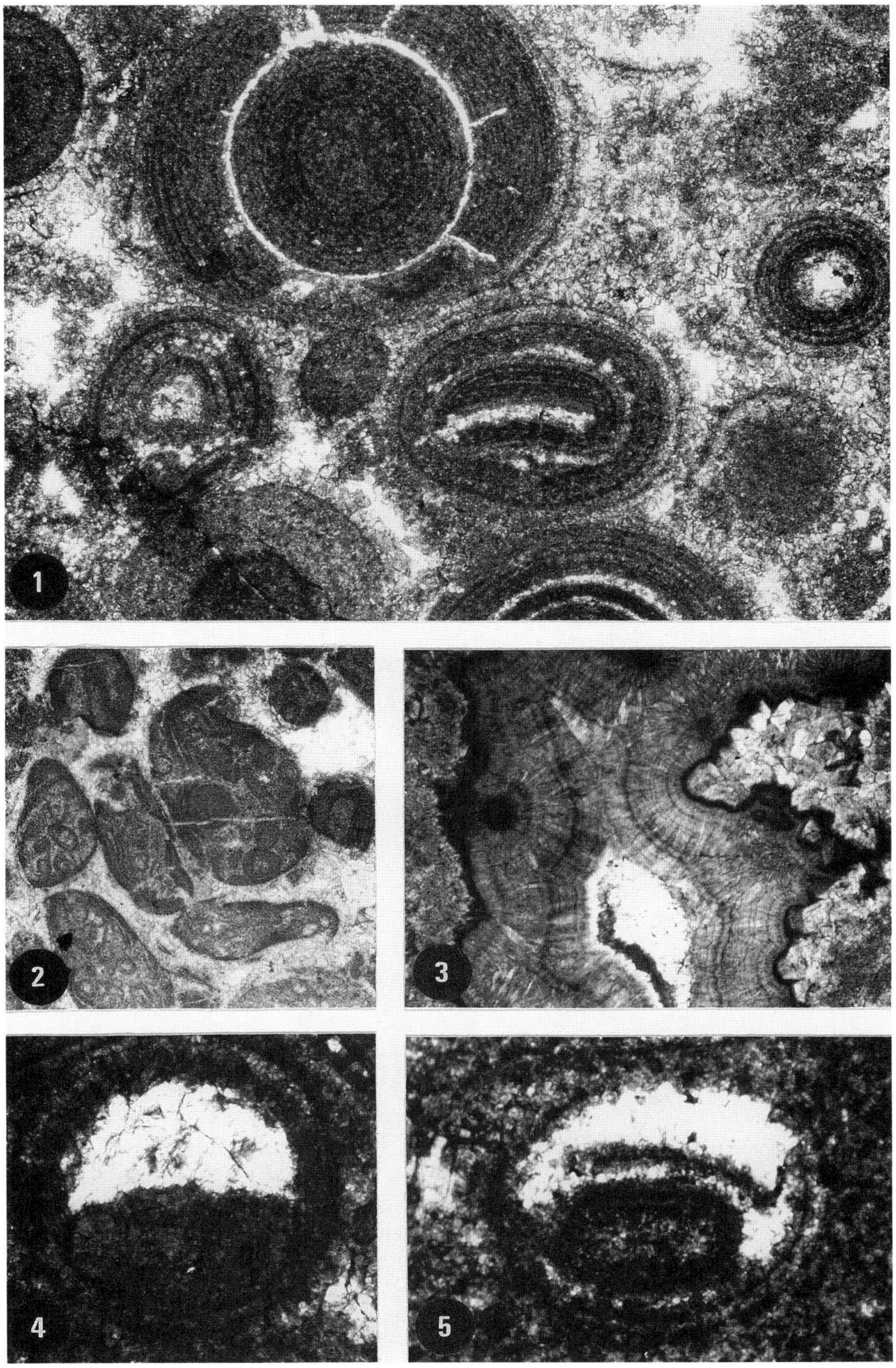
1
2
3
4
5

PLATE 3

Photomicrographs of limestone horizon at the top of the Heritage Group: (Minaret Formation). Figs. 1–4: Webers Peaks (Springer Peak); Figs. 5–6: Yochelson Ridge

(All photomicrographs taken in plain light.)

Figure 1. Trilobite-echinoderm-biopelsparite. Thin section Ant 79-3 × 7.

Figure 2. Oncosparite. Trilobites common as nuclei. Thin section Ant 79-12 × 7.

Figure 3. Echinoderm fragment with rim cement. Thin section Ant 79-5 × 15.

Figure 4. Trilobite pelsparite. Thin section Ant 79-8 × 15.

Figure 5. Oosparite, silty. Ooids very elongated by foliation (cf. Pl. 4, Fig. 3). Bedding not at all visible in thin sections. Thin section Ant 79-28 × 7.

Figure 6. Siltstone with single ooids and coated grains. Thin section Ant 79-29 × 15.

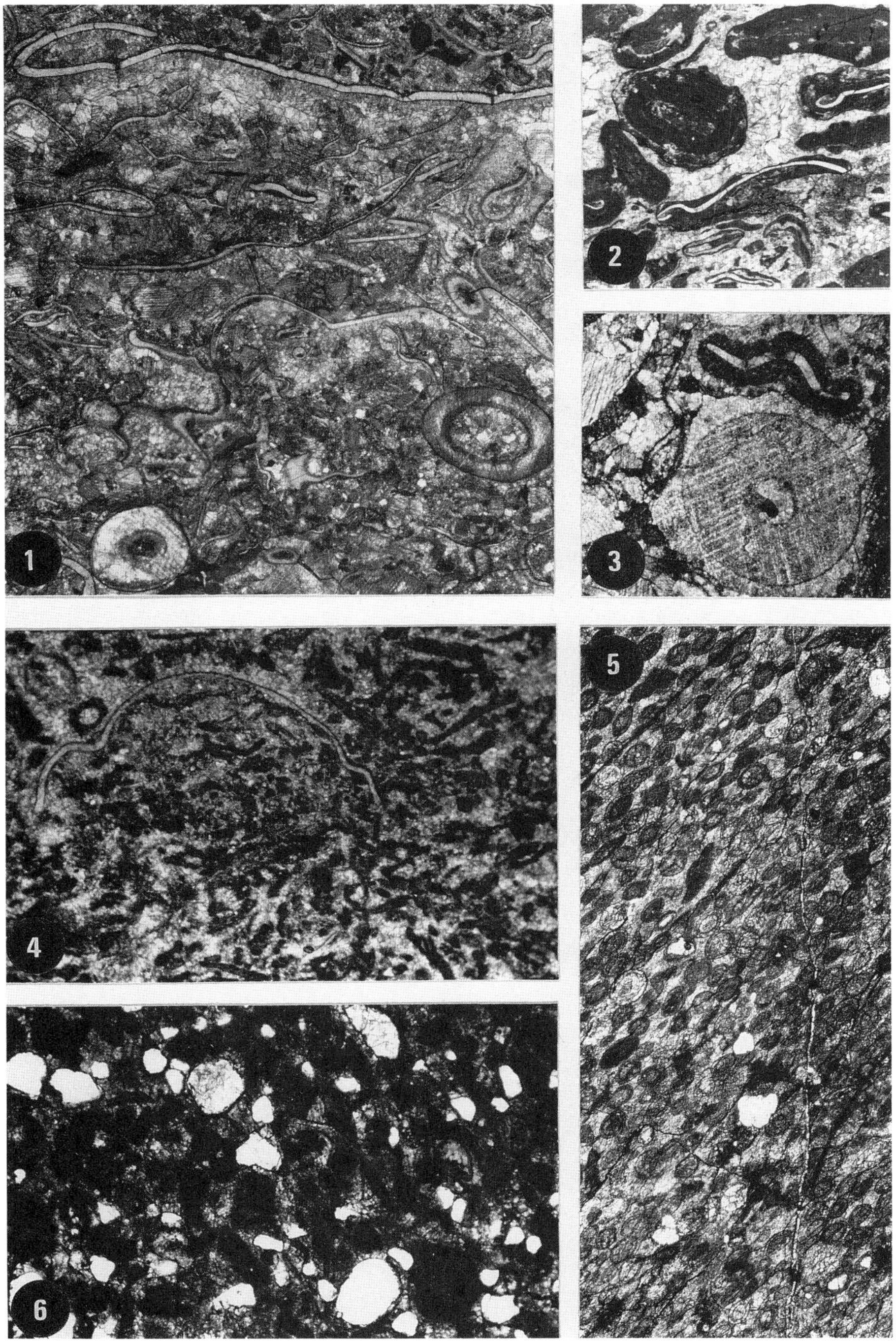
1
2
3
4
5
6

PLATE 4

Photomicrographs of deformation and "metamorphism" of ooids and oncoids (Heritage Range from N to S, Minaret Formation)

(All photomicrographs taken in plain light.)

Figure 1. Oncolite. Fossils as nuclei are still visible. Oncoids partly recrystallized (white) and elongated. Bingham Peak. Polished slab. Specimen Ant 79-48 × 1, 2.

Figure 2. Oolite, extensively recrystallized. Concentric structures almost completely destroyed. Crumbly texture. S Mt. Dolence. Thin section Ant 79-159 × 7.

Figure 3. Oolite. Concentric structure is still preserved; the ooids are elongated tectonically (long axis = bisector of the two foliations). Primary foliation is distinct in the prekinematic sparite cement. A second foliation is present. Yochelson Ridge. Thin section Ant 79-30 × 15.

Figure 4. Oolite, recrystallized into radial aggregates. Patriot Hills. Thin section Ant 79-208 × 15.

Figure 5. Recrystallized carbonate rock. Only echinoderms with rim cements are preserved. Marble Hills. Thin section Ant 79-231 × 15.

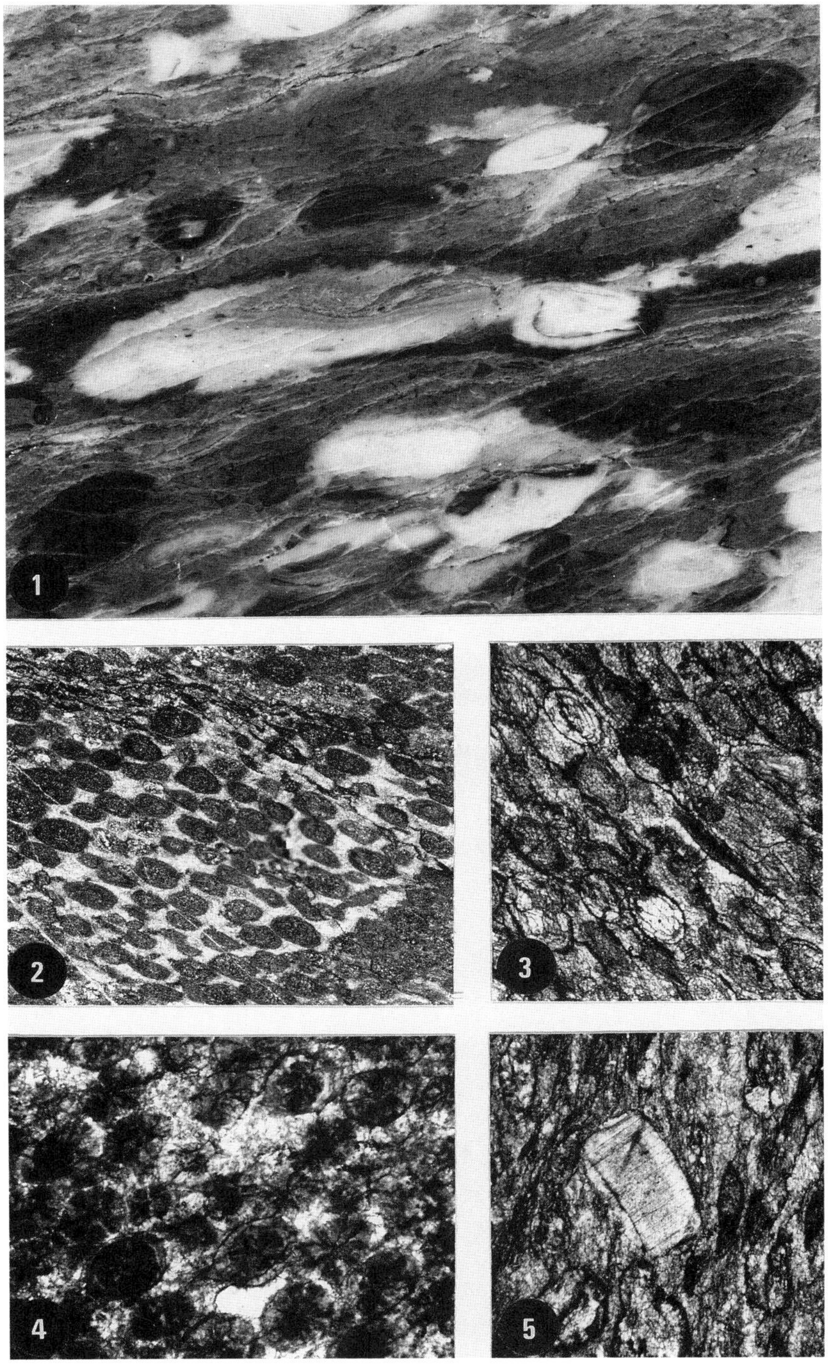
1
2
3
4
5

PLATE 5

Photomicrographs of carbonates of the Minaret Formation (Figs. 1 to 3). Photomicrographs of carbonate clasts from the Permo-Carboniferous diamictite (Whiteout Conglomerate) (Figs. 4, 5).

Figure 1. Intrasparite or pelsparite with shells (trilobite) indicating a Paleozoic age for the limestone. Liberty Hills. Thin section Ant 89-A14 × 30.

Figure 2. Biomicrite, very deformed. Marble Hills. Thin section Ant 89-224 × 3.

Figure 3. Trilobite cross section, very deformed. Marble Hills. Thin section Ant 89-A-72 × 7.

Figure 4. Archaeocyathan-stromatolite bindstone. Interior cemented by dog-tooth cement. Remaining cavity filled with broken cement silt and micrite. Meyer Hills. Thin section Ant 89-92 × 3.

Figure 5. *Epiphyton* sp. Mt. Lymburner. Thin section Ant 89-204 × 80.

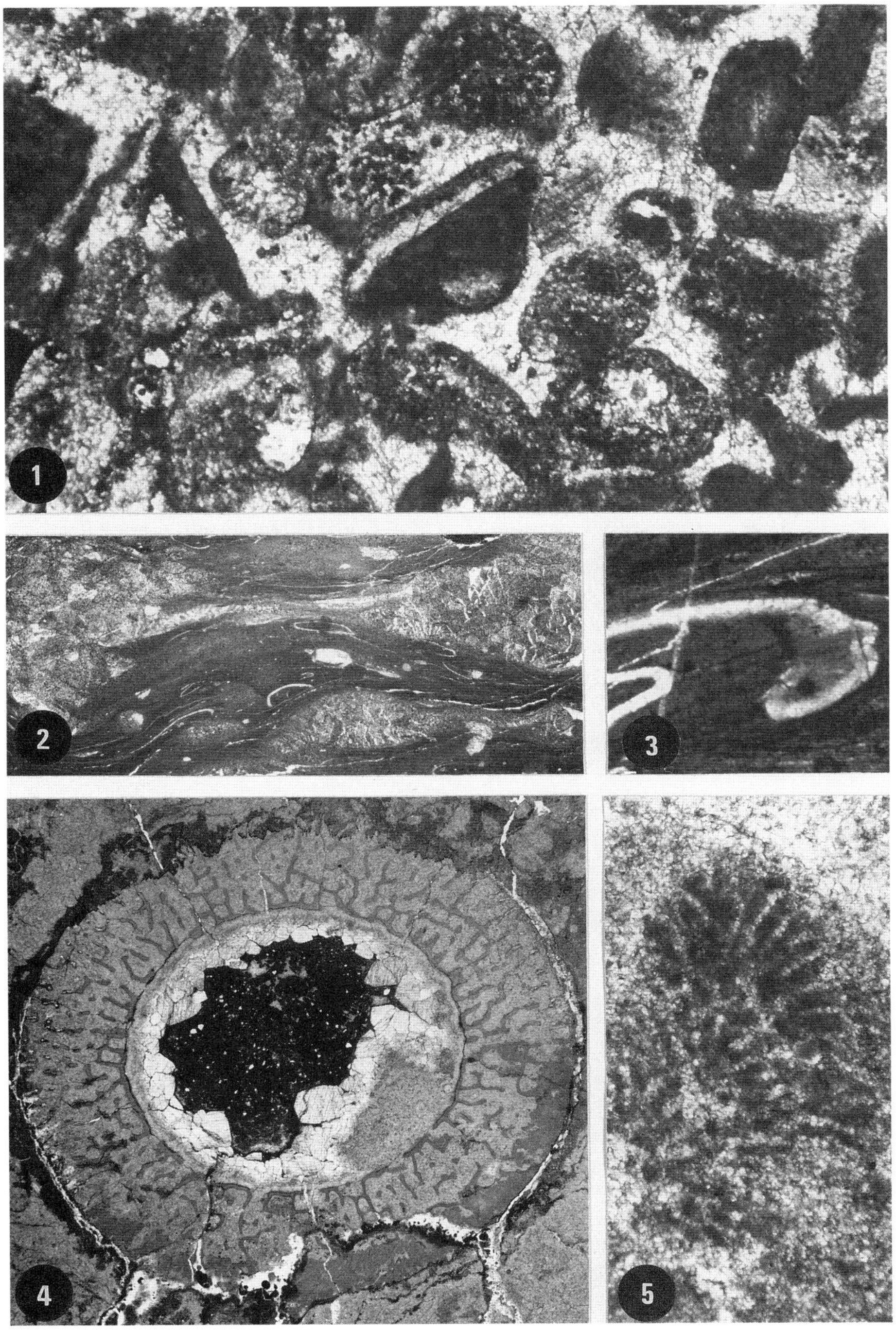
1
2
3
4
5

REFERENCES CITED

Buggisch, W., 1982, Conodonten aus den Ellsworth Mountains (Oberkambrium/Westantarktis): Zeitschrift der Deutschen Geologischen Gesellschaft (Hannover), v. 133, p. 493–507.

Carozzi, A. V., 1963, Half-moon oolites: Journal of Sedimentary Petrology, v. 33, p. 633–645.

Craddock, C., 1969, Geology of the Ellsworth Mountains, *in* Bushnell, V. C., and Craddock, C., eds., Geologic maps of Antarctica: American Geographical Society, Folio 12, sheet 4.

de Wit, M. J., 1977, The evolution of the Scotia Arc as a key to the reconstruction of southwestern Gondwana: Tectonophysics, v. 37, p. 53–82.

Flügel, E., 1978, Methods of microfacial investigations in limestones: Berlin, Springer-Verlag, 454 p.

Hill, D., 1965, Archaeocyatha from Antarctica and a review of the phylum: Transantarctic Expedition 1955–1958 Scientific Reports, v. 10, 151 p.

—— , 1972, Archaeocyatha, *in* Teichert, C., ed., Treatise on invertebrate paleontology: Geological Society of America, part E, v. 1, 158 p.

Kolsov, P. N., 1979, On time appearance of cyanophyta, widely distributed in the Cambrian: Bulletin des Centres de Recherches Exploration-Production Elf-Aquitaine, v. 3, p. 665–667.

Logan, B. W., Rezak, R., and Ginsburg, R. N., 1964, Classification and environmental significance of algal stromatolites: Journal of Geology, v. 72, p. 68–83.

Radwanski, A., and Birkenmajer, K., 1977, Oolitic/pisolitic dolostones from the late Precambrian of south Spitsbergen; Their sedimentary environments and diagenesis: Acta Geologica Polonica, v. 27, p. 1–39.

Schmidt, D. L., and Ford, A. B., 1969, Geology of the Pensacola and Thiel Mountains, *in* Bushnell, V. C., and Craddock, C., eds., Geologic maps of Antarctica: American Geographical Society, Folio 12, Sheet, 5.

Voronova, L. G., 1979, Calcitizated algae of the Precambrian and the early Cambrian: Bulletin des Centres de Recherches Exploration-Production Elf-Aquitaine, v. 3, p. 867–871.

Watts, D. R., and Bramall, A. M., 1980, Paleomagnetic investigation in the Ellsworth Mountains: Antarctic Journal of the United States, v. 15, no. 5, p. 34–36.

—— , 1981, Palaeomagnetic evidence for a displaced terrain in western Antarctica: Nature, 239, no. 5834, p. 638–641.

Webers, G. F., 1972, Unusual upper Cambrian fauna from West Antarctica, *in* Adie, R. J., ed., Antarctic geology and geophysics: Oslo, Universitetsforlaget, p. 235–237.

—— , 1981, Cambrian rocks of the Ellsworth Mountains, West Antarctica, *in* Taylor, M. E., ed., Short Papers for the 2nd International Symposium on the Cambrian System, Golden, Colorado, August 1981: U.S. Geological Survey Open-File Report 81-743, p 236–238.

Yochelson, E. L., Flower, R. H., and Webers, G. F., 1973, The bearing of the new late Cambrian monoplacophoran genus *Knightoconus* upon the origin of the Cephalopoda: Lethaia, v. 6, p. 275–309.

Manuscript Accepted by the Society March 1, 1989

Printed in U.S.A.

Geological Society of America
Memoir 170
1992

Chapter 7

Middle Cambrian trilobites from the Ellsworth Mountains, West Antarctica

J. B. Jago
Department of Applied Geology, South Australian Institute of Technology, The Levels, South Australia 5095, Australia
G. F. Webers
Department of Geology, Macalester College, St. Paul, Minnesota 55105

ABSTRACT

Middle Cambrian or probably Middle Cambrian trilobites were collected at six localities in the Heritage Range of the Ellsworth Mountains, including Yochelson Ridge, Drake Icefall area, Edson Hills, Liberty Hills (two locations), and Marble Hills. The total fauna includes 14 genera (2 new) and 32 species (5 new). Due to original preservation and/or deformation, a number of the forms are not assignable to specific taxa. The new taxa described herein are *Peronopsis deons* sp. nov., *Pagetia edsonensis* sp. nov., *Sohopleura drakensis* gen. et sp. nov., *Pseudobergeronites spinosa* gen. et sp. nov., and *Blountia perplexa* sp. nov. The trilobites were found in formations of the upper Heritage Group (Springer Peak Formation, Drake Icefall Formation, and Liberty Hills Formation). The trilobite faunas described show affinities with faunas from northern Victoria Land (Antarctica), Tasmania, Queensland, China, Kazakhstan, and North America. All faunas are probably Middle Cambrian in age (Templetonian and Boomerangian on the Australian biochronological scale).

INTRODUCTION

The first collection of Middle Cambrian trilobites in the Ellsworth Mountains was made by Robert Rutford in 1964 in the Drake Icefall area of Soholt Peaks. Rutford, then a member of a University of Minnesota expedition under the direction of Campbell Craddock, collected a small number of specimens of a single taxon (Webers, 1977). No subsequent discoveries were made until the 1979–1980 Ellsworth Mountains expedition, when the above fauna was recollected and five additional faunas were found.

LITHO- AND BIOSTRATIGRAPHY

Middle Cambrian or probably Middle Cambrian trilobites were collected in the Heritage Range of the Ellsworth Mountains at Yochelson Ridge, Drake Icefall area, Edson Hills, Liberty Hills, and Marble Hills (Fig. 1). Two collections from the Liberty Hills are from loose limestone boulders that were probably derived from the nearby Minaret Formation. Specimens from all other localities were collected in situ. Figure 2 shows the stratigraphic relations among the Middle Cambrian localities. Facies changes in the Minaret Formation and especially the upper Heritage Group are complex, and a detailed discussion of these units can be found in Buggisch and Webers (this volume) and in Webers and others (this volume). The Middle Cambrian faunas described here show affinities with faunas from northern Victoria Land, Tasmania, Queensland, China, Kazakhstan, and North America. They are correlated with the Australian Middle Cambrian biochronological scale of Öpik (1979). The ages of the six faunas represented in this paper are discussed separately below, along with a brief description of the fossiliferous localities.

The Edson Hills have yielded a well-preserved Middle Cambrian fauna (W79-126) at the upper end of the valley of the Hyde Glacier (about 79°49′S,83°54′W). The trilobites are found in black calcareous concretions within the black shales near the base of the Drake Icefall Formation. This locality, together with the Drake Icefall area locality, has been instrumental in correlating thick stratigraphic sections of the Heritage Group. No fossil groups other than trilobites were found. The fauna

Jago, J. B., and Webers, G. F., 1992, Middle Cambrian trilobites from the Ellsworth Mountains, West Antarctica, *in* Webers, G. F., Craddock, C., and Splettstoesser, J. F., Geology and Paleontology of the Ellsworth Mountains, West Antarctica: Boulder, Colorado, Geological Society of America Memoir 170.

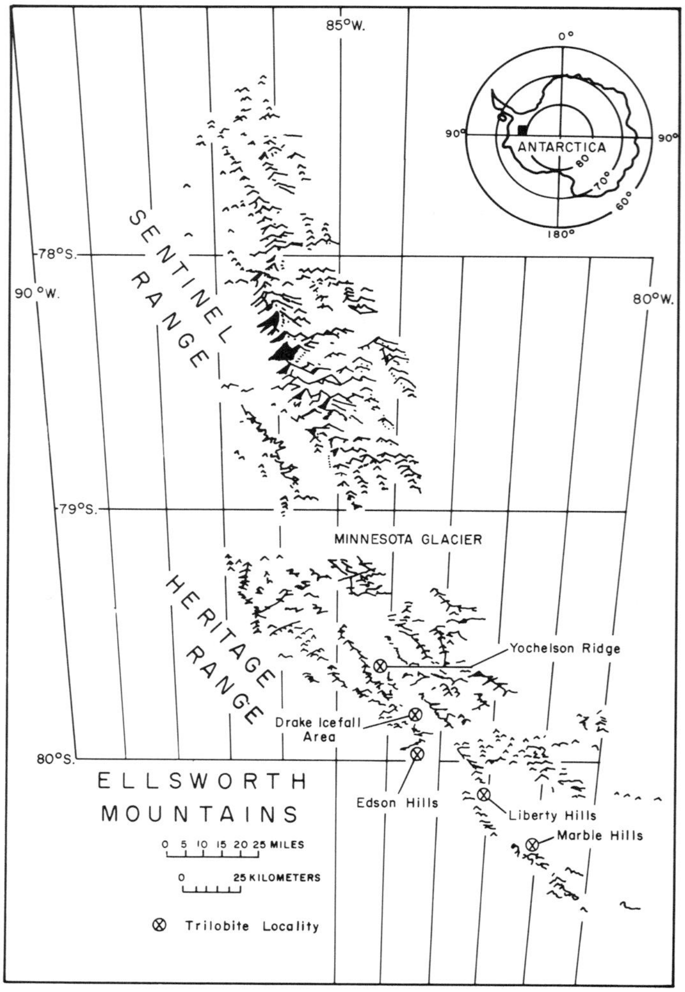

Figure 1. Ellsworth Mountains showing Middle Cambrian trilobite localities.

consists of three new species—*Peronopsis deons, Pagetia edsonensis,* and *Sohopleura drakensis*—plus an unassigned librigena. *Pagetia edsonensis* is close to *P. silicunda* and *P. macrommatia* of Jell (1975). As noted by Jell (1975, p. 20), the number of pygidial axial rings in *Pagetia* decreases with time; *P. edsonensis* has five axial rings plus a terminus. Of the Australian Middle Cambrian species of *Pagetia* tabulated by Jell (1975, Fig. 10, p. 27), all those with five or more pygidial axial rings occur in the *Ptychagnostus gibbus, Xystridura templetonensis,* and *Redlichia chinensis* zones, thus suggesting that the Edson Hills fauna is no younger than the *P. gibbus* Zone. *Peronopsis deons* compares closely with *P. interstricta,* which in North America spans several zones starting with the *Ptychagnostus gibbus* Zone (Robison, 1982). Hence the Edson Hills fauna is placed tentatively in either the *Ptychagnostus gibbus* or *Xystridura templetonensis* zones, i.e., Templetonian in the Australian Middle Cambrian zonal scheme.

The Drake Icefall area has yielded Middle Cambrian trilobites (M64-51) at about 79°44′S,83°58′W. The fauna was found near the center of the 1,000-m-thick Drake Icefall Formation in black shaly limestone stringers. The Drake Icefall Formation at this locality is predominantly a highly deformed black shale with shaly limestone stringers near the center. Parasitic folding precludes an exact position of the fauna above the Union Glacier Formation. The fauna consists of only one species, *Sohopleura drakensis* gen. et sp. nov., which also occurs in the Edson Hills fauna. Stratigraphically it occurs a short distance above the Templetonian Edson Hills fauna and well below the Boomerangian Yochelson Ridge fauna. The presence of *Sohopleura drakensis* in both the Drake Icefall area and the Edson Hills faunas suggests that the Drake Icefall area fauna is not much younger than the Edson Hills fauna. It is probably of either Templetonian or Floran age on the Australian biochronological scale.

The Liberty Hills has yielded two loose boulders containing Middle Cambrian trilobites. One of these (PT79-1) is in a valley exposing beds of the Liberty Hills Formation just northeast of Mt. Rosenthal (about 80°02′S,83°10′W). No other fossils were found with the trilobites, although an adjacent boulder yielded forms probably assignable to the fossil mollusk *Orthotheca*(?). It is probable that the boulders came from the nearby Minaret Formation. Orthothecids have been found in place about 50 m above the base of the Minaret Formation at Mt. Rosenthal. The only trilobite present in this fauna is a cranidium of a pagodiid related to *Lisania,* which occurs in the late Middle Cambrian of China. Hence a similar age is suggested for this specimen. The second boulder (PA79-1) is from a valley just east of Kelley Peak in the southern Liberty Hills (about 80°10′S,82°43′W). As at Mt. Rosenthal, the loose limestone boulder was in an area exposing beds of the Liberty Hills Formation. The boulder is similar in lithology to that of the nearby Minaret Formation, from which it is probably derived. The fauna consists of the following trilobites: *Ptychagnostus* sp. 2; *Hypagnostus* sp. 1; *Fuchouia* sp.; Dameselliidae, gen. et sp. indet.; Asaphiscidae, gen. et sp. indet.; pygidium, gen. et sp. indet. no. 3. This fauna is clearly of late Middle Cambrian age. However, because no previously described species are identifiable, an exact age cannot be determined. A Boomerangian age is suggested.

A small, poorly preserved, Middle Cambrian trilobite fauna (P79-5) was found in the Minaret Formation of the Marble Hills at about 80°17′S,82°09′W. The Minaret Formation is as much as 600 m thick here and is strongly deformed. Only four trilobite specimens were found along with six specimens of the fossil mollusk *Latouchella*(?). The trilobite fauna consists of the following: Kingstoniidae, gen. et sp. indet.; Dorypygidae, gen. et sp. indet. Because both trilobite species represented here are in open nomenclature, an exact age cannot be assigned to the fauna. The combination of a dorypygid and a kingstoniid trilobite, as well as the inferred stratigraphic position of the fauna, suggests a late Middle Cambrian age. The presence of the mollusk *Latouchella*(?) is consistent with this age.

Yochelson Ridge (centered about 79°38′S,84°18′W) is a north-south ridge extending northward from the Soholt Peaks in the west-central Heritage Range. The Middle Cambrian trilobites

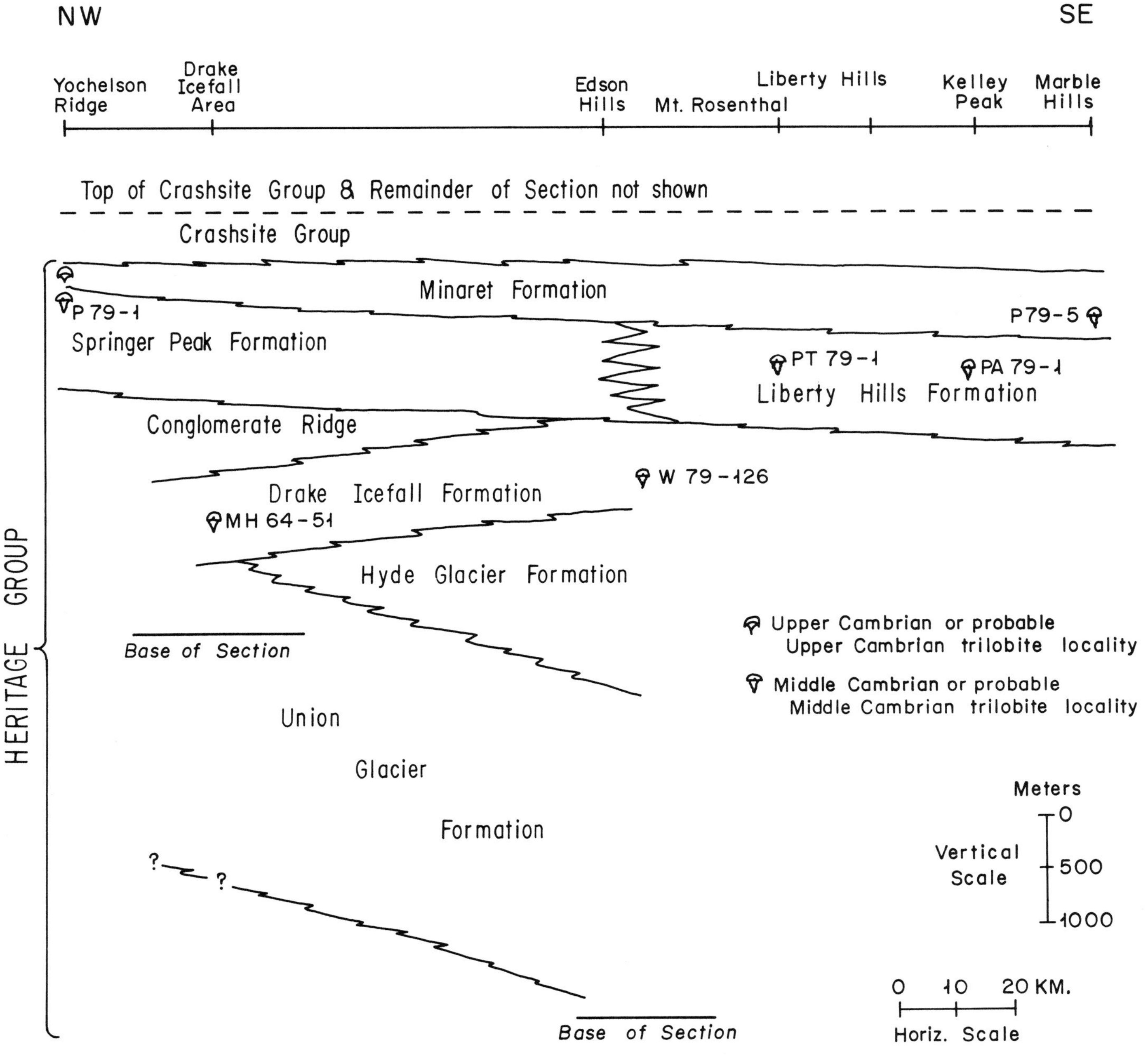

Figure 2. Stratigraphic relationships of rock units with horizons indicated for localities yielding Middle Cambrian or possible Middle Cambrian trilobite faunas. Distance between the Drake Icefall area and the Edson Hills has been enlarged six times to show stratigraphic detail. The Heritage Group comprises the Union Glacier, Hyde Glacier, Drake Icefall, Conglomerate Ridge, Springer Peak, and Liberty Hills Formations.

(P79-1) occur in a black shaly limestone in the upper beds of the Springer Peak Formation about 26 m below the contact with the overlying Minaret Formation. The upper beds of the Springer Peak Formation at this locality consist of argillites, black calcareous shales, graywackes, and black shaly limestone stringers. Late Cambrian trilobites occur 7 m below, and Late Cambrian trilobites, monoplacophorans, algae, gastropods, archaeocyathids, pelmatozoans, and brachiopods occur in the lower 10 m above the Springer Peak–Minaret contact. The Late Cambrian trilobites are described by Shergold and Webers (this volume).

The following trilobites make up the fauna from locality P79-1:

Ptychagnostus aculeatus (Angelin)
Ptychagnostus sp. 1
Hypagnostus sp. 2
Tomagnostella sp.
Baltagnostus(?) sp.
Lejopyge sp.
Grandagnostus sp.
cf. *Grandagnostus* sp.
Clavagnostus sp. 1
Clavagnostus sp. 2
Valenagnostus sp.
Agnostoid, gen. et ap. indet. no. 1
Agnostoid, gen. et sp. indet. no. 2
Agnostoid, gen. et sp. indet. no. 3
Pseudobergeronites spinosa gen. et sp. nov.
Blountia perplexa sp. nov.
cranidium, gen. et sp. indet. no. 1
pygidium, gen. et sp. indet. no. 1
pygidium, gen. et sp. indet. no. 2
pygidium, gen. et sp. indet. no. 4

The only previously described species listed above, *Ptychagnostus aculeatus,* is known from the Boomerangian *Lejopyge laevigata* I Zone of Queensland (Öpik, 1979), the *Lejopyge calva* Zone of the Great Basin (Robison, 1976), and the *Solenopleura brachymetopa* Zone of Sweden (Westergård, 1946). Although *P. aculeatus* is known only from a partial cephalon in the fauna described here, the other trilobites in the P79-1 fauna are consistent with the proposed Boomerangian, *Lejopyge laevigata* I Zone age.

SYSTEMATIC PALEONTOLOGY

The terminology used for agnostoid trilobites is essentially that of Robison (1982); that used in the description of *Pagetia edsonensis* is after Jell (1975); that used for the polymeroid trilobites is after Harrington and others (1959). Classification of the Agnostoidea follows that of Öpik (1967), while that of the polymeroid trilobites follows Harrington and others (1959). Catalog numbers are those of the U.S. National Museum (USNM).

Class TRILOBITA Walch, 1771
Order MIOMERA Jaekel, 1909
Superfamily AGNOSTOIDEA McCoy, 1849
Family AGNOSTIDAE McCoy, 1849
Subfamily QUADRAGNOSTINAE Howell, 1835

Genus *Peronopsis* Corda, 1847

Synonymy. See Robison *in* Jell and Robison, 1978, p. 8.
Type Species. Battus integer Beyrich 1845, p. 44, Plate 1, Fig. 19.
Diagnosis. See Robison *in* Jell and Robison, 1978, p. 8.

Peronopsis deons sp. nov.
(Plate 1, Figs. 1-7)

Material. Several reasonably well-preserved cephala and pygidia are available from the W79-126 fauna.
Holotype. Pygidium, USNM334086, pl. 1, fig. 2.
Diagnosis. Subquadrate cephalon, with unconstricted acrolobe, about as wide as long. Narrow, shallow border furrow; narrow border widens slightly to anterior. Smooth genae. Small, simple basal lobes. Axial furrows shallow markedly to anterior. Length of glabella about 0.70 to 0.75 that of cephalon. Posteroglabella is widest at posterior where it is strongly convex. Deep, wide transverse glabellar furrow curved gently to posterior. Strongly convex subquadrate pygidium about as wide as long. Wide, shallow border furrow; border widens markedly to the posterior. Pygidial border swellings placed a little forward of posterior end of unconstricted acrolobe. Smooth pleural fields. Axis does not extend full length of acrolobe: deep wide axial furrows. Axial length 0.70 to 0.75 that of pygidium; strongly convex axis slightly constricted at second lobe, which bears a prominent elongated node.
Description. Moderately convex subquadrate cephalon with unconstricted acrolobe is about as wide as long. Narrow, shallow border furrow; narrow border widens slightly to anterior. In some specimens (e.g., Plate 1, Figs. 1, 6) there is a slight posterior deflection of border at anterior of cephalon. Smooth genae; small, simple basal lobes. Axial furrows moderately deep posteriorly. They shallow markedly forward. Glabella widest at anterior end of basal lobes from where it tapers gently to transverse glabellar furrow; widest part of anteroglabella slightly wider than anterior end of posteroglabella. Both glabellar rear and front broadly rounded. Glabella length 0.70 to 0.75 that of cephalon. Length of posteroglabella about 0.6 that of glabella. Posteroglabella strongly convex posteriorly; it flattens markedly forward. Pair of faint, shallow lateral glabellar furrows placed well forward on posteroglabella. No node on posteroglabella. Deep, wide transverse glabellar furrow curved gently to posterior.

Strongly convex subquadrate pygidium with unconstricted acrolobe about as wide as long. Wide, shallow border furrow widens slightly to posterior; border quite narrow anteriorly but widens considerably to posterior. A little forward of posterior end of acrolobe are pygidial border swellings rather than true spines. Axis does not extend full length of acrolobe. Smooth pleural fields separated behind the axis by short postaxial median furrow in some specimens. Axis outlined by wide deep axial furrows; it has a length about 0.70 to 0.75 that of pygidium. Strongly convex axis comprises three lobes; it is slightly constricted at second axial lobe. First and second lobes separated by shallow anterior lateral furrow that arches gently to anterior. Second lobe separated from posteroaxis by a furrow that is evident only by a pair of lateral indentations in axis. Prominent elongated node on second axial lobe. Length of posteroaxis about half that of axis. Bluntly rounded axial rear.
Discussion. P. deons is quite close to *P. interstricta* (White). The only difference in the cephala is that the posterior part of the posteroglabella is more convex in *deons* than in *interstricta*. The pygidium of *P. interstricta* is markedly wider than is long (Robison, 1964, p. 530), whereas that of *P. deons* has approximately equal length and width. Not all pygidia of *deons* have a postaxial median furrow, whereas that in *interstricta* is well marked. As noted by Robison (1982, p. 156), the pygidial axis of *interstricta* is semilanceolate; that of *deons* is slightly ogival rather than lanceolate.

P. deons is similar to *P. egenus* (Resser and Endo), as illustrated by Yin (1978, p. 390, Plate 144, Figs. 21, 22), although the pygidial axis of *egenus* is longer than that of *deons*. Incidentally, it is not clear if *P. egenus* as illustrated by Yin is the same species as *Agnostus egenus* as illustrated by Resser and Endo (1937, Plate 30, Figs. 12–14) and reillustrated by Lu and others (1965, Plate 5, Figs. 12, 13), because of the very poorly preserved nature of the pygidium originally described by Resser and Endo.

P. deons also resembles the specimens figured by Jegorova and others (1982, Plate 52, Figs. 1, 2; Plate 56, Figs. 7, 8) as *Peronopsis* ex. gr. *scutalis* (Salter). However, neither of the cephala illustrated by Jegorova and others shows the slight posterior deflection of the anterior border seen in some specimens of *deons*. The pygidium illustrated by Jegorova and others (Plate 52, Fig. 2) has a pair of posterior border swellings, but these are placed farther forward than those of *deons*. The pygidium illustrated by Jegorova and others (Plate 56, Fig. 8) has no posterior border swellings.

Genus *Valenagnostus* Jago 1976

Type Species. Agnostus nudus Beyrich var. *marginata* Brögger, 1878, p. 73, Plate 6, Fig. 3.
Diagnosis. See Jago, 1976a, p. 142.

Valenagnostus sp.
(Plate 1, Figs. 9–10)

One poorly preserved partial pygidium (USNM334095) from the P79-1 fauna with a wide border and essentially effaced acrolobe probably belongs in *Valenagnostus*. The poorly preserved, essentially effaced cephalon (USNM334094) with a faintly outlined glabella may also belong in *Valenagnostus*.

Genus *Grandagnostus* Howell, 1935

Synonymy. See Jago, 1976a, p. 141. To this should be added:
Grandagnostus Rushton, 1978, p. 255; Rushton, 1979, p. 52.
Phalacroma Khairulina, 1973, p. 48.
Phaldagnostus Ivshin *in* Jegorova and others, 1960, p. 167;

Romanenko, 1977, p. 168; Ergaliev, 1980, p. 81.
Phalagnostus Rosova, 1964, p. 17.
Type Species. Grandagnostus vermontensis Howell, 1935, p. 221, Plate 22, Figs. 8–11.
Diagnosis. See Jago, 1976a, p. 141.
Discussion. The concept of *Grandagnostus* has been discussed recently by Jago (1976a) and Rushton (1978); Ivshin (*in* Jegorova and others, 1960) erected the genus *Phaldagnostus* with type species *P. orbiformis,* which was illustrated with only a line diagram. Subsequent illustrations of *orbiformis* (e.g., Romanenko, 1977, Plate 23, Fig. 34; Ergaliev, 1980, Plate 12, Figs. 5, 6) suggest that *Phaldagnostus* should be placed in synonymy with *Grandagnostus. Phalacroma rabutensis* as illustrated by Khairulina (1973, Plate 3, Fig. 16) also appears to belong in *Grandagnostus.*

Grandagnostus sp.
(Plate 1, Figs. 13, 14, 18, 19)

Material. One cephalon and four pygidia from the P79-1 fauna are assigned to *Grandagnostus* sp.
Remarks. The poor preservation of the available specimens precludes assignment to species, although the figured specimens clearly belong in *Grandagnostus.*

cf. **_Grandagnostus_ sp.**
(Plate 1, Figs. 8, 11, 12, 15, 16)

Material. Several cephala and pygidia from the P79-1 fauna are placed in this species.
Description. Flatly convex cephalon about as wide as long. Narrow, shallow border furrow; narrow gently convex rim that widens anteriorly. Completely effaced acrolobe shows no sign of axial furrows, basal lobes or nodes.

Moderately convex pygidium more convex than cephalon. Narrow, shallow border furrow; flat slightly elevated border is widest about half way around pygidium. Prominent node on centroanterior part of pygidium. Apart from the node and faint indications of wide pygidial axis, acrolobe is entirely effaced.
Discussion. The combination of a flatly convex cephalon with an entirely effaced acrolobe and anteriorly widening border with an almost effaced moderately convex pygidium with centrally widened border is unknown in any previously described agnostoid genus. However, the erection of a new genus of effaced agnostoids based on the available material is not warranted.

Cf. *Grandagnostus* differs from *Grandagnostus* in that the latter has a very narrow or undeveloped cephalic border and a pygidial border that widens uniformly to the posterior. It differs from *Phalagnostus* in that the latter has a transverse furrow that extends across the anterolateral corner of the pygidium. Cf. *Grandagnostus* differs from *Valenagnostus* in that the latter has a much wider pygidial border and a distinctive pygidial axis.

The pygidium is similar to that of the species described by Westergård (1946, p. 91, Figs. 10–14) as *Phoidagnostus bituberculatus* in that it has a centrally widened pygidial border. However, the cephalic border of *bituberculatus* is faint or absent, and this species has been placed in *Grandagnostus* by Rushton (1978, p. 255).

Phoidagnostus augustiformus, as illustrated by Jegorova and others (1982, Plate 20, Figs. 9–10; Plate 25, Fig. 2), may belong in the same genus as cf. *Grandagnostus* sp. However, the species illustrated by Jegorova and others (1982) appears to be different from *augustiformis* as illustrated by Fedyanina (1977, Plate 20, Figs. 3, 4, 7, 10). Rushton placed *augustiformis* in *Grandagnostus.*

Genus _Hypagnostus_ Jaekel 1909

Synonymy. See Jago, 1976a, p. 140 for pre-1976 synonymy, but as noted below, *Tomagnostella* is now regarded as a separate genus. To the synonymy of Jago (1976a) should be added:
Hypagnostus Rushton, 1979, p. 52; Öpik, 1979, p. 65.
Type species. Agnostus parvifrons Linnarsson, 1869, p. 82, Plate 2, Figs. 56, 57.
Diagnosis. See Öpik, 1979, p. 65.

Hypagnostus sp. 1
(Plate 1, Figs. 20, 23; Plate 2, Fig. 27)

Material. Two cephala; one reasonably well preserved and the other poorly preserved. Both come from the PA79-1 fauna.
Description. Cephalon probably a little wider than long allowing for distortion. Narrow, shallow border furrow; narrow border widens very slightly to the anterior. Smooth genae. Only posteroglabella present; it is outlined by narrow, shallow axial furrows and an anteriorly curved transverse glabellar furrow; it has a length a little over half that of cephalon. Small node placed well forward on posteroglabella. Basal lobes are small, simple, and connected by narrow occipital band.
Discussion. In the specimen illustrated in Plate 1, Fig. 20, the ratio of the length of the posteroglabella to that of the cephalon is greater than in described species of *Hypagnostus.* However, this ratio may have been exaggerated by the tectonic distortion.

Hypagnostus sp. 2
(Plate 1, Fig. 22)

Material. One poorly preserved cephalon from the P79-1 fauna.
Description. Moderately convex specimen has narrow border furrow and narrow border. Smooth genae. Poorly preserved single lobed glabella about half length of cephalon. It is outlined by axial furrows that shallow anteriorly. Small basal lobes.
Discussion. This specimen is too poorly preserved to be placed in any particular species of *Hypagnostus.* It is possible that the pygidium described as Agnostoid, gen. et sp. indet. no. 3 (Plate 2, Fig. 6) belongs in the same species as *Hypagnostus* sp. 2.

Genus _Tomagnostella_ Kobayashi, 1939

Synonymy. Tomagnostella Kobayashi, 1939, p. 150; Howell, 1959, p. 186; Öpik, 1979, p. 17.
Type Species. Agnostus exsculptus Angelin, 1851, p. 7, Plate 6, Fig. 8.
Diagnosis. See Öpik, 1979, p. 71.
Discussion. Öpik (1979, p. 65, 71) suggested that *Hypagnostus* and *Tomagnostella,* which have been placed in synonymy by Westergård (1946) and later workers, should in fact be separate genera. We support Öpik's view.

Tomagnostella sp.
(Plate 2, Fig. 1)

Material. Two partial cranidia from the P79-1 fauna, the better one of which, USNM334118, is figured.
Description. Shallow border furrow; narrow border. Single lobed glabella about 0.5 length of cephalon; it is separated from genae by a straight transverse glabellar furrow, which is deepest at either end. Two pairs of short lateral glabellar furrows evenly spaced along glabella. Small node centrally placed opposite anterior pair of furrows. Small, simple basal lobes connected by narrow occipital band. Genae moderately scrobiculate with almost straight scrobicules extending from border to 0.50 to 0.75 the distance to glabella. Short, transversely elongated pit extends from either end of transverse glabellar furrow.
Discussion. The two cephala are more scrobiculate than any of the other species of *Tomagnostella* listed by Öpik (1979, p. 72), with the exception of *T. exsculptus* (Angelin), although *T. varicosus* (Öpik) is almost as scrobiculate. However, the two partial cephala do not warrant the erection of a new species.

Subfamily PTYGAGNOSTINAE Kobayashi 1939

Genus *Ptychagnostus* Jaekel, 1909

Synonymy. See Robison, 1982, p. 135; Robison, 1984, p. 12.
Type Species. Agnostus punctuosus Angelin, 1851, p. 8.
Diagnosis. See Robison, 1982, p. 135.
Discussion. Robison (1982, 1984) thoroughly reviewed *Ptychagnostus* and related genera. Robison (1982) placed such genera as *Goniagnostus* and *Onymagnostus* in synonymy with *Ptychagnostus,* but after further consideration he decided (Robison 1984, p. 12) that *Goniagnostus* (*sensu* Öpik, 1979) and *Onymagnostus* should remain as separate genera. The writers prefer Robison's 1982 view to that of 1984.

Ptychagnostus aculeatus (Angelin)
(Plate 1, Fig. 17)

Synonymy. The pre-1973 synonymy is given by Khairulina 1973, p. 34. To this should be added
Ptychagnostus aculeatus Lu and others, 1965, p. 37, Plate 3, Figs. 11–12 and synonymy therein.
Ptychagnostus (Ptychagnostus) aculeatus Palmer, 1968, p. 28, Plate 6, Fig. 20.
Ptychagnostus (Ptychagnostus) cf. *aculeatus* Jago, 1976a, p. 152, Plate 23, Fig. 12.
Ptychagnostus aculeatus Kindle, 1982, Plate 1–2, Figs. 16, 19, 23.
Ptychagnostus aculeatus Xiang and others, 1981, Plate 7, Figs. 6, 7.
Material. One poorly preserved partial cephalon (USNM334105) from the P79-1 fauna.
Discussion. As noted by Jago (1976a) and Robison (1976), *Ptychagnostus aculeatus* is a cosmopolitan species whose biostratigraphic usefulness is limited by the fact that it is uncommon or rare at all known localities.

Ptychagnostus sp. 1
(Plate 1, Fig. 24)

Material. One almost complete pygidium from the P79-1 fauna.
Description. Gently convex pygidium, with unconstricted acrolobe slightly wider than long. Moderately wide pygidial border widens posteriorly, narrow, shallow border furrow. No sign of posterolateral spines. Narrow, shallow shoulder furrows; narrow convex shoulders. Smooth pleural fields separated behind axis by short postaxial median furrow that shallows posteriorly. Length of pygidial axis about 0.75 that of pygidium. Narrow pygidial axis outlined by narrow, moderately deep axial furrows; it tapers posteriorly to sharply rounded posterior. Axis widest close to anterior. Details of pygidial axis poorly preserved, but it appears to be a quadrilobate axis similar to *Ptychagnostus nathorsti* and related species. Elongated node on second axial lobe.
Discussion. This pygidium is too poorly preserved for specific identification. However, the narrow quadrilobate nature of the axis clearly indicates affiliation with *Ptychagnostus nathorsti* and related species. Öpik (1961), in erecting his subgenera *Ptychagnostus (Ptychagnostus)* and *P. (Goniagnostus),* considered the distinguishing features of *P. (Goniagnostus)* to be the quadrilobate nature of the pygidial axis and the presence of pygidial spines. The quadrilobate pygidium under discussion has no such spines, thus supporting Robison's (1982) contention that *Goniagnostus* is a junior synonym of *Ptychagnostus.*

Ptychagnostus sp. 2
(Plate 1, Fig. 21)

Material. One moderately well preserved partial pygidium from the PA79-1 fauna.
Description. Narrow, shallow border furrow; narrow, flat border widens slightly to posterior. Posterolateral spines absent. Smooth pleural fields. Narrow axis extends almost full length of acrolobe; it is outlined by narrow, shallow, axial furrows. Axis tripartite with a constriction at second axial lobe on which is a distinct low node.
Discussion. The narrow pygidial axis, narrow, shallow axial and border furrows, and narrow border are similar to species of *Ptychagnostus* that Öpik (1979) included in his genus *Onymagnostus.* However, the one available specimen is insufficiently preserved to be placed in any existing species or be the basis of a new species.

Genus *Lejopyge* Hawle and Corda, 1847

Synonymy. See Robison, 1984, p. 36.
Phoidagnostus Whitehouse, 1936, p. 93.
Type Species. Battus laevigatus Dalman, 1828, p. 136.
Diagnosis: See Robison, 1984, p. 36.
Discussion: Lejopyge, its species, and its subspecies have been discussed in detail by Daily and Jago (1975), Jago (1976b), Öpik (1979), and Robison (1984). Öpik and Robison have considerably widened the concept of *Lejopyge* in that they have included species with ordinary ptychagnostid dorsal furrows, whereas previous authors have generally included only effaced species within the genus.

Öpik (1979) described three new species of *Lejopyge* that have wider pygidial borders than previously described species. The further broadening of the generic concept was supported by Robison (1984), who placed *Pseudophalacroma* Pokrovskaya 1958 in synonymy with *Lejopyge,* a view supported by the authors. The specimens figured by Poulsen and Anderson (1975, Plate 2, Figs. 1–9) as *Lejopyge laevigata* belong in *Grandagnostus* as is indicated by the shape of the pygidium.

Lejopyge sp.
(Plate 1, Figs. 26, 27)

Material. Several partial cephala and pygidia from the P79-1 fauna.
Remarks. The figured specimens are too poorly preserved to warrant formal description. They show no scrobiculation and probably belong in *Lejopyge laevigata.*

Family CLAVAGNOSTIDAE Howell 1937
Subfamily CLAVAGNOSTINAE Howell 1937

Genus *Clavagnostus* Howell 1937

Synonymy. See Jago and Daily, 1974, p. 97.
Type Species. Agnostus repandus Westergård *in* Holm and Westergård, 1930, p. 13, Plate 4, Figs. 11, 12 (only).
Diagnosis. See Jago and Daily, 1974, p. 97.
The concept of *Clavagnostus* outlined by Jago and Daily (1974) is followed here with the following amendment. In the diagnosis given by Jago and Daily it is stated that the glabella of *Clavagnostus* has no transverse glabellar furrow. This should be amended to allow for a poorly defined (see Plate 2, Fig. 3, herein) or a vestigial transverse glabellar furrow as indeed is shown by *C. milli* Jago and Daily (1974, Plate 11, Fig. 12).

Öpik (1967), in describing *C. bisectus,* figured three cephala and a pygidium. The cephala come from the early Late Cambrian *Glyptagnostus stolidotus* Zone and although from two different localities appear to belong in the one species. However, the pygidium figured by Öpik (1967, Plate 55, Fig. 9) was from a third locality and from the late Middle Cambrian *Erediaspis eretes* Zone. Hence, it must be doubtful that this pygidium belongs in the same species as the three cranidia.

Although poorly preserved, *C. dentatus* Khairulina (1973, Plate 1, Figs. 14–20) appears to belong in the *C. repandus* species group as described by Jago and Daily. *C. trispinus* Zhou and Yang (p. 17, Plate 1, Fig. 2 *in* Yang, 1978) is known only from one pygidium and may also belong in the *C. repandus* species group but is characterized by the presence of a third pygidial spine. It is doubtful if the pygidium figured by Yang (1978, Plate 1, Fig. 3) as *C. sulcatus* is that species because the posterior part of the axis is much too narrow.

Qian (1982) erected the genus *Acanthagnostus* on the basis of two cephala that appear similar to *Clavagnostus.* These cephala have a faint, anteriorly placed transverse glabellar furrow, similar to those in *C. milli* and *C.* sp. 1 described below. We regard *Acanthagnostus* as a possible synonym of *Clavagnostus,* but until pygidia of *Acanthagnostus* are found a firm decision cannot be made.

Clavagnostus sp. 1
(Plate 2, Figs. 1-4, 9)

Material. Two cephala and two pygidia, none of which is particularly well preserved; all have been slightly distorted. All are from the P79-1 fauna.
Description. Cephalon about as wide as long; narrow, convex, elevated border: narrow, shallow, border furrow widens anteriorly. Median preglabellar furrow absent. Narrow spines, moderate length. Glabellar length and width about 0.67 and 0.33 that of cephalon, respectively. Elliptical glabella bounded by shallow axial furrows. Basal lobes somewhat elongated. Details of connecting ring not visible. Faint transverse glabellar furrow. No lateral axial furrows on posteroglabella. Genae smooth.

Pygidium about as wide as long. Narrow convex lateral borders; wider posterior border widest at center. Short posterolateral spines. Smooth pleural fields. Long tapering axis extends entire length of acrolobe; it is outlined by narrow, shallow axial furrow. Transverse axial furrows absent. Anterior part of axis bears an elongated ridge. Pits occur about 0.65 of distance from anterior to posterior of axis.
Discussion. The specimens belong in the *C. repandus* species group of Jago and Daily (1974); however, they are too poorly preserved for specific identification.

Clavagnostus sp. 2
(Plate 2, Fig. 7)

Material. One poorly preserved internal mold of a cephalon, USNM334125, from the P79-1 fauna clearly belongs in *Clavagnostus.* It has an elliptical glabella much shorter than that of *Clavagnostus* sp. 1. It appears to belong in the *C. repandus* group of species as defined by Jago and Daily (1974).

Clavagnostus(?) sp.
(Plate 2, Fig. 5)

Material. One reasonably well preserved pygidium from the P79-1 fauna. It has been widened tectonically.
Description. Allowing for distortion the pygidium is probably slightly wider than long. Acrolobe unconstricted. Narrow shallow border furrow; narrow border widens slightly posteriorly. On the available specimen it is not certain that the apparent right posterolateral spine is genuine. Pleural fields distinctly pitted with pits being arranged in two fairly distinct rows on each pleural field. One row of 4 to 5 pits close to and aligned parallel to acrolobe margin. The other less distinct row of 3 to 4 shallower pits subparallel with axis, with the most posterior pits merging with axial furrows in centroposterior part of posteroaxis. Axis outlined by narrow, shallow axial furrows; it extends almost full length of acrolobe. Very short postaxial median furrow. Axis comprises three lobes. At anterior it is about 0.4 width of acrolobe; it is widest at anterior of anterior axial lobe and constricted at anterior of posteroaxis before tapering gently to bluntly rounded posterior. Anterior lobe tripartite with axial and lateral lobules of equal size. Anterior lobe separated from second axial lobe by furrow that is arched gently forward. Second axial lobe bears prominent elongated node that extends on to anterior or posteroaxis. Centroposterior part of posteroaxis bears pair of longitudinally elongated pits.
Discussion. The longitudinally elongated pits on the pygidial axis are similar to those in *Clavagnostus* and related genera. However, the posteroaxis of this specimen is broader and with a more rounded posterior end than in previously described species of *Clavagnostus.* The pits in the pleural field are much more prominent than those in *C.*(?) *rawlingi* Jago and Daily, the only previously described possible species of *Clavagnostus* with such pits. However, it should be noted that *C. sulcatus* Westergård has very shallow depressions on the pleural fields (see Jago and Daily, 1974, Plate 11, Figs. 5, 6). Hence, although it is quite possible that the pygidium described above represents a new species of *Clavagnostus,* we believe that the erection of new species on the basis of a single specimen is unwise.

Family DIPLAGNOSTIDAE Whitehouse 1936
Subfamily DIPLAGNOSTINAE Whitehouse 1936

Genus *Baltagnostus* Lochman, 1944

Synonymy. See Robison *in* Jell and Robison, 1978, p. 4. To this should be added Öpik, 1979, p. 47.
Type Species. Proagnostus? centerensis Resser, 1938, p. 48.
Diagnosis. See Robison *in* Jell and Robison, 1978, p. 4.

***Baltagnostus*(?) sp.**
(Plate 1, Figs. 25, 28, 29)

Material. Several cephala and pygidia from the P79-1 fauna.
Description. Subrectangular cephalon a little wider than long. Wide, shallow border furrow; narrow elevated border. Smooth genae separated by well defined median preglabellar furrow. Glabella outlined by wide deep axial furrows; glabella has a length about two-thirds that of cephalon. Anteroglabella separated from posteroglabella by shallow, straight transverse furrow, which is deepest laterally. At its widest, just to posterior of transverse glabellar furrow, glabellar width is a little over 0.3 that of cephalon. Pair of prominent lateral glabellar furrows placed at a position about two-thirds of distance from glabellar posterior to transverse glabellar furrow. Small, simple basal lobes.

No available pygidium is well preserved. Wide shallow border furrow; convex elevated border. Smooth pleural fields. Wide pygidial axis, with almost effaced lateral axial furrows, extends almost to posterior border. It is outlined by axial furrows that shallow posteriorly. Prominent anteriorly placed axial node. Presence or absence of posterolateral spines cannot be determined on available material.
Discussion. It is not certain that the cephalon and pygidia described above belong either in a single species or in *Baltagnostus.* However, the features of the pygidium that suggest affiliation with *Baltagnostus* are the wide long pygidial axis and the almost effaced lateral axial furrows. The transversely subrectangular nature of the cephalon also suggest affiliation with *Baltagnostus.* However, the median preglabellar furrow in the species described herein is better developed than in any described species of *Baltagnostus.*

Family and Subfamily unknown

Agnostoid, gen. et sp. indet. no. 1
(Plate 2, Fig. 10)

A poorly preserved partial pygidium with an expanded posteroaxis that reaches the posterior border and a prominent elongated node on the second axial lobe cannot be placed with certainty into any genus. The expanded nature and length of the posteroaxis is similar to that in *Oedorhachis* Resser. However, *Oedorhachis* has a zonate pygidium, and although the central part of the posterior border is not preserved on the available specimen, as far as can be seen it is not zonate. Agnostoid, gen. et sp. indet. no. 1 comes from the P79-1 fauna.

Agnostoid, gen. et sp. indet. no. 2
(Plate 2, Fig. 8)

Material. One almost complete pygidium from the P79-1 fauna.
Description. Pygidium about as wide as long. Shallow border furrow; narrow, elevated border widens posteriorly. Wide axis outlined by shallow axial furrows that shallow posteriorly. Axis may reach posterior border but posterior of axis merges with smooth pleural fields. Axis has width about half that of pygidium at anterior; it tapers only slightly to the posterior. No sign of transverse axial furrows. Low node on anterior part of axis.
Discussion. This pygidium is similar to that described by Rushton (1978, p. 254, pl. 24, fig. 5) as *Hypagnostus* (or *Grandagnostus*?) sp. The present specimen differs from that of Rushton in having less well defined axial furrows and a comparatively wider rim.

Agnostoid, gen. et sp. indet. no. 3
(Plate 2, Fig. 6)

A poorly preserved internal mold of a pygidium (USNM334124) from the P79-1 fauna has a tapering axis that stops just short of the posterior border. There is a prominent node on second axial lobe; the border is wide and flat; there is a narrow, shallow border furrow; the border widens slightly immediately to the posterior of the axis.

This pygidium may belong in *Hypagnostus* and could be the pygidium of *Hypagnostus* sp. 2, which also occurs in the P79-1 fauna.

Superfamily EODISCOIDEA Richter, 1932
Family EODISCIDAE Raymond, 1913

Genus *Pagetia* Walcott, 1916

Synonymy. See Jell, 1975, p. 30; Palmer and Halley, 1979, p. 77.
Type Species. Pagetia bootes Walcott, 1916, p. 408, Plate 67, Figs. 1, 1a–f.
Diagnosis. See Jell, 1975, p. 30.

***Pagetia edsonensis* sp. nov.**
(Plate 2, Figs. 11–19)

Material. Several cephala and pygidia from the W79-126 fauna.
Holotype. Cephalon, USNM334129a and b (Plate 2, Figs. 11, 12).
Diagnosis. Semi-elliptical cephalon with smooth surface; moderately convex glabella with sharply rounded anterior. Occipital furrow strongly impressed laterally. Wide, deep axial furrow. Small bacculae. Short (sag.), long (tr.), shallow preglabellar field. Gently arcuate low eye ridges almost straight close to glabella but at abaxial extremities are sharply curved to the posterior. Shallow, border furrow deepens posteriorly. Wide, posteriorly narrowing border depressed in front of preglabellar field. Semicircular, sharply convex pygidium with axis of five rings and a long terminus with long narrow spine. Second axial ring is the longest (sag.). Narrow, shallow border furrow; very narrow border. Almost smooth pleural areas with up to two shallow pleural furrows.
Description. Semielliptical cephalon has a moderately convex glabella with a weak transverse glabellar furrow evident in some specimens (e.g., Plate 2, Fig. 18). No lateral glabellar furrows. Occipital furrow strongly impressed laterally. Base of cranidial spine seen on several specimens but spine nowhere preserved. Sharply rounded glabellar front. At its widest, just anterior of occpital furrows, glabella has a width about 0.35 that of cephalon; length of glabella about 0.7 that of cephalon. Wide deep axial

furrow. Short (sag.), long (tr.), shallow preglabellar field. Small bacculae placed close to ends of occipital ring. Fixed cheeks posterior of eye ridge gently convex; anterior of eye ridge the fixed cheeks slope evenly to border furrow. Gently arcuate, low eye ridges arise just forward of the transverse glabellar furrow and are directed slightly to the posterior near the axis but at their abaxial extremities are curved sharply to the posterior. Low, indistinct palpebral lobes. No palpebral furrows. Details of facial sutures nowhere clearly preserved. Free cheeks not preserved. Shallow border furrow merges with the fixed cheek near preglabellar field but becomes more distinct posteriorly. Wide border slightly depressed in front of glabellar field; border narrows posteriorly. The number of shallow border scrobicules cannot be determined from available material, but there are at least eleven. Scrobicules end in distinct pits well inside margin of cephalon.

Strongly convex semicircular pygidium. Axis tapers gently to posterior; it stands out well above pleural areas. At second axial ring axis has width about 0.3 that of pygidium. Axis comprises five axial rings plus a long terminus that bears a long narrow spine. First axial ring distinctly shorter than others; the second is longest (sag.) with third, fourth, and fifth rings successively slightly shorter. Articulating half-ring markedly narrower (tr.) than first axial ring, but at its longest (sag.) it is a little longer than first axial ring. Transaxial furrows straight, moderately impressed at either extremity but very shallow centrally. Transaxial furrows become progressively less well impressed posteriorly, with furrows between fourth and fifth rings and between fifth ring and terminus almost effaced in most specimens. Anterior three axial rings bear very low tubercles that are not visible on all specimens. Narrow, shallow border furrow; very narrow, gently convex border. Pleural areas almost smooth but with up to two shallow pleural furrows visible. Anterior pleural strip of first pygidial segment quite long (exsag.) at fulcral point, which occurs midway between axial furrow and anterolateral corner of pygidium.

Discussion. Of previously described species of *Pagetia, P. edsonensis* is best compared with *P. silicunda* and *P. macrommatia* as recognized by Jell (1975), and *P. ellsi* and *P. quebecensis* as figured by Rasetti (1966). All these species have a glabella with a faint transverse glabellar furrow, faint or no signs of lateral glabellar furrows, and a pygidium with five axial rings plus a terminus that bears a long narrow axial spine (where preserved).

P. edsonensis is closest to *P. silicunda* Jell. The pygidium of *edsonensis* is very similar to that of *silicunda,* although the axial tubercles are more prominent and the pygidial axis narrower in *silicunda* than in *edsonensis.* The glabella of *silicunda* extends closer to the border than that of *edsonensis;* the eye ridges of *silicunda* are fainter than those of *edsonensis.*

The axial furrow of the cephalon of *P. macrommatia* is deeper and wider than that of *edsonensis;* the bacculae of *edsonensis* are smaller than those of *macrommatia.* The pygidial axis of *edsonensis* is wider than that of *macrommatia;* the axial tubercles and the transaxial furrows are more prominent in *macrommatia* than in *edsonensis.*

P. edsonensis differs from *P. ellsi* in that there are faint but distinct signs of lateral glabellar furrows in *P. ellsi;* the pygidial axis of *ellsi* is much narrower than that of *edsonensis.* The glabella of *quebecensis* reaches farther forward than does that of *edsonensis;* the eye ridges of *edsonensis* are better defined than those of *quebecensis.*

POLYMEROID TRILOBITES
Order CORYNEXOCHIDA Kobayashi 1935
Family DORYPYGIDAE Kobayashi 1935

Genus and species undetermined
(Plate 2, Figs 20, 21)

A partial cephalon and a partial pygidium, both of which are poorly preserved and come from the P79-5 fauna, probably belong in the same species of Dorypygidae. However, their preservation is such that no description is warranted.

Family DOLICHOMETOPIDAE Walcott, 1916

Genus *Fuchouia* Resser and Endo, 1937

Fuchouia Resser and Endo, 1937, p. 225; Lu and others, 1965, p. 116; Yin, 1978, p. 466, Öpik, 1982, p. 26.
Amphoton (Fuchouia) Kobayashi, 1942, p. 162; Harrington and others, 1959, p. 222; Öpik, 1961, p. 139.
Type Species. Bathyuriscus manchuriensis Walcott, 1911, p. 97, Plate 16, Fig. 4.
Diagnosis. See Öpik, 1982, p. 26.

Fuchouia sp.
(Plate 2, Figs 26, 27)

Material. One markedly distorted cranidial internal mold and two internal molds of hypostomes from the PA79-1 fauna.
Remarks. The cranidium is not well enough preserved to warrant formal description. The palpebral lobe has a length just under half that of the glabella, and hence it would fall into species Group A of Öpik (1982, p. 27).

Order PTYCHOPARIIDA Swinnerton 1915
Family SOLENOPLEURIDAE Angelin, 1854

Genus *Sohopleura* nov.

Type Species. S. drakensis sp. nov.
Diagnosis. Member of the Solenopleuridae with no surface ornament. Cranidium about 1.4 times wider than long. Strongly convex gently tapering glabella extends almost to shallow anterior border furrow; between the palpebral lobes glabella has width 0.40 to 0.45 that of cranidium. Gently impressed axial furrows shallow forward. Prominent convex occipital ring bears small centrally placed node. Lateral glabellar furrows either absent or very gently impressed. Low eye ridges extend around broadly rounded glabellar front to form a parafrontal band. Fixigenae flat except at anterior and posterior margins where they slope down to border furrow. Preocular sections of facial suture diverge gently and then converge sharply forward of border furrow; postocular sections of facial suture diverge gently. Almost flat librigena with genal spine. Thorax of at least nine segments; moderately impressed pleural furrows stop just short of rounded pleural extremities.
Discussion. When compared with previously described members of the Solenopleuridae, *Sohopleura* appears closest to *Solenoparia* Kobayashi, and the writers originally considered placing *Sohopleura drakensis* as described below in *Solenoparia.* However, in all well documented spe-

cies of *Solenoparia,* including the type species *Solenoparia toxeus,* the glabella tapers more and extends farther forward than does that of *drakensis.* The anterior part of the axial furrow of most species of *Solenoparia* is much more impressed than that of *drakensis.*

Sohopleura differs from *Parasolenopleura* Westergård in that the glabella of the latter tapers forward more, is relatively wider, and is not as long as that of *Sohopleura. Parasolenopleura* has shallow but distinct lateral glabellar furrows, whereas the glabella of *Sohopleura* is almost effaced.

Sohopleura differs from *Solenopleura* in that the glabella of *Solenopleura* tapers forward more than that of *Sohopleura.* Lateral glabellar furrows are evident on *Solenopleura* but not on *Sohopleura.* The surface of *Solenopleura* is faintly to clearly granulose, whereas that of *Sohopleura* is smooth. The glabella of *drakensis* reaches farther forward than it does in most species of *Solenopleura,* although the glabella of *Solenopleura holometopa* has a length similar to that of *Sohopleura drakensis.*

It should be noted that Snajdr (1958) placed *Parasolenopleura* in synonymy with *Solenopleurina* Ruzicka. The questionn of the validity and relationship of such genera as *Parasolenopleura, Solenopleurina, Solenoparia,* and *Pseudosolenoparia* Zhou is reserved for future discussion. Zhang and Jell (1987) have thoroughly reviewed many of the Chinese species of the Solenopleuridae.

Sohopleura drakensis sp. nov.
(Plate 2, Figs. 22–24; Plate 3, Figs. 1–4)

Material. Several cranidia in various states of preservation; one partially enrolled specimen comprising a cranidium and parts of nine thoracic segments. The specimens come from the W79-126 and MH64-51 faunas.
Holotype. Cranidium, USNM334137, Plate 2, Fig. 22.
Diagnosis. See generic diagnosis.
Description. Surface ornament lacking on all available specimens. Cranidium about 1.4 times wide than long. Length of strongly convex glabella (including occipital ring) about 0.75 times that of cranidium; between palpebral lobes glabellar width is 0.40 to 0.45 that of cranidium. Glabella narrows forward only slightly to broadly rounded anterior. Axial furrow gently impressed posteriorly, shallows forward. No distinct anterior axial furrow; anterior of glabella slopes steeply down and merges with very short (sag.) preglabellar field, in turn merging with anterior border furrow. Moderately impressed occipital furrow widens adaxially. Prominent convex occipital ring longest (sag.) at center; at either end it continues across axial furrow as a low ridge into the fixigena. Small median occipital node. Forward of occipital furrow glabella is almost entirely effaced with only faint signs of lateral glabellar furrows on some specimens. Broad anterior border furrow merges into broad gently convex anterior border. Short (exsag.) slightly elevated, gently curved palpebral lobes opposite centro-anterior part of glabella. From anterior of palpebral lobes low eye ridges are directed forward to anterior of glabella and in some specimens continue around front of glabella as a parafrontal band. Very shallow palpebral furrows.

Preocular areas of fixigenae slope steeply to border furrow. Palpebral and posterior areas of fixigenae almost flat except at posterior where fixigenae slope down to broad moderately impressed posterior border furrow that widens adaxially. Posterior border furrow meets posterior margin near ends of occipital ring. Preocular sections of facial suture diverge slightly to anterior of border furrow, then converge sharply forward. Postocular sections of facial suture diverge gently.

The only available thorax is incomplete; it comprises nine segments, each about eleven times wider than long. Axis has width about 0.3 that of each segment; it stands well above pleural areas. Moderately impressed pleural furrows stop just short of rounded pleural extremities.

Hypostome and pygidium not known.

Family KINGSTONIIDAE Kobayashi, 1933

Genus and species indet.
(Plate 3, Figs. 5–6)

Material. One small specimen from the P79-5 fauna comprising a cranidium and parts of seven thoracic segments.
Descriptions. Semicircular moderately convex almost effaced cranidium about 2 mm long and 2 mm wide. Occipital ring and furrow both prominent. Posterior margin of occipital ring curved markedly to posterior. Just forward of occipital furrow the low glabella has width about 0.5 that of cranidium. Axial furrow is moderately impressed at occipital ring but shallows markedly anteriorly and merges with rest of cranidium about halfway from posterior to anterior margin of cranidium. There is some suggestion of a broadly rounded glabellar anterior with a glabellar length about 0.75 that of cranidium. Border area very narrow. No sign of a palpebral lobe nor any sign of a change in direction of facial suture either anterior or posterior of expected position of palpebral lobe.

Preserved part of thorax comprises parts of seven segments. Axis elevated well above pleural areas. Shallow, narrow axial furrows. Axis has width about 0.35 to 0.4 of each segment. Where preserved pleural furrows are shallow. Adaxial halves of pleural areas are almost flat; the abaxial halves of pleural areas are curved sharply downward. Details of pleural extremities not preserved.
Discussion. This specimen if correctly assigned is one of the oldest representatives of the family, which is mainly of Late Cambrian age, and the first to reveal the thoracic structure. The thorax is similar to that of the Cambrian illaenurid trilobite *Parakoldinia kureiskaya* Rosova (Rosova, 1977, Plate 6, Fig. 1).

Family ASAPHISCIDAE Raymond 1924
Subfamily BLOUNTINAE Lochman 1944

Genus *Blountia* Walcott 1916

Type Species. Blountia mimula Walcott 1916, p. 399, Plate 61, Figs. 4–4c.
Differential diagnosis. The cranidium is indistinguishable from that of *Blountia* as diagnosed by Palmer (1962, p. 22). The pygidium differs from all other members of *Blountia* in possessing a long posterior spine.

Blountia perplexa sp. nov.
(Plate 3, Figs. 7–15)

Material. Several cranidia, one librigena, and three pygidia from the P79-1 fauna.
Holotype. Pygidium, USNM334149, Plate 3, Figs. 9–10.
Description. Smooth cranidium; glabella only slightly elevated above fixigenae. Length of glabella about 0.70 to 0.75 that of cranidium. Glabellar front broadly rounded. Axial and preglabellar furrows weakly impressed. Occipital ring gently convex; gently impressed occipital furrow curved anteriorly at either extremity where it shallows markedly. Lateral glabellar furrows almost effaced; 1p furrows at about midlength of glabella are broad depressions rather than distinct furrows. In some specimens there is a suggestion of a pair of depressions where 2p furrows would be expected (e.g., Plate 3, Fig. 7). Short (sag.) almost flat preglabellar field. Broad gently depressed anterior border furrow merges into gently upsloping anterior border. Short palpebral lobes extend forward from opposite 1p furrows; shallow palpebral furrows; low eye ridges are directed anteriorly from anterior ends of palpebral furrows. Preocular areas of fixigenae slope evenly down to anterior border furrow. Small

palpebral areas almost flat. Posterior areas of fixigenae slope gently and evenly away from glabella. Broad gently impressed posterior border furrows meet axial furrow near posterior of occipital ring. Gently convex posterior borders widen abaxially. Preocular sections of facial suture diverge gently up to anterior border furrow from where they converge gently. Gently curved postocular sections of facial suture diverge markedly to posterior.

Only available librigena poorly preserved. It is almost flat with a wide gently impressed border furrow and almost flat wide lateral border.

Semicircular smooth pygidium possesses long posterior spine. Gently impressed axial furrows shallow posteriorly. Axial length about 0.75 that of pygidium. Short (sag.) articulating half ring. Moderately tapering narrow axis has five axial rings plus terminus. Axial rings become successively shorter to posterior. Transaxial furrows gently impressed. Four pleural furrows on pleural areas; they become successively shorter and less well impressed to posterior. Broad shallow border furrow merges with flat slightly upsloping border.

Asaphiscidae, gen. et sp. indet.
(Plate 3, Figs. 16–18; Plate 4, Figs. 1–2)

Material. Several reasonably well preserved large pygidia are known from the PA79-1 fauna.
Description. Smooth moderately convex transversely elliptical pygidium. Axial furrow gently impressed. Axis slightly more convex than pleural areas; it narrows gently to posterior where it reaches narrow shallow border furrow. Articulating half-ring very short (sag.). Axis is almost effaced, and the number of axial rings cannot be determined. Pleural fields without pleural furrows. Shallow border furrow merges with wide border. Large articulating facets. Fulcra placed a little closer to axial furrow than to anterolateral corners.
Discussion. This pygidium appears to belong in the Asaphiscidae. However, there are no cranidia in the PA79-1 fauna with which it can be linked. Hence it is left in open nomenclature.

Family PAGODIIDAE Kobayashi, 1935

Pagodiidae gen. et sp. indet.
(Plate 4, Fig. 3)

Material. An almost complete, moderately well preserved internal mold of a cranidium is the only known specimen from the P79-1 fauna.
Description. Smooth cranidium about as wide as long. Moderately convex glabella extends to anterior border. Between the palpebral lobes, glabella has width about 0.5 that of cranidium. Glabella tapers slightly to broadly rounded anterior. Moderately impressed axial furrow; gently impressed preglabellar furrow. Anterior border furrow very gently impressed; border almost flat and slightly downsloping. Occipital ring almost flat; occipital furrow gently impressed abaxially, but at center hardly impressed at all. Gently impressed 1p furrows directed markedly to posterior; 2p furrows very gently impressed and directed only slightly to posterior. No other lateral glabellar furrows visible. Slightly elevated palpebral lobes extend from opposite 1p furrows to just forward of 2p furrows; they are curved gently abaxially. Moderately impressed palpebral furrows. Prominent eye ridges meet axial furrow a little to posterior of anterior border furrow. Preocular areas of fixigenae slope gently down to anterior border furrow; palpebral areas slightly convex; posterior areas of fixigenae slope markedly downward. Details of posterior border area poorly preserved. Preocular sections of facial suture diverge gently and then converge gently forward; postocular sections of facial suture diverge markedly to posterior.
Discussion. This cranidium appears to belong in the Pagodiidae, probably in a genus related to *Lisania.* However, there is insufficient material available to make a definite generic or specific assignment.

Family DAMESELLIDAE Kobayashi, 1935
Subfamily DREPANURINAE Hupé, 1953

Genus *Pseudobergeronites* nov.

Type Species. P. spinosa sp. nov.
Diagnosis. Granulose damesellid cranidium with moderately convex forward tapering glabella with truncated anterior. Deeply impressed axial furrow. Glabellar length about 0.8 that of cranidium. Strongly convex, abaxially expanded occipital ring with low central node. Abaxially expanding and deepening occipital furrow. Four pairs of lateral glabellar furrows; 1p furrows bifurcate with deep posterior branch and shallow anterior branch. Swollen 1p lateral glabellar lobes. The 2p, 3p, and 4p lateral glabellar furrows become successively well impressed and less posteriorly directed. Flat preglabellar field. Gently sinusoidal palpebral lobes; moderately impressed palpebral furrow; low eye ridges. Preocular sections of facial suture converge gently; postocular sections diverge markedly.

Large smooth pygidium with seven pairs of prominent marginal spines; seven pairs of pleural ribs defined by clearly outlined pleural furrows are pitted where they meet the gently concave border. All spines strongly directed to the posterior; anterior pair extends farther to posterior than any other pair of spines; second, third, fourth, and fifth pairs of spines of about equal size; sixth pair of spines larger than second to fifth pairs; seventh pair of spines quite short. Axis of five axial rings and long terminus that just reaches border.
Discussion. Of previously described genera *Pseudobergeronites* seems to be most closely related to *Bergeronites* and hence is tentatively placed in the Drepanurinae. The pygidium of *Pseudobergeronites* is similar to that of *Bergeronites,* particularly to that of *Bergeronites* aff. *dissidens* (Öpik, 1967, p. 343, Figs. 1–31). However, the cranidium of *P. spinosa* is quite different from that of either *Bergeronites* or *Drepanura* in that in these two genera the glabella reaches the anterior border, whereas the glabella of *P. spinosa* stops well short of the border.

***Pseudobergeronites spinosa* sp. nov.**
(Plate 4, Figs. 5–11)

Material. Several partial cranidia and pygidia from the P79-1 fauna.
Holotype. Cranidium, USNM334163, Plate 4, Fig. 7.
Diagnosis. See generic diagnosis.
Description. Moderately convex cranidium with surface covered with coarse granules except in furrows. Moderately convex glabella tapers forward to truncated anterior; outlined by deeply impressed axial furrow. Length of glabella about 0.8 that of cranidium. Strongly convex occipital ring expands abaxially and bears low central node. Occipital furrow wide and shallow at center; deep and narrow at extremities. Four pairs of lateral glabellar furrows. Deeply impressed 1p furrows bifurcate with deeper posterior branch directed strongly to posterior; shallow anterior branch leaves posterior branch at about its midpoint and is directed slightly to anterior. The 1p lateral glabellar lobes are quite swollen. Gently impressed 2p furrows directed inward and slightly to posterior; 3p furrows short, gently impressed, and directed very slightly to posterior; short 4p furrows are very faint. Flat preglabellar field. Details of anterior border not clear on any available specimen.

Gently sinusoidal palpebral lobes extend from opposite middle of posterior lateral glabellar lobes to opposite 2p furrows. Moderately impressed palpebral furrows. Low eye ridges extend from near 3p furrows to anterior of palpebral lobes. Preocular sections of facial suture converge gently; postocular sections diverge markedly. Preocular areas of fixigenae slope down to anterior border area. Palpebral areas of fixigenae almost flat. Posterior areas of fixigenae slope downward markedly abaxially. Wide (tr.) posterolateral limbs with deeply impressed posterior border furrow and narrow, convex posterior border.

Large, smooth surfaced pygidium with seven pairs of prominent marginal spines corresponding to number of well defined pleural ribs. Anterior pleural furrows deeply impressed; posterior pleural furrows gently impressed. Near each pleural furrow where it meets the border is a shallow pit that appears to be an extension of pleural furrows, although Öpik (1967, p. 342) has interpreted similar structures in *Bergeronites dissidens* as the distal ends of interpleural furrows. However, there are no signs of interpleural furrows on available specimens of *spinosa*. The pits in *spinosa* are elongated and extend about halfway across border area. Spines are large flat blades directed posteriorly. Anterior pair extends farther to posterior than do any other spines. Second, third, fourth, and fifth pairs of spines of approximately equal length and also directed markedly to posterior; sixth pair a little longer than second to fifth pairs; seventh pair quite short. Flatly concave pygidial border. Moderately elevated pygidial axis has length about 0.8 that of pygidium; axial furrows well impressed shallowing posteriorly. Axis of five axial rings plus a long terminus that just reaches border. The four anterior axial rings separated by well defined transaxial furrows. Transaxial furrow separating fifth axial segment from terminus poorly impressed.

Damesellidae, gen. et sp. indet.
(Plate 4, Fig. 16)

Material. One internal mold of a partial cranidium from the P79-1 fauna.
Description. The smooth cranidium has an occipital furrow that is deeply impressed abaxially and gently impressed adaxially, where it is curved forward. Axial furrow moderately impressed. 1p furrows moderately impressed abaxially and shallow adaxially; directed markedly to posterior and almost meet occipital furrow. 2p furrows gently impressed and directed inward and slightly to posterior. Palpebral areas of fixigenae narrow (tr.); palpebral furrow shallow; palpebral lobe narrow. No other visible details.
Remarks. This cranidium belongs in the Damesellidae, but is insufficiently preserved to place it in any described genus or species.

Cranidium, gen. et sp. indet. no. 1
(Plate 4, Fig. 4)

*Remarks.*One very poorly preserved partial cranidium, USNM334161 is figured simply for the record. It comes from the P79-1 fauna.

Librigena, gen. et sp. indet. no. 1
(Plate 2, Fig. 25)

Remarks. This librigena from the W79-126 fauna is too poorly preserved to be assigned to any genus or species.

Pygidium, gen. et sp. indet. no. 1
(Plate 4, Fig. 12)

Material. One reasonably well preserved external mold of a partial pygidium from the P79-1 fauna.
Description. Gently convex pygidium with a smooth surface is about twice as wide as long. Gently impressed axial furrows shallow posteriorly. Axial length about 0.75 that of pygidium; short (sag.) articulating half-ring. Gently tapering axis of five axial rings plus long (sag.) slightly expanded terminus. Short (sag.) axial rings become successively shorter to posterior. Gently impressed transaxial furrows curved slightly to anterior; they become successively shallower to posterior. Pleural areas with three pleural ribs separated by gently impressed pleural furrows that shallow abaxially. There is no distinct border furrow, and border merges with pleural areas.
Discussion. Although this pygidium shows similarities to those of some members of the Dolichometopidae, it cannot be assigned to any particular genus or species.

Pygidium, gen. et sp. indet. no. 2
(Plate 4, Fig. 15)

Remarks. The very poorly preserved internal mold figured in Plate 4, Fig. 15 is too poorly preserved to warrant description or assignation to any genus or species. It comes from the P79-1 fauna.

Pygidium, gen. et sp. indet. no. 3
(Plate 4, Fig. 14)

Material. One poorly preserved internal mold of a partial pygidium, USNM334170; from the PA79-1 fauna.
Description. Pygidial axis extends almost to posterior border; shallow transaxial furrows; other axial details not visible. Six gently impressed wide pleural furrows and five narrow interpleural furrows on pleural areas. All furrows stop short of pygidial margin. Border merges with pleural area; no distinct border furrow.
Remarks. This pygidium is unsufficiently preserved to be placed in any described genus or species, although it shows some resemblance to *Bathyuriscus* (e.g., see Robison, 1964, Plate 83, Figs. 7, 9.

Pygidium, gen. et sp. indet. no. 4
(Plate 4, Fig. 13)

Material. One reasonably well preserved external mold of a partial pygidium from the P79-1 fauna.
Description. Pygidial axis, outlined by gently impressed axial furrows, comprises seven short (sag.) axial rings plus a short terminus. Axial rings become shorter posteriorly. Axis just reaches border furrow. Transaxial furrows become successively narrower to posterior. Six pleural furrows become successively shorter and less clearly outlined to posterior. Original segmentation of pleural areas is outlined between both first and second and second and third pleural segments, with a distinct drop from first to second segment and a smaller drop from second to third segment. Narrow, shallow border furrow. Moderately wide border narrows slightly to posterior except immediately behind pygidial axis where margin of pygidium is deflected slightly to anterior.
Remarks. There is insufficient material to allow assignation to a described genus or species.

ACKNOWLEDGMENTS

We wish to thank Dr. J. H. Shergold (Bureau of Mineral Resources, Canberra), Dr. P. A. Jell (Queensland Museum, Brisbane), and Professor R. A. Robison (University of Kansas, Lawrence) for valuable comments on an earlier version of this manuscript. Of the several people who assisted in the collection of the trilobites, we wish to thank particularly Dr. J. Pojeta and Dr. E. Yochelson of the U.S. Geological Survey.

PLATE 1

Figures 1–7. *Peronopsis deons* sp. nov.

Figure 1. USNM334085, internal mold of cephalon, locality W79-126, ×8.5.

Figure 2. USNM334086, internal mold of holotype pygidium, locality W79-126, ×9.

Figure 3. USNM334087, internal mold of pygidium, locality W79-126, ×8.

Figure 4. USNM334088, external mold of cephalon, locality W79-126, ×9. USNM334089, external mold of pygidium, locality W79-126, ×9.

Figure 5. USNM334093, external mold of pygidium, locality W79-126, ×8.

Figure 6. USNM334090, external mold of pygidium, locality W79-126, ×7. USNM334091, external mold of cephalon, locality W79-126, ×7.

Figure 7 (bottom). USNM334092, external mold of cephalon, locality W79-126, ×6.5.

Figure 7 (top). *Pagetia edsonensis* sp. nov. USNM334097, external mold of pygidium also figured in pl. 2, fig. 17, locality W79-126, ×6.5. USNM334098, external mold of cranidium, locality W79-126, ×6.5.

Figures 9–10. *Valenagnostus* sp.

Figure 9. USNM334094, external mold of cephalon tentatively placed in *Valenagnostus* sp., locality P79-1, ×11.

Figure 10. USNM334095, internal mold of pygidium, locality P79-1, ×8.

Figures 8, 11, 12, 15, 16. cf. *Grandagnostus* sp.

Figure 8. USNM334096, internal mold of cephalon, locality P79-1, ×4.5.

Figure 11. USNM334099, internal mold of cephalon, locality P79-1, ×5.3.

Figure 12. USNM334100, internal mold of cephalon, locality P79-1, ×6.

Figure 15. USNM334103, internal mold of pygidium, locality P79-1, ×4.5.

Figure 16. USNM334104, external mold of pygidium, locality P79-1, ×6.

Figures 13, 14, 18, 19. *Grandagnostus* sp.

Figure 13. USNM334101, internal mold of cephalon, locality P79-1, ×5.5.

Figure 14. USNM334102, partly exfoliated pygidium, locality P79-1, ×7.

Figure 18. USNM334106, internal mold of pygidium, locality P79-1, ×6.5.

Figure 19. USNM334107, internal mold of pygidium, locality P79-1, ×11.

Figure 17. *Ptychagnostus aculeatus* (Angelin), USNM334105, external mold of partial cephalon, locality P79-1, ×7.

Figure 20. *Hypagnostus* sp. 1, USNM334108, internal mold of cephalon, locality PA79-1, ×9.

Figure 21. *Ptychagnostus* sp. 2, USNM334111, internal mold of pygidium, locality PA79-1, ×7.5.

Figure 22. *Hypagnostus* sp. 2, USNM334109, internal mold of cephalon, locality P79-1, ×6.5.

Figure 23. *Hypagnostus* sp. 1, USNM334110, internal mold of cephalon, locality PA79-1, ×10.

Figure 24. *Ptychagnostus* sp. 1, USNM334112, internal mold of pygidium, locality P79-1, ×12.

Figures 25, 28, 29. *Baltagnostus*(?) sp.

Figure 25. USNM334113, internal mold of cephalon, locality P79-1, ×6.5.

Figure 28. USNM334114, external mold of pygidium, locality P79-1, ×10.

Figure 29. USNM334115, internal mold of pygidium, locality P79-1, ×7.

Figures 26, 27. *Lejopyge* sp.

Figure 26. USNM334116, internal mold of cephalon, locality P79-1, ×10.

Figure 27. USNM334117, internal mold of cephalon and partial pygidium, locality P79-1, ×3.

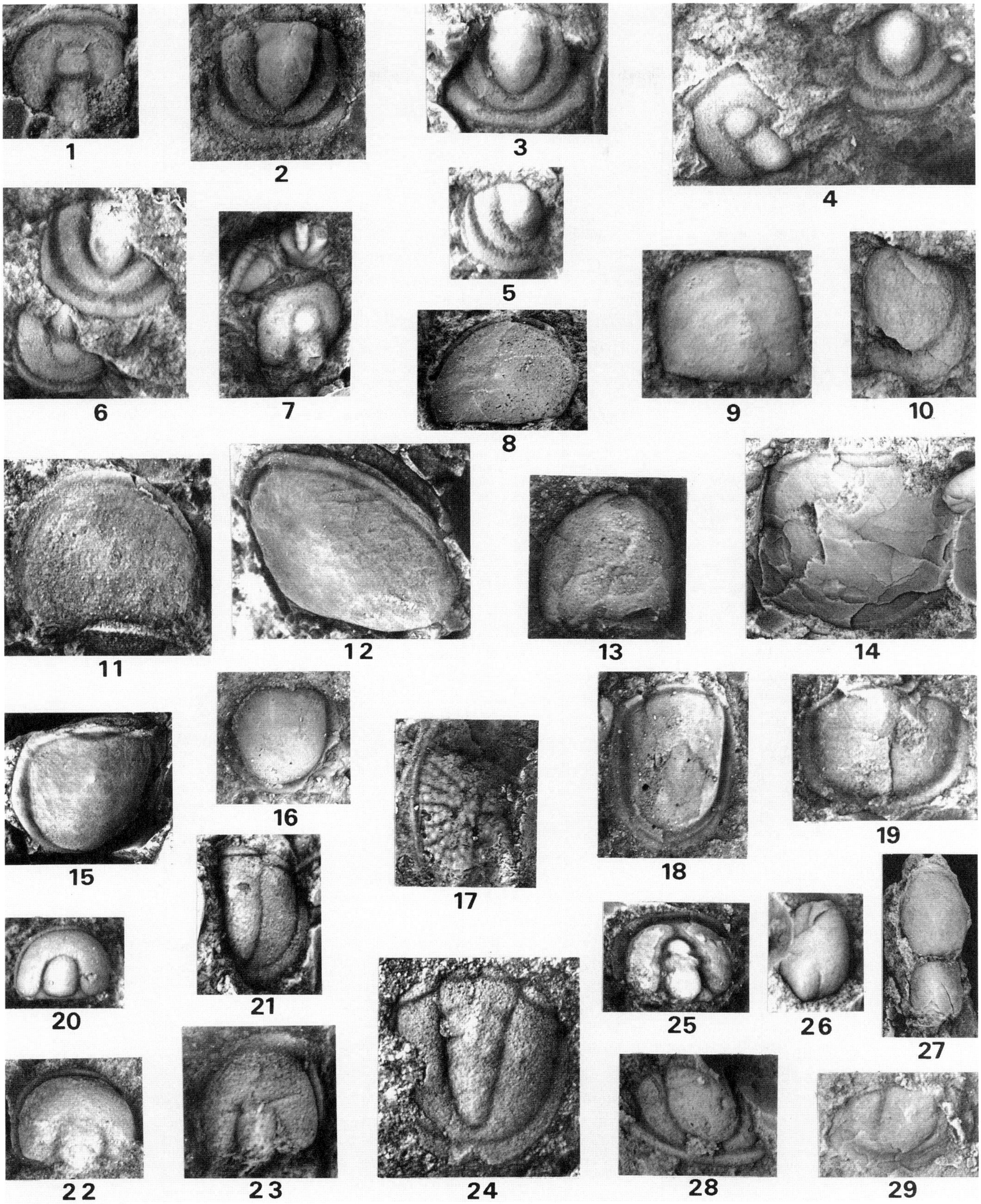
1
2
3
4
5
6
7
8
9
10
11
12
13
14
15
16
17
18
19
20
21
22
23
24
25
26
27
28
29

PLATE 2

Figure 1 (top). *Tomagnostella* sp., USNM334118, internal mold of cephalon, locality P79-1, ×8.

Figures 1–4, 9. *Clavagnostus* sp. 1.

Figure 1 (bottom). USNM334119, internal mold of partial cephalon hidden by cephalon of *Tomagnostella,* locality P79-1, ×8.

Figure 2. USNM334120, internal mold of cephalon, locality P79-1, ×8.

Figure 3. USNM334121, internal mold of pygidium, locality P79-1, ×11.

Figure 4. USNM334122, external mold of pygidium, locality P79-1, ×7.

Figure 9. USNM334127, external mold of pygidium, locality P79-1, ×8.5.

Figure 5. *Clavagnostus*(?) sp., USNM334123, external mold of pygidium, locality P79-1, ×9.

Figure 6. Agnostoid, gen. et sp. indet. 3., USNM334124, internal mold of pygidium, locality P79-1, ×8.

Figure 7. *Clavagnostus* sp. 2, USNM334125, internal mold of cephalon, locality P79-1, ×10.

Figure 8. Agnostoid, gen. et sp. indet. no. 2, USNM334126, internal mold of pygidium, locality P79-1, ×13.

Figure 10. Agnostoid, gen. et sp. indet. no. 1, USNM334128, internal mold of pygidium, locality P79-1, ×8.

Figures 11–19. *Pagetia edsonensis* sp. nov.

Figure 11. USNM334129a, internal mold of holotype cranidium, locality W79-126, ×14.

Figure 12. USNM334129b, partly exfoliated holotype cranidium, locality W79-126, ×8.

Figure 13. USNM334130, partly exfoliated cranidium, locality W79-126, ×12.

Figure 14. USNM334131, partly exfoliated pygidium, locality W79-126, ×12.

Figure 15. USNM334132, internal mold of pygidium, locality W79-126, ×16.

Figure 16. USNM334133, internal mold of pygidium showing anterior part of terminal spine, locality W79-126, ×9.5.

Figure 17. USNM334097, external mold of pygidium showing terminal spine, locality W79-126, ×12. Specimen also figured in pl. 1, fig. 7.

Figure 18. USNM334134, poorly preserved external mold of cranidium, locality W79-126, ×8.

Figure 19. USNM334135, partly exfoliated pygidium, locality W79-126, ×12.

Figures 20, 21. Dorypygidae, gen. et sp. indet.

Figure 20. USNM334136, internal mold of partial cranidium, locality P79-5, ×6.

Figure 21. USNM334142, internal mold of partial pygidium, locality P79-5, ×3.

Figures 22–24. *Sohopleura drakensis* gen. et sp. nov.

Figure 22. USNM334137, partly exfoliated holotype cranidium, locality MH64-51, ×3.5.

Figure 23. USNM334140, partly exfoliated cranidium, locality W79-126, ×4.

Figure 24. USNM334138, partly exfoliated cranidium, locality MH64-51, ×3.4.

Figure 25. Librigena, gen. et sp. indet., no. 1. USNM 334139, internal mold of right librigena, locality W79-126, ×5.5

Figures 26, 27. *Fuchouia* sp.

Figure 26. USNM334141, internal mold of cranidium, locality PA79-1, ×3.

Figure 27. USNM334143, internal mold of hypostome associated with cephalon of *Hypagnostus* sp. 1 (see also pl. 1, fig. 20) and pygidium of *Ptychagnostus* sp. 2 (see also pl. 1, fig. 21), locality PA79-1, ×5.5.

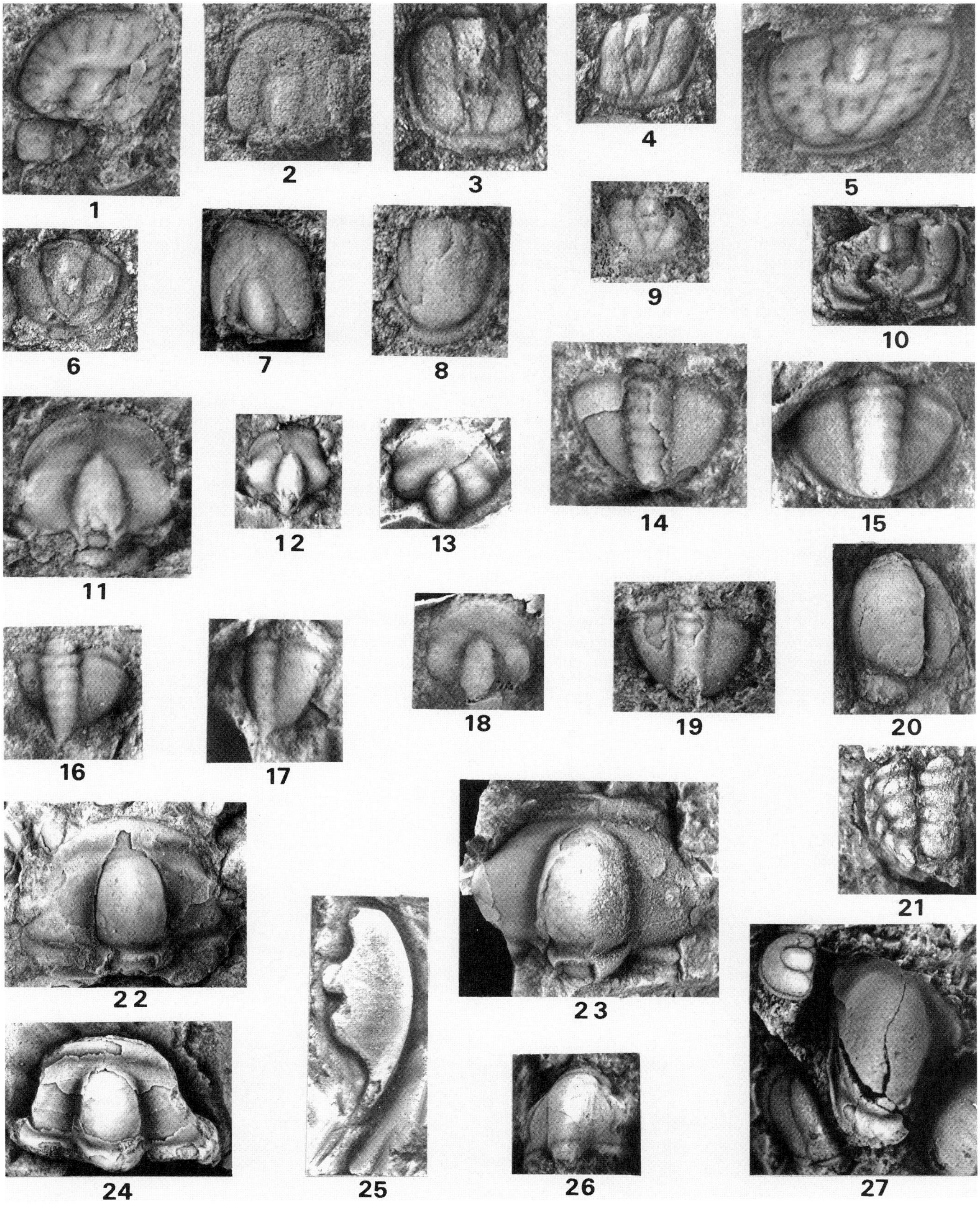
1
2
3
4
5
6
7
8
9
10
11
12
13
14
15
16
17
18
19
20
21
22
23
24
25
26
27

PLATE 3

Figures 1–4. *Sohopleura drakensis* gen. et sp. nov. USNM334144a, partly exfoliated cranidium and partial thorax, locality MH64-51. Fig. 1, partly exfoliated cranidium, ×6. Fig. 2, partial thorax, ×4.75. Fig. 3, side view, ×5.5. Fig. 4, oblique view, ×4.

Figures 5–6. Kingstoniidae, gen. et sp. indet. USNM334145, cranidium and parts of seven thoracic segments, locality P79-5.

Figure 5. Fig. 5, cranidium, ×11; Fig. 6, cranidium plus thoracic segments, ×11.

Figures 7–15. *Blountia perplexa* sp. nov.

Figure 7. USNM334146, internal mold of partial cranidium, locality P79-1, ×2.35.

Figure 8. USNM334147, internal mold of pygidium with broken spine. USNM334148, internal mold of partial cranidium; locality P79-1, ×3.67.

Figures 9–10. USNM334149, internal mold of holotype pygidium, locality P79-1; Fig. 9, ×4; Fig. 10, ×3.

Figure 11. USNM334150, internal mold of left librigena, locality P79-1, ×2.

Figure 12. USNM334151, internal mold of cranidium, locality P79-1, ×2.

Figure 13. USNM334152, external mold of cranidium, locality P79-1, ×5.

Figure 14. USNM334153, internal mold of pygidium with a broken spine, locality P79-1, ×4.

Figure 15. USNM334154, internal mold of cranidium, locality P79-1, ×2.5.

Figure 16–18. Asaphiscidae, gen. et sp. indet.

Figure 16. USNM334155, internal mold of pygidium, locality PA79-1, ×3.

Figure 17. USNM334156a, external mold of pygidium, locality PA79-1, ×2.

Figure 18. USNM334157, partly exfoliated pygidium, locality PA79-1, ×2.67.

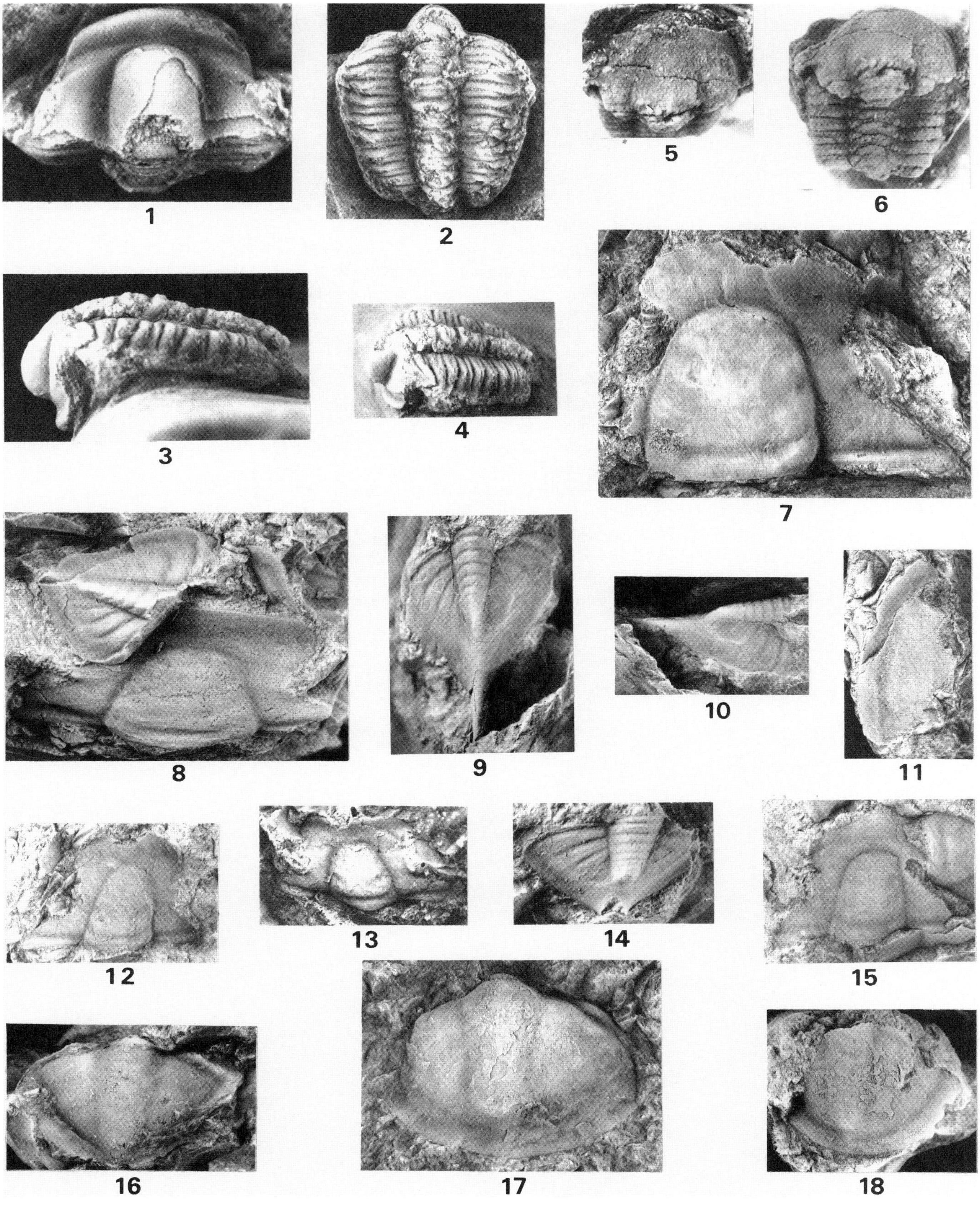
1
2
5
6
3
4
7
8
9
10
11
12
13
14
15
16
17
18

PLATE 4

Figures 1–2. Asaphiscidae, gen. et sp. indet.

Figure 1. USNM334158, partly exfoliated pygidium, locality PA79-1, ×3.

Figure 2. USNM334159, partly exfoliated pygidium, locality PA79-1, ×3.

Figure 3. Pagodiidae, gen. et sp. indet. USNM334160, internal mold of cranidium, locality PT79-1, ×5.5.

Figure 4. Cranidium, gen. et sp. indet. no. 1 USNM334161, partly exfoliated partial cranidium, locality P79-1, ×2.5.

Figures 5–11. *Pseudobergeronites spinosa* gen. et sp. nov.

Figure 5. USNM334162, almost completely exfoliated cranidium, locality P79-1, ×3.

Figure 6. Different view of specimen shown in Fig. 5, ×3.

Figure 7. USNM334163, partly exfoliated holotype cranidium, locality P79-1, ×3.

Figure 8. USNM334164, internal mold of partial pygidium, locality P79-1, ×2.

Figure 9. USNM334165, partly exfoliated partial cranidium, locality P79-1, ×2.5.

Figure 10. USNM334166, internal mold of pygidium, locality P79-1, ×1.67.

Figure 11. USNM334167, external mold of pygidium, locality P79-1, ×2.

Figure 12. Pygidium, gen. et sp. indet. no. 1, USNM334168, external mold of pygidium, locality P79-1, ×2.3.

Figure 13. Pygidium, gen. et sp. indet. no. 4, USNM334169, external mold of pygidium, locality P79-1, ×3.

Figure 14. Pygidium, gen. et sp. indet. no. 3. USNM334170, internal mold of pygidium, locality PA79-1, ×4.5.

Figure 15. Pygidium, gen. et sp. indet. no. 2. USNM 334171, internal mold of pygidium locality P79-1, ×4.5.

Figure 16. Damesellidae, gen. et sp. indet. USNM334172, internal mold of a partial cranidium, locality PA79-1, ×4.5.

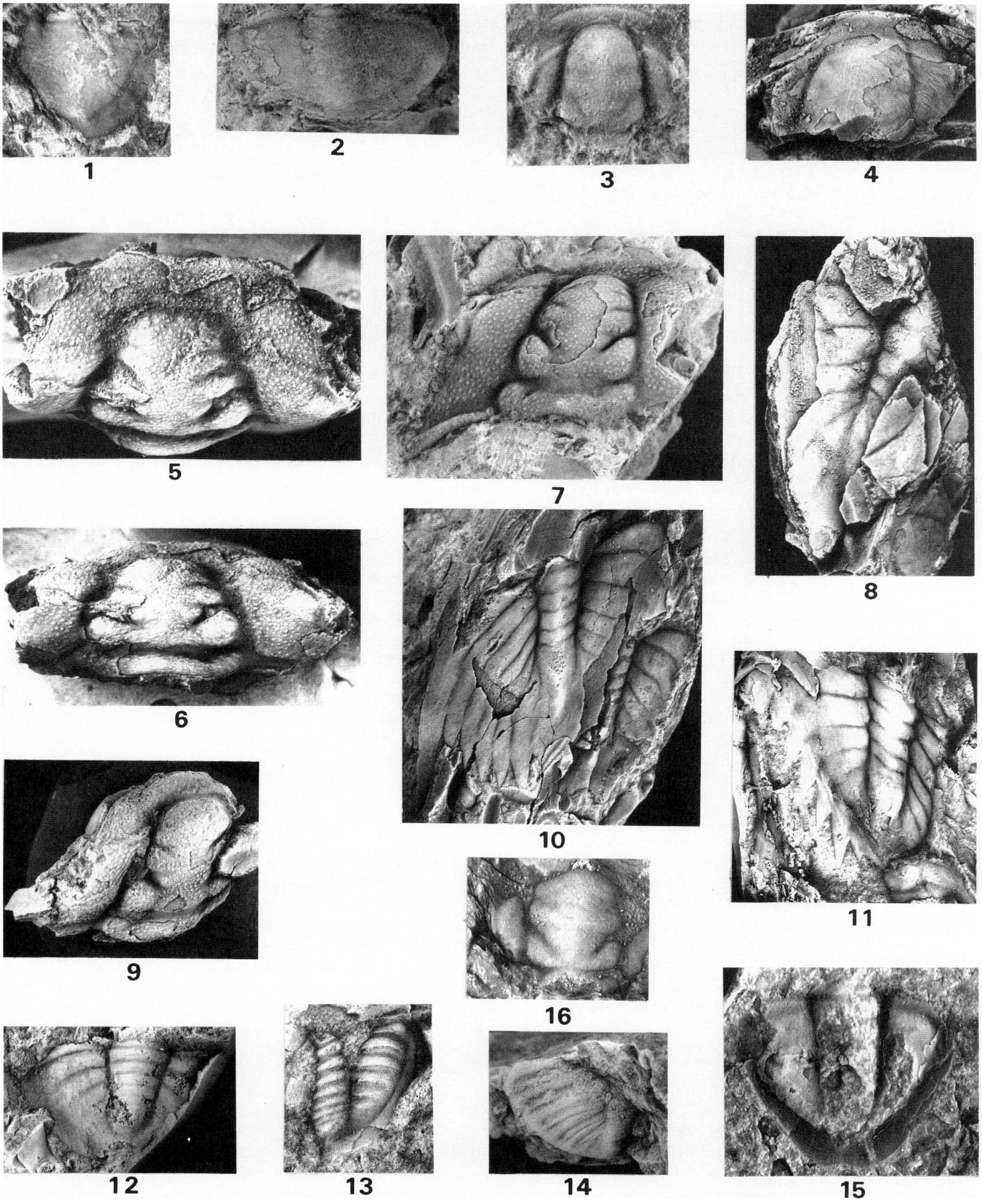

FIGURED SPECIMENS

USNM No.	Locality	Details
334085	W79-126	Cephalon of *Peronopsis deons* sp. nov. Plate 1, fig. 1
334086	W79-126	Holotype pygidium of *Peronopsis deons* sp. nov., Plate 1, fig. 2
334087	W79-126	Pygidium of *Peronopsis deons* sp. nov., Plate 1, fig. 3
334088	W79-126	Cephalon of *Peronopsis deons* sp. nov., Plate 1, fig. 4
334089	W79-126	Pygidium of *Peronopsis deons* sp. nov., Plate 1, fig. 4
334090	W79-126	Pygidium of *Peronopsis deons* sp. nov., Plate 1, fig. 6
334091	W79-126	Cephalon of *Peronopsis deons* sp. nov., Plate 1, fig. 6
334092	W79-126	Cephalon of *Peronopsis deons* sp. nov., Plate 1, fig. 7
334093	W79-126	Pygidium of *Peronopsis deons* sp. nov., Plate 1, fig. 5
334094	P79-1	Cephalon tentatively placed in *Valenagnostus* sp., Plate 1, fig. 9
334095	P79-1	Pygidium of *Valenagnostus* sp., Plate 1, fig. 10
334096	P79-1	Cephalon of cf. *Grandagnostus* sp., Plate 1, fig. 8
334097	W79-126	Pygidium of *Pagetia edsonensis* sp. nov., Plate 1, fig. 7; Plate 2, fig. 17
334098	W79-126	Cranidium of *Pagetia edsonensis* sp. nov., Plate 1, fig. 7
334099	P79-1	Cephalon of cf. *Grandagnostus* sp., Plate 1, fig. 11
334100	P79-1	Cephalon of cf. *Grandagnostus* sp., Plate 1, fig. 12
334101	P79-1	Cephalon of *Grandagnostus* sp., Plate 1, fig. 13
334102	P79-1	Pygidium of *Grandagnostus* sp. Plate 1, fig. 14
334103	P79-1	Pygidium of cf. *Grandagnostus* sp., Plate 1, fig. 15
334104	P79-1	Pygidium of cf. *Grandagnostus* sp., Plate 1, fig. 16
334105	P79-1	Partial cephalon of *Ptychagnostus aculeatus* (Angelin), Plate 1, fig 17
334106	P79-1	Pygidium of *Grandagnostus* sp., Plate 1, fig. 18
334107	P79-1	Pygidium of *Grandagnostus* sp. Plate 1, fig. 19
334108	PA79-1	Cephalon of *Hypagnostus* sp. 1, Plate 1, fig. 20; Plate 2, fig. 27
334109	P79-1	Cephalon of *Hypagnostus* sp. 2, Plate 1, fig. 22
334110	PA79-1	Cephalon of *Hypagnostus* sp. 3, Plate 1, fig. 23
334111	PA79-1	Pygidium of *Ptychagnostus* sp. 2, Plate 1, fig. 21; Plate 2, fig. 27
334112	P79-1	Pygidium of *Ptychagnostus* sp. 1, Plate 1, fig. 24
334113	P79-1	Cephalon of *Baltagnostus*(?) sp., Plate 1, fig. 25
334114	P79-1	Pygidium of *Baltagnostus*(?) sp., Plate 1, fig. 28
334115	P79-1	Pygidium of *Baltagnostus*(?) sp., Plate 1, fig. 29
334116	P79-1	Cephalon of *Lejopyge* sp. Plate 1, fig. 26
334117	P79-1	Cephalon and partial pygidium of *Lejopyge* sp., Plate 1, fig. 27
334118	P79-1	Cephalon of *Tomagnostella* sp., Plate 2, fig. 1
334119	P79-1	Partial cephalon of *Clavagnostus* sp. 1, Plate 2, fig. 1
334120	P79-1	Cephalon of *Clavagnostus* sp. 1, Plate 2, fig. 2
334121	P79-1	Pygidium of *Clavagnostus* sp. 1, Plate 2, fig. 3
334122	P79-1	Pygidium of *Clavagnostus* sp. 1, Plate 2, fig. 4
334123	P79-1	Pygidium of *Clavagnostus*(?) sp., Plate 2, fig. 5
334124	P79-1	Pygidium of Agnostid, gen et sp. indet. no. 1, Plate 2, fig. 6
334125	P79-1	Cephalon of *Clavagnostus* sp. 2, Plate 2, fig. 7
334126	P79-1	Pygidium of Agnostid, gen et sp. indet. no. 2, Plate 2, fig. 8
334127	P79-1	Pygidium of *Clavagnostus* sp. 1, Plate 2, fig. 9
334128	P79-1	Pygidium of Agnostid, gen et sp. indet. no. 3, Plate 2, fig. 10
334129a	W79-126	Internal mold of holotype cranidium of *Pagetia edsonensis* sp. nov., Plate 2, Fig. 11
334129b	W79-126	External mold of holotype cranidium of *Pagetia edsonensis* sp. nov., Plate 2, fig. 12
334130	W79-126	Cranidium of *Pagetia edsonensis* sp. nov., Plate 2, fig. 13
334131	W79-126	Pygidium of *Pagetia edsonensis* sp. nov., Plate 2, fig. 14
334132	W79-126	Pygidium of *Pagetia edsonensis* sp. nov., Plate 2, fig. 15
334133	W79-126	Pygidium of *Pagetia edsonensis* sp. nov., Plate 2, fig. 16
334134	W79-126	Cranidium of *Pagetia edsonensis* sp. nov., Plate 2, fig. 18
334135	W79-126	Pygidium of *Pagetia edsonensis* sp. nov., Plate 2, fig. 19
334136	P79-5	Partial cranidium of *Dorypygidae*, gen et sp. indet., Plate 2, fig. 20
334137	MH64-51	Holotype cranidium of *Sohopleura drakensis* gen et sp. nov., Plate 2, fig. 22
334138	MH64-51	Cranidium of *Sohopleura drakensis* gen et. sp. nov., Plate 2, fig. 24

FIGURED SPECIMENS (continued)

USNM No.	Locality	Details
334139	MH64-51	Right librigena of *Sohopleura drakensis* gen et sp. nov., Plate 2, fig. 25
334140	W79-126	Cranidium of *Sohopleura drakensis* gen et sp. nov., Plate 2, fig. 23
334141	PA79-1	Cranidium of *Fuchouia* sp, Plate 2, fig. 26
334142	P79-5	Partial pygidium of Dorypygidae gen et sp. indet., Plate 2, fig. 21
334143	PA79-1	Hypostome of *Fuchouia* sp., Plate 2, fig. 27
334144b	MH64-51	Cranidium and partial thorax of *Sohopleura drakensis* gen et sp. nov., Plate 3, figs. 1-4
334144a	MH64-51	Counterpart of above
334145	P79-5	Cranidium and parts of seven thoracic segments, Kingstonidae, gen et sp. indet. Plate 3, figs. 5-6
334146	P79-1	Cranidium of *Blountia perplexa* sp. nov., Plate 3, fig. 7
334147	P79-1	Pygidium of *Blountia perplexa,* sp. nov., Plate 3, fig. 8
334148	P79-1	Cranidium of *Blountia perplexa,* sp. nov., Plate 3, fig. 8
334149	P79-1	Holotype pygidium of *Blountia perplexa,* subgen et sp. nov., Plate 3, figs. 9-10
334150	P79-1	Left librigena of *Blountia perplexa* sp. nov., Plate 3, fig. 11
334151	P79-1	Cranidium of *Blountia perplexa,* et sp. nov., Plate 3, fig. 12
334152	P79-1	Cranidium of *Blountia perplexa,* Plate 3, fig. 13
334153	P79-1	Pygidium of *Blountia perplexa,* sp. nov., Plate 3, fig. 14
334154	P79-1	Cranidium of *Blountia perplexa,* sp. nov., Plate 3, fig. 15
334155	PA79-1	Pygidium of Asaphiscidae gen et sp. indet., Plate 3, fig. 16
334156a	PA79-1	Pygidium of Asaphiscidae, gen et sp. indet., Plate 3, fig. 17
334156b	PA79-1	Counterpart of above
334157	PA79-1	Pygidium of Asaphiscidae gen et sp. indet., Plate 3, fig. 18
334158	PA79-1	Pygidium of Asaphiscidae, gen et sp. indet., Plate 4, fig. 1

USNM No.	Locality	Detail
334159	PA79-1	Pygidium of Asaphiscidae, gen et sp indet., Plate 4, fig. 2
334160	PT79-1	Cranidium of Pagodiidae, gen et sp. indet., Plate 4, fig. 3
334161	P79-1	Cranidium, gen et sp. indet no. 1, Plate 4, fig. 4
334162	P79-1	Cranidium of *Pseudobergeronites spinos* sp. nov., Plate 4, figs. 5-6
334163	P79-1	Holotype cranidium of *Pseudobergeronites spinosa* sp. nov., Plate 4, fig. 7
334164	P79-1	Pygidium of *Pseudobergeronites spinosa* sp. nov., Plate 4, fig. 8
334165	P79-1	Cranidium of *Pseudobergeronites spinosa* sp. nov., Plate 4, fig. 9
334166	P79-1	Pygidium of *Pseudobergeronites spinosa* sp. nov., Plate 4, fig. 10
334167	P79-1	Pygidium of *Pseudobergeronites spinosa* sp. nov., Plate 4, fig. 11
334168	P79-1	Pygidium, gen et sp. indet no. 1, Plate 4, fig. 12
334169	P79-1	Pygidium, gen et sp. indet. no. 4, Plate 4, fig. 13
334170	PA79-1	Pygidium, gen et sp. indet. no. 3, Plate 4, fig. 14
334171	P79-1	Pygidium, gen et sp. indet. no. 2, Plate 4, fig. 15
334172	PA79-1	Cranidium; Damesellidae, gen et sp. indet., Plate 4, fig. 16

UNFIGURED SPECIMENS
(i.e., bulk specimens)

USNM No.	Locality	Detail
334173	W79-126	*Peronopsis deons* sp. nov.
334174	W79-126	*Pagetia edsonensis* sp. nov.
334175	P79-1	*Pseudobergeronites spinosa* sp. nov.
334176	P79-1	*Grandagnostus* sp.
334177	MH64-51	*Sohopleura drakensis* sp. nov.
334178	P79-1	*Blountia perplexa* sp. nov.
334179	PA79-1	Asaphiscidae, gen et sp. indet.
334180	PA79-1	*Fuchouia* sp.

REFERENCES CITED

Angelin, N. P., 1851, Palaeontologica Scandinavica; Pars 1, Crustacea formationis transitionis; Academy Regiae Scientarium Suecanae, p. 1–24.

Beyrich, E., 1845, Über einige böhmische Trilobiten: Berlin, Reimer, 47 p.

Brögger, W. C., 1878, Om paradoxideskifrene ved krekling: Nyt Magazin fur Naturvidenskaberne, v. 24, p. 18–88.

Daily, B., and Jago, J. B., 1975, The trilobite *Lejopyge* Hawle and Corda and the middle–upper Cambrian boundary: Palaeontology, v. 18, p. 527–550.

Dalman, J. W., 1828, Arsberättelse om nyare zoologiska arbeten och upptäckter: Stockholm, Vetenskapsakad. Arsberätt.

Ergaliev, G. K., 1980, The Middle and Upper Cambrian trilobites of the Maly Karatau: Academy Nauk Kazakhstan SSR, Alma-Ata, 211 p. (in Russian).

Fedyanina, E. S., 1977, Trilobites of the Orlinogorskaya series (Mount orlinaya, northeast Salair): Novosibirsk, U.S.S.R., Trudy Institut of Geology and Geophysics, v. 313, p. 145–152 (in Russian).

Harrington, H. J., and others, 1959, Arthropoda 1, *in* Moore, R. C., ed., Treatise on invertebrate paleontology: Geological Society of America, Part O, 560 p.

Holm, G., and Westergård, A. H., 1930, A Middle Cambrian fauna from Bennett Island: U.R.S.S., Memoirs of the Academy of Sciences, v. 21, no. 8, 25 p.

Howell, B. F., 1935, New Middle Cambrian agnostian trilobites from Vermont: Journal of Paleontology, v. 9, p. 218–221.

——, 1959, Suborder Agnostina, *in* Moore, R. C., ed., Treatise on invertebrate paleontology: Geological Society of America, Part O, p. 172–186.

Jago, J. B., 1976a, Late Middle Cambrian agnostid trilobites from northwestern Tasmania: Palaeontology, v. 19, p. 133–172.

——, 1976b, Late Middle Cambrian agnostid trilobites from the Gunns Plains area, northwestern Tasmania: Papers and Proceedings of the Royal Society of Tasmania, v. 110, p. 1–18.

Jago, J. B., and Daily, B., 1974, The trilobite *Clavagnostus* Howell from the Cambrian of Tasmania: Palaeontology, v. 17, p. 95–109.

Jegorova, L. I., and others, 1960, The Arthropoda, *in* Khalfin, L. L., ed., Palaeozoic biostratigraphy of the Sayan–Altai Mountain region; Volume 1, The lower Palaeozoic: Sibir, Nauchno-issled Inst. Geologii, Geofiziki i Mineral 'nogo Syr'ya, Trudy, v. 19, p. 152–253 (in Russian).

Jegorova, L. I., Pegel, T. B., and Chernysheva, N. E., 1982, Trilobite descriptions, *in* Sokolov, B. S., ed., The Mayan Stage of the type locality (Middle Cambrian of the Siberian Platform): U.S.S.R., Transactions of the Academy of Sciences, Ministry of Geology, v. 8, p. 57–141 (in Russian).

Jell, P. A., 1975, Australian Middle Cambrian eodiscoids with a review of the superfamily: Palaeontographica A, v. 150, p. 1–97.

Jell, P. A., and Robison, R. A., 1978, Revision of a late Middle Cambrian trilobite faunule from northwestern Queensland: Lawrence, University of Kansas Paleontological Contributions Paper 90, p. 1–21.

Khairulina, T. I., 1973, Biostratigraphy and trilobites of the Miasky stage of the Middle Cambrian of the Turkestan Range: Tashkent, Uzbekistan S.S.R., Ministry of Geology, 112, p.

Kindle, C. H., 1982, Middle Cambrian to Lower Ordovician trilobites from the Cow Head Group, western Newfoundland; The C. H. Kindle collection: Geological Survey of Canada Paper 82-1C, p 1–17.

Kobayashi, T., 1939, On the Agnostids, Part 1: Tokyo, Japan, University of Tokyo Journal of the Faculty of Science, section 2, v. 5, p. 69–198.

——, 1942, On the Dolichometopinae: Tokyo, Japan, University of Tokyo Journal of the Faculty of Science, section 2, v. 5, p. 143–206.

Linnarsson, J.G.O., 1869, Om Vestergötlands Cambriska och Siluriska aflagringar: Stockholm, K. Svenska Ventenskapsakademiens, Handl, v. 8, no. 2, p. 1–89.

Lu, Y.-H., Chang, W.-T., Chu, C.-L., Chien, Y.-Y., and Hsiang, L., 1965, Fossils of each group of China: Chinese trilobites: Peking, Science Publishing Co., 2 volumes, 766 p. (in Chinese).

Öpik, A. A., 1961, Cambrian geology and palaeontology of the headwaters of the Burke River, Queensland: Australia Bureau of Mineral Resources Bulletin 53, 249 p.

——, 1967, The Mindyallan fauna of northwestern Queensland: Australia Bureau of Mineral Resources Bulletin 74, 2 volumes, 571 p.

——, 1979, Middle Cambrian agnostids; Systematics and biostratigraphy: Australia Bureau of Mineral Resources Bulletin 172, 2 volumes, 188 p., 67 pl.

——, 1982, Dolichometopid trilobites of Queensland, Northern Territory and New South Wales: Australia Bureau of Mineral Resources Bulletin 175, 85 p.

Palmer, A. R., 1962, *Glyptagnostus* and associated trilobites in the United States: U.S. Geological Survey Professional Paper 374-A, 49 p.

——, 1968, Cambrian trilobites of east-central Alaska: U.S. Geological Survey Professional Paper 559-B, 115 p.

Palmer, A. R., and Halley, R. B., 1979, Physical stratigraphy and trilobite biostratigraphy of the Carrara formation (Lower and Middle Cambrian) in the southern Great Basin: U.S. Geological Survey Professional Paper 1047, 131 p.

Pokrovskaya, N. V., 1958, Middle Cambrian agnostids of Yakutia, Part 1: Akademiia Nauk S.S.S.R., Geological Institute, Trudy, v. 16, 96 p. (in Russian).

Poulsen, V., and Anderson, M. M., 1975, The middle Upper Cambrian transition in southeastern Newfoundland, Canada: Canadian Journal of Earth Sciences, v. 12, p. 2065–2079.

Qian, Y.-Y., 1982, Ontogeny of *Pseudagnostus benxiensis* sp. nov. (Trilobita): Acta Palaeontologica Sinica, v. 21, p. 632–644 (in Chinese with English summary).

Rasetti, F., 1966, Revision of the North American species of the Cambrian trilobite genus *Pagetia:* Journal of Paleontology, v. 40, p. 502–522.

Resser, C. E., 1938, Cambrian System (restricted) of the southern Appalachians: Geological Society of America Special Paper 14, 140 p.

Resser, C. E., and Endo, R., 1937, Description of the fossils, *in* Endo, R., and Resser, C. E., eds., The Sinian and Cambrian formations and fossils of southern Manchoukuo: Manchurian Science Museum Bulletin 1, 474 p.

Robison, R. A., 1964, Late Middle Cambrian faunas from western Utah: Journal of Paleontology, v. 38, p. 510–566.

——, 1976, Middle Cambrian biostratigraphy of the Great Basin: Provo, Utah, Brigham Young University Geology Studies, v. 23, p. 93–109.

——, 1982, Some Middle Cambrian agnostoid trilobites from western North America: Journal of Paleontology, v. 56, p. 132–160.

——, 1984, Cambrian Agnostida of North America and Greenland; Part 1, Ptychagnostidae: Lawrence, University of Kansas Paleontological Contributions Paper 109, p. 1–59.

Romanenko, E. V., 1977, Cambrian trilobites from the section on the Bol'shaya Isha River (northeast Altai): Novosibirsk, U.S.S.R., Trudy Institute of Geology and Geophysics, v. 313, p. 161–184 (in Russian).

Rosova, A. V., 1964, Biostratigraphy and description of the trilobites of the Middle and Upper Cambrian of the northwestern Siberian Platform: Akademiia Nauk S.S.R. Sibirskoe Otdelenie, Institut geologii i geofiziki, 147 p. (in Russian).

——, 1977, Some Upper Cambrian and Lower Ordovician trilobites from the Rybnaya, Khantajka, Kurejka, and Letnyaya River basins: Novosibirsk, U.S.S.R., Trudy Institute of Geology and Geophysics, v. 313, p. 53–84.

Rushton, A.W.A., 1978, Fossils from the middle Upper Cambrian transition in the Nuneaton district: Palaeontology, v. 21, p. 245–283.

——, 1979, A review of the Middle Cambrian Agnostida from the Abbey Slates, England: Alcheringa, v. 3, p. 43–61.

Snajdr, M., 1958, Trilobiti ceskeho stredniho kambria: Rozpravy Ustredniho ustavu geologickeho, v. 24, p. 1–280 (in Czech with English summary).

Walcott, C. D., 1911, Cambrian geology and paleontology; 2, Cambrian faunas of China: Smithsonian Miscellaneous Collections, v. 57, no. 4, p. 69–108.

——, 1916, Cambrian geology and paleontology; 3, Cambrian trilobites: Smithsonian Miscellaneous Collections, v. 64, no. 5, p. 303–456.

Webers, G. F., 1977, Recent paleontological investigations in the Ellsworth Mountains of West Antarctica: Antarctic Journal of the United States, v. 12, no. 4, p. 120–121.

Westergård, A. H., 1946, Agnostidea of the Middle Cambrian of Sweden: Sveriges Geologiska Undersökning, ser. C., no. 477, Arsbok 40, no. 1, 140 p.

Whitehouse, F. W., 1936, The Cambrian faunas of northeastern Australia, parts 1 and 2: Queensland Museum Memoirs, v. 11, p. 59–112.

Xiang, L., and others, 1981, The Cambrian system of China: Beijing, China, Geological Publishing House, 210 p. (in Chinese).

Yang, J.-L., 1978, Middle and Upper Cambrian trilobites of western Hunan and eastern Guizhou: Beijing, China, Geological Publishing House Professional Papers of Stratigraphy and Palaeontology, v. 4, p. 1–74 (in Chinese with English summary).

Yin, G., 1978, A handbook of palaeontology of southwest China; Part 1, Kewichou, Beijing, China, Geological Publishing House, 843 p. (in Chinese).

Zhang Wentang and Jell, P. A., 1987, Cambrian trilobites of North China: Beijing, China, Science Press, pp. 1–459.

Manuscript Accepted by the Society March 1, 1989

Printed in U.S.A.

Geological Society of America
Memoir 170
1992

Chapter 8

Late Dresbachian (Idamean) and other trilobite faunas from the Heritage Range, Ellsworth Mountains, West Antarctica

J. H. Shergold
Bureau of Mineral Resources, P.O. Box 378, Canberra, ACT 2601, Australia
G. F. Webers
Department of Geology, Macalester College, St. Paul, Minnesota 55105

ABSTRACT

Late Cambrian (Late Dresbachian; Idamean) trilobite faunas have been recovered from four localities (Inferno Ridge, Yochelson Ridge, Springer Peak, and the Windy Peak–Pipe Peak area) in the Heritage Range of the Ellsworth Mountains, West Antarctica. Trilobites are most abundant, best preserved, and most diverse in the Minaret Formation at Springer Peak, where they are associated with monoplacophorans, gastropods, rostroconchs, archaeocyathids, articulate and inarticulate brachiopods, pelmatozoans, conodonts and algae.

The Heritage Range material includes taxa with North American, Australian, Chinese, and southern Russian affinities and represents three distinct faunal assemblages. Trilobites from the base of the Minaret Formation at Yochelson Ridge represent the *Aphelaspis* Zone, and trilobites from the Minaret Formation at Springer Peak represent the *Dunderbergia* Zone. A single parabolinoidid trilobite species (gen. and sp. indet.) from the Windy Peak–Pipe Peak area may represent the *Elvinia* Zone. Deformed solenopleuracean trilobites stratigraphically below the other three faunas (gen. and sp. indet.) were recovered from Inferno Ridge.

From the four localities 15 genera, 1 new subgenus, and 24 species (8 new) are described. New taxa include *Idolagnostus (Obelagnostus) imitor* n. subgen., n. sp., *Eugonocare? nebulosum* n. sp., *Changshanocephalus? suspicor* n. sp., *Bathyholcus? conifrons* n. sp., *Protemnites magnificans* n. sp., *Onchopeltis variabilis* n. sp., *?O. acis* n. sp., and *O.? neutra* n. sp.

INTRODUCTION

Late Cambrian trilobite faunas have been recovered from four localities (Inferno Ridge, Yochelson Ridge, Springer Peak, and the Windy Peak–Pipe Peak area) in the Heritage Range of the Ellsworth Mountains, West Antarctica (Fig. 1). Small collections from Springer Peak and Inferno Ridge were obtained during 1962–1963 and 1963–1964 expeditions to the Ellsworth Mountains under the direction of C. Craddock; a greater amount of material was collected during the 1979–1980 expedition to the Ellsworth Mountains under the direction of G. Webers.

LITHOSTRATIGRAPHY AND LOCALITY DESCRIPTION

Figure 2 shows the columnar section of the Ellsworth Mountains, and Figure 3 shows the stratigraphic relations of the four localities. Details of the stratigraphy of the involved forma-

Shergold, J. H., and Webers, G. F., 1992, Late Dresbachian (Idamean) and other trilobite faunas from the Heritage Range, Ellsworth Mountains, West Antarctica, *in* Webers, G. F., Craddock, C., and Splettstoesser, J. F., Geology and Paleontology of the Ellsworth Mountains, West Antarctica: Boulder, Colorado, Geological Society of America Memoir 170.

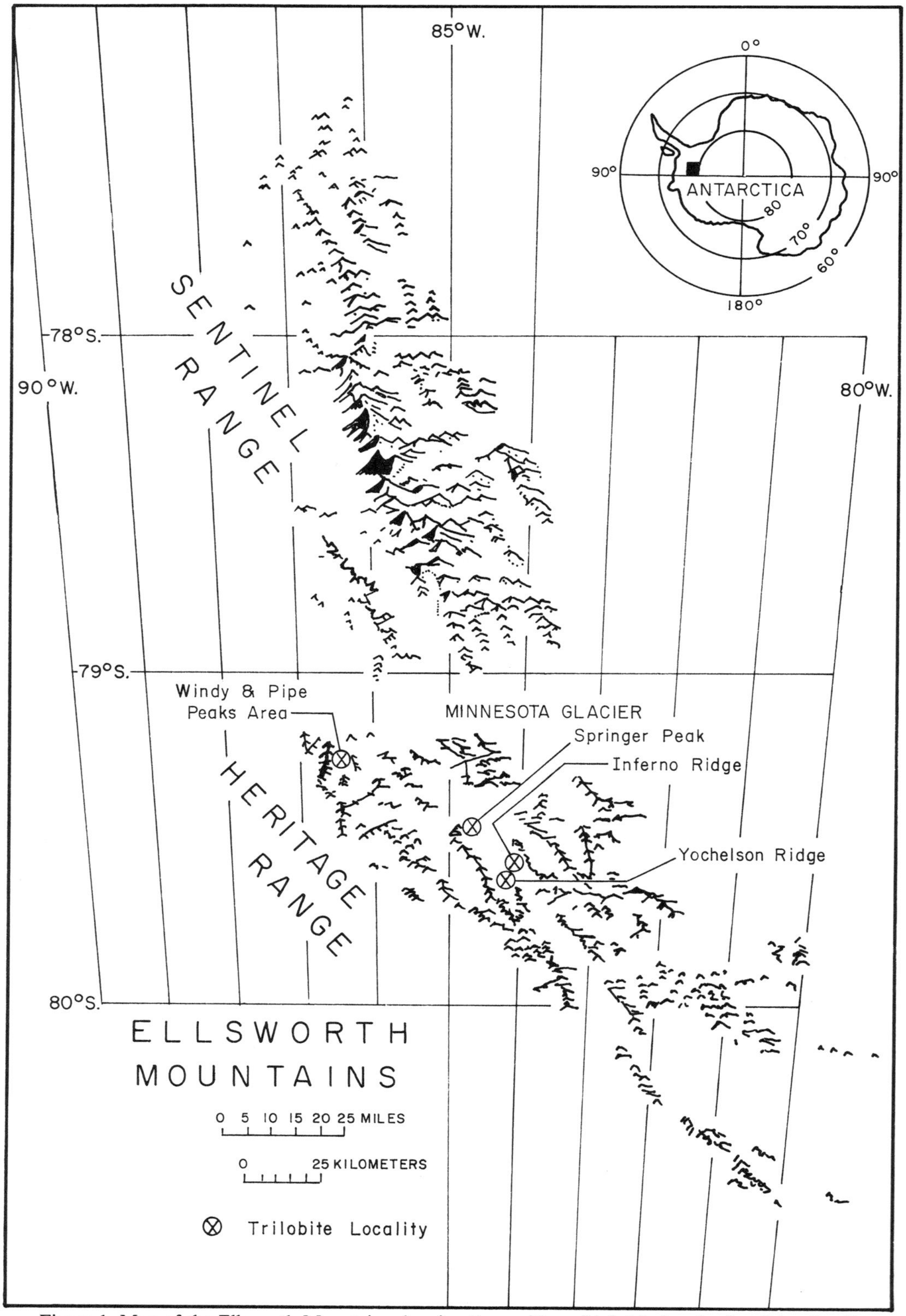

Figure 1. Map of the Ellsworth Mountains showing the position of trilobite-bearing localities. Small inset shows location of the Ellsworth Mountains within Antarctica.

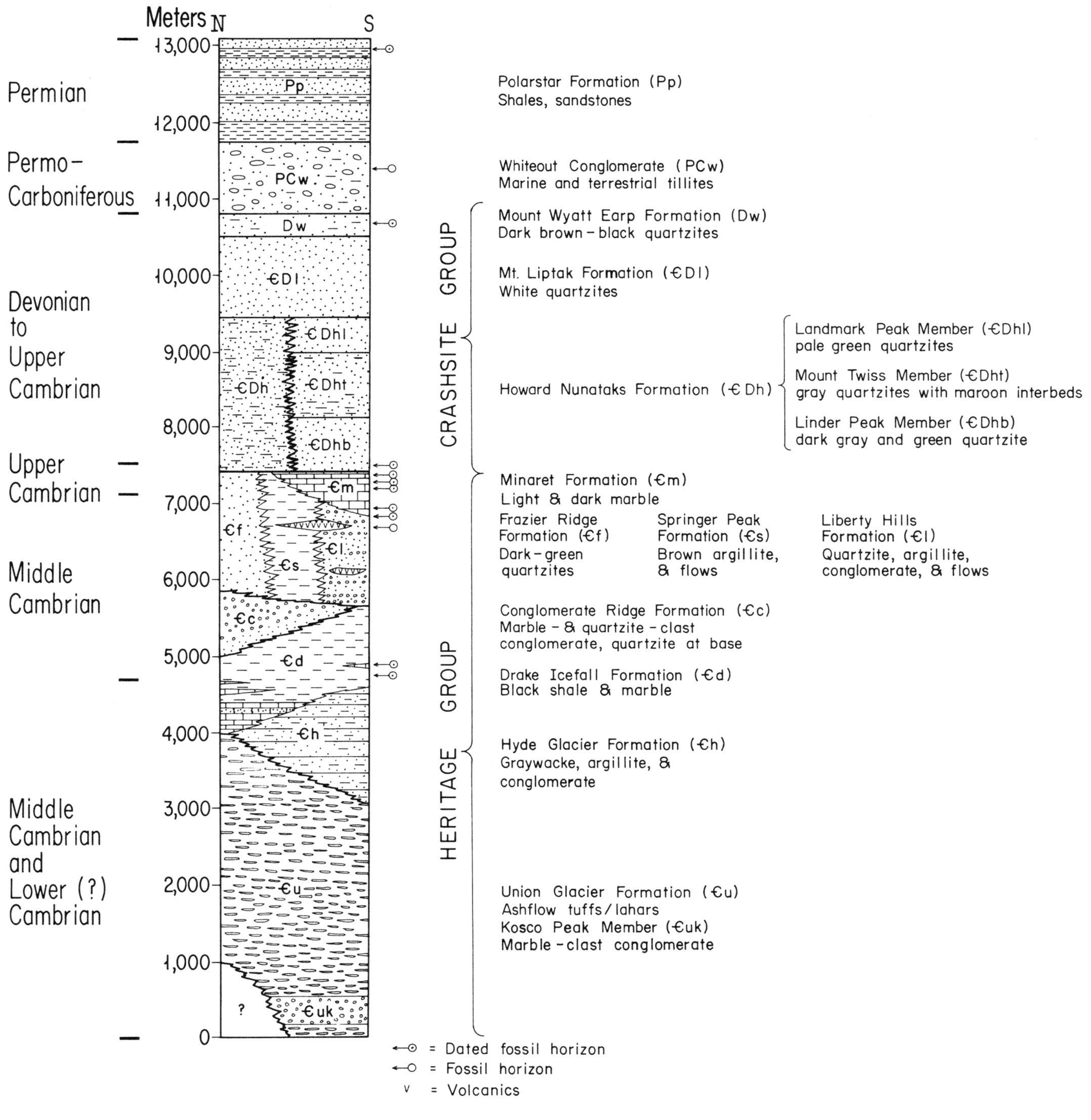

Figure 2. Columnar section, Ellsworth Mountains, Antarctica.

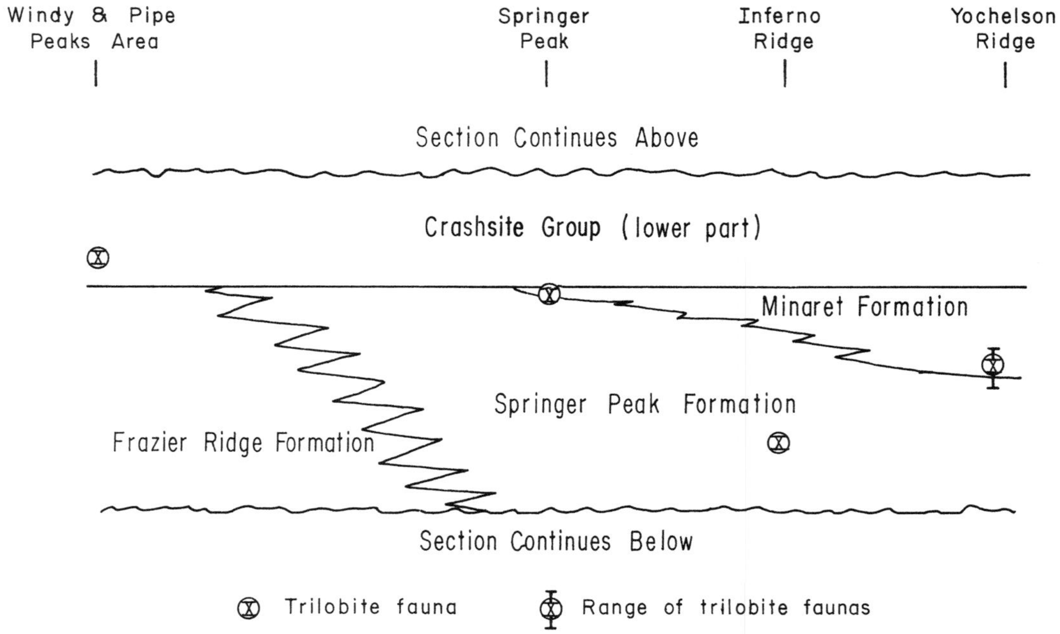

Figure 3. Stratigraphic relations of trilobite-bearing localities.

tions can be found in Webers and others (this volume). Localities yielding late Cambrian trilobites are described below.

Inferno Ridge

Inferno Ridge (centered about 79°25′S, 84°15′W) is an elongate north-south ridge exposing the Springer Peak Formation. Exposed sedimentary rocks consist of black shale, argillite, and graywacke. The Springer Peak Formation is estimated to be 1,000 m thick and the nodular black shales are thought to occur about midway through the stratigraphic sequence. Exact stratigraphic position is difficult to determine because of the limited stratigraphic thickness of exposed beds at Inferno Ridge, the similarity of stratigraphy throughout the formation, and the marked development of parasitic folding and fracture cleavage.

The fauna consists almost entirely of trilobites encased in nodules within the black shale beds. The trilobites are usually entire specimens, but they are strongly deformed. Solopleuracean trilobites (gen. and sp. indet.), probably representing a single species, make up the trilobite fauna. A conodont cluster (Buggisch and others, this volume) completes the faunal picture. The age of this fauna could not be determined because of poor preservation and could be as old as Middle Cambrian. It is included here as a matter of convenience.

Yochelson Ridge

Yochelson Ridge (centered about 79°38′S, 84°18′W) is a north-south ridge extending north from the Soholt Peaks in the west-central Heritage Range. Trilobite faunas are found in the northern tip of the ridge where it splits into east and west subridges. Both subridges expose the upper beds of the Springer Peak Formation and the lower beds of the Minaret Formation. The upper beds of the Springer Peak Formation at this locality consist of argillite, black calcareous shale, graywacke, and black shaly limestone stringers. Late Cambrian trilobites described in this chapter from this locality are from the *Aphelaspis* Zone (see Table 1). Monoplacophorans, gastropods, pelmatozoans, archaeocyathids, and brachiopods are associated with the Late Cambrian trilobites. Middle Cambrian trilobites are also present at Yochelson Ridge in the Springer Peak Formation and are described by Jago and Webers (this volume).

The following is a brief description of the fossiliferous strata of the east subridge:

P79-3. Upper Springer Peak Formation. Probably within 10 m of the contact with the overlying Minaret Formation. The fauna occurs in a light to gray shaly limestone.

YP79-13. Upper Springer Peak Formation. The fauna occurs in a 1-m-thick black shaly limestone within black shale about 7 m below the contact with the Minaret Formation.

YP79-8. Minaret Formation. The fauna occurs in a black nodular limestone in the lower 1.5 m of the Minaret Formation.

YP79-9. Minaret Formation. The fauna occurs in a dark gray limestone about 3 m above the base of the formation.

YP79-11. Minaret Formation. The fauna occurs in a dark gray limestone estimated to be about 3 m above the base of the formation. This limestone is at about the same level as YP79-9 but to the south. It is kept separate because the two collections are

TABLE 1. DISTRIBUTION OF TAXA IN ELLSWORTH MOUNTAINS TRILOBITE COLLECTIONS

	YP79-8	P79-3	YP79-13	YP79-9	YP79-11	YP79-2	W63c	YP79-4
Glyptagnostus reticulatus	X							
Aphelaspis cf. *subdita*	X	X						
Aphelaspis cf. *australis*	X	X	X	X	X			
Idolagnostus (Obelagnostus) imitor	X		X					
Aphelaspis cf. *walcotti*		X						
Prochuangia sp. indet.		X						
Ammangnostus? sp. indet.		X	X					
Pseudagnostus cf. *idalis*			X			X		
Eugonocare? nebulosum			X				X	
Aphelaspis cf. *lata*			X	X	X			
Changshanocephalus? suspicor			X	X	X			
Bathyholcus? conifrons						X	X	
Protemnites? magnificans						X	X	
Onchopeltis variabilis						X	X	
Onchopeltis? acis						X	X	
Onchopeltis? neutra						X	X	
Kormagnostella cf. *minuta*						X	X	
Pseudagnostus cf. *vastulus*						X	X	
Homognostus cf. *ultraobesus*						X	X	
Protemnites aff. *P. elegans*							X	
Erixanium cf. indet.							X	
Stigmatoa sp. indet.							X	
Parabolinoidid? gen. and sp. indet.								X
			Aphelaspis Zone				*Dunderbergia* Zone	?

on opposite sides of a small fault, and the base of the Minaret is not exposed at YP79-11.

Springer Peak

Springer Peak (about 79°25′S, 84°52′W) exposes about 2,500 m of the Crashsite Group, as much as 30 m of the Minaret Formation, and about 700 m of the Springer Peak Formation. The Minaret Formation is at least 600 m thick in the southern Heritage Range and thins northward to a thickness of 8 m at the northernmost exposures at Springer Peak. The Minaret Formation has not been found northwest of Springer Peak.

The fauna occurs in the light- to dark-gray limestone of the Minaret Formation. The fauna is unusually well preserved and generally undeformed. The limestone is a trilobite coquina associated with monoplacophorans, gastropods, rostroconchs, archaeocyathids, articulate and inarticulate brachiopods, pelmatozoans, conodonts, and algae. A similar but less diverse fauna is found at Bingham Peak (about 4 km to the south), where the preservation is relatively poor.

The two samples (W63c and YP79-2) were collected at different times and represent the entire stratigraphic thickness of the Minaret Formation at Springer Peak.

Windy Peak–Pipe Peak

The Windy Peak–Pipe Peak area (centered about 79°10′S, 86°05′ W) is about 30 km northwest of Springer Peak in the northwest Heritage Range. The Minaret Formation does not occur in this area, but its stratigraphic position is indicated by the lowest of a series of alternating black shale and green-to-buff orthoquartzite. Some of the orthoquartzites are calcareous. These beds represent the lowermost beds of the Linder Peak Member of the Howard Nunataks Formation of the Crashsite Group. Trilobites were collected from a narrow stratigraphic interval at four locations in this area: Pipe Peak, Matney Peak, Windy Peak, and the Reuther Nunataks. They are represented by a single sample—YP79-4. The fauna consists of a single parabolinoidid species that may represent the *Elvinia* Zone. The fauna is included in this paper as a matter of convenience.

BIOSTRATIGRAPHY

The trilobites described here are of late Dresbachian and possibly initial Franconian ages when assessed against the standard North American Late Cambrian biochronologic scale. In terms of other regional scales, they are Idamean and immediately post-Idamean in Australia, early Changshanian in China, and Sakian in the USSR (Kazakhstan) and equate with the *Olenus* Zone and possibly *Parabolina spinulosa* Zone faunas of northern Europe. The North American zonal scheme is used here even though the subdivisions of these zones cannot be recognized. Alternatively, the Australian terms early and late Idamean could be applied, but these have a lesser degree of biochronological resolution. The Australian biostratigraphic scheme, proposed by Öpik (1963) and Henderson (1976), is considered too highly resolved for application to the Ellsworth faunas, although correlations can be suggested.

The Heritage Range material includes taxa with North American, Australian, Chinese, and southern Russian affinities and represents three distinct faunal assemblages. The earliest of these—occurring at and within the base of the Minaret Formation at Yochelson Ridge—is represented by collections P79-3, YP79-8, YP79-9, YP79-11 and YP79-13. The distribution of taxa among these collections is shown in Table 1.

This assemblage is characterized by aphelaspidinids and is considered here to represent the early part, at least, of the *Aphelaspis* Zone in the United States and its correlatives elsewhere (Table 1). Confirming the presence of the initial *Aphelaspis* Zone is the occurrence in sample P79-3 of *Aphelaspis* cf. *walcotti,* a species occurring in the early *Aphelaspis* Zone of Texas (Palmer, 1955) and Tennessee (Rasetti, 1965), and *A.* cf. *subdita,* which occurs with *Glyptagnostus reticulatus* in Nevada (Palmer, 1962).

With the exception of *Glyptagnostus reticulatus,* which has a cosmopolitan distribution, the *Aphelaspis* Zone assemblage in Antarctica represents a mingling of North American and Australo-Sinian biofacies. As in North America, aphelaspidinid morphology is much more varied than in the contemporaneous Australian biofacies in which species of *Olenus* are more frequently associated with *G. reticulatus* than species of *Aphelaspis,* although other aphelaspidines like *Eugonocare* provide a link. *Proceratopyge,* of common occurrence in Australian and Chinese outer carbonate belt environments, is not recorded in the Heritage Range, but its niche may be occupied here by *Changhanocephalus?*. The associated agnostid genera are all known from northern Australia, but two of them, *Idolagnostus* and *Ammagnostus?,* are common there in the *Glyptagnostus stolidotus* Zone, which predates that based on *G. reticulatus* and *Aphelaspis.* However, *Idolagnostus* is represented by its new subgenus *Obelagnostus,* currently not known in Australia. Species of *Innitagnostus,* common in *Aphelaspis* Zone equivalents in Australia, are not represented in the Heritage Range assemblages. The presence of *Prochuangia* and *Changshanocephalus?* provides a further link with China and South Korea.

The second faunal assemblage occurs at the top of the Minaret Formation in the vicinity of Springer Peak, some 25 km north-northwest of Yochelson Ridge, and is represented by collections YP79-2 and W63c. This is the assemblage noted earlier by Webers (1972), containing *Onchopeltis,* which, as in Quebec, is associated with a form similar to *Bathyholcus.* The assemblage also contains species of *Erixanium* and *Stigmatoa,* originally described from Australia (Öpik, 1963) but now known to have a wider distribution. *Erixanium* also occurs in the early *Dunderbergia* Zone of Nevada and Utah (Palmer, 1965), and in Kazakhstan (Ergaliev, 1980, 1981), permitting correlation of the Australian late Idamean zones of *Erixanium sentum* and *Stigmatoa diloma* (of Henderson, 1976) with the *Dunderbergia* Zone of North America, a correlation supported by the determination of *Pseudagnostus* cf. *vastulus.* While the North American/Australian relations established during the preceding *Aphelaspis* Zone continue into that of *Dunderbergia,* relations are also evident with the southern USSR. *Bathyholcus?* closely resembles *Didwudina* from central Kazakhstan; *Eugonocare* occurs in southern Kazakhstan; *Kormagnostella* was originally described from the Sayan-Altai, which has also yielded species of *Stigmatoa; Homagnostus* is represented by a species similar to one originally described from the Kuznetsk Alatau; and the species complexes referred here to *Onchopeltis* and *Protemnites* have morphological counterparts in Novaya Zemlya (*Pesaia*) and central Kazakhstan (*Olentella*), respectively.

The third, and presumed youngest, assemblage occurs in green calcareous siltstone at Windy Peak and Pipe Peak, a further 30 km to the northwest of Springer Peak, at locality YP79-4. This assemblage contains a single species of possible parabolinoidid trilobite that may suggest an initial Franconian (*Elvinia* Zone or even younger) age. Since it has not been possible to accurately classify this trilobite, however, biostratigraphic analysis remains speculative.

SYSTEMATIC PALEONTOLOGY

The descriptive terminology used in this chapter is based on that defined in the Treatise on Invertebrate Paleontology, Part O, Arthropoda 1 (Harrington and others, 1959, p. 117–126). Additionally, we use terminology proposed and used mainly for miomerids by Öpik (1961a, 1961b, 1963, 1967) and Shergold (1972, 1975, 1977, 1980, 1982).

Symbols used in the text for measured parameters have been defined previously (e.g., Shergold, 1972, 1975, 1980). Those used here are:

L_c maximum length (sag.) of cephalon or cranidium
L_b length (sag.) of anterior cranidial border, cephalic border, or pygidial border
G length (sag.) of glabella
G_n length (sag.) of glabella plus occipital ring
L_{p1} maximum pygidial length (sag.), including articulating half-ring
L_{p2} pygidial length (sag.), excluding articulating half-ring

All material described in this paper is deposited in the collections of the U.S. National Museum, Washington, D.C., and is prefixed USNM.

Order MIOMERA Jaekel, 1909
Suborder AGNOSTINA Salter, 1864
Family AGNOSTIDAE M'Coy, 1849
Subfamily AGNOSTINAE M'Coy, 1849

Genus *Homagnostus* Howell, 1935

Type species. Agnostus pisiformis Lin. var. *obesus* Belt (1867, p. 295, Plate xii, Figs. 4a–d); Late Cambrian; "Lower Lingula Flags," River Mawddach, North Wales; designated Howell (1935, p. 15).
Comments. Belt's (1867) figures of *Homagnostus obesus* have long been considered uninterpretable, and the concepts of the species and of the genus *Homagnostus* have rested on two complete but slightly different specimens figured by Westergård (1922, Plate 1, Figs. 4–5). Recently, however, Rushton (in Allen and others, 1981, Plate 16, Fig. 2) has designated as the lectotype of *H. obesus* a more or less complete specimen from among Belt's cotypes. Though useful, this specimen has a foreshortened pygidium that has lost its posterior margin so that the extent of the axis must be estimated, and there is some distortion of the cephalon to the extent that the anterior border has been pushed over the acrolobe, and a proper determination of the degree of deliquiation of the marginal furrows is not possible. The lectotype does show, however, the presence of a median preglabellar furrow, suggesting that Westergård's (1922) Figure 4 is likely to be conspecific and that his (1922) Figure 5, which has no such furrow, must be regarded as a separate taxon. Combining the morphology of the lectotype and Westergård's Figure 4, *H. obesus* is seen to be an en grande tenue species with a subquadrate to subcircular, strongly deliquiate cephalon and a pygidium that has narrower (sag.) posterior borders, is less deliquiate, has a long axis extending close to but not contacting the posterior marginal furrow, and has retral spines placed in front of the rear of the termination of the axis. Rushton (1978, Plate 25, Fig. 4; 1983, Plate 14, Figs. 1–10) has considered many nondeliquiate specimens to be conspecific with *H. obesus.*

Homagnostus cf. *ultraobesus* Lermontova, 1940
(Plate 1, Figs. 1–6)

Synonymy.
1922. *Agnostus pisiformis obesus* Belt; Westergård, 1922, p. 116, Plate 1, Fig. 5, *non* Figs. 4, 6.
cf. 1940. *Homagnostus ultraobesus* Lerm.; Lermontova, 1940, p. 124, Plate 49, Figs. 9, 9a.
Material. Twelve cephala, USNM 333936–333938 and 333942a–i, ranging in length between 1.20 and 3.10 mm; and fifteen pygidia, USNM 333939–333941 and 333943a–1, ranging (L_{p2}) between 1.25 and 2.50 mm.
Occurrence. Minaret Formation, Springer Peak, localities YP79-2 and W63c.
Age. Late Cambrian, late Dresbachian, *Dunderbergia* Zone.
Description. Cephalon subcircular, subovoid, or subquadrate according to preservation with maximum width (tr.) at anterior glabellar furrow; strongly deliquiate, borders 8 to 14 percent of cephalic length (sag.), anteriorly tapering acrolobe undivided by median preglabellar furrow. Glabella 67 to 77 percent cephalic length, with semicircular anterior lobe and rectilinear or gently posteriorly curved, well-incised anterior furrow; long, relatively wide (tr.) posterior lobe rounded at rear; anterolateral lobe sometimes faintly indicated; node situated just under half-way along glabella from rear; basal lobes prominent, divided into swollen posterior, and elongate (exsag.) anterior portions. Exoskeletal surface faintly textured on some specimens.

Pygidium subovoid to subquadrate according to direction of deformation, with maximum transverse width just behind second axial furrow when undeformed, with narrower borders than in cephalon, but equally prominent, deliquiate marginal furrows. Axis elevated, swollen posteriorly, 73 to 90 percent of pygidial length (L_{p2}), plethoid, extending almost to marginal furrows but always separated by narrow band of acrolobe; gently constricted acrolobes; posterior lobe occupying more than half the total axial length; retral posterolateral spines, the bases of which lie in advance of transverse line across rear of posterior axial lobe.
Comments. Antarctic cephala cannot be adequately distinguished from that figured by Lermontova (1940, Plate 49, Fig. 9) as *Homagnostus ultraobesus.* The type pygidium (Lermontova, 1940, Fig. 9a) is laterally compressed but nevertheless appears to fall within the morphological variation illustrated herein. The confer has been added to the Antarctic determinations to accommodate the degree of uncertainty that has resulted from comparing a paradigm of some 27 specimens with two indifferent illustrations.

Homagnostus cf. *ultraobesus* Lermontova very closely resembles one of Westergård's (1922, Plate 1, Fig. 5) specimens of *Agnostus pisiformis obesus* Belt from the *Olenus* Zone of Skåne, southern Sweden, in the degree of deliquiation of the marginal furrows in both cephalon and pygidium and in its faintly constricted pygidial acrolobe. Westergård's specimen is slightly crushed so that there is an apparent difference in its shape and the extent of the postaxial acrolobe. This specimen was nominated as the holotype of *H. obesus laevis* by Westergård (1947, p. 4). If synonymy is proved, then this name will have priority over *ultraobesus.*

H. cf. *ultraobesus* also resembles *H. hoiformis* Kobayashi (1933, Plate 10, Figs. 1–3) from the early Changshanian, *Chuangia* Zone of the Wuhutsui Basin, Liaoning, northern China, the latter differing only in retaining traces of a median preglabellar furrow on the cephalon as in *H. obesus* (Belt). Generally speaking, *H. tumidosus* (Hall and Whitfield, 1877), as refigured by Palmer (1955, Plate 19, Figs. 3–4) from the Dunderberg Shale of Nevada, has narrower marginal furrows, its glabella has a greater anterior taper, the posterior axial lobe of the pygidium is possibly less swollen, and the specimen retains a median preglabellar furrow proximally. However, laterally compressed forms of *H. tumidosus,* e.g., in Palmer (1968, Plate 7, Figs. 3, 8) have marginal furrows not readily distinguished from those of *H. ultraobesus. H. comptus* Palmer (1962, Plate 1, Figs. 12–15), from the Lower Dunderberg Formation, also differs by degree of deliquiation.

Genus *Idolagnostus* Öpik, 1967

Type species. Idolagnostus agrestis Öpik (1967, p. 104–106, Plate 59, Figs. 9, 10; Plate 60, Figs. 1–2; Plate 63, Figs 10), O'Hara Shale; Late Cambrian, Mindyallan, *Glyptagnostus stolidotus* Zone, northern Australia; by original designation.

Subgenus *Obelagnostus* subgen. nov.

Name. Gk. *obelos,* pointed pillar, prefixing existing generic name *Agnostus* and referring to the prominent glabellar format.
Type species. Here designated, *Idolagnostus* (*Obelagnostus*) *imitor* sp. nov.; Minaret Formation; Late Cambrian, late Dresbachian, *Aphelaspis* Zone, Ellsworth Mountains, Antarctica; monotypical.
Diagnosis. Obelagnostus is a subgenus of *Idolagnostus* Öpik characterized by the total effacement of the anterior transverse glabellar furrow

and thus formation of a composite anterior lobe and almost total effacement of pygidial axial features.

Comments. *Obelagnostus* is regarded as a subgenus of *Idolagnostus* by loss of the anterior furrows of the glabella and accessory furrows of the pygidium. Such a classification, rather than the erection of a new generic taxon, emphasizes the overall similarity between *Obelagnostus* and *Idolagnostus* and probable genetic relation. Precedent for such a low-level taxonomic procedure presently exists among the species groups recognized within *Neoagnostus* (see Shergold, 1977, Plate 16), wherein a significant degree of homeomorphy exists in the relations between the *bilobus* and *araneavelatus* species groups and the subgenera *Idolagnostus* (*Idolagnostus*) and *I.* (*Obelagnostus*). Agnostids with a trilobate glabella are not uncommon (Öpik, 1967, p. 104), but the glabellar similarity between species of *I.* (*Obelagnostus*) and neoagnosti of the *N. bilobus* group, particularly *N. aspidoides* Kobayashi, the type species of *Neoagnostus,* is remarkable. Equally remarkable is the degree of glabellar similarity between species of the *N. araneavelatus* group and *I.* (*Obelagnostus*), for in both of them effacement of the anterior glabellar furrows has produced an essentially similar composite anterior lobe. Significant also is the effacement of the accessory furrows of the pygidium in both taxonomic groups, Evidently, *Neoagnostus,* considered to be derived from a diplagnostid source (Pseudagnostinae), evolved at the time of the Cambrian-Ordovician transition through the same processes of furrow elimination as did the agnostid-derived (Agnostinae) *Idolagnostus* earlier in the late Cambrian.

The pygidium of *Obelagnostus,* by virtue of its degree of effacement, resembles that of the diplagnostoid *Pseudagnostina* (cf., Palmer, 1962, Plate 2, Figs. 18, 23–25; Öpik, 1967, Plate 63, Fig. 10), but the latter may be distinguished by posteriorly converging pygidial axial furrows. *Xestagnostus* (Diplagnostidae, Pseudagnostinae) is also pygidially similar: it possesses similarly oriented axial furrows but quite different border morphology.

Idolagnostus (*Idolagnostus*) is a predominantly Mindyallan (early Dresbachian) form described only from northern Australia: *I.* (*I.*) *agrestis* Öpik and *I.* (*I.*) *dryas* Öpik occur in the *Erediaspis eretes, Cyclagnostus quasivespa,* and *Glyptagnostus stolidotus* Zones in western Queensland. Somewhat younger, and comparable in age with *I.* (*Obelagnostus*) *imitor* sp. nov., is the semieffaced *Idolagnostus* sp. [indet]., known by a single cephalon from the late Idamean *Stigmatoa diloma* Zone, also in western Queensland (Shergold, 1982, p. 19–20).

Idolagnostus (Obelagnostus) imitor subgen. and sp. nov.
(Plate 1, Figs. 7–13)

Name. L., *imitor,* imitator, alluding to homeomorphy with earlier described species of the *Neoagnostus araneavelatus* group, as indicated above.

Holotype. USNM 333946, cephalon illustrated here on Plate 1, Fig. 8.

Material. Thirteen cephala, USNM 333944–333946 and 333951a–d, ranging in length between 1.30 and 2.40 mm; and eight pygidia, USNM 333947–333950 and 333952a–d, measuring (L_{p2}) 1.20 to 2.10 mm.

Occurrence. Basal Minaret Formation, Yochelson Ridge, YP79-13, YP79-8.

Age. Late Cambrian, late Dresbachian, *Aphelaspis* Zone.

Diagnosis. See subgeneric diagnosis.

Description. Cephalon subquadrate to subrectangular depending on preservation, expanding in width anteriorly, widest across front of anterior glabellar lobe; with narrow (sag.) borders, less than 10 percent of cephalic length (sag.), bearing nondeliquiate marginal furrows; acrolobe subovoid to subquadrate divided sagittally by faint median preglabellar furrow, unconstricted. Glabella occupying 63 to 74 percent of total cephalic length, anteriorly sharply pointed, laterally constricted at level of median lateral furrows, posteriorly bluntly terminated; anterior furrow presumably totally effaced to give composite frontal lobe, separated from posterior lobe by transverse or anteriorly forward bowed furrow equivalent to that faint furrow normally separating anterolateral and posterior lobes; axial node lying approximately one-third glabellar length from rear, according to preservation, immediately behind this furrow; large undivided triangular basal lobes.

Pygidium transversely subovoid to subrectangular, with length (L_{p2}) between 75 to 80 percent of maximum width, which lies close to transverse mid-line of pygidium; with wider borders than in cephalon, up to 15 percent pygidial length (L_{p2}); nondeliquiate marginal furrows; retral posterolateral spines with bases lying slightly in front of transverse line across rear of acrolobe. Acrolobe subcircular to subovoid, unconstricted laterally. Axial furrows parallel, faint; transverse axial furrows generally effaced; accessory furrows effaced; weakly defined axial node; terminal node present at posterior extremity of acrolobe.

Subfamily GLYPTAGNOSTINAE Whitehouse, 1936

The familial classification of Glyptagnostinae has been discussed at length previously (Shergold, 1982, p. 23). It is here regarded as a subfamily of Agnostidae. Generic content of Glyptagnostinae is also discussed by Shergold (1982).

Genus *Glyptagnostus* Whitehouse, 1936

Type species. *Glyptagnostus toreuma* Whitehouse (1936, p. 101–103, Plate 9, Figs. 17–20), Late Cambrian, Idamean; Georgina Limestone, western Queensland, Australia; synonymized with *Glyptagnostus reticulatus* (Angelin) by Westergård (1947, p. 5).

Other species. See Kobayashi (1949), Öpik (1961a, p. 428; 1967, p. 167), Palmer (1962, p. 15–18), and Shergold (1982, p. 23).

Glyptagnostus reticulatus Angelin (1851) sensu lato
(Plate 2, Figs. 13–16)

Material. Five cephala, USNM 333956–333959, with lengths between 2.50 and 4.10 mm; and a single meraspid pygidium, USNM 333960, measuring (L_{p1}) 0.85 mm.

Occurrence. Basal Minaret Formation, Yochelson Ridge, locality YP79-8.

Age. Late Cambrian, late Dresbachian, early *Aphelaspis* Zone.

Comments. The concept of *Glyptagnostus reticulatus,* and discussion of its subspecies and synonymy, has been reviewed by Shergold (1982, p. 23–24). The material from the Heritage Range is referred here to *G. reticulatus* (Angelin) in a broad sense: insufficient material, and particularly the lack of holaspid pygidia, is available to allocate a subspecific name with certainty. It seems probable that all specimens found so far can be accommodated, however, in *G. reticulatus reticulatus.*

Degree of scrobiculation differentiates *G. reticulatus* from *G. stolidotus* Öpik, 1961a, Plate 70; Palmer, 1962, Plate 2), but meraspids of the former are similar to holaspids of the latter, indicating a probable neotenous derivation of *reticulatus* from *stolidotus.* The meraspid pygidium illustrated here (Plate 2, Fig. 13) is entirely morphologically compatible with that of the same size associated with *G. reticulatus reticulatus* in northern Australia (Shergold, 1982, Plate 4) but shows the terminal node to greater advantage. In both specimens the axial lobe is simple, and the pleural zone bears four pairs of primary scrobicules, together with a faint postaxial one. Holaspid pygidia of *G. stolidotus* also have a small number of primary pleural scrobicules, but secondary and tertiary ones appear to be inserted at the periphery of the acrolobe during morphogenesis.

Intensity of scrobiculation differentiates the subspecies of *G. reticulatus* in which the simple radial *stolidotus* pattern is overlain by a concentric reticulation that terminates in the nodulose condition of the subspecies *G. reticulatus nodulosus* Westergård. The morphogenesis of northern Australian samples of *G. reticulatus reticulatus* (see Shergold, 1982, p. 24) indicates an increase in the number and density of scrobicular bifurcations with size. The Antarctic cephala, size for size, appear to fall largely into a pattern similar to the Australian material.

Subfamily INCERTAE SEDIS
Genus *Kormagnostella* Romanenko, 1967

Type species. *Kormagnostella glabrata* E. Romanenko (*in* Romanenko and Romanenko, 1967, p. 75, Plate 1, Figs. 22–23), Late Cambrian, Kulbitch Suite, Kulbitch Spring, Gorniya Altai, southern Siberian Platform, USSR; by original designation.
Comments. *Kormagnostella* has been reviewed by Shergold (1982, p. 24–25), who regards *Litagnostoides* Schrank (1975, p. 593–594, Plate 1, Figs. 4–10) from the *Lotosoides quadriceps* fauna of Saimaji, Liaoning, northern China, as a synonym. Two other species besides the type are known with certainty: *K. minuta* (Schrank, 1975) and *K. inventa* Shergold (1982, p. 25–26, Plate 6, Figs. 12–17) from the *Stigmatoa diloma* Zone in the Pomegranate Limestone of the Burke River area, northwestern Queensland. *K. longa* Ergaliev (1980, p. 76, Plate 4, Fig. 20), known from a single pygidium from the early Late Cambrian *Kormagnostus simplex* Zone of southern Karatau, is not readily acceptable in *Kormagnostella.*

***Kormagnostella* cf. *minuta* (Schrank, 1975)**
(Plate 1, Figs. 14–15)

Synonymy.
cf. 1975 *Litagnostoides minutus* n. sp.; Schrank, 1975, p. 593–594, Plate 1, Figs. 4–10.
Material. One cephalon, USNM 333953, length 1.20 mm; and two pygidia, USNM 333954–333955, with lengths (L_{p2}) of 1.40 and 1.70 mm.
Occurrence. Minaret Formation, Springer Peak, localities W63c and YP79-2.
Age. Late Cambrian, late Dresbachian, *Dunderbergia* Zone.
Comments. The cephalon is complete and well preserved and is characterized by a transverse subcircular shape, narrow borders, and nondeliquiate marginal furrows. There is no median preglabellar furrow, and only a faint trace of the anterior glabellar lobe is discernible. The axial glabellar node lies well back on the posterior lobe. The pygidium is featureless, apart from its wide, flat borders; wide, deliquiate marginal furrows; and prominent axial node. The size and orientation of the posterolateral spines is not known.

The Antarctic material is more similar to that described as *Litagnostoides minutus* Schrank from northern China than to those species described from Siberia or northern Australia. Cephala described by Schrank (1975) are less transverse than that illustrated herein, but in other respects identical, and the pygidia cannot be distinguished. The confer is added accordingly to acknowledge the slight difference in cephalic shape.

No comparison can be made with the cephalon of the type species from southern Siberia (Romanenko, 1967, Plate 1, Fig. 23), which is incomplete and obliquely illustrated. The pygidium of *K. glabrata* Romanenko has narrower (tr., sag.) pygidial borders and traces of axial furrows. The Australian *K. inventa* Shergold (1982, Plate 6, Figs. 12–17) has a much more ovoid cephalic outline and deliquiate marginal furrows in both cephalon and pygidium. The latter also possesses well-defined axial furrows, demonstrably setting it apart from the Antarctic species.

Agnostid gen. and sp. indet.
(Plate 2, Fig. 6)

Material. Two cephala, USNM 333970–333971, with lengths of 2.10 and 3.00 mm respectively.
Occurrence. Minaret Formation, Springer Peak, locality W63c.
Age. Late Cambrian, late Dresbachian, *Dunderbergia* Zone.
Comments. These cephala are subovoid or subrectangular, with deliquiate marginal furrows, unconstricted acrolobes, no median preglabellar furrow, and a slender glabella with prominent anterior lobe and transverse anterior furrow, indications of both median lateral and anterior lateral segments, a posterior lobe that is angulate at the rear, and large triangular basal lobes. The axial glabellar node lies between the indicated anterior and median lateral segments.

Cephala with similar glabellar proportions and segmentation have been previously illustrated from the Georgina Basin of northern Australia as *Innitagnostus?* sp. (Shergold, 1982, p. 21, Plate 5, Figs. 7–8), but these have a different shape and have a posteriorly angulate glabella and distinct median preglabellar furrow. The basal lobes are similarly large, however, and the axial node similarly situated.

Also similar is the cephalon assigned by Palmer (1962, p. 20, Plate 2, Fig. 9) to *Acmarhachis* sp. This cephalon, from Alabama, has comparable glabellar proportions, axial node, basal lobes, and undivided acrolobe, and it is also deliquiate. It differs in having a more square anterior glabellar lobe, and it has no indications of median lateral lobes.

A paucity of material prevents us from confidently determining this species, which is accordingly left under open nomenclature.

Family DIPLAGNOSTIDAE Whitehouse, 1936 emend. Öpik, 1967
Subfamily PSEUDAGNOSTINAE Whitehouse, 1936

Genus *Pseudagnostus* Jaekel, 1909
Subgenus *Pseudagnostus* Jaekel, 1909

Type species. *Agnostus cyclopyge* Tullberg (1880, p. 26, Plate 2, Figs. 15a, 15c), Zones of *Parabolina spinulosa* with *Orusia lenticularis* and *Olenus,* Andrarum, Skåne, Sweden (*fide* Westergård, 1922, p. 116–117); designated Jaekel (1909, p. 400).
Comments. The concept of *Pseudagnostus* has been reviewed previously (Shergold, 1975, 1977), and other species referred to the genus up to 1976 have been listed and assigned to species groups and subgeneric taxa (Shergold, 1977). The pseudagnosti described here are all spectaculate and fall within the subgenus *Pseudagnostus* and species groups based on *P.* (*P.*) *communis* (Hall and Whitfield) and *P.* (*P.*) *cyclopyge* (Tullberg). They are morphologically most similar to species described previously from northern Australia.

***Pseudagnostus (Pseudagnostus)* cf. *vastulus* Whitehouse, 1936**
(Plate 2, Figs. 1–5)

Synonymy.
1936. *Pseudagnostus vastulus* sp. nov.; Whitehouse, 1936, p. 99–100, Plate 10, Figs. 3–4.
1936. *Pseudagnostus nuperus* sp. nov.; Whitehouse, 1936, p. 100, Plate 10, Fig. 5, ?6, *non* Fig. 7 (indet.).
1971. *Pseudagnostus vastulus* Whitehouse, 1936; Hill and others, 1971, Plate Cm XII, Figs. 8–9.

1976. *Pseudagnostus curtare* sp. nov.; Henderson, 1976, p. 330–331, Plate 47, Figs. 1–5.
1982. *Pseudagnostus (Pseudagnostus) vastulus* Whitehouse, 1936; Shergold, 1982, p. 28–30, Plate 1, Figs. 1–14.
non 1963. *Pseudagnostus vastulus; sensu* Öpik, 1963, p. 50–53.
non 1976. *Pseudagnostus vastulus* Whitehouse; *sensu* Henderson, 1976, p. 328, 330, Plate 47, Figs. 10–12 [= *Pseudagnostus* cf. *idalis* Öpik, 1967].
Material. Nine cephala, USNM 333961–333963 and 333966a–f, ranging in length between 2.20 and 3.35 mm; and five pygidia, USNM 333964–333965 and 333967a–b, with lengths (L_{p2}) between 1.65 and 2.80 mm.
Occurrence. Minaret Formation, Springer Peak, localities W63c and YP79-2.
Age. Late Cambrian, late Dresbachian, *Dunderbergia* Zone.
Comments. Full discussion of the problems associated with the synonymy of *Pseudagnostus (P.) vastulus* Whitehouse, and detailed description, have been given previously (Shergold, 1982, p. 28–30).

Antarctic cephala assigned to *P. (P.)* cf. *vastulus* are very similar to those from northern Australia (Shergold, 1982, Plate 1, Figs. 1–14). They have the same subcircular shape, wider (tr.) than long (sag.) when undeformed; have similar narrow borders and deeply incised subdeliquiate marginal furrows; and bear a faintly impressed median preglabellar furrow. Pygidia are similarly subcircular and have a similar degree of effacement and deliquiation. Small holaspides have the transversely ovoid deuterolobe that characterizes the early morphogenesis of the Australian taxon, but later holaspides have a more regular flask-shaped deuterolobe similar to that of *P. (P.) communis* (Hall and Whitfield) *sensu* Palmer (1955, Plate 19, Figs. 20–21). The confer is added on account of this characteristic.

Comparisons of *P. (P.) vastulus* with other similar species have been published elsewhere (Shergold, 1982) and are equally applicable for the Antarctic taxon. *P. (P.)* cf. *vastulus* differs from *P. (P.) communis,* from the Dunderberg Shale of Nevada, only by its less elongate cephalic shape and degrees of effacement and deliquiation. In fact, laterally compressed cephala of *P. (P.)* cf. *vastulus* could easily be mistaken for the American species.

The Antarctic specimens differ from *P. (P.) cyclopyge* (Tullberg) in possessing transversely proportionately wider and rounded rather than elongate cephala and pygidia. Characteristically *P. (P.) vastulus* possesses a transversely ovoid deuterolobe that contrasts with the elongate one of *cyclopyge.*

P. (P.) cf. *vastulus* is readily distinguished from the previously illustrated *Pseudagnostus* sp. from northern Victoria Land, Antarctica, by its degree of effacement and less elongate shield shape (see Shergold and others, 1976, Plate 41, Figs. 9–11).

Pseudagnostus (Pseudagnostus) cf. *idalis* Öpik, 1967
(Plate 2, Figs. 7–8)

Synonymy.
cf. 1967 *Pseudagnostus idalis* sp. nov.; Öpik, 1967, p. 153–154, Plate 62, Figs. 8–9; Plate 63, Fig. 1, *non* Fig. 3 [= *P. (P.) idalis sagittus* Shergold, 1982].
cf. 1976 *Pseudagnostus vastulus* Whitehouse; Henderson, 1976, p. 328–330, Plate 47, Figs. 10–12.
cf. 1982 *Pseudagnostus idalis idalis* Öpik; Shergold, 1982, p. 26–27, Plate 2, Figs. 1–13.
Material. Two cephala, USNM 333968–333969, with lengths of 2.50 and 3.00 mm respectively.
Occurrence. Minaret Formation, Springer Peak, locality YP79-2, and Yochelson Ridge, locality YP79-13.
Age. Late Cambrian, late Dresbachian, *Dunderbergia* Zone.
Comments. *Pseudagnostus idalis* Öpik has similarly shaped cephala and pygidia to *P. (P.) communis* (Hall and Whitfield). The latter, however, is a semieffaced species, where *P. idalis* has en grande tenue cephala and pygidia that have strongly deliquiate marginal furrows. The cephalon has a spectaculate glabella and a varyingly impressed median preglabellar furrow. The pygidium has a plethoid, ampullate deuterolobe and retral posterolateral spines situated a little in front of a transverse line drawn across the rear of the deuterolobe.

The cephala at hand are exfoliated, and it is not possible to assess degree of effacement, but they are deliquiate, and have a rather weakly developed median preglabellar furrow. In general, the axial glabellar node lies a little farther rearward on the posterior lobe than in those taxa listed in the synonymy, hence the addition of the confer to the determination. The position of the glabellar node appears to migrate forward during the development of pseudagnostinid phylogenetic lineages. If this holds true for the *P. idalis* complex, then the Antarctic taxon may fall into the earlier part of the *idalis* lineage that extends, at least in northern Australia, throughout the Idamean Stage as redefined by Shergold (1982). Specimens previously illustrated by Henderson (1976) as *P. vastulus* and by Shergold (1982) are all from the later part of the lineage—from the *Stigmatoa diloma* Zone.

Subfamily AMMAGNOSTINAE Öpik, 1967

Genus *Ammagnostus* Öpik, 1967

Type species. *Ammagnostus psammius* Öpik (1967, p. 139–141, Plate 55, Fig. 3; Plate 66, Figs. 1–4), O'Hara Shale, Late Cambrian, Mindyallan, Zone of *Glyptagnostus stolidotus,* northern Australia, by original designation.
Comments. Öpik (1967) described three further species from the Mindyallan of northern Australia: *Ammagnostus integriceps* (Öpik, 1967, p. 141–143, Plate 66, Figs. 5–8) and *A. euryaxis* (p. 143–144, Plate 66, Fig. 9), also from the O'Hara Shale, *G. stolidotus* Zone; and *A. mitis* (p. 144–145, Plate 66, Figs. 10–11) from the Mungerebar Limestone, *Cyclagnostus quasivespa* Zone. Subsequently, Schrank (1975, p. 604, Plate 1, Figs. 2–3) has described *Ammagnostus* sp. from the *Kaolishania quadriceps* assemblage at Saimaji, northern China, and Ergaliev (1980, p. 85–86, Plate 3, Fig. 12; Plate 5, Figs. 8–9) has described *A. integriceps* Öpik from the Ajusockanian *Kormagnostella simplex* and *Glyptagnostus stolidotus* Zones of Lesser Karatau, southern Kazakhstan (see also Ergaliev, 1981, p. 83).

Ammagnostus species have an en grande tenue cephalon and variably effaced pygidium. They have relatively narrow cephalic, and slightly wider pygidial borders, constricted pygidial acrolobes, and a large, transversely expanded posterior axial lobe that contacts the posterior marginal furrow and bears a terminal node situated some way in front of its observed termination.

The Antarctic specimens referred here to *Ammagnostus?* are younger than all previously described species, except perhaps that from northern China (Schrank, 1975). Nevertheless, they seem to possess the essential diagnostic characteristics of the genus, particularly those of the pygidium. They differ from other ammagnostines, however, in having constricted cephalic and pygidial borders. Although their cephalic structures do not resemble contemporaneous genera like *Homagnostus, Connagnostus sensu* Shergold (1975, 1980) or *Agnostascus,* and their pygidial marginal furrows are insufficiently wide for classification with other Ammagnostinae, e.g., *Agnostoglossa* Öpik, the characteristics noted above are considered insufficient grounds for the recognition of a new generic taxon, and they are questionably referred to *Ammagnostus* as the closest existing genus.

Ammagnostus? sp. indet.
(Plate 2, Figs. 9–12)

Material. Four cephala, USNM 333972 and 333976a–c, ranging in length between 1.00 and 2.65 mm; and seven pygidia, USNM 333973–333975 and 333977a–e, with lengths (L_{p2}) between 1.60 and 2.25 mm.
Occurrence. Basal Minaret Formation, Yochelson Ridge, localities P79-3, YP79-13.
Descriptive notes. Cephalon en grande tenue; subrectangular, almost as long as wide, with maximum width at the level of the anterior glabellar furrow; with wide (sag.) borders, about 14 percent of the cephalic length, bearing shallow deliquiate marginal furrows. Cephalic acrolobe constricted laterally, undivided by median preglabellar furrow sagittally. Glabella between two-thirds and three-quarters the cephalic length, cylindrical, with rounded anterior lobe separated by prominent transverse anterior furrow from poorly defined anterolateral lobes; posterior lobe elevated, bluntly rounded at the rear, bearing axial node that is situated half-way along glabella from rear; simple, triangular basal lobes of variable prominence.

Pygidium subcircular to subquadrate, decidedly wider (tr.) than long (sag.), length varying between 73 to 88 percent of maximum width, which lies across middle of posterior axial lobe according to preservation; with borders up to 14 percent the pygidial length (L_{p2}) bearing subdeliquiate marginal furrows, and posterolateral spines lying in advance of a transverse line across rear of posterior lobe; faintly constricted lateral acrolobes. Axis long, extending to marginal furrows, characterized by an expanded posterior lobe longer (sag.) than remainder of axis, and three anterior segments. First transverse furrows confluent sagittally define narrow (exsag.) anterior axial lobe; second transverse furrows very faint, not contacting axial furrows abaxially, divided sagittally by posterior portion of axial node that overlaps anterior edge of posterior axial lobe; latter bearing terminal node in advance of rear of axis.

Parietal morphology is unknown.
Comments. In terms of cephalic shape *Ammagnostus?* sp. indeterminate differs from all previously described species (Öpik, 1967; Ergaliev, 1980) of Mindyallan or equivalent age and is most similar to that species from the *Lotosoides quadriceps* fauna described by Schrank (1975) from Saimaji, Liaoning Province, China, as *Ammagnostus* sp. However, the wide borders and constricted acrolobes appear to distinguish *Ammagnostus* sp. from cephala of all described species of *Ammagnostus.* Pygidially it closely resembles the type species, *A. psammius* Öpik, even to the possession of a narrow third pair of lobes in the anterior portion of the axis. Its narrower pygidial borders set it aside from *A. integriceps* Öpik, *A. euryaxis* Öpik, and *A. mitis* Öpik. The last, and *Ammagnostus* sp. *sensu* Schrank, are further distinguished by narrower (tr.) posterior axial lobes.

Order PTYCHOPARIIDA Swinnerton, 1915
Suborder PTYCHOPARIINA Swinnerton, 1915
Superfamily SOLENOPLEURACEA Angelin, 1854
Family EULOMIDAE Kobayashi, 1955
Subfamily EULOMINAE Kobayashi, 1955

Genus *Stigmatoa* Öpik, 1963

Type species. Stigmatoa diloma Öpik (1963, p. 89–90, Plate 4, Fig. 2), Late Cambrian, Idamean, *Stigmatoa diloma* Zone (Henderson, 1976); Georgina Limestone, Pomegranate Limestone, western Queensland, Australia; designated Öpik (1963, p. 87).
Comments. Stigmatoa has been diagnosed by both Öpik (1963, p. 87–89) and Henderson (1976, p. 352–353) from northern Australian material. The genus also occurs in southern Siberia, in the Gorniya Altai, where two species have been described by Romanenko (*in* Zhuravleva and Rozova, 1977, p. 179–181), and has also been previously reported from northern Victoria Land, Antarctica, by Shergold and others (1976, p. 265–266). The previously named species are listed in Shergold and others (1976, p. 265) and Shergold (1982, p. 31). For discussion of the familial nomenclature, see Shergold (1980, p. 43).

Stigmatoa sp. indet.
(Plate 5, Fig. 11)

Material. A single cranidium, USNM 334006, measuring (L_c) 5.20 mm, excluding the occipital spine.
Occurrence. Minaret Formation, Springer Peak, locality W63c.
Age. Late Cambrian, late Dresbachian, *Dunderbergia* Zone.
Comments. The characteristic occipital spine is broken off, and the palpebral lobes are largely destroyed. Nevertheless, the shape of the glabella and its relationship to the preglabellar area strongly suggest classification with *Stigmatoa.* In general, this specimen appears to have palpebral lobes situated closer to the glabella than in all described species except *S. tysoni* Öpik (1963, Plate 4, Fig. 3; Henderson, 1976, Plate 51, Fig. 8), but that species has a thicker, reflected anterior cranidial border whereas the present cranidium has a preglabellar morphology more like that of *S. diloma* Öpik (1963, Plate 4, Fig. 2; Henderson, 1976, Plate 51, Figs. 5–6; Shergold, 1982, Plate 7, Fig. 1). It is differentiated from previous species by its mostly effaced glabellar furrows: Possibly the degree of effacement is similar to that exhibited by *S. reticulata* Romanenko (*in* Zhuravleva and Rozova, 1977, Plate 25, Figs. 2–3) from southern Siberia, but the Russian species is severely deformed, preventing definitive comparison.

Stigmatoa sp. indet. has a very finely granulose prosopon like the specimen described previously (Shergold and others, 1976, Plate 41, Fig. 1) as *Stigmatoa* sp. from northern Victoria Land. It is not possible to establish a clear relation between the two Antarctic specimens because that from Victoria Land has lost its preglabellar area. The strong possibility that the two may represent a single taxon can only be confirmed by the discovery of more material from either locality.

Family and Subfamily INCERTAE SEDIS

Genus *Bathyholcus* Rasetti, 1961

Type species. Bathyholcus gaspensis Rasetti (1961, p. 110, Plate 23, Figs. 18–23), pebbles in Ordovician Conglomerate, Grosses Roches, Matane, Quebec; and Frederick Limestone, Maryland; Late Cambrian, *Dunderbergia* Zone; by original designation.
Other species. Bathyholcus sulcatus Rasetti (1961, p. 110, Plate 23, Figs. 24–26), Lévis Conglomerate, Lévis, Quebec, Canada.
Comments. Bathyholcus was erected for strongly convex (sag.) cranidia with deeply incised, narrow dorsal furrows. Other characteristics noted by Rasetti (1961, p. 109) are not here considered relevant to generic diagnosis.

The most morphologically comparable genus is *Didwudina* Ivshin (1962, p. 185–186), which occurs with *Irvingella* in the Selety Gorizont of the Kulbitch Yarus in central Kazakhstan and is represented by two species: *D. strigosa* Ivshin and *D. longa* Ivshin. *Didwudina* has a comparable convexity (sag.) and similar glabellar shape and occipital characteristics. Its palpebral morphology is also similar, but the lobes are situated closer to the glabella; and the anterior cranidial marginal furrow is more definite than in the type species of *Bathyholcus.* Nevertheless, the similarities are such that there would be some justification for regarding these genera as synonymous.

The material described below is assigned to *Bathyholcus* with some reservation since Rasetti (1961, p. 110) places great emphasis on the depth and narrowness of the axial furrows, the uniformity of which is no doubt a reflection of the degree of inflation of the glabella, and the clarity of which may be a result of Rasetti's photographic techniques (blackening before whitening?). Nevertheless, the Antarctic material has the same overall cranidial plan, degree of effacement, and palpebral morphology as the type species. It has a more conical glabella, the anterior cranidial marginal furrow is slightly less effaced, and there is a greater degree of vaulting when viewed anteriorly. Ivshin's (1962) species of *Didwudina* seem to have a lower convexity than either Canadian or Antarctic material. *D. longa* has an elongate glabella. Should *Didwudina* become an accepted synonym of *Bathyholcus,* then the morphological details indicated here would lie within the limits of an expanded concept of *Bathyholcus* and be of specific value. In the meantime, the generic determination is here qualified.

Bathyholcus? conifrons sp. nov.
(Plate 3, Fig. 1–9)

Name. L. *conifrons,* conical front, referring to the shape of the glabella.
Types. Holotype, cranidium, USNM 333983, illustrated on Plate 3, Figs. 6–8; paratypes, USNM 333978–333982, 333984–333985a–r.
Material. Twenty-four cranidia ranging in length between 2.40 and 7.90 mm; and a single librigena (USNM 333984).
Occurrence. Minaret Formation, Springer Peak, localities W63c, YP79-2.
Age. Late Cambrian, late Dresbachian, *Dunderbergia* Zone.
Diagnosis. A probable species of *Bathyholcus* Rasetti with anteriorly tapering conical glabella, with effaced furrows; short preglabellar field; and sharply incised anterior cranidial marginal furrow.
Description. The cranidium is convex (sag., tr.), subtriangular in shape, being widest across the posterolateral limbs and apparently narrowest (tr.) across the preglabellar area; gently vaulted dorsoventrally.

Glabella long (sag.), 86 to 93 percent of total cranidial length, widest (tr.) at preoccipital lobes, tapering gently forward, and anteriorly rounded; appreciably convex when viewed laterally (Plate 3, Fig. 7). Glabellar furrows effaced on exoskeleton; occasionally preserved parietally when preoccipital furrows are seen to be sigmoidal and compound (Plate 3, Fig. 4); median lateral furrows are short, transverse, slightly anteriorly curved; anterior lateral furrows very faint, even parietally, transverse, curvature following that of glabella at this level; all furrows contacting axial furrows.

Occipital furrow wide (sag.) and relatively deep. Occipital ring raised above preoccipital glabellar lobes in profile; narrow (sag.); as wide as (tr.) or even wider in some specimens than maximum preoccipital glabellar width; abaxial ends of occipital lobe anterolaterally projected; median occipital node present.

Preocular sections of facial suture moderately divergent, enclosing short, steeply convex (sag., exsag.) preglabellar field, and narrow (sag.), anteriorly downsloping border, 6 to 10 percent of total cranidial length (sag.), separated by prominent, narrow (sag.) anterior marginal furrow, more deeply incised laterally than sagittally. Preglabellar field apparently becomes shorter and more convex (sag.) during morphogenesis. Postocular sections of facial suture widely divergent, enclosing long (tr.), bladelike, posterolateral limbs bearing prominent deeply incised posterior cranidial marginal furrows and distally thickening (exsag.) posterior cranidial margins.

Palpebral lobes arcuate, 40 to 60 percent of glabellar length (G) (35 to 48 percent Gn), depending on degree of morphogenesis; situated close to axial furrows, palpebral width (tr.) approximately one-third of glabellar width at median lateral lobes, with mid-points close to mid-length of glabella (G); anteriorly connected to shallow area in floor of axial furrows at anterolateral corners of glabella by very faint posterolaterally directed ocular ridges.

Librigena with prominent, convex (tr.) lateral border bearing deeply incised marginal furrow; short posterior border, terminated adaxially by convex forward posterior marginal furrow; lateral and posterior marginal furrows meet at distinct angle; no genal spine, but instead a distinctive point at genal angle; gently convex (tr., exsag.) genal field, surmounted by sharply defined subocular groove and narrow eye socle.

External cranidial and librigenal prosopon very finely granulose; parietal surfaces smooth. Doublure below anterior cranidial border possesses transverse terrace lines, as does leading edge of lateral librigenal border.
Comments. *Bathyholcus? conifrons* sp. nov. has a more anteriorly tapered glabella than that of the type species, *B. gaspensis* Rasetti (1961, Plate 23, Figs. 18–23). Its anterior cranidial marginal furrow is more deeply incised and more reminiscent of *B. sulcatus* Rasetti (1961, Plate 23, Figs. 24–26) from which *B? conifrons* is also differentiated by its more anteriorly tapered glabella and additionally its elevated convexity.

Species of *Didwudina* described by Ivshin (1962) differ in having their palpebral lobes situated closer to the axial furrows than in any species of *Bathyholcus.* Species of *Delleana* Ivshin (1962) have a glabellar shape similar to those of *Bathyholcus* and have similarly situated palpebral lobes. They are distinguished, however, by a more extensive (sag.) anterior border and preglabellar field.

Solenopleuracean gen. and sp. indet.
(Plate 9, Figs. 9–10)

Material. Ten articulated cephalo-thoraces or thoraco-pygidia are known and assumed to belong to a single taxon. Figured specimens are USNM 334063–334064; unfigured material is USNM 334065a–h.
Occurrence. Inferno Ridge, locality R64-1NF.
Age. Except Cambrian, not known in detail.
Comments. All specimens are to some degree deformed preventing detailed identification. The cranidium appears to have a faintly furrowed anteriorly rounded glabella and small anteriorly situated palpebral lobes. It may also have a relatively wide (sag.) anterior cranidial border, but all parameters are distorted. Eleven thoracic segments are preserved with pointed, but not spinose, tips. A nonspinose pygidium has three or four pleural segments and six axial ones and may have had a well-defined border. Trilobites with such morphology may have a Middle or Late Cambrian age. Since the latter may apply they are included here for the sake of completeness.

Superfamily OLENACEA Burmeister, 1843
Family ELVINIIDAE Kobayshi, 1935
Subfamily INCERTAE SEDIS

Samples W63c and YP79-2 contain a large number of elvinioid cranidia with intergrading morphological characteristics. End members of graded morphological sequences have been referred previously to genera like *Pesaia* Walcott and Resser, 1924; *Protemnites* Whitehouse, 1939; *Onchopeltis* Rasetti, 1944; *Olentella* Ivshin, 1956; and *Prismenaspis* Henderson, 1976. Here, they are assigned to two genera: *Onchopeltis* (three species, two qualified) and *Protemnites* (two species). *Onchopeltis,* as conceived here, is a possible senior synonym of *Tatulaspis* Ivshin, 1955, a poorly known cranidial genus from Kazakhstan of approximately similar age. There is a rather wide variation exhibited by species assigned here to *Onchopeltis* and uncertainty as to the correct matching of their tagmata. Specific morphological variation in *Onchopeltis* is expressed in modification to the anterior cranidial border, which may be considerably widened (as in *Pesaia*) or linguate and anteriorly pointed. There is also some modification of the abaxial portions of the anterior cranidial marginal furrow, which may become deep and narrow (exsag.) at the expense of the sagittal portion, which remains elevated as a bridge between the border and the preglabellar field (also a characteristic of *Pesaia*). In the forms referred to *Protemnites* the anterior border remains simple, but there is a tendency to form a low boss sagittally on the preglabellar field, and the anterior cranidial marginal furrow is bowed forward. Such forms resemble some species of *Olentella* (particularly *O. shidertensis* Ivshin, 1956) and *Prismenaspis*. Among all taxa recognized here there is a great deal of variability in glabellar shape, in the size and position of the palpebral lobes and degree of obliquity of the ocular ridges, and prosopon. It is impossible to say with certainty which pygidia and librigenae in the collections can be matched with the illustrated cranidia. Those figured are assigned on the basis of size range and prosopon.

The familial relationships of the generic group indicated remain in doubt and can only be resolved by revision from as much as possible of all the genera concerned. It is likely, from palpebral and preglabellar morphology, that they represent a new subfamilial taxon related to Elviniinae or Dokimocephalinae as construed by Palmer (1965).

Genus *Onchopeltis* Rasetti, 1944

Type species. Onchopeltis spectabilis Rasetti (1944, p. 250, Plate 39, Figs. 1–5), Late Cambrian, Dresbachian; Lévis Conglomerate (Boulder 51), Lévis, Quebec, Canada; by original designation.
Other species. Onchopeltis curcunchensis Rusconi (1955b, p. 34–35, Plate 3, Fig. 5), Late Cambrian; San Isidro region, Mendoza, Argentina. ?*Onchopeltis platycephalus* Rusconi (1955a, p. 3; 1955b, p. 35), from the same area, is a *nomen nudum*.
Comments. The presence of possible species of *Onchopeltis* in the late Cambrian of the Heritage Range has been noted previously by Webers (1972). The new species *variabilis* is referred to *Onchopeltis* with some confidence as a result of the similarity shown with cranidia of the type species, *O. spectabilis*, illustrated by Rasetti, and of examination of nonillustrated paratype material deposited in the British Museum of Natural History, London. Other species described here are referred to *Onchopeltis* with varying degrees of uncertainty, and these determinations are queried.

The generic diagnosis, based on *O. spectabilis* Rasetti, 1944, is adequate. However, the pygidium originally assigned to this species may not be correct in light of the possible association indicated by the Heritage Range material and possible classification with Elviniidae. An alternative nonillustrated paratype pygidium in the British Museum collections is distinctly more elviniid.

Onchopeltis spectabilis was originally described from Boulder 51 in the Lévis Conglomerate at Lévis. Subsequently, the species was reported to occur in Pebble G-11 in Ordovician conglomerate at Grosses Roches, Matane County, Gaspe Peninsula, Quebec (Rasetti, 1961, p. 108), where it is associated with *Bathyholcus gaspensis* Rasetti, 1961. An association of *Onchopeltis* and *Bathyholcus*-like morphologies is also reported here from Springer Peak.

Onchopeltis variabilis sp. nov.
(Plate 4, Figs. 1–10)

Name. L., *variabilis*, variable, referring to the variable proportions of the preglabellar field.
Types. Holotype, cranidium, USNM 333986, illustrated on Plate 4, Fig. 1; paratypes, USNM 333987–333996.
Material. Nineteen cranidia, USNM 333986–333991 and 333994, ranging in length between 7.50 and 33.90 mm; six librigenae, USNM 333992, 333995; and eight pygidia, USNM 333993, with length (L_{p2}) between 3 and 12.10 mm.
Occurrence. Minaret Formation, Springer Peak, localities W63c and YP79-2.
Age. Late Cambrian, late Dresbachian, *Dunderbergia* Zone.
Diagnosis. A species referred to *Onchopeltis* with proportionately long, flat preglabellar field (sag.) shortening during morphogenesis, simple anterior cranidial marginal furrow, not interrupted sagittally; simple, gently convex, evenly rounded anterior cranidial border; small palpebral lobes, lying in advance of the mid-point of the glabella; highly granulose cranidial and librigenal prosopon.
Description. Cranidium elongate (sag.), anteriorly gently rounded, vaulted in anterior profile.

Glabella proportionately long (G:Lc 60 to 66 percent) and narrow (tr.), anteriorly tapering, variably rounded at front, moderately convex (sag.) in lateral profile. Glabellar furrows very faint: preoccipital ones sigmoidal almost meeting sagittally; median lateral furrows very faint, transverse, gently curved; anterior lateral furrows generally effaced.

Occipital furrow deep and wide, completely isolating glabella and occipital ring. Occipital ring very slightly wider (tr.) than preoccipital glabellar lobes; narrow (sag.).

Palpebral lobes small (A:G 31 to 38 percent; A:Gn 24 to 31 percent), arcuate, situated anterior to mid-point of glabella (G), close to axial furrows; palpebral areas strongly sloping into axial furrows; palpebral furrows well defined. Palpebral lobes connected to parafrontal band (Plate 4, Fig. 2) by long, posterolaterally sloping ocular ridges.

Preocular sections of facial suture run direct to cranidial margin without appreciable divergence, enclosing preocular areas that slope strongly anterolaterally; an anteriorly sloping, flat preglabellar field of variable length; and gently convex (sag.) anterior cranidial border lying in horizontal plane. Anterior cranidial marginal furrow narrow (sag.), more deeply incised abaxially, on some specimens quite shallow sagittally. Anterior cranidial border simple, anteriorly rounded, vaulted in anterior profile, 11 to 15 percent of the total cranidial length (sag.). Postocular sections of facial suture diverge strongly to posterior cranidial margin enclosing broad-based, triangular posterolateral limbs; posterior marginal furrow widens distally.

Associated librigena with broad lateral and posterior margins drawn into short, laterally deflected spine at genal angle; posterior border short (tr.). Lateral border drawn into short prong anterolaterally. Lateral and posterior marginal furrows shallow, meeting at genal angle. Genal field broad (tr.), gently convex (tr., sag.), bearing narrow eye socle and shallow subocular groove.

Associated pygidium subtrapezoidal or semicircular, slightly vaulted posteriorly, nonspinose, with length (L_{p2}) 32 to 36 percent of maximum width. Axis long (sag.) reaching to posterior marginal furrow, comprising four rings separated by well-defined transverse furrows and a terminal piece. No postaxial ridge. Pleural zone contains three, possibly four, pleural segments indicated by wide (exsag.), shallow pleural furrows and faint interpleural furrows. Marginal furrow continuous, shallow. Geniculation weak, wide spaced from axial furrows; prominent articulating facets; simple crescentic articulating half-ring.

Both external and parietal surfaces of cranidium and librigena heavily granulose; pygidium faintly granulose or, more generally, smooth. Anterior facing edge of cranidial border and lateral margin of librigenal border striate. Some specimens display very fine caeca crossing lateral librigenal marginal furrow and anterior cranidial marginal furrow.
Comments. Cranidia of *Onchopeltis variabilis* sp. nov. closely resemble those illustrated by Rasetti (1944, Plate 39, Figs. 1–3) for the type species, *O. spectabilis* Rasetti. The Antarctic species appears to have a less convex (sag.) preglabellar field, but the dimensions of the latter do appear to decrease with increasing size as indicated in Plate 4. The palpebral lobes of *O. variabilis* may also be more slighty advanced with

respect to the mid-point of the glabella. The librigena illustrated by Rasetti is also very similar to that assigned here to *O. variabilis.* They differ mainly in the angle of emergence of the genal spine, which is distinctly longer in *spectabilis* and very slightly advanced in *variabilis,.* Pygidia assigned to *O. spectabilis* and *O. variabilis,* however, are quite different. If *Onchopeltis* is an elviniid genus, the pygidium assigned to *O. variabilis* here is more likely to represent the genus than the one figured by Rasetti, and although wider (tr.), is more comparable to an unfigured paratype in the Rasetti collection held by the British Museum of Natural History, London.

Early holaspid cranidia of *O. variabilis* also show some resemblance to some of those figured as *Tatulaspis princeps* by Ivshin (1955, 1956) from central Kazakhstan. The latter are not well illustrated but seem to have wider palpebral lobes and distinctive pear-shaped glabellae, more similar perhaps to some Housiinae.

***Onchopeltis? neutra* sp. nov.**
(Plate 5, Figs. 7–10)

Name. L., *neuter,* neutral, neither one nor the other, referring to the morphological position of this species, which lies between those described above.
Types. Holotype, cranidium, USNM 334001, illustrated on Plate 5, Figs. 7–8; paratypes, USNM 334002–334005.
Material. Thirty-eight cranidia with lengths between 4.40 and 16.30 mm.
Occurrence. Minaret Formation, Springer Peak, localities W63c and YP79-2.
Age. Late Cambrian, late Dresbachian, *Dunderbergia* Zone.
Diagnosis. A species referred questionably to *Onchopeltis* with wide (sag.) often transversely furrowed, anterior cranidial border; incipient preglabellar boss; and smooth to lightly granulose prosopon.
Description. Cranidium transverse, anteriorly rounded. Glabella proportionately short (sag.) (G:Lc 52 to 62%), and wide (tr.), subparallel-sided, obtusely rounded anteriorly. Glabellar furrows faint, both exoskeletally and parietally; preoccipital furrows sigmoidal posterolaterally directed; median lateral furrows transverse, gently anteriorly curved; anterior lateral furrows generally effaced. Glabella with moderately strong convexity (sag.) in lateral profile.

Occipital furrow deep, sagittally anteriorly bowed, effectively separating glabella and occipital ring; occipital ring sagittally wide, lateral ends narrow (exsag.) and turned anteriorly; median node present sagittally, in anterior half of occipital ring.

Palpebral lobes arcuate, proportionately long (A:G 38 to 55 percent), extending from opposite anterior lateral glabellar furrows to preoccipital furrows, with mid-points bracketing glabellar mid-point; connected anteriorly to ill-defined parafrontal band by posterolaterally sloping or even approximately transverse, duplicated, ocular ridges. Palpebral furrows well defined; palpebral areas gently convex (tr.), sloping to axial furrows.

Preocular sections of facial suture directed with minimum divergence to anterolateral cranidial margins, circumscribing an anteriorly curved cranidial border, 14 to 24 percent of the cranidial length, which may be undivided and gently convex in a horizontal plane or be transversely divided and slightly reflected (Plate 5, Fig. 7). Preocular and preglabellar areas convex (exsag., sag.). Anterior cranidial marginal furrow deep abaxially, appreciably shallow sagittally, often bowed forward at sagittal line by presence of incipient preglabellar boss. Postocular facial sutures enclose triangular posterolateral limbs.

Exoskeletal prosopon consists of low-density granulation; parietal surfaces invariably smooth. Leading edge of anterior cranidial border striate.
Comments. Onchopeltis? neutra sp. nov. falls morphologically between *O. variabilis* and *O.? acis.* Its anterior cranidial border is anteriorly rounded, as in *O. variabilis,* but is widened (sag., exsag.) and sometimes transversely furrowed. Palpebral morphology, however, more closely resembles *O.? acis,* having longer palpebral lobes and more transverse ocular ridges than *O. variabilis.* The preglabellar field is convex, as in *O.? acis,* and the granulose prosopon is not dense. *O.? neutra* differs from both species in possessing a proportionately shorter (sag.) glabella and preglabellar field developing an incipient boss. Since the size range of *O.? neutra* overlaps that of *O. variabilis,* it is possible that the former represents early morphogenetic stages of the latter, but it seems difficult to reconcile the characteristic differences outlined to obtain compatibility.

The holotype cranidium of *O.? neutra* has similarity with that of *Pesaia exsculpta* Walcott and Resser (1924, p. 9, Plate 2, Fig. 12) from Novaya Zemlya, although the latter has an anteriorly truncated glabella and more steeply ridged preglabellar field. There is also some resemblance to *Olentella shidertensis* Ivshin (1956, p. 68–69, Plate 6, Figs. 1–11; Plate 7, Figs. 1–8) from central Kazakhstan, particularly in the relation of the anteriorly diverted cranidial marginal furrow to the anterior border and preglabellar field. *O.? neutra* is distinguished from *Olentella shidertensis* by its glabellar shape and extent of its anterior cranidial border.

Cranidia described from the Idamean Cupala Creek Formation of northwestern New South Wales, Australia, referred to *Prismenaspis* sp. nov. (Jell, *in* Powell and others, 1982, Fig. 10/9–12), have a similarly long (sag.), transversely furrowed anterior cranidial border to that of *O.? neutra.* All of Jell's specimens are deformed, so that an accurate comparison is not possible. Their more posteriorly oblique ocular ridges may indicate different species, but generically they must be classified with *Onchopeltis? neutra.*

***Onchopeltis? acis* sp. nov.**
(Plate 5, Figs. 1–6)

Name. Gk., *akis,* f., point or beak, and referring to the pointed beaklike form of the anterior cranidial border.
Types. Holotype, cranidium, USNM 333997, illustrated on Plate 5, Fig. 1; paratype cranidia, USNM 333998–333999, 334000.
Material. Eleven cranidia with lengths between 5.60 and 15.00 mm.
Occurrence. Minaret Formation, Springer Peak, localities W63c and YP79-2.
Age. Late Cambrian, late Dresbachian, *Dunderbergia* Zone.
Diagnosis. A species questionably referred to *Onchopeltis* with its anterior cranidial border drawn out into an adventrally directed point.
Description. Cranidium convex (tr., sag.), anteriorly distinctively pointed.

Glabella (G) 59 to 62 percent of cranidial length (sag.), anteriorly tapering and obtusely rounded. Glabellar furrows externally effaced, faint even parietally; where visible, preoccipital furrows sigmoidal, posterosagittally directed; median lateral furrows transverse, anteriorly curved; anterior lateral furrows with similar orientation. Glabella appreciably convex (sag.) in lateral profile.

Occipital furrow wide (sag.), completely separating glabella and occipital ring. Occipital ring as wide (tr.) as preoccipital glabellar lobes; raised slightly above them in lateral profile; bearing a median node (sag.) anteriorly situated.

Palpebral lobes arcuate, moderately long (exsag.) (A:G 46 to 49 percent; A:G 38 percent for three measured specimens), extending from opposite anterior part of median lateral glabellar lobes; connected to anterolateral corners of glabella by posterolaterally directed, often curved, duplicated ocular ridges. Palpebral furrows well defined; palpebral areas sloping to axial furrows, gently convex (tr.).

Preocular sections of facial suture directed to anterolateral cranidial margin with little divergence, meeting sagittally at a distinct point to enclose appreciably convex (exsag., sag.) preocular and preglabellar areas; and convex (sag.) adventrally directed deltoid anterior cranidial border (Plate 5, Figs. 3–5) of variable length. Anterior cranidial marginal deeply incised abaxially, very shallow sagittally. Postocular sections of facial suture diverge rapidly to enclose long (tr.), broad-based, triangular posterolateral limbs.

Low-density granulosity characterizes the prosopon of both exoskeletal and parietal surfaces. The leading edge of the anterior border appears to be striate.

Comments. This species has the general cranidial format of *Onchopeltis variabilis* but differs on several significant characteristics. Most prominent is the modification of the anterior cranidial border, which is developed into a downsloping prong with diminished lateral dimensions. Furthermore, on many parietal surfaces, the anterior marginal furrow is interrupted sagittally (Plate 5, Fig. 6), which has not been observed in *O. variabilis. O.? acis* sp. nov. has a convex preglabellar field and more convex (exsag.) preocular areas. It also has a shorter glabella; longer (exsag.) palpebral lobes, centered more closely around the mid-point of the glabella; and accordingly less oblique ocular ridges. The general convexity of the cranidium is greater, and its granulose prosopon is less dense.

It is impossible to assign a librigena or pygidium to *O.? acis* with any degree of confidence.

Genus *Protemnites* Whitehouse, 1939

Type species. By original designation, *Protemnites elegans* Whitehouse (1939, p. 210, Plate 22, Figs. 12, a,b, *non* Fig. 13 = pagodiid pygidium), exact stratigraphic horizon and locality unknown, but from the Late Cambrian, Georgina Limestone, Glenormiston area, western Queensland, Australia.

Comments. This genus has been revived, reviewed, and rediagnosed by Shergold (1982, p. 35), who has also refigured the holotype cranidium of the type species (Shergold, 1982, Plate 9, Fig. 1). Other species, all Australian, are *Protemnites brownensis* Henderson (1976, p. 351, Plate 50, Figs. 14–19) from the late Idamean, *Stigmatoa diloma* Assemblage-Zone, Georgina Limestone, Glenormiston area, western Queensland; *Protemnites burkensis* Shergold (1982, p. 35–36, Plate 9, Figs. 3–7) from the post-Idamean *Irvingella tropica* Zone, Mount Murray, Burke River Structural Belt, western Queensland; and *Protemnites* sp. indet. (Shergold, 1982, p. 36, Plate 9, Figs. 8–9) from the *Stigmatoa diloma* Assemblage-Zone at the same locality.

Cranidially, *Protemnites* intergrades morphologically with both *Onchopeltis* Rasetti and *Olentella* Ivshin and could be regarded as a senior subjective synonym of the latter. The genera particularly resemble each other in the structure of the preglabellar area, in glabellar shape and degree of effacement of the glabellar furrows, and in palpebral morphology.

Protemnites sp. aff. *P. elegans* Whitehouse, 1939
(Plate 6, Figs. 7–11)

Synonymy.
aff. 1939. *Protemnites elegans* sp. nov.; Whitehouse, 1939, p. 210, Plate 22, Figs. 12a,b, *non* Fig. 13 = pagodiid species.
aff. 1982. *Protemnites elegans* Whitehouse; Shergold, 1982, p. 35, Plate 9, Fig. 1 (refigured holotype).
Material. Ten cranidia ranging in length between 6.10 and 14.30 mm. Specimens USNM 334007–334011 are illustrated; 334012a–e are non-illustrated supplementary materials.
Occurrence. Minaret Formation, Springer Peak, locality W63c.
Age. Late Cambrian, late Dresbachian, *Dunderbergia* Zone.
Comments. The ten cranidia separated under the name *Protemnites* aff. *P. elegans* Whitehouse have a transverse rather than elongate cranidium with proportionately shorter (sag.), anteriorly rounded glabella, preglabellar and anterior cranidial border. The glabellar furrows are faint but are still visible on the exoskeletal surface, and the ocular ridges are almost transverse. A finely granulose prosopon is visible when the exoskeleton is preserved. These characteristics are shared with the holotype cranidium of *Protemnites elegans* (see Shergold, 1982, Plate 9, Fig. 1). Specimens like USNM 334010 (Plate 6, Fig. 10) are morphologically very close to the holotype. Others, however, have a less tapered glabella and wider (tr.) preglabellar areas.

Protemnites aff. *P. elegans* also closely resembles *Olentella shidertensis* Ivshin (1955, p. 68–69, Plate 6, Figs. 1–11; Plate 7, Figs. 1–8) from central Kazakhstan. Specimens can only be differentiated by glabellar shape, those from Kazakhstan being more truncate. Nevertheless, it seems possible that *O. shidertensis* may well represent *Protemnites* in the Soviet Union. Similarly, specimens from southern Kazakhstan (Lesser Karatau), illustrated by Ergaliev (1980, Plate 11, Figs. 12–14) as *Prismenaspis trisulcatus* Ergaliev, would also seem more appropriately classified with *Protemnites.*

Protemnites? magnificans sp. nov.
(Plate 6, Figs. 1–6)

Name. L., *magnificans,* noble, eminent, referring to the impressive format of this species.
Types. Holotype, cranidium, USNM 334013, illustrated on Plate 6, Figs. 1, 5–6; paratypes, USNM 334014–334017.
Material. Eleven cranidia ranging in length between 8.80 and 20.50 mm.
Occurrence. Minaret Formation, Springer Peak, localities W63c and YP79-2.
Age. Late Cambrian, late Dresbachian, *Dunderbergia* Zone.
Diagnosis. A species questionably referred to *Protemnites* Whitehouse with elongate cranidium, wide (sag.) anterior cranidial border; anteriorly truncate glabella; effaced glabellar furrows; and oblique ocular ridges.
Description. Cranidial format elongate (sag.) with low degree of vaulting (tr.) but moderate degree of sagittal convexity.

Glabella (G) long, (sag.), 69 to 73 percent of cranidial length, narrow (tr.), anteriorly tapering, truncate across frontal lobe. Glabellar furrows generally effaced on exoskeleton, very faint parietally.When visible preoccipital furrows complex, sigmoidal, posterosagittally orientated; median lateral furrows, transverse, gently curved; anterior lateral furrows also transverse, barely curved. Glabella with low to moderate convexity (sag.) in lateral profile.

Occipital ring marginally wider (tr.) than preoccipital glabellar lobes, separated from glabella by sharply incised occipital furrow; narrow (sag.), with faint median node anteriorly sited; depressed to dorsal level of preoccipital glabellar lobes in lateral profile.

Palpebral lobes arcuate; narrow (tr.); length varying according to size (A:G 36 to 47 percent; A:Gn 29 to 37 percent); extending from opposite anterior lateral glabellar furrows to front of median lateral glabellar lobes; connected to parafrontal band by delicate, curved, oblique ocular ridges. Palpebral areas convex (tr.), as wide (tr.) as length (exsag.) of palpebral lobes at their mid-points. Palpebral furrows well impressed.

Preocular sections of facial suture run direct to anterolateral cranidial margin, thence arch forward to enclose a sagittally elongated, convex (sag.) anterior cranidial border that bears traces of shallow transverse bisecting furrow; convex (exsag.) preocular areas; flat, anteriorly sloping preglabellar field; and anterior cranidial marginal furrow that is sagittally shallow and anteriorly deflected on the sagittal line. Postocular sections of facial suture diverge rapidly to enclose long (tr.), triangular posterolateral limbs bearing shallow, distally expanding posterior marginal furrows.

Cranidial prosopon finely granulose overall; preglabellar field and preocular areas faintly caecate on both exoskeletons and parietal surfaces.
Comments. This species is classified with *Protemnites* because its palpebral morphology and the general relation of the elements of the preglabellar area are similar. *P.? magnificans* sp. nov. has a longer (sag.) glabella than that of the the type species, *P. elegans* Whitehouse but is similarly tapered. However, its furrows are more effaced, and it is anteriorly more truncate. *P.? magnificans* also has an apparently longer (sag.) anterior cranidial border. The species have a similar prosopon. *P.? magnificans* differs from *P. burkensis* by its more elongate cranidial form and its proportionately longer (sag.) preglabellar field. In *P. burkensis* the palpebral lobes are situated considerably farther from the axial furrows.

Family PTEROCEPHALIIDAE Kobayashi, 1935
Subfamily APHELASPIDINAE Palmer, 1960

Genus *Aphelaspis* Resser, 1935

Type species. Aphelaspis walcotti Resser (1938, p. 59, Plate 13, Fig. 14), Late Cambrian, Nolichucky Formation, Virginia, United States; designated Palmer (1953, p. 157).
Other species. For American species assigned to *Aphelaspis* see Palmer (1962) and Rasetti (1965), and for species of this genus outside North America see Shergold (1982, p. 37).
Comments. This genus has been revised and rediagnosed by Palmer (1953, 1954, 1960, 1962, 1965), who has also restricted the number of species erected by Resser. Comments on the characteristics used for specific discrimination among *Aphelaspis* have been published by Rasetti (1965, p. 73–75). The present morphological variation within the genus is such that it is not considered warranted to establish new taxa for the Antarctic material: most observed morphological variation can be encompassed within existing species, four of which are recognized here.

Species referable to *Aphelaspis* in the Heritage Range form a distinct complex of interrelated morphologies, secondarily complicated by deformation and distortion that makes much of the material indeterminate since the main characteristics used in discriminating them, dimensions of the preglabellar area, width of interocular areas, orientation of the palpebral ridges, glabellar shape, etc., have been frequently exaggerated. In the Heritage Range, as in North America, species of *Aphelaspis* occur most commonly in association with *Glyptagnostus reticulatus* and predate species of *Erixanium.* In northern Victoria Land and northern Australia, however, *Aphelaspis* also occurs in younger biozones (Shergold *in* Shergold and others, 1976; Öpik, 1963), which may also be the situation in central Kazakhstan (Ivshin, 1956) where *G. reticulatus* is unrecorded. Due to difficulty in their interpretation no comparison is made here with the species of *Aphelaspis* described from Kazakhstan by Ivshin (1956), all of which have longer (exsag.) or more posteriorly sited palpebral lobes.

Aphelaspis cf. *walcotti* Resser, 1938
(Plate 9, Figs. 1–3)

Synonymy.
See Palmer (1953, p. 157; 1954, p. 746) and Rasetti (1965, p. 76).
Materials. Two cranidia, USNM 334018–334019, with lengths of 6.40 and 6.60 mm; and a single librigena, USNM 334020.
Occurrence. Basal Minaret Formation, Yochelson Ridge, P79-3.
Age. Late Cambrian, late Dresbachian, *Aphelaspis* Zone (associated with *Glyptagnostus reticulatus*).
Comments. Cranidia referred to *Aphelaspis* cf. *walcotti* Resser from West Antarctica are characterized by a subrectangular anteriorly truncate poorly furrowed glabella; partially effaced occipital furrow; moderately large palpebral lobes, between 50 and 55 percent of the glabellar length (sag.), situated less than the length of one palpebral lobe away from the axial furrows; gently sloping ocular ridges; and a broad (tr.) preglabellar area one-third the cranidial length (sag.), which comprises a border almost as long (sag.) as the preglabellar field. In these characteristics the Antarctic material closely resembles cranidia of *A. walcotti*, particularly those described by Rasetti (1965, p. 76–77, Plate 18, Figs. 10–11, 13–16) from the Nolichucky Formation of northeastern Tennessee. Some specimens previously figured from the Riley Formation, central Texas (Palmer 1954, p. 746–748, Plate 84, Figs. 2, 4–8), have shorter palpebral lobes and others a more anteriorly tapering and rounded glabella or more prominent glabellar furrows. The Texas and Tennessee librigenae both have a longer posterior margin than that illustrated here but in other respects are similar.

Aphelaspis cf. *subdita* Palmer, 1962
(Plate 7, Figs. 1–4)

Synonymy.
cf. 1962 *Aphelaspis subditus* n. sp.; Palmer, 1962, p. 35–36, Plate 4, Figs. 20–22, 25.
cf. 1965 *Aphelaspis subditus* Palmer; Palmer, 1965, p. 60, Plate 8, Figs. 22–26.
Material. Four cranidia, USNM 334021–334022 and 334025, ranging in length between 5.5 and 8.60 mm; one hypostoma, USNM 334027; one librigena, USNM 334023; and three pygidia, USNM 334024 and 334025, between 2.10 and 2.50 mm long (L_{p2}).
Occurrence Basal Minaret Formation, Yochelson Ridge, localities YP79-8, P79-3.
Age. Late Cambrian, late Dresbachian, *Dunderbergia* Zone, associated with *Glyptagnostus reticulatus.*
Comments. Cranidia are characteristically elongate (sag.) with a subrectangular, anteriorly truncate, poorly furrowed glabella 58 to 60 percent of the cranidial length; a partially effaced occipital furrow; relatively small palpebral lobes (A:G about 40 percent; A:Gn 31 to 33 percent on two measured specimens), situated in advance of the mid-point of the glabella (G), the length of one palpebral lobe away from the axial furrows; transverse ocular ridges; long (sag.) preglabellar area with width (tr.) approximating interocular width (tr.), comprising a narrow anterior cranidial border about half as long (sag.) as the preglabellar field. The associated librigena has a prominent eye socle and short laterally deflected genal spine. The pygidium is ellipsoidal, lacking marginal spines and a differentiated border, containing three axial segments and three poorly defined pleural ones.

The narrow-bordered cranidium with faintly furrowed glabella and anteriorly sited palpebral lobes resembles specimens illustrated as *Aphelaspis subditus* by Palmer (1962, p. 35, Plate 4, Fig. 20, in particular) from the Dunderberg Shale, Mt. Hamilton district, Nevada. Subsequently illustrated material of this species (Palmer, 1965, p. 60, Plate 8, Figs. 22–26), from Mt. Hamilton and McGill, Nevada, has longer (exsag.) palpebral lobes and more anteriorly rounded glabellae. Librigenae and pygidia referred to *A. subditus* by Palmer rather closely resemble material from the Heritage Range (cf., Palmer, 1965, Plate 8, Figs. 7, 20), and the hypostoma (unfigured here) is essentially the same as that figured as *A. brachyphasis* Palmer (Palmer, 1965, Fig. 21).

The anteriorly truncate glabella, transverse ocular ridges, and small, anteriorly sited palpebral lobes also resemble *Aphelaspis buttsi* (Kobayashi) (see Palmer, 1962, Plate 4, Fig. 31), but its pygidium, known from a complete exoskeleton, has lateral borders and one more axial segment, at least. On the balance of characteristics, classification with *A. subdita* Palmer is preferred.

A previously illustrated cranidial fragment from northern Victoria

Land, described as *Aphelaspid* sp. 1 (Shergold and others, 1976, p. 272–273, Plate 41, Fig. 4), may also belong with *Aphelaspis* cf. *subdita* Palmer, but if so, it is the youngest known specimen referable to this species.

Late Dresbachian aphelaspids associated with *Glyptagnostus*are not easily discriminated from some early olenids. *Olenus rarus* Orlowski (1968, p. 268–269, Plate 4, Figs. 6–19), from the Holy Cross Mountains in southern Poland, is cranidially comparable in characteristics associated with the glabella and preglabellar area. On some specimens of the Polish species, the palpebral lobes are more posteriorly situated than on others and, together with the variation in glabellar shape, may indicate that *O. rarus* is a composite species. *Olenus alpha* Henningsmoen (1957, p. 100–102, Plate 9, Figs. 1–6), from the *Agnostus pisiformis* Zone in Norway, is also essentially similar, but its cranidium has a shorter preglabellar area and possesses glabellar furrows. The pygidium of *O. alpha* (Henningsmoen, 1957, Fig. 6) also resembles that of *Aphelaspis* cf. *subdita* in both shape and segmentation, differing only by its less effaced furrows.

Aphelaspis cf. *lata* Rasetti, 1965
(Plate 7, Figs. 5–8)

Synonymy.
cf. 1965 *Aphelaspis lata* Rasetti, new species; Rasetti, 1965, p. 87–88, Figs. 8–20.
Material. Thirteen cranidia, USNM 334028–334029 and 334032a–k, ranging in length between 2.55 and 9.00 mm; three librigenae, USNM 334030 and 334033a–b; and one pygidium, USNM 334031, measuring (L_{p2}) 2.40 mm.
Occurrence. Uppermost Heritage Group, Yochelson Ridge, locality YP79-13; basal Minaret Formation, same area, localities YP79-9 and YP79-11.
Age. Late Cambrian, late Dresbachian, *Aphelaspis* Zone.
Comments. Cranidia of this species closely resemble those described above as *Aphelaspis* cf. *subdita* Palmer in degree of effacement of the glabellar furrows and glabellar shape; the transverse ocular ridges; and proportionate dimensions of the anterior cranidial border, which is narrow, and preglabellar field. The cranidium of *A.* cf. *lata,* however, is transverse rather than elongate (sag.), the interocular width (tr.) is significantly greater, and there is a definite tendency to form an incipient preglabellar boss. Some specimens (not figured) also have a forward-arched anterior cranidial marginal furrow that imparts to the preglabella area an appearance similar to that of the elviniid genus *Dunderbergia,* especially when material is deformed in the sagittal dimension. The librigena found associated has a relatively long (tr.) posterior margin and short genal spine. The associated pygidium is transverse, elliptical, lacking marginal spines and well-defined borders, and has three axial segments and two or three pleural ones. Essentially it is indistinguishable from that assigned here to *Aphelaspis* cf. *subdita* Palmer.

The Heritage Range material is compared with *Aphelaspis lata* Rasetti because of its wide palpebral areas; small, anteriorly sited palpebral lobes, transverse ocular ridges; and general appearance of the preglabellar area. Possibly the glabella is narrower (tr.) and more anteriorly tapered on cranidia figured by Rasetti (1965, Plate 16, Figs. 8, 15–16). The pygidium has similar shape and segmentation, although a narrow border is indicated on specimens illustrated by Rasetti (see 1965, Plate 16, Figs. 12–14, 17–18).

Also cranidially similar is *Aphelaspis buttsi* (Kobayashi), which is based on dorso-ventrally flattened material similar to that at locality YP79-13. *A. buttsi,* however, has a more posteriorly pointed pygidium, and its librigena has a long genal spine (see Palmer, 1962, Plate 4, Fig. 31).

In *Aphelaspis washburnensis* Rasetti (1965, p. 85–86, Plate 17, Figs. 15–21), from the Nolichucky Formation of northeastern Tennessee, there is also a tendency to produce an incipient preglabellar boss, and this species has in addition similar anteriorly situated palpebral lobes, transverse ocular ridges, and wide (tr.) palpebral areas. There is, however, greater pygidial distinction than with *Aphelaspis lata* Rasetti in that *A. washburnensis* has proportionately broad flat borders, and this fact has influenced the determination made here.

Aphelaspis cf. *australis* Henderson, 1976
(Plate 7, Figs. 9–13)

Synonymy.
cf. 1976 *Aphelaspis australis* sp. nov.; Henderson 1976, p. 342–343, Plate 49, Figs. 5–7.
Material. Twenty-five cranidia, USNM 334034–334036 and 334040a–v, with lengths between 2.80 and 12.10 mm; one hypostoma, USNM 334041; four librigenae, USNM 334037 and 334042a–b; and eleven pygidia, USNM 334038–334039 and 334043a–i, measuring (L_{p2}) 2.50 to 4.30 mm.
Occurrence. Uppermost Heritage Group and basal Minaret Formation, Yochelson Ridge, localities P79-3, YP79-8, YP79-9, YP79-11, and YP79-13.
Age. Late Cambrian, late Dresbachian, *Aphelaspis* Zone.
Comments. Many specimens are varyingly deformed and are placed in this taxon with a degree of uncertainty. Deformation prevents an adequate quantitative assessment of much of the material.

Aphelaspis australis Henderson is diagnosed (Henderson 1976, p. 342) as having a short preglabellar area, elongate palpebral lobes, and a pygidium that is twice as long as wide. In addition, it has an anteriorly rounded glabella with distinct furrows and transverse or only very gently sloping ocular ridges. The pygidium is transverse and has a distinct border.

Material from the Heritage Range possesses the diagnostic characteristics of the type material. However, its glabellar furrows tend toward effacement, and the palpebral lobes may be slightly shorter (exsag.) than in the Australian material. Glabellar shape, position of the palpebral lobes adjacent to the anterior half of the glabella; and proportions of the preglabellar area are essentially comparable. Henderson's illustrated pygidium (1976, Plate 49, Fig. 7) is possibly more transverse than those illustrated here, and, although it contains a similar number of axial segments, only three pleural segments can be discerned, whereas there is a faint fourth on the Antarctic pygidia.

Cranidia from the Heritage Range also resemble those illustrated by Rasetti (1965, Plate 16, Figs. 1, 6, 7) as *Aphelaspis buttsi* (Kobayashi), but the latter may be differentiated by their stronger glabellar furrows. Pygidia referred to *A. buttsi* by Rasetti (1965, Figs. 2–4) have more prominent pleural and interpleural furrows, the former cutting across the distal ends of the segmental opisthopleura whereas the latter continue straight to the marginal furrow, the effect being to produce an isolated lobate opisthopleuron. A hint of this structure is seen on Henderson's specimen and on the specimen of *Aphelaspis* cf. *australis* figured here on Plate 7, Fig. 12. It is typical of many olenid pygidia (see Henningsmoen, 1957).

Aphelaspis cf. *australis* Henderson has a typical *Aphelaspis* hypostoma (not figured) comparable to that previously illustrated for *A. brachyphasis* Palmer (see Palmer, 1962, p. 4, Fig. 19; 1965, Plate 8, Fig. 21). The librigena closely resembles that referred by Palmer (1968, Plate 9, Fig. 18) to *Aphelaspis? haguei,* having a slightly sigmoidal posterior marginal furrow that is more deeply incised than the lateral marginal furrow.

Aphelaspis cf. *australis* has a similar time span in the Heritage Range as in the Georgina Limestone of northern Australia, where, according to Henderson (1976, Table 2), it ranges from the *Glyptagnostus reticulatus* Zone into that of *Proceratopyge cryptica*.

Genus *Eugonocare* Whitehouse, 1939

Type species. *Eugonocare tessellatum* Whitehouse (1939, p. 226, Plate 23, Figs. 15, 17, *non* Figs. 16, 18; Plate 25, Fig. 7b *fide* Henderson, 1976), Late Cambrian, Idamean, Georgina Limestone, western Queensland, Australia.
Other species. See Shergold (1982, p. 38). Material referable to this genus has been described from Australia (Whitehouse, 1939; Henderson, 1976; Shergold, 1982), China (eastern Guizhou, Lu, 1956), northwestern Siberia (as *Aphelaspis? buttsi* [Kobayashi], by Rozova, 1977), and southern Kazakhstan (Ergaliev, 1980).
Comments. *Eugonocare* is a pterocephaliid genus, presently referred to the Subfamily Aphelaspidinae, the cranidial morphology of which compares very closely with that of *Olenaspella* Wilson, 1956. In fact, cranidial characteristics of established species of *Eugonocare* intergrade with those of *Olenaspella* as recognized by Palmer (1962, p. 36–39; 1965, p. 36–65), Ivshin (1962, p. 68–70), and Ergaliev (1980, p. 123–125). The genera are usually distinguished on the basis of their pygidia: those of *Eugonocare* are apparently nonspinose throughout morphogenesis (Shergold, 1982, Plate 10), whereas those of *Olenaspella* have a variably spinose margin. Palmer (1965, p. 64) has drawn attention to a third closely related aphelaspidine genus, described from Yakutia in northern Siberia by Lermontova (1940, Plate 49, Figs. 4, 4a, 4b) as *Crepicephalus borealis*. This has a *Eugonocare* or *Olenaspella*-like cranidium matched with a *Crepicephalus*-like spinose pygidium, and this combination has been verified by Palmer on the basis of completely articulated exoskeletons in the collections of the Geological Institute, Moscow. *Notoaphelaspis* Jell, 1982 also resembles *Eugonocare* in preglabellar and ocular characteristics but has glabellar features reminiscent of *Erixanium* in addition. Its pygidium is nonspinose.

Since no pygidium matching any of these taxa has yet been found in the present Antarctic collections, the cranidia illustrated here on Plate 8 could belong to any except *Notoaphelaspis*. On the balance of characteristics a tentative assignment to *Eugonocare* is favored, taking into consideration the tendency to effacement of the glabellar furrows that are much less strongly impressed than those of *Olenaspella*, the relative width (sag., tr.) of the anterior cranidial border, and arrangement and convexity (sag.) of the preglabellar area in general. It should be emphasized, however, that this determination is temporary: there are some morphological differences from both *Eugonocare* and *Olenaspella* that may eventually prove significant, among them the tendency toward an anteriorly rounded glabella (not evident on all specimens), inclined ocular ridges, and palpebral lobes that may be closer to the axial furrows. A new generic taxon is not contemplated, however, in the absence of librigenal and pygidial information.

Eugonocare? nebulosum sp. nov.
(Plate 8, Figs. 3–10)

Name. L., *nebulosum* n., indefinite; referring to the mostly effaced glabellar furrows.
Types. Holotype, cranidium USNM 334051, illustrated on Plate 8, Fig. 8; paratypes USNM 334047–334050, 334052–334054a–i.
Material. Sixteen cranidia ranging in length between 1.55 and 14.10 mm.
Occurrence. Minaret Formation, Yochelson Ridge, YP79-13; Springer Peak, W63c.
Age. Late Cambrian, late Dresbachian, *Aphelaspis* and *Dunderbergia* Zones.
Diagnosis. A species tentatively assigned to *Eugonocare* with largely effaced glabellar furrows, wide (sag., tr.) anterior cranidial border, and generally anteriorly rounded glabella; posteriorly inclined ocular ridges and palpebral lobes relatively close to the axial furrows.
Description. The material has been variably deformed, and most measured parameters show a wide variation. Relatively undeformed cranidia have a broadly arcuate anterior contour and low lateral and anterior profiles. Transverse width across preglabellar area closely approximates that across posterolateral limbs.

Glabella parallel-sided, anteriorly gently rounded, even pointed in some cases, 51 to 65 percent of the cranidial length (sag.); glabellar furrows in general poorly defined on exoskeleton, but when visible preoccipital furrows are sigmoidal with a tendency to join across sagittal line; median lateral furrows posterosagittally directed, not connected sagittally; anterior lateral furrows not seen. Occipital furrow generally more strongly impressed than glabellar furrows, generally reaching axial furrows laterally, but not in all specimens.

Palpebral lobes of moderate length (A:G, 41 to 63 percent; A:Gn, 32 to 52 percent), situated about mid-length of glabella (G); separated from palpebral areas by significant palpebral furrows; anteriorly merging into prominent narrow, posterolaterally directed, ocular ridges; width (tr.) posterior palpebral area less than half preoccipital glabellar width (tr.).

Preocular facial sutures widely divergent, enclosing broad (tr.) preglabellar area composed of variably wide (sag.), forward sloping, anterior cranidial border, 10 to 19 percent of cranidial length (sag.), and correspondingly variable, gently downsloping preglabellar field, caecate when exfoliated; anterior cranidial marginal furrow well defined, pitted in caecate specimens. Postocular facial sutures enclose short (tr.), narrow (exsag.), triangular posterolateral limbs.

Librigenae, hypostomata, thoracic segments, and pygidia not yet identified.
Comments. Compared with *Eugonocare tessellatum* Whitehouse, *E.? nebulosum* sp. nov. has in general a less anteriorly truncate glabella, more severely effaced glabellar furrows, more steeply inclined ocular ridges, and probably a proportionately wider (sag.) anterior cranidial border. The cranidia assigned to *Crepicephalus borealis* Lermontova are more similar in respect of glabellar shape and degree of effacement. In general species of *Olenaspella*, like *Eugonocare*, have a more truncate glabella but with more strongly developed furrows. Some species of *Olenaspella*, like the type species *Parabolinella? evansi* Kobayashi (1936, p. 92–93, Plate 15, Figs. 7–10), are very much more olenid than aphelaspidinid, in their narrow anterior cranidial borders and relatively short palpebral lobes situated close to the glabella. Subsequently figured *Parabolinella evansi* figured by Kobayashi (e.g., 1938, Plate 16, Fig. 11) are in fact more like *Olenaspella* as conceived by Palmer (1962, 1965).

Subfamily Erixaniinae Öpik, 1963

Öpik (1963, p. 77) regarded *Erixanium* as the sole genus within his new family Erixaniidae, at that time known only from northern Australia. Öpik (1963) considered Erixaniidae to be Ptychopariacea with 12 thoracic segments; a cranidium with subrectangular glabella, wide (sag.) preglabellar field, and narrow anterior border, large palpebral lobes, and large rostral shield; and a pygidium having a short axis, narrow border and doublure. Since the publication of this diagnosis, Palmer (1965) has demonstrated that Aphelaspidinae (Pterocephaliidae) can have either 12 or 13 thoracic segments, so that the number of thoracic segments can no longer be regarded as diagnostic. Öpik (1963) himself drew cranidial comparison with *Litocephalus* (Aphelaspidinae), and other aphelaspidines, e.g., *Eugonocare*, also have similar preglabellar areas. Nevertheless, the combination of long palpebral lobes, subrectangular glabella, and large rostral shield seems diagnostic of *Erixanium* and sets it apart. Similarly, its pygidium, having a generally short axis and correspondingly long postaxial area, often carinate, and narrow borders and doublure, is characteristic. Such characteristics, however, could be derived readily from a pterocephaliine ancestor, possibly from *Pterocephalia* itself, which calls into question the familial affinity of *Erixanium*. Although the genus could be accommodated in Pterocephaliidae without necessitating too much change to that family's diagnosis (Palmer, 1965, p. 57), it does possess characteristics that cut across present subfamilial divisions: Notably it has aphelaspidine cranidial features combined with pterocephaliine pygidia. In effect, only the large rostral shield currently diagnoses the proposed family Erixaniidae. Pending further investigation of pterocephaliid ventral morphology, it is suggested here that *Erixanium* be regarded as a pterocephaliid genus and that rostral differences, in combination with the modified pterocephaliine pygidium, be regarded as subfamilial characteristics, thereby bringing a subfamily Erixaniinae into line with the existing pterocephaliid subfamilies Aphelaspidinae, Pterocephaliinae, and Housiinae.

Genus *Erixanium* Öpik, 1963

Type species. By original designation (Öpik, 1963, p. 77), *Erixanium sentum (Öpik* (1963, p. 78–81, Plate 8, Figs. 1–4, 7–8; Plate 9, Figs. 1–5); Late Cambrian Idamean, *Erixanium sentum* Zone; Pomegranate and Georgina limestones; western Queensland, Australia.
Comments. Öpik (1963) described two other species of *Erixanium* from western Queensland, *E. strabum* (1963, p. 81–82, Plate 8, Figs. 5–6) and *E. alienum* (1963, p. 82–84, text-figs. 28–29). The genus also occurs in North America, where it has been described by Palmer (1960, 1965) from the late Dresbachian *Dunderbergia* Zone of Nevada and Utah: *viz. E. carinatum* Palmer (1965, p. 49, Plate 17, Figs. 19–21), *E. multisegmentum* Palmer (1965, p. 49–50, Plate 17, Figs. 17–18), *E.? brachyaspis* Palmer (1965, p. 50, Plate 17, Figs. 14–16), and *Erixanium* sp. (Palmer, 1965, Plate 17, Fig. 22). Ergaliev (1980, p. 128, Plate 12, Figs. 14–15) has recently described *E. carinatum* Palmer from the *Homagnostus longiformis* Zone of the Saks Yarus in the Lesser Karatau Range, southern Kazakhstan; and Palmer (1965, p. 49) has quoted the presence of the genus in Yakutia, northern Siberia.

Species of *Erixanium* are differentiated mainly on pygidial characteristics and on the shape of the glabella and whether or not its furrows open into the axial furrows. The last is, however, a very variable characteristic, as shown by Öpik's illustrations of *E. sentum,* and likely to depend on mode of preservation, fainter glabellar furrows being expected if shell is preserved.

***Erixanium* cf. *sentum* Öpik, 1963**
(Plate 8, Figs. 1–2)

Synonymy.
cf. 1963 *Erixanium sentum* sp. nov.; Öpik, 1963, p. 78–81, Plate 8, Figs. 1–4, 7–8; Plate 9, Figs. 1–5.
Material. Three cranidia, USNM 334044–334046, one preserved with shell, one partly exfoliated, and the third as a parietal surface; two measurable specimens having lengths of 3.55 and 4.20 mm. The third is considerably larger and is approximately 6.50 mm long.
Occurrence. Minaret Formation, Springer Peak, locality W63c.
Age. Late Cambrian, late Dresbachian, *Dunderbergia* Zone.
Comments. The specimens at hand are certainly representative of *Erixanium,* but in the absence of pygidia the specific determination remains questionable. They are all referred here to a single taxon even though they exhibit variation in glabellar format in various combinations. Two specimens, the smallest and the largest, have anteriorly gently tapered glabellae. The largest, being a parietal surface, displays three pairs of transverse anteriorly gently curved glabellar furrows, none of which definitely open out into the axial furrows. The smallest specimen is partly exfoliated and shows similar furrows. These specimens resemble cranidia that Öpik (1963, Plates 8–9) has referred to *Erixanium sentum:* The furrows are not sufficiently strongly defined for inclusion in *E. carinatum* Palmer of *E. multisegmentum* Palmer (see Palmer, 1965, Plates 17, Figs. 19 and 17 respectively). The third (USNM 334044) specimen is preserved with shell and has a rectangular, parallel-sided glabella, the furrows of which are obscure. This specimen closely resembles the cranidium Öpik illustrated (1963, Fig. 28a) as *Erixanium alienum,* or that which Ergaliev (1980, Plate 12, Fig. 14) has referred to *E. carinatum* Palmer. However, obscure furrows are also present on some of the paratype specimens of *E. sentum* (e.g., Öpik, 1963, Plate 8, Fig. 3). Preglabellar and the palpebral characteristics as far as they can be determined for all three specimens are similar and comparable with those of *E. sentum.* Accordingly, all specimens are regarded as being within the morphological variation of that taxon.

Superfamily DAMESELLACEA Kobayashi, 1935
Family CHANGSHANIIDAE Kobayashi, 1935

Genus *Changshanocephalus* Sun, 1935

Type species. By original designation, *Changshanocephalus reedi* Sun (1935, p. 41, Plate 1, Fig. 25), Late Cambrian, Changshanian, *Chuangia tawenkouensis* Zone, Shantung, China.
Other species. The following species have been placed in *Changshanocephalus* and are listed here regardless of possible erroneous assignment. *Changshanocephalus?* sp. Sun (1935, p. 41, Plate 1, Fig. 26), age and location as for type species. Kobayashi (1960, p. 395) and Lu and others (1965, p. 174–175) also include *Yokusenia conica* Endo (1944, p. 76, Plate 8, Figs. 7–8), from the Paishan Formation, Fengtien, Liaoning, and questionably *Ptychoparia* (*Proampyx*) *burea* Walcott (1905, p. 86; 1913, p. 145, Plate 14, Fig. 3), from the upper Kiulung Group of Shantung, China. Subsequently, Kobayashi (1962, p. 88) assigned *Anomocarella baucis* Walcott (1905, p. 55; 1913, p. 196–197, Plate 20, Figs. 2, 2a) from the upper Chaumitien Limestone of Changhia, Shantung. Schrank (1975, p. 598–599), following Kobayashi (1962), synonymized *Anomocare majus* Dames (1883, p. 18, Plate 2, Fig. 15), from Saimaji, Liaoning, northern China, and also referred the species *majus* to *Changshanocephalus.*
Comments. Pygidia from the Heritage Range illustrated herein are remarkably similar to those illustrated by Dames (1883, p. 17–18, Plate 1, Fig. 19; Plate 2, Fig. 15) as *Anomocare majus* and *A. subcostatum* respectively. These were referred to *Anomocarella* by Kobayashi (1937) and Lu and others (1965), even though Walcott (1905, p. 54) diagnosed this genus only on the basis of cranidial characteristics. Most species previously referred to *Anomocarella* are of Middle Cambrian age (see Lu and others, 1965, p. 316–329), with the exception of *A. baucis* Walcott, 1905 and *A. bergioni* Walcott, 1905, of Late Cambrian age, which have been referred to *Changshanocephalus* and *Changshania* (or *Maladioidella*) respectively by Kobayashi (1962, p. 88). More recently, Schrank (1975, p. 599) has classified *Anomocare majus* Dames (= *A. subcostatus* Dames) with *Changshanocephalus.* Additionally, he has figured a cranidium with a long (sag.), concave preglabellar area (Schrank, 1975, Plate 6, Fig. 3), which closely resembles those illustrated herein from the Heritage Range.

Although the synonymy of *Anomocare majus* can be accepted readily, the reference of this species to *Changshanocephalus* can only be accepted with reservation. The genus is based on a single cranidium of the type species, *C. reedi* Sun (1935, p. 41, Plate 1, Fig. 25, refigured in Lu and others, 1965, Plate 29, Fig. 8), which cannot be adequately interpreted. *Changshanocephalus* is accordingly not well established, and its concept is not clear. As diagnosed by Sun (1935, p. 40), *Changshanocephalus* has a broad, conical, and moderately convex glabella, large palpebral lobes, and a distinct, broad, concave frontal brim. However, if Lu and others (1965, p. 174), following Kobayashi (1962, p. 88), are correct in assigning *Yokusenia conica* Endo to *Changshanocephalus,* then the genus also accommodates cranidia with a preglabellar area not unlike that of *Protemnites* or *Olentella.*

Since it is not possible to resolve the concept of *Changshanocephalus* in this chapter, we use the generic name in the sense used by

Schrank with some degree of reservation, bearing in mind Sun's (1935) diagnosis and Schrank's (1975) reillustrations of Dames's (1883) material.

The lack of a definite generic concept also obscures the familial classification of *Changshanocephalus.* Mostly, large-eyed Ptychopariida have been classified among Anomocaridae or Proasaphiscidae, but these families have a well-differentiated preglabellar field and anterior cranidial border and are not appropriate for the classification of *Changshanocephalus,* as construed here. Others, with long palpebral lobes and broad, concave preglabellar areas, have been classified with Ceratopygacea (Asaphida): Ceratopygids, however, generally have semicircular palpebral lobes rather than arcuate ones, and their pygidial morphology is quite different from that observed here for *Changshanocephalus.*

Lu and others (1965) regard *Changshanocephalus* as an Asian pterocephaliid, but the palpebral morphology does not seem to support this view. The family Changshaniidae Kobayashi, 1935, however, which includes trilobites with a narrow, forward-tapering glabella, long, arcuate palpebral lobes, and a variably concave preglabellar area, could well accommodate this genus. *Parachangshania,* seemingly related to *Changshanocephalus,* has been classified by Lu and others (1965) as a changshaniid, and *Mecophrys* Shergold (1982) may represent another genus of the family.

Changshanocephalus? suspicor sp. nov.
(Plate 10, Figs. 1–10)

Name. L., *suspicor,* surmise, conjective, referring to the classificatory position of this taxon.
Types. Holotype, cranidium, USNM 334069a, the counterpart of which is illustrated on Plate 10, Fig. 1; paratypes, USNM 334070–334081a–k.
Material. Eighteen cranidia, USNM 334069–334072 and 335079a–d, ranging in length between 5.60 and 16.00 mm; three librigenae USNM 334073, 335080a–b; and 25 pygidia, USNM 334074–334078, 334081a–k, measuring (L_{p2}) 1.80 to 13.00 mm.
Occurrence. Uppermost Heritage Group and basal Minaret Formation, Yochelson Ridge, localities YP79-9, YP79-11, and YP79-13.
Age. Late Cambrian, late Dresbachian, *Aphelaspis* Zone.
Diagnosis. A species, questionably assigned to *Changshanocephalus,* distinguished by the following combination of characteristics: narrow (tr.) forward-tapering glabella; long, arcuate palpebral lobes close to axial furrows; long (sag.), broadly rounded, concave preglabellar area without perceptible differentiation into preglabellar field and border; nonspinose, semicircular pygidium with only first pleural furrows deeply incised and distally confluent with marginal furrows, separating narrow (exsag.) propleuron of first segment from remainder of tagma.
Description. Material variably compressed but indicating an elongate cranidium, anteriorly broadly rounded, and with low convexity (tr. sag.).

Glabella (G) moderately long (sag.), 55 to 62 percent of cranidial length (sag.), straight-sided, anteriorly tapering and broadly rounded; with low convexity (sag.). Glabellar furrows effaced on both exoskeleton and parietal surfaces.

Occipital furrow mostly effaced on exoskeletons, present only as ill-defined depression sagittally; better defined parietally as distinct furrow, widest sagittally. Occipital ring narrow (sag.), slightly wider (tr.) than preoccipital glabellar lobes, raised in lateral profile very slightly above dorsal surface of glabella; bearing median node.

Palpebral lobes long (A:G 54 to 68 percent; A:Gn 44 to 55 percent), arcuate, situated posteriorly with mid-points behind glabellar midpoint and close to axial furrows; connected anteriorly to faint, very oblique ocular ridges. Palpebral furrows sharply defined; palpebral areas narrow (tr.), flat or gently convex (tr.).

Preocular sections of facial suture sigmoidal, diverging strongly to anterolateral cranidial margins, meeting sagittally in wide arc to enclose basically concave preglabellar area in which preglabellar field, anterior cranidial border, and preocular areas are poorly differentiated unless specimens deformed when distinction becomes more apparent (cf., Plate 10, Figs. 2, 4). Postocular facial sutures diverge rapidly to enclose short (tr.), subrectangular posterolateral limbs.

Librigena quite characteristic with large eye socle and short posterolateral border to accommodate reduced posterolateral limb. Lateral marginal furrow wide and very shallow, not merging with posterior marginal furrow, which terminates in a pit before reaching genal angle. Genal field with low convexity (tr.), ill-defined subocular groove, and narrow eye socle. Genal angle drawn into short, broad-based, stout spine.

Pygidium semicircular, length (L_{p2}) 50 to 60 percent of width, nonspinose, characteristically with effaced external furrows on both axis and pleural zone with exception of first pair pleural furrows, which are sharply incised and distally confluent with poorly defined, shallow marginal furrows. Parietally five or six axial segments are separated by faint transverse furrows and up to five pairs of pleural segments indicated by even fainter pleural furrows. First pleural furrow isolates narrow (exsag.) propleuron of first pleural segment, which merges distally with lateral pygidial borders.

Anterior portion of preglabellar area, particularly along anterior cranidial margin; lateral and posterior borders of librigena; and all borders of pygidium and its doublure bear terrace-lined prosopon.
Comments. Changshanocephalus? suspicor sp. nov. cannot adequately be compared with the type species *C. reedi* Sun. The latter appears to have a better differentiated preglabellar area and a less elongate glabella, bearing furrows. *Changshanocephalus?* sp. (Sun, 1935, Plate 1, Fig. 26) has the appropriate extended preglabellar area, but its palpebral lobes are crescentic, and it may represent a ceratopygacean genus. *Yokusenia conica* Endo (1944, Plate 8, Fig. 7) has a short glabella, middling palpebral lobes, and a well-differentiated preglabellar area with an upturned anterior cranidial border quite distinct from specimens referred here to *Changshanocephalus. Ptychoparia burea* Walcott (1905, p. 86) is similar.

Pygidia of *C.? suspicor* resemble those assigned to *Anomocarella baucis* Walcott (1905) and *Anomocare majus* Dames (1883). Pygidia of the latter figured by Schrank (1975, Plate 6) are virtually indistinguishable in shape and general organization from those described here. They do, however, possess an additional axial segment. Similarly, one of the cranidia figured by Schrank (1975, Plate 6, Fig. 3) is very close to the Heritage Range material, differing only in its possession of glabellar furrows. Other cranidia, however (Plate 5, Figs. 7–9), are rather more distinctive with well-differentiated preglabellar areas and palpebral lobes more distantly spaced from the axial furrows. Librigenae referred to *Changshanocephalus majus* by Schrank (Plate 6, Figs. 5, 7) are essentially similar to those associated herein.

Superfamily LEIOSTEGIACEA Bradley, 1925
Family LEIOSTEGIIDAE Bradley, 1925
Subfamily PAGODIINAE Kobayashi, 1935

Genus *Prochuangia* Kobayashi, 1935

Type species. By original designation, *Prochuangia mansuyi* Kobayashi (1935, p. 186–187, Plate 8, Fig. 8; Plate 10, Figs. 1–7), Late Cambrian, Saisho-ri, South Korea.
Other species. Species assigned to *Prochuangia* have been listed by Shergold (*in* Shergold and others, 1976, p. 277–278; 1980, p. 66).
Comments. As information has accrued recently, it has become evident that there is little morphological distinction between *Prochuangia* and *Lotosoides* Shergold, 1975, which was originally conceived as a subgenus of *Pagodia* Walcott, 1905. Thorough revision of material described prior to 1975 is required to establish possible synonymy. In the meantime, *Prochuangia* is used here to classify pagodiines with spinose pygidia.

Prochuangia sp. indet.
(Plate 10, Figs. 11–12)

Material. One cranidium, USNM 334066, measuring 8.80 mm long; and two pygidia, USNM 334067 and 334068, with lengths (L_{p2}) of 4.10 and 6.15 mm.
Occurrence. Basal Minaret Formation, Yochelson Ridge, locality P79-3.
Age. Late Cambrian, late Dresbachian, *Aphelaspis* Zone (associated with *Glyptagnostus reticulatus*).
Comments. Insufficient material is available to accurately determine this species of *Prochuangia.* The cranidium is mostly exfoliated; elongate (sag.); with apparently effaced glabellar furrows; long, posterolaterally oblique simple ocular ridges; and small (exsag.) palpebral lobes situated to the rear of the mid-point of the glabella, distant from the axial furrows. The exoskeletal prosopon appears to be smooth. The pygidia are essentially semicircular with a pair of stout posterolaterally directed spines derived from the opisthopleuron of the first and propleuron of the second segments. The axis contains up to five segments and extends close to the posterior pygidial margin. Only the first segmental pleural furrows are defined. Like the cranidium, the exoskeletal prosopon appears to have been smooth.

Only the smooth prosopon distinguishes this species from material earlier described from northern Victoria Land (see Shergold and others, 1976, p. 277–281, Plate 40, Figs. 1–6). The same characteristic separates the Heritage Range material from the pagodiine described by Shrank (1975, p. 597, Plate 4, Figs. 1–9) from Saimaji, Liaoning, China, as *Kaolishania? quadriceps* (Dames). The Chinese material has a low-density granulosity but is otherwise compatible. *Prochuangia glabella* Shergold (1980, p. 67–69, Plate 23, Figs. 1–8) is also smooth, but this species has a short pygidial axis, and its pygidial spines are drawn from the whole of the first pleural segment. *Prochuangia mansuyi* Kobayashi (1935, p. 186, Plate 10, Figs. 1–7) may also be smooth but has quite different cranidial proportions.

Superfamily Incertae Sedis
Family PARABOLINOIDIDAE Lochman, 1956

Parabolinoidid? gen. and sp. indet.
(Plate 9, Figs. 4–8)

Material. Two cephalo-thoraces up to 58.5 mm long (sag.), USNM 334059, 334062; six disarticulated cranidia, USNM, 334055–334057, and 334060a–c, measuring 6.60 to 14.10 mm; and two pygidia, USNM 334058 and 334061, one of which has a length (L_{p2}) of 4.80 mm.
Occurrence. Uppermost Minaret Formation, Pipe Peak and Windy Peak, horizon YP79-4.
Age. Late Cambrian, stage and zone uncertain.
Description. All tagmata preserved as flattened molds and casts, completely decalcified.

Glabella conical with gently curved anteriorly tapering sides and truncate frontal lobe; 52 to 61 percent of cranidial length (sag.); glabellar furrows mostly effaced but where preserved preoccipital ones gently sigmoidal posterosagitally directed; median lateral furrows transverse, faint, gently curved forward, anterior lateral furrows always effaced.

Occipital furrow narrow (sag.), transverse, isolating occipital ring that is sagittally narrow but transversely wider than preoccipital glabellar lobes.

Palpebral lobes small (A:G 27 to 37 percent, A:Gn 21 to 28 percent), situated adjacent to frontal glabellar lobe well in front of mid-point of glabella, about one palpebral lobe length removed from axial furrows; ocular ridges more or less transverse, faint; palpebral furrows ill defined.

Preocular sections of facial suture not widely divergent, meeting in broad arch sagitally; enclosing forward-sloping preglabellar field, separated from narrow (sag.) flat-lying anterior cranidial border by shallow, ill-defined marginal furrow. Postocular sections enclosing broad-based triangular posterolateral limbs of considerable transverse width; some specimens bacculate (e.g., Plate 9, Fig. 4).

Thorax contains 13 segments, all with wide (tr.), narrow (sag.) axial rings, and pleurae with wide (exsag.) pleural furrows; pleural tips pointed probably not spinose; weakly geniculate.

Pygidium small, subtriangular, posteriorly slightly pointed, twice as wide as long; axis bluntly terminated posteriorly, comprising four well-defined segments; four pleural segments with prominent pleural furrows orientated parallel to interpleural furrows; restricted marginal furrow; margins entire, nonspinose.

One cranidium tentatively referred to this species is granulose; prosopon of remainder unknown.
Comments. The available material may be classified among several family groups. The truncato-conical glabella combined with small, anteriorly situated palpebral lobes and small subtriangular pygidium suggests reference either to Parabolinoididae, Olenidae, or Shirakiellidae. Classification with the first is tentatively suggested here.

Parabolinoidid? genus and species indeterminate bears reasonably close resemblance to some species of *Parabolinoides,* particularly those with a downsloping preglabellar field and flat anterior cranidial border like *P. expansa* Nelson (1951, p. 776, Plate 107, Figs. 1, 3) and *P. hedrus* Kurtz (*in* Bell and others, 1952, p. 186, Plate 32, Figs. 2a–c), both from the *Conaspis* Zone of the Upper Mississippi Valley, United States. Generally, however, species of *Parabolinoides* have palpebral lobes situated closer to the axial furrows, and their pygidia, where known, are spinose. Parabolinoidids are known with granulose prosopon but are not generally known in a bacculate condition. The exception may be *Bernia obtusa* (Frederickson, 1949), which shows an apparent baccular convexity. Apart from its stronger glabellar furrows and anteriorly rounded glabella, this species has an overall similar cranidial format.

The nature of the palpebral lobes, ocular ridges, narrow anterior cranidial border, and pygidial shape are characteristics resembling those of some olenids that develop a long (sag.) preglabellar area, e.g., *Olenus* and *Parabolinella. Olenus transversus* Linnarsson (see Westergård, 1922, Plate 3, Figs. 11–17; Plate 5, Figs. 16–17) has an appropriate glabellar shape and transverse subtriangular pygidium. The almost complete exoskeleton illustrated by Westergård (1922, Plate 3, Fig. 11) is remarkably close to that illustrated here. Notwithstanding such similarities, olenids are not usually as robust as the Antarctic material, and they are not normally bacculate; nor are they granulose.

Shirakiellidae are also characterized by a truncato-conical glabella with effaced furrows and small anteriorly situated palpebral lobes, but they lack a well-differentiated preglabellar area (see *S. lateconvexa* Kobayashi, 1935, p. 323, Plate 7, Figs. 15–16), as do certain genera classified with Parabolinoididae (e.g., *Croixana* Nelson, 1951). The pygidial characteristics of shirakiellids are unknown.

On the balance of characteristics, classification with Parabolinoididae is accordingly preferred. Paraboliniodidids have been recorded elsewhere in the southern hemisphere—by Öpik (1969) in the Bonaparte Gulf Basin of northwestern Australia—but these have not yet been described.

PLATE 1

Figures 1–6. *Homagnostus* cf. *ultraobesus* Lermontova, 1940. 1, USNM 333936, cephalon exfoliated, with characteristic shape and degree of deliquiation; length 2.00 mm, locality W63c, ×16. 2, USNM 333937, cephalon, exfoliated partially compressed; length 1.70 mm, locality W63c, ×16. 3, USNM 333938, cephalon, partially exfoliated, with very wide marginal furrows; length 2.80 mm, locality W63c, ×12. 4, USNM 333939, pygidium, exfoliated, undeformed, showing characteristic axial length; length (L_{p2}) 2.15 mm, locality W63c, ×12. 5, USNM 333940, pygidium, laterally compressed, partially exfoliated; length (L_{p2}) 2.30 mm, locality W63c, ×12. 6, USNM 333941, pygidium, incomplete, mostly exfoliated; length (L_{p2}) 2.50 mm (estimated), locality W63c, ×12.

Figures 7–13. *Idolagnostus (Obelagnostus) imitor* subgen. et sp. nov. 7, USNM 333944, cephalon, incomplete; length 1.95 mm, locality YP79-13, ×16. 8, USNM 333946, holotype, cephalon, showing median preglabellar furrow; length 1.40 mm, locality YP79-13, ×20. 9, USNM 333945, cephalon, slightly foreshortened, lacking median preglabellar furrow; length 1.65 mm, locality YP79-13, ×16. 10, USNM 333947, pygidium showing typical degree of effacement; length (L_{p2}) 2.00 mm, locality YP79-13, ×12. 11, USNM 333948, pygidium as above; length (L_{p2}) 2.20 mm, locality YP79-8, ×16. 12, USNM 333949, pygidium as above; length (L_{p2}) 1.70 mm, locality YP79-13, ×16. 13, USNM 333950, latex replica of small pygidium; length (L_{p2}) 1.20 mm, locality YP 79-13, ×16.

Figures 14–15. *Kormagnostella* cf. *minuta* (Schrank, 1975). 14, USNM 333953, well-preserved cephalon retaining trace of anterior glabellar lobe; length 1.20 mm, locality W63c, ×20. 15, USNM 333954, incomplete, largely effaced, broad-bordered pygidium; length (L_{p2}) 1.70 mm, locality YP79-2, ×16.

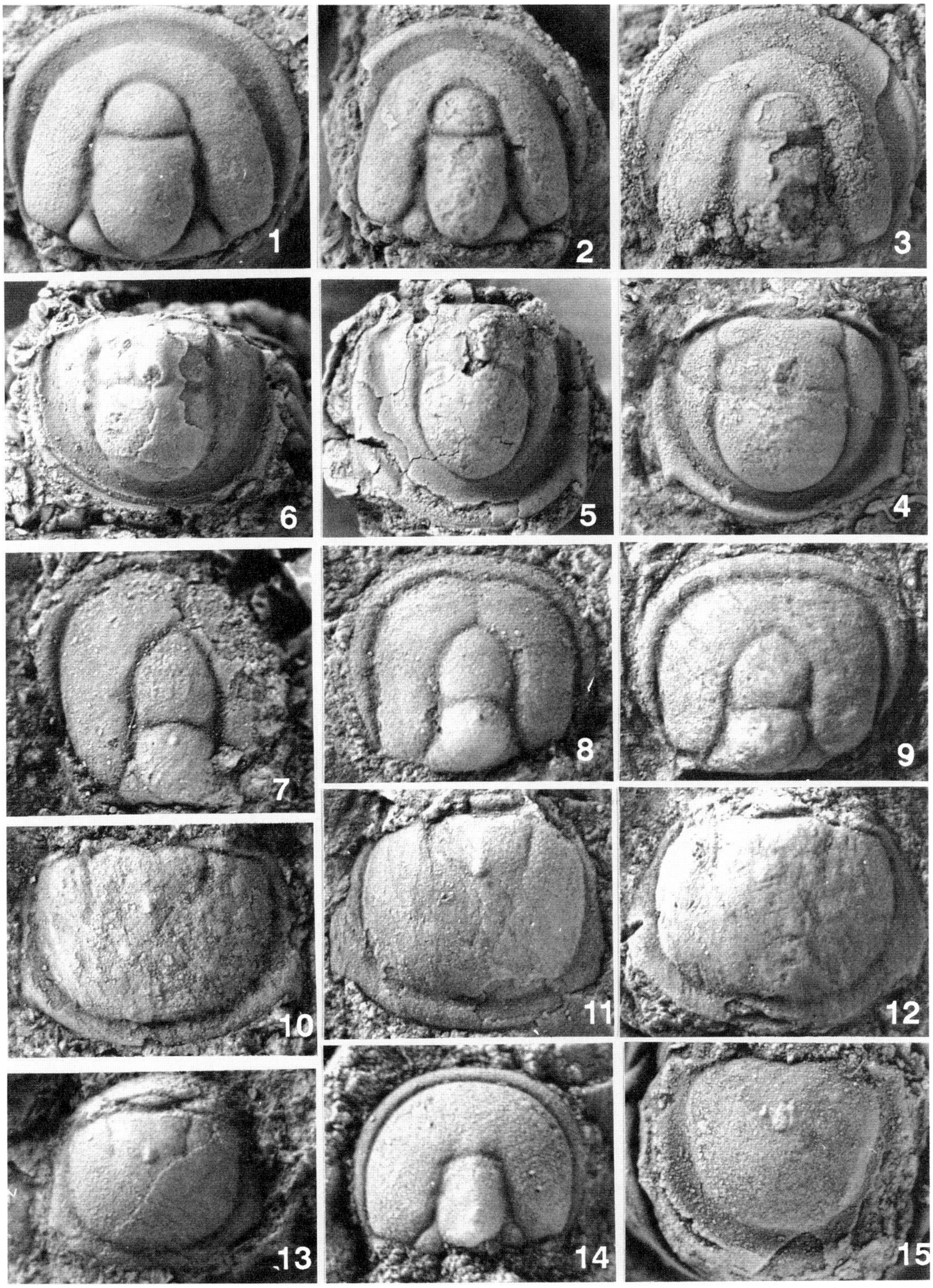
1
2
3
6
5
4
7
8
9
10
11
12
13
14
15

PLATE 2

Figures 1–5. *Pseudagnostus (Pseudagnostus)* cf. *vastulus* Whitehouse, 1936. 1, USNM 333961, cephalon, partially effaced; length 2.20 mm, locality W63c, ×12. 2, USNM 333962, as above; length 3.20 mm, locality W63c, ×8. 3, USNM 333963, as above; length 3.30 mm, locality W63c, ×8. 4, USNM 333964, pygidium, partially exfoliated; length (L_{p2}) 2.30 mm, locality W63c×12. 5, USNM 333965, as above, incomplete, with less extensive deuterolobe; length (L_{p2}) 2.50 mm, locality W63c, ×12.

Figure 6. Agnostid gen. and sp. indeterminate. USNM 333970, cephalon, exfoliate, with narrow (tr.) glabella, and traces of lateral lobes; length 3.0 mm, locality W63c, ×12.

Figures 7–8. *Pseudagnostus (Pseudagnostus)* cf. *idalis* Öpik, 1967. 7, USNM 333969, cephalon, exfoliated; length 3.00 mm, locality YP79-2, ×10. 8, USNM 333968, as above; length 2.50 mm, locality YP79-2, ×12.

Figures 9–12. *Ammagnostus?* sp. indeterminate. 9, USNM 333972, poorly preserved cephalon with wide borders; length 2.65 mm, locality YP79-13, ×12. 10, USNM 333973, latex replica of pygidial fragment with well-defined axial characteristics; length (Lp_2) 2.25 mm, locality YP79-13, ×12. 11, USNM 333974, latex replica of pygidium with wide borders; length (L_{p2}) 1.75 mm, locality YP79-13, ×16. 12, USNM 333975, slightly deformed pygidium; length (L_{p2}) 2.20 mm, locality YP79-13, ×12.

Figures 13–16. *Glyptagnostus reticulatus* (Angelin, 1851) *sensu lato*. 13, USNM 333959, meraspid pygidium with only four pairs of lateral scrobicules and prominent terminal axial node; length (L_{p2}) 0.85 mm; locality YP79-8, ×24. 14, USNM 333956, sagitally deformed large cephalon with full complement of scrobiculations; length 4.00 mm, locality YP79-8, ×6. 15, USNM 333957, transversely deformed cephalon; length (estimated) 2.50 mm, locality YP79-8, ×12. 16, USNM 333958, obliquely deformed cephalon; length 3.40 mm, locality YP79-8, ×8.

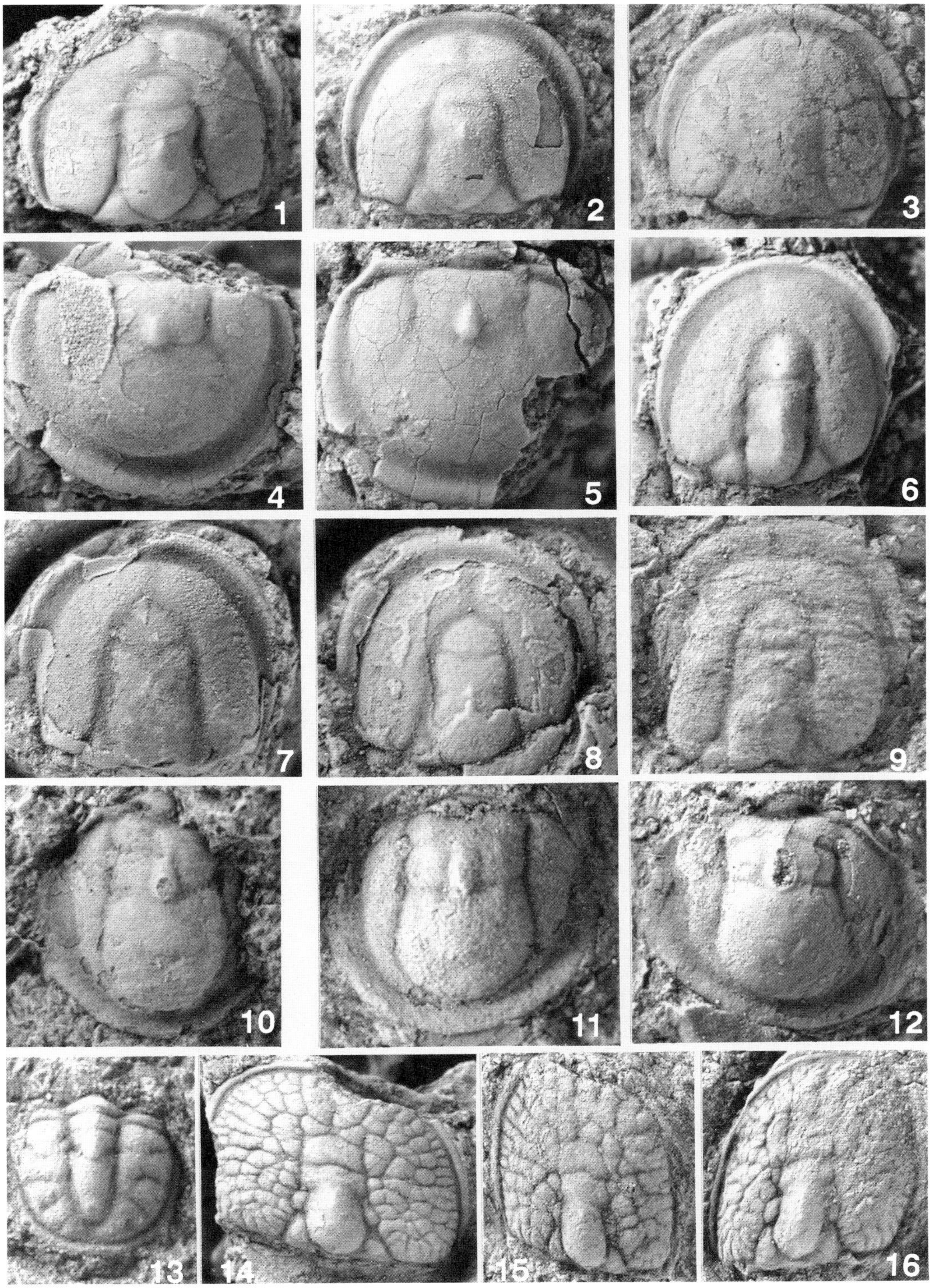
1
2
3
4
5
6
7
8
9
10
11
12
13
14
15
16

PLATE 3

Figures 1–9. *Bathyholcus? conifrons* sp. nov. 1, USNM 333978, cranidium showing degree of effacement and fine granulosity of exoskeletal surface; length 3.00 mm, locality W63c, ×12. 2, USNM 333979, early holaspid cranidial exoskeleton showing extent of palpebral lobes; length 2.4 mm, locality W63c, ×12. 3, USNM 333980, as above, partially exfoliated; length 3.70 mm, locality W63c, ×8. 4, USNM 333987, exfoliated late holaspid cranidium exhibiting complex glabellar muscle scar impressions; length 7.90 mm, locality W63c, ×4. 5, USNM 333982, partially exfoliated cranidium with relatively long proglabellar field; length 5.50 mm, locality W63c, ×6. 6, USNM 333983, holotype cranidium; length 5.60 mm, locality W63c, ×6. 7, USNM 333983, holotype, lateral aspect, ×6. 8, USNM 333983, holotype, anterior view, ×6. 9, USNM 333984, librigena assigned to this species with decided genal angle, and marginal terrace lines; locality W63c, ×12.

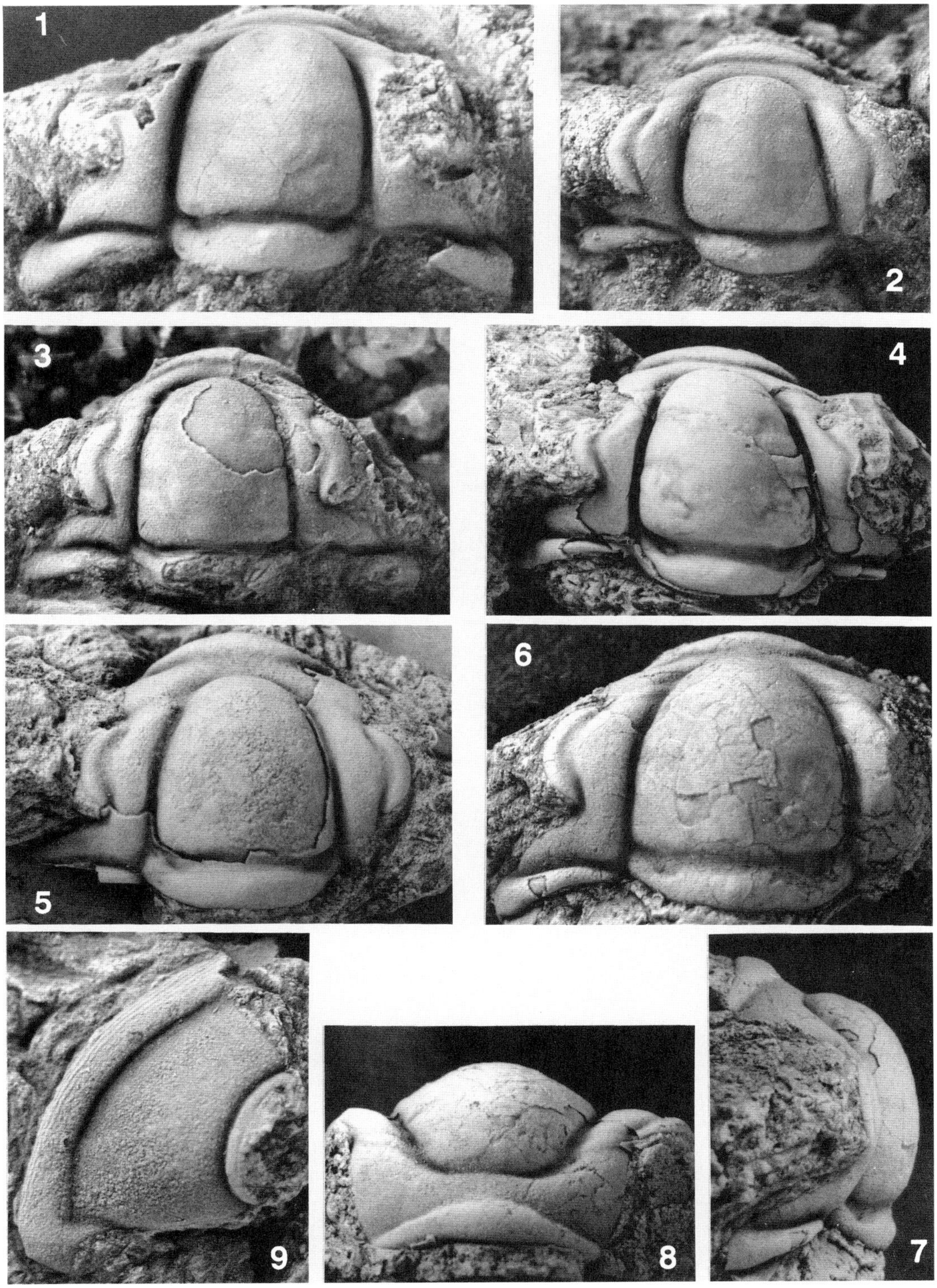
1
2
3
4
5
6
9
8
7

PLATE 4

Figures 1–10. *Onchopeltis variabilis* sp. nov. 1, USNM 333986, holotype cranidium showing typical late holaspid morphology; length 15.00 mm, locality W63c, ×3. 2, USNM 333987, large late holaspid cranidium showing traces of glabellar furrows; length 26.65 mm, locality W63c, ×2. 3, USNM 333988, late holaspid cranidium; length 11.30 mm, locality W63c, ×3. 4, USNM 333989, partially exfoliated late holaspid cranidium, highly granulose; length 17.00 mm, locality W63c, ×2. 5, USNM 333990, exfoliated cranidium; length 12.30 mm, locality W63c, ×2. 6, USNM 333990, as above, lateral view, ×3. 7, USNM 333990, as above, anterior view, ×3. 8, USNM 333991, partially exfoliated cranidium; length 7.50 mm, locality W63c, ×4. 9, USNM 333992, librigenal exoskeleton showing marginal terrace lines, short genal spine, and caeca crossing the shallow marginal furrow; locality W63c, ×3. 10, USNM 333993, exfoliated pygidium possibly belonging to this species; length (L_{p2}) 6.70 mm, locality YP79-2, ×3.

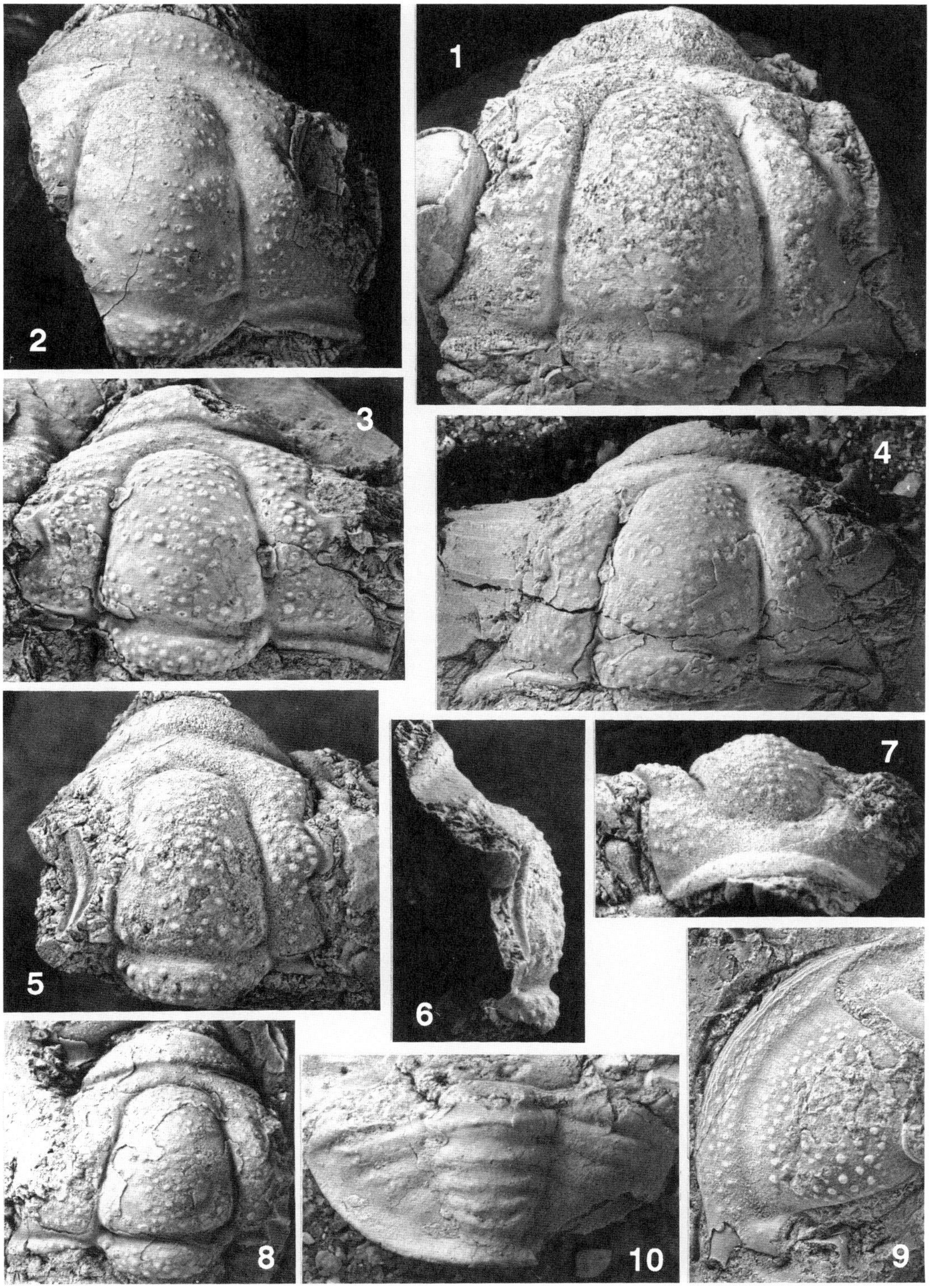
1
2
3
4
5
6
7
8
10
9

PLATE 5

Figures 1–6. *Onchopeltis? acis* sp. nov. 1, USNM 333997, holotype, cranidium, partially exfoliated; length 10.80 mm, locality W63c, ×3. 2, USNM 333998, cranidium, incompletely exfoliated; length 10.40 mm, locality W63c, ×3.3, USNM 333998, left lateral profile, ×3. 4, USNM 333998, right lateral profile, ×3. 5, USNM 333998, anterior view, ×3. 6, USNM 333999, cranidial fragment showing morphology of preglabellar area to advantage.

Figures 7–10. *Onchopeltis? neutra* sp. nov. 7, USNM 334001, holotype cranidium with transversely furrowed anterior cranidial border; length 5.20 mm, locality W63c, ×6. 8, USNM 334001, lateral aspect of holotype, ×6. 9, USNM 334002, exfoliated cranidium; length (estimated) 8.80 mm, locality W63c, ×4. 10, USNM 334003, as above; length 7.80, locality W63c, ×4.

Fig. 11. *Stigmatoa* sp. indeterminate. USNM 334006, exfoliated cranidium, lacking occipital spine; length 5.20 mm, locality W63c, ×8.

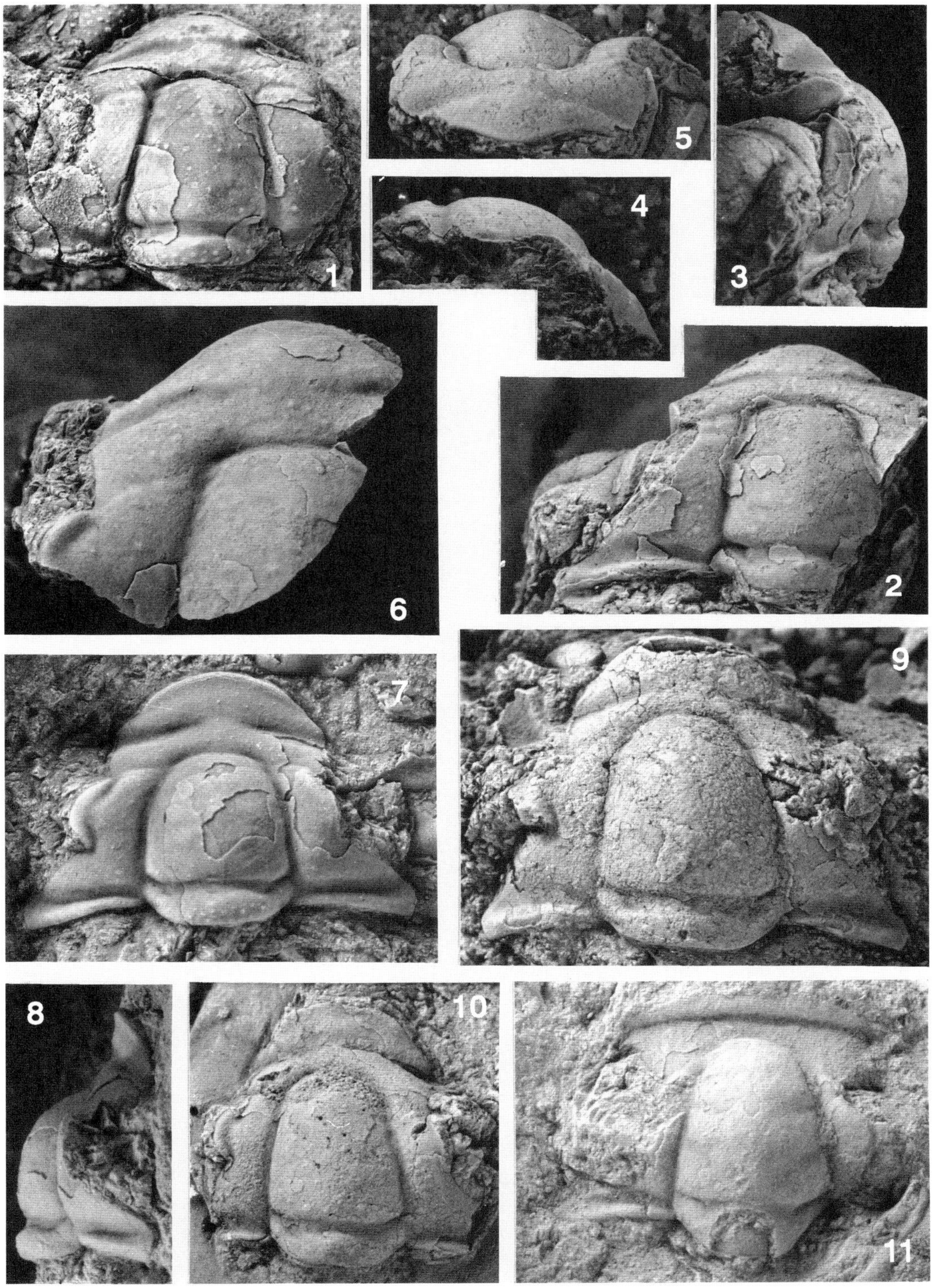
1
2
3
4
5
6
7
8
9
10
11

PLATE 6

Figures 1–6. *Protemnites? magnificans* sp. nov. 1, USNM 334013, holotype cranidium, exfoliated, showing characteristically long preglabellar area; length (estimated) 18.70 mm, locality W63c, ×2. 2, USNM 334014, partially exfoliated cranidium showing extent of posterolateral limbs; length of glabella (G) 5.80 mm, locality W63c, ×4. 3, USNM 334015, partially exfoliated cranidium showing parietal muscle scar impressions; length 8.80 mm, locality W63c, ×4. 4, USNM 334016, late holaspid cranidium; length 17.10 mm, locality W63c, ×2. 5, USNM 334013, holotype, left lateral aspect, ×2. 6, USNM 334013, holotype, anterior view, ×2.

Figures 7–11. *Protemnites* sp. aff. *P. elegans* Whitehouse, 1939. 7, USNM 334007, exfoliated incomplete cranidium; length 8.10 mm, locality W63c, ×4. 8, USNM 334008, partially exfoliated cranidium showing glabellar muscle scar impressions and caecal system of preglabellar field; length 12.40 mm, locality W63c, ×3. 9, USNM 334009, lateral view of exfoliated cranidium; length 11.90 mm, locality W63c, ×3. 10, USNM 334010, partially exfoliated cranidium most similar to Whitehouse's holotype of *P. elegans*; length 6.10 mm, locality W63c, ×6. 11, USNM 334011, mostly exfoliated cranidium; length 11.20 mm, locality W63c, ×3.

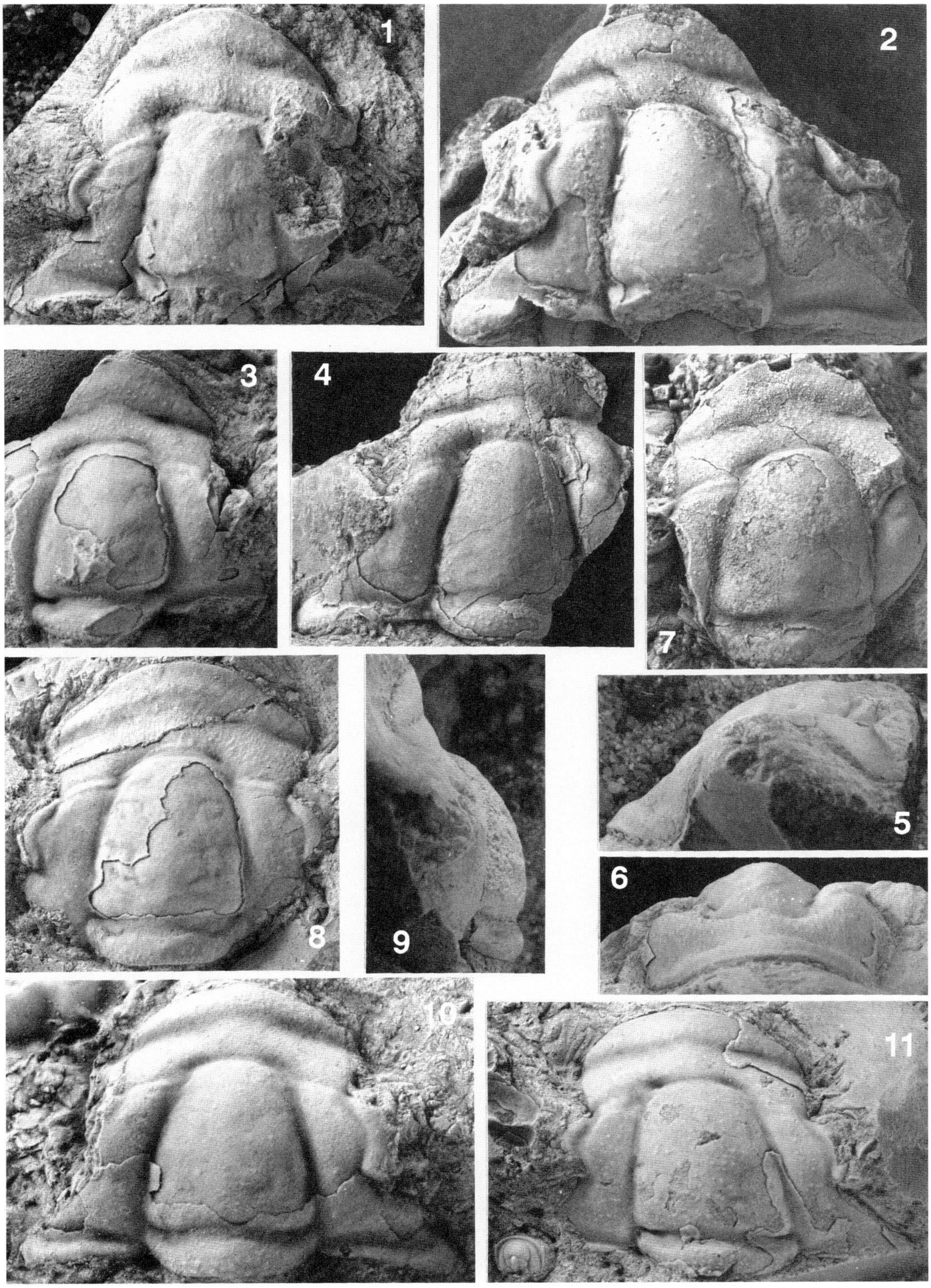
1
2
3
4
7
5
6
8
9
10
11

PLATE 7

Figures 1–4. *Aphelaspis* cf. *subdita* Palmer, 1962. 1, USNM 334021, partially exfoliated cranidium; length 5.50 mm, locality YP79-8, ×6. 2, USNM 334022, exfoliated cranidium with typical truncate glabella, narrow cranidial border, and more or less transverse ocular ridges; length 8.60 mm, locality P79-3, ×4. 3, USNM 334023, librigena with pronounced eye socle; locality P79-3, ×6. 4. USNM 334024, pygidium with one unliberated segment, lacking borders; length (L_{p2}) 2.10 mm, locality P79-3, ×6.

Figures 5–8. *Aphelaspis* cf. *lata* Rasetti, 1965. 5, USNM 334031, pygidium, lacking borders; length (L_{p2}) 2.40 mm, locality YP79-13, ×6. 6, USNM 334028, cranidium with wide-spaced palpebral lobes, anteriorly truncate glabella, and wide preglabellar area; length 7.70 mm, locality YP79-13, ×4.7, USNM 334030, associated librigena; locality YP79-13, ×4. 8, USNM 334029, cranidium with incipient preglabellar boss; length 7.60 mm, locality YP79-13, ×4.

Figures 9–13. *Aphelaspis* cf. *australis* Henderson, 1976. 9, USNM 334034, partially exfoliated, punctate cranidium, associated with librigena (USNM 334037) tentatively assigned to this species; length 7.80 mm, locality YP79-8, ×4. 10, USNM 334035, slightly deformed cranidium; length 10.10 mm, locality YP79-13, ×3. 11, USNM 334036, latex replica of partially exfoliated cranidium; length 5.70 mm, locality P79-3, ×6. 12, USNM 334038, pygidium with eroded borders; length (L_{p2}) 3.35 mm, locality YP79-13, ×6. 13, USNM 334039, latex replica of pygidium with well-preserved border; length (L_{p2}) 2.50 mm, locality P79-3, ×8.

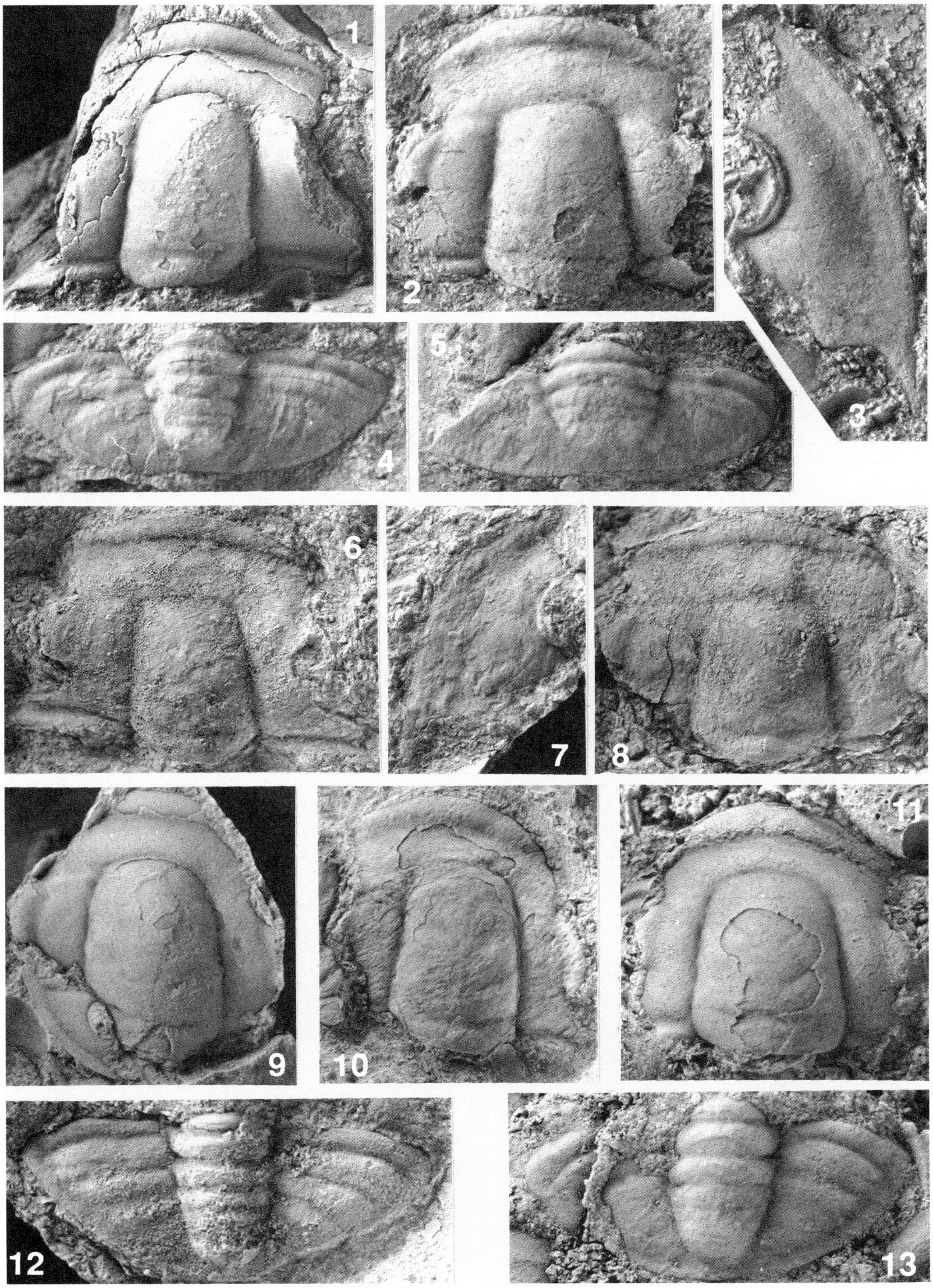
1
2
3
4
5
6
7
8
9
10
11
12
13

PLATE 8

Figures 1–2. *Erixanium* cf. *sentum* Öpik, 1963. 1, USNM 334044, partially exfoliated cranidium; estimated length 4.20 mm; locality W63c, ×8. 2, USNM 334045, poorly preserved cranidium, length 3.55 mm, locality W63c, ×8.

Figures 3–10. *Eugonocare? nebulosum* sp. nov. 3, USNM 334047, small partially exfoliated cranidium with three pairs of glabellar furrows; length 2.00 mm, locality W63c, ×8. 4, USNM 334048, as above, furrows effaced; length 2.40 mm; locality W63c, ×12. 5, USNM 334049, partially exfoliated cranidium with sagitally continuous preoccipital glabellar furrows; length 14.10 mm, locality W63c, ×2. 6, USNM 334050, deformed cranidium; length 5.30 mm, locality W63c, ×6. 7, USNM 334050, as above, lateral aspect, ×6. 8, USNM 334051, holotype cranidium retaining most of exoskeleton; length 4.80 mm, locality W63c, ×6. 9, USNM 334052, exfoliated, effaced cranidium with anteriorly rounded glabella; length 6.30 mm, locality W63c, ×6. 10, USNM 334053a, typically exfoliated cranidium; length 6.00 mm, locality W63c, ×6.

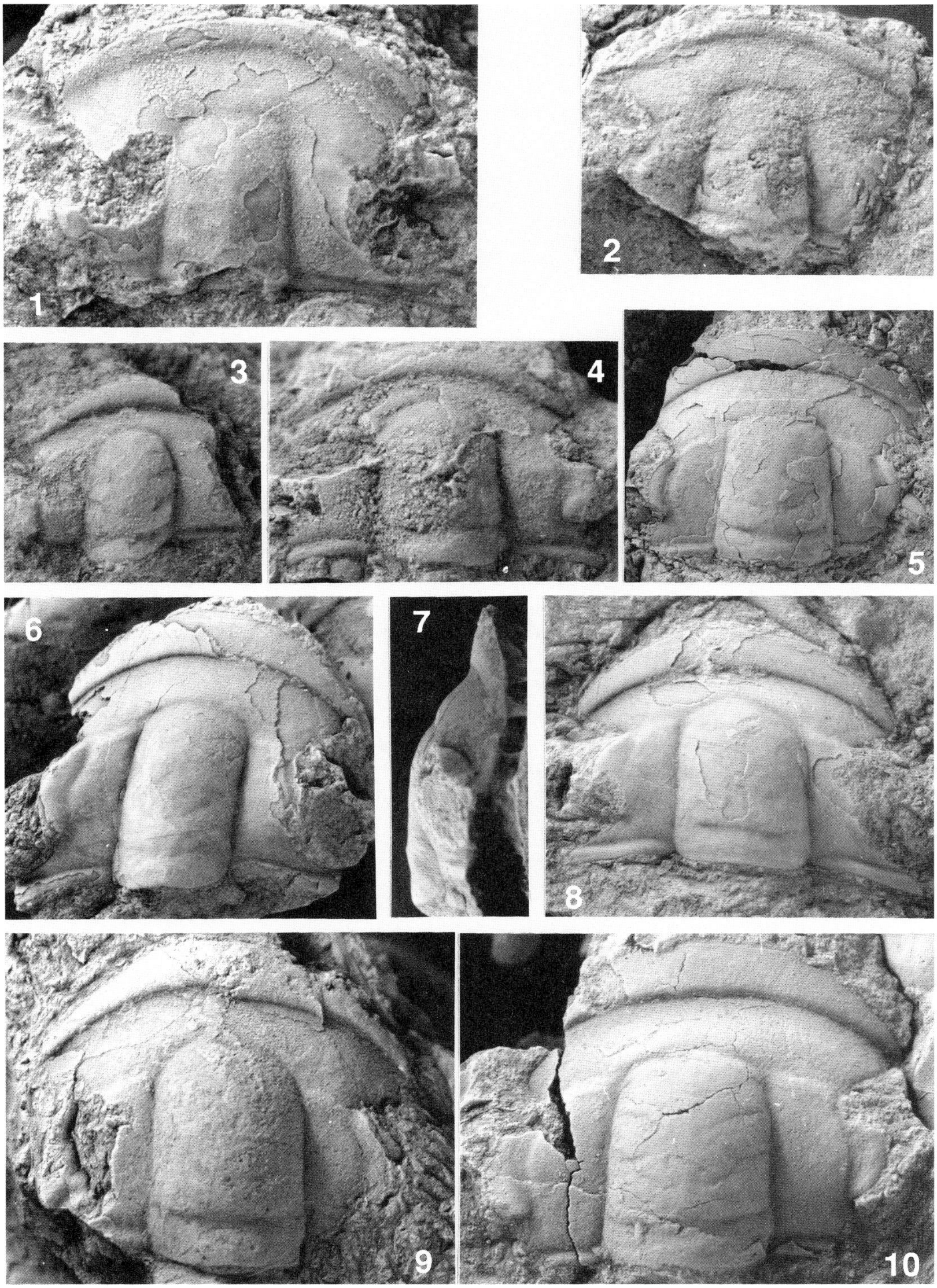
1
2
3
4
5
6
7
8
9
10

PLATE 9

Figures 1–3. *Aphelaspis* cf. *walcotti* Resser, 1938. 1, USNM 334020, typical aphelaspid librigena with short genal spine and posterior border; locality P79-3, ×3. 2, USNM 334019, exfoliated cranidium with narrow glabella; length 6.40 mm, locality P79-3, ×5. 3, USNM 334018, complete effaced cranidium; length 6.60 mm, locality P79-3, ×5.

Figures 4–8. Parabolinoididid? gen. and sp. indeterminate. 4, USNM 334055, exfoliated cranidium with truncate glabella and bacculae; length 14.10 mm; locality YP79-4, ×2. 5, USNM 334056, as above, with small anteriorly situated palpebral lobes; length 12.00 mm, locality YP79-4, ×2. 6, USNM 334057, as above, replica with traces of glabellar furrows; length 8.80 mm, locality YP79-4, ×4. 7, USNM 334058, replica of exfoliated pygidium, with posteriorly angled margin; length (L_{p2}) 4.80 mm, locality YP79-4, ×4. 8, USNM 334059, replica of more or less complete exoskeleton, with anteriorly rounded glabella; total length about 58.50 mm, locality YP79-4, ×1.

Figures 9–10. Solenopleuracean trilobite indeterminate. 9, USNM 334063, latex replica of relatively undeformed cephalo-thorax preserving librigena, with small anteriorly situated palpebral lobes; locality R64-INF, ×2. 10, USNM 334064, laterally compressed cephalo-thorax; locality R64-INF, ×1.5.

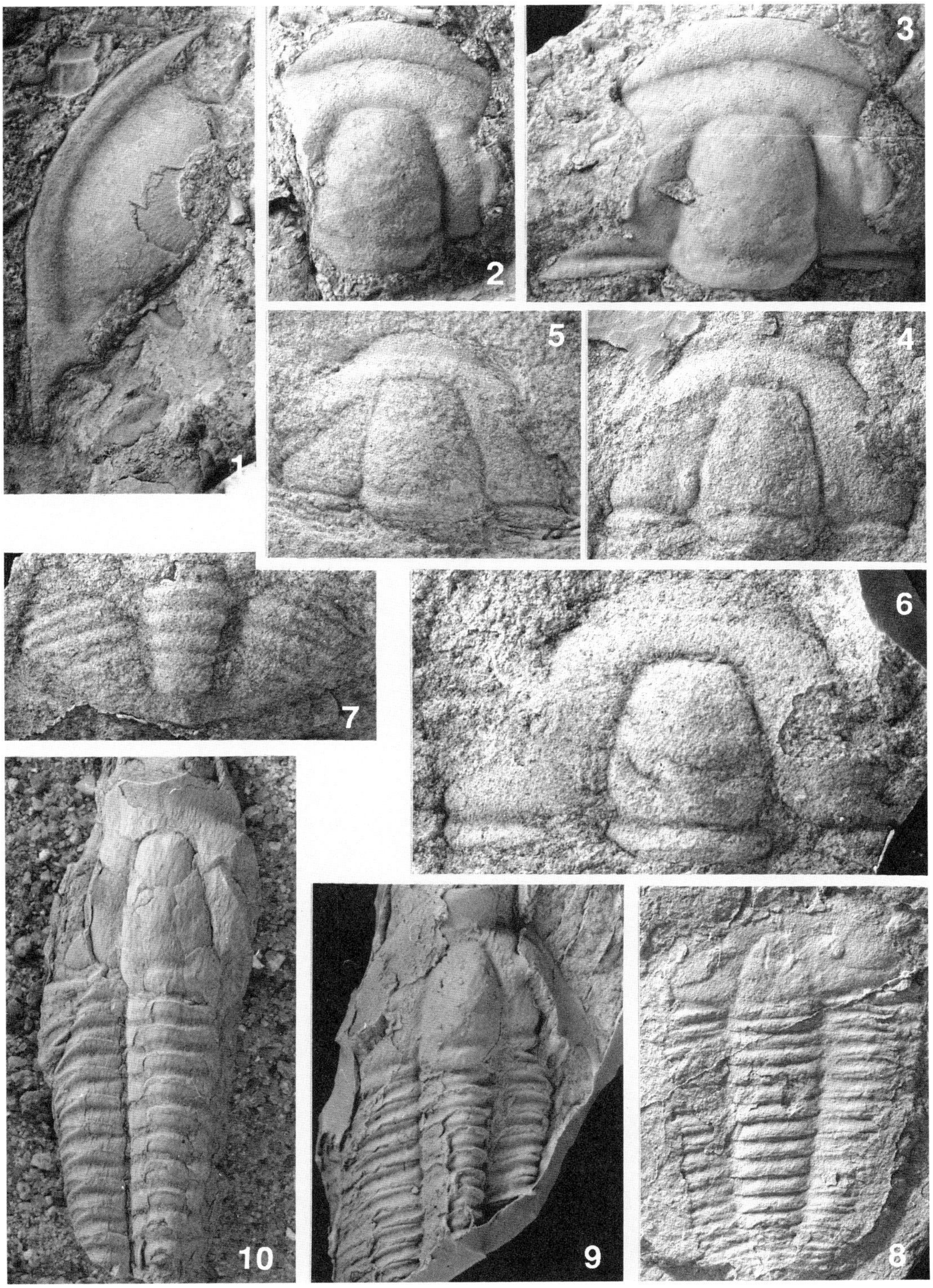
1
2
3
5
4
7
6
10
9
8

PLATE 10

Figures 1–10. *Changshanocephalus? suspicor* sp. nov. 1, USNM 334069b, latex replica from counterpart of holotype cranidium; length 11.90 mm, locality YP79-13, ×3. 2, USNM 334070, laterally compressed cranidium showing extent of palpebral lobes; length 5.60 mm, locality YP79-13, ×6. 3, USNM 334071, laterally compressed cranidium showing full extent of preglabellar area; length 8.20 mm, locality YP79-13, ×4. 4, USNM 334072, sagitally compressed cranidium illustrated for contrast; estimated length 10.25 mm, locality YP79-13, ×3. 5, USNM 334073, librigena with narrow eye socle, associated with pygidium USNM 334074; locality YP79-13, ×6. 6, USNM 334075, mostly exfoliated pygidium; estimated length (L_{p2}) 5.10 mm, locality YP79-13, ×4. 7, USNM 334074, partly exfoliated pygidium; estimated length (L_{p2}) 3.00 mm, locality YP79-13, ×8. 8, USNM 334076, partially exfoliated pygidium; length (L_{p2}) 10.00 mm, locality YP79-13, ×2. 9, USNM 334077, partially exfoliated pygidium showing extent of lateral doublure; estimated length (L_{p2}) 4.00 mm, locality YP79-13, ×6. 10, USNM 334078, pygidium showing degree of external effacement; length (L_{p2}) 7.00 mm, locality YP79-13, ×4.

Figures 11–12. *Prochuangia* sp. indeterminate. 11, USNM 334066, mostly exfoliated, slightly deformed cranidium; length 8.80 mm, locality P79-3, ×4. 12, USNM 334067, exfoliated pygidium; length (L_{p2}) 6.15 mm, locality P79-3, ×4.

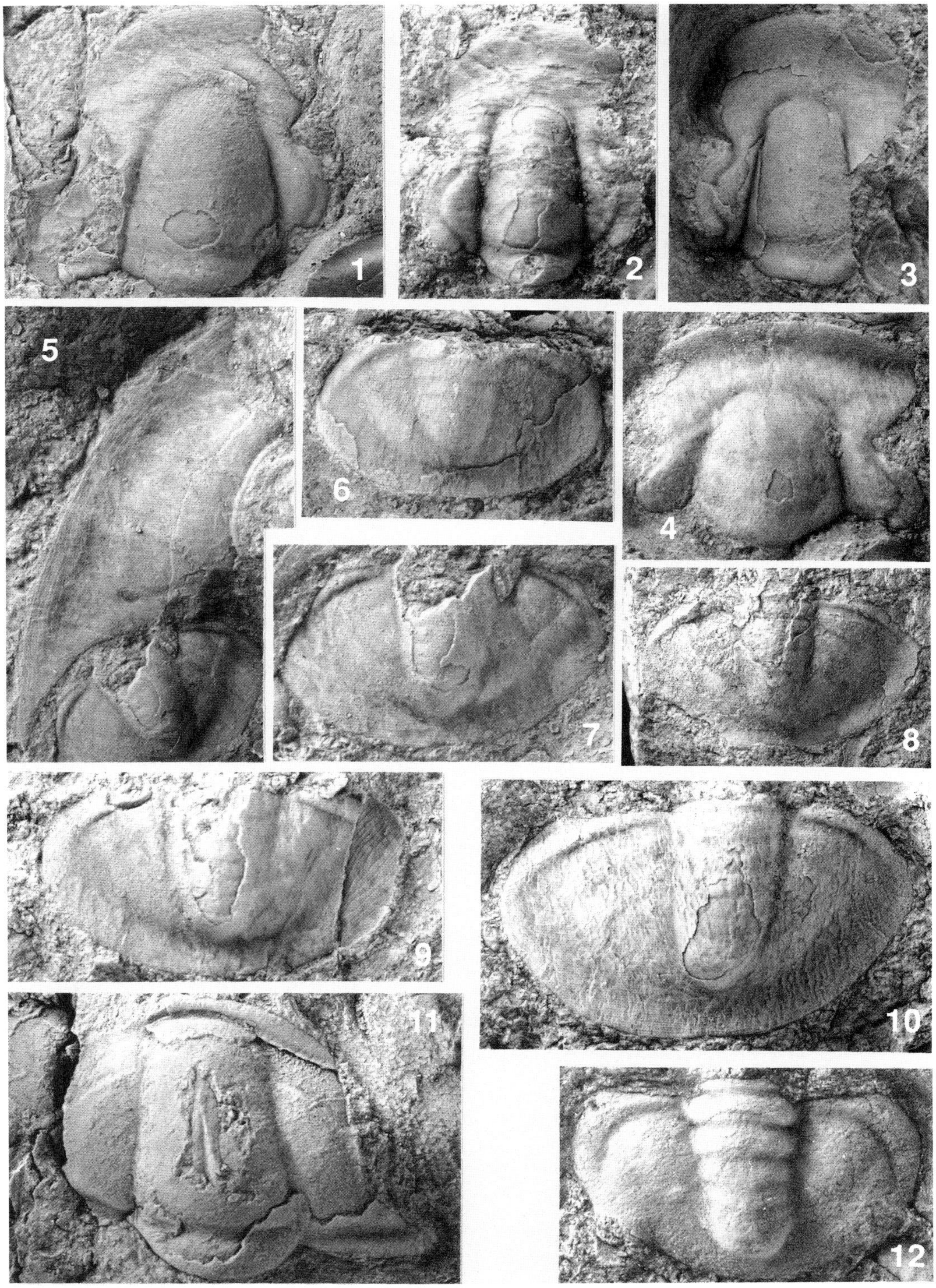
1
2
3
5
6
4
7
8
9
10
11
12

REFERENCES CITED

Allen, P. M., Jackson, A. A., and Rushton, A.W.A., 1981, The stratigraphy of the Mawddach Group in the Cambrian sequence of North Wales: Yorkshire Geological Society Proceedings, v. 43, no. 3, p. 295–329.

Angelin, N. P., 1851–1854, Palaeontologia Scandinavica, Pars 1, Crustacea formationis transitionis, p. 1–24, pl. 1–24 [1851]; Pars 2, no separate titles p. 1–ix5, 21–92, pl 25–41 [1854]: Stockholm, Academia Regiae Scientarium Suecanae (Holmiae).

Bell, W. C., Feniak, O. W., and Kurtz, V. E., 1952, Trilobites of the Franconia Formation, southeast Minnesota: Journal of Paleontology, v. 26, no. 2, p. 175–198.

Belt, T., 1867, On some new trilobites from the Upper Cambrian rocks of North Wales: Geological Magazine, v. 4, p. 294–295.

Bradley, J. H., 1925, Trilobites of the Beekmantown in the Philipsburg region of Quebec: Canadian Field Naturalist, v. 39, p. 7–8.

Burmeister, H., 1843, Die Organization der Trilobiten: Berlin, G. Reimer, 148 p.

Dames, W., 1883, Cambrische Trilobiten von Liautung, *in* von Richthofen, F. F., ed., China; Ergebnisse einiger Reisen und darauf gegrundeter Studien 4: Berlin, Palaeontologisches Teil, 288 p.

Endo, R., 1944, Restudies on the Cambrian formations and fossils of southern Manchukuo: Central National Museum of Manchukuo Bulletin, v. 7, p. 1–100.

Ergaliev, G. K., 1980, Middle and Upper Cambrian trilobites of Lesser Karatau: Kazakhstan SSR Academy of Sciences, K. I., Satpaev Institute of Geological Sciences, Alma-Ata, Science Press, 211 p. (in Russian).

—— , 1981, Upper Cambrian biostratigraphy of the Kyrshabakty Section, Maly Karatau, southern Kazakhstan, *in* Taylor, M. E., ed., Short papers for the 2nd International Symposium on the Cambrian System, 1981: U.S. Geological Survey Open-File Report 81-743, p. 82–88.

Frederickson, E. A., 1949, Trilobite fauna of the Upper Cambrian Honey Creek Formation: Journal of Paleontology, v. 23, no. 4, p. 341–363.

Hall, J., and Whitfield, R. P. 1877, Palaeontology, Part 2, *in* King, C., ed., Report of the geological exploration of the 40th Parallel, 4: U.S. Army Professional Paper of the Engineer Department, p. 198–302.

Harrington, J. H., Moore, R. C., and Stubblefield, C. J., 1959, Morphological terms applied to Trilobita, *in* Moore, R. C., ed., Treatise on invertebrate paleontology; Part O, Arthropoda 1: Lawrence, University of Kansas Press, p. 117–126.

Henderson, R. A., 1976, Upper Cambrian (Idamean) trilobites from western Queensland: Palaeontology, v. 19, no. 2, p. 325–364.

Henningsmoen, G., 1957, The trilobite family Olenidae, with description of Norwegian material and remarks on the Olenid and Tremadocian series: Norske Videnskaps-Akademi i Oslo, Årsbok; Avhandlinger, Matematisk-naturvidenskapileg klasse, Skrifter, 303 p.

Hill, D., Playford, G., and Woods, J. T., eds., 1971, Cambrian fossils of Queensland: Brisbane, Queensland Palaeontographical Society, 32 p.

Howell, B. F., 1935, Some New Brunswick Cambrian agnostians: Wagner Free Institute of Science Bulletin, v. 10, no. 2, p. 13–16.

Ivshin, N. K., 1955, Eight new genera of trilobites from the Upper Cambrian of central Kazakhstan: Izvestiya Akademii Nauk Kazakhoskoy SSR, Seriya Geologischeskaya Biulleteny 21, p. 106–123.

—— , 1956, Upper Cambrian trilobites of Kazakhstan, 1: Alma-Ata, Kazakhstan SSR Academy of Sciences Transactions of the Institute of Geological Sciences, 98 p. (in Russian).

—— , 1962, Upper Cambrian trilobites of Kazakhstan, 2: Alma-Ata, Kazakhstan SSR Academy of Sciences Transactions of the Institute of Geological Sciences, 412 p. (in Russian).

Jaekel, O., 1909, Über die Agnostiden: Zeitschrift der Deutschen Geologischen gesellschaft, v. 61, p. 380–401.

Kobayashi, T., 1933, Upper Cambrian of the Wuhutsui Basin, Liaotung, with special reference to the limit of the Chaumitien (or Upper Cambrian) of eastern Asia, and its subdivision: Journal of Japanese Geology and Geography, v. 11, no. 1-2, p. 55–155.

—— , 1935, The Cambro-Ordovician formations and faunas of South Chosen, Palaeontology; Part 3, Cambrian faunas of South Chosen with special study on the Cambrian trilobite genera and families: Journal of the Faculty of Science Imperial University of Tokyo, series 2, v. 4, no. 2, p. 49–344.

—— , 1936, On the *Parabolinella* fauna from Province Jujuy, Argentina, with a note on the Olenidae: Journal of Japanese Geology and Geography, v. 13, no. 1-2, p. 85–102.

—— , 1937, Restudy of the Dames' types of the Cambrian trilobites from Liaotung: Transactions and Proceedings of the Palaeontological Society of Japan, v. 12, no. 7, p. 70–86.

—— , 1938, Upper Cambrian fossils from British Columbia with a discussion on the isolated occurrence of the so-called "Olenus" Beds of Mount Jubilee: Journal of Japanese Geology and Geography, v. 15, p. 149–192.

—— , 1949, The *Glyptagnostus* hemera, the oldest world-instant: Journal of Japanese Geology and Geography, v. 21, no. 1-4, p. 1–6.

—— , 1955, The Ordovician fossils in the McKay Group in British Columbia, western Canada, with a note on the early Ordovician palaeogeography: Journal of the Faculty of Science University of Tokyo, series 2, v. 9, no. 3, p. 355–493.

—— , 1960, The Cambro-Ordovician formations and faunas of South Korea; Part 7, Palaeontology; Part 6, Supplement to the Cambrian faunas of Tsuibon Zone with notes on some trilobite genera and families: Journal of the Faculty of Science University of Tokyo, series 2, v. 12, no. 2, p. 329–420.

—— , 1962, The Cambro-Ordovician formations and faunas of South Korea; Part 9, Palaeontology; Part 8, The Machari fauna: Journal of the Faculty of Science University of Tokyo, series 2, v. 14, no. 1, p. 1–152.

Lermontova, E. V., 1940, Arthropoda, *in* Vologdin, A. G., ed., Atlas of the leading forms of fossil faunas of the USSR; Vol. 1, Cambrian: Moscow, Leningrad, Council of Peoples Commissars of the USSR and Geological Committee of the All-Union Scientific Prospecting Institute (VSEGEI), State Editorial Office for Geological Literature, p. 112–157.

Lochman, C., 1956, The evolution of some Upper Cambrian and Lower Ordovician trilobite families: Journal of Paleontology, v. 30, no. 3, p. 445–462.

Lu, Y.-H., 1956, An Upper Cambrian trilobite faunule from Kueichow: Acta Palaeontologica Sinica, v. 4, no. 3, p. 365–380.

Lu, Y.-H., Chang, W. T., Chu, C.-L., Chien, Y.-Y., and Hsiang, L.-W., 1965, Chinese fossils of all groups; Trilobita: Peking, Science Publishing Co., v. 1, 362 p., v. 2, p. 363–766 (in Chinese).

M'Coy, F., 1849, On the classification of British fossil Crustacea, with notices of new forms in the University collection at Cambridge: Annals of the Magazine of Natural History, series 2, v. 4, p. 161–179, 392–414.

Nelson, C. A., 1951, Cambrian trilobites from the St. Croix Valley: Journal of Paleontology, v. 25, no. 6, p. 765–784.

Öpik, A. A., 1961a, Alimentary caeca of agnostids and other trilobites: Palaeontology, v. 3, no. 4, p. 410–438.

—— , 1961b, The geology and palaeontology of the headwaters of the Burke River, Queensland: Bureau of Mineral Resources Geology and Geophysics of Australia Bulletin 53, 249 p.

—— , 1963, Early Upper Cambrian fossils from Queensland: Bureau of Mineral Resources Geology and Geophysics of Australia Bulletin 64, 133 p.

—— , 1967, The Mindyallan fauna of northwestern Queensland: Bureau of Mineral Resources Geology and Geophysics of Australia Bulletin 74, v. 1, 404 p.; v. 2, 166 p.

—— , 1969, Appendix 3, The Cambrian and Ordovician sequence, Cambridge Gulf area, *in* Kaulbach, J. A., and Veevers, J. J., 1969, The Cambrian and Ordovician geology of the southern part of the Bonaparte Gulf Basin, and the Cambrian and Devonian geology of the outliers, Western Australia: Bureau of Mineral Resources Geology and Geophysics of Australia Report

109, p. 74–77.
Orłowski, S., 1968, Upper Cambrian fauna of the Holy Cross Mountains: Acta Geologica Polonica, v. 28, no. 2, p. 257–291.
Palmer, A. R., 1953, *Aphelaspis* Resser and its genotype: Journal of Paleontology, v. 27, p. 157.
——, 1954, The fauna of the Riley Formation in central Texas: Journal of Paleontology, v. 28, no. 6, p. 709–786.
——, 1955, Upper Cambrian Agnostidae of the Eureka District, Nevada: Journal of Paleontology, v. 29, no. 1, p. 86–101.
——, 1960, Trilobites of the Upper Cambrian Dunderberg Shale, Eureka district, Nevada: U.S. Geological Survey Professional Paper 334-C, 109 p.
——, 1962, *Glyptagnostus* and associated trilobites in the United States: U.S. Geological Survey Professional Paper 374-F, 49 p.
——, 1965, Trilobites of the Late Cambrian pterocephaliid biomere in the Great Basin, United States: U.S. Geological Survey Professional Paper 493, 105 p.
——, 1968, Cambrian trilobites of east-central Alaska: U.S. Geological Survey Professional Paper 559-B, 115 p.
Powell, C. McA., Neef, G., Crane, D., Jell, P. A., and Percival, I. G., 1982, Timing of deformation indicated by Late Cambrian (Idamean) fossils in the Cupala Creek Formation, northwestern New South Wales: Proceedings of the Linnean Society of New South Wales, v. 106, no. 2, p. 127–150.
Rasetti, F., 1944, Upper Cambrian trilobites from the Levis Conglomerate: Journal of Paleontology, v. 18, no. 3, p. 229–258.
——, 1961, Dresbachian and Franconian trilobites of the Conococheague and Frederick Limestone of the central Appalachians: Journal of Paleontology, v. 35, no. 1, p. 104–124.
——, 1965, Upper Cambrian trilobite faunas of northeastern Tennessee: Smithsonian Miscellaneous Collections, v. 148, no. 3, 127 p.
Resser, C. E., 1935, Nomenclature of some Cambrian trilobites: Smithsonian Miscellaneous Collections, v. 93, no. 5, p. 1–46.
——, 1938, Cambrian System (restricted) of the southern Appalachians: Geological Society of America Special Paper 15, 140 p.
Romanenko, E. V., 1967, Description of Cambrian trilobites, *in* Romanenko, E. V., and Romanenko, M. F., Some problems of palaeogeography and Cambrian trilobites of the Altai Mountains: Proceedings of the All-Union Geographical Society, Altai Division, v. 8, p. 62–96 (in Russian).
——, 1977, Cambrian trilobites from the region of the Bolyshoy Ishe River, northeast Altay, *in* Zhuravleva, I. T., and Rozova, A. V., eds., 1977, Biostratigraphy and fauna of the Upper Cambrian boundary beds (new data from the Asiatic regions of the USSR): Novosibirsk, Academy of Sciences of the USSR, Siberian Division, Transactions of the Institute of Geology and Geophysics, v. 313, p. 161–184 (in Russian).
Romanenko, E. V., and Romanenko, M. F., 1967, Some problems of palaeogeography and Cambrian trilobites of the Altai Mountains: Proceedings of the All-Union Geographical Society, Altai Division, v. 8, p. 62–93 (in Russian).
Rozova, A. V., 1977, Some Upper Cambrian and Lower Ordovician trilobites from the Rybnoy, Khantayki, Kureyki and Lerney River basins, *in* Zhuravleva, I. T., and Rozova, A. V., eds., Biostratigraphy and fauna of the Upper Cambrian boundary beds (new data from the Asiatic regions of the USSR): Novosibirsk, Academy of Sciences of the USSR, Siberian Division, Transactions of the Institute of Geology and Geophysics, v. 313, p. 54–84 (in Russian).
Rusconi, C., 1955a, Mas fosiles Cambricos y Ordovicios de San Isidro, Mendoza: Boletin Paleontologico Buenos Aires, v. 31, p. 1–4.
——, 1955b, Fosiles Cambricos y Ordovicios al oeste de San Isidro, Mendoza: Revista del Museo de Historia Natural de Mendoza, v. 7, no. 1-4, p. 3–63.
Rushton, A.W.A., 1978, Fossils from the middle Upper Cambrian transition in the Nuneaton district: Palaeontology, v. 21, p. 245–283.
——, 1983, Trilobites from the Upper Cambrian Olenus Zone in central England: Special Papers in Palaeontology, v. 30, p. 107–139.
Schrank, E., 1975, Kambrische Trilobiten der China-kollektion von Richthofen; 2, Die fauna mit *Kaolishania? quadriceps* von Saimaki: Zeitschrift für Geologische Wissenschaften, v. 3, no. 5, p. 591–619.
Shergold, J. H., 1972, Late Upper Cambrian trilobites from the Gola Beds, western Queensland: Bureau of Mineral Resources Geology and Geophysics of Australia Bulletin 112, 126 p.
——, 1975, Late Cambrian and early Ordovician trilobites from the Burke River Structural Belt, western Queensland, Australia: Bureau of Mineral Resources Geology and Geophysics of Australia Bulletin 153, v. 1, 251 p.; v. 2, 58 plates.
——, 1977, Classification of the trilobite *Pseudagnostus:* Palaeontology, v. 20, no. 1, p. 69–100.
——, 1980, Late Cambrian trilobites from the Chatsworth Limestone, western Queensland: Bureau of Mineral Resources Geology and Geophysics of Australia Bulletin 186, 111 p.
——, 1982, Idamean (Late Cambrian) trilobites, Burke River Structural Belt, western Queensland: Bureau of Mineral Resources Geology and Geophysics of Australia Bulletin 187, 69 p.
Shergold, J. H., Cooper, R. A., MacKinnon, D. I., and Yochelson, E. L., 1976, Late Cambrian Brachiopoda, Mollusca, and Trilobita from northern Victoria Land, Antarctica: Palaeontology, v. 19, no. 2, p. 247–291.
Sun, Y.-C., 1935, The Upper Cambrian trilobite faunas of north China: Palaeontologia Sinica, series B., v. 2, no. 2, p. 1–69.
Swinnerton, H. H., 1915, Suggestions for the revised classification of trilobites: Geological Magazine, Decade 6, v. 2, p. 487–496, 538–545.
Tullberg, S. A., 1880, Om *Agnostus*-arterna i de Kambriska afflagringarne vid Andrarum: Sveriges Geologiska Undersökning, series C, v. 42, 37 p.
Walcott, C. D., 1905, Cambrian faunas of China: Proceedings of the U.S. National Museum, v. 29, p. 1–106.
——, 1913, The Cambrian faunas of China, *in* Research in China, v. 3: Publications of the Carnegie Institution,v . 54, 375 p.
Walcott, D. C., and Resser, C. E., 1924, Trilobites from the Ozarkian sandstones of the island of Novaya Zemlya, *in* Report of the scientific results of the Norwegian expedition to Novaya-Zemlya 1921: Kristiania, Society of Arts and Science, v. 24, p. 3–14.
Webers, G. F., 1972, Unusual Upper Cambrian fauna from West Antarctica, *in* Adie, R. J., ed., 1972, Antarctic geology and geophysics: Oslo, Universitetsforlaget, p. 235–237.
Westergård, A. H., 1922, Sveriges Olendiskiffer: Sveriges Geologiska Undersökning, series Ca, v. 18, p. 1–188 (in Swedish); p. 189–205 (in English).
——, 1947, Supplementary notes on the Upper Cambrian trilobites of Sweden: Sveriges Geologiska Undersökning, series C, Årsbok 41, no. 8, p. 3–34.
Whitehouse, F. W., 1936, The Cambrian faunas of northeastern Australia; Part 1, Stratigraphic outline; Part 2, Trilobita (Miomera): Memoirs of the Queensland Museum, v. 11, no. 1, p. 59–112.
——, 1939, The Cambrian faunas of northeastern Australia; Part 3, The polymerid trilobites: Memoirs of the Queensland Museum, v. 11, no. 3, p. 179–282.
Zhuravleva, I. T., and Rozova, A. V., eds., 1977, Biostratigraphy and fauna of the Upper Cambrian boundary beds (new data from the Asiatic regions of the USSR): Novosibirsk, Academy of Sciences of the USSR, Siberian Division, Transactions of the Institute of Geology and Geophysics, v. 313, 355 p. (in Russian).

Manuscript Accepted by the Society June 18, 1989

ADDENDUM

Between the time this paper was originally submitted (1983) and its going to press, descriptions of trilobites from some of our localities have been published by Soloviev and others (1984) from material collected by Samsonov while accompanying the 1979–1980 Ellsworth Mountains Expedition. Specifically, material from localities YP79-3, 79-8, 79-9, 79-11, and 79-13 has been published by Soloviev and others (1984), who, contrary to this account, recognize both Early and Late Dresbachian assemblages. It is apparent that large collections such as those described herein were not available to the Soviet authors, who have adopted a rather different approach to the systematic palaeontology. In our opinion the preservation and degree of deformation of the material warrants a cautious systematic treatment. Soloviev and others (1984) have classified some of this material to subspecific rank, apparently without taking preservational and deformational differences into account. Furthermore, they have misidentified taxa, thus creating a fictitious and misleading biostratigraphy.

Much of the material described by Soloviev and others (1984) can either be directly synonymized or, where in our opinion erroneously identified, reclassified. Due to poor illustration we are not able to recognize all of the taxa described, as indicated in the suggested correlation of the materials tabulated below.

Soloviev and others (1984)	This Chapter
Hypagnostus cf. *correctus* Öpik	Not recognized.
Ammagnostus integriceps Öpik	*Ammagnostus*? sp.
Agnostoglossa bassa Öpik	*Ammagnostus*? sp.
Homagnostus aff. *fecundus* Pokrovskaya & Ergaliev	*Ammagnostus*? sp.
Idolagnostus dryas Öpik	*Idolagnostus* (*Obelagnostus*) *imitor* subgen. et sp. nov.
Glyptagnostus (*Glyptagnostus*) ?*reticulatus angelini* Resser	*Glyptagnostus reticulatus* (Angelin) *sensu lato*
Pseudagnostus (*Pseudagnostus*) aff. *bulgosus* Öpik	*Pseudagnostus* (*Pseudagnostus*) cf. *idalis* Öpik
Pseudagnostus (*Pseudagnostina*) cf. *contracta* Palmer (pygidia only)	*Idolagnostus* (*Obelagnostus*) *imitor* subgen. et sp. nov.
Norwoodia quadrangularis Palmer *longa* Solovjev	Not recognized.
Coosia cf. *modesta* Lochman	Not recognized.
Coosia coreanica Kobayashi *soholti* Solovjev	*Changshanocephalus*? *suspicor* sp. nov.
Coosia coreanica Kobayashi *sulca* Solovjev	*Changshanocephalus*? *suspicor* sp. nov.
Coosia coreanica Kobayashi *angulata* Solovjev	*Changshanocephalus*? *suspicor* sp. nov.
?*Coosia ignota* Solovjev	Not recognized.
?*Leichneyella* sp. indet.	Not recognized.
Blountia (*Mindycrusta*) *advena Öpikantarctica* Solovjev	Not recognized.
Blountia cf. *glabra* (Walcott)	*Changshanocephalus*? *suspicor* sp. nov.
Blandicephalus heritidji Solovjev	*Changshanocephalus*? *suspicor* sp. nov. (cranidia only)
Aphelaspis (*Labiostria*) *antarctica antarctica* Solovjev	*Aphelaspis* cf. *australis* Henderson
Aphelaspis (*Labiostria*) *antarctica angulata* Solovjev	*Aphelaspis* cf. *australis* Henderson
Aphelaspis (*Proaulacopleura*) *cylindrica* Solovjev	*Aphelaspis* cf. *subdita* Palmer
Aphelaspis (*Proaulacopleura*) cf. *buttsi* (Kobayashi)	*Aphelaspis* cf. *australis* Henderson
Aphelaspis (*Eugonocare*) *soholti* Solovjev	*Aphelaspis* cf. *walcotti* Resser
Aphelaspis (*Eugonocare*) cf. *tesselatum* (Whitehouse)	*Aphelaspis* cf. *lata* Rasetti
Aphelaspis sp. indet. A	Not recognized.
Aphelaspis sp. indet. B.	Not recognized.
Pterocephalia cf. *concava* Palmer	Not recognized.
?*Prismenaspis* sp. indet.	Not recognized.
Densonella sp.	Not recognized.
Rhyssometopus (*Rostrifinis*) cf. *rostrifinis* Öpik	Not recognized.

REFERENCE:

Soloviev, I. A., Popov, L. E., and Samsonov, V. V., 1984, New data on Upper Cambrian fauna of Ellsworth and Pensacola Mountains (West Antarctica): Moscow, USSR Academy of Sciences, The Antarctic, The Committee Reports, v. 23, p. 46–71, 8 pls. (in Russian, but republished in German by the Bundesanstalt für Geowissenschaften und Rohstoffe, Hannover, in 1986).

Printed in U.S.A.

Geological Society of America
Memoir 170
1992

Chapter 9

Cambrian conodonts from the Springer Peak and Minaret Formations, Ellsworth Mountains, West Antarctica

Werner Buggisch
Institut für Geologie, Universität Erlangen, D-8500 Erlangen, Federal Republic of Germany
Gerald F. Webers
Department of Geology, Macalester College, St. Paul, Minnesota 55105
John E. Repetski
U.S. Geological Survey, 970 National Center, Reston, Virginia 22092
Linda Glenister
2030A South Hannibal Way, Aurora, Colorado 80013

ABSTRACT

Cambrian rocks of the Springer Peak and Minaret formations of the upper Heritage Group contain conodont faunas. The upper Middle Cambrian Springer Peak Formation has yielded a conodont cluster referable to *Phakelodus,* and the Upper Cambrian Minaret Formation has yielded conodonts of the genera *Furnishina, Proacodus, Phakelodus,* and *Westergaardodina.* Three species each of *Furnishina* and *Westergaardodina* are known worldwide; other specimens of *Westergaardodina* may represent new species, but low numbers and generally poor preservation preclude reliable diagnosis. All of the identified paraconodont and protoconodont taxa are long ranging, but they are consistent with the Cambrian age of the host rocks as determined by the other contained fossils.

INTRODUCTION

Cambrian conodonts were described first by Müller (1959) from Europe and North America. Since then more than 100 papers concerning or mentioning Cambrian conodonts from Europe, North America, and Asia have been published. However, Cambrian conodonts were unknown from Antarctica until 1981 when Buggisch reported them from the Ellsworth Mountains. Those specimens were found in the Upper Cambrian Minaret Formation. Almost at the same time, Webers, Repetski, and Glenister discovered conodonts in samples collected by Webers and others from the same formation and locality as Buggisch's material. Interestingly, although the conodont faunas originated from the same locality, the single sample collected by Buggisch contained only the genera *Furnishina* and *Proacodus*(?), whereas the other samples contained only *Westergaardodina* and *Phakelodus.*

During the GANOVEX III expedition (1982/83), two research groups collected limestones in the Ross Orogen of northern Victoria Land that yielded Cambrian-Ordovician conodonts. Burrett and Findlay (1984) reported *Proconodontus, Prooneotodus, Prosagittodontus, Iapetognathus,* and *Westergaardodina,* and Wright, Ross, and Repetski (1984) reported species of *"Prooneotodus"* (= *Phakelodus*), *Hirsutodontus,* and *Teridontus.* In 1987, Buggisch and Repetski reported and described these and additional taxa from the same locality. The conodonts of the Ellsworth Mountains are described here. All of the taxa recovered are either protoconodonts (*Phakelodus*) or paraconodonts, and all have relatively long stratigraphic ranges. Discovery and documentation of this sparse fauna broadens the known geographic ranges of these taxa. To date, Cambrian conodonts remain unknown from Africa and South America.

Buggisch, W., Webers, G. F., Repetski, J. E., and Glenister, L., 1992, Cambrian conodonts from the Springer Peak and Minaret Formations, Ellsworth Mountains, West Antarctica, *in* Webers, G. F., Craddock, C., and Splettstoesser, J. F., Geology and Paleontology of the Ellsworth Mountains, West Antarctica: Boulder, Colorado, Geological Society of America Memoir 170.

DESCRIPTION OF LOCALITIES

Two localities—Inferno Ridge and Springer Peak—within the Heritage Range of the Ellsworth Mountains (Fig. 1) have yielded conodonts.

Inferno Ridge (Springer Peak Formation)

Inferno Ridge, in the central Heritage Range, exposes beds of the Springer Peak Formation. This formation (Webers and others, this volume) consists of at least 1,000 m of black shale, deltaic graywacke, and argillite. The thickness and exact stratigraphic position of the beds within the Springer Peak Formation at Inferno Ridge are difficult to estimate because of extensive parasitic folding. The formation is of late Middle Cambrian age, with some possibility that the uppermost beds reach up to the early Late Cambrian. The exposures at Inferno Ridge, from which conodonts were collected, are considered to be of late Middle Cambrian age.

The black shales at Inferno Ridge contain abundant carbonate nodules. These nodules contain a number of complete but highly deformed, unidentifiable trilobites and have yielded a single poorly preserved conodont cluster. Preservation of the material only permits assignment to *Phakelodus?* sp.

Springer Peak (Minaret Formation)

In the northern part of Webers Peaks the limestones of the Minaret Formation thin to a feather edge at Springer Peak (Webers and others, this volume). Here the limestone is about 8 m thick and contains a rich fauna of trilobites and mollusks. Minor elements of the biota include one species of archaeocyath, brachiopods, echinoderm fragments, hyoliths, and algae. The trilobites are dated as Late Cambrian (late Dresbachian; Idamean) by Shergold and Webers (this volume). The presence of oncoids, intraclasts, pelsparites, biosparites, coquinas, and stromatactis is evidence that these limestones represent shallow-water facies (Buggisch and Webers, 1982).

The limestones, which wedge out to the northwest, can be traced with increasing thicknesses in a southerly direction from Springer Peak to the southern Heritage Range, where they are as thick as 700 m in the Marble Hills (Buggisch and Webers, this volume). The conodont-bearing samples were discovered in outcrops east of Springer Peak. Most of the conodonts are highly altered and recrystallized. Nevertheless, some of the specimens, particularly those of *Westergaardodina,* are much better preserved than are specimens of the other taxa, although all elements were collected at the same locality.

DESCRIPTION OF THE CONODONT TAXA

Phylum CONODONTA Pander, 1856
Class CONODONTATA Pander, 1856

Genus *Furnishina* Müller 1959
Type species: *Furnishina furnishi* Müller 1959

***Furnishina asymmetrica?* Müller 1959**
Plate 1, Figs. 8–10

*1959 *Furnishina asymmetrica* n. sp., Müller, p. 451, Plate 11, Figs. 16, 19.
1966 *Furnishina asymmetrica* Müller. Nogami, p. 354, Plate 9, Figs. 1, 2.
1971 *Furnishina asymmetrica* Müller. Müller, p. 14, Plate 1, Figs. 13, 16, Text-fig. 1b.
1973 *Furnishina asymmetrica* Müller. Müller, p. 39, Plate 1, Figs. 6, 8, 9.
1975 *Furnishina asymmetrica* Müller. Lee, p. 79, Plate 1, Fig. 1, Text-fig. 2A.
1976 *Furnishina asymmetrica* Müller. Miller and Paden, p. 595, Plate 1, Figs. 13, 14.
1978 *Furnishina asymmetrica* Müller. Abaimova, p. 78, Plate 7, Fig. 1.
1979 *Furnishina asymmetrica* Müller. Bednarczyk, p. 427, Plate 3, Figs. 17–19, 22, 23.
1980 *Furnishina asymmetrica* Müller. Landing and others, p. 27, Fig. 7j–m.
1982 *Furnishina asymmetrica* Müller? Buggisch, p. 498, Plate 2, Figs. 12–14.
1982 *Furnishina asymmetrica* Müller. An, p. 131, Plate 2, Figs. 10, 11.
1983 *Furnishina asymmetrica* Müller. An and others, p. 98, Plate 2, Figs. 13, 14.

Remarks. The margin of the basal cavity of this specimen is damaged. The entire specimen is highly altered and considerably recrystallized. Nevertheless, the three sharp-edged carinae, the triangular cross section nearly up to the tip, and the distinct asymmetry of the denticle are close to the original description of Müller.
Material. One specimen.
Locality. Minaret Formation (Upper Cambrian) at Springer Peak.
Repository. Geol. Inst. Erlangen: M 6-1.

***Furnishina furnishi* Müller 1959**
Plate 1, Fig. 3–7

*1959 *Furnishina furnishi* n. sp. Müller, p. 452, Plate 11, Figs. 5, 6, 8, 9, 11–15, 17, 18?; Plate 12, Figs. 1, 2, 6D, 6E.
1966 *Furnishina furnishi* Müller. Nogami, p. 354, Plate 9, Figs. 5–7.
1969 *Furnishina furnishi* Müller. Miller, p. 430.
1969 *Furnishina furnishi* Müller. Clark and Robison, p. 1045, Fig. 1b.
1971 *Sagittodontus furnishi* (Müller). Druce and Jones, p. 87, Plate 9, Figs. 1a–4c; Fig. 28c, d.
1971 *Furnishina furnishi* Müller. Müller, Plate 1, Figs. 9, 12, 14, 15.
1973 *Furnishina furnishi* Müller. Müller, p. 39, Plate 1, Figs. 4, 5a, 7, 10.
1975 *Furnishina furnishi* Müller. Abaimova and Ergaliev, p. 391, Plate 14, Figs. 1, 2.
1976 *Furnishina furnishi* Müller. Miller and Paden, p. 595, Plate 1, Figs. 8–12.
1977 *Furnishina furnishi* Müller. Gedik, Plate 2, Fig. 4.
1978 *Furnishina furnishi* Müller. Abaimova, Plate 7, Fig. 2.
1979 *Furnishina furnishi* Müller. Bednarczyk, p. 427, Plate 1, Figs. 2–9, 12; Plate 3, Figs. 14, 20, 21.

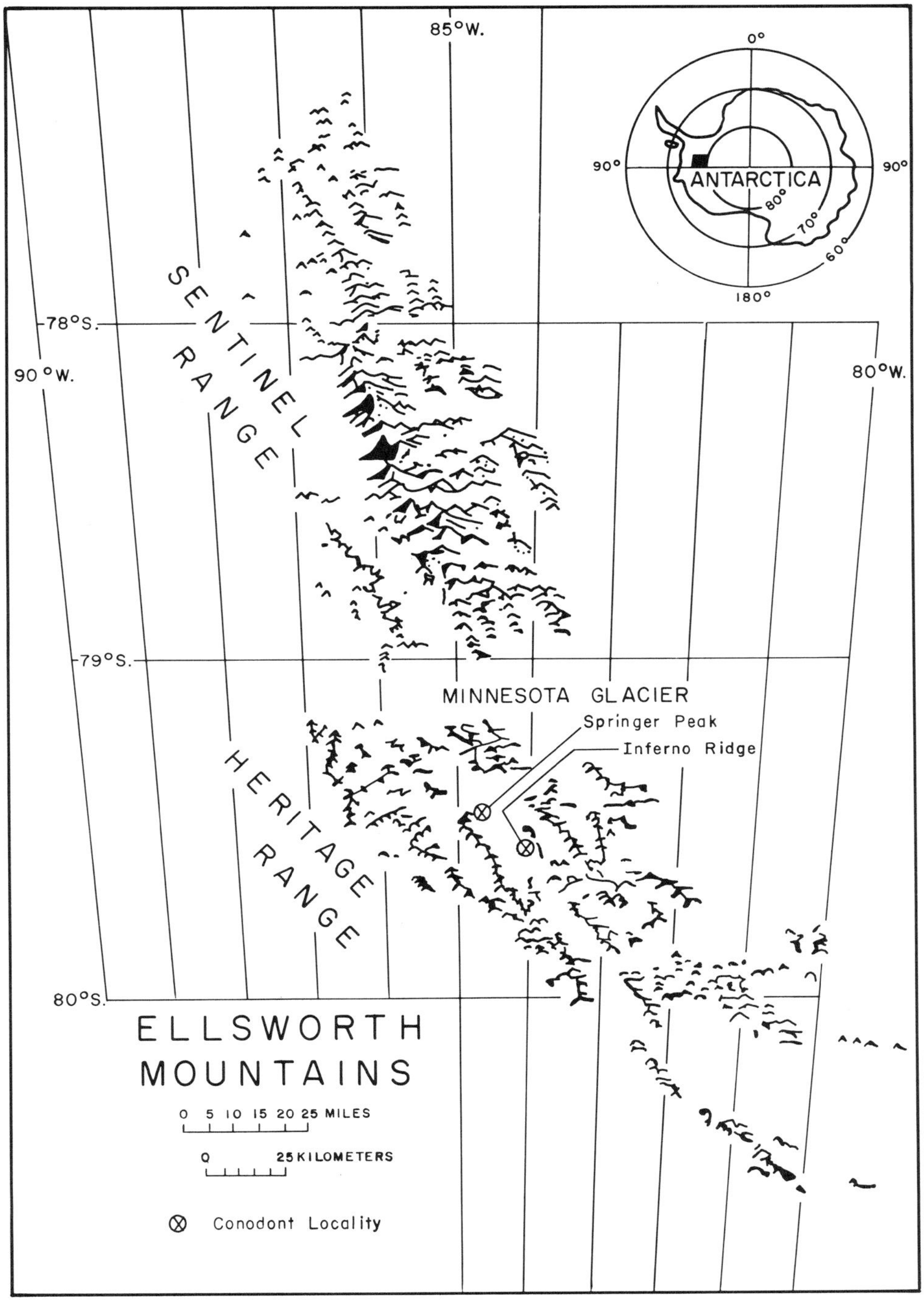

Figure 1. Map showing location of conodont collecting localities at Springer Peak and Inferno Ridge in Heritage Range, Ellsworth Mountains, West Antarctica.

1982 *Furnishina furnishi* Müller. Buggisch, p. 498–499, Plate 2, Figs. 4–6, 9–11, 19–22.
1982 *Furnishina furnishi* Müller. An, Plate 1, Figs. 14, 15; Plate 2, Figs. 3–4, 6–9; Plate 3, Fig. 13.
1983 *Furnishina furnishi* Müller. Wang, Plate 2, Fig. 13.
1983 *Furnishina furnishi* Müller. An and others, [in part], p. 99, Plate 2, Figs. 5, 7, 8, 10 only.

Remarks. *Furnishina furnishi* is the most common species of the genus, and it shows considerable morphologic variation. Elongation of the anterior side (A side *sensu* Müller, 1959) may be diagnostic. The cross section is round at the tip and becomes more or less pronouncedly triangular toward the base. The specimens at hand are highly recrystallized.
Material. Five specimens.
Locality. Minaret Formation (Upper Cambrian) at Springer Peak.
Repository. Geol. Inst. Erlangen: M6-2 through M 6-6.

Furnishina quadrata Müller 1959
Plate 1, Figs. 11–14.

*1959 *Furnishina quadrata* n. sp. Müller, p. 453, Plate 12, Figs. 2, 4, 9, 6c.
1966 *Furnishina quadrata* Müller. Nogami, p. 356, Plate 9, Figs. 3, 4.
1975 *Furnishina quadrata* Müller. Abaimova and Ergaliev, p. 392, Plate 4, Figs. 3, 4, 6, 7.
1979 *Furnishina quadrata* Müller. Bednarczyk, p. 428, Plate 1, Figs. 15, 16.
1978 *Furnishina quadrata* Müller. Landing and others, p. 76, Fig. 2E.
1982 *Furnishina quadrata* Müller. Buggisch, p. 499, Plate 2, Figs. 15–18, 23, 24.

Remarks. The typical subquadrate cross section at the basal margin of these specimens corresponds to the description of Müller (1959). The tip of the altered denticle of the specimens at hand is replaced by a single apatite crystal.
Material. Two specimens.
Locality. Minaret Formation (Upper Cambrian) at Springer Peak.
Repository. Geol. Inst. Erlangen: M 6-7, M 6-8.

Genus _Proacodus_ Müller 1959
Type species: _Proacodus obliquus_ Müller 1959

Proacodus sp.?
Plate 1, Figs. 1, 2

1982 *Proacodus* sp.? Buggisch, p. 499–500, Plate 2, Figs. 1–3.

Remarks. A single denticle with an extremely expanded basal cavity and a weakly undulatory basal margin. Except for the only weakly developed asymmetry of this specimen, it is close to the shape of *Proacodus obliquus* Müller. A firm identification of this specimen is not possible because the upper part of the denticle is completely recrystallized into a single apatite crystal.
Material. One specimen.
Locality. Minaret Formation (Upper Cambrian) at Springer Peak.
Repository. Geol. Inst. Erlangen: M 6-9.

Genus _Phakelodus_ Miller 1984
Type species: _Oneotodus tenuis_ Müller 1959

Phakelodus tenuis (Müller), 1959
Plate 2, Figs. 19, 20, 22

*1959 *Oneotodus tenuis* n. sp. Müller, p. 457, Plate 13, Figs. 13, 14, 20.
1966 *Oneotodus tenuis* Müller. Nogami, p. 356, Plate 9, Figs. 11, 12.
1971 *Oneotodus tenuis* Müller. Druce and Jones, p. 83.
1971 *Oneotodus tenuis* Müller. Müller, Plate 1, Figs. 1, 4–6.
1973 *Prooneotodus tenuis* (Müller). Müller, p. 45, Plate 1, Figs. 1, 3.
1975 *Prooneotodus tenuis* (Müller). Lee, Plate 1, Figs. 14, 15, 17.
1975 *Oneodontus tenuis* (Müller). Abaimova and Ergaliev, p. 394, Plate 14, Figs. 9, 10.
1976 *Prooneotodus tenuis* (Müller). Miller and Paden, Plate 1, Figs. 20–23.
1977 *"Prooneotodus" tenuis* (Müller). Landing, Plate 1, Figs. 1–9, Text-fig. 1.
1978 *Prooneotodus tenuis* (Müller). Abaimova, Plate 8, Figs. 2, 4, 9.
1979 *"Prooneotodus" tenuis* (Müller). Tipnis and others, Plate 1, Fig. 6.
1980 *"Prooneotodus" tenuis* (Müller). Landing and others, Figs. 8M, 8N.
1982 *Prooneotodus*? *tenuis* (Müller). Szaniawski, p. 807, Text-fig. 1.
1982 *Prooneotodus tenuis* (Müller). An, p. 145, Plate 1, Fig. 1.
1983 *Prooneotodus tenuis* (Müller). Wang, Plate 4, Figs. 10, 11.
1983 *Prooneotodus tenuis* (Müller). An and others, p. 130, Plate 5, Figs. 2, 3.

Remarks. The specimens do not differ significantly from the type material of Müller. Longitudinal striation as shown by Müller (1973, Plate Fig. 3a) is not visible, but the outcropping of growth lamellae can be demonstrated on the upper surface of at least one of the specimens from the Ellsworth Mountains (Plate 2, Fig. 22).
Material. Three specimens, more or less complete, plus two fragments possibly of *P. tenuis* elements.
Locality. Minaret Formation (Upper Cambrian) at Springer Peak.
Repository. U.S. National Museum (USNM) 409283, 409284, and 409285.

Phakelodus? sp.
Plate 1, Fig. 15

Remarks. Clusters of several elements that are nearly identical to *Phakelodus tenuis* (Müller) are preserved rather commonly in Upper Cambrian strata. The condition of preservation of the cluster at hand from Inferno Ridge (Middle Cambrian) is not sufficient for determination of a species, but overall element shape and configuration of the cluster suggest *P. tenuis.*
Material. One cluster and some cluster fragments.
Locality. Springer Peak Formation (upper Middle Cambrian) at Inferno Ridge.
Repository. Geol. Inst. Erlangen: M6-10.

Genus _Westergaardodina_ Müller 1959
Type species: _Westergaardodina bicuspidata_ Müller 1959
Westergaardodina bicuspidata Müller 1959
Plate 2, Figs. 1, 2, 13, 14, 17, 18, ?11, ?12

*1959 *Westergaardodina bicuspidata* n. sp. Müller 1959, Figs. 1, 4, 7, 9, 10, 14.
1966 *Westergaardodina bicuspidata* Müller. Hamar, Plate 6, Fig. 1, Text-figs. 2, 3.

1971 *Westergaardodina bicuspidata* Müller. Druce and Jones, Plate 7, Figs. 1–4, Text-fig. 32.
1973 *Westergaardodina bicuspidata* Müller. Müller, Plate 2, Figs. 8, 9.
1978 *Westergaardodina bicuspidata* Müller. Landing and others, Text-fig. 2-c.
1982 *Westergaardodina bicuspidata* Müller. An, Plate 7, Figs. 6–8.
1983 *Westergaardodina bicuspidata* Müller. Wang, Plate 2, Figs. 16–19.
1984 *Westergaardodina bicuspidata* Müller. Borovko and others, p. 57, Plate 1, Fig. 3.

Remarks. In his original description, Müller (1959) considered possession of a distinct median denticle to be a primitive character for this species. All of the specimens from Antarctica assigned to this species show a more or less distinct median denticle. The condition of the material studied varies from well preserved to almost completely recrystallized. In addition, one specimen (Plate 2, Figs. 11, 12) could be identified only tentatively because of poor preservational state.
Material. Four specimens, including one assigned with query.
Locality. Minaret Formation (Upper Cambrian) at Springer Peak.
Repository. USNM 409286, 409287, 409288, and 409289.

Westergaardodina moessebergensis Müller 1959
Plate 2, Figs. 3, 4, ?7

*1959 *Westergaardodina moessebergensis* n. sp. Müller, Plate 14, Figs. 11, 12, 15.
1966 *Westergaardodina moessebergensis* Müller. Nogami, Plate 10, Figs. 1, 2.
1971 *Westergaardodina moessebergensis* Müller. Müller, Plate 2, Fig. 5.
1973 *Westergaardodina moessebergensis* Müller. Müller, Plate 2, Fig. 6.
1984 *Westergaardodina moessebergensis* Müller. Borovko and others, p. 57, Plate 1, Fig. 1.

Remarks. The bicuspidate Antarctic specimens are like the type material of the species described from Vastergotland, Sweden. The Antarctic specimens show a reduced but distinct median denticle that is observed only rarely in a few specimens of the original material of *W. moessebergensis.* The large basal cavity that continues from one lateral denticle to the other is diagnostic for this species.
Material. Two specimens, one fragmentary and identified only with question.
Locality. Minaret Formation (Upper Cambrian) at Springer Peak.
Repository. USNM 409290, 409291.

Westergaardodina tricuspidata Müller 1959
Plate 2, Figs. 9, 10, ?8, ?21

*1959 *Westergaardodina tricuspidata* n. sp. Müller, Plate 15, Figs. 3, 5–6.
1966 ? *Westergaardodina tricuspidata* Müller. Nogami, Plate 10, Fig. 5.
1971 *Westergaardodina tricuspidata* Müller. Müller, Plate 2, Fig. 12.
1982 *Westergaardodina tricuspidata* Müller. An, Plate 7, Fig. 10; Plate 8, Figs. 6, 9.

Remarks. The middle denticle of this W-shaped *Westergaardodina* is strongly recrystallized and is broken distally. Nevertheless, a sharp posterior edge is well preserved. The lateral cavities are large; they extend the distal two-thirds of the length of the outer denticles and reach the tips.

Two of the specimens (Plate 2, Figs. 8, 21) are preserved too poorly for positive identification but appear to have the features diagnostic for *W. tricuspidata.*
Material. Three specimens.
Locality. Minaret Formation (Upper Cambrian) at Springer Peak.
Repository. USNM 409292, 409293, 409294.

Westergaardodina n. sp. A
Plate 2, Figs. 15, 16

Remarks. The contour of this new species is similar to that of *W. bicuspidata.* A small median denticle is developed, comparable to the primitive forms of *W. bicuspidata* in our Antarctic material. In contrast to *W. bicuspidata,* however, the basal cavity of the new species is very large and extends the entire length of the outer margin of the element as does the cavity for *W. moessebergensis.* The specimen at hand has a more basally pointed basal margin than is typical for *W. moessebergensis*; the basal margin of the latter is more uniformly curved and U-shaped.
Material. One specimen.
Locality. Minaret Formation (Upper Cambrian) at Springer Peak.
Repository. USNM 409295.

Westergaardodina n. sp. B
Plate 2, Figs. 5, 6

Remarks. This new species shows some similarity to *W. tricuspidata.* The element is symmetrical and has three distinct denticles; a sharp edge is developed along the posterior of the median denticle (compare with *W. tricuspidata,* Plate 2, Fig. 10). In contrast to *W. tricuspidata,* the new species has a very large basal cavity that extends along the entire outer margin.
Material. One specimen.
Locality. Minaret Formation (Upper Cambrian) at Springer Peak.
Repository. UNSM 409296.

PLATE 1

Figures 1, 2. *Proacodus* sp.? Lateral view (1) and lower side (2), Springer Peak, ×100. Geol. Inst. Erlangen M6-9.

Figures 3–5. *Furnishina furnishi* MÜLLER. Lateral view (3), lower side (4), B-side oblique from below (5), Springer Peak, ×100. Geol. Inst. Erlangen M6-2.

Figure 6. *Furnishina furnishi* MÜLLER. B-side oblique from below, Springer Peak, ×100, Geol. Inst. Erlangen M6-3.

Figure 7. *Furnishina furnishi* MÜLLER. Lateral view, Springer Peak, ×100, Geol. Inst. Erlangen M 6-4.

Figures 8–10. *Furnishina asymmetrica* MÜLLER. Lateral view (8), upper view (9), basal view (10), Springer Peak, ×100, Geol. Inst. Erlangen M 6-1.

Figures 11, 12. *Furnishina quadrata* MÜLLER. Lateral view (11), basal view (12), Springer Peak, ×100, Geol. Inst. Erlangen M 6-7.

Figures 13, 14. *Furnishina quadrata* MÜLLER. Upper view (13), lateral view (14), Springer Peak, ×100, Geol. Inst. Erlangen M 6-8.

Figure 15. Cluster of *Phakelodus*? sp. Inferno Ridge, ×75, Geol. Inst. Erlangen M 6-10.

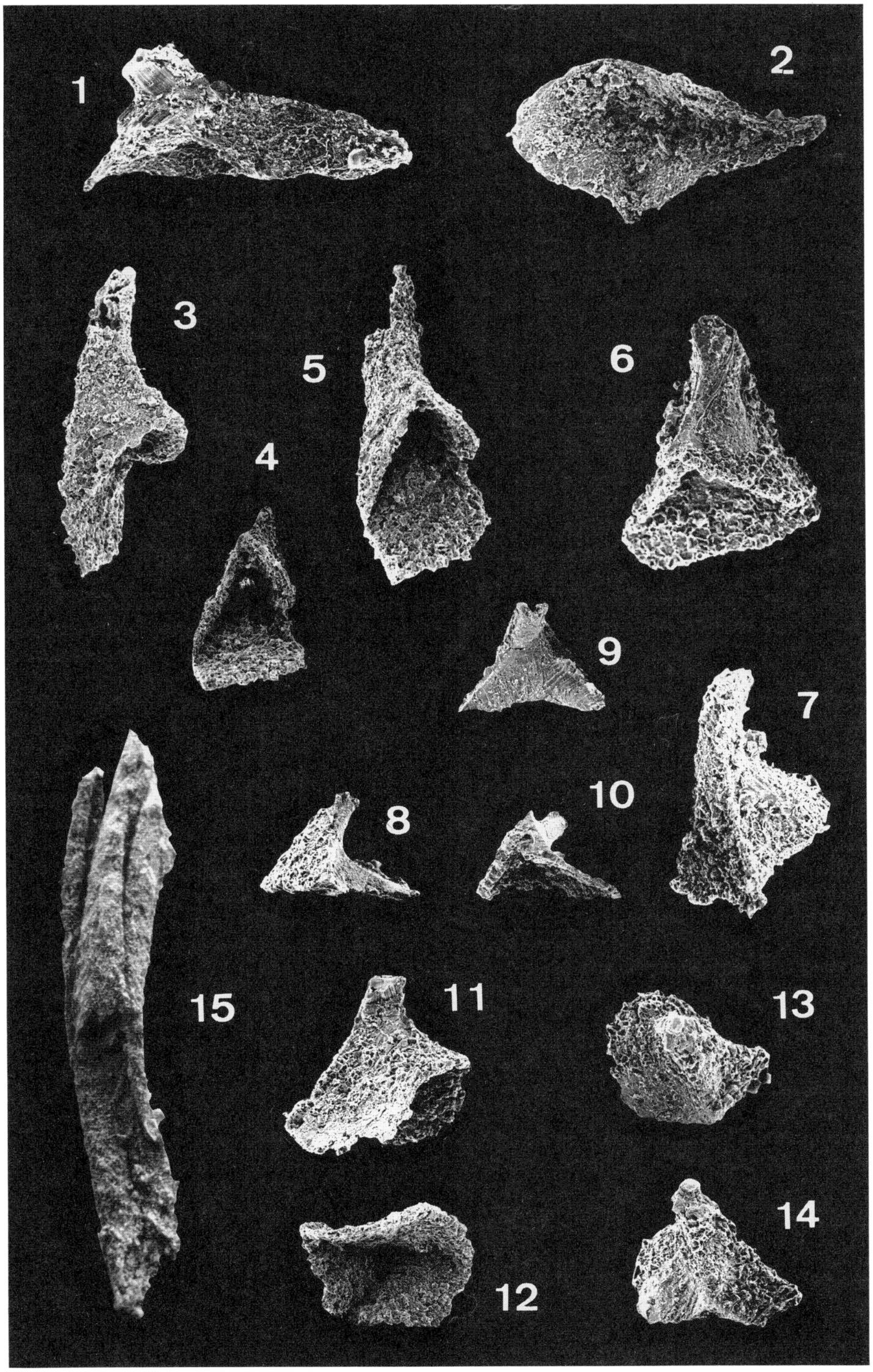
1
2
3
5
6
4
9
7
10
8
15
11
13
12
14

PLATE 2

All specimens from Springer Peak locality. Elements photographed using SEM and carbon coating technique of Repetski and Brown (1982); collection reposited in collections of Paleobiology Department, U.S. National Museum of Natural History, Washington, D.C., U.S.A.

Figures 1, 2, 13, 14, 17, 18. *Westergaardodina bicuspidata* MÜLLER. Anterior (1, 13, 17) and posterior (2, 14, 18) views of three specimens; USNM 409286, ×43, USNM 409287, ×38, and USNM 409288, ×23, respectively.

Figures 3, 4. *Westergaardodina moessebergensis* MÜLLER. Posterior (3) and anterior (4) views of specimen USNM 409290, ×25.

Figures 5, 6. *Westergaardodina* n. sp. B. Anterior (5) and posterior (6) views of specimen USNM 409296, ×40.

Figure 7. *Westergaardodina moessebergensis* MÜLLER? Posterior view of fragmentary specimen USNM 409291, ×35.

Figures 8, 21. *Westergaardodina tricuspidata* MÜLLER? Posterior view of poorly preserved and broken elements; 8: USNM 409293, ×33; 21: USNM 409294, ×50.

Figures 9, 10. *Westergaardodina tricuspidata* MÜLLER. Anterior and posterior views, respectively, of specimen USNM 409292, ×43.

Figures 11, 12. *Westergaardodina bicuspidata* MÜLLER? Posterior and anterior views, respectively, of specimen USNM 409289, ×58.

Figures 15, 16. *Westergaardodina* n. sp. A. Posterior and anterior views, respectively, of specimen USNM 409295, ×33.

Figures 19, 20, 22. *Phakelodus tenuis* (MÜLLER). Left (19, 20) and right (22) lateral views of USNM 409283, ×38, USNM 409284, ×30, and USNM 409285, ×30, respectively.

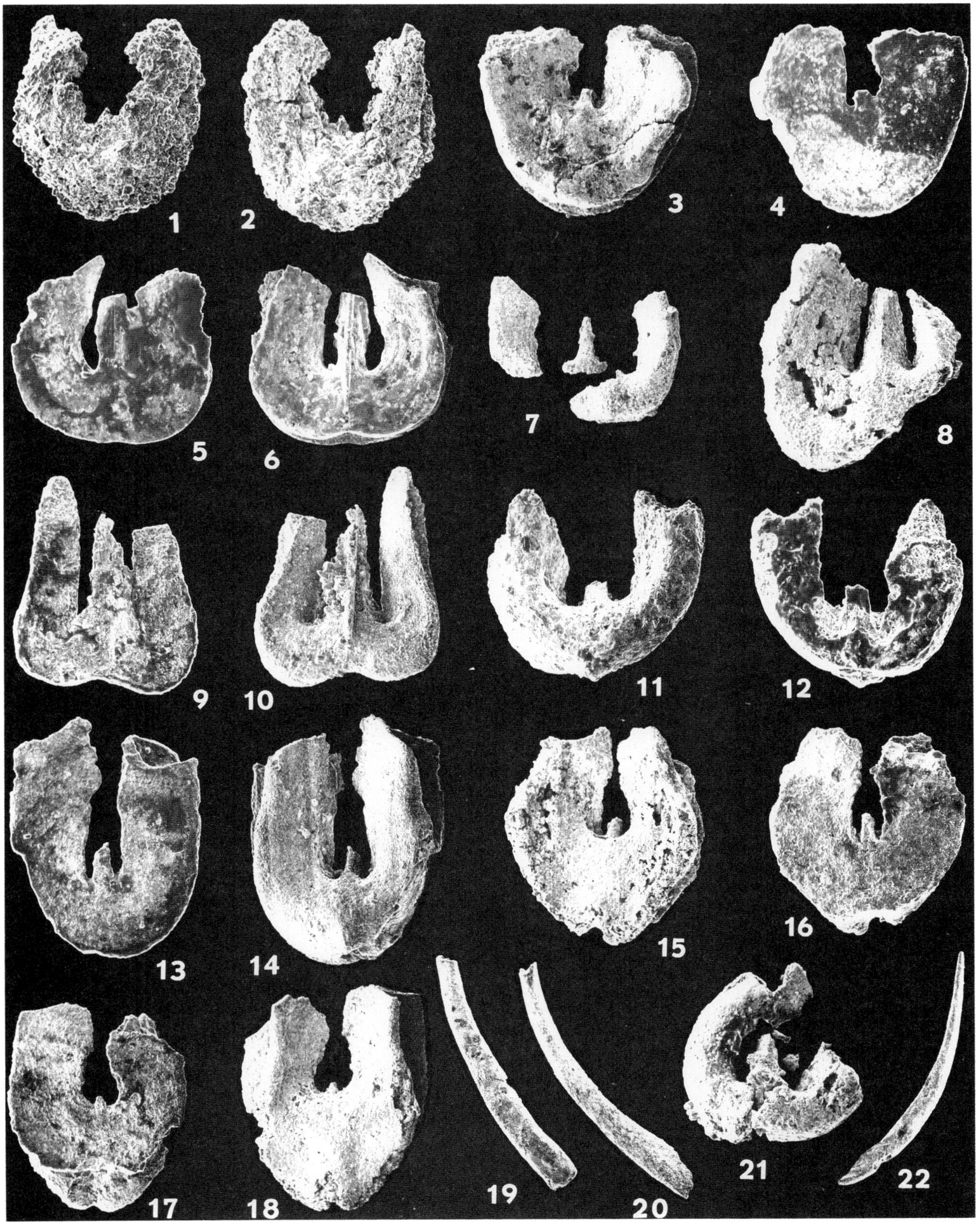
1
2
3
4
5
6
7
8
9
10
11
12
13
14
15
16
17
18
19
20
21
22

REFERENCES CITED

Abaimova, G. P., 1978, First Cambrian conodonts from the central region of Kazakhstan: Paleontological Journal, v. 1978, no. 4, p. 77–87.

——, 1980, Apparati kembrijskich konodontov iz Kazakhstana: Paleont. Zhurnal. Akad. Nauk. SSSR, v. 1980, p. 143–146.

Abaimova, G. P., and Ergaliev, G. K., 1975, The discovery of conodonts in the Middle and Upper Cambrian of the Lesser Karatau: Akademii Nauk SSSR, Sibir, Otdel., Inst. Geologii i Geofiz., Trudy no. 333, Moskva, p. 390–394, 399–400, p. 14.

An, T., 1982, Study on the Cambrian conodonts from north and northeast China: Sci. Rep., Inst. Geosci., Univ. Tsukuba, Sec. B., v. 3, p. 113–159.

An, T., and Yang, C., 1980, Cambro-Ordovician conodonts in China, with special reference to the boundary between the Cambrian and Ordovician Systems, *in* Science Papers on Geology for International Exchange: Paris, 26th International Geological Congress, v. 4, p. 14–23.

An, T., and 10 others, 1983, The conodonts of North China and the adjacent regions: Beijing, Science Press, 223 p.

Azmi, R. J., 1980, Cambrian conodonts from the Parahio Valley of Spiti, Himchal Pradesh [abs.]: Him. Geo. Se., Delhi, Abstr. Programme, p. 4.

Azmi, R. J., Joshi, M. N., and Juyal, K. P., 1980, Discovery of the Cambro-Ordovician conodonts from the Mussoorie Tal Phosphorite and its significance in correlation of the Lesser Himalaya [abs.]: Him. Geol. Sem, Delhi, Abstr. Programme, p. 4–5.

Barnes, C. R., Rexroad, C. B., and Miller, J. F., 1973, Lower Paleozoic conodont provincialism, *in* Rhodes, F.H.T., ed., Conodont paleozoology: Geological Society of America Special Paper 141, p. 157–190.

Bednarczyk, W., 1979, Upper Cambrian to Lower Ordovician conodonts of Leba Elevation, NW Poland, and their stratigraphic significance: Acta geol. Polonica, v. 29, p. 409–442.

Bengtson, S., 1976, The structure of some Middle Cambrian conodonts and the early evolution of conodont structure and function: Lethaia, v. 9, p. 185–206.

Borovko, N. G., and six others, 1984, Opornyii razres pogranichnykh otlozhenii Kembrijai i Ordovika severo-zapada Russkoi platy (r. izhora): Moscova, Izvestija Akademii Nauk SSSR, Seria Geologicheskaja.

Buggisch, W., 1981, Erste paläozoische Conodonten aus der Antarktis (Oberkambrium: Ellsworth Mountains) [abs.]: Jahresversammlung Programm u. Exkursionsführer, Erlangen, 28–30 September 1981, Paläontologische Gesellschaft, p. 51.

——, 1982, Conodonten aus den Ellsworth Mountains (Oberkambrium/Westantarktis): Zeitschrift deutsch. geol. Gesellschaft, v. 133, p. 493–507.

Buggisch, W., and Webers, G. F., 1982, Zur Facies der Karbonatgesteine in den Ellsworth Mountains (Paläozoikum, Westantarktis): Facies, v. 7, p. 199–228.

Burrett, C. F., and Findlay, R. H., 1984, Cambrian and Ordovician conodonts from the Robertson Bay Group, Antarctica, and their tectonic significance: Nature, v. 307, p. 723–726.

Clark, D. L., and Miller, J. F., 1969, Early evolution of conodonts: Geological Society of America Bulletin, v. 80, p. 125–134.

Clark, D. L., and Robison, R. A., 1969, Oldest conodonts of North America: Journal of Paleontology, v. 43, p. 1044–1046.

Craddock, C., 1969, Geology of the Ellsworth Mountains, *in* Bushnell, V. C., and Craddock, C., eds., Geologic maps of Antarctica: New York, American Geographical Society Antarctic Map Folio Series, Folio 12, Plate 4, scale 1:1,000,000.

Craddock, C., Bastien, T. W., Rutford, R. H., and Anderson, J. J., 1955, Glossopteris discovered in West Antarctica: Science, v. 148, p. 634–637.

Druce, E. C., and Jones, P. J., 1968, Stratigraphical significance of conodonts in the Upper Cambrian and Lower Ordovician sequences of the Boulia Region, western Queensland: Australian Journal of Science, v. 31, p. 88.

——, 1971, Cambro-Ordovician conodonts from the Burke River structural belt, Queensland: Australian Bureau of Mineral Resources Geological and Geophysical Bulletin, v. 110, p. 1–158.

Dutro, J. T., Jr., Palmer, A. R., Repetski, J. E., and Brosge, W. P., 1984, Middle Cambrian fossils from the Doonerak anticlinorium, central Brooks Range, Alaska: Journal of Paleontology, v. 48, p. 1364–1371.

Fahraeus, L. E., and Nowlan, G. S., 1978, Franconian (Late Cambrian) to early Champlainian (Middle Ordovician) conodonts from the Cow Head Group, western Newfoundland: Journal of Paleontology, v. 52, p. 444–471.

Gedik, I., 1977, Conodont biostratigraphy in the Middle Taurus: Bulletin of the Geological Society of Turkey, v. 20, p. 35–48.

Goodwin, P. W., 1961, Late Cambrian and Early Ordovician conodonts from the Bighorn Mountains, Wyoming, *in* Geological Society of America abstracts for 1961: Geological Society of America Special Paper 68, p. 183.

Grant, R. E., 1965, Faunas and stratigraphy of the Snowy Range Formation (Upper Cambrian) in southwestern Montana and northwestern Wyoming: Geological Society of America Memoir 96, 171 p.

Hamar, G., 1966, The Middle Ordovician of the Oslo region, Norway, 22nd preliminary report on conodonts from the Oslo-Asker and Ringerike districts: Norsk Geologisk Tidsskrift, v. 46, p. 27–83.

Jones, P. J., 1961, Discovery of conodonts in the Upper Cambrian of Queensland: Australian Journal of Science, v. 24, p. 143.

Jones, P. J., Shergold, J. H., and Druce, E. C., 1971, Late Cambrian and Early Ordovician stages in western Queensland: Geological Society of Australia Journal, v. 18, p. 1–32.

Koucky, F. L., Cygan, N. C., and Rhodes, F.H.T., 1961, Conodonts from the eastern flank of the central part of the Bighorn Mountains, Wyoming: Journal of Paleontology, v. 35, p. 877–879.

Koucky, F. L., Rhodes, F.H.T., and Chung, H. H., 1967, Upper Cambrian–Lower Ordovician conodont sequences of the Bighorn Mountains, Wyoming: Geological Society of America 1967 Annual Meeting Program, p. 123–124.

Kurtz, V. E., 1978, Late Cambrian and Early Ordovician conodont biostratigraphy from the Bighorn Mountains, Wyoming and Montana: Geological Society of America Abstracts with Programs, v. 10, p. 219.

——, 1980, Conodonts from the upper Eminence and lower Gasconade Formations and their bearing on the position of the Cambrian-Ordovician boundary in Missouri: Geological Society of America Abstracts with Programs, v. 12, p. 232.

Landing, E., 1974, Early and Middle Cambrian conodonts from the Taconic Allochthon, eastern New York: Journal of Paleontology, v. 48, p. 1241–1248.

——, 1977, *"Prooneotodus" tenuis* (Müller, 1959) apparatuese from the Taconic Allochthon, eastern New York; Construction, taphonomy and the protoconodont "supertooth" model: Journal of Paleontology, v. 51, p. 1072–1084.

——, 1979, Conodonts and biostratigraphy of the Hoyt limestone (Late Cambrian, Tremealeauan), eastern New York: Journal of Paleontology, v. 53, p. 1023–1029.

——, 1980, Late Cambrian–Early Ordovician macrofaunas and phosphatic microfaunas, St. John Group, New Brunswick: Journal of Paleontology, v. 54, p. 752–761.

Landing, E., Taylor, M. E., and Erdtmann, B. D., 1978, Correlation of the Cambrian-Ordovician boundary between the Acado-Baltic and the North American faunal provinces: Geology, v. 6, p. 75–78.

Landing, E., Ludvigsen, R., and von Bitter, P. H., 1980, Upper Cambrian to Lower Ordovician conodont biostratigraphy and biofacies, Rabbitkettle Formation, District of Mackenzie: Royal Ontario Museum Life Science Contribution 126, p. 1–42.

Lee, H.-Y., 1975, Conodonts from the Upper Cambrian formations, Kangweon-Do, South Korea, and its stratigraphical significance: Yonsei-Nonchong, v. 12, p. 71–89.

Lochmann, C., 1964, Upper Cambrian faunas from the subsurface Deadwood Formation, Williston basin, Montana: Journal of Paleontology, v. 38, p. 33–60.

Miller, J. F., 1969, Conodont fauna of the Notch Peak Limestone (Cambro-Ordovician), House Range, Utah: Journal of Paleontology, v. 43, p. 413–439.

——, 1970, Conodont zonation of the uppermost Cambrian and lowest Ordovi-

cian: Geological Society of America Abstracts with Programs, v. 7, p. 624.
——, 1973, Natural conodont assemblages from the Upper Cambrian of Warwickshire, Great Britain: Geological Society of America Abstracts with Programs, v. 5, p. 338–339.
——, 1975, Conodont faunas from the Cambrian and lowest Ordovician of western America: Geological Society of America Abstracts with Programs, v. 7, p. 1200–1201.
——, 1977, Conodont biostratigraphy and intercontinental correlations across the Cambrian-Ordovician Boundary: Sydney, 25th International Geological Congress, v. 1, p. 274.
——, 1980, Taxonomic revision of some Upper Cambrian and Lower Ordovician conodonts with comments on their evolution: Lawrence, University of Kansas Paleontological Contribution 99, p. 1–39.
——, 1984, Cambrian and earliest Ordovician conodont evolution, biofacies, and provincialism, *in* Clark, D. L., ed., Conodont biofacies and provincialism: Geological Society of America Special Paper 196, p. 43–68.
Miller, J. F., and Kurtz, V. E., 1975, Evolution and biostratigraphy of the conodont family Clavohamulidae Lindstrom, 1970, Cambro-Ordovician: Geological Society of America Abstracts with Programs, v. 7, p. 823.
——, 1979, Reassignment of the Dolomite Point Formation of east Greenland from the Middle Cambrian to the Lower Ordovician based on conodonts: Geological Society of America Abstracts with Programs, v. 11, p. 480.
Miller, J. F., and Melby, J. H., 1971, Trempealeauan conodonts, *in* Clark, D. L., ed., Conodonts and biostratigraphy of the Wisconsin Paleozoic: Wisconsin Geological Survey Information Circular 19, p. 4–9.
Miller, R., and Paden, E. A., 1976, Upper Cambrian stratigraphy and conodonts from eastern California: Journal of Paleontology, v. 50, p. 590–597.
Müller, K. J., 1959, Kambrische Conodonten: Z. dt. geol. Ges., v. 111, p. 434–485.
——, 1971, Cambrian conodont faunas, *in* Sweet, W. C., and Bergstrom, S. M., eds., Symposium on conodont biostratigraphy: Geological Society of America Memoir 127, p. 5–20.
——, 1973, Late Cambrian and Early Ordovician conodonts from northern Iran: Geological Survey of Iran Report 30, 77 p.
Müller, K. J., and Andres, D., 1976, Eine Conodontengruppe von *Prooneotodus tenuis* (Muller, 1959) in naturlichem Zusammenhang aus dem overen Kambrium von Schweden: Palaont. Z., v. 50, p. 193–200.
Müller, K. J., and Nogami, Y., 1971, Uber den Feinbau der Conodonten: Mem. Fac. Sci., Kyoto Univ. Ser. Geol. Min. v. 38, p. 1–87.
Nogami, Y., 1966, Kambrische Conodonten von China; Teil 1, Conodonten aus den overkambrischen Kushan-Schichten: Mem. College Sci., Univ. Kyoto, Ser. B., v. 32, p. 351–366.
——, 1967, Kambrische Conodonten von China; Teil 2, Conodonten aus den hoch oberkamberischen Kushan-Schichten: Mem. College Sci., Univ. Kyoto, Ser. B., v. 33, p. 211–218.
Nowlan, G. S., and Barnes, C. R., 1976, Late Cambrian and Early Ordovician conodont biostratigraphy of the eastern Canadian Arctic Islands: Geological Society of America Abstracts with Programs, v. 8, p. 501–502.
Ozgul, N., and Gedik, I., 1973, New data on the conodont faunas of the Caltepe Limestone and Seydisehir Formation, lower Paleozoic of central Tauras Range (Turkey): Geological Society of Turkey Bulletin, v. 16, p. 39–52.
Pantoja-Alor, J., and Robinson, R. A., 1967, Paleozoic sedimentary rocks in Oaxaca, Mexico: Science, v. 157, p. 1033–1035.
Poulsen, V., 1966, Early Cambrian distacodontid conodonts from Bornholm: Biol. Medd. Danske Vidensk. Selsk., v. 23, p. 1–12.
Repetski, J. E., and Brown, W. R., 1982, On illustrating conodont type specimens using the scanning electron microscope; New techniques and a recommendation: Journal of Paleontology, v. 56, p. 908–911.
Szaniawski, H., 1971, New species of Upper Cambrian conodonts from Poland: Acta Palaeont. Polonica, v. 16, p. 401–412.
——, 1982, Chaetognath grasping spines recognized among Cambrian protoconodonts: Journal of Paleontology, v. 56, p. 806–810.
Tipnis, R. S., and Cecile, M. P., 1980, Early Paleozoic conodont biostratigraphy and paleogeography of northwestern Canada [abs.]: American Association of Petroleum Geologists Bulletin, v. 64, p. 793.
Tipnis, R. S., Chatterton, B.D.E., and Ludvigsen, R., 1979, Ordovician conodont biostratigraphy of the southern district of Mackenzie, Canada: Geological Association of Canada Special Paper 18, p. 39–91.
van Wamel, W. A., 1974, Conodont biostratigraphy of the Upper Cambrian and Lower Ordovician of northwestern Oland, south-eastern Sweden: Utrecht Micropaleontological Bulletin, v. 10, p. 1–22.
Wang, Z., 1983, Outline of uppermost Cambrian and lowermost Ordovician conodonts in north and northeast China with some suggestions to the Cambrian-Ordovician boundary, *in* Papers for the symposium on the Cambrian-Ordovician and Ordovician-Silurian boundaries, Nanjing, China, October 1983: Nanjing Institute of Geology and Palaeontology, Academia Sinica, p. 31–39, 158–161.
Webers, G. F., 1972, Unusual Upper Cambrian fauna from West Antarctica, *in* Adie, R., ed., Antarctica geology and geophysics: Oslo, Universitetsforlaget, p. 235–237.
Wright, T. O., Ross, R. J., Jr., and Repetski, J. E., 1984, Newly discovered youngest Cambrian or oldest Ordovician fossils from the Robertson Bay terrane (formerly Precambrian), northern Victoria Land, Antarctica: Geology, v. 12, p. 301–305.

Manuscript Accepted by the Society March 1, 1989

Printed in U.S.A.

Geological Society of America
Memoir 170
1992

Chapter 10

Cambrian mollusca from the Minaret Formation, Ellsworth Mountains, West Antarctica

Gerald F. Webers
Department of Geology, Macalester College, St. Paul, Minnesota 55105
John Pojeta, Jr., and Ellis L. Yochelson
U.S. Geological Survey, 970 National Center, Reston, Virginia 22092

ABSTRACT

Cambrian mollusks are known from four localities in the limestone of the Minaret Formation, Heritage Range, Ellsworth Mountains, West Antarctica. The most diverse and best-preserved specimens are from the coquina at the feather edge of the Minaret Formation, on the northeastern side of Springer Peak, Webers Peaks. This locality has provided one of the finest Upper Cambrian mollusk faunas in the world. The mollusks indicate a Dresbachian to Franconian age. The trilobites associated with the mollusks define the age of the rocks at Springer Peak as late Dresbachian (Idamean). These rocks were first thought to be Precambrian in age.

From the four localities, 19 genera (4 new) and 20 species (12 new) are described; there are 7 species of monoplacophorans placed in 7 genera, 6 species of gastropods placed in 6 genera, 3 species of hyoliths placed in 3 genera, and 3 species of rostroconchs placed in 3 genera. One calcareous tubular organism is described under the hyoliths as Orthothecida? species indeterminate. The higher taxa are presented in the order of decreasing abundance of specimens in the coquina at Springer Peak. Mollusks make up about 5 percent of the coquina, which at this locality is as much as 8 m thick. The remainder of the coquina is almost entirely trilobite fragments; minor elements of the biota are archaeocyaths, inarticulate and articulate brachiopods, echinoderm fragments, conodonts, and algae. The fossiliferous beds at Springer Peak are interpreted as having been deposited in a medium- to high-energy, nearshore environment under normal marine conditions. The less fossiliferous limestone beds above and below the coquina are laminated, and some contain pisoliths; this evidence of algal activity suggests a low-energy environment.

Some of the Upper Cambrian species of mollusks found at Springer Peak occur farther south in the Minaret Formation limestones at Bingham Peak and Yochelson Ridge. One species of helcionellacean mollusk, not found elsewhere, was recovered from the Minaret Formation at its type locality in the Marble Hills. This species is classified as *Latouchella*? species indeterminate; it shows that the Minaret Formation is Cambrian in age throughout its thickness and outcrop area. Various of the genera of mollusks known from the Minaret Formation, as well as one species, are geographically widespread in rocks of Late Cambrian age in Australia, northeastern China, and the upper Mississippi

Webers, G. F., Pojeta, J., Jr., and Yochelson, E. L., 1992, Cambrian mollusca from the Minaret Formation, Ellsworth Mountains, West Antarctica, *in* Webers, G. F., Craddock, C., and Splettstoesser, J. F., Geology and Paleontology of the Ellsworth Mountains, West Antarctica: Boulder, Colorado, Geological Society of America Memoir 170.

River Valley and Ozark Dome regions of the United States. Geographic distributions are discussed under each taxon.

The new taxa of mollusks are: (1) monoplacophorans—*Cosminoconella runnegari* n. gen., n. sp.; *Ellsworthoconus andersoni* n. gen., n. sp.; *Kirengella pyramidalis* n. sp.; *Proconus incertis* n. gen., n. sp.; and *Proplina rutfordi* n. sp.; (2) gastropods—*Aremellia batteni* n. gen., n. sp.; *Euomphalopsis splettstoesseri* n. sp.; *Kobayashiella? heritagensis* n. sp.; *"Maclurites" thomsoni* n. sp.; and *Matherella antarctica* n. sp.; (3) hyoliths—*Linevitus? springerensis* n. sp.; and (4) rostroconchs—*Apoptopegma craddocki* n. sp.

INTRODUCTION

In 1963, G. F. Webers, as a member of the University of Minnesota Expedition to the Ellsworth Mountains, found a trilobite-rich coquina at Springer Peak near the northern end of the Heritage Range (Fig. 1; Plate 19, Fig. 2). The fauna from this coquina became the key to dating the limestones of the Heritage Range. Robert Rutford, University of Texas at Dallas, collected additional material from the Springer Peak locality in 1964. Recollection and description of the Springer Peak fauna were two of the goals of the 1979–80 U.S. Macalester College expedition to the Ellsworth Mountains. The Upper Cambrian mollusk fauna from the Ellsworth Mountains is preserved as well as, or better than, those described from Australia (Pojeta and others, 1977), China (Chen and Teichert, 1983), and the upper Mississippi Valley, U.S. (Berkey, 1898).

Yochelson and others (1973) described *Knightoconus antarcticus* from Springer Peak, but they did not consider in any detail other mollusks known to be present. Preliminary reports about mollusks from Springer Peak, other than *Knightoconus antarcticus,* were published by Webers (1972, 1977, 1982), Pojeta and others (1981), and Webers and others (1982). The information, conclusions, and taxonomic assignments given here supersede those preliminary reports. Our conclusion on the age of the Springer Peak fauna has remained the same as that in earlier reports, but some generic assignments have changed. Our understanding of paleobiogeographic distributions has improved considerably. Most of the specimens illustrated came from Springer Peak, and a few came from less mollusk rich outcrops at Yochelson Ridge and near Minaret Peak; no specimens are illustrated of the mollusks from Bingham Peak (Fig. 1; Plate 19).

This chapter contains descriptions of hyoliths, which are here included in the Mollusca as a matter of convenience and compromise. Pojeta regards the Hyolitha as a separate phylum only distantly related to Mollusca (Runnegar and others, 1975; Pojeta, 1989). As a further compromise, a helcionellacean, *Latouchella*? species indeterminate, is assigned to the Monoplacophora. Yochelson (1978) did not consider helcionellaceans to be monoplacophorans. Proposed subphylum divisions of Mollusca (Runnegar and Pojeta, 1974) are omitted by preference of Yochelson. Other differences in interpretation among the authors are mentioned in the appropriate places in the body of the text. Some poorly preserved calcareous tubes from the Minaret Formation at Mount Rosenthal (Fig. 1), Liberty Hills, are described as Orthothecida? species indeterminate under Hyolitha, because we could think of nowhere else to describe them; we do not regard them as hyoliths.

All three of us participated in the U.S./Macalester College 1979–80 expedition, which was organized by G. F. Webers, and we collected most of the specimens described here. We have all analyzed the specimens. In the main, Webers contributed the mechanical preparation, sorting, and descriptions of the material collected in 1963 and 1964. Most preparation of the 1979–80 collections and all photography of the specimens and outcrops were done by Pojeta. He also did the heating and quenching preparation of the specimens described below. Pojeta prepared the plates and he prepared drafts of the plate explanations, the rostroconch section, and the description of *Latouchella*? species indeterminate. Yochelson prepared drafts of the sections of hyoliths, gastropods, and monoplacophorans. The writing of the introductory sections was a joint effort. All of us examined the fossils in detail and are jointly responsible for the lower-level taxonomic descriptions and for the new taxa.

STRATIGRAPHY

The lithostratigraphy of the Minaret Formation is described elsewhere in this volume by Webers and others. Here we give a capsule summary of the lithostratigraphy and a stratigraphic column for the Ellsworth Mountains (Fig. 2). The Minaret Formation is largely a limestone unit deposited under normal marine conditions in low- to high-energy nearshore regimes. The formation is thickest in the Marble Hills at and south of, Minaret Peak (Fig. 1; Plate 19, Fig. 6), where as much as 600 m is exposed; it thins to the north and northeast. Field data suggest that north of the Marble Hills the lower part of the limestone is the most fossiliferous. At Yochelson Ridge, it has thinned to 85 m (Fig. 1; Plate 19, Fig. 5). At Springer Peak, in about 1.5 km, the Minaret thins from slightly more than 20 m thick at the southern exposure (Fig. 1; Plate 19, Fig. 1) to less than 8 m at the northernmost feather-edge exposure (Plate 19, Fig. 2). Most of the fossils described here were obtained from this 8-m-thick coquina.

The unusually good preservation of the fauna at the Springer Peak locality may be due, in part, to the thin feather-edge limestone of the Minaret Formation lying between two incompetent argillite units, which probably absorbed most of the tectonic and

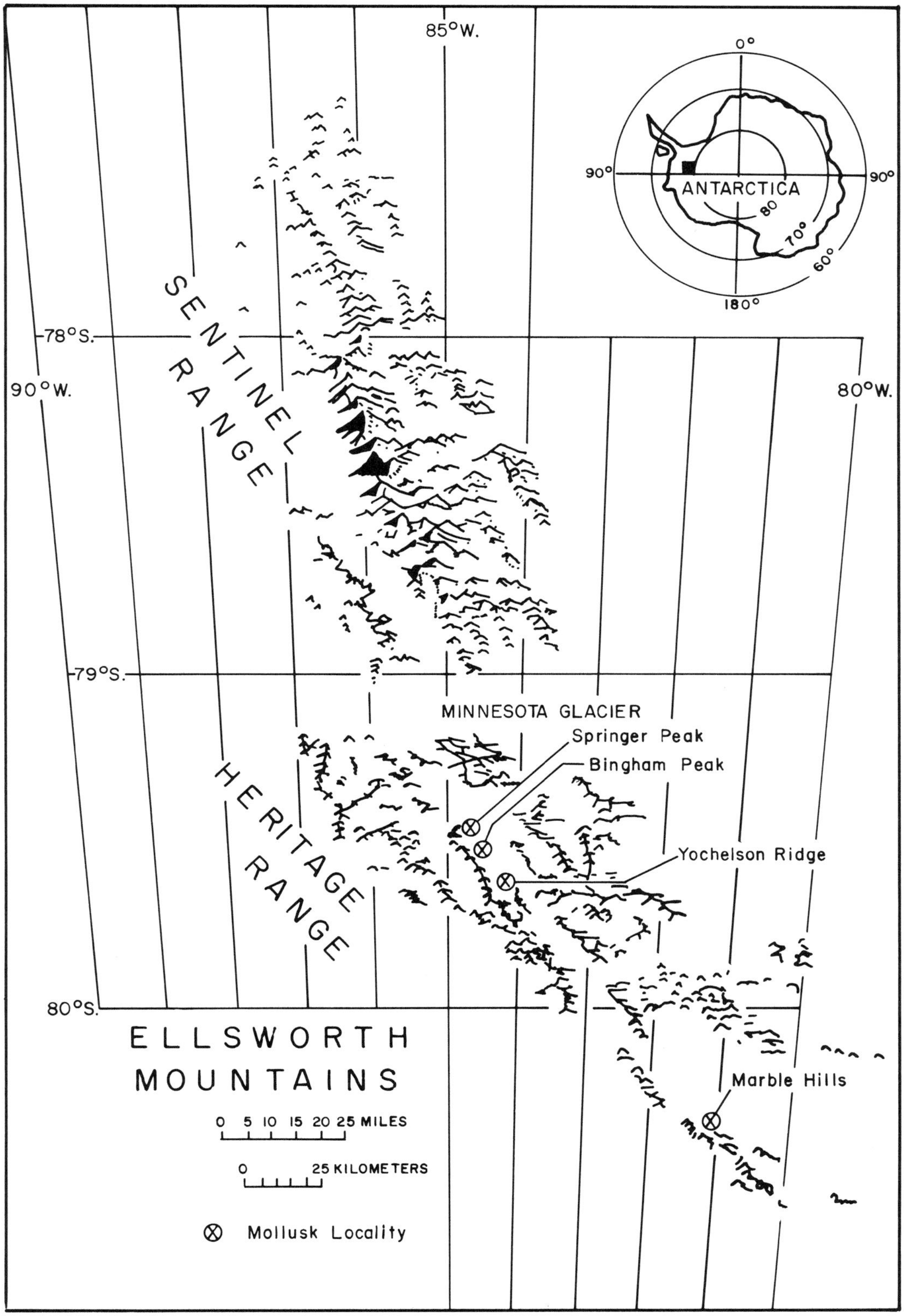

Figure 1. Locality map showing the places where Cambrian mollusks described in this chapter were collected from limestone of the Minaret Formation, Heritage Range, Ellsworth Mountains, West Antarctica.

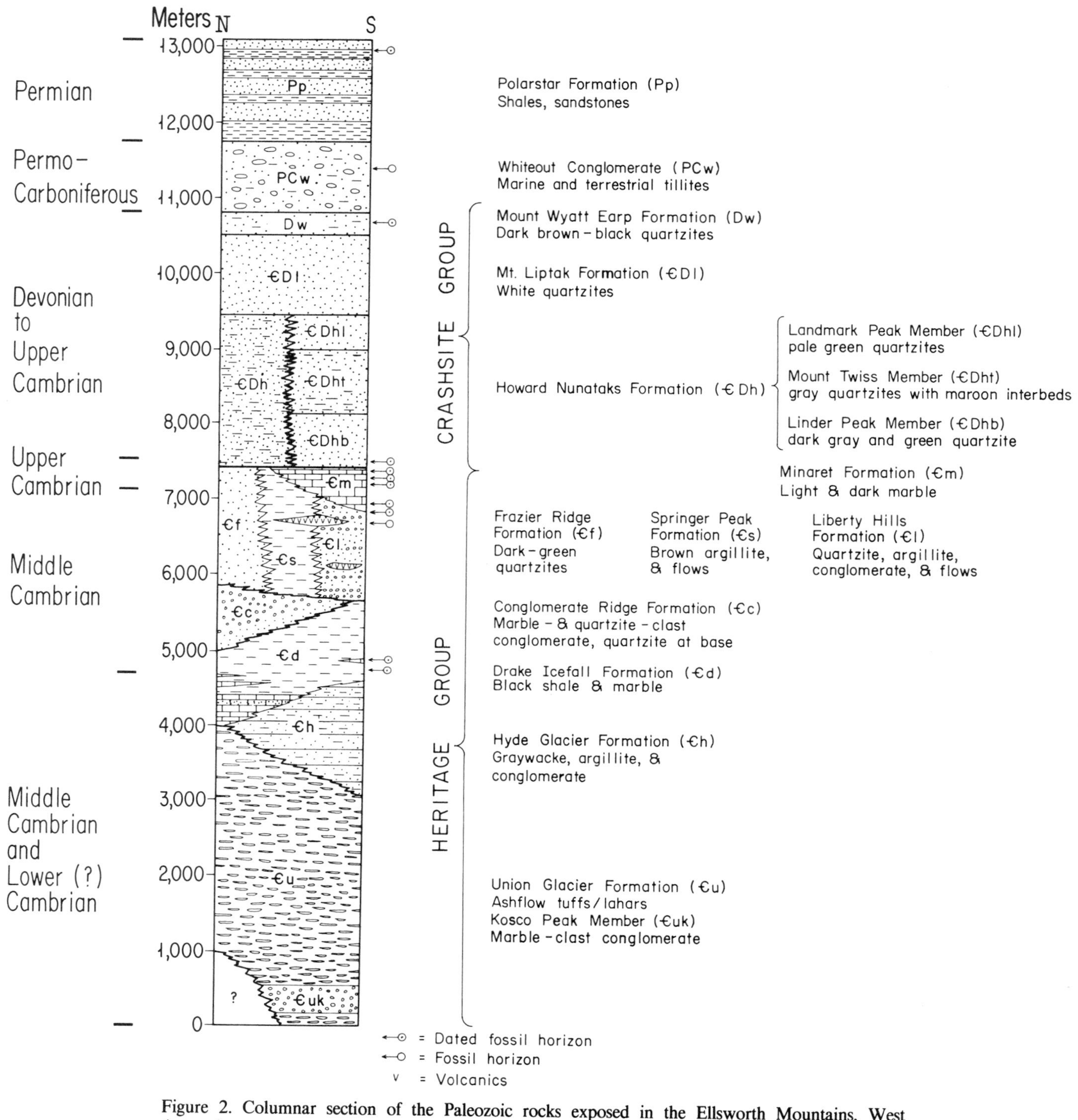

Figure 2. Columnar section of the Paleozoic rocks exposed in the Ellsworth Mountains, West Antarctica.

orogenic forces during the Jurassic folding of the rocks of the Ellsworth Mountains (Plate 19, Figs. 1, 2). As the Minaret thickens to the south, the limestone and the contained fossils become progressively more deformed and the fossils less abundant. This deformation can even be seen along the exposures at the base of Springer Peak. Here fossils are abundant and well preserved in the 8-m-thick coquina at the feather-edge, but they are less abundant and are not well preserved 1.5 km to the south, where the limestone is 20 m thick In the Marble Hills, exposures of the Minaret Formation are so deformed that we were surprised when the helcionellaceans and polymerid trilobites were found.

Our specimens are from five localities (I–V), which are discussed next.

(I) *The Springer Peak locality* is the northernmost locality of the Minaret Formation and is just northeast of Springer Peak (Fig. 1; Plate 19, Fig. 2) at the north end of Webers Peaks; this is the locality here called the feather-edge of the formation where the limestone is an 8-m-thick coquina. The coquina is composed largely of trilobite fragments, which make up considerably more than 90 percent of the specimens found; only two small enrolled articulated trilobites were found. The trilobites have been placed in 13 genera and 19 species (Shergold and Webers, this volume); they date the beds as Dresbachian/Idamean (Late Cambrian).

Mollusks at Springer Peak make up about 5 percent of the fauna and are represented in decreasing abundance by monoplacophorans, gastropods, hyoliths, and rostroconchs. In the field, emphasis was given to obtaining well-preserved mollusks at the expense of trilobites, and it is possible that the actual percentage of specimens of mollusks is less than 5. The mollusks from here represent one of the most diverse noncephalopod Upper Cambrian faunas known (Webers, 1972, 1977, 1982; Yochelson and others, 1973; Pojeta and others, 1981; Webers and others, 1982).

In relative abundance, the molluscan fauna is dominated by monoplacophorans. The species *Cosminoconella runnegari* and *Knightoconus antarcticus* are the two most common mollusk taxa at Springer Peak. In addition to trilobites and mollusks, the biota at Springer Peak contains archaeocyaths, inarticulate and articulate brachiopods, echinoderm fragments, conodonts, and algae; most of these are described elsewhere in this volume by various authors. Part of the reason for the abundance of mollusks at Springer Peak is the probable nearshore medium- to high-energy environment in which the limestone at the feather edge of the Minaret was deposited. In Phanerozoic rocks, such environments commonly have abundant mollusks. About as many hours were spent collecting at the feather-edge outcrop at Springer Peak as were spent collecting in the Liberty Hills and at Yochelson Ridge.

Paleoecological and petrological interpretations of the limestone of the Minaret Formation are presented by Buggisch and Webers (this volume). Here we describe the gross petrology of hand specimens of the rocks from which mollusks were obtained at each of our localities. At the feather-edge locality, the unweathered limestone is dense, and spar fills most of the voids between the shells. The rocks are fossiliferous packstone to poorly washed grainstone, and these two lithologies are interlayered; the color is medium gray (N5) on the rock color chart of Goddard and others (1948). This type of rock suggests medium- to high-energy conditions. Most of the spar was chemically precipitated, possibly by fresh phreatic ground water, and there is only minor recrystallization of the carbonate mud.

We accept the age of the Minaret Formation at Springer Peak as determined from trilobites as late Dresbachian (Idamean) (Shergold and Webers, this volume). The mollusks confirm this age assignment but are less restrictive in age range. The univalved taxa are most like those described by Berkey (1898) from rocks of Dresbachian-Franconian age in Wisconsin and Minnesota, U.S. The pseudobivalved rostroconch *Ribeiria australiensis* is known elsewhere only from rocks of Mindyallan (early Dresbachian) age in northern Australia (Pojeta and Runnegar, 1976). The hyolith *Tcharatheca webersi* is known elsewhere only from Idamean-age rocks in northern Victoria Land, Antarctica (Yochelson, 1976).

The mollusks known from Springer Peak are placed in 17 genera and 17 species as follows:

Monoplacophora

(1) *Cosminoconella runnegari,* n. gen., n. sp.
(2) *Ellsworthoconus andersoni,* n. gen., n. sp.
(3) *Kirengella pyramidalis,* n. sp.
(4) *Knightoconus antarcticus,* Yochelson, Flower, and Webers
(5) *Proconus incertis,* n. gen., n. sp.
(6) *Proplina rutfordi,* n. sp.

Gastropoda

(7) *Aremellia batteni,* n. gen., n. sp.
(8) *Euomphalopsis splettstoesseri,* n. sp.
(9) *Kobayashiella*? *heritagensis,* n. sp.
(10) *"Maclurites" thomsoni,* n. sp.
(11) *Matherella antarctica,* n. sp.

Hyolitha

(12) *Hyolithes*? species indeterminate
(13) *Linevitus*? *springerensis,* n. sp.
(14) *Tcharatheca webersi,* (Yochelson)

Rostroconchia

(15) *Apoptopegma craddocki,* n. sp.
(16) *Ribeiria australiensis,* Pojeta and Runnegar
(17) *Wanwania* species A

(II) *The Bingham Peak locality* is about 4.8 km south-southeast of Springer Peak (Fig 1; Plate 19, Figs. 3, 4). No attempt was made to collect extensively from this locality, and it is not as rich in fossils as the feather-edge locality (I). Sufficient time was devoted to collecting here to determine that the trilobites and mollusks are closely comparable to those at the northern end of Springer Peak.

Hand specimens of the rocks at Bingham Peak, from which the mollusks were obtained, are interlayered wackestone to localized packstone containing peloids; fossils are concentrated along distinct layers forming a muddy packstone. Large, whole fossils are not present. The color is dark gray (N3). This type of rock suggests current conditions alternating between quiet water and periods of higher energy, with better washing and orientation of fossil debris. All species of mollusks known from Bingham Peak also occur at Springer Peak. *Cosminoconella runnegari* and *Knightoconus antarcticus,* the most abundant species of mollusks at Springer Peak, are absent at Bingham Peak. The changing energy conditions and periodic deposition of fossil debris in the beds sampled at Bingham Peak may account for this difference. They may also explain the abundance of *Hyolithes*? species indeterminate in the small collection from Bingham Peak; here 12 specimens were found, whereas only two specimens were found at Springer Peak.

The mollusks known from Bingham Peak are placed in seven genera and seven species as follow:

Monoplacophora

(1) *Ellsworthoconus andersoni,* n. gen., n. sp.
(2) *Proplina rutfordi,* n. sp.

Gastropoda

(3) *Aremellia batteni,* n. gen., n. sp.
(4) *Euomphalopsis splettstoesseri,* n. sp.
(5) *Kobayashiella*? *heritagensis,* n. sp.

Hyolitha

(6) *Hyolithes*? species indeterminate
(7) *Linevitus*? *springerensis,* n. sp.

(III) *The Yochelson Ridge locality* is about 16 km southeast of Bingham Peak at the northern end of the Soholt Hills (Fig. 1; Plate 19, Fig. 5). It is an area of extremely complex geology, which exposes a number of disconnected outcrops of the Minaret Formation limestone. The only way one outcrop can be related to another is by the contained fossils. All the fossils collected from the limestones indicate a late Dresbachian (Idamean) age. The trilobites are described by Shergold and Webers (this volume).

At Yochelson Ridge, hand specimens of the unweathered limestone show that it is a dense fossiliferous grainstone, which is poorly washed in places, to a fossiliferous packstone containing numerous rounded intraclasts to peloids. Color ranges from medium light gray (N6) to medium gray (N5). This type of rock suggests medium- to high-energy conditions. All species of mollusks known from Yochelson Ridge, except *Scaevogyra*? species indeterminate, also occur at Springer Peak. The most abundant taxon is *Proplina rutfordi,* which is represented by 13 deformed specimens.

The mollusks known from Yochelson Ridge are placed in five genera and five species as follow.

Monoplacophora

(1) *Cosminoconella runnegari,* n. gen., n. sp.
(2) *Knightoconus antarcticus,* Yochelson, Flower, and Webers
(3) *Proplina rutfordi,* n. sp.

Gastropoda

(4) *Scaevogyra*? species indeterminate

Rostroconchia

(5) *Apoptopegma craddocki,* n. sp.

(IV) *The Mount Rosenthal locality* (Fig. 1), about 51 km southeast of the northern end of Yochelson Ridge, yielded some simple, nearly straight calcareous tubes. For convenience, we have described these tubes under the name Orthothecida? species indeterminate. However, we do not regard the tubes as hyoliths. The tubes do show that in the northern Liberty Hills, the Minaret Formation is Phanerozoic in age. Elsewhere in the Liberty Hills, the Minaret Formation did not yield any fossil mollusks or hyoliths.

The tubes occur in sandy fossiliferous limey packstone to grainstone. The quartz grains are very fine sand to coarse silt in size. The packstone is peloidal and is fairly well washed in places. The quartz-rich areas and the fossils are concentrated in distinct layers; the fossils are generally horizontal. The color of the rock is medium dark gray (N4). This type of rock suggests moderate, though consistent, current conditions that winnowed away the mud and sorted the sediment fairly well. The small size of the quartz grains and the presence of peloids suggest moderate-energy conditions.

(V) *The Minaret Peak locality,* in the northern Marble Hills, is the southernmost locality at which fossils were found in the carbonate rocks of the Minaret Formation (Fig. 1; Plate 19, Fig. 6). This locality is the type section for the Minaret Formation and is located on the U.S. Geological Survey Liberty Hills topographic map (1:250,000, SU 16-20/2*, 1967). The locality is to the right of the word *Marble,* on the east side of the middle closed 1,400 m contour. The coordinates at 77°S and 85°W are S8000-W7700/1x8.

At this locality, six mollusks, identified as *Latouchella*? species indeterminate, and a few poorly preserved polymerid trilobites were found. The trilobites are not identifiable to genus or species but have what is judged to be a Middle Cambrian aspect. *Latouchella*-like helcionellaceans range throughout the Cambrian, but they are most common in Lower and lower Middle Cambrian rocks.

Even though the specimens are poorly preserved and not fully identifiable, they show that at its type section the Minaret Formation is Cambrian in age. They suggest that at this locality it may be older than Late Cambrian. Previously, the metamorphosed limestones in the Marble Hills were thought to be Proterozoic in age (Craddock, 1969). Reinterpretation of the age of the Minaret Formation at its type locality is one of the major findings of the 1979–80 expedition.

The rock in which the fossils occur is conglomeratic marble. The clasts are subangular and in the size range of coarse sand to small pebbles. Most of the matrix and the clasts are recrystallized carbonate mud.

Solovjev, in Solovjev and others (1984), had described two species of Upper Cambrian mollusks from the Neptune Range, Antarctica, which are called *Scenella* sp. and *Pollicina* sp. Neither is well illustrated. *Pollicina* sp. is elongated like *Knightoconus antarcticus.*

COMPARISON WITH OTHER MOLLUSCAN FAUNAS

Runnegar (1981a) provided brief summaries of the geographic distribution and diversity of major Upper Cambrian mollusk faunas of the world outside Antarctica. He noted that important Upper Cambrian mollusk faunas are known from northeastern China (Fengshanian), Australia (Mindyallan-Payntonian), and North America (Dresbachian-Trempealeauan). The Chinese faunas (Chen and others, 1983; Chen and Teichert, 1983) occur in the highest Cambrian, in rocks that had long been thought to be earliest Ordovician in age, the stromatolitic Wanwangou (= Wanwankou of earlier literature) Limestone and equivalents. The fauna is rich in cephalopods, rostroconchs, and gastropods. At least 120 species of cephalopods and 20 species of rostroconchs are known from the Wanwangou. Gastropods and monoplacophorans have been little studied since Kobayashi (1933) described 11 species of the two groups. Various nonpaleontological literature suggests that gastropods are abundant in the Wanwangou but monoplacophorans do not seem to be common.

The Upper Cambrian molluscan faunas of Australia are dominated by rostroconchs (Pojeta and others, 1977) and gastropods (Öpik, 1967; Runnegar, under study). Cephalopods from the Upper Cambrian of Australia are under study by Mary Wade (Queensland Museum), and monoplacophorans seem to be rare. Ten species of rostroconchs are known, which locally can be abundant enough to form coquinas. Most of the rostroconchs occur in nearshore deposits of latest Cambrian age (Payntonian), but some rostroconchs and gastropods occur in limestone as old as Mindyallan.

In North America, Upper Cambrian molluscan faunas are dominated by large monoplacophorans, gastropods, and *Matthevia;* rostroconchs and cephalopods are rare (Whitfield, 1882, Berkey, 1898; Yochelson and others, 1973; Pojeta and Runnegar, 1976; Runnegar and others, 1979). *Matthevia* is not known outside of North America, where it is confined to stromatolitic limestone largely of Trempealeauan age. Ludvigsen and Westrop (1983) suggested that the type species, *M. variabilis* Walcott, may be Franconian in age at the type locality in New York. Most North American monoplacophorans and gastropods occur in the sandy Upper Cambrian rocks (Franconian-Dresbachian) of the upper Mississippi Valley and in carbonate rocks in the Ozark Plateau and Appalachian Mountains.

In general terms, the Minaret fauna is most like that of North America because it is dominated by monoplacophorans and gastropods associated with rare rostroconchs. Cephalopods have not been found in the Minaret fauna. *Knightoconus antarcticus* and *Proplina rutfordi* from the Minaret are very similar to species placed in *Hypseloconus* Berkey and *Tryblidium* Lindström by Berkey (1898). A significant difference in preservation is that the material described by Berkey is from sandstone, whereas the Antarctic material is from limestone; both rock types, however, were deposited near the shore. Yochelson (1976) described a small fauna of Upper Cambrian mollusks (including hyoliths) from the Bowers Group, northern Victoria Land, Antarctica. That fauna has the hyolith *Tcharatheca webersi,* and perhaps the gastropod *Scaevogyra,* in common with the Minaret fauna. The only other species of mollusk known to occur both in the Minaret and elsewhere is *Ribeiria australiensis,* which occurs in Mindyallan rocks in Queensland, Australia. In the systematic section, we point out geographic similarities to various regions as appropriate. Although greatest similarities seem to be with North America, this may be because we have more comparative material and literature available for that continent.

Recently, Upper Cambrian mollusks have received considerable study (Pojeta and Runnegar, 1976; Pojeta and others, 1977; Runnegar and others, 1979; Chen and Teichert, 1983; Chen and others, 1983, among others). Most of these recent studies have dealt with cephalopods, rostroconchs, and *Matthevia.* Gastropods and monoplacophorans of the Upper Cambrian have largely been described in older works, usually as incidental parts of studies that dealt primarily with trilobites. An exception to this is the study of *Knightoconus antarcticus* by Yochelson and others (1973).

Enough information is now available to show that Upper Cambrian mollusks are of considerable biostratigraphic value. Runnegar and Pojeta (1989) have assembled a range chart of all known genera except the cephalopods, which were dealt with by Chen and Teichert (1983). Elements of the Upper Cambrian fauna of the Ellsworth Mountains occur worldwide. We anticipate that future studies will show even more similarities and that Upper Cambrian mollusks will prove to be useful paleobiogeographic tools. We record similarities to Siberian, Chinese, North American, and Australian faunas. No arrangement of land masses in the Late Cambrian places all occurrences of these mollusks in close proximity. If, as we suspect, many Upper Cambrian mollusks do not show provinciality, then they should be of considerable use in establishing correlations between the various provinces dominated by trilobites. At the very least, they will reinforce the correlations based on trilobites.

At this stage in our knowledge, it seems likely to us that many Upper Cambrian mollusks were essentially worldwide in distribution. At all localities where we have been able to obtain data on the occurrences of Upper Cambrian mollusks, shallow-water nearshore conditions are indicated. We suggest that Upper Cambrian mollusks may have had long-lived larvae in order to explain the wide distributions. For the mollusks from the Ellsworth Mountains, in which we have been able to observe the earliest preserved part of the shell, this presumed larval stage is always simple and relatively large.

PREPARATION OF SPECIMENS

At the principal locality, northeast of Springer Peak, limestone of the Minaret Formation crops out with a nearly vertical dip. It trends north-south across the landscape as a spine-like outcrop standing above adjacent low-lying argillites to either side

(Plate 19, Figs. 1, 2). Many limestone blocks form a rubble at the foot of the slope at the northern feather-edge outcrop (Plate 19, Fig. 2). This rubble is probably the result of intense frost wedging. Many of the blocks in the rubble are friable, and excellent float specimens can be obtained from such blocks by tapping the blocks with a hammer. However, many specimens are incomplete and delicate because of breakage resulting from the weathering.

In marked contrast, many of the blocks of rubble and the material from the outcrop at Bingham Peak and elsewhere are not friable. The coquina is slightly argillaceous dense limestone with most of its pore space filled with spar. It is difficult to extract specimens mechanically from such limestone. Ordinarily, specimens are not freed from the rock when it is struck with a hammer; the fracture goes through both the fossils and the matrix, and one is left with sections of fossils to study. At present, such sections provide little useful taxonomic or biostratigraphic information. A few specimens were prepared from such rocks using grinding wheels and needles; however, this procedure is time consuming and produces only a few specimens for study. It is particularly difficult to obtain useful specimens of small species using mechanical preparation.

In an effort to obtain additional specimens, Pojeta used a heating and quenching technique. Most specimens were obtained from the dense limestone after it was alternately heated and then quenched in cold water. This method is limited by the size of the source of heat. In Pojeta's laboratory, the source of heat was a Meker-type high-temperature burner, which produces a short wide flame of uniform intensity with almost no gold cone. This type of burner can be used in place of a laboratory blast furnace for many purposes. Rock specimens were placed on a flame shield or wire gauze squares (with or without asbestos centers) of appropriate size on a tripod stand, which is 24 cm high and 15 cm in diameter. The rock specimens were covered with a metal cylinder (1,360-g coffee can) 15 cm in diameter and 17 cm high, which was closed at one end. The closed end had a number of holes made with a three-penny nail (about 30 mm long). The cylinder gives both uniform heat and protection against rock explosions. Heat was applied directly to the rock specimen for about 1 hour. The specimen was then removed with tongs and dropped into the coldest running water available so that the rock was completely immersed. The rock was left in the cold running water until it reached the same temperature as the water. This procedure was repeated as many times as necessary, on different sides of the rock, until the rock could be broken apart with the fingers. A small chisel and a small hammer were also useful, but most of the specimens of fossils described and figured here were literally recovered by hand.

All the heating and quenching procedures should be carried out in a laboratory fume hood with appropriate safety equipment such as gloves, apron, goggles, etc. We recommend starting with small samples that do not show any obviously important specimens until the procedures are mastered.

Heating and quenching allow one to separate rock matrix from the fossils by taking advantage of the anisotropy between them. Many more specimens become available for study than can be obtained by the use of grinding wheels and needles. Small specimens have been recovered, including enrolled trilobites as small as 5 mm and rostroconchs as small as 7 mm long. A drawback to the method is that often the shells of mollusks and articulate brachiopods will separate from the internal molds (steinkerns), and the shell itself may fracture into many small pieces. This separation does not always happen, and our plates show that in almost all taxa a few specimens retain some to most of the shell.

The heating and quenching method may produce fractures in some of the larger fossils, such as those shown on Plate 7, Figures 5–7; Plate 8, Figure 12; Plate 9, Figure 13; Plate 10, Figure 7; and Plate 14, Figure 12. However, the cause of the fractures is not certain, because some specimens that were prepared mechanically or were obtained from outcrop are also highly fractured, such as those shown on Plate 6, Figures 5–8; Plate 10, Figures 1–4; and Plate 8, Figure 3. It may be that taphonomy, weathering, tectonism, and orogeny affected the Ellsworth Mountains specimens more than did our laboratory procedures. In general, small specimens are less fractured than large specimens. Compare the specimens shown on Plate 11 with those mentioned above. The mollusks shown on Plate 11, Figures 1, 2, and 9, were prepared mechanically; all others illustrated on Plate 11 were prepared by heating and quenching. There is a slight difference in color between the heated and quenched specimens and those collected on the outcrop or prepared mechanically, but this is not seen when the fossils are coated with NH_4Cl.

Delicate features, such as those shown on Plate 2, Figures 5–11; Plate 5; Plate 6, Figures 2, 3; and Plate 15, Figures 2, 3, can be exposed by the heating and quenching technique. We suggest that the heating and quenching technique be used more widely for dense, slightly argillaceous limestones, in which fractures break through fossil specimens instead of around them in fresh rock. The technique is easy to use, and the equipment needed is inexpensive.

We are not entirely certain what happens chemically and physically to the rock in the heating and quenching technique. Clearly some CO_2 is driven off because where the flame touches the rock there is calcining, as is evidenced by the presence of white, powdery CaO. However, this is a minor happening. Water may be driven out of the rock, and this may alter some clay minerals. However, most of the effect seems to be physical thermal shock and the creation of steam when the hot rocks are dropped into the cold running water. The thermal shock and steam may enlarge any small joints and help separate shell from matrix and internal mold (steinkern).

Standard 35-mm photographic techniques of ammonium chloride–whitened specimens were used for most of the figures on the plates. Specimens shown on Plate 2, Figures 16, 17; Plate 3, Figures 1–3; Plate 5, Figure 9; and Plate 17, Figures 4–8, and all the specimens shown on Plate 18, were photographed with a scanning-electron microscope. These specimens were first coated with carbon and then with gold-palladium. The specimens photographed with the SEM do not necessarily have the conventional apparent highlight from the upper left.

All specimens studied are in the collections of the Department of Paleobiology, U.S. National Museum of Natural History (USNM). For each new species, we figure the holotype Of the new taxa, only *Matherella antarctica* is based on a single specimen. All additional illustrated specimens of new taxa are paratypes. For some taxa, better specimens that are not illustrated have also been selected as unfigured paratypes and have been assigned a single lot number. Remaining specimens, which contribute nothing additional to the concept of the taxon, have been placed in the general collections. Hypotypes are assigned individual numbers for each specimen of previously described species in accordance with general practice; a few well-preserved unfigured individuals have also been given lot numbers.

Except as noted in the text or plate explanations, all illustrated material is from northeast of Springer Peak, the outcrop at the feather-edge of the Minaret Formation (Fig. 1; Plate 19, Fig. 2). The mollusks that document the age of the Minaret at other localities within the Heritage Range have been assigned museum numbers and are considered to be either unfigured paratypes or hypotypes, as appropriate.

SYSTEMATIC PALEONTOLOGY

Class HYOLITHA Marek

Introduction

Hyolitha has been used as a group name for an extinct phylum (Runnegar, 1980; Pojeta, 1989), as an extinct class of Mollusca (Marek and Yochelson, 1976, Yochelson, 1979; and earlier papers by Marek), and as a lower-level taxon within the Monoplacrophora (Dzik, 1978, 1980). Earlier systematic assignments of the hyoliths (see Fisher, 1962) are of historical interest only. The material from the Ellsworth Mountains provides no data bearing on the first two alternatives listed above, but for convenience these fossils are described along with the Mollusca. We all concur that the interpretation by Dzik is a retrograde step in systematics that does not need to be discussed here.

When treated as a class, the Hyolitha is commonly divided into two orders, the Hyolithida and Orthothecida. Peel and Yochelson (1980; 1984) noted a third order, of late Paleozoic age, not represented in the Cambrian. Four other orders—Globorilida, Camerothecida, Diplothecida, and Hyolithellida—have been proposed, all of which contain a few genera exclusively Cambrian in age. The Hyolithellida are now generally considered not to belong to the Hyolitha. The Diplothecida and Camerothecida have been questioned by Marek (1967, p. 64–65), for they are both based on poorly preserved material and may not be appropriate biological subdivisions; the Globorilida is also poorly founded. We recognize two orders within the Ellsworth Mountains material—Hyolithida and Orthothecida.

Order HYOLITHIDA Matthew

Discussion. To the best of our knowledge, there is no generally accepted use of superfamilies within the order Hyolithida, and none are used here.

Family HYOLITHIDAE Nicholson

The family Sulcavidae was erected by Syssoiev in 1958 for hyoliths that have "cuts" near the base of the aperture. These areas are the openings from which the supports, or helens, protrude between the shell and the operculum. To the best of our knowledge, all members of the Hyolithida have such helens, and the aperture must accommodate them in some way. Because we are uncertain that the family grouping is appropriate, we have transferred *Linevitus* to the conventional family Hyolithidae.

Genus *Linevitus* Syssoiev, 1958
Linevitus? springerensis [n. sp.]

Plate 2, Fig. 18; Plate 3, Figs. 4, 14, 15; Plate 4, Figs. 1–6, 8–12

1981. *Hyolithes* sp., Pojeta and others, U.S. Geological Survey Open File Report 81-743, p. 168.
1982. *Hyolithes* sp. A, *Hyolithes* sp. B. Webers, Antarctic geoscience, p. 636, Fig. 78.2-14, 19, 20.

Description. Hyolithid having a prominent rounded dorsal median carina and distinct lateral carinae. Ligula short, uniformly curved at the apertural margin, extending through about one-third the circumference of a circle. Aperture oxygonal, the growth lines straight, inclined about 30° from vertical from the dorsum almost to lateral carinae, abruptly turning toward vertical just before reaching the carina and continuing smoothly, convex forward across the base. Shell moderately narrow, expanding at an angle near 17°; dorsally convex, showing indication of slight logarithmic curvature throughout its length. Basal surface in cross section very gently inflated, arched convexly downward. Shell relatively low. Upper surface composed and flattened rather than arched, joining at an obtuse angle of approximately 120°. Juncture of sides and base sharp and accentuated by a distinct flangelike carina extending outward from the basal surface. Dorsum marked by a broad, distinctly raised, prominent, rounded carina. Ornament confined to prominent, closely spaced growth lines stronger on the lateral slopes than on the base. Shell of two layers, the outer layer thinner but greatly thickened at the dorsum and at the junction of sides and base.

Remarks. The morphologic details available are sufficient to place this form within the family Hyolithidae. The type species of the genus is from the Middle Ordovician of Sweden and has prominent lateral carina. However, it lacks a prominent median carina, and for this reason we question the generic placement of this new species. No operculum is known, and according to Marek (1967, p. 66), the operculum is a key feature for assignment both at the generic and family level. The literature is overburdened by generic names based on poorly defined species; although our material is adequate to formally define a genus, we see no need to add yet another name concept at this time.

This species shows some variation in the height of the median carina, much of which seems to be related to preservation. On steinkerns, the lateral carinae are not so prominent, and they show no great elevation at the dorsal crest although all three carinae are prominent in cross section (Plate 3, Figs. 14, 15). On specimens retaining the inner layer, the dorsum is only slightly more elevated than the anticipated height of the dorsum (Plate 4, Fig. 4).

In the presence of a prominent raised median carina (Plate 4, Fig. 1), the Antarctic form is similar to both *Hyolithes hospe* Holm and *H.* species number 12 of Holm (1893) from the late Early Ordovician sequence of Öland, Sweden. We are unable to find any species of *Hyolithes* described from northern China by Walcott (1913) or by Endo (1937) that have such a prominent carina, although admittedly some of their taxa are known only from the basal surface. The species is definitely not the form that is most common in the Great Lakes region of North America.

So far as is known, representatives of the Hyolithida have not been described previously from Antarctica.

Types. Holotype, USNM 385144, figured paratypes, USNM 385136, 385142, 385143, 385145–385149; unfigured paratypes USNM 388311, 388312.

Type locality. From the feather-edge of the Minaret Formation on the northern side of Springer Peak, Webers Peaks, Heritage Range (Fig. 1).

Distribution. Springer Peak, Bingham Peak (USNM 388312).

Etymology. Springer Peak has produced a wealth of fossils, and it is appropriate that this locality be noted by the species name.

Genus *Hyolithes* Eichwald, 1840
***Hyolithes*? [sp.] indeterminate**
Plate 3, Figs. 11–13; Plate 4, Fig. 7

1982. *Orthotheca* sp., Webers, Antarctic geoscience, p. 636, Fig. 78.2-14.

Remarks. Several incomplete specimens of hyoliths clearly do not belong to *L.*? *springerensis.* Insofar as they are preserved, the specimens fit in the concept of *Hyolithes.* The Middle Ordovician genus from the Baltic region has tended to become a catchall for a variety of poorly known species. The Ellsworth material has a simple, nearly semicircular cross section (Plate 3, Fig. 12), lacking all evidence of a median crest and lateral flanges. The steinkern is smooth; although the crest and flanges are subdued on the steinkern of *L?. springerensis,* they can be easily discerned. The lateral slopes of *Hyolithes*? sp. indet. form a slightly smaller obtuse angle than the slopes in *L?. springerensis.* One individual has broadly rounded lateral carinae (Plate 3, Figs. 11, 12). The ligula is simple and small (Plate 3, Fig. 11) and is covered with fine growth lines. No other details are available.

Material. Figured specimen, USNM 385141; unfigured specimens USNM 386587, 388193.

Locality of figured material. From the feather-edge of the Minaret Formation on the northern side of Springer Peak, Webers Peaks, Heritage Range (Fig. 1).

Distribution. Springer Peak, Bingham Peak (USNM 386587).

Order ORTHOTHECIDA Marek
Superfamily ORTHOTHECACEA Syssoiev

Discussion. In his monographic work on hyoliths, Syssoiev (1972, p. 47) erected the superfamily Orthothecoidea. Presumably he followed Recommendation 29A of the International Code of Zoological Nomenclature in regard to the suffix. However, traditionally the suffix "-acea" has been used within the Mollusca and we have made this modification, although Pojeta does not regard hyoliths as mollusks.

Syssoiev (1972, p. 47) divided the Orthothecimorpha, equivalent to the Orthothecida, into three superfamilies, the Circothecoidea, Orthothecoidea, and Exilithecoidea. On the basis of observations of other Cambrian faunas, at least some of the taxa within the Circothecoidea might as well be assigned to other tube-dwelling organisms. The Orthothecoidea even more so than the Hyolithida seem to contain a number of fossils whose assignment to the group may be spurious. Without having undertaken any detailed study of the order, we nevertheless suggest that the morphologic limits of the Orthothecoidea are somewhat narrower than the current literature indicates.

During the course of this study, considerable difficulty was encountered in distinguishing fragmentary orthothecoids from the broken tips of elongate genal spines of trilobites. Conway Morris (1982) noted that a spine of *Wiwaxia* was described originally as an *Orthotheca.* Because of observations of the superficial similarity noted above, we are convinced that some of the species of Orthothecoidea described in the literature are arthropod fragments.

Genus *Tcharatheca* Syssoiev 1972
***Tcharatheca webersi* (Yochelson), 1976**
Plate 3, Figs. 5–9

1976. *Contitheca webersi* Yochelson, *in* Shergold and others, p. 260–261, Plate 39, Figs. 1–9, 22.
1982. *Contitheca* sp., Webers, Antarctic geoscience, p. 636, Fig. 78.2-16.

Remarks. This species was first described from northern Victoria Land, Antarctica. The identification of this species in the Ellsworth Mountains is based on less than a dozen specimens. Nothing of significance can be added to the original description; no operculum is yet known for the species. The illustrations of this new material show the salient features of a base that is concave downward, combined with the dorsal outline nearly semicircular in profile. The result is a reniform or "kidney-bean" cross section (Plate 3, Figs. 5–8). Abundant, uniformly spaced, longitudinal threads on the base and covering the dorsum can also be seen (Plate 3, Fig. 9). Most specimens are incomplete in the apical region. We have no information on whether septa are present, but if they do occur, they are confined to the juvenile part of the shell.

Curvature of the base and on the dorsum is slight and is not as great as that in some specimens in the type lot. Curvature of the dorsum and the identation of the base may have been emphasized by tectonism in the northern Victoria Land specimens. The new material from the Heritage Range indicates that the species should be transferred to *Tcharatheca,* for that genus has a smooth arched dorsum. In *Contitheca,* the median line is more clearly defined by the presence of a change in curvature of the profile.

It may be hazardous to apply the same specific name to specimens from areas as widely separated geographically as northern Victoria Land and West Antarctica, but the specimens are morphologically close, and we have no basis for applying a different specific name. At this stage of our knowledge of the Cambrian faunas of Antarctica, we prefer to err on the side of too few specific names, rather than too many. Precise correlations are uncertain, but at least both stratigraphic units yielding this fossil are Late Cambrian in age.

Runnegar and others (1975, p. 184, Fig. 3) illustrated the silicified gut of an orthothecid from the Middle Cambrian Nelson Limestone, from Mount Dover, Neptune Range, Antarctica. Because of differences in preservation, this material cannot be compared on the specific level with the Late Cambrian form.

Figured specimens. USNM 385137–385140.

Locality. From the feather-edge of the Minaret Formation at the northern side of Springer Peak, Webers Peaks, Heritage Range (Fig. 1).

***Orthothecida*? sp. indet.**
Plate 3, Fig. 10; Plate 5, Figs. 14, 15; Plate 16, Fig. 12

Remarks. The designation Orthothecida? species indeterminate is used for calcareous tubes that are approximately circular in cross section. The tubes expand only slightly in width during growth (Plate 3, Fig. 10). The tubes seem to be straight rather than curved. No growth lines are evident. In cross section, adjacent tubes are varyingly compressed (Plate 5, Fig. 14; Plate 16, Fig. 12). These tubes are not mollusks for there are no features, apart from a calcareous hard part, that in any way link them to the mollusks. Presumably, within the Orthothecacea some similar tube-like fossils have been named, but we prefer not to assign any generic

name to the Antarctic material. Some of the specimens, at first glance, appear more complex, but this is a result of imbrication of tubes (Plate 5, Fig. 15). Such imbrication is not uncommon (Yochelson, 1984a) and may require only modest water movement to effect the telescoping and stacking of specimens.
Material. Figured hypotypes, USNM 385172–385174, 386586; unfigured specimens 388313.
Locality. Minaret Formation, Mount Rosenthal, Liberty Hills (Fig. 1).

Class MONOPLACOPHORA Wenz

Introduction

As originally conceived by Wenz (1940) and elaborated upon by Knight (1952), the class Monoplacophora was defined for Paleozoic fossil mollusks having a bilaterally symmetrical shell. The soft parts secreting the shell were assumed to be bilaterally symmetrical, as demonstrated by the presence of multiple paired muscle scars within the shell; both internal and external bilaterality were embodied in the taxonomic concept. The concept of a bilaterally symmetrical mollusk in a single bilaterally symmetrical shell was dramatically verified by Lemche (1957), who described the living monoplacophoran *Neopilina galathaea.* Currently, the Monoplacophora is generally accepted as a class of the Mollusca. The two authorities who defined the class differed on its contents, particularly in connection with the Bellerophontacea, a controversy that continues to this day; the Upper Cambrian monoplacophorans described below provide no new data on this issue.

Quite apart from this controversy, from a practical standpoint there are difficulties in assigning fossils to the Monoplacophora. Some living gastropods develop a secondarily bilaterally symmetrical shell. One of the earliest patelliform mollusks to be named from rocks of Paleozoic age was *Metoptoma* Phillips, from the Lower Carboniferous of England. That genus is characterized by a horseshoe-shaped muscle scar on the shell interior, and because of this feature, *Metoptoma* is now assigned to the gastropod superfamily Patellacea. Because of the similarity of *Metoptoma,* in general shape, to shells from older rocks, the name was used in the early literature for almost any Paleozoic patelliform shell. Although Paleozoic monoplacophoran shells are bilaterally symmetrical, not all Paleozoic bilaterally symmetrical shells are regarded as monoplacophorans.

Following the work of Lindström (1884) on the Silurian of Gotland, the generic names *Tryblidium* and *Pilina* were introduced into the literature for what were thought to be fossil patellacean gastropods. *Pilina* has not been used by many subsequent workers, but *Tryblidium* was repeatedly used, and many species were assigned to this genus. The type species of both genera display the bilaterally symmetrical, multiple paired muscle scars that are the prime characteristic of fossils placed in the class. However, many fossil taxa that are generally accepted as having monoplacophoran affinities do not contain a single specimen that shows multiple muscle scars, and the shells of modern monoplacophorans do not show muscle scars. There is some indication that presence of these scars reflects, in part, possession of a thick shell and, in part, life in a specialized clinging environment.

In spite of a careful search, we have not found any trace of discrete paired muscle scars on any specimen of the Upper Cambrian bilaterally symmetrical mollusks from the Ellsworth Mountains that we assign to the Monoplacophora. They all have in common a moderately thin shell. As indicated, assignment of species of fossils to Monoplacophora in the absence of paired muscle scars requires judgment and interpretation. However, if one does not proceed from the conchological features of general shape, little is likely to be accomplished in our understanding of the diversity of the class. We are as confident of assignment of various taxa to the Monoplacophora as one can be from the study of exterior shell details and internal molds (steinkerns).

The descriptive term *cap-shaped shell* has been applied by authors to many monoplacophoran shells. Our impression is that this was originally used to indicate a resemblance to the liberty cap of the American and French Revolutions, but it has taken on a variety of meanings. In a liberty cap, the apex hangs over the edge, but in an American baseball cap, it does not; the word *cap* means only a covering for the head, and caps have many shapes. We suggest that the term *cap-shaped* be abandoned.

For Paleozoic gastropods and pelecypods, descriptive terms are based on common genera; for gastropods, terms such as murchisoniiform or holopeiform are in common usage. We recommend a similar scheme for monoplacophorans and suggest that terms such as piliniform or tryblidiform would better convey the concept of overall shape.

Although there is as yet no consensus concerning the principal features to be used in classification of monoplacophorans, there seems to be general agreement that the position of the apex relative to the apertural margin is important. Thus, taxa having the apex situated more or less subcentrally above the aperture are classified differently from those having a prominent apex extending past the vertical projection of the apertural margin. This difference is quite evident and useful, regardless of the issue of whether the apex is to be oriented as anterior or posterior (Fig. 3).

Monoplacophorans are abundant and show a diversity of shell forms in the Ellsworth Mountains. Knowledge of them poses some interesting questions. *Knightoconus antarcticus* is a septate species, but it is associated with other shells that do not have such internal partitions. The function, or use, of the septa to the organism when first evolved remains unclear.

In Yochelson's notion of the class limits, by Middle Ordovician time, and perhaps earlier, monoplacophorans were uncommon organisms and showed far less diversity than they showed in the Cambrian and Early Ordovician. The factors leading to this restriction are not evident. However, other definitions of the Monoplacophora, which include the Bellerophontacea (Runnegar and Jell, 1976; Runnegar, 1983), do not support this interpretation.

Distinguishing taxa within the Monoplacophora is far more difficult than among other groups of mollusks. Some of the species of Monoplacophora seem to show more individual variation than do the gastropods or the rostroconchs from the Ellsworth Mountains. Even though we have no quantitative way of expressing the degree of individual variation within a species, we are confident of this observation. The amount of variation in some of the monoplacophorans suggests that special caution should be exercised in the naming of new species. We do not yet understand why they are so variable.

Divisions of the Monoplacophora

As proposed by Knight and Yochelson (1960), the Monoplacophora contained two orders, and a third was questionably assigned. In a quarter of a century, some genera that were originally placed in the class have been removed, and, conversely, additional monoplacophoran genera have been named. We realize that ordinal divisions within the class are not well founded, but rather than attempt a major revision of the class at this time, we use herein the ordinal and family-level taxa used by Knight and Yochelson (1960). Horný (1965) coined the terms *tergomyan* and *cyclomyan,* but these refer to morphologic states of musculature and are not formal systematic terms. Pedal retractor muscle scars are not known unequivocally in any of the Antarctic species.

According to Horný (1965), cyclomyans have a more or less complete ring of pedal retractor muscle scars, commonly fused to form a continuous band, as in *Archinacella* Ulrich and Scofield; they have the

Figure 3. Orientation of, and descriptive terms applied to, proplinid monoplacophorans. 1, 4, 7, 9: *Ellsworthoconus andersoni* new genus, new species (compare with Plate 14, Figs. 3-6). In this species, the apex projects little, or not at all, forward of the anterior apertural margin. 2, 3, 5, 6, 8, 10: *Proplina rutfordi* new species (compare with Plate 12, Figs. 11, 12, 15–18). In this species, the apex projects well forward of the anterior apertural margin. 1–3 are anterior views; 4 and 6 are dorsal views; 5, 9, and 10 are left-lateral views; and 7 and 8 are right right-lateral views. (A) = apex (beak); (Am) = apertural margin; (At) = anterior; (D) = dorsal; (L) = left; (La) = lateral area (arch); (P) = posterior; (R) = right; (Saa) = subapical area; and (V) = ventral. Length is the greatest anterior-posterior dimension; height is the greatest dorsal-ventral dimension; and width is the greatest right-left dimension, when the aperture is held horizontally. Rozov (1969) also proposed a system of terminology and measurements for monoplacophorans.

apex of the shell inside the muscle scar zone. Tergomyans have discrete pedal retractor muscle scars arranged in a more or less complete circle located toward the apex and with the apex overhanging the circle of muscle scars, as in *Pilina* Koken. Harper and Rollins (1982) defined cyclomyans as having a wide variety of coiled and uncoiled shell forms in which the axis of the shell apex intersects, or lies within, the plane of the muscle field. They defined tergomyans as cap-shaped or spoon-shaped shells having the apical axis lying outside the plane of the muscle field.

Order TRYBLIDIOIDEA Lemche
Superfamily TRYBLIDIACEA Pilsbry *in* Zittel-Eastman

Discussion. As used by Knight and Yochelson (1960), the superfamily Tryblidiacea contained the typical family and the Palaeacmaeidae. Within the Palaeacmaeidae, only the type genus of Late Cambrian age and, questionably, *Scenella* Billings of Early Cambrian age were included.

Yochelson and Stanley (1982) indicated that the Lower Ordovician genus *Palaeolophacmaea* Donaldson, originally thought to be a patelliform gastropod or a monoplacophoran, is better interpreted as a chitinoid float of a siphonophorate coelenterate, and they suggested that *Palaeacmaea* Hall also may be a similar coelenterate. *Scenella* Billings was described on the basis of a Lower Cambrian species, but it did not come into more general usage until after the publication of Ulrich and Scofield's (1987) work, in which *Archinacella* was also named. Both these names have been used widely for Upper Cambrian and younger symmetrical shells. Yochelson and Cid (1984) described *Scenella morenoensis* and assigned it to the coelenterates. They indicated that the type species, *S. reticulata,* should also be reassigned; however, some of the species later assigned to *Scenella* are accepted as mollusks.

With the transfer of the type species of *Scenella* and *Palaeacmaea,* in neither of which muscle scars are known, from the Mollusca by Yochelson and Stanley (1982) and Yochelson and Cid (1984), the superfamily contains only the genera in the Tryblidiidae. The Tryblidiidae is here defined as relatively flat shells having the apex at, or slightly overhanging, the anterior margin and having clearly developed symmetrical pedal retractor muscle scars arranged in a nearly circular to fusiform pattern open at the apical end.

In addition to the Silurian *Tryblidium* Lindström, this family would include at least *Pilina* Lindström, now known to range down into the Upper Ordovician (Peel, 1977), the Lower Ordovician *Bipulvina* Yochelson, and the Devonian *Drahomira* Horný. *Tryblidium simplex* (Billings) form the Upper Cambrian of eastern Canada has a shape near the limits of definition of the family in terms of height of the shell.

Family PROPLINIDAE Knight and Yochelson

Discussion. As used by Knight and Yochelson (1960), the Tryblidiidae included four subfamilies for Paleozoic fossils and one for living monoplacophorans. Additional information from the new Antarctic material, and reinterpretation of the original basis for the classification, prompts us to herein raise Proplininae to family rank.

When originally proposed, the Proplininae contained *Proplina* and questionably *Vallatotheca* Foerste, a rare genus described from the Upper Ordovician. Examination of the general shape and muscle scars in a topotype of the type species, *V. manitoulini* Foerste (unfortunately broken during preparation), suggests that the genus is more closely allied to *Archaeophialia* and may even by a synonym of that genus. It bears little relationship to *Proplina.* Runnegar and Jell (1976) assigned small Middle Cambrian specimens to *Vallatotheca*; muscle scars are unknown in those forms.

Proplina is widely misunderstood, partly because of work by Yochelson (1958). Few reports of muscle scars in pre-Silurian species were known at that time. In the paper cited, Yochelson reported multiple paired scars in a shell from the lower Lower Ordovician of Missouri, United States, which he assigned to *Proplina cornutiformis* (Walcott), the type species of the genus. This report of muscle scars was reproduced by Knight and Yochelson (1960, Fig. 46[3b]) as indicative of the genus and has since come into the literature as typifying the taxon.

Re-examination of the type lot and additional topotype specimens of *P. cornutiformis* convinces us that a reassignment of the Missouri specimen is in order. No muscle scars are known for the type species or for any of the other species assigned to the genus. We suggest that the material figured by Yochelson (1958) be transferred to *"Tryblidium"* species; perhaps a new genus is needed to accommodate it.

The Proplinidae are defined herein as strongly curved shells having the apex overhanging the presumed anterior and extending distinctly away from the apertural margin; the shells are thin and strongly arched laterally and have a widely oval aperture. Although the shell profile is convexo-concave in lateral view, it is far lower and far more strongly curved on the dorsal slope than any of the taxa referred to the Hypseloconidae.

Genus *Proplina* Kobayashi, 1933

Description. Strongly curved and strongly arched shells, bilaterally symmetrical. Protoconch unknown. Dorsum very well arched, completing about a quarter of a whorl. Apex extending anterior of margin, shell area below it well curved so that early growth stages protrude a significant distance beyond margin. Lateral slopes well rounded, whorl profile nearly hemispherical. Aperture all in one place, oval in outline, greatest width near center of shell.

Discussion. The name *Proplina* was coined to indicate similarity and affinity to the Ordovician-Silurian *Pilina* Lindström. *Pilina* is a very low, broad patelliform shell. The apex does overhang the margin but by a very small amount. In addition, the aperture is slightly geniculate, curving out of a plane. In contrast, *Proplina* is an elongate, anteriorly narrow shell, having the cross section well rounded (Fig. 3-2). The apex is extremely prominent and extends for some distance past the apertural margin (Figs. 3-5, 8, 10). The area of shell between the apex and the margin is significantly large. This region below the apex is curved and is strongly inclined subparallel to the curvature of the dorsum, so that the protrusion of the apex is accentuated. The apertural margin lies in a plane.

Muscle scars are conspicuous in *Pilina,* but are unknown in *Proplina.* On the basis of the muscle scars available in *Pilina, Tryblidium,* and a few other genera, the apex in low monoplacophorans is considered to be anterior. For purposes of description, we follow this convention in *Proplina.*

Proplina rutfordi n. sp
Plate 12, Figs. 6–8, 11–18

1982. N. gen., n. sp., Webers, Antarctic geosciences, p. 636, fig. 78.2-3.

Description. Linguliform shells having anterior area below apex divided into three obscurely distinguished segments. Protoconch unknown, early growth stages hook shaped. Shell very elongate, apex extending over apertural margin for almost one-fourth of total length of shell; periphery near anterior, its width about one-third of total aperture length; curvature constant at all growth stages. Lateral parts of shell region below apex flattened and inclined anteriorward, separated by a central moderately well rounded region. Dorsal part of shell arched in profile, rather than uniformly rounded. Apertural margin in a plane. Shell thin. Interior

irregularly rugose. Growth lines closely spaced, protruding slightly above interspaces.
Remarks. P. rutfordi differs from *P. cornutiformis* by its much lower shell and correspondingly more gently rounded dorsal outline. In both species, growth lines are fairly prominent, closely spaced, and very slightly irregular (Plate 12, Figs. 13, 16); the pattern of growth lines is a distinctive feature of the genus. *Proplina ampla* Kobayashi (1933, p. 264) from the Upper Cambrian Wanwangou Formation of northeastern China is similar to *P. rutfordi* in shell height but seems to have a shorter area below the apex and a correspondingly broader shell; its surface texture is unknown. *Proplina*(?) sp. (Kobayashi, 1933, p. 265), from the same interval, is shorter and more strongly curved dorsally than *P. rutfordi. Proplina* sp. (Cloud and Barnes, 1948, Plate 40, Fig. 21) from the Lower Ordovician Tanyard Formation of Texas is also more strongly curved dorsally. In *P. rutfordi,* the area below the apex (Plate 12, Fig. 17) is modified into two lateral areas on each side of a more rounded subapical area. The lateral slopes are flattened (Figs. 3-2, 3). Growth lines are prominent throughout ontogeny and are more closely spaced with age (Plate 12, Figs. 14, 16). In contrast, *Proconus incertis* has periodic growth rugae (Plate 13, Figs. 6–9).

F. Rasetti (oral communication to Yochelson, 1968) collected specimens from isolated blocks at Levis, Quebec, Canada, that have a blunted apex on the steinkerns, suggestive of an inner septum or apical thickening (see Yochelson and others, 1973, p. 289). Rasetti considered the Levis specimens to be Late Cambrian in age. Those specimens are similar to *P. cornutiformis* (Walcott). We see evidence of apical blunting on one steinkern of *P. rutfordi* (Plate 12, Figs. 6, 7). This is probably a thickening within the apex because there is no indication of any septation on the small specimen shown on Plate 12, Figures 13, 14.
Types. Holotype USNM 385208; figured paratypes USNM 385204, 385206, 385207; unfigured paratypes, USNM 388315, 388316.
Type locality. From the feather edge of the Minaret Formation on the northeastern side of Springer Peak, Webers Peaks, Heritage Range (Fig. 1).
Distribution. Springer Peak; Bingham Peak (USNM 388315); Yochelson Ridge (USNM 388316).
Etymology. Robert Rutford, University of Texas, Dallas, United States, provided additional materal from an earlier expedition to the Ellsworth Mountains and also participated in the 1979–80 expedition.

Genus *Proconus* n. gen.

Type species. Proconus incertis n. sp.
Description. Low to moderately low shells, having a distinct apex. Aperture broadly oval. Dorsum with slight logarithmic curvature. Area below apex nearly straight. Muscle scars unknown.
Discussion. This genus may be considered as allied to *Proplina,* but the angle of logarithmic curvature of the shell is much lower, producing a relatively longer shell. Whereas *Proplina* at maturity completes perhaps one quarter of a whorl, *Proconus* shows hardly any obvious curvature in lateral outline. It appears in profile like a scalene triangle. The degree of extension of the apex away from the aperture is remarkable.

It is unfortunate that the logarithmic spiral presents such formidable difficulties in measurement, particularly when the rate of expansion is low. In theory, the differences between *Proplina* and *Proconus* could be easily expressed mathematically, but in practice this is difficult to do. Modeling by computer may be the next step toward clarifying understanding of the geometry of monoplacophoran shells.

Even measuring the rate of expansion of width presents problems. In *Proplina,* the shell becomes relatively wide after the early growth stages (Plate 12, Fig. 14); in *Proconus,* the shell widens much more gradually (Plate 13, Fig. 3). However, the differences constitute only a few degrees of divergence on either side of the plane of bilateral symmetry and are too small to measure reliably.

The periodic thickening of the aperture in *Proconus* is comparable to that of *Tryblidium* but probably does not indicate any relationship. *Tryblidium* is a flat shell bearing prominent lamellae on the exterior; its closest ancestor is probably the Lower Ordovician *Bipulvina* Yochelson from Missouri, United States. Fragmentary specimens of *Proconus* may be distinguished from *Proplina* because they show essentially no growth lines between the periodic faint varices.

Proconus incertis n. sp.

Plate 12, Figs. 9, 10, 19–22; Plate 13, Figs. 1–12

Description. Relatively low shell with apex protruding far beyond anteroventral margin. Apex strongly overhanging aperture, extending nearly one-third of total length of shell. Aperture elongate oval, the greatest width near the center of the shell. Area below apex nearly straight and flattened laterally. Shell with distinct rugosities of growth. Dorsum slightly curved, lateral slopes very little arched.
Remarks. This shell has an unusual shape. The extension of the apex is so great that the shell appears in danger of tipping forward (Plate 13, Figs. 4, 8, 12). Naturally, the shell would not tip over when the soft parts were within the shell and it was buoyed up by water. This problem of seeming overextension of the apex is compounded by one large steinkern assigned to the species, in which a seemingly blunted apex gives a suggestion of shell thickening in the apical area (Plate 13, Figs. 1, 2). Perhaps *Proconus incertis* was sedentary.

Specimens of *P. incertis* have prominent comarginal growth rugae (Plate 12, Figs. 9, 10). These appear to be spaced periodically, with the separation between them gradually increasing with size. These varices are indicated on the interior (Plate 12, Figs. 19, 22) by expansion and constriction. The steinkerns of large specimens of *Proplina rutfordi* are more arched and many are relatively smoother.

Metoptoma retrosa Whitfield from the Upper Cambrian of Wisconsin, United States, is assigned to this genus. It differs from *P. incertis* in being a more distinctly curved shell laterally and in having relatively less protrusion of the apex.
Types. Holotype, USNM 386554; figured paratypes, USNM 385205, 385209–385214, 386555; unfigured paratypes USNM 388317.
Type locality. From the feather edge of the Minaret Formation on the northern side of Springer Peak, Webers Peaks, Heritage Range (Fig. 1).
Etymology. Proconus is an apt name for a cone that is presumed to be forwardly inclined. The specific name indicates the uncertainty that surrounds all attempts to distinguish species of monoplacophorans.

Genus *Ellsworthoconus* n. gen.

Type species. Ellsworthoconus andersoni n. sp.
Description. Bilaterally symmetrical shells, having distinct logarithmic curvature of both anterior and dorsal slopes. Apex assumed anterior. Protoconch unknown, but early and late growth stages in comparable pattern throughout growth. Apex at, or just overhanging, apertural margin. Anterior and dorsal slopes both expanding slowly so that shells become relatively high. As size increases lateral slopes approach vertical. Retractor muscle scars unknown.
Discussion. During the last two decades, high curved shells have been traditionally assigned to *Hypseloconus* and low curved shells to *Proplina.* As noted, we judge *Proplina* to be a relatively specialized form having the apex extended relatively far away from the vertical projection of the apertural plane. Because *Proplina* has stronger curvature and is lower, it has relatively little shell area below the apex. Other species, which are

placed in *Proplina,* have the area below the apex almost as strongly curved as the lateral profile of the dorsum so that the apex is never far removed from the anterior margin of the aperture. Another way to express this same distinction is to consider the empty shell resting on a flat surface. If a vertical plane from the apex nearly touches the margin of the anterior aperture, the shell is stable. In *Proplina,* the apex is distinctly in front of the aperture; in *Proconus,* it is so far forward that the empty shell is nearly unstable.

Ellsworthoconus is an extremely stable shell, when viewed in this way. At most, the apex extends slightly beyond the apertural margin at all growth stages (Figs. 3–4, 7, 9). The curvature is relatively slight after the initial part of the shell so that although the individual gets much higher, it does not get appreciably wider. Thus, the apex remains close to the margin of the aperture. In further contrast to *Proplina,* which has a well-arched anterior profile (Fig. 3–2), large specimens of *Ellsworthoconus* have the lateral slopes very steeply inclined and even nearly vertical (Fig. 3–1).

Ellsworthoconus and *Hypseloconus* differ primarily in terms of length, width, and height (Fig. 3–2, 7, 10). The Hypseloconidae are high so that the vertical height from apex to aperture is significantly greater than the length of the aperture (Plate 16, Fig. 1). In *Ellsworthoconus,* the length of the aperture is less than twice the height. In addition, *Ellsworthoconus* curves consistently in the same direction during growth, whereas the hypseloconid *Knightoconus* reverses curvature during growth (Plate 15, Figs. 1–4).

Stinchcomb (1980, p. 45, text Fig. 1D) figured, but did not name, from the Upper Cambrian Potosi Dolomite of Missouri, United States, a monoplacophoran that is a representative of *Ellsworthoconus.* It is possible that the poorly known *Micropileus* Wilson, from the Middle Ordovician of Canada, is related to *Ellsworthoconus.*

Ellsworthoconus andersoni n. sp

Plate 13, figures 13–18; Plate 14

1981. *Proplina* sp., Pojeta and others, U.S. Geological Survey Open-File Report 81–743, p. 167.

1982. *Proplina* n. sp., Webers, Antarctic geoscience, p. 636, fig. 78.2–13.

Description. Moderately large, very strongly arched, bilaterally symmetrical shells. Protoconch unknown, early growth stages expanding uniformly, later growth stages having more posteriorward than lateral expansion. Subapical area showing distinct logarithmic curvature in early growth stages, much less inclined with increasing size. Anterior margin and adjacent slopes well rounded. Dorsum showing a distinct logarithmic curving completing about one-quarter whorl. Dorsal region slightly inclined laterally, juncture with lateral slopes well rounded but distinct. Lateral slopes having essentially no curvature after early growth stages, becoming increasingly steep with maturity. Shell thin, thickening slightly with maturity. Growth lines faint and closely spaced. Retractor muscle scars unknown.

Remarks. This species is somewhat variable in shape in the early growth stages. Viewing the dorsal outline laterally, in some specimens it can be seen that the first part of the logarithmic spiral appears to expand slowly (Plate 13, Fig. 16) so that there is obvious curvature near the apex. Others expand more rapidly (Plate 14, Fig. 3) so that the dorsum has far less curvature. There is also some difference in the dorsal view, one individual being somewhat flattened dorsally (Plate 13, Fig. 14). However, attempts to distinguish two species within the available material have not been successful.

Some of the individual variation may be a consequence of minor deformation. Some may be a consequence of comparing steinkerns with shelled material. Even though the shell is thin, it is markedly thicker than the shells of other monoplacophorans; the increase in shell thickness with size may be seen on steinkerns that retain patches of the shell (Plate 13, Fig. 17).

No retractor muscle scars have been seen. However, one unusual feature is present internally (Plate 14, Fig. 12). The lateral slopes posterior to the apex do not expand uniformly but are constricted. This constriction is represented on the steinkern by a relatively deep depression. Although only the largest steinkern shows this depression, smaller specimens do show a constriction in the shell surface in this region. This feature is not a muscle scar, for a scar would be shown on the internal filling by a protrusion rather than a depression. There is no obvious function for the slight lateral indentation near the apex. It may be a pathologic feature comparable to a blister pearl formed in some pelecypods when there is an irritation between the mantle and the inner surface of the shell. Various elongate markings on the lateral slopes are also best seen on the largest specimen; these might be impressions of radial mantle muscles.

Ellsworthoconus andersoni differs from *Metoptoma barabooensis* Whitfield from the Upper Cambrian of Wisconsin, United States, here transferred to *Ellsworthoconus,* in being lower and less inflated. *Proplina bridgei* Kobayashi (1933, p. 263–265) from the Upper Cambrian Wanwangou Dolomite of northern China, here also transferred to this genus, is lower and broader anteriorward than the type species.

Metoptoma similis Whitfield and *M. perovalis* Whitfield, from the Upper Cambrian of Wisconsin, United States, are both quite low. *Proplina* sp. of Cloud and Barnes (1948, Plate 41, Fig. 23) from the Lower Ordovician Gorman Formation of Texas, United States, is similar in shape. All three are transferred to *Ellsworthoconus* but may eventually be found to be distinct enough to be placed in a separate genus, if the philosophy of classification suggested here is followed.

Types. Holotype, USNM 386900; figured paratypes, USNM 386556–386558, 386560–386567; unfigured paratypes, USNM 388318, 388319.

Type locality. From the feather edge of the Minaret Formation on the northeastern side of Springer Peak, Webers Peaks, Heritage Range (Fig. 1).

Distribution. Springer Peak; Bingham Peak (USNM 388319).

Etymology. Among those who participated in the Ellsworth Mountains Expedition, John Anderson, Victoria University, Wellington, New Zealand, one of the younger Old Antarctic Explorers, was a constant source of knowledge and assistance in the field.

Order KIRENGELLIDA Rozov

Discussion. "The shell is cone-shaped' sometimes there is a slight tendency toward planispiral coiling, but not more than one-quarter to one-third of a volution. The apex usually is not central, although this is not impossible. Characteristically the apex is shifted forward and even may overhang the anterior margin. Muscle scars (6 to 8 pairs) are placed concentrically around the apex." (Rozov, 1975, p. 42, translated by M. Balanc.)

In defining the order, Kirengellida Rozov (1975, p. 42) included three families. He raised the Archaeophialinae of Knight and Yochelson to a family. As originally conceived, it contained the type genus of Middle and Late Ordovician age, questionably the poorly known Middle Ordovician genus *Micropileus* Wilson, and equally questionably the Lower Devonian genus *Calloconus* Perner. Horný (1963, p. 58–59) transferred *Calloconus* to the Patellacea (Gastropoda); *Micropileus* remains poorly known but need not be allied to *Archaeophialia.* In addition to *Archaeophialia,* we would include *Vallatotheca* Foerste, as noted in the discussion of the Proplinidae, in the family. The Romaniellidae are discussed below in connection with the Hypseloconidae.

Family KIRENGELLIDAE Starabogatov

Discussion. As considered by Rozov (1975, p. 42) the family Kirengellidae contained *Kirengella* Rozov, *Moyerokania* Rozov, and *Scenella* Billings. As noted above, some species that have been assigned to *Scenella* may not be monoplacophorans. The Lower Ordovician *Yochelsoniella* Flower may belong in this family; the form is laterally compressed so that the aperture is nearly lozenge shaped. Assigned genera have in common a moderately high form, intermediate between *Hypseloconus* and *Pilina.* They are all more or less subconical and have the apex distinctly above the aperture, rather than overhanging it. Muscle scars are known in the type species of *Kirengella* (Rozov, 1968) from Siberia. Similar scars were described by Stinchcomb (1980) in *Hypseloconus* cf. *H. simplex* Berkey, from Missouri, United States, a species that we here assign to *Kirengella.*

Not too much emphasis should be placed on the number or pattern of muscle scars at this stage of our understanding of the monoplacophorans. Nevertheless, there is some general similarity between those of *Kirengella* and those of *Archaeophialia,* and the two families seem to be related. Muscle scars are not known for *Yochelsoniella.*

Genus *Kirengella* Rozov, 1968

Description. "Shell of medium size and medium height with wide elliptic base and apex displaced slightly anteriorly. Anterior, posterior and lateral slopes straight. Surface of the shell having abundant small concentric furrows and vaguely expressed concentric folds; elements of radial structure are absent. Six pairs of muscle impressions, of which the three anterior pairs are situated individually, the impressions of 4th and 5th pairs are in relation to each side and the impressions of the 6th pair are represented as two tubercles on the posterior slope." (Rozov, 1968, p. 1428, translated by M. Balanc.)
Discussion. As discussed by Yochelson and others (1973), the question of anterior and posterior in some monoplacophorans, even with the muscle scars preserved, is open to different interpretations. They suggested that if the largest scar of *Kirengella* is homologized with that of *Pilina* or *Tryblidium,* the apex is posterior rather than anterior. For purposes of description, we will follow the convention that the apex is anterior. Small specimens from Antarctica suggest that the early growth stages of this genus may be a simple high cone that abruptly changes its rate of expansion to the characteristic more gentle slope seen on larger shells.

Some species that have been assigned to *Hypseloconus* are better placed in *Kirengella.* In effect, those forms that are not concavo-convex in lateral profile are reassigned there. Among others, these include *Metoptoma augusta* Billings, *Hypseloconus stabilis* Berkey, *H. washingtonense* Stinchcomb, and aff. *Proplina* of Cloud and Barnes (1948, Plate 41, Fig. 20). *Metoptoma recurvum* Whitfield from the Upper Cambrian of Wisconsin, United States, is here transferred to *Kirengella* because the aperture is more elongate than that of typical species of *Hypseloconus.*

Yochelsoniella may be distinguished from *Kirengella* by its exceedingly narrow width. The aperture of *Yochelsoniella* is more lozenge shaped, whereas in *Kirengella* the outline of the aperture approaches an egg shape. Billings (see Billings, 1865, p. 86–87) described several species that he assigned to *Metoptoma* from "limestone No. 2" at Point Levis, Quebec, Canada. Presumably those are of Early Ordovician age; few of the species have been illustrated. Apparently *M. melissa* has an apex that does not overhang the margin. The same is reported for *M. orphyne,* and it is possible that both are species of *Kirengella* or *Yochelsoniella. Metoptoma augusta* is noted by Billings as possibly not being symmetrical, and the species may not be a monoplacophoran.

Knight and Yochelson (1958, p. 44, Plate 5, Figs. 1–3) referred a specimen from the Lower Ordovician of Missouri, United States, to *Hypseloconus* species. This should now be assigned to *Kirengella.* At the time, they suggested that a raised circular band on the steinkern was a muscle scar. In light of the muscle scars known for *Kirengella* and those described by Stinchcomb (1980) for *Hypseloconus* cf. *H. stabilis* Berkey, this now seems unlikely. The species figured by Stinchcomb (1980) is here placed in *Kirengella.*

Kirengella pyramidalis n. sp.

Plate 10, Figs. 1–7

Description. Conical shells having a broadly oval aperture and the apex toward the presumed anterior. Protoconch unknown, shell seemingly with uniform shape at all growth stages. Apex within area of aperture, located strongly anteriorward. Lateral slopes just below and anterior of apex flattened, joining anteriorly at a rounded juncture so that this part of the shell is distinctly triangular, the sides diverging at about 45 degrees from the plane of bilateral symmetry. Juncture of lateral slopes with anterior slopes moderately distinct but not set off by any ridges or angulations. Middorsum well rounded so that there is a continuous curve from the one side of the aperture through the dorsum to the other. Aperture distinctly oval, the greatest width near midlength.
Remarks. In the type species, *K. ayaktchica* Rozov, from the Upper Cambrian of Siberia, U.S.S.R., muscle scars are conspicuous and the apex is subcentral. In *K. pyramidalis,* the apex is distinctly toward one end of the shell, so that the presumed anterior slope is quite short. Lateral and posterior slopes are subequal. No muscle scars are known, but only a few specimens are available, and most of these retain the shell.
Types. Holotype, USNM 385188; figured paratypes, USNM 385189, 385190; unfigured paratypes, USNM 388320.
Type locality. From the feather edge of the Minaret Formation on the northeastern side of Springer Peak, Webers Peaks, Heritage Range (Fig. 1).
Etymology. The specific name is given in token of the overall pyramidal shape of this species.

Family HYPSELOCONIDAE Knight

Discussion. The family Hypseloconidae was placed by Knight and Yochelson (1960) in the Order Archinacelloidea. At that time, no muscle scars were known, and a ring-shaped impression was interpreted by Knight and Yochelson (1958) as a muscle scar, a view long since abandoned by Yochelson. As proposed, the family contained the type genus *Hypseloconus* Berkey and two other genera questionably assigned to it. *Pollicina* Holzapfel, from the Middle Ordovician of the Baltic region, is based on few specimens. Fortunately, topotypes are available in the Natural History Museum in Stockholm, Sweden, and examination of them shows that this is a slightly asymmetrical form; it is an open-coiled euomphalacean, much like *Lytospira* Koken (Yochelson, unpublished data), rather than a monoplacophoran. *Ozarkoconus* Heller continues to be poorly known, but it also may not be a monoplacophoran; its placement is uncertain.

Rozov (1975, p. 42) proposed the family Romaniellidae and included in it *Romaniella* Doguzhaeva from the Upper Cambrian and Upper Ordovician of Siberia, U.S.S.R., *Nyuella* Rozov from the same region, but of Early Ordovician age, and *Hypseloconus.* The taxon Hypseloconidae has priority.

We assign to the Hypseloconidae the two Upper Cambrian genera *Shelbyoceras* Ulrich, Foerste, and Miller and *Knightoconus* Yochelson, Flower, and Webers. All have in common a distinctly curved quite high shell. Specimens assigned to *Hypseloconus* that are lower and have the shell apex more or less subcentral are better assigned to *Kirengella* or one of its allies. On the basis of this reassignment, the extremely high curved cones that are placed in the Monoplacophora are then confined to the Upper Cambrian.

Some of the species of hypseloconids such as *Metoptoma*? *peracuta* Walcott are very poorly known. However, *Hypseloconus bonneterrense* Stinchcomb and *H. elongatus* Berkey, the type species, are well preserved, and the absence of septa can be demonstrated by steinkerns that end at a tiny apical area.

Berkey (1898) distinguished seven species or "varieties" of *Hypseloconus,* four of which are from the Franconian-age sandstone at Taylors Falls, Minnesota, United States. For several of these, the apical condition is unknown. Some poorly preserved specimens suggest that septate forms may be present, and septate cones have been collected from the locality (R. Sloan, oral communication, 1982). Much confusion surrounds the details of the type species of *Hypseloconus,* but for the present, *Hypseloconus* is considered nonseptate; *Shelbyoceras* and *Knightoconus* are both septate.

The reversal of curvature during early growth is well documented for *Knightoconus.* It has also been observed in several specimens of *H. elongatus* Berkey. Possibly this change in shape during early growth is a feature of the family, but observations on more species are needed to confirm this suggestion.

Genus *Knightoconus* Yochelson, Flower, and Webers, 1973

Discussion. This genus is the only member of the Ellsworth molluscan fauna that was formally described previously. This high conical curved form falls readily within the Hypseloconidae by consequence of its general shape. It is distinguished from the type genus by the presence of multiple internal septae in the smaller parts of the shell. In this regard, it is most similar to *Shelbyoceras* Ulrich, Foerste, and Miller; that genus was originally considered to be a cephalopod, but it has been reinterpreted as a septate monoplacophoran (Stinchcomb and Echols, 1966; Stinchcomb, 1980). *Knightoconus* differs from *Shelbyoceras* in being slightly more arched and expanding at a slightly greater rate on the concave, presumably posterior, part of the shell. *Shelbyoceras* gives a general appearance of being slightly more elongate and in that sense is more like the earliest cephalopod *Plectronoceras.*

When Yochelson and others (1973) described *Knightoconus,* they emphasized that its septate condition suggested that the genus might have been a precursor of the cephalopods. This speculation has not been generally considered in the literature but has been accepted, with modification, by at least one author (Bandel, 1982, p. 83–84).

Knightoconus antarcticus Yochelson, Flower, and Webers, 1973
Plate 15; Plate 16, Figs. 1–11; Plate 17

1973. *Knightoconus antarcticus* Yochelson, Flower, and Webers, Lethaia, v. 6, p. 290, Fig. 9a–c, p. 292, Fig. 10a–b.
1981. *Knightoconus antarcticus* Pojeta, Webers, and Yochelson, U.S. Geological Survey Open-File Report 81-743, p. 167.
1982. *Knightoconus antarcticus* Webers, Antarctic geoscience, p. 636, Fig. 78.2-4.

Remarks. A long description and a number of figures of this species were given in the original publication. There is no need to repeat the description, although some points are discussed below and supplemental illustrations are given.

The previous evidence of multiple septa has been confirmed by examination of a large number of new specimens (Plate 15, Fig. 6; Plate 16, Figs. 1–8). The septa of *K. antarcticus* are not portions of spheroidal surfaces (Plate 16, Fig. 8) like septa in some late Paleozoic orthoconic cephalopods. Rather, they appear to be little arched, except near the junction of the shell wall and the septum itself. Arching is more evident on larger specimens, but this may be a function of larger size, making observation easier. Examination of many specimens also shows that the plane of the septum is at an angle to the plane of the growth lines, which marks the earlier position of the aperture (Plate 16, Fig. 9). Because the septa were laid down later than the external shell, this angle probably is nothing more than a consequence of the curvature of the soft parts as the shell increased in both length and volume.

If one takes the convex side as anterior, the aperture is consistently narrower at this end (Plate 15, Fig. 11). The aperture itself is very nearly symmetrical, but at almost all growth stages, the aperture is oval rather than elliptical. The narrower anterior may be a useful criterion for orienting conical shells that lack muscle scars. The relatively long expanse of shell posteriorward (Plate 16, Fig. 2) was termed by Knight (1952, p. 26) "a trainlike hood," but this seems an inappropriate morphologic term. On this part of the shell, Bandel (1982, Fig. 58) restored an operculum to *Plectronoceras,* the earliest known cephalopod, but we have no evidence that a calcareous operculum was ever present in *Knightoconus,* and no strong basis exists for this aspect of his reconstruction.

Our evidence indicates that the early growth of this species is marked by a change in curvature of the shell (Plate 15, Figs. 1–4). After a few millimeters of growth, the shell then reversed and thereafter is concave posteriorward. There is no trace of a coiled or strongly curved protoconch. This hook-shaped apex and the reversal of curvature are confirmed in several specimens.

Pojeta and Runnegar (1976, Plate 15) illustrated a generally similar nontorted apex in a Middle Ordovician specimen of ?*Macroscenella.* That specimen has no reversal of curvature. Thus, attempting to distinguish anterior from posterior in the Monoplacophora will continue to present uncertainties.

Material. Figured topotypes USNM 386568–386585, 386588, 386589; unfigured specimens USNM 388321, 388322.

Distribution. Springer Peak, Yochelson Ridge (USNM 388322) (Fig. 1).

?Order ARCHINACELLOIDEA Knight and Yochelson

Discussion. The order Archinacelloidea was based on the premise of a complete or incomplete ring-shaped muscle. Included taxa ranged from very low and broad shells to quite high and narrow cones. There is now considerable question of the validity of the premise for the order. The type family contained only two genera, and one of these, *Ptychopeltis* Perner, has been reassigned to the inarticulate Brachiopoda (Horný, 1963, p. 61–64). The only other family, the Hypseloconidae, was associated because of a presumed ring-shaped muscle scar. As mentioned under *Kirengella,* this feature is probably not a muscle scar.

Archinacella is reasonably well known from its Middle Ordovician type species, but its placement remains uncertain. A horseshoe-shaped muscle scar is present below the apex; Knight (1941) gave details. Knight and Yochelson (1958, Plate 5, Fig. 4) illustrated a specimen having two paired muscle spots on the dorsum and used this as confirmatory data in assigning the genus to the Monoplacophora. At the time of that work, prevailing theory was that the Patellacea did not originate until the mid-Paleozoic and that all patelliform shells in the early Paleozoic necessarily fit into the Monoplacophora. The then-recent discovery of a living monoplacophoran had a profound impact on discussions of the importance and diversity of the monoplacophorans.

On the basis of new information and further study of early and mid-Paleozoic patelliform mollusks, E.L.Y. thinks that the Patellacea may have begun earlier than the mid-Paleozoic. An undescribed Middle Ordovician patelliform specimen from Utah (Yochelson, unpublished data) is circular in apertural view, rather than oval, and has a muscle scar similar to the scar in the type species of *Archinacella.* It also contains other internal markings and is convincingly interpreted as a patellacean gastropod. The assignment of *Archinacella* must remain uncertain until that material is described and evaluated.

Superfamily UNCERTAIN

Discussion. The new genus described below has as one of its features a ringlike sulcus in the apical region. Possibly this links the form to the Archinacellacea. In other features, this taxon is different enough from other shells described in this work that any high-level placement must be uncertain. Radial ornament is uncommon on most monoplacophorans but is found in *Archaeophialia* and *Vallatotheca.* The new genus could be a gastropod as it is like some of the capuliform taxa living today. The limpet shape may be regarded as a convergent end member of several unrelated lineages within the gastropods, and in that sense, the appearance of this shell form in the Late Cambrian would be extremely surprising. The shell appears symmetrical, and on balance, it probably is better assigned to the Monoplacophora.

Genus *Cosminoconella* new genus

Type species. Cosminoconella runnegari new species
Description. Strongly curved, capuliform mollusks bearing abundant lirate ornament. Shell strongly arched, completing about one-quarter of a whorl. Aperture having an oval shape, posterior wider than presumed anterior and lateral margins nearly straight. Lateral slopes little inflated, curving smoothly to join dorsum. Greatest width strongly posteriorward. Shell moderately thin, ornamented by uniformly spaced abundant spiral lirae.
Discussion. The morphology is deceptively simple in being a rapidly expanding, strongly curved cone. In general terms, the shape is more like that of *Proplina,* as defined herein, than that of any other early Paleozoic molluscan genus. However, the aperture is relatively much shorter, and the beak does not strongly overhang the anterior apertural margin. When viewed from above, the flattening of the dorsal shell segment is pronounced. The apex is more hooklike, with a proportionally shorter aperture, than that of *Ellsworthoconus.*

Although the radial ornament is vaguely suggestive of the Upper Ordovician presumed monoplacophoran *Helcionopsis,* in that genus, the ornament is incised, rather than raised, and growth lines are fairly prominent. The compressed curved form of *Cosminoconella* is quite different from the broad low shell of *Helcionopsis.* The Middle Cambrian specimens assigned by Runnegar and Jell (1976) to *Vallatotheca* show prominent radial ornament but differ in being rugose.

Most specimens are obviously bilaterally symmetrical, although in some of the material there is a hint of asymmetry, as evidenced by a slight inclination of the dorsal surface. This inclination may be entirely the result of diagenetic or tectonic effects and has been discounted, but the possibility remains that this could be a fundamental feature of the organism. The only previous report of asymmetry in a propliniform monoplacophoran shell is *Cyrtonellopsis* Yochelson from the Lower Ordovician of Missouri, United States. This is known from a single species, represented by three specimens. The shell expands rapidly and is "hook shaped," but on each of the three individuals the apex has a different orientation relative to the main axis of the shell. In that taxon, there is no question that the asymmetry is really present. None of the Antarctic specimens complete as much as half a whorl.

Cosminoconella runnegari n. sp.
Plate 11; Plate 12, figures 1–5

1982. N. gen., n. sp., Webers, Antarctic geoscience, p. 636, Figs. 78.2-6, 9.

Description. Strongly curved, rapidly expanding mollusks having an ovoid aperture. Protoconch unknown. Early growth stages and all succeeding growth uniform. Strongly curved, mature specimens completing about one-fourth of a whorl. Expanding rapidly; width of aperture approximately same as length. Greatest width close to presumed posterior. Area below apex little curved. Lateral slopes only gently arched, the juncture with anterior and posterior slopes smooth. Aperture in a single plane. Growth lines weak, seemingly not interrupting the radial ornament. Ornament of several dozen, raised rounded lirae, extending from the apex, uniformly spaced with interspaces five to six times width of lirae. Shell moderately thin, of two layers; at maturity, the ornament reflected on the steinkern.
Remarks. This species is characterized by its wide, very rapidly expanding shape. As the species is used here, there is some variation in the rate of shell expansion, some specimens being about two-thirds the width of others, at comparable shell height. However, all attempts to separate these into clear morphotypes have been unsuccessful.

The shell is moderately thin but is thicker than that of other monoplacophorans in the fauna that are the same size (Plate 11, Figs. 3, 5). Ornament is most prominent on the outer layer (Plate 11, Figs. 1, 18). Where the outer layer is lost, the ornament is much reduced (Plate 11, Fig. 10). To avoid the possibility that steinkerns might be confused with those of *Proplina rutfordi,* we illustrate only (1) specimens for which a counterpart is known showing that the spiral lirae are preserved, or (2) specimens that show the ringlike sulcus (Plate 11, Figs. 3, 13, 15; Plate 12, Figs. 1, 5).

The most interesting feature is seen on the steinkern. Below the apex, a relatively wide sulcus is present (Plate 11, Fig. 4). It is not present on all specimens (Plate 11, Fig. 19), but in the examples where it is present, it is prominent. The sulcus shallows anteriorward (Plate 11, Fig. 13; Plate 12, Fig. 2) and may not be present on the anterior slope of the steinkern. Essentially no indication of this feature appears on the exterior of the shell, in contrast to *Ellsworthoconus andersoni* where lateral indentations are moderately prominent. The sulcus is apparently the result of internal thickening of the shell, but its function is not obvious.

For the moment, we prefer to consider the distinct ornament of *Cosminoconella* as a key generic feature and relegate the internal feature to a specific distinction. As our knowledge of the Upper Cambrian monoplacophorans increases, the peculiar internal feature may take on greater significance in classification.
Types. Holotype, USNM 385191; figured paratypes USNM 385192–385203; unfigured paratypes 388324.
Type locality. From the feather edge of the Minaret Formation, on the northeastern side of Springer Peak, Webers Peaks, Heritage Range (Fig. 1).
Distribution. Springer Peak; Yochelson Ridge (USNM 388324).
Etymology. This small ornately ornamented conical shell is named for Bruce Runnegar, a productive and thoughtful worker who is also involved in unraveling the early history of the Mollusca. *Cosminoconella* is derived from the Greek words meaning ornamented cone.

Order CYRTONELLIDA Horný
Superfamily HELCIONELLACEA Wenz
Family HELCIONELLIDAE Wenz

Genus *Latouchella* Cobbold, 1921

Discussion. In his study of the type species of what were then considered Paleozoic gastropods and are now known to include the Monoplacophora, Knight (1941) overlooked this genus. As a consequence, the type species, *L. costata* Cobbold, is not well known, and, unfortunately, the type specimens are not particularly well preserved (Yochelson, unpublished data). They add little to the original description and line drawings.

Runnegar and Jell (1976) distinguished among the three genera of rugose helcionellids currently in the literature. *Helcionella* Grabau and Shimer has a broad cross section and a large aperture, which is about subcircular in outline. *Latouchella* has a narrow, compressed cross section and, correspondingly, an elliptical to subrectangular aperture. Coreospira Saito has an even more rectangular aperture. All three genera have comarginal rugae and grew by adding wedge-shaped increments.

Latouchella? species indeterminate
Plate 2, Figs. 16, 17; Plate 3, Figs. 1–3

Remarks. Poorly preserved helcionellaceans were found at the type locality of the Minaret Formation in the northern Marble Hills. They show the key features of rugose comarginal ornament with fine spiral ornament on the shell between these rugae (Plate 3, Figs. 1, 2) and wedge-shaped growth increments (Plate 2, Figs. 16, 17; Plate 3, Fig. 3). These features clearly place the material within the Helcionellidae. Both *Helcionella* and *Latouchella* bear fine spiral ornament. Even though the aperture cannot be observed, the shell overall is relatively compressed. Accordingly, we have assigned the material to the latter genus, but with question.

As described in the literature, *Latouchella* has a stratigraphic range from the Lower Cambrian (Tommotian) to the Upper Cambrian (Changshanian) (Dresbachian). However, the genus is far more common in pre-Dresbachian rocks. *Helcionella* also has a long range. It occurs predominantly in the Lower Cambrian (Tommotian-Ordian, Runnegar and Pojeta, 1989).

"Helcionella" was reported by Öpik (1967, p. 47) from the Upper Cambrian (Mindyallan) of Queensland, Australia. Palmer (1982) figured a specimen of what is probably *Helcionella,* which he identified as *Pelagiella* species from the Dresbachian in the subsurface of Indiana, United States. In spite of these occurrences, *Helcionella* also is predominantly a pre-Dresbachian form. Even though we are not certain of the generic assignment, we are confident that this form indicates beds of Cambrian age, probably older than Late Cambrian.

Figured specimens. Hypotypes, USNM 385129, 385130; unfigured specimens cataloged as USNM 390102.

Distribution. Northern Marble Hills, south-southeast of Minaret Peak, (Fig. 1).

Class GASTROPODA Cuvier

Introduction

The class Gastropoda is based on living organisms that have undergone torsion—that is, twisting of the soft parts—or are judged to be descended from torted ancestors. In the absence of soft parts, assignment of fossils to the Gastropoda carries a degree of uncertainty. Commonly the shell is coiled in three dimensions, but many living species have shells that are secondarily bilaterally symmetrical; some bilaterally symmetrical fossil shells have also been assigned to the gastropods by comparison to living forms.

Not all coiled calcareous forms are gastropods—witness the common annelid worm *Spirorbis.* Without elaborating on the problem of lack of data on the soft parts, we need only say that we think the species described below are very likely to be gastropods. Various forms in the Lower and Middle Cambrian have been assigned to the Gastropoda. Of particular interest is *Yuwenia* Runnegar (1981b, p. 315) from the Lower Cambrian of Australia. This small form coils sinistrally in conventional orientation and shows general similarity of shape to some of the Upper Cambrian taxa described below. The ancestry and origin of the class, as well as its oldest known representative, are vigorously debated.

Recently, Linsley and Kier (1984) proposed the concept of a class Paragastropoda. This is based on the assumption of a nontorted mollusk in a coiled shell. Without passing judgment on this concept, we do not use it here for purposes of classification.

Gastropods from the Minaret Formation bring to light two significant findings. First, if one applies conventional orientation, all the Antarctic species are left-handed. In modern faunas, left-handed gastropods are rare, and this rarity has been present since Ordovician time. Second, all the Ellsworth Mountains gastropods possess a simple margin. However, in the Ordovician, and throughout most of the Paleozoic, species having a distinct slit in the margin predominate. Most malacologists would accept the Paleozoic species having a slit in the margin as directly ancestral to modern gastropods. Yet in this Upper Cambrian assemblage only species having a simple margin are present. How these Cambrian species are related to Ordovician and younger species having slits is a topic under consideration.

Order ARCHAEOGASTROPODA Thiele
Suborder MACLURITINA Cox and Knight

Discussion. The suborder Macluritina as defined by Knight and others (1960) included the Macluritacea and the Euomphalacea. Yochelson (1984b) suggested that these superfamilies are not related. For convenience, we continue to follow the conventional definition.

Superfamily MACLURITACEA Fischer

Discussion. As used by Knight and others (1960), the superfamily Macluritacea was divided into two families, the Macluritidae and the Onychochilidae. There are genera in the Ellsworth fauna that may fit into both groups. Since the work of Knight (1952) on fossil monoplacophorans and early gastropods, a great deal of attention has been directed toward these fossils. In a section entitled "*Maclurites* and its allies" Knight (1952, p. 36–40) proposed that hyperstrophy occurred in several early and mid-Paleozoic taxa. In left-handed or sinistral gastropods, the general shell form and soft anatomy are mirror images of that of the more common dextral gastropods; a few species among living forms have both dextral and sinistral individuals in the same population. In hyperstrophy, a rare condition observed in a few living forms, the soft parts are twisted in a direction opposite to that of the soft parts of orthostrophic gastropods; the shell of hyperstrophic species is coiled "left-handed" in the conventional orientation. However, if the spire is interpreted as being depressed, rather than rising above the aperture, the shell is right-handed, as are the soft parts. Along with the reverse coiling of the shell, the soft anatomy of the hyperstrophic gastropod is also modified.

Although hyperstrophy had been known for more than a century as a phenomenon among living gastropods and had even been suggested for the common and widespread Ordovician *Maclurites* LeSueur, until the work of Knight (1952) it had not figured in classification of Paleozoic gastropods. In general, one can divide the gastropods listed by Knight into (1) coiled forms having a more or less flat base (= depressed spire in hyperstrophic coiling), such as *Maclurites,* and (2) coiled species that were clearly higher spired though always to the "left," such as *Matherella* Walcott. This conchological division was formally established in Knight and others (1960) with the use of the families Macluritidae and Onychochilidae.

Knight (1952) did not include all Paleozoic sinistral-appearing gastropods within his assemblages of hyperstrophic taxa; however, only a few such taxa were excluded. He gave no criteria by which one could distinguish a sinistral orthostrophic from a hyperstrophic gastropod, apart from a ridge around the pseudoumbilicus of the latter. In discussing hyperstrophy in early Paleozoic gastropods, Knight (1952, p. 38) indicated that "one line of collateral evidence is that supplied by the peculiar angulation on the 'base' of these early Paleozoic shells. If we regard these shells as hyperstrophic, the angulation is no longer anomalous. It becomes the trace of the dorsal anal emargination." This is an interesting observation, but it need not indicate hyperstrophy. Some of the presumed hyperstrophic forms that Knight discussed have no basal angulation.

In the final analysis, in fossils, because we lack direct knowledge of the soft parts, the concept of hyperstrophy depends on an interpretation of the operculum. In the modern fauna, the shell in orthostrophic species coils in one direction and the operculum in another. In hyperstrophic species, when placed in conventional orientation, both shell and operculum coil in the same direction. The operculum of *Maclurites* is well known and has been interpreted as coiling in the same direction as the shell, when placed in orthostrophic orientation. Other genera are linked to *Maclurites* by similarity of form. It is by no means clear that Cambrian forms that have been referred to *Maclurites* belong to that genus. One negative piece of information is that no Cambrian opercula have been reported in the literature or are known in collections. In spite of careful search on outcrops and in the laboratory, no one has been able to find any. Thus, we consider the issue of hyperstrophy among Cambrian gastropods to be unresolved and open to further interpretations. Because the significance of hyperstrophy for the classification of early Paleozoic gastropods is under scrutiny, here, we describe all taxa as sinistral when placed in orthostrophic orientation.

Family MACLURITIDAE Fischer

Discussion. The outline of the Macluritidae as it has been used for the last two decades was given by Knight (1952, p. 36–37) and formalized in the classification of Knight and others (1960, p. I-188–I-189). They assigned seven taxa to the family as genera or subgenera.

Since 1960, one genus has been added, the late Early Ordovician age *Teiichispira* Yochelson and Jones (1968). Other genera have been removed. *Omphalocirrus* Ryckholt was considered a euomphalacean by Yochelson (1966). More recently, *Lecanospira* Butts and, by inference, *Barnesella* Bridge and Cloud were transferred to the Pleurotomariacea *sensu lato* (Yochelson, 1984b) because of the shape of the aperture. In this work, *Macluritella* Kirk and *Euomphalopsis* Ulrich and Bridge *in* Bridge are moved to the Euomphalacea, in part because of their open coiling.

As emended, the family Macluritidae now contains only three genera—*Maclurites* LeSueur, *Palliseria* Wilson, and *Teiichispira* Yochelson and Jones—all of which are most common in the upper Lower Ordovician or the Middle Ordovician. For each, a calcified operculum is known, although that of *Palliseria* has not as yet been described (Yochelson, 1989). Each of these opercula is somewhat different in shape and in thickness.

In the classification of 1960, the Onychochilidae were presented ahead of the Macluritidae; at that time, the oldest known Macluritidae were in rocks of late Early Ordovician age, whereas presumed Onychochilidae were known from the Cambrian. As a result of this investigation, they are now both known to occur in Upper Cambrian rocks. As will be discussed below, the Onychochilidae are subject to even more reinterpretation than Macluritidae. It may well be that combining the two family groups into one superfamily will prove to be an inadequate concept, but for the time being, it remains convenient to separate the extremely low-spired macluritid taxa from the others.

Genus *Maclurites* LeSueur, 1818

Discussion. This genus is based on *Maclurites magna* LeSueur, a Middle Ordovician species. Because its shape is so different from the conventional dextral trochiform gastropod, *Maclurites* has interested malacologists since it was first described. For nearly 150 years, the organism has been interpreted as hyperstrophic, that is, "pseudosinistral" (=ultradextral) because the operculum appears to coil to the left (see Knight, 1952, for details). The operculum is a flat plate and can also be interpreted as one that expands by accretion on one edge but does not coil. R. M. Linsley (oral communication, 1983) has suggested that *Maclurites* lived a sedentary existence as a filter feeder on reef flats. This is a plausible suggestion, because consolidated reef deposits are the rock type in which so many Middle Ordovician specimens are found. Whether *Maclurites* was indeed hyperstrophic is a different question. Possibly the shell morphology should be correlated with a sessile life habit and not with hyperstrophy.

Considerable emphasis was placed by Knight (1952) on the presence of an angulation surrounding the pseudoumbilicus in *Maclurites* and other macluritaceans. This angulation is not present in the Cambrian material from the Ellsworth Mountains, which has a well-rounded transition from the outer whorl surface to the inner. The material from Antarctica may be adequate to propose a new genus. Growth lines are not well known on most of the shell surface, although the shape of the aperture is clear. Entirely for the sake of expediency, this material is placed in *Maclurites* with quotation marks.

Scaevogyra Whitfield, first described from North America, is a common Upper Cambrian gastropod, but it remains poorly known in some details. It differs from the Middle Ordovician *Maclurites* in having a distinct trochiform spire, rather than a flattened base. The characters of the umbilicus in the type species are not well known, and an angulation may exist around the umbilicus (or pseudoumbilicus) as in *Maclurites.* Knight (1952) thought that both genera were fairly closely related. In the species of *"Maclurites"* from Antarctica described below, the base is nearly flat so that the inner whorls hardly protrude when a specimen is viewed from the side. On the other hand, the simple aperture and its apparent flare are similar to these features in *Scaevogyra.* It is certainly possible that this new species of *"Maclurites"* is better assigned to *Scaevogyra.* For the time being, the difference between a flat or nearly flat base and a protruding spire may be used to distinguish the two concepts. Should it turn out that the presumed umbilical angulation of *Scaevogyra* is actually a well-rounded segment of the whorl profile, the new species described below would fit as an exceedingly low-spired member of *Scaevogyra.*

"Maclurites" thomsoni n. sp

Plate 6, Figs. 1–8; Plate 7, Figs. 1–7

1982. N. gen., n. sp., Webers, Antarctic geoscience, p. 636, fig. 78.2-15.

Description. Extremely low-spired gastropods with impressed sutures and well-rounded outer whorl face. All known whorls coiled in same direction, sinistral in orthostrophic orientation. Basal surface protruding to a slight extent in early stages but generally flattened; basal suture strongly and distinctly impressed; whorl area well rounded adjacent to suture, flattened for about half of its width beyond the sutural region; juncture of basal and outer whorl faces following the arc of a small circle; outer whorl surface gently curved but with upper part of whorl again strongly inclined inward, following the arc of a circle of smaller diameter than that below; pseudoumbilical wall inclined steeply downward and gently inward, very gently curved so that in cross section this part of the whorl overhangs that which impinges on the penultimate whorl. Pseudoumbilical sutures impressed. Shell has about four slowly expanding whorls in contact at all growth stages. Shell surface smooth, growth lines extremely faint, their course generally orthocline on basal surface, and probably less than 5 degrees from vertical on the outer whorl face but not known in detail. Operculum unknown.

Remarks. Although this species is superficially similar to *Euomphalopsis splettstoesseri,* the latter has a nearly circular whorl profile, whereas in *"M." thomsoni,* the profile is distinctly asymmetrical (Plate 7, Fig. 3). The body whorl of *"M." thomsoni* impinges strongly on the penultimate whorl at all growth stages, and the shell is also more obviously asymmetrical. It is the second least common of the gastropods collected in the Ellsworth Mountains, whereas *E. splettstoesseri* is relatively abundant.

Although *"M." thomsoni* is asymmetrical, in conventional sinistral orientation the base is nearly flat (Plate 6, Fig. 8); protrusion of the spire is slight. Another important feature is the well-rounded whorl section leading from the outer face into the pseudoumbilicus (Plate 6, Figs. 6–8). The shell in *Maclurites* is relatively thick, and features of the steinkern do not always accurately portray the external form. In contrast, this species seems to have an exceedingly thin shell. Finally, the larger specimens show irregularities on the body whorl (Plate 6, Figs. 6–8), and the largest specimen shows either the growing edge of the shell or a slight apertural flare (Plate 7, Fig. 7).

"Maclurites" thomsoni shows some similarity to species of *Maclurites* that do not have a flat base. Billings (1865) described several species of *"Maclurea"* from the upper Lower Ordovician rocks of Newfoundland, Canada, that were never illustrated. The descriptions of these indicate that *Maclurea rotundatus* seems to be most similar to *"M." thomsoni,* differing in having the outer whorl face more strongly inclined.

Types. Holotype, USNM 385166; figured paratypes 385161–385165, 385167; unfigured paratypes USNM 388325.

Type locality. From the feather edge of the Minaret Formation north of Springer Peak, Webers Peaks, Heritage Range (Fig. 1).

Etymology. M.R.A. Thomson, a member of the British Antarctic Survey for many years and a paleontologist who has a distinguished record in unraveling the geologic history of Antarctica, assisted in collection of fossils in the Ellsworth Mountains. It is fitting that the species be named in his honor.

?Family SCAEVOGYRIDAE Wenz

Discussion. As used by Knight and others (1960, the Onychochilidae was subdivided into the typical subfamily and the Scaevogyrinae. The Onychochilinae were distinguished because they had the basal "spire" high, whereas the Scaevogyrinae contained lower and broader shells. The type species of the various genera that Knight (1952) considered to be in this family were described and illustrated by him earlier (Knight, 1941), and the reader is referred to that work for more details. The post-Cambrian taxa placed in the Onychochilidae have a variety of shell forms. Almost all taxa involved were rare and poorly known, and in general, 40 years later they continue to be both rare and poorly known. Because of the considerable general interest in the question of hyperstrophic coiling, we comment on what was known of these taxa until 1960 and then follow the scanty literature since then.

Koken (1925) originally used the Onychochilidae as a family that contained four genera—*Onychochilus, Laeogyra, Helicotis,* and *Mimospira*; all are of Middle Ordovician or younger age. As presented in the classification of Knight and others (1960, p. 187), the subfamily Onychochilinae contained only two of the genera given by Koken—*Laeogyra* and *Onychochilus. Palaeopupa* Foerste was placed in subjective synonymy with *Onychochilus.* Three additional genera, two of Late Cambrian age (*Matherella* Walcott and *Matherellina* Kobayashi) and one of Early Devonian age (*Sinistracirsa* Cossmann), were added. *Mimospira* was moved to the Clisospiracea. *Helicotis* Koken was placed in the Scaevogyrinae with question.

We begin with the geologically youngest genus. *Sinistracirsa* Cossmann is known only from its type species; it was described from the Lower Devonian of Czechoslovakia from two incomplete specimens, one of which has been lost (Knight, 1941, p. 318–319). To the best of our knowledge, it has not been reported subsequently. Because of its high spire, this genus might equally well be interpreted as a sinistral loxonematacean, rather than an onychochilid.

The type species of *Onychochilus* Linström, *O. physa,* was described from two shells from the Middle Silurian of Gotland, Sweden; its subjective synonym, *Palaeopupa* Foerste, is also known from a single species based on two specimens from beds of equivalent age in Ohio, United States (Knight, 1941, p. 215–216, 230). In the more than 90 and 80 years, respectively, since each was described, no additional specimens have been found. A second species that Lindström assigned to the genus is a regularly expanding trochiform gastropod, and the third is a high-spired form much like *Sinistracirsa* Cossmann. It is probably correct to treat *Onychochilus* and *Palaeopupa* as synonyms, though irregular expansion of whorls is much more evident in the latter. The aperture of *O. physa* is elongated and in a sense slitlike, but there is no real basal "angulation," for there is no umbilical depression. Wenz (1938, p. 367) placed the genus in its own subfamily, in part because of the direction of coiling, but judged that it fit easily within the Subulitidae. This genus could certainly be transferred to the Subulitacea. Apart from the direction of coiling, it has the characteristic shape of taxa of that superfamily.

Laeogyra bohemica, the type species of *Laeogyra* Perner, from the Middle Ordovician of Czechoslovakia (Knight, 1941, p. 167–168), was described from three specimens. So little was known of the genus that it was not even illustrated by Knight and others (1960). The base of the shell is unknown, and thus we have no information on the junction of the outer face with the inner part of the shell. Apart from the "hyperstrophic" coiling, no objective data indicate that *Laeogyra* is closely related to the other genera discussed.

In our view all the onychochilid genera discussed above should be put aside and not considered further in the question of hyperstrophic coiling. They are either poorly known or as easily considered to be sinistral.

The remaining two genera are Late Cambrian in age and do appear to be related to each other. *Matherella walcotti,* the type species of *Matherellina* Kobayashi, was based on three specimens (Knight, 1941, p. 189) from northern China. It is perhaps somewhat better known than *Laeogyra,* for the slightly sunken juncture of outer face and base may be seen. The available faunal evidence from beds that yielded the three specimens has been reinterpreted recently and suggests that this form occurs in beds equivalent to the Trempealeauan, that is, latest Cambrian in the current American sense. So far as we know, only the type species has been described. The remaining genus in the subfamily as of 1960 was *Matherella* Walcott from the Upper Cambrian Hoyt Limestone in eastern New York, United States (Knight, 1941, p. 187–189), which is also known only from the type species, *Billingsia saratogensis* Miller. In recent years, *Matherella* has been found to be widely distributed in the Upper Cambrian of the western United States.

The subfamily Scaevogyrinae Wenz, 1938 (p. 238–280) was treated by its author as a family under the Trochonematacea, a characteristically dextral group. *Scaevogyra* Whitfield itself was divided into the nominate subgenus and *S. (Versispira)* Perner. Another genus divided into subgenera was *Matherella* Walcott; it, *Matherellina* Kobayashi, and, questionably, *Palaeopupa* Foerste were treated as subgenera. *Laeogyra* Perner was treated in the text as a full genus, but in the figure caption, it was used as a subgenus under *Scaevogyra.* Finally, *Antispira* Perner was included but not illustrated.

Within the classification of Knight and others (1960), five genera were placed in the subfamily Scaevogyrinae. The Upper Cambrian *Scaevogyra* Whitfield is reasonably abundant and well known, although most specimens are not well preserved. *Kobayashiella* Endo, the other Upper Cambrian genus in the subfamily is known from a single species and that from a single specimen from northern China (Knight, 1941, p. 165); this genus was not included in the classification of Wenz. The Ordovician *Antispira* Perner was originally described on the basis of a species founded in a single specimen, which has subsequently been lost (Knight, 1941, p. 41). *Versispira* Perner is also known from a single species based on a single specimen (Knight, 1941, p. 381–382) from the same locality as *Antispira*. The differences between these two generic concepts is not at all evident. The locality for both, Trubska, near Beraun, Czechoslovakia, is judged to be Middle Ordovician Llandeilian (Horný, 1963). Finally, the Middle Ordovician *Helicotis* Koken was questionably included. It too is known from a type species described from a single specimen about which Knight (1941, p. 144) wrote "it is unfortunate that a genus should be based on a species founded on such poor materials"; the genus was not illustrated by Knight and others (1960).

One may summarize by noting that none of the Middle Ordovician and younger taxa placed in the family in 1960 are well enough known to be placed anywhere in a classification with any reasonable degree of confidence. Since 1960, one work, by Wängberg-Eriksson (1979), described a diverse fauna of moderately to very high-spired "Macluritacean gastropods from the Upper Ordovician and Silurian of Sweden." It sheds some light on Silurian taxa from Gotland described by Lindström (1884) and describes a number of new forms from the Upper Ordovician. These are all unusual if considered to be hyperstrophic, but if they are interpreted as sinistral gastropods, they are not that remarkable in form.

Under the circumstances, we suggest returning the Scaevogyrinae to family rank and provisionally linking it to the Macluritidae, within the Macluritacea, pending further clarification of the concept of hyperstrophy. The Onychochilidae can be used as a "form taxon" for the Middle Ordovician and younger "sinistral" taxa that cannot be placed in other families, and this family can be assigned to Gastropoda Incertae Sedis. The redefined Scaevogyridae includes *Scaevogyra, Matherella, Matherellina,* and *Kobayashiella*. The family is defined as prominently trochiform gastropods that are sinistral in the orthostrophic position and that have a moderately well rounded whorl profile, in part as a result of impressed sutures and in part as a result of a distinct umbilicus, and a straight outer lip, steeply inclined.

Genus *Matherella* Walcott, 1912

Discussion. This genus is the most elongate, or high-spired, of the Scaevogyridae. Because the shape of the aperture is linked to the overall form, it is also the one that has the growth lines nearest to vertical and the smallest umbilicus.

Matherellina is distinguished from *Matherella* primarily in that it is reported to have a sinus near the juncture of the outer face and umbilicus. This sinus is based on undulations seen on one of the steinkerns in the type lot (see Knight, 1941). At the time of the description of *Matherellina* from northern China, the Wanwangou Dolomite was considered to be Early Ordovician in age. It has been redefined as Late Cambrian, and no stratigraphic separation between these two genera now exists. When *Matherellina* is better known it may be placed in the synonymy of *Matherella*.

Scaevogyra elevata Whitfield from Wisconsin, United States, is relatively higher than the type species of *Matherella*. For the moment it can be assigned to *Scaevogyra*; when it is better known it might serve as the type species for another allied genus. *Scaevogyra elevata* is also known from the Wilberns Formation of north-central Texas, United States (Ulrich and Bridge *in* Dake and Bridge, 1932, Plate 12, Fig. 1; Cloud and Barnes, 1948, p. 138, Figs. 44, 49, 50).

Matherella antarctica n. sp.
Plate 4, Figs. 13–15

1981. *Matherella* sp., Pojeta, Webers, and Yochelson, U.S. Geological Survey Open-File Report 81-743, p. 168.
1982. N. gen., n. sp., Webers, Antarctic geoscience, p. 636, Fig. 78.2-11.

Description. Elongate, smooth, slowly expanding, trochiform possibly sinistral gastropods. Shell of few whorls, sinistral in orthostrophic orientation; high spired. Sutures distinctly impressed; outer whorl face very little arched, little flattened between the sutural area and the base. Base apparently anomphalus, the juncture of outer face and basal surfaces well rounded, following the arc of a small circle. Whorls embracing progressively lower on the face of each succeeding whorl. Growth lines prosocline, inclined less than 15 degrees from the axis of coiling. Shell smooth with exceedingly faint growth lines.
Remarks. This species, quite frankly, is based in part on the geographic distance between North America and Antarctica. Because the shell is so simple, it is difficult to find characteristics of significance, apart from general shape. *Matherella antarctica* seems to be a bit more slender than *M. saratogensis* (Miller), the type species from the Upper Cambrian Hoyt Limestone of New York, United States. *Matherella antarctica* lacks an umbilicus (Plate 4, Fig. 13), and the early whorls do not protrude as high as those in the type species. The closeness of morphology between *M. antarctica* and *M. saratogensis* reinforces our views on the widespread distribution of Upper Cambrian mollusks.

Matherella antarctica is similar to *Kobayashiella*? *heritagensis* in overall form; it differs in being slightly narrower at the same growth stage and in having no umbilicus. Juveniles of *K.*? *heritagensis* that lack rugosities on the steinkern and on which the base is hidden by matrix cannot be distinguished from the early growth stages of *M. antarctica*. In larger specimens, the higher spire of *M. antarctica* is evident. We have considered the possibility that some specimens assigned to *K.*? *heritagensis* have been elongated by tectonic deformation but have rejected it; deformation would affect both the width and the inclination of the sutures of earlier whorls, and we saw no evidence of this. If one sets aside small specimens that might be assigned to either genus, the unique specimen of *Matherella* is only about one hundredth as common in the Ellsworth fauna as is *Kobayashiella*? and is the rarest of the species assigned to the Gastropoda.
Types. Holotype, USNM 385150.
Type locality. From the feather edge of the Minaret Formation, north of Springer Peak, Webers Peaks, Heritage Range (Fig. 1).
Etymology. The species *Matherella antarctica* is named for the continent where it was first found.

Genus *Scaevogyra* Whitfield, 1878

Discussion. As currently understood, *Scaevogyra* is characterized by a relatively large shell, having a distinct but moderately low spire and whorls that expand rapidly so that the body whorl is much wider than high. The type species, *S. sweezyei* Whitfield, is umbilicate.

Scaevogyra? species indeterminate
Plate 4, Fig. 16

Remarks. One specimen has whorls that are relatively wide so that the specimen expands rapidly. The spire is significantly more pronounced than in any of the specimens of *"Maclurites" thomsoni*. This elongation may be a result of tectonic deformation, but that seems unlikely as a smaller, even less well preserved, specimen on the same piece of matrix has the same whorl shape. The entire base is not exposed, but what can be seen is smoothly convex. If an umbilicus were present, it would have to be small, but the available data suggest that the base is anomphalus.

Yochelson (*in* Shergold and others, 1976, p. 262) identified *Scaevogyra* sp. indet. from the Upper Cambrian of northern Victoria Land, Antarctica, and noted that the material seemed to have a convex base and to lack an umbilicus. In general terms, the northern Victoria Land collection is comparable to the material from the Ellsworth Mountains,

though both lots are inadequately preserved. If this form lacks an umbilicus, it cannot be placed in *Scaevogyra*.
Material. Figured specimen, USNM 385151.
Locality. Minaret Formation, Yochelson Ridge, Heritage Range (Fig. 1).

Genus *Kobayashiella* Endo, 1937

Discussion. Generic assignment of the Antarctic material poses a problem. *Kobayashiella* is known from a single species. The nomenclatural history of the genus, and an excellent description and a good photograph of the holotype of the type species, *K. circe* (Walcott), were provided by Knight (1941, p. 165). In essence, it is a low, very rapidly expanding form, having a rounded juncture of the outer face with the basal surface and bearing closely spaced strong growth increments, suggesting varices on the exterior of the shell. However, the holotype is a tiny fragment and clearly only a small part of a larger individual. Knight (1941) noted a small sinus in the inner trace of the varices of *Kobayashiella* near the juncture of base and outer whorl face. Our information is insufficient to confirm this feature in the new species, in spite of the large number of specimens available. For this reason, we assign it to the genus with question.

The Upper Cambrian *Scaevogyra obliqua* Whitfield, a poorly known species from Wisconsin, United States, may be congeneric with *Kobayashiella*. The type species of *Scaevogyra, S. sweezyei* Whitfield (Knight, 1941, p. 306), is known from relatively large, poorly preserved steinkerns that occur in beds of Trempealeauan age. Satisfactory comparison between them and the type species of *Kobayashiella* is virtually impossible. The shape of the largest specimen of *K.*? *heritagensis* from the Ellsworth Mountains is more or less comparable to that of *Scaevogyra,* and the two generic concepts may be the same. In order to confirm this, more needs to be known of the early growth stages of *Scaevogyra*. Should well-preserved specimens be found that show occasional thickenings or rugosities in the early growth stages, *Kobayashiella* could then be placed in synonymy. Until such information is available, the wisest course again seems to be to use the generic assignment with question.

Kobayashiella? *heritagensis* n. sp.

Plate 5, Figs. 1–13; Plate 18, Figs. 1–5, 7, 8.

1981. *Scaevogyra* sp., Pojeta, Webers, and Yochelson, U.S. Geological Survey Open-File Report 81-743, p. 168.
1982. *Scaevogyra* n. sp., Webers, Antarctic geoscience, p. 636, Fig. 78.2-8.

Description. Small, very low-spired gastropods. Shell sinistral in orthostrophic orientation, extremely low spired, the apical angle being more than 150 degrees. Protoconch simple, vermiform; all whorls observed uniform in coiling. Whorl profile moderately well rounded from impressed suture but following the arc of a large circle so that at the middle area the outer whorl face appears slightly flattened; periphery near midwhorl. At base of outer whorl face, profile turning strongly inward forming nearly a semicircle until joining the base. Basal walls inclined steeply inward, seemingly nearly straight. Shell thin, details unknown. Growth lines poorly known, straight or nearly straight from suture to base with an oblique inclination. Steinkerns show some variation from smooth to rugose at irregular intervals, the rugosities a result of apertural thickenings.
Remarks. Kobayashiella? *heritagensis* differs from *K. circe* (Walcott), the type species, in lacking prominent ridges on the steinkern. Walcott (1905) described *K. circe* from the Upper Cambrian of central China, on the basis of a single specimen. Endo (1937, p. 314–315) identified two equally tiny specimens of that species from the Upper Cambrian of Manchuria.

Of particular importance in this species is the individual variation in regard to the presence of rugosities parallel to the growth lines (Plate 5, Figs. 1, 13). Less than a quarter of the specimens possess them, even though we have chosen to emphasize them in illustration for they show the course of the growth lines. In those specimens in which the rugosities are preserved, there is variation in the number and in the spacing (Plate 5, Figs. 1, 11–13). Many large specimens lack the rugosities so they are not a feature of ontogeny.

Another morphologic feature of considerable interest is the early whorl, which is elongate and vermiform in appearance (Plate 18, Figs. 1–5, 7, 8). Although the preservation is relatively coarse, this may be the protoconch; we see no evidence of any coiled protoconch preceding it.
Types. Holotype, USNM 385152; figured paratypes, USNM 385153–385160; 386191; unfigured paratypes, USNM 388326, 388327.
Type locality. From the feather edge of the Minaret Formation at the northern side of Springer Peak, Webers Peaks, Heritage Range (Fig. 1).
Distribution. Springer Peak, Bingham Peak (USNM 388327).
Etymology. The species is named for the Heritage Range, Ellsworth Mountains.

Superfamily EUOMPHALACEA deKoninck

Discussion. In the classification of Knight and others (1960, p. I-189), *Euomophalopsis* was placed as a subgenus under *Macluritella* Kirk, and both were assigned to the Macluritacea. Restudy of the type lot of *Macluritella stantoni* Kirk, the type species, has shown it to have a more pronounced triangular cross section than *Euomphalopsis involuta* Ulrich and Bridge, the type species. We judge that the two again should be ranked as distinct genera. *Macluritella stantoni* is from the Manitou Dolomite of Early Ordovician age in Colorado, United States. Several attempts have been made to collect additional material from the type locality of that genus, but to no avail.

Knight (1952) emphasized the importance of a sinus on the dorsum as an indication of relationship. Probably too much has been made of this feature (Yochelson, 1984b). Neither genus shows a pronounced pseudoumbilical carina, and the whorl profile has much in common with that of younger euomphalaceans. The open coiling, slow rate of whorl expansion, and low height-to-width ratio of the whorl profile are additional features that distinguish both genera from *Maclurites*. Accordingly, the two genera are transferred out of the Macluritacea.

Family EUOMPHALIDAE deKoninck

Discussion. Even though the genus *Euomphalus* is characterized by a distinct angulation of the whorl profile, some genera within the family lack it. In particular, the nearly circular whorl profile of the late Paleozoic *Straparollus* is so close to that of *Euomphalopsis* that fragments of each cannot be distinguished.

Genus *Euomphalopsis* Ulrich and Bridge, *in* Bridge, 1931

Description. Nearly planispiral, open-coiled gastropods with a suboval to nearly circular cross section, open coiled at maturity. Early growth stages and most of mature whorl in contact; part of mature body whorl free. Early whorls are within the area of the expanding aperture so that the shell is nearly bilaterally symmetrical. Whorl profile approaching oval in shape, the basal surface very slightly flattened. Aperture simple, so steeply inclined as to be nearly orthocline; growth lines closely spaced and extremely faint. Shell of at least two layers. Ornament lacking.
Discussion. Euomphalopsis involuta, the type species, is of early Early Ordovician age, as it was first described from the Gasconade Formation of Missouri, United States. The assignment of a new species of Late

Cambrian age to the genus constitutes a downward extension of its range. However, Butts (1926, Plate 14, Fig. 10) illustrated a specimen from near Talladega, Alabama, United States, that he assigned to *Pelagiella* sp.? This is associated with *Sinuopea* and *Scaevogyra,* and Butts suggested that it might be of Cambrian age. The specimen is a *Euomphalopsis* but is too poor to compare with either the type species or the new material from Antarctica.

Knight (1941, p. 120, Plate 72, Fig. 46) described and illustrated a very slight sinus on the dorsal surface of the holotype of *E. involuta.* It is present, but how significant this shallow indentation may be is quite another question. We have examined a large suite of topotypes of this species and specimens of what may be undescribed species of the genus. Most are poorly preserved, but those that show growth lines show no indication of a dorsal sinus.

Euomphalopsis splettstoesseri n. sp.

Plate 7, Figs. 8–13; Plate 8, Figs. 1–12; Plate 9, Figs. 1, 2

1982. N. gen. n . sp., Webers, Antarctic geoscience, p. 636, Fig. 78.2-18.

Description. Rapidly expanding gastropods having aperture nearly circular and simple. Protoconch unknown; earliest observed whorl large. Early youth stages in contact but impinging only slightly; later whorls in contact but not impinging; most of mature body whorl free. Profile nearly circular except for a slight basal flattening. Cross section expanding rapidly. Asymmetrically coiled, very slightly sinistral. Shell smooth. Growth lines exceedingly faint, prosocline on outer whorl surface, inclined less than 10 degrees from vertical. Shell with thin outer layer and thick inner layer of uniform width around circumference.

Remarks. Fragmentary specimens that have lost the earlier coiled part of the shell (Plate 9, Figs. 1, 2) appear at first glance to be high, slightly curved monoplacophorans, broken near the apical area. Although we cannot cite any examples from the literature, we are reasonably certain that during the past 20 years such shells would have been assigned to the Monoplacophora. The large "height" of the curved shell relative to the size of the aperture should serve to avoid most such pitfalls in assignment. So far as we know, no curved monoplacophorans have the apex overhanging the shell margin by a great distance; were the partial body whorl to be restored as a monoplacophoran, it would certainly overhang the margin. The nearly circular aperture (Plate 8, Figs. 4, 6) is unlike that of a monoplacophoran. Upon close examination, the asymmetry may be seen (Plate 8, Figs. 6, 11).

Euomphalopsis splettstoesseri is an unusual shell because its whorl cross section expands rapidly (Plate 8, Fig. 4), but the logarithmic spiral expands even more rapidly. The consequence is a large, only slightly curved body whorl (Plate 8, Figs. 1, 2). It differs from *E. involuta* Ulrich *in* Butts (1926) in having a greater rate of whorl expansion and thus a higher whorl profile and less curvature at a comparable size. Although the two taxa appear greatly different, only minor changes in basic geometry are needed to change one species to the other. In *E. splettstoesseri,* the point where open coiling begins varies. In most specimens, this begins relatively soon after the initial coil (Plate 8, Fig. 4), whereas others appear to be much larger before the last part of the body whorl becomes free (Plate 8, Fig. 8); in fact the difference in size is probably less than one-eighth of a whorl between the smallest and largest specimens having the whorls still in contact.

The earliest observed whorl (Plate 8, Fig. 5), which is certainly close to the protoconch, is relatively wide. Its width suggests that if an earlier larval shell existed it was simple and large.

Examination of many steinkerns has not revealed any features suggestive of muscle scars or any other change in shell thickness.

Although *E. splettstoesseri* and *"Maclurites" thomsoni* have a superficial similarity, the latter species is not open coiled and the body whorl definitely impinges on earlier whorls, resulting in a fundamental difference in whorl profile. These differences are in addition to the prominent pseudoumbilicus of *"M." thomsoni* compared to the nearly symmetrical character of *E. splettstoesseri.*

Euomphalus strongi variety *sinistrocirca* Berkey from the Upper Cambrian of Wisconsin, United States, is based on poorly preserved material. Examination of the type lot at the American Museum of Natural History, New York City, suggests that it should be assigned to *Euomphalopsis,* though the quality of preservation creates a great deal of uncertainty. The types do not reveal any significant details on the species level.

Types. Holotype, USNM 385181; figured paratypes, 385168–385171, 385175–385180, 385182; unfigured paratypes, USNM 388228, 388229.

Type locality. From the feather edge of the Minaret Formation north of Springer Peak, Webers Peaks, Heritage Range (Fig. 1).

Distribution. Springer Peak, Bingham Peak (USNM 388229).

Etymology. John Splettstoesser was a member of the first field party to set foot in the Ellsworth Mountains, in the 1961–62 field season, and was logistics coordinator in the 1979–80 season. It is appropriate that "S plus thirteen" be honored by having a species from the Ellsworth Mountains named for him.

?Superfamily ORIOSTOMATACEA Wenz

Discussion. The Oriostomatacea have many features in common with the Euomphalacea, such as a steeply inclined aperture and a low, rapidly expanding shell. However, oriostomataceans are readily distinguished from them by prominent spiral ornament. In the classification of Knight and others (1960), these two groups were widely separated, for in that work the Euomphalacea were closely linked to the Macluritacea in the Suborder Macluritina. This association of superfamilies is now judged to be in error (Yochelson, 1984b). In the 1960 scheme, the Euomphalacea were judged as an ancient group, whereas the Oriostomatacea were considered primarily a Silurian and Early Devonian offshoot of the Pleurotomariacea.

If the genus described below is correctly assigned, both the Oriostomatacea (*Aremellia*) and the Euomphalacea (*Euomphalopsis*) are equally old. This implies that loss of one gill may have occurred independently a number of times in various gastropod lineages. Thus, one may consider both as distinct superfamilies but having many genera converging in shell form. The presence of spiral ornament in the Oriostomatacea implies a large mantle, which was thrown into wrinkles when expanded, each wrinkle being the site of deposition of a revolving element on the exterior of the shell.

For reasons still unknown, but perhaps in some way associated with this presumably large mantle, the Oriostomatacea favored life in a reefal environment. The abundance of oriostomataceans in the Silurian thus illustrates ecological preference rather than the explosive evolution of a newly derived stock. As partial support for this interpretation, there is similarity in lithology between the various "reefy" limestones of Silurian age that contain oriostomataceans in abundance and both the Lower Devonian of Czechoslovakia and Upper Ordovician of Sweden, from which they have also been reported.

The discovery by Wängberg-Eriksson (1964) that the Upper Ordovician *Isospira* Koken from Sweden is really an open-coiled oriostomatacean has interesting implications. Other open-coiled genera are known in the superfamily. Conventional wisdom is that reef or reefal environments are rich in food resources. Speculation that open-coiling inhibits movement of gastropods (Yochelson, 1971) and that, in effect, food comes to them, rather than *vice versa,* definitely is consistent with their occurrence in reefal areas.

Assignment of a new Upper Cambrian form to the Oriostomatacea is certainly influenced by our interpretation that the molluscan-rich Upper Cambrian limestone of the Ellsworth Mountains was deposited in shallow water. The algal-like banding of the rocks is suggestive of shallow water. This interpretation of the depositional environment was suggested long before these presumed oriostomataceans were identified. Thus, the presence of presumed oriostomataceans may be taken as further corroboration of the environment of deposition.

The early record of the Oriostomatacea is extremely sparse. The Ordovician *Isospira* was the oldest known heretofore. The enigmatic *Ozarkoconus* Heller from the middle Lower Ordovician of Missouri, United States, may be similar to *Aremellia* described below and thus could serve to link these two taxa temporally. Possibly Heller's genus is a senior synonym of *Aremellia,* described below, but it is so rare that only the type species is known, and the type specimen is poorly known in several features. Seemingly, it is distinguished in part from other oriostomatid taxa by finer spiral lirae. However, it was originally described as a monoplacophoran, and it may be bilaterally symmetrical.

Study of the Middle Ordovician slightly curved euomphalacean *Lytospira* Koken (see Yochelson, 1963, p. 179–181) suggests that open-coiled forms might be more common than anticipated. Further support that this new form may be a gastropod rather than a monoplacophoran is based on two previously unpublished observations by ELY. First, examination of specimens of the Middle Ordovician *Pollicina* Koken, housed in the Riks Natural History Museum, Stockholm, Sweden, indicates that this genus is a partially coiled euomphalacean whose cross section is oval rather than triangular like *Lytospira* (Yochelson, unpublished data). *Pollicina* is not a high-curved monoplacophoran as assigned by Knight and Yochelson (1960). Open coiling as a shell form was both more common and earlier than is noted in the literature.

Second is the occurrence of the poorly known *Spirodentalium* Walcott, 1891 in the Upper Cambrian of Wisconsin, United States. Examination of the type lot suggests that this elongate curved form is not a scaphopod, but rather it is comparable to *Lytospira*; other Upper Cambrian open-coiled euomphalaceans are known from Missouri (Yochelson, unpublished data). *Spirodentalium* is from the Upper Cambrian (Franconian) of Wisconsin and thus is only slightly younger than the molluscan-rich beds in the Ellsworth Mountains.

Family UNCERTAIN

Aremellia n. gen.

Type species. Aremellia batteni n. sp.
Description. Very slightly asymmetrical curved shells, bearing spiral ornament. Shell completing about one-quarter whorl, whorl width expanding fairly rapidly. Cross section asymmetrically oval with one side flattened. Shell bearing numerous spiral lira.
Discussion. The placement of this genus is uncertain. If it is accepted as a monoplacophoran, the shell is a relatively high, distinct curved cone, but one that does not expand as rapidly as *Knightoconus.* It differs from other monoplacophoran genera in having an anterior that is flattened rather than oval. When oriented with the apex anterior, the shell is slightly asymmetrical. If, on the other hand, this form is viewed as an open-coiled gastropod lying on its side, the asymmetry and the spiral ornament are readily interpreted. However, the rate at which the shell expands is exceptionally large compared to the rates of other known open-coiled taxa. There is thus a real dilemma as to where to assign this genus.

Although in the discussion of the superfamily, considerable emphasis was placed on open coiling, this, too, must be qualified. Forms such as *Spirodentalium* or *Lytospira* are quite elongate, are little curved, and expand the whorl slowly. *Pollicina* is more strongly curved but also has a slow rate of whorl expansion. In contrast, *Aremellia* expands very rapidly and, thus, more closely resembles high-curved cones generally accepted as monoplacophorans, such as *Hypseloconus* or *Knightoconus.* From what one may infer about the relationship of the foot size to the shell, it might have been possible for *Aremellia* to hold the shell erect and to crawl from time to time. Like any other hypothesis about the soft parts of extinct organisms that have no close living descendant, this interpretation must be viewed as speculation. On the other hand, we have no reason to assume that an organism lived the same way all the time or that it lived the same way during its early life as it did at maturity. One may envision *Aremellia* as living most of its mature life with one side down on the substrate, although it may have been more active when it was smaller.

Aremellia batteni n. sp

Plate 9, Figs. 3–13; Plate 18, Fig. 6

1982. New genus, new species, Webers, Antarctic geoscience, p. 636, Fig. 78.2-2.

Description. Partially coiled, rapidly expanding shell, bearing strong spiral lirae. Protoconch seemingly simple, flattened. Shell strongly curved, mature specimens extending for about a quarter of a whorl at maximum. Shell expanding rapidly so that aperture is approximately twice as wide as high. Shell very slightly asymmetrical; whorl profile gently rounded on the upper whorl surface, curving gradually onto more arched outer whorl face and this in turn curving slightly more abruptly into flattened basal surface; inner (concave) surface strongly curved and not as wide as outer whorl surface. Growth lines simple, prosocline, inclined about 15 degrees from vertical on outer face. Surface having between 40 and 45 uniformly spaced, rounded, spiral lirae, interspaces about four times width of lira. Shell of two layers, lirae obscure on inner one.
Remarks. So far as we are aware, there are no species in the literature to which this new species should be compared. We are also unaware of any previously described species that should be assigned to the genus.

The whorl profile is slightly asymmetrical (Plate 9, Fig. 4, 9). This asymmetry of the shell, following conventional orientation, is in a sinistral direction (Plate 9, Figs. 9–12). For the moment, this is judged to be only a feature of specific rank, though it should be reevaluated when more species of the genus are described.

The earliest observed whorl is hookshaped and simple (Plate 18, Fig. 6). Allowing for this, one may suggest a maximum coiling of half a whorl. No protoconch was observed. This is generally a moderately small species. The largest specimen (Plate 9, Fig. 13) is about twice the size of most individuals.

Steinkerns of *A. batteni* are difficult to distinguish from those of high conical monoplacophorans. Some show a clear distinction in curvature between the base and the outer whorl face, but most do not, particularly the smaller individuals. The spiral ornament commonly is reflected on the steinkern (Plate 9, Figs. 5, 6, 10).
Types. Holotype, USNM 385186; figured paratypes, USNM 385183–385185, 385187; unfigured paratypes, USNM 388230, 388231.
Type locality. From the feather edge of the Minaret Formation north of Springer Peak, Webers Peaks, Heritage Range (Fig. 1).
Distribution. Springer Peak, Bingham Peak (USNM 388231).
Etymology. Enigmatic forms deserve enigmatic names. The generic name chosen is based on the initials RML for a fellow worker, Robert M. Linsley, who has added to our understanding of form and function of gastropods. In regard to the specific name, it is only appropriate that one of the most productive workers on American Paleozoic gastropods, Roger L. Batten, also be linked to this unusual shell.

Class ROSTROCONCHIA Pojeta, Runnegar, Morris, and Newell

Introduction

Discussion. All the known diasome mollusks from the Minaret Formation belong to the class Rostroconchia. Pojeta and Runnegar (1989) provided the most recent definition and evaluation of the concept Diasoma and showed the central position that rostroconchs occupy in understanding that concept. Major modern summaries of rostroconchs are by Pojeta and Runnegar (1976), Pojeta and others (1977), and Runnegar (1978).

At present, two genera of Lower Cambrian rostroconchs are known, and they range through most of the epoch. *Heraultipegma* Pojeta and Runnegar occurs in Siberia, U.S.S.R. (Matthews and Missarzhevsky, 1975), China (Qian and others, 1979), France (Pojeta and Runnegar, 1976), Newfoundland, Canada (Bengtson and Fletcher, 1983), and South Australia (Daily and others, 1976). *Watsonella* Grabau occurs in Massachusetts, United States (Pojeta and Runnegar, 1976), and China (Pojeta and Runnegar, 1989). Middle Cambrian rostroconchs are not yet well known; *Ribeiria* Sharpe occurs in the Middle Cambrian of New Brunswick, Canada (Pojeta and Runnegar, 1989), and an as yet undescribed ribeirioid has been found in the Middle Cambrian of Queensland, Australia (Runnegar, 1983). All Lower and Middle Cambrian rostroconchs belong to the order Ribeirioida.

Thirteen genera of rostroconchs have been identified from the Upper Cambrian rocks of the world. Eight of these have a Cambrian record limited to the uppermost rocks of the System—*Cymatoopegma* Pojeta, Gilbert-Tomlinson, and Shergold, *Kimopegma* Pojeta, Gilbert-Tomlinson, and Shergold, and *Pinnocaris* Etheridge are known from Australia; *Eoischyrinia* Kobayashi, *Eopteria* Billings, *Pseudotechnophorus* Kobayashi, *Wanwanella* Kobayashi, and *Wanwanoidea* Kobayashi are known from northeastern China.

The other five genera are known from older Upper Cambrian rocks—*Apoptopegma* Pojeta, Gilbert-Tomlinson, and Shergold (Antarctica and Australia), *Oepikila* Runnegar and Pojeta (Australia), *Pleuropegma* Pojeta, Gilbert-Tomlinson, and Shergold (Australia), *Ribeiria* (Antarctica and Australia), and *Wanwania* Kobayashi (Antarctica). In the uppermost Cambrian, *Ribeiria* is known from Australia, China, and North America, and *Wanwania* is known from China; the other three genera are not known in uppermost Cambrian rocks. Nine of the Upper Cambrian genera are placed in the Ribeirioida; the other four belong to the order Conocardioida.

The range chart published for Cambrian rostroconchs by Pojeta (1979) has been updated by Runnegar and Pojeta (1989), who showed the known ranges of all genera of Cambrian mollusks, except cephalopods. A geographic overlay on their chart indicates that western Pacific Upper Cambrian rostroconchs should prove useful for helping to establish paleogeographies in China, Australia, and Antarctica.

Unlike the univalves described above, no new genera of the pseudobivalved rostroconchs have been found in the Heritage Range, Ellsworth Mountains. Three species of ribeirioid rostroconchs, placed in three genera, are known from the Minaret Formation—*Ribeiria australiensis* Pojeta and Runnegar, *Apoptopegma craddocki* n. sp., and *Wanwania* sp. A. These are the first occurrences of these genera reported from the Antarctic. This is also the oldest known occurrence of *Wanwania*.

Order RIBEIRIOIDA Kobayashi
Family RIBEIRIIDAE Kobayashi

Genus *Ribeiria* Sharpe, 1853

Diagnosis. Posteriorly elongated ribeirioids in which dorsal and ventral margins are not subparallel, beak is not the anteriormost point of shell, and rugose comarginal ornament is lacking; dorsum carinate or not.

Stratigraphic range. Middle Cambrian (*Paradoxides davidis* Zone) through Upper Ordovician (Ashgillian).

Remarks. In the Cambrian, *Ribeiria* is known from New Brunswick, Canada (*Paradoxides davidis* Zone, Pojeta and Runnegar, 1989); Australia (Mindyallan-Payntonian, Pojeta and others, 1977); China (Fengshanian; Chen and Teichert, 1983, p. 11); New York, United States (Trempealeauan, Pojeta and Runnegar, 1976); and the Dresbachian (Idamean) of Antarctica (herein).

Ribeiria australiensis Pojeta and Runnegar, 1976
Plate 1, Figs. 4–12

1976. *Ribeiria australiensis* Pojeta and Runnegar, U.S. Geological Survey Professional Paper 968, p. 50, Plate 4, Figs. 26–29.
1977. *Ribeiria australiensis* Pojeta, Gilbert-Tomlinson, and Shergold, Australian Bureau of Mineral Resources Bull. 171, p. 13, Plate 1, Figs. 1–20.
1977. *Ribeiria* sp., Webers, Antarctic Journal of the United States, v. 12, p. 120.
1981. *Ribeiria* n. sp., Pojeta, Webers, and Yochelson, U.S. Geological Survey Open-File Report 81-743, p. 168.
1982. *Ribeiria* sp. Webers, Antarctic geoscience, p. 636, Fig. 78.2-5.

Description. Small ovate *Ribeiria* with nearly straight dorsum bearing carina; umbo projecting little, or not at all, above rest of dorsal margin; dorsal carina extending posteriorly from beak, pegma small.

Material and locality. In the Ellsworth Mountains, *R. australiensis* is known from five internal molds (USNM 385115–385119), of which the two best preserved are figured. All the Ellsworth Mountains specimens are from the feather edge of the Minaret Formation, on the northeastern side of Springer Peak, Webers Peaks, Heritage Range (Fig. 1).

Comparison. The holotype and two paratypes of *R. australiensis* (Plate 1, Figs. 7–10) are figured for comparison with the Antarctic material. These specimens are from the Mungerebar Limestone (Mindyallan), Queensland, Australia (Pojeta and others, 1977). The Antarctic specimens are not well preserved, whereas the Australian specimens are silicified replicas. They are similar in the following ways: (1) the beak and umbo are not prominent structures, and the dorsal margin is straight or nearly so in lateral view (compare Plate 1, Figs. 5, 12, with Plate 1, Figs. 9, 10); (2) the cross-section shape is subcircular (compare Plate 1, Fig. 6, with Plate 1, Fig. 7) and not elliptical as in *Apoptopegma craddocki* (Plate 2, Fig. 6); (3) both are small, obese shells with dorsal carinas (compare Plate 1, Figs. 4, 11, with Plate 1, Fig. 8); and (4) there are no prominent constrictions in dorsal profile, such as are seen in *A. craddocki* (Plate 1, Fig. 17; Plate 2, Fig. 4). On the average, Antarctic specimens of *Ribeiria australiensis* are 25 to 30 percent smaller than adult-sized Australian specimens; 7 to 7.5 mm long, versus about 10 mm long. The Antarctic specimens also have a more prominent pegma than do the Australian specimens.

Genus Apoptopegma Pojeta, Gilbert-Tomlinson, and Shergold, 1977

Diagnosis. Small posteriorly elongated ribeiriids in which beak is anteriormost projection of shell; margin of shell below beak slopes posteroventrally and not projecting anteriorly; dorsal margin carinate.

Stratigraphic range. Upper Cambrian (Mindyallan) through Lower Ordovician (Datsonian).

Remarks. Apoptopegma is known only from Antarctica and Queensland, Australia (Pojeta and others, 1977). In Antarctica, only the new species *A. craddocki* is known; it occurs in the Heritage Range, Ellsworth Mountains. Two or three species are known from the Datsonian part of the Ninmaroo Formation of the Georgina Basin, Queensland. Runnegar and Pojeta (1989) note the likely occurrence of a species of *Apoptopegma* in rocks of Mindyallan age in Australia, but this species has yet to be described.

Apoptopegma craddocki n. sp.
Plate 1, Figs. 13–18; Plate 2, Figs. 1–15

1977. N. gen., n. sp., Webers, Antarctic Journal of the United States, v. 12, p. 120.
1981. *Apoptopegma* n. sp., Pojeta, Webers, and Yochelson, U.S. Geological Survey Open-File Report 81-743, p. 168.
1982. N. gen., n. sp., Webers, Antarctic geoscience, p. 636, Fig. 78.2-1

Description. *Apoptopegma* having arcuate to sigmoidal dorsal margin, posterior part of dorsal margin curves slightly upward upon reaching posterior gape; dorsal carina strong; anteroventral margin oblique to dorsal margin, sigmoidal; ventral margin arcuate; posterior margin oblique to erect. Beak forming prominent ventral hook. Ornament comarginal, poorly known. Anterior shell gape is funnel-shaped aperture immediately below beak; posterior shell gape small dorsal slit; ventral gape probably not present; between anterior funnel-shaped gape and umbo, lateral sulcus present, which is especially prominent in internal molds.

Internal features not well known; pegma a slightly to strongly oblique structure between anterior funnel-shaped gape and beak. Anterior and ventral margins having marginal denticles.

Types. *Apoptopegma craddocki* is known from the holotype (USNM 385120, Plate 1, Figs. 15–17), seven figured paratypes (USNM 385121–385127, Plate 1, Figs. 13, 14, 18; Plate 2, Figs. 1–15), and 11 unfigured paratypes (USNM 385128). All known specimens are internal molds, some of which have minor amounts of shell adhering. The holotype is 11.3 mm long, 9.2 mm high, and 4.1 mm wide (both valves).

Type locality. All specimens, except USNM 385127, are from the feather edge of the Minaret Formation on the northeastern side of Springer Peak, Webers Peaks, Heritage Range (Fig. 1).

Distribution. Specimen USNM 385127 is from the Minaret Formation, on the northwestern limb of Yochelson Ridge. Here the limestone of the Minaret overlies argillite and is broken by normal faults. However, along the eastern face of the limestone exposure is an unfaulted section. At the base of this section, a small collection, containing this species, was made. Limestone of the Minaret Formation at Yochelson Ridge has yielded trilobites that indicate a Dresbachian (Idamean) age, and the finding of *Apoptopegma craddocki* there is consistent with that age determination.

Etymology. The species is named for Campbell Craddock, University of Wisconsin at Madison, United States, a long-time geologist and explorer of Antarctica, especially the Ellsworth Mountains.

Remarks and comparisons. *Apoptopegma craddocki* is unique among ribeirioid rostroconchs in having an expanded funnel-shaped anterior gape (Plate 2, Figs. 6–8). Similar structures occur in various conocardioids, including the Upper Cambrian genus *Pseudotechnophorus* Kobayashi (Pojeta and Runnegar, 1976, Plate 20; Runnegar, 1978, Fig. 12[k]) and in all members of the family Conocardiidae (Pojeta and Runnegar, 1976, Plates 38, 43). Anterior feeding tubes are also known in the univalved Cambrian helcionellaceans *Yochelcionella* Runnegar and Pojeta and *Eotebenna* Runnegar and Jell (Pojeta and Runnegar, 1976, Plate 1; Runnegar and Jell, 1976, Fig. 11; Runnegar, 1978, Plate 2, Figs. 19–21). At the present time, all these anterior tubes are interpreted, by Pojeta, as convergent homeomorphic structures.

The probable absence of a ventral gape is an interpretation based both on functional analysis and the method by which most of the figured specimens were prepared. Ribeiriids may or may not have ventral gapes. For *Apoptopegma craddocki,* it seems unlikely that an animal with the anterior and posterior gapes limited to the dorsal margin of the shell would have the ventral gape of primitive or unspecialized ribeirioids. *A. craddocki* is already specialized for conducting water currents through the body in a way that species like *Ribeiria australiensis* are not (compare Plate 1, Fig. 7, with Plate 2, Figs. 5, 6–8). As noted above, Pojeta regards the occurrence of anterior tubes as an iterative specialization in various rostroconchs and helcionellaceans. Like most other rostroconchs, *A. craddocki* was probably infaunal (Runnegar, 1978, Fig. 1); restriction of the size of the shell gapes would allow a more efficient flow of water through the animal. There is nothing that strongly suggests that *A. craddocki* had anterior protrusible feeding structures like those that probably occurred in the conocardioids *Eopteria, Pseudotechnophorus,* and *Hippocardia* Brown (Runnegar, 1978, Fig. 12); such structures were also likely present in conocardiids (Pojeta and Runnegar, 1976, p. 19). However, the muscle scars of *A. craddocki* are not known.

The heating and quenching of the rock broke down the boundary between specimen and rock, and it also tended to remove shell from the internal mold and to make the feather edges of the molds friable so that they broke back slightly. We believe this is what happened in the specimen shown on Plate 2, Figure 1.

A. craddocki differs from all other species placed in *Apoptopegma* in having the anterior funnel-shaped aperture and in having the arcuate to sigmoidal dorsum.

Genus *Wanwania* Kobayashi, 1933

Diagnosis. Dorsoventrally elongated ribeiriids in which shell is higher than long or is subquadrate; shell not elongated posteriorly.

Stratigraphic range. Upper Cambrian (Dresbachian/Idamean)–Lower Ordovician (Datsonian).

Remarks. The specimen of *Wanwania* sp. A described here is the oldest known representative of the genus (Dresbachian/Idamean). Latest Cambrian (Fengshanian) species are known from northeastern China (Chen and Teichert, 1983, p. 11), and an earliest Ordovician (Datsonian) species has been described from the Ninmaroo Formation of Queensland, Australia (Pojeta and others, 1977).

Wanwania sp. A
Plate 1, Figs. 1–3

1981. *Wanwania* sp. Pojeta, Webers, and Yochelson, U.S. Geological Survey Open-File Report 81-743, p. 168.

Discussion. The only specimen collected (USNM 385129a) has the characteristic subquadrate shape and sharply posteriorly sloping anteroventral margin of the type species of the genus, *W. cambrica* Kobayashi (Pojeta and Runnegar, 1976, Plate 3, Figs. 11–14). These same features can be seen in the Lower Ordovician species *W. drucei* Pojeta and others (1977, Plate 5).

The Antarctic specimen is an internal mold with some shell adhering. It is missing the entire upper umbonal area, and height cannot be measured accurately. Measurements are: length, 11 mm; height, 9.4 mm; and width, 5 mm (both valves). The major difference between *W. cambrica* and *Wanwania* sp. A is that in the former the pegma is oblique, whereas in the latter the pegma is almost horizontal; in *W. drucei,* the pegma is almost erect. In the other two known species of *Wanwania, W. compressa* Kobayashi and *W. ambonychiformis* Kobayashi (Pojeta and Runnegar, 1976, Plate 3, Figs. 6–10), the shell is higher than it is long and is not subquadrate; these two species occur in the Fengshanian of northeastern China.

Material and locality. *Wanwania* species A is known from one incomplete specimen (USNM 385129) from the feather edge of the Minaret Formation on the northeastern side of Springer Peak, Webers Peaks, Heritage Range (Fig. 1).

PLATE 1

Figures 1–3. *Wanwania* sp. A. Right-lateral, left-lateral, and dorsal views of an incomplete internal mold showing subquadrate profile and impression of pegma (arrows) (×5). USNM 385129a.

Figures 4–12. *Ribeiria australiensis* Pojeta and Runnegar, 1976.

4–6. Dorsal, right-lateral, and anterior views of nearly complete internal mold showing dorsal carina (black arrow), impression of pegma (white arrows), and cross-section shape (×10). USNM 385115.

7, 8. Anterior and dorsal views of silicified *holotype* showing cross-section shape and dorsal carina (arrow) (×4). Mungerebar Limestone (Mindyallan), Queensland, Australia. Specimen at the Australian Bureau of Mineral Resources (BMR) CPC14670.

9. Right-lateral view of a silicified paratype showing nearly straight dorsal margin (×4). Horizon and locality the same as for the specimens shown in Figures 7, 8 above. BMR CPC14671.

10. Left-lateral view of a silicified paratype showing nearly straight dorsal margin (×4). Horizon and locality the same as for Figures 7, 8 above. BMR CPC14672.

11, 12. Dorsal and left-lateral views of an incomplete internal mold showing dorsal carina (arrow) (×10). USNM 385116.

Figures 13–18. *Apoptopegma craddocki* n. sp.

13, 14. Dorsal and left-lateral views of an internal mold, which is incomplete anteroventrally, showing hooked beak and marginal denticles (arrow) (×10). USNM 385121, paratype.

15–17. Right-lateral view of an internal mold with some shell at the extreme anterior end showing marginal denticles (arrow). (×4). Left-lateral view showing marginal denticles (black arrow), pegma (white arrow), and impression of the comarginal ornament posteriorly (×10). Dorsal view showing carina (arrow) and profile for comparison with *Ribeiria australiensis* in Figures 8 and 11 above (×4). USNM 385120, *holotype.*

18. Right-lateral view of incomplete internal mold showing hooked beak, marginal denticles, and impression of pegma (×4). USNM 385122, paratype.

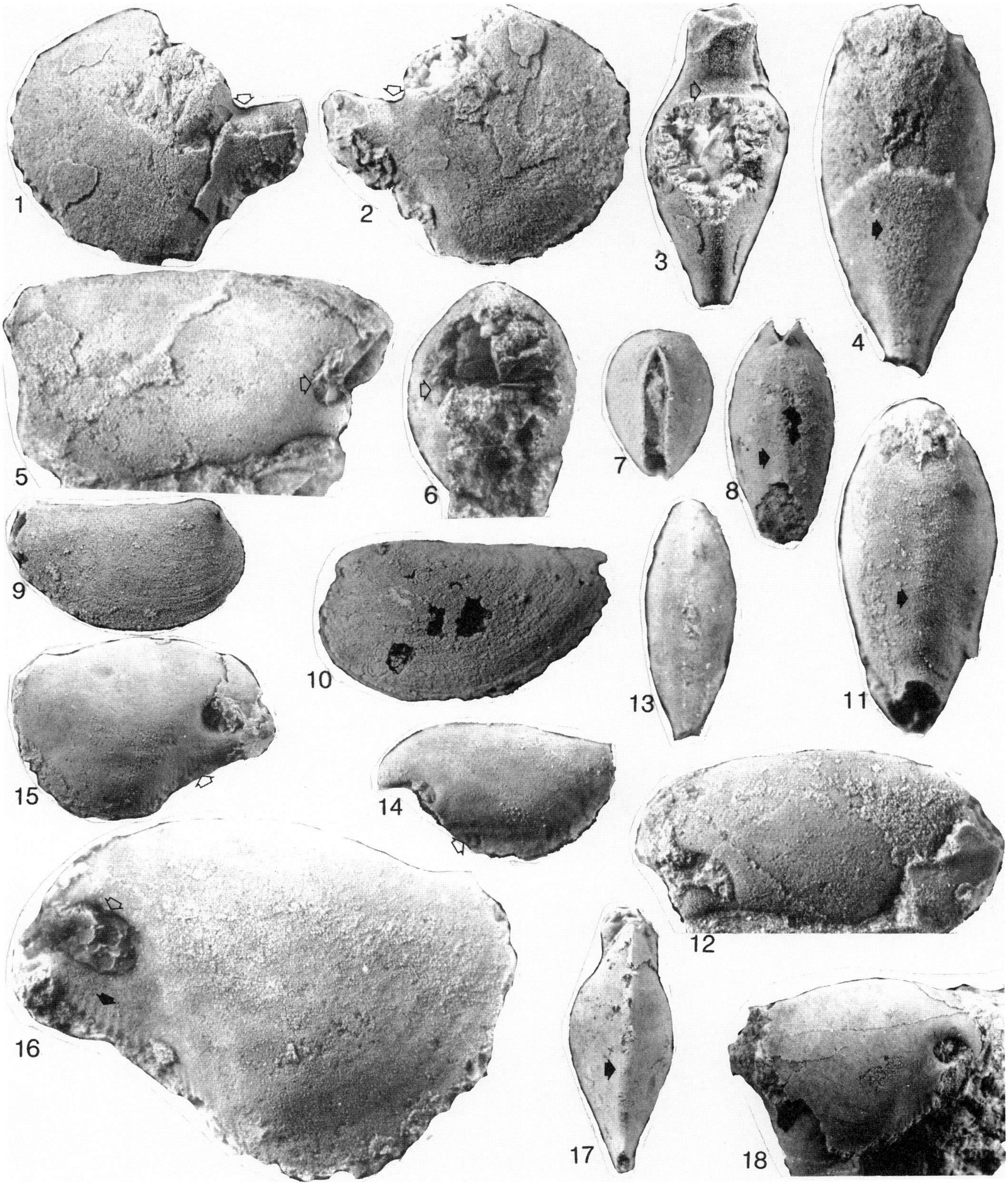
1
2
3
4
5
6
7
8
9
10
11
12
13
14
15
16
17
18

PLATE 2

Figures 1–15. *Apoptopegma craddocki* [n. sp.]

1, 2. Ventral and right-lateral views of an internal mold showing broken ventral margin simulating a gape and lateral profile with anteroventral sulcus and pegma (×4). USNM 385123, paratype.

3–6. Left-lateral and dorsal views (×4) and posterior and anterior views showing shell gapes (arrows) (×8) of a partially shelled specimen. USNM 385124, paratype.

7. Oblique anterior view showing carina, hooked beak, pegma, and funnel-shaped anterior gape (arrow) (×8). USNM 385125, paratype.

8–11. Partially shelled specimen, anterior view showing funnel-shaped gape (×8); dorsal view (×4); enlargement of anterior end of dorsal view showing flaring of anterior gape (arrow) on right side, left side broken off (×8); and left-lateral view showing anteroventral sulcus (×4). USNM 385126, paratype.

12–15. Internal mold with anteroventral portion missing, ventral view showing impression of pegma (arrow), dorsal view showing carina, right-lateral, and left-lateral views (×5). This specimen is from the Minaret Formation on the northwestern limb of Yochelson Ridge. USNM 385127, paratype.

Figures 16, 17. *Latouchella*? [sp. indet.] Composite mold (?), right-lateral view showing rugose, wedge-shaped growth increments (×20), and oblique dorsal view of lower wedge-shaped rugae (×60). Specimen from the Minaret Formation, south-southeast of Minaret Peak, northern Marble Hills. USNM 385129.

Figure 18. *Linevitus*? *springerensis* n. sp. Dorsal view of shelled specimen showing growth lines at right angle to carina (arrow) (×6). USNM 385136, paratype.

1
2
3
4
5
6
7
8
9
10
11
12
13
14
15
16
17
18

PLATE 3

Figures 1–3. *Latouchella*? sp. indet.

1, 2. Dorsal view of composite mold (?) showing rugose comarginal ornament and fine spiral threads between rugae (×45), and enlargement of central and lower part of Figure 1 (×110). USNM 385130.

3. Oblique lateral view of right side showing wedge-shaped rugae (×25). Same specimen as is shown in Plate 2, Figure 16, 17. USNM 358129. Both specimens from the Minaret Formation, south-southeast of Minaret Peak, northern Marble Hills.

Figures 4, 14, 15. *Linevitus*? *springerensis* n. sp.

4. Anterior ventral side (ligula?) of a shelled specimen showing arcuate growth lines (×5). USNM 385136a, paratype.

14, 15. Two views showing triangular cross section and dorsal and lateral carinas (×5, ×7). USNM 385142, 385143, paratypes.

Figures 5–9. *Tcharatheca webersi* (Yochelson), 1976.

5, 6. Views of the convex dorsum and concave venter of an incomplete internal mold (×5). USNM 385137.

7. Internal mold showing reniform cross section (×6). USNM 385138.

8. View of concave vector of an incomplete internal mold (×6). USNM 385139.

9. Oblique dorsal view of incomplete partially shelled specimen showing faint longitudinal costellae (×6). USNM 385140.

Figure 10. Orthothecida? sp. indet. View of the length of an incomplete tube of uncertain affinities (×3). Specimen from the Minaret Formation, Mount Rosenthal, Liberty Hills. USNM 385172.

Figure 11-13. *Hyolithes*? sp. indet. Oblique left-lateral, cross-section, and dorsal views of an incomplete, partially shelled specimen, showing broad lateral carina on left side, semicircular cross section, lack of a dorsal carina, and some of the ligula (arrow) (×5). USNM 385141.

1
2
3
4
5
6
7
8
9
10
11
12
13
14
15

PLATE 4

Figures 1–6, 8–12. *Linevitus*? *springerensis* n. sp.

1, 2. Incomplete shelled specimen, dorsal view showing growth lines at right angles to carina, and oblique left-lateral view showing slightly upturned posterior end and sigmoidal growth lines (×6). USNM 385144, *holotype.*

3, 4. Incomplete shelled specimen, left-lateral view showing sigmoidal growth lines and dorsal view showing growth lines at right angle to carina. (×6). USNM 385145, paratype.

5, 6. Incomplete largely decorticated specimen, dorsal view showing carina and view of broadly arcuate venter (×6). USNM 385146, paratype.

8–10. Incomplete shelled specimen, oblique right-lateral view, dorsal view, and ventral view showing broadly arcuate growth lines (×6). USNM 385147, paratype.

11, 12. Two specimens showing the triangular cross section (×7.5). USNM 385148, 385149, paratypes.

Figure 7. *Hyolithes*? sp. indet. Slightly oblique right-lateral view of the specimen shown on Plate 3, Figures 11–13, showing lack of preservation of a broad carina (×6). USNM 385141.

Figures 13–15. *Matherella antarctica* n. sp. Apertural (×6), back (×3), and oblique back (×3) views of an internal mold showing the large body whorl and prominent sutures. Specimen placed in the orthostrophic position. USNM 385150, *holotype.*

Figure 16. *Scaevogyra*? sp. indet. Back view of an internal mold (×3). Specimen placed in the orthostrophic position; from the Minaret Formation at Yochelson Ridge. USNM 385151.

PLATE 5

Figures 1–13. *Kobayashiella*? *heritagensis* n. sp. (All specimens photographed in orthostrophic position.)

1, 2. Back and apical views of partially shelled specimen which shows the course of the growth rugae, in the lateral view, near the aperture and on the penultimate whorl (×10). USNM 385152, *holotype.*

3, 4. Basal and apical views of a partially shelled specimen; the rounded whorl surface curving into the umbilical area is clearly seen on the base (×12). USNM 385153, paratype.

5. Back view of a moderately well preserved specimen (×10). USNM 385154, paratype.

6, 7. Apical and oblique back view of a steinkern showing the elongate smooth early whorls and the beginning of growth rugae in the later whorls (×10). USNM 385155, paratype.

8, 9. Basal (×7.5) and oblique back (×10) views. USNM 385156, paratype.

10. Oblique apical view of a steinkern showing the deep sutures between the whorls (×5). USNM 385157, paratype.

11. Back view of a partially shelled specimen showing three whorls and growth rugae near the aperture (×10). USNM 385158, paratype.

Figures 1–13. *Kobayashiella*? *heritagensis* n. sp.

12. Back view of an incomplete specimen showing prominent growth rugae on the body whorl (×10). USNM 385159, paratype.

13. Slightly oblique back view of a steinkern showing growth rugae on the body whorl (×10). USNM 385160, paratype.

Figures 14, 15. Orthothecida? sp. indet. Both specimens from the Minaret Formation, Mount Rosenthal, Liberty Hills.

14. Section through rock showing variation in cross-section shape of tubes caused by compression; photographed under water (×3). USNM 385173.

15. Naturally weathered surface showing imbrication of tubes (arrow) (×4). USNM 385174.

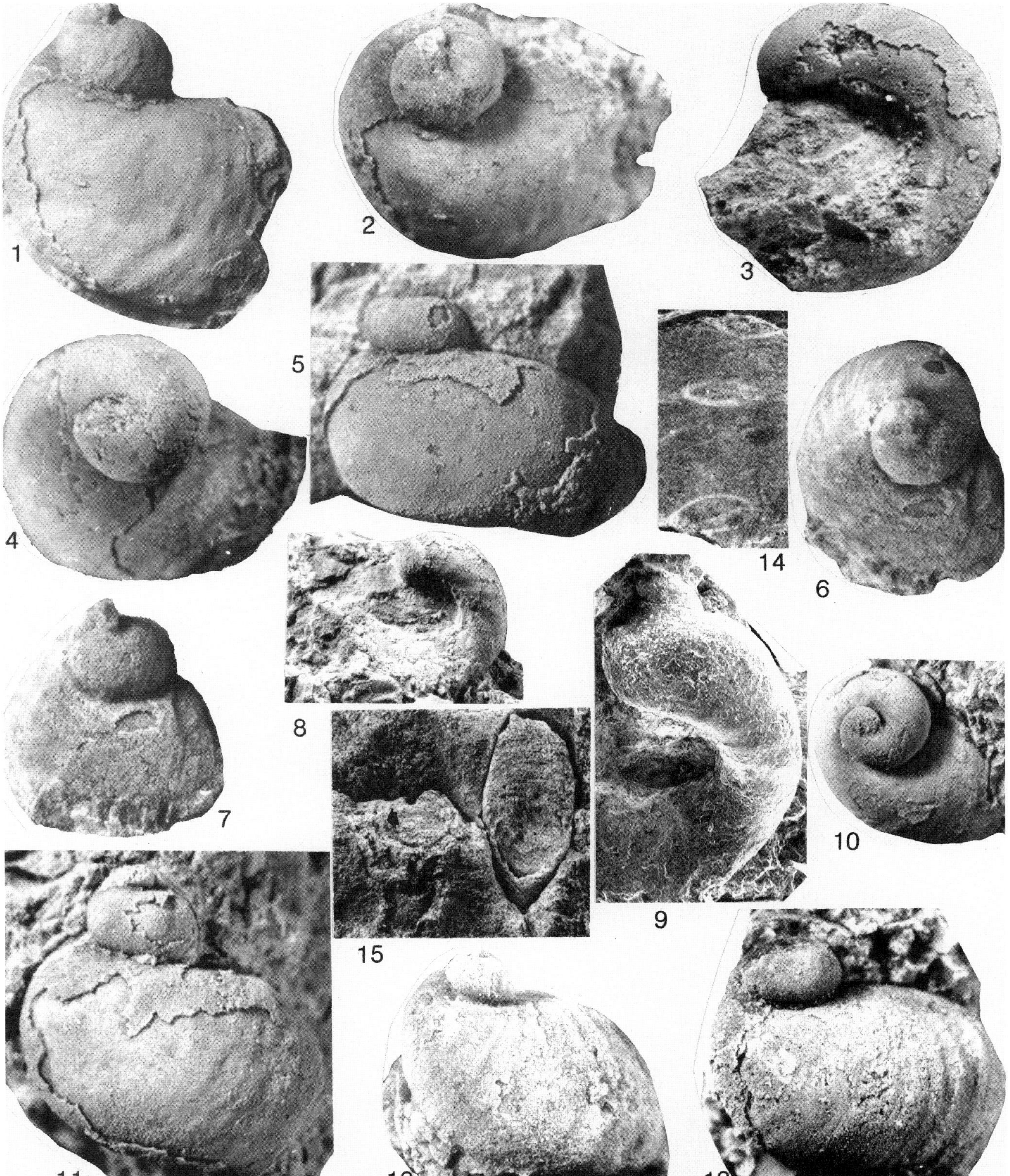
1
2
3
4
5
6
7
8
9
10
11
12
13
14
15

PLATE 6

Figures 1–8. *"Maclurites" thomsoni* n. sp. (All specimens photographed in orthostrophic position.)

1. Basal view of steinkern (×3). USNM 385161, paratype.

2, 3. Slightly oblique views of a small distorted largely decorticated specimen showing the large protoconch (arrows) (×3, ×6). USNM 385162, paratype.

4. Slightly oblique top view of a largely shelled specimen showing faint growth lines and the pseudoumbilicus largely filled with matrix (×3). USNM 385163, paratype.

5. Basal view of a largely decorticated specimen showing the deep suture (×3). USNM 385164, paratype.

6–8. Basal, top, and oblique back views of a largely decorticated specimen (×3). The simple shell margin is well shown in Figure 8 by two prominent comarginal rugae. USNM 385165, paratype.

1
2
3
4
5
6
7
8

PLATE 7

Figures 1–7. *"Maclurites" thomsoni* n. sp.

1–4. Mostly decorticated shell, basal, top, and apertural views (×3) and enlargement of basal view showing the large protoconch (arrow) (×6). USNM 385166, *holotype.*

5–7. Lateral, back, and top views of a large distorted steinkern (×2). USNM 385167, paratype.

Figures 8–13. *Euomphalopsis splettstoesseri* n. sp.

8, 9. Lateral and basal views of a large steinkern lacking the early whorls (×3). USNM 385168, paratype.

10. Lateral view of a small steinkern (×3). USNM 385169, paratype.

11, 12. Basal and lateral views of a composite (?) mold showing faint growth lines near the aperture (×3). USNM 385170, paratype.

13. Fragmentary steinkern of the open-coiled body whorl (×2). USNM 385171, paratype.

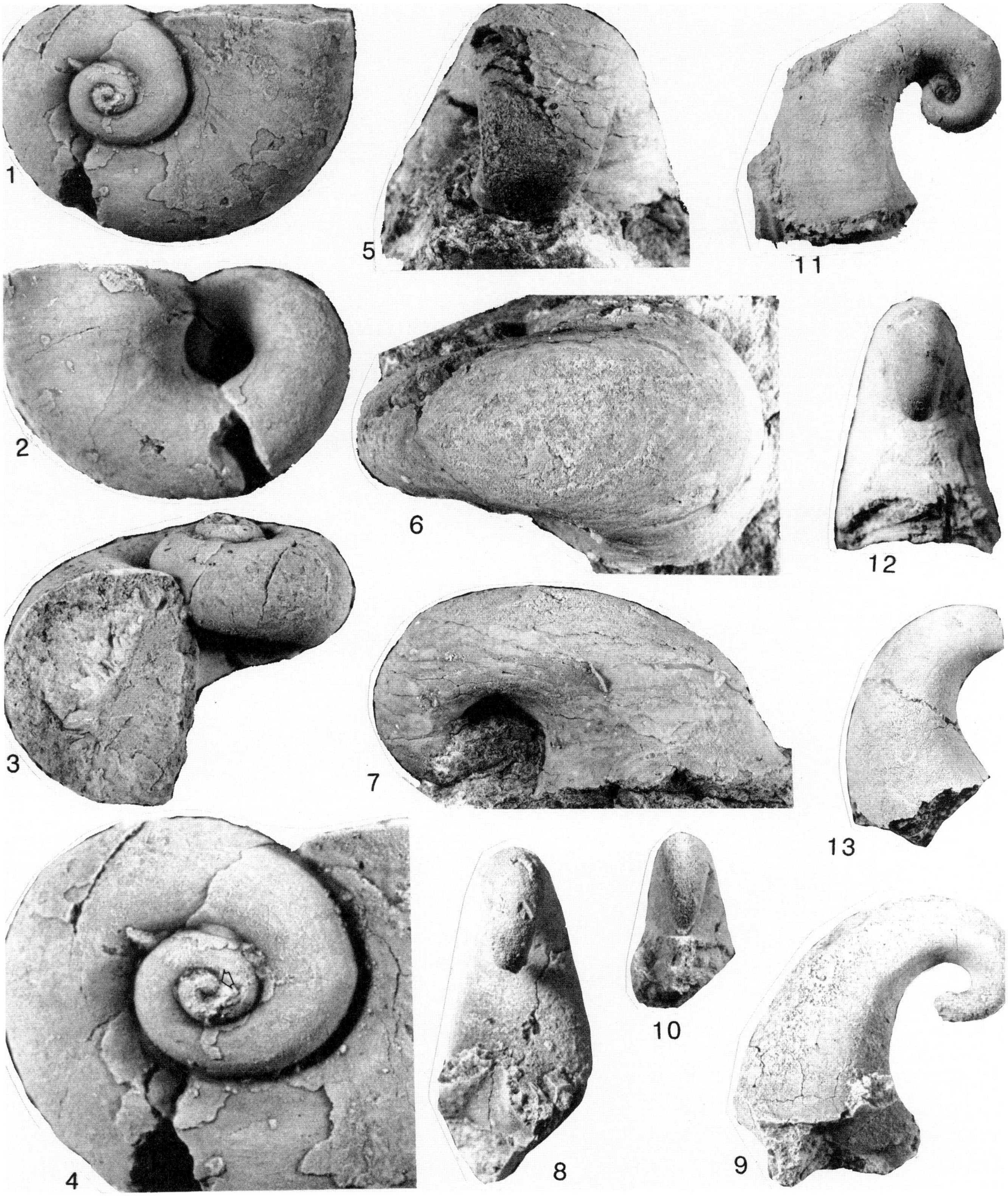
1
2
3
4
5
6
7
8
9
10
11
12
13

PLATE 8

Figures 1–12. *Euomphalopsis splettstoesseri* n. sp.

1, 2. Basal and lateral views of a steinkern lacking the early whorls (×3). USNM 385175, paratype.

3. Basal view of a weathered specimen in matrix (×3). USNM 385176, paratype.

4, 5. Basal view of a small specimen preserving the early whorls (×3) and enlargement of early whorls (×6). USNM 385177, paratype.

6, 7. Lateral view showing asymmetry and basal view showing the open coiling of the shell after the initial whorls; small patches of the thin shell are present on the left side (×3). USNM 385178, paratype.

8. Basal view of a small, largely decorticated specimen showing the unbroken straight apertural margin (×3). USNM 385179, paratype.

9. Basal view of a small steinkern retaining shell near the suture (×3). USNM 385180, paratype.

10–12. Basal, lateral, and upper views of a specimen showing some growth lines near the base and the early whorls (×3). USNM 385181, *holotype*.

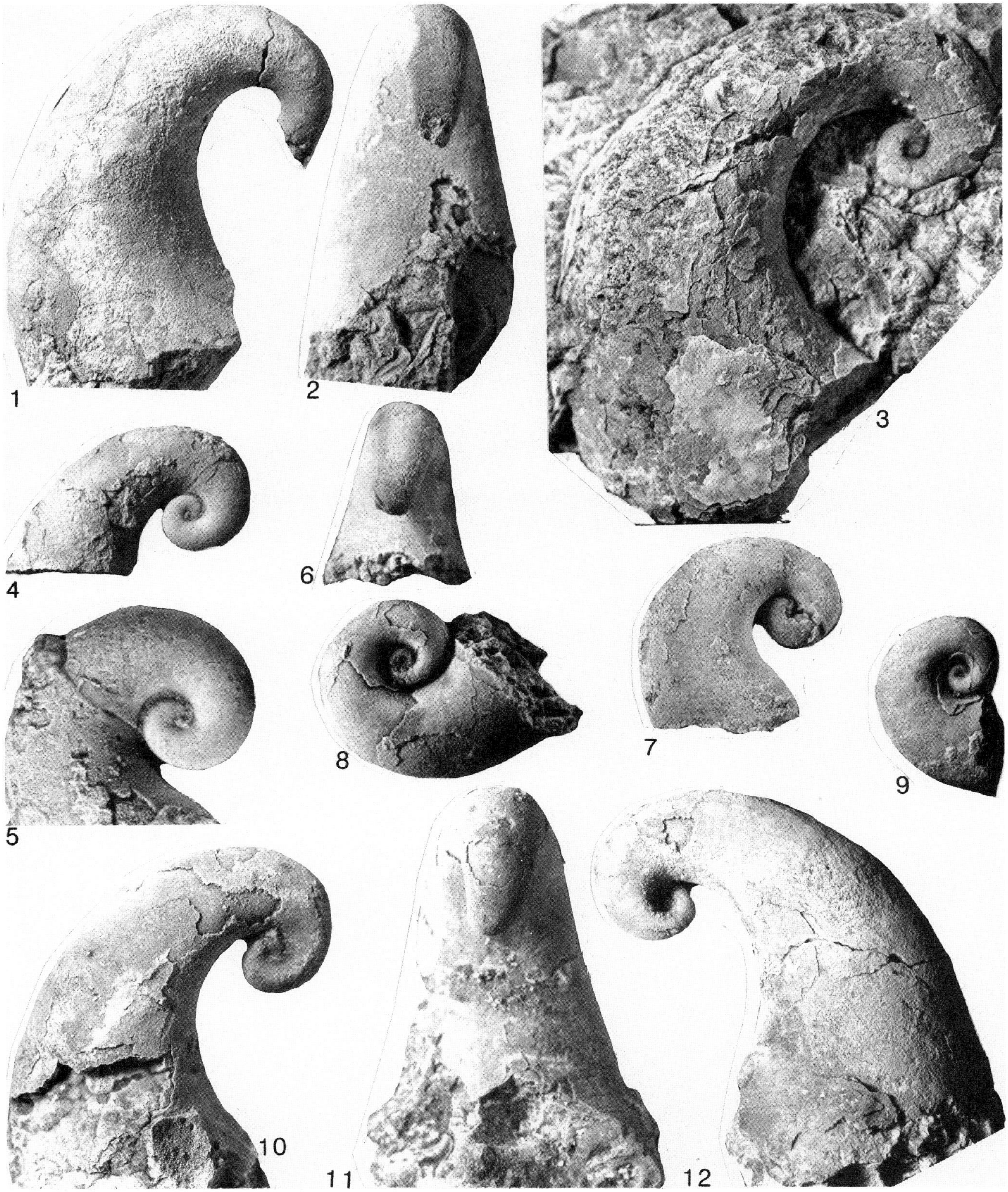
1
2
3
4
6
5
8
7
9
10
11
12

PLATE 9

Figures 1, 2. *Euomphalopsis splettstoesseri* n. sp. Basal and lateral views; this mostly decorticated steinkern is the largest specimen known (×3). USNM 385182, paratype.

Figures 3–13. *Aremellia batteni* n. sp.

3, 4. Basal and back views of a largely decorticated specimen (×5). USNM 385183, paratype.

5, 6. Basal and back views of a steinkern lacking the apex and showing well the slight sinistral asymmetry (×6). USNM 385184, paratype.

7, 8. Apical and back views of a steinkern showing cross-section shape and spiral ornament (×6). USNM 385185, paratype.

9–12. Apical, upper, basal, and back views of a largely decorticated specimen, which shows the slight sinistral asymmetry well and the spiral ribs near the aperture on the basal side (×6). USNM 385186, *holotype.*

13. Upper view of the largest known specimen (×6). USNM 385187, paratype.

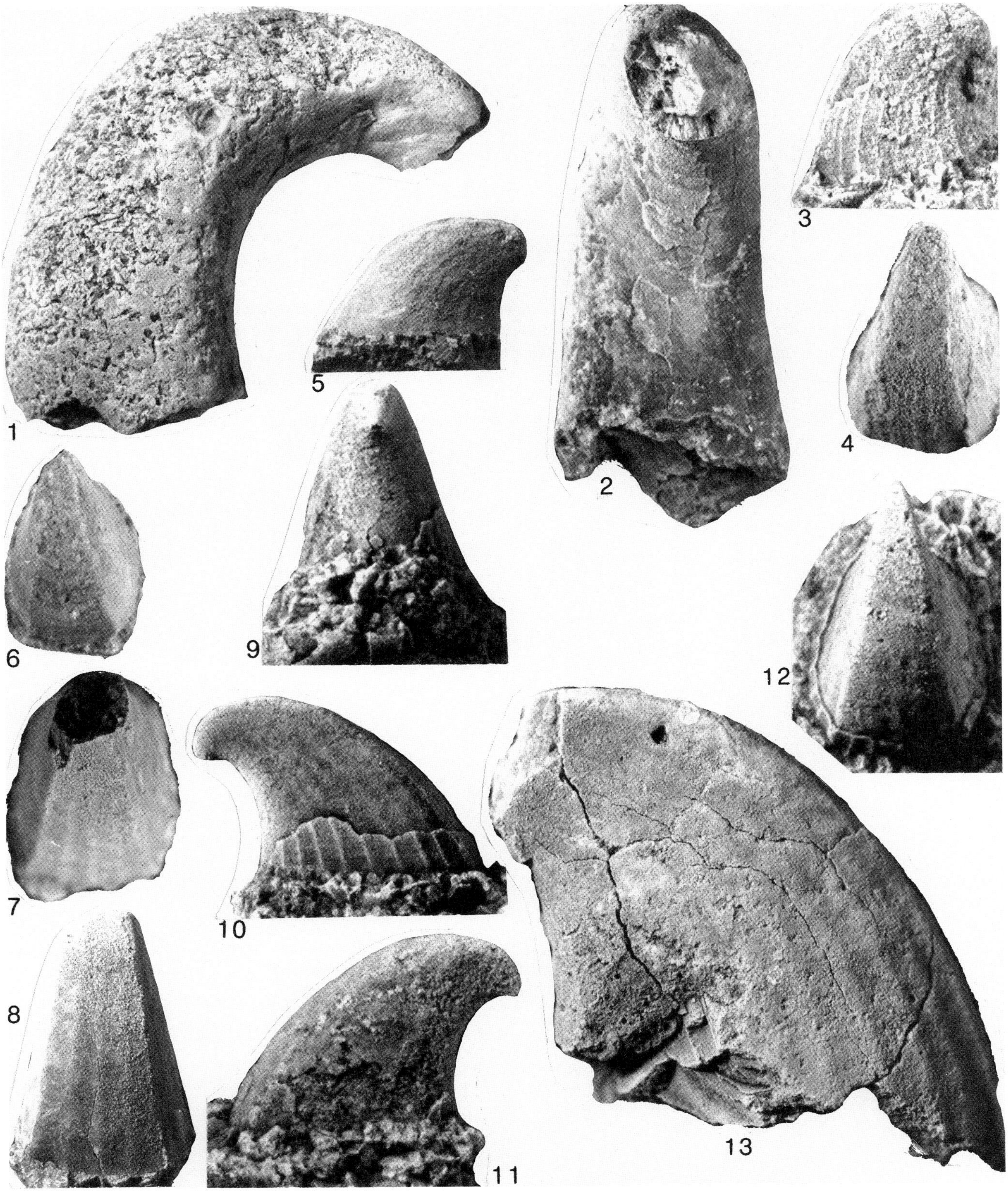
1
2
3
4
5
6
7
8
9
10
11
12
13

PLATE 10

Figures 1–7. *Kirengella pyramidalis* n. sp.

1–4. Anterior, dorsal, left-lateral, and right-lateral views showing profiles and closely spaced growth lines (×2). USNM 385188, *holotype.*

5, 6. Anterior and right-lateral views of a largely decorticated specimen showing growth lines and rugae, incomplete dorsally and posteriorly (×2). USNM 385189, paratype.

7. Right-lateral view of a highly fractured steinkern; the profile of this specimen suggests that the apex was more posterior than the apex on the specimen shown in Figures 1–4 above (×2). USNM 385190, paratype.

1
2
3
4
5
6
7

PLATE 11

Figures 1–19. *Cosminoconella runnegari* n. gen., n. sp.

1, 2. Left-lateral and dorsal views of a largely shelled specimen showing profiles and ornament (×6). USNM 385191, *holotype.*

3–5. Right-lateral, dorsal, and left-lateral views of a largely decorticated specimen showing profiles, ornament, and prominent subcomarginal sulcus about one-third of the height from the apex (×6). USNM 385192, paratype.

6, 7. Left-lateral and dorsal views of a largely shelled specimen (×6). USNM 385193, paratype.

8. Right-lateral view of a largely decorticated specimen (×6). USNM 385194, paratype.

9. Dorsal view of a shelled specimen showing profile and simple spiral ribs (×6). USNM 385195, paratype.

10, 11. Left-lateral and dorsal views showing profiles and ornament (×6). USNM 385196, paratype.

12. Left-lateral view (×6). USNM 385197, paratype.

13. Left-lateral view of internal mold with faint spiral rib impressions and the subcomarginal sulcus (×6). USNM 385198, paratype.

14. Dorsal view (×6). USNM 385199, paratype.

15, 16. Left-lateral and right-lateral views of internal mold (×6). USNM 385200, paratype.

17, 18. Right-lateral and oblique dorsal views of a largely decorticated specimen (×6). USNM 385201, paratype.

19. Right-lateral view of internal mold with faint impression of spiral ribs (×6). USNM 385202, paratype. Note the variation in length/height ratio from specimens such as those shown in Figures 1 and 15, to Figure 10, to Figures 5, 12, 13, to Figure 19.

1
2
3
4
5
6
7
8
9
10
11
12
13
14
15
16
17
18
19

PLATE 12

Figures 1–5. *Cosminoconella runnegari* n. gen., n. sp.

1–4. Dorsal, right-lateral, anterior, and oblique anterior views of an internal mold showing subcomarginal sulci and ridges (×6). USNM 385202a, paratype.

5. Dorsal view of internal mold showing faint subcomarginal sulcus (×6). USNM 385203, paratype.

Figures 6–8, 11–18. *Proplina rutfordi* n. sp.

6–8. Right-lateral, dorsal, and anterior views of a partially decorticated specimen showing profiles (×3). USNM 385204, paratype.

11, 12. Left-lateral and anterior views of a shelled specimen showing profiles (×3). USNM 385206, paratype.

13, 14. Right-lateral and dorsal views of a small shelled specimen showing profiles and ornament (×10). USNM 385207, paratype.

15–18. Left-lateral, dorsal, anterior, and right-lateral views of a partially shelled specimen showing profiles and rugose comarginal ornament on internal mold but finer growth increments on the shell (×3). USNM 385208, *holotype.*

Figures 9, 10, 19–22. *Proconus incertis* n. gen., n. sp.

9, 10. Left-lateral and dorsal views of a shelled specimen showing profiles and rugose comarginal ornament (×3). USNM 385205, paratype.

19. Dorsal view of a small partially shelled specimen (×3). USNM 385209, paratype.

20, 21. Left-lateral and dorsal views of a composite mold (?) showing lateral and dorsal profiles and rugae (×3). USNM 385210, paratype.

22. Dorsal view of a partially shelled specimen showing that on the internal mold (upper right) only the coarse growth increments are preserved (×6). USNM 385211, paratype.

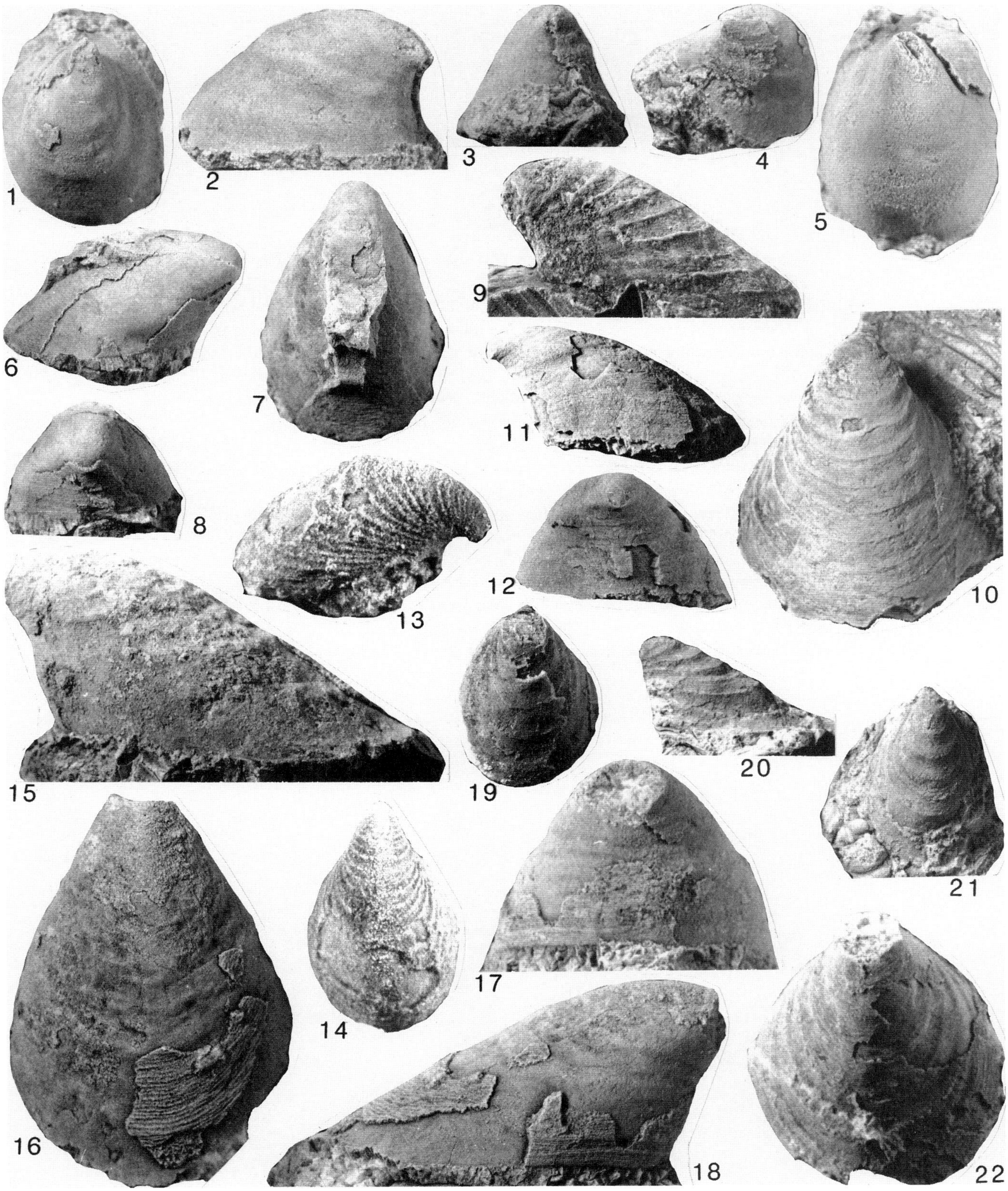
1
2
3
4
5
6
7
8
9
10
11
12
13
14
15
16
17
18
19
20
21
22

PLATE 13

Figures 1–12. *Proconus incertis* n. gen., n. sp.

1, 2. Anterior and left-lateral views of large internal mold showing profiles (×1). USNM 385212, paratype.

3, 4. Dorsal and right-lateral views of internal mold showing ornament and profiles (×6). USNM 385213, paratype.

5. Right-lateral view of a largely decorticated specimen showing profile (×6). USNM 385214, paratype.

6–9. Anterior, right-lateral, left-lateral, and dorsal views of incomplete internal mold showing ornament and profiles (×3). USNM 386554, *holotype.*

10–12. Anterior, dorsal, and left-lateral views of incomplete internal mold showing profiles and ornament (×3). USNM 386555, paratype.

Figures 13–18. *Ellsworthoconus andersoni* n. gen., n. sp.

13. Left-lateral view of a partially shelled specimen showing ornament and profile (×2). USNM 386556, paratype.

14–16. Dorsal, anterior, and right-lateral views of a partially shelled specimen showing profiles and ornament (×2). USNM 386557, paratype.

17. Left-lateral view of a specimen with little shell preserved showing the profile (×2). USNM 386558, paratype.

18. Right-lateral view of an internal mold showing profile (×2). USNM 386559, paratype.

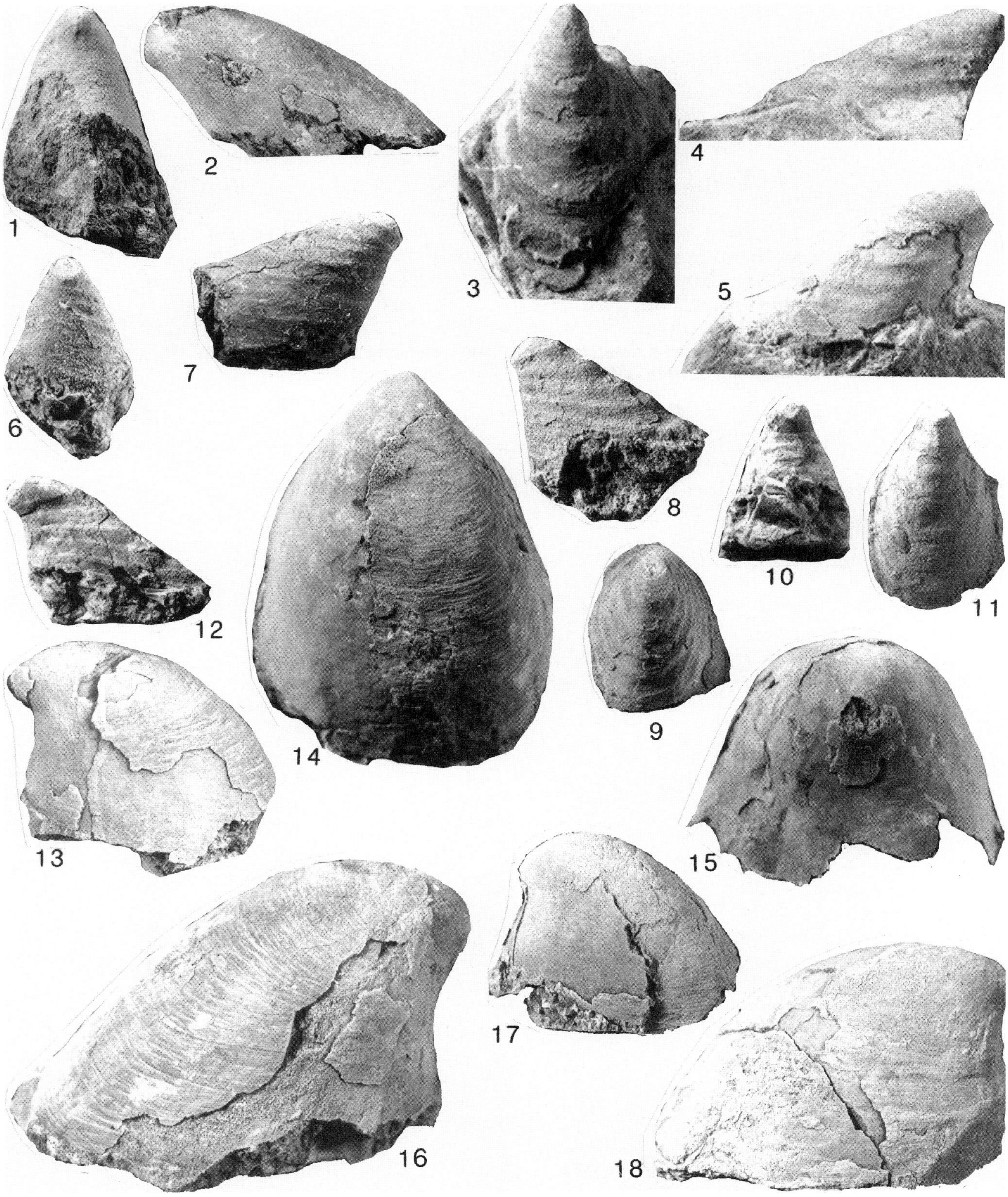
1
2
3
4
5
6
7
8
9
10
11
12
13
14
15
16
17
18

PLATE 14

Figures 1–12. *Ellsworthoconus andersoni* n. gen., n. sp. Figures 1–3, 8–11 are arranged in a growth series (×2) of left-lateral views of *holotype* (Fig. 3; USNM 386900) and six paratypes (Figs. 1, 2, 8–11; USNM 386560–386565). Figures 4–6 are dorsal (×2), anterior (×2), and right-lateral (×3) views of the holotype. Figure 7 is a right-lateral view (×3) of a paratype (USNM 386566). Figure 12 is a left-lateral view of an internal mold showing impressions of probable radial mantle muscles and a blister-pearllike impression (arrow) (×2) (USNM 386567) paratype.

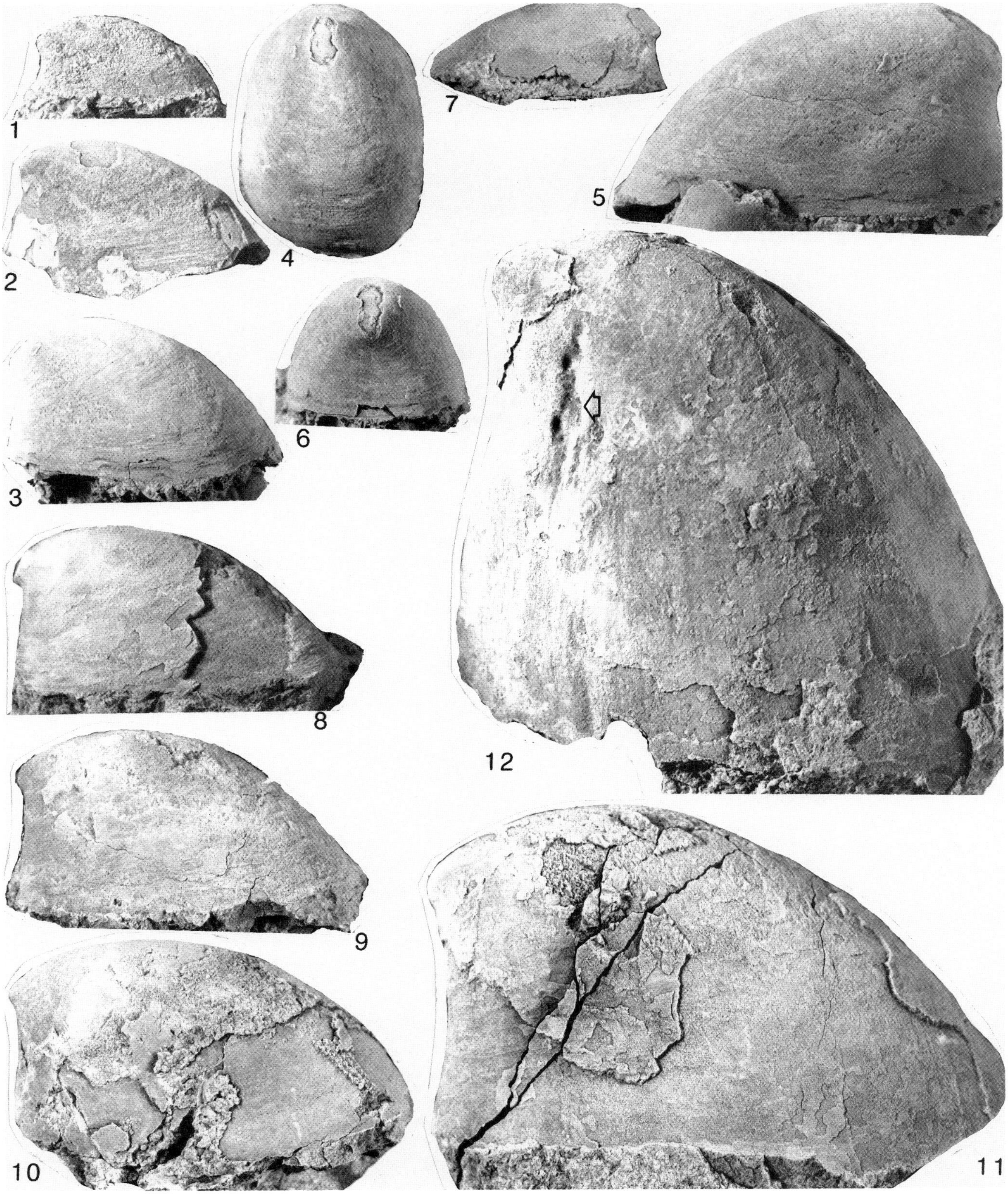
1
2
3
4
5
6
7
8
9
10
11
12

PLATE 15

Figures 1–12. *Knightoconus antarcticus* Yochelson, Flower, and Webers, 1973 (All specimens are topotypes.)

1. Right-lateral view of a small partially shelled specimen showing reversal of curvature between juvenile and adult shell (arrow) (×10). USNM 386568.

2, 3. Left-lateral views of a small internal mold showing reversal of curvature between juvenile and adult shell at a septum (white arrows) and prominent trail on adult shell (black arrow) (×3, ×12). USNM 386569.

4. Right-lateral view of a small internal mold showing reversal of curvature between juvenile and adult shell (arrow) (×10). USNM 386570.

5. Oblique left-lateral view of an incomplete shelled specimen showing ornament and septum (arrow) (×3). USNM 386571.

6. Oblique right-lateral view of an incomplete internal mold showing molds of two septa (arrows) the upper one of which is incompletely preserved (×3). USNM 386572.

7, 8. Anterior and left-lateral views of an internal mold (×3). USNM 386573.

9, 10. Left-lateral and posterior views of an incomplete internal mold showing trace of a septum (arrows) (×3). USNM 386574.

11, 12. Dorsal and right-lateral views of an internal mold showing profiles, trail, and trace of a septum (arrow) (×3). USNM 386575.

1
2
3
4
5
6
7
8
9
10
11
12

PLATE 16

Figures 1–11. *Knightoconus antarcticus* Yochelson, Flower, and Webers, 1973. (All specimens are topotypes.)

1–7. Left-lateral views of seven specimens, mostly internal molds, showing variation in curvature and position of septa (arrows). Figure 2 shows ornament and trail. Figure 1 (×6), all others (×3). USNM 386576–386581, 386573.

8. Oblique posterior view of internal mold showing septum (arrow), trail, and ornament (×3). USNM 386582.

9. Right-lateral view of anterior end of a specimen preserving little shell (black arrow) and showing that the ornament can be reflected on the internal mold, two closely spaced septa (white arrows), and planar aperture (×3). USNM 386583.

10. Left-lateral view of a small internal mold showing reversal of curvature between juvenile and adult shell (×12). USNM 386584.

11. Right-lateral view of an internal mold; the line in the upper half of the specimen is a fracture and not a septum (×3). USNM 386585.

Figure 12. Orthothecida? sp. indet. Section showing three cross sections compressed to varying degrees (×3). Specimen from the Minaret Formation, Mount Rosenthal, Liberty Hills. USNM 386586.

1
2
3
4
5
6
7
8
9
10
11
12

PLATE 17

Figures 1–8. *Knightoconus antarcticus* Yochelson, Flower, and Webers, 1973. (All specimens are topotypes.)

1. Left-lateral view of internal mold showing ornament, trail, and septum (arrow) (×3). USNM 386582.
2. Left-lateral view of a partially shelled specimen showing ornament; the line in the upper half of the specimen is a fracture and not a septum (×3). USNM 386588.

3–8. This specimen was found at the apex of a crushed *Knightoconus antarcticus* and is the smallest known juvenile specimen of the species. The very earliest shell is missing. Note that the curvature is in the same direction as the juvenile part of the shell seen on Plate 15, Figures 1–4, and Plate 16, Figure 10. Figure 3 (×12) is a light photograph of the right side of the specimen. All other figures are SEM photographs of the posterior (×30), left side (×40), right side (×35), and anterior (×35). Figure 8 (×85) is an enlargement of the anterodorsal part of the right side, which shows vague radial markings. We are not sure if these indicate a different ornament for the larval shell, or if they were somehow caused by taphonomy or tectonism. USNM 386589.

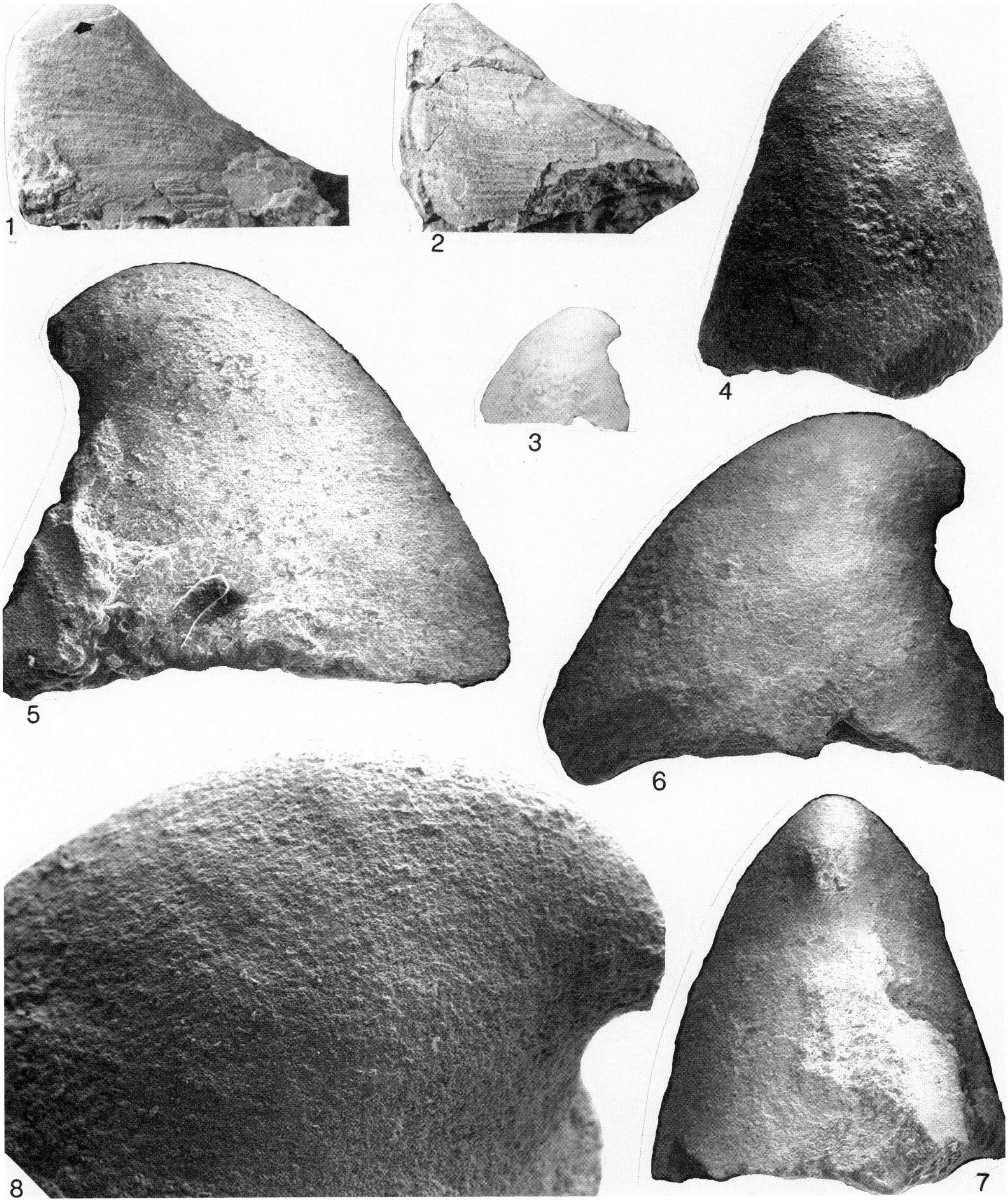
1
2
3
4
5
6
7
8

PLATE 18

Figures 1–5, 7, 8. *Kobayashiella*? *heritagensis* n. sp. Early whorls of specimens.

1, 2. Apical and side views of the specimen shown on 5, Figure 11 (×35; ×20). USNM 385158, paratype.

3, 4. Apical and side views (×22). USNM 386191, paratype.

5. Side view of the specimen shown on Plate 5, Figures 8, 9 (×30). USNM 385156, paratype.

7, 8. Apical and basal views of the specimen shown on Plate 5, Figure 5 (×35). USNM 385154, paratype.

Figure 6. *Aremellia batteni* n. gen., n. sp. Outer part of base of specimen shown on Plate 9, Figures 3, 4 (×15). USNM 385183, paratype.

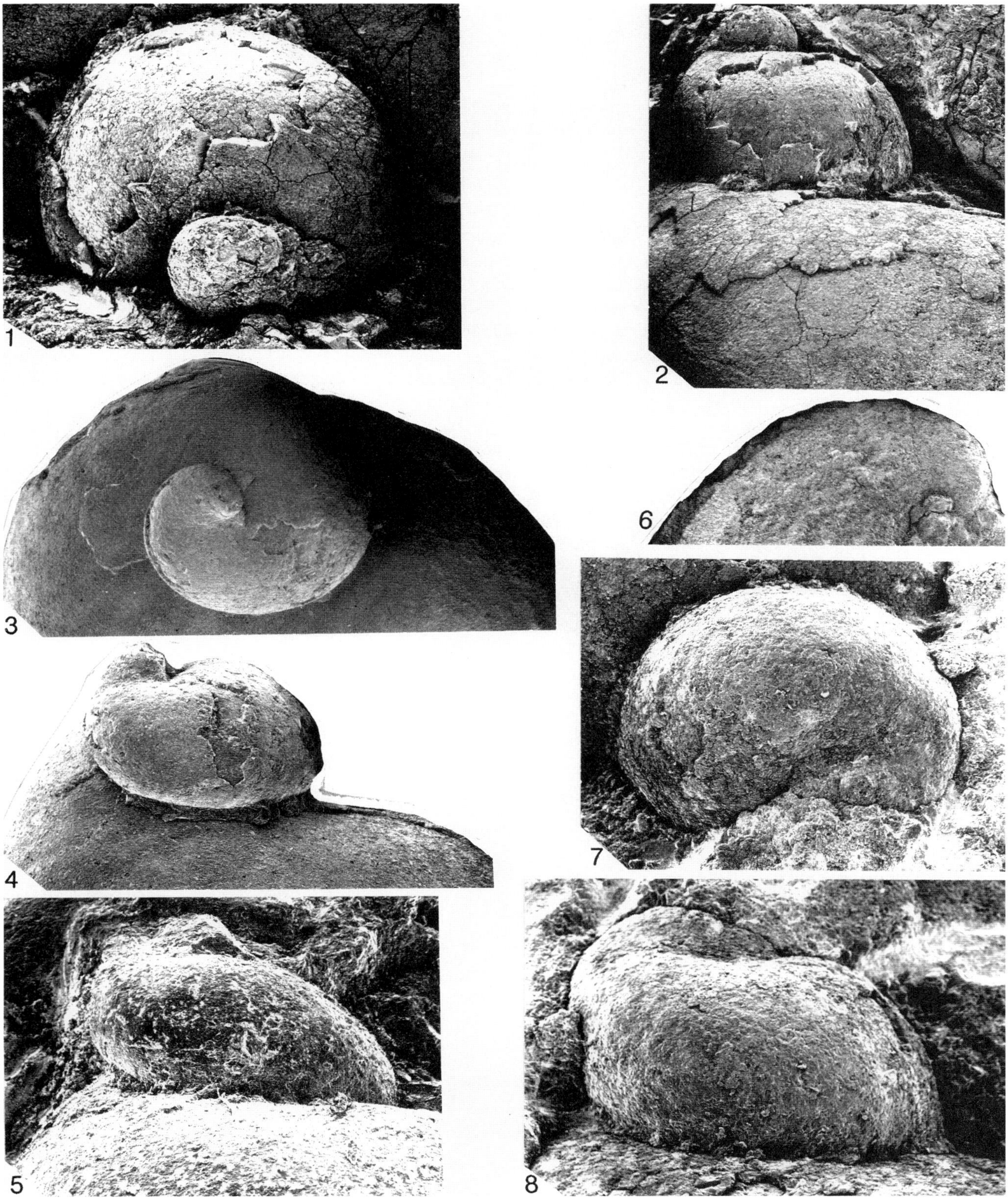

PLATE 19

(Major collecting localities for mollusks)

Figures 1, 2. Springer Peak: (1) Outcrop of Minaret Formation limestone (arrows) looking north; less resistant argillites to either side. The base of Springer Peak (1,460 m high) is to the left. Springer Peak is composed of Crashsite Quartzite. (2) Feather edge of the Minaret Formation limestone (arrows) looking south from the junction of Webers Peaks and the Splettstoesser Glacier.

Figures 3, 4. Bingham Peak: (3) Outcrop of Minaret Formation limestone (arrows) looking west from the Balish Glacier. Bingham Peak (1,540 m high), in the center of the photograph, is composed of Crashsite Quartzite. (4) Closer view of the northern end of the outcrop seen in Figure 3; arrow points to a man for scale.

Figure 5. Fault selvage in the limestone of the Minaret Formation at Yochelson Ridge. The collection made below and left of the fault yielded trilobites and the mollusks *Apoptopegma craddocki* and *Proplina rutfordi.*

Figure 6. Looking south at Minaret Peak (arrow), which is 1,613 m high. Minaret Peak is composed of limestone of the Minaret Formation.

1
2
3
4
5
6

ACKNOWLEDGMENTS

We gratefully acknowledge the field support of a number of geologists who are not paleontologists and of several nongeologists, all of whom were instrumental in acquiring a large collection of fossils. Notable among these in 1963 were Campbell Craddock, Jerry Dolence, and Benjamin Drake; the topographic engineers Rob Collier and Dean Edson were enthusiastic collectors. The additional material recovered by Robert Rutford in 1964 played a significant role in demonstrating the need for further work in the region. During the 1979–80 expedition, as a result of the transitional nature of local parties, almost every geologist present contributed to the search of the Minaret Formation for fossils or to collecting at Springer Peak. Lawrence Rosen and John Anderson found the first fossils from the Minaret Formation, at its type locality, in the Marble Hills.

Marija Balanc and Susan Goda, U.S. Geological Survey, opaqued the photographic prints. Balanc printed all photographs and did the final mounting of them on the montages of the plates. R. T. Lierman, U.S. Geological Survey, provided the gross petrographic descriptions from hand samples.

Fieldwork by all of us, and study of the fossils by Webers, were supported by grants DPP 7821720 and DPP 8214212 from the National Science Foundation, Division of Polar Programs, to Macalester College.

REFERENCES CITED

Bandel, K., 1982, Morphologie und Bildung der fruhontogenetischen Gehause bei conchiferen Mollusken: Facies, v. 7, p. 1–198.

Bengtson, S., and Fletcher, T. P., 1983, The oldest sequence of skeletal fossils in the lower Cambrian of southeastern Newfoundland: Canadian Journal of Earth Sciences, v. 20, p. 525–536.

Berkey, C. P., 1898, Geology of the St. Croix Dalles; 3, Paleontology: American Geologist, v. 21, p. 270–294.

Billings, E., 1865, Palaeozoic fossils, v. l: Montreal, Geological Survey of Canada, 426 p.

Bridge, J., 1931, Geology of the Eminence and Cardareva Quadrangles: Missouri Bureau of Geology and Mines, 2nd series, v. 24, 228 p.

Butts, C., 1926, Geology of Alabama; The Paleozoic rocks: Alabama Geological Survey Special Report 14, p. 40–230.

Chen, J., and Teichert, C., 1983, Cambrian cephalopoda of China: Palaeontographica, abt. A, v. 181, 102 p.

Chen, J., and 5 others, 1983, Faunal sequence across the Cambrian-Ordovician boundary in northern China and its international correlation: Geologica et Palaeontologica, v. 17, p. 1–15.

Cloud, P. E., Jr., and Barnes, V. E., 1948, The Ellenburger Group of central Texas: Austin, University of Texas Publication 4621, 473 p.

Cobbold, E. S., 1921, The Cambrian horizons of Comley, and their brachiopoda, pteropoda, and gastropoda: Quarterly Journal of the Geological Society of London, v. 76, no. 4, p. 325–386.

Conway Morris, S., 1982, *Wiwaxia corrugata* (Matthew), a problematic Middle Cambrian animal from the Burgess Shale of British Columbia, *in* Proceedings, North American Paleontological Convention 3: Lawrence, Kansas, Allen Press, v. 1, p. 93–98.

Craddock, C., 1969, Geology of the Ellsworth Mountains, *in* Bushnell, V. C., and Craddock, C., eds., Geologic maps of Antarctica: New York, American Geographical Society, Antarctic Map Folio Series, Folio 12, Plate 4, 1:1,000,000.

Daily, B., Firman, J. B., Forbes, B. G., and Lindsay, J. M., 1976, Geology, *in* Twindle, C. R., Tyler, M. J., and Webb, B. P., eds., Natural history of the Adelaide region: Royal Society of South Australia, Inc., p. 5–42.

Dake, C. L., and Bridge, J., 1932, Faunal correlation of the Ellenburger Limestone of Texas: Geological Society of America Bulletin, v. 43, p. 725–748.

Dzik, J., 1978, Larval development of hyoliths: Lethaia, v. 11, p. 293–299.

—— , 1980, Ontogeny of *Bactrotheca* and related hyoliths: Geologiska Foreningens i Stockholm Forhandlingar, v. 102-223-233.

Eichwald, C. E., von, 1840, Ueber das silurische Schichtenssystem in Esthland: Zeitschrift der Natur und Heilkunde der medizinischen Academie zu St. Petersburg, heft 1, p. 114; heft 2, p. 115–210.

Endo, R., 1937, Addenda to parts 1 and 2, *in* Endo, R., and Resser, C. E., eds., The Sinian and Cambrian formations and fossils of southern Manchoukou: Manchurian Science Museum Bulletin, v. 1, p. 302–369.

Fisher, D., 1962, Small conoidal shells of uncertain affinities, *in* Moore, R. C., ed., Treatise on invertebrate paleontology; v. W, Miscellanea: Lawrence, Kansas, University of Kansas Press, p. W98–W143.

Goddard, E. N., and others, 1948, Rock color chart: Geological Society of America, 16 p.

Harper, J. A., and Rollins, H. B., 1982, Recognition of Monoplacophora and Gastropoda in the fossil record; A functional morphological look at the bellerophont controversy, *in* Proceedings, 3rd North American Paleontological Convention: Lawrence, Kansas, Allen Press, v. 1, p. 227–236.

Holm, G., 1893, Sveriges Kambrisk-Siluriska Hyolithidae och Conulariidae: Sveriges Geologiska Undersokning C., v. 112, 173 p.

Horný R. J., 1963, Lower Paleozoic Monoplacophora and patellid Gastropoda (Mollusca) of Bohemia: Sbornik Ustredniho Ustavu Geologikeho, oddil paleontologicky, v. 28, p. 17–81.

—— , 1965, *Cyrtolites* Conrad, 1838, and its position among the Monoplacophora (Mollusca): Sbornik Narodiniho Muzea v Praze, v. 21B, no. 2, p. 57–70.

Knight, J. B., 1941, Paleozoic gastropod genotypes: Geological Society of America Special Paper 32, 510 p.

—— , 1952, Primitive fossil gastropods and their bearing on gastropod classification: Smithsonian Miscellaneous Collections, v. 117, no. 3, p. 1–56.

Knight, J. B., and Yochelson, E. L., 1958, A reconsideration of the relationships of the Monoplacophora and the primitive Gastropoda: Malacological Society of London Proceedings, v. 33, no. 1, p. 37–48.

—— , 1960, Monoplacophora, *in* Moore, R. C., ed., Treatise on invertebrate paleontology; v. I, Mollusca 1: Lawrence, University of Kansas Press, p. I77–I83.

Knight, J. B., Batten, R. L., and Yochelson, E. L., 1960, Descriptions of Paleozoic gastropods, *in* Moore, R. C., ed., Treatise on invertebrate paleontology; v. I, Mollusca 1: Lawrence, University of Kansas Press, selected pages.

Kobayashi, T., 1933, Faunal studies of the Wanwanian (Basal Ordovician) Series with special notes on the Ribeiridae and the Ellesmeroceroids: Imperial University of Tokyo Faculty of Science Journal, v. 3, no. 7, p. 249–328.

Koken, E., 1925, Die Gastropoden des baltischen Untersilurs, edited by J. Perner: Leningrad, Academie des Sciences de Russie Memores, 8th series, Classe Physico-Mathematique, v. 38, no. 1, 326 p.

Lemche, H., 1957, A new deep-sea mollusc of the Cambro-Devonian class Monoplacophora: Nature, v. 179, p. 413–416.

LeSueur, C. A., 1818, Observations on a new genus of fossil shells: Academy of Natural Sciences of Philadelphia Journal, v. 1, p. 310–313.

Lindström, G., 1884, On the Silurian Gastropoda and Pteropoda of Gotland:

Kungliga Svenska Vetenskapsakademiens Handlinger, v. 19, no. 6, 250 p.
Linsley, R. M., and Kier, W. M., 1984, The Paragastropoda; A proposal for a new class of Paleozoic Mollusca: Malacologica, v. 25, p. 241–254.
Ludvigsen, R., and Westrop, S. R., 1983, Franconian trilobites of New York state: New York State Museum Memoir 23, 82 p.
Marek, L., 1967, The class Hyolitha in the Caradoc of Bohemia: Sbornik Geologickych Ved, Paleontologie, v. 9, p. 51–114.
Marek, L., and Yochelson, E. L., 1976, Aspects of the biology of Hyolitha (Mollusca): Lethaia, v. 9, p. 65–82.
Matthews, S. C., and Missarzhevsky, V. V., 1975, Small shelly fossils of late Precambrian and Early Cambrian age: A review of recent work: Geological Society of London Journal, v. 131, p. 289–304.
Öpik, A. A., 1967, The Mindyallan fauna of northwestern Queensland: Australian Bureau of Mineral Resources Bulletin 74, 404 p.
Palmer, A. R., 1982, Fossils of Dresbachian and Franconian (Cambrian) age from the subsurface of west-central Indiana: Indiana Department of Natural Resources Geological Survey Special Report 29, 12 p.
Peel, J. S., 1977, Relationship and internal structure of a new *Pilina* (Monoplacophora) from the Late Ordovician of Oklahoma: Journal of Paleontology, v. 51, p. 116–122.
Peel, J. S., and Yochelson, E. L., 1980, Giant Mollusca (Hyolitha) from the Permian of East Greenland: Gronlands geologiske Undersogelse Rapport, v. 101, p. 65–67.
——, 1984, Permian Toxeumorphida from Greenland; An appraisal of the molluscan class Xenochonchia: Lethaia, v. 17, p. 211–221.
Pojeta, J., Jr., 1979, Geographic distribution of Cambrian and Ordovician rostrochonch mollusks, *in* Gray, J., and Boucot, A. J., eds., Historical biogeography, plate tectonics, and the changing environment: Corvallis: Oregon State University Press, p. 27–36.
Pojeta, J., Jr., Gilbert-Tomlinson, J., and Shergold, J H., 1977, Cambrian and Ordovician rostroconch mollusks from northern Australia: Australian Bureau of Mineral Resources Bulletin 171, 54 p.
——, 1989, The early evolution of some diasome mollusks, *in* Trueman, E. R., ed., Evolution of the mollusca: New York, Academic Press (in press).
Pojeta, J. Jr., and Runnegar, B., 1976, The Paleontology of rostroconch mollusks and the early history of the Phylum Mollusca: U.S. Geological Survey Professional Paper 968, 88 p.
——, 1989, The early evolution of some diasome mollusks, in; Trueman, E. R., ed., Evolution of the mollusca: New York, Academic Press (in press).
Pojeta, J., Jr., Webers, G. F., and Yochelson, E. L., 1981, Upper Cambrian mollusks from the Ellsworth Mountains, West Antarctica, *in* Taylor, M. E., ed., Short papers for the 2nd International Symposium on the Cambrian System: U.S. Geological Survey Open-File Report 81-743, p. 167–168.
Qian, Y., Chen, M., and Chen, Y., 1979, Hyolithids and other small shelly fossils from the Lower Cambrian Huangshandon Formation in the eastern part of the Yangtze Gorge: Acta Palaeontologica Sinica, v. 18, p. 207–230 (in Chinese).
Rozov, S. N., 1968, New genus of Late Cambrian molluska from the class of Monoplacophora, southern part of Siberian platform: U.S.S.R. Academy of Sciences, Doklady, v. 183, no. 6, p. 1427–1430 (in Russian).
——, 1969, Morphology and terminology of Monoplacophora: Paleontological Journal 1969, no. 4, p. 107–110 (in Russian).
——, 1975, A new order of Monoplacophora: Palaeontological Journal, 1975, no. 1, p. 41–45 (in Russian).
Runnegar, B., 1978, Origin and evolution of the class Rostroconchia: Philosophical Transactions of the Royal Society of London, series B., v. 284, p. 319–333.
——, 1980, Hyolitha; Status of the phylum: Lethaia, v. 13, p. 21–25.
——, 1981a, Biostratigraphy of Cambrian molluska, *in* Taylor, M. E., ed., Short papers for the 2nd International Symposium on the Cambrian System: U.S. Geological Survey Open-File Report 81-743, p. 198–202.
——, 1981b, Muscle scars, shell form, and torsion in Cambrian and Ordovician univalved mollusks: Lethaia, v. 14, no. 4, p. 311–322.
——, 1983, Molluscan phylogeny revisited: Association of Australasian Palaeontologists Memoir 1, p. 121–144.
Runnegar, B., and Jell, P. A., 1976, Australian Middle Cambrian molluscs and their bearing on early molluscan evolution: Alcheringa, v. 1, p. 109–191.
Runnegar, B., and Pojeta, J., Jr., 1974, Molluscan phylogeny: The paleontological viewpoint: Science, v. 186, p. 311–317.
——, 1989, Origin and diversification of the Mollusca, *in* Trueman, E. R., ed., Evolution of the mollusca: New York, Academic Press (in press).
Runnegar, B., and 5 others, 1975, Biology of the Hyolitha: Lethaia, v. 8, p. 181–191.
Runnegar, B., Pojeta, J., Jr., Taylor, M. E., and Collins, D., 1979, New species of the Cambrian and Ordovician chitons *Matthevia* and *Chelodes* from Wisconsin and Queensland; Evidence for the early history of polyplacophoran mollusks: Journal of Paleontology, v. 53, p. 1374–1394.
Sharpe, D., 1853, Description of new species of Zoophyta and Mollusca, Appendix B *in* Riberio, C., Carboniferous and Silurian formations of the neighborhood of Bussaco in Portugal: Quarterly Journal of the Geological Society of London, v. 9, p. 146–158.
Shergold, J. H., Cooper, R. A., MacKinnon, D. I., and Yochelson, E. L., 1976, Late Cambrian Brachiopoda, Mollusca, and Trilobita from northern Victoria Land, Antarctica: Palaeontology, v. 19, p. 247–291.
Solovjev, I. A., Popov, L. E., and Samsonov, V. V., 1984, New data about Upper Cambrian faunas from the Ellsworth Mountains and Pensacola Mountains, western Antarctica: U.S.S.R. Academy of Science Antarctic Committee Report 23, p. 65–66 (in Russian).
Stinchcomb, B. L., 1980, New information on Late Cambrian Monoplacophora *Hypseloconus* and *Shelbyoceras:* Journal of Paleontology, v. 54, p. 45–49.
Stinchcomb, B. L., and Echols, D. J., 1966, Missouri Upper Cambrian Monoplacophora previously considered cephalopods: Journal of Paleontology, v. 40, p. 647–650.
Syssoiev, V. A., 1958, Order Hyolithida, *in* Orlov, A., ed., Osnovy Paleontologii, Mollyuski go lovonogie, P. Ammonoidei, Teseratity i anmonity: Vnatrira Kovinnye, Prilozhenie, Konikonkhii Gosgeoltekhizdat, p. 187–188 (in Russian).
——, 1972, Biostratigraphy and hyolith orthotecimorphs from the Lower Cambrian of the Siberian platform: Moscow, Publishing House Nauka, 152 p. (in Russian).
Ulrich, E. O., and Scofield, W. F., 1897, The Lower Silurian gastropoda of Minnesota, *in* Geology of Minnesota; Final Report: v. 3, no. 2, p. 183–1081.
Walcott, C. D., 1905, Cambrian faunas of China: Proceedings of the United States National Museum, v. 29, p. 1–106.
——, 1912, Cambrian geology and paleontology; 2, no. 9, New York Potsdam-Hoyt fauna: Smithsonian Miscellaneous Collections, v. 57, no. 9, p. 252–294.
——, 1913, The Cambrian faunas of China, *in* Research in China: Carnegie Institution of Washington Publication 54, p. 1–228.
Wängberg-Eriksson, K., 1964, *Isospira reticulata* n. sp. from the Upper Ordovician Boda Limestone, Sweden: Geologiska Foreningens i Stockholm Forhandlingar, v. 86, p. 229–237.
——, 1979, Macluritacean gastropods from the Ordovician and Silurian of Sweden: Sveriges Geologiska Undersokning C, v. 758, p. 33.
Webers, G. F., 1972, Unusual Upper Cambrian fauna from West Antarctica, *in* Adie, R. J., ed., Antarctic geology and geophysics: Oslo, Universitetsforlaget, p. 235–237.
——, 1977, Paleontological investigations in the Ellsworth Mountains, West Antarctica: Antarctic Journal of the United States, v. 12, no. 4, p. 120–121.
——, 1982, Upper Cambrian molluscs from the Ellsworth Mountains, *in* Craddock, C., ed., Antarctic geoscience: Madison, University of Wisconsin Press, p. 635–638.
Webers, G. F. Pojeta, J., Jr., and Yochelson, E. L., 1982, Cambrian Mollusca from the Ellsworth Mountains Antarctica: Geological Society of America Abstracts with Programs, v. 14, p. 643.
Wenz, W., 1938, Gastropoden, 1 and 2, *in* Schindewolf, O., ed., Handbuch der Palaozoologie: Berlin, Allgemeiner Teil und Prosobranchia (pars), 480 p.
——, 1940, Ursprung und fruhe Stammes-geschichte der Gastropoden: Archives

Molluskunde, v. 72, p. 1–10.
Whitfield, R. P., 1878, Preliminary descriptions of new species of fossils from the lower geological formations of Wisconsin: Wisconsin Geological Survey Annual Report for 1877, p. 50–89.
—— , 1882, Paleontology, *in* Geology of Wisconsin, 4: Wisconsin Geological Survey, p. 163–363.
Yochelson, E. L., 1958, Some Lower Ordovician monoplacophoran molluska from Missouri: Washington Academy of Sciences Journal, v. 48, p. 8–14.
—— , 1963, The Middle Ordovician of the Oslo region, Norway; 15, Monoplacophora and gastropoda: Norsk Geologisk Tidsskrift, v. 43, no. 2, p. 133–213.
—— , 1966, A reinvestigation of the Middle Devonian gastropods *Arctomphalus* and *Omphalocirrus:* Norsk Polarinstitutt Arbok, 1965, p. 37–48.
—— , 1971, A new Late Devonian gastropod and its bearing on problems of open coiling and septation, *in* Dutro, J. T., Jr., ed., Paleozoic perspectives; A paleontological tribute to G. Arthur Cooper: Smithsonian Contributions to Paleobiology, v. 3, p. 231–241.
—— , 1976, Phylum mollusca, *in* Shergold, J. H., Cooper R. A., MacKinnon, D. I., and Yochelson, E. L., Late Cambrian brachiopoda, mollusca, and trilobita from northern Victoria Land, Antarctica: Palaeontology, v. 19, p. 258–263.
—— , 1978, An alternative approach to the interpretation of the phylogeny of ancient mollusks: Malacologia, v. 17, p. 165–191.
—— , 1979, Early radiation of Mollusca and mollusc-like groups, *in* House, M. R., ed., The origin of major invertebrate groups; Systematics Association Special Volume 12: London, Academic Press, p. 323–358.
—— , 1984a, Speculative functional morphology and morphology that could not function; The example of *Hyolithes* and *Biconulites:* Malacologia, v. 25, no. 1, p. 253–264.
—— , 1984b, Historic and current considerations for revision of Paleozoic gastropod classification: Journal of Paleontology, v. 59, p. 259–269.
—— , 1989, The operculum of the early Middle Ordovician gastropod *Palliseria robusta* Wilson: Journal of Paleontology (in press).
Yochelson, E. L., and Cid, G. L., 1984, *Scenella morenaensis* new species, from the Cambrian of Spain and its bearing on the systematic position of *Scenella:* Lethaia, v. 17, p. 331–340.
Yochelson, E. L., and Jones, C. R., 1968, *Teiichispira,* a new Early Ordovician gastropod genus: U.S. Geological Survey Professional Paper 613-B, 15 p.
Yochelson, E. L., and Stanley, G. D., Jr., 1982, An early Ordovician patelliform gastropod *Palaeolophacmaea* reinterpreted as a coelenterate: Lethaia, v. 15, p. 323–330.
Yochelson, E. L., Flower, R. H., and Webers, G. F., 1973, The bearing of the new Late Cambrian monoplacophoran genus *Knightoconus* upon the origin of the Cephalopoda: Lethaia, v. 6, p. 275–309.

Manuscript Accepted by the Society March 1, 1989

Printed in U.S.A.

Geological Society of America
Memoir 170
1992

Chapter 11

Brachiopods, archaeocyathids, and Pelmatozoa from the Minaret Formation of the Ellsworth Mountains, West Antarctica

Robert A. Henderson
Department of Geology, James Cook University of North Queensland, Post Office, James Cook University, Queensland, Australia 4811
Francoise Debrenne
ER 154 CNRS, Institut de Paleontologie, 8 Rue Buffon, Paris, France 75005
A. J. Rowell
Department of Geology, University of Kansas, Lawrence, Kansas 66045
G. F. Webers
Department of Geology, Macalester College, St. Paul, Minnesota 55105

ABSTRACT

A diverse Late Cambrian fossil fauna was recovered from the Minaret Formation, Springer Peak, Heritage Range, Ellsworth Mountains, West Antarctica. Fossils include trilobites, molluscs, conodonts, brachiopods, archaeocyathids, and Pelmatozoa. The articulate and inarticulate brachiopods, archaeocyathids, and Pelmatozoa are described here; the remainder of the fauna is described elsewhere in this volume.

In articulate brachiopods make up a small percentage of the Springer Peak fauna but are abundant enough to be present in most hand specimens. Preservation is poor. Genera present include *Dactylotreta, Lingulella, Micromitra, Quadrisonia, Schizambon, Treptotreta, Zhanatella,* and *Angulotreta,* which is represented by a new species *A. ellsworthensis.* The assemblage has affinities with those of Australia, suggesting an early Late Cambrian (upper Idamean) age.

Articulate brachiopods are rare in the Springer Peak fauna, and a single species resembling *Billingsella borukaevi* is present. Archaeocyathids make up a small percentage of the fauna but are abundant on some bedding surfaces. They are represented by a single species referred to *Antarcticocyathus webersi.* Pelmatozoa are rare, but abundant columnals, probably representing eocrinoids, are present on some bedding surfaces.

INTRODUCTION, STRATIGRAPHY, AND PELMATOZOA

G. F. Webers

INTRODUCTION AND STRATIGRAPHY

A well-preserved fossil fauna was recovered from the Minaret Formation at Springer Peak, Heritage Range, Ellsworth Mountains, West Antarctica (Fig. 1). The trilobite-mollusc coquina has yielded 20 species of molluscs including monoplacophorans, gastropods, rostroconchs, and hyoliths (Webers, Pojeta, and Yochelson, this volume); 22 species of trilobites (Shergold and Webers, this volume); and the articulate and inarticulate brachiopods, archaeocyathids, and pelmatozoans described here.

The Minaret Group was first described by Craddock (1969) and considered to be of probable Precambrian age. Splettstoesser and Webers (1980) reported the age to be Late Cambrian on the

Henderson, R. A., Debrenne, F., Rowell, A. J., and Webers, G. F., 1992, Brachiopods, archaeocyathids, and Pelmatozoa from the Minaret Formation of the Ellsworth Mountains, West Antarctica, *in* Webers, G. F., Craddock, C., and Splettstoesser, J. F., Geology and Paleontology of the Ellsworth Mountains, West Antarctica: Boulder, Colorado, Geological Society of America Memoir 170.

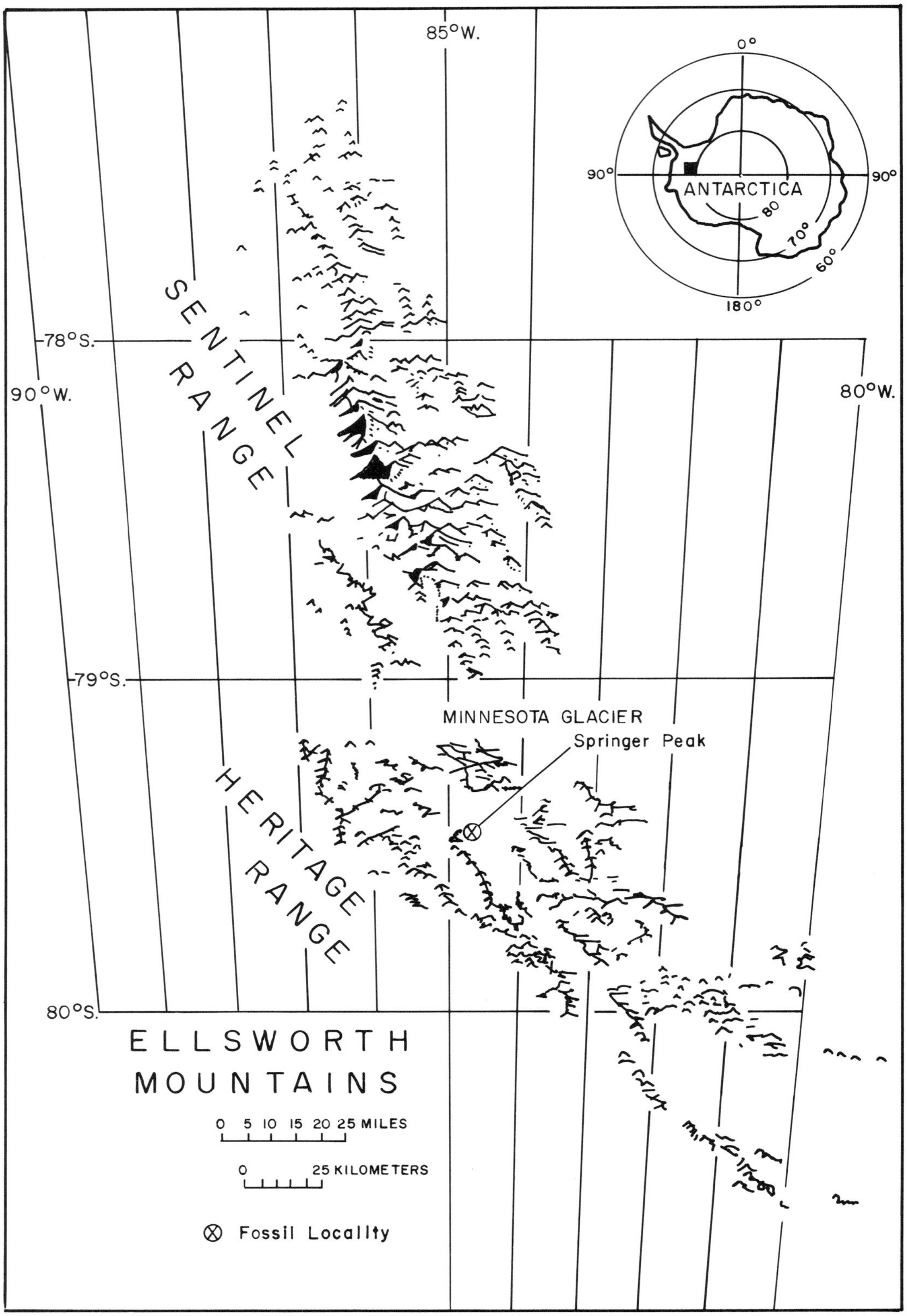

Figure 1. Map of the Ellsworth Mountains showing the location of Springer Peak.

basis of trilobite faunas, and Webers and Spörli (1983) downgraded the group to formational status.

The Minaret Formation is the highest formation in the Heritage Group and is composed almost entirely of white to gray marble (Webers and others, this volume). It includes a variety of carbonate rocks, including biosparite, oncosparite, pelsparite, oolite, biomicrite, and dolomicrite, representing a variety of environments from high-energy normal marine to low-energy restricted marine (Buggisch and Webers, this volume).

The type area of the Minaret is the Marble Hills in the southern Heritage Range where it reaches its maximum thickness of at least 600 m and could be as thick as 700 m. Sedimentary structures include oolites, oncolites, flat-pebble conglomerates, and a few ripples and cross-beds.

The Minaret thins almost uniformly to the north. At Springer Peak, some 115 km to the north in the northern Webers Peaks, the Minaret thins to a feather edge with a thickness of only 8 m.

The Minaret is typically highly sheared, recrystallized, and unfossiliferous. Vague forms suggestive of fossils are present at many localities exposing the Minaret, but their authenticity at most localities could not be established. A small collection of deformed trilobites (Shergold and Webers, this volume) and molluscs (Webers, Pojeta, and Yochelson, this volume) were recovered from sheared limestone beds in the Marble Hills. Deformed fossils including articulate brachiopods, a trilobite, and an archaeocyathid were collected near Huggler Peak in the Anderson Massif but could not be identified in detail. Their presence is consistent with correlation with the fauna at Springer Peak. A fauna generally similar to the Springer Peak fauna but more restricted and more poorly preserved was recovered from Bingham Peak just 8 km south of Springer Peak. All specimens described in this paper are from Springer Peak.

Sedimentary structures indicate that the deposition of Minaret sediments was for the most part in open, shallow-marine, agitated environments. Sedimentation of the Minaret probably began in the late Middle Cambrian in the Marble Hills area and continued into the Late Cambrian. At Springer Peak, sedimentation did not begin until the beginning of the Late Cambrian. Paleomagnetic studies and the faunal diversity of the Minaret Formation (Webers, 1972) suggest semitropical to tropical latitudes.

Specimens referred to in this chapter are curated by the U.S. National Museum of Natural History, Washington, D.C.

PELMATOZOA

Pelmatozoan debris is generally rare in the Minaret Formation, although rare bedding surfaces (Plate 4, Fig. 7) from Springer Peak show pelmatozoan debris making up as much as 50 percent of the surface material. The debris consists of scattered columnals and, rarely, sections of the pelma. Columnals are round, with a range in diameter from 1 to 4 mm and an average of 1.5 mm. Columnal thickness ranges between 0.25 and 0.30 mm. Axial canals are round and relatively large and typically account for 50 percent of the columnal diameter. Sections of the pelma as large as 17 mm have been recovered. No definite thecal plates have been found, although Plate 4, Figure 6 may represent such a plate.

Recrystallization and the lack of thecal plates preclude identification to taxonomic levels lower than Pelmatozoa. The large axial canals and the age of the strata, however, suggest that the pelmatozoans are probably eocrinoids.

REFERENCES CITED

Craddock, C., 1969, Geology of the Ellsworth Mountains, *in* Bushnell, V. C., and Craddock, C., eds., Geologic maps of Antarctica: New York, American Geographical Society Antarctic Map Folio Series, Folio 12, Plate 4.

Splettstoesser, J. F., and Webers, G. F., 1980, Geological investigations and logistics in the Ellsworth Mountains, 1979–80: Antarctic Journal of the United States, v. 15, no. 5, p. 36–39.

Webers, G. F., 1972, Unusual Upper Cambrian fauna from West Antarctica, *in* Adie, R. J., ed., Antarctic geology and geophysics: Oslo, Universitetsforlaget, p. 235–237.

Webers, G. F., and Spörli, K. B., 1983, Palaeontological and stratigraphic investigations in the Ellsworth Mountains, West Antarctica, *in* Oliver, R. L., James, P. R., and Jago, J. B., eds., Antarctic earth science: Canberra, Australian Academy of Science, p. 261–264.

INARTICULATE BRACHIOPODA

R. A. Henderson

ABSTRACT

A poorly preserved inarticulate brachiopod fauna from limestone of the Minaret Formation, Ellsworth Mountains, Antarctica, contains the genera *Dactylotreta, Lingulella, Micromitra, Quadrisonia, Schizambon, Treptotreta, Zhanatella,* and *Angulotreta,* represented by a new species *A. ellsworthensis.* The assemblage has affinities with those of Australia, and Australian range data on its component taxa place it as early Upper Cambrian, probably within the upper part of the Idamean Stage.

The Cambrian Minaret Formation from the Ellsworth Mountains, Antarctica, contains a diverse fauna of invertebrate fossils (Webers, 1970, 1972, and this volume; Solov'ev and others, 1984), including an inarticulate brachiopod assemblage. The specimens described here were obtained by L. Glenister by acid dissolution of limestone. In all, more than 800 valves are represented, but almost all are broken and most have been distorted, presumably as a result of strain imposed during folding of the enclosing strata. As a result, measurements are possible for only a few valves. The collection contains several new taxa, but the unsatisfactory mode of preservation has precluded the formal naming of any except the dominant species, which makes up over 80 percent of the fauna.

Morphological terminology is from Williams and Rowell (1965), and Australian biostratigraphic terminology is from Shergold (1975, 1980), Henderson (1976), and Öpik (1979).

SYSTEMATIC DESCRIPTIONS

Family Acrotretidae Schuchert 1893
Subfamily Acrotretinae Schuchert 1893

Genus *Angulotreta* Palmer 1955

Type Species. Angulotreta triangularis Palmer, 1955.
Diagnosis. Procline, conical Acrotretinae lacking a strong intertrough. Apical process a ridge extending along the posterior valve slope to the apical region, tapering anteriorly, with its crest open as an elongate, subtriangular apical cavity. It may continue a short distance onto the anterior valve slope as a laterally expanding, poorly defined apron. The internal pedicle tube diverges slightly from the posterior slope with the internal pedicle opening lying within the posterior sector of the apical process. Dorsal valve weakly convex with a well-developed pseudointerarea.
Remarks. Angulotreta is distinguished only by details of its apical process from the Middle and early Upper Cambrian genus *Treptotreta* Henderson and MacKinnon, 1981. In *Angulotreta* the apical process is largely restricted to the posterior slope and apex *sensu stricto* with a small, poorly defined apron anterior to the apex. In *Treptotreta* the posterior sector of the apical process is less ridge-like, with the apical cavity less elongate, and the anterior apron is more expansive and more clearly defined. In addition, the internal pedicle tube lies against the posterior valve slope in *Treptotreta,* whereas it is slightly divergent in *Angulotreta.*

The shape of the apical process is quite variable in species of both genera, and extreme variants of *Treptotreta* are indistinguishable from *Angulotreta* and vice versa. Populations of specimens, however, are readily separable. The distinction between *Angulotreta* and *Treptotreta,* although very slight, is maintained on the basis that the Australasian populations of *Treptotreta,* now known from about 100 samples from different horizons throughout the Middle and Upper Cambrian, are consistently different from Upper Cambrian North American *Angulotreta.*

Physotreta Rowell, 1966 is a contemporary, allied genus, also best discriminated on details of the apical process, which is strongly developed as a ridge or expanding apron in the apex and on the anterior valve slope but lacks a clear extension onto the posterior valve slope. *Discinia microscopica* Shumard, referred to *Angulotreta* by Bell and Ellinwood (1962) is best placed in *Physotreta.*

Prototreta Bell 1938 of Middle Cambrian age is also comparable to *Angulotreta* and its allies but is distinguished by possessing a deep, narrow intertrough on the ventral pseudointerarea across which growth lines cannot be traced as discussed by Rowell (1966, p. 3).
Range. Angulotreta is known to range throughout the Upper Cambrian from Dresbachian to Trempealeauan.

***Angulotreta ellsworthensis* sp. nov.**
(Plate 1, Figs. 1–5, 9–14)

1984 Angulotreta? sp. Popov in Solov'ev and others, p. 67, pl. 8, Figs. 8–14, text-fig. 4a.
Material. Holotype USNM 333907, paratypes USNM 333908–333913, over 600 additional specimens, most of them fragmentary.
Diagnosis. Ventral valve a low cone with valve height less than half the valve width. Dorsal valve with a strongly bladed median septum.
Description. Shell conical, up to 2.5 mm in diameter. Commissural surface oval with length 78 to 84 percent of width. Protegular ornamented with fine pits whereas later growth stages bear variably spaced growth lines.

Ventral valve. Procline, with the apex situated 15 to 35 percent of valve length from the posterior margin. Valve height is 40 to 43 percent of width. Posterior slope straight, rarely weakly convex; anterior slope is straight or slightly convex with the lateral apical angle varying between 85° and 95°. Lateral slopes weakly convex with the posterior apical angle measuring some 100°. Protegulum declined posteriorly with respect to the commissural plane of mature growth stages so that the valve apex appears truncated. Pseudointerarea poorly discriminated with a weakly depressed central zone across which growth lines are deflected dorsally. Apical process a well-defined ridge with its crest broken open to form an elongate, triangular apical cavity. It extends along the posterior valve slope, narrowing to the apex, which it occludes, and it typically continues anterior of the apex for a short distance as a low, expanding sheet that passes smoothly into confluence with the anterior valve slope. Poorly defined to obscure *vascula lateralia* border the apical process. An internal foramen is located in its posterior sector, leading to an internal pedicle tube that diverges slightly from the valve wall. Apical pits lie posterior to the internal foramen, at the margin of the apical process. A pair of poorly defined cardinal muscle scars are located on the posterior valve wall.

Dorsal valve. Gently convex, with distinct protegular nodes. Pseudointerarea well developed with a concave medium plate and broad, flattened propareas. Median septum strong, subtriangular, projecting beyond the commissural plane and extending anteriorly to a point representing about 70 percent of the valve length. Cardinal muscle scars clear with the distance between their distal extremities measuring about 50 percent of the valve width. Central muscle scars are absent.
Remarks. Like other *Angulotreta* species, *A. ellsworthensis* is quite variable. The apical process is more strongly developed and more extensive anteriorly on some specimens than others, and for some its ridged portion does not taper. The external shape of the pedicle valve as seen in profile shows some variety, and the median septum is stronger in some dorsal valves compared to others.

Typical specimens compare closely with the type species, *A. triangularis* Palmer, 1955 particularly in the form of the apical process. However, *A. triangularis,* like *A. postapicalis* Palmer, 1955 and *A. tectonensis* (Walcott) (redescribed by Grant, 1960), is distinguished by the height of its ventral valves, which exceeds half the width. *Angulotreta catheta* Grant, 1960 resembles *A. ellsworthensis* in general shape, but its dorsal valve lacks a blade-like median septum, and the same distinction applies to *A. glabra* Grant, 1960 and *A. vescula* Grant, 1960. The specimens described by Popov in Solov'ev and others (1984) from the Minaret Formation as *Augulotreta*(?) sp. almost certainly belong to the present species.

Genus *Dactylotreta* Rowell and Henderson, 1978
***Dactylotreta* sp.**
(Plate 2, Figs. 5–10)

Material. Nine fragmentary dorsal valves and three ventral valve fragments.
Description. Ventral valve represented by an acutely conical fragment showing a large apical process that must have completely occluded the apical region of the interior. The apical process surface is marked by a narrow ridge extending from the posterior valve slope and margined by prominent apical pits. Anteriorly the ridge broadens and becomes less obvious; here it is bordered by narrow, deeply impressed *vascula lateralia.* Internal foramen circular located within the apical process immediately posterior of its midlength.

Dorsal valve flat or very gently concave, with clear protegular nodes. Its posterolateral margins are almost straight and converge so that the valve posterior is bluntly pointed. Pseudointerarea large, occupying about 20 percent of the valve length. It has a large, shallowly excavated median plate and extensive, flat propareas. Median septum a very low blade rising to an apex near its posterior margin and extending for approximately two-thirds of the valve length. Cardinal muscle scars rather narrowly set with the distance between their distal extremities less than half the valve width.

Remarks. *Dactylotreta* is known from five Australian species and one from the United States, ranging through much of the Late Cambrian, but of these only the type species *D. redunca* Rowell and Henderson, 1978 has been described. The Ellsworth material is closely comparable to, and probably conspecific with, *D. redunca,* which is known in Australia from the Upper Idamean Stage (Zones of *Stigmatoa diloma* and *Irvingella tropica*). *Dactylotreta solitaria* Popov, recently described in Solov'ev and others (1984) from Late Cambrian Nelson Formation of the Neptune Range, appears to be distinct from the specimens described here. The posterolateral margins of its dorsal valve are not as straight and the dorsal valve posterior is, in consequence, distinctly less pointed.

Genus *Quadrisonia* Rowell and Henderson, 1978
***Quadrisonia* sp. nov.**
(Plate 2, Figs. 1–4)

Material. 10 ventral valves and 8 dorsal valves.

Description. Shell low, subconical, slightly wider than long and up to 2 mm in diameter.

Ventral valve procline with height 30 to 35 percent of width and the apex located 10 to 25 percent of the valve length from the posterior margin. Anterior slope straight or weakly convex, posterior slope straight or weakly concave, lateral apical angle about 90°. Lateral slopes are straight to very slightly convex, and the posterior apical angle measures about 100°. Pseudointerarea flattened with a very weak median concavity across which growth lines are continuous and show a dorsal deflection. Apical process extends as a ridge from the posterior valve slope across the apex to the anterior valve slope where it expands as an apron of variable extent, reaching to the valve midlength on some specimens but barely showing on others. A circular internal foramen lies within the apical process at its posterior extremity and is typically margined by, or confluent with, a small apical cavity. *Vascula lateralia* and cardinal muscle scars are clearly marked on some valves, obscure on others. Apical pits are small to obscure, located immediately posterior of the internal pedicle opening.

Dorsal valve convex with clear protegular nodes. Pseudointerarea small with a prominent median plate bordered by small propareas. Cardinal muscle scars small, closely set with the distance between their distal extremities slightly less than half the valve width. Median septum a low, subtriangular blade, its tip projecting just beyond the commissural plane.

Remarks. In addition to the type species, *Q. minor* Rowell and Henderson, 1978, *Quadrisonia* is represented by two undescribed species in Australia and one in the United States. Of these, the Ellsworth specimens most closely resembles an Australian species that ranges throughout the Idamean stage. Ventral valves are comparable, but dorsal valves of the Ellsworth suite are a little less convex and show a greater posterior extent of the median septum than the Australian comparative. These distinctions are small, however, and the two suites are probably conspecific.

Genus *Treptotreta* Henderson and MacKinnon, 1981
***Treptotreta* sp. nov.**
(Plate 1, Figs. 6–8)

Material. Seven ventral valves.

Description. Valves conical, up to 1.6 mm in diameter, a little wider than long with valve height about 70 percent of valve width. Both lateral and posterior apical angles are 70° to 80° except for initial growth stages, up to a valve width of some 0.4 mm, which may be more acutely conical so that the apical region has a pointed appearance. The protegulum is declined anteriorly with respect to the commissural plane of advanced growth stages so that the axis of maximum height extends to the anterior lip of the external foramen. Pseudointerarea slightly flattened with a medial, shallow concavity across which growth lines are continuous. Apical process well developed, extending as a ridge from the anterior valve slope to the apex and continuing a short distance along the posterior slope. Internal pedicle opening at the posterior of the apical process, with an oval or triangular apical cavity located adjacent to it.

Remarks. Dorsal valves are probably represented in the collection but given the homomorphy of *Angulotreta* and *Treptotreta* dorsal valves and the indifferent preservation of the Ellsworth assemblage, they would not be recognized.

These Ellsworth specimens are indistinguishable from an undescribed Australian species that ranges through much of the Idamean Stage, from the Zone of *Glyptagnostus reticulatus* to the Zone of *Stigmatoa diloma.*

The specimens of Idamean age from the Bowers Group, northern Victoria Land, Antarctica, described by Shergold and others (1976) as *Protetreta* may well belong to this species also. As noted above, *Prototreta* is distinguished from *Treptotreta* only by possessing a deep, narrow intertrough on the pedicle pseudointerarea, across which growth lines are not continuous. Whether the Victoria Land specimens possess such a structure is conjectural (see Shergold and others, 1976, Plate 38, Fig. 9).

Family Siphonotretidae Kutorga 1848
Genus *Schizambon* Walcott 1884
***Schizambon* sp.**
(Plate 2, Figs. 13–14)

Material. A single abraded ventral valve fragment.

Description. The fragment shows a bluntly pointed apex overhanging a catacline pseudointerarea. An impressed pedicle track with anteriorly concave growth lines and straight, slightly divergent margins begins at the apex and extends to the margin of the fragment. Traces of concentric growth lines of a slightly beaded or nodular character are apparent.

Remarks. Although a poorly preserved fragment, the distinctive pedicle track and character of the ornament show this specimen to represent *Schizambon.* In Australia this genus first appears in poorly fossiliferous strata underlying the *Wentsuia iota/Rhaptagnostus apsis* Assemblage Zone (see Shergold, 1980) and thought to be immediately post-Idamean in age. In Victoria Land, Antarctica, *Schizambon* ranges a little lower, being known from the Bowers Group as part of an assemblage assigned to the Idamean Stage by Shergold and others (1976).

Family Zhanatellidae Koneva 1986
Genus *Zhanatella* Koneva 1986
***Zhanatella* sp.**
(Plate 2, Figs. 15–16; Plate 3, Figs. 7–8, 10)

Material. Five ventral valves and three dorsal valves, all fragmentary.

Description. Shell weakly biconvex with width equal to or perhaps exceeding length, ornamented with fine, closely spaced growth lines.

Ventral valve with a well-rounded posterior margin, cut by a distinct notch, the emarginatura of Koneva (1986), on the median line. Pseudointerarea consists of two orthocline propareas separated by a very short, deep pedicle groove. An impressed, oval muscle field lies posterior to the valve center, with a pair of small, widely separated muscle scars immediately behind.

Dorsal valve with a rounded posterior margin and a crescentic, orthocline pseudointerarea. An elongate muscle field with a raised mar-

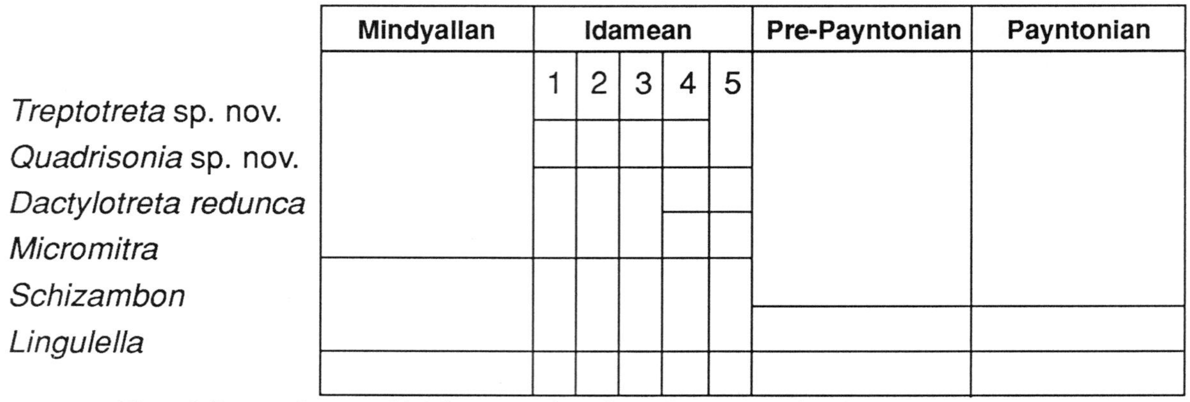

Mindyallan
Idamean
Pre-Payntonian
Payntonian
1
2
3
4
5
Treptotreta sp. nov.
Quadrisonia sp. nov.
Dactylotreta redunca
Micromitra
Schizambon
Lingulella

Figure 2. Ranges of comparable Minaret taxa from Australia, plotted against Australian state divisions. The Idamean zonation is from Henderson (1976): 1 - Zone of *Glyptagnostus reticulatus,* 2 - Zone of *Proceratopyge cryptica,* 3 - Zone of *Erixanium sentum,* 4 - Zone of *Stigmatoa diloma,* 5 - Zone of *Irvingella tropica.*

gin lies along the posterior valve midline, divided by a weak median ridge. Two small transversely elongate muscle scars are located adjacent to the lateral margins of the pseudointerarea, each marking the posterior termination of narrow, shallowly inscribed *vascula lateralia.*

Remarks. These fragmentary specimens are sufficiently distinctive in their morphology to be placed with confidence in *Zhanatella* Koneva 1986, which has recently been described from the Upper Cambrian Sakian and Aksayan Stages of Kazakhstan. Although only the type species, *Z. rotunda* Koneva has as yet been described, Koneva (1986) regarded the genus as ranging through much of the Cambrian Period in the U.S.S.R.

Subfamily Lingulellinae Schuchert 1893
Genus *Lingulella* Salter 1866
***Lingulella* sp. A**
(Plate 3, Figs. 12–17)

Material. Eight ventral and four dorsal valves.

Description. Shell elongate, gently biconvex, with the axis of maximum width lying anterior to the valve center. Valve width to valve length ratio is approximately 0.67. External ornament of concentric fila.

Ventral valve has a pointed apex with flattened posterolateral valve margins subtending an angle of about 90°. Pseudointerarea with a prominent, elongate pedicle groove that tapers posteriorly. Propareas elongate, subtriangular, flattened; divided into two subequal fields by a median longitudinal flexure line. Both fields bear growth lines parallel to their free margins, but those of the proximal field are more conspicuous.

Dorsal valve with a rounded posterior margin. Pseudointerarea a crescentic zone of thickened shell raised slightly at the extremities and with a very faint central concavity, essentially following the contours of the inner shell surface. Its growth lines are deflected anteriorly on the midline. The interior of the best-preserved specimen shows three very faint, slightly divergent ridges orientated longitudinally in the mid-valve region.

Remarks. Specimens from the Minaret Formation referred to *Lingulella* by Solov'ev and others (1984) are probably conspecific with specimens described here. *Lingulella* of this general morphology, perhaps even conspecific with the Ellsworth specimens, occur throughout the Middle and Late Cambrian in Australia, from the Floran Stage (early middle Cambrian) to the Payntonian Stage (latest Cambrian). Very similar forms also occur in the Lower Ordovician of North America, as shown by Krause and Rowell (1975).

***Lingulella* sp. B**
(Plate 2, Figs. 11–12)

Material. Three dorsal valves.

Description. Valve convex, elongate and with a rounded posterior margin. Pseudointerarea a lip of thickened shell with a distinct central concavity. A short longitudinal ridge extends from the pseudointerarea to an impressed, oval muscle field located immediately posterior of the valve center and divided medially by a faint longitudinal ridge.

Remarks. The valve convexity and internal morphology set these valves apart from those of *Lingulella* sp. A.

Family Paterinidae Schuchert 1893
Genus *Micromitra* Meek 1873
***Micromitra* sp.**
(Plate 3, Figs. 1–6, 9, 11)

Material. Six ventral valves and eighteen dorsal valves.

Description. Shell up to 3.2 mm in diameter with length some two-thirds of width, ornamented by regular concentric growth lines and irregular, weakly developed radial costellae.

Ventral valve a low, steeply procline cone with height approximately half of width. The axis of maximum height lies at the umbo or immediately anterior to it. Posterior valve surface slightly flattened and cut by a prominent delthyrium, the margins of which diverge at an angle of 70° to 80°. The proximal edge of the delthyrium bears a thin anteriorly projecting tip of shell, and the homeodeltidium is very small or not developed.

Dorsal valve convex with a long, almost straight posterior margin. Notothyrium broad, its margins subtending an angle of approximately 150° and lacking a homeochilidium. The edge of the notothyrium bears a thin, orthocline pseudointeraea that broadens toward the midline where it follows the contour of the notothyrium apex to form a concave median plate.

Remarks. Ornamentation is variable. On some specimens the costellae are well developed and the shell surface has a reticulate appearance, but on others the costellae are weak and growth lines dominate. The well-developed dorsal pseudointerarea resembles that of the Middle Cambrian *Micromitra* species figured by Henderson and MacKinnon (1981, Fig. 12K). Contemporary Australian forms are set apart from the Ellsworth species by possessing a well-developed homeodeltidium. The range of *Micromitra* in Australia is from the early Middle Cambrian Floran Stage to the early Upper Cambrian Idamean Stage.

DISCUSSION

Australian affinities with the Minaret Formation assemblage allow it to be compared to the biostratigraphic succession of inarticulate brachiopods established for the Georgina Province, northeastern Australia, by Rowell and Henderson (1978), Henderson and MacKinnon (1981), and Henderson (unpublished). The ranges of comparable Australian taxa are shown in Figure 2. The biostratigraphic data thus place the Minaret assemblage as probably of late Idamean age.

Almost all the specimens are black or almost so and in this respect resemble those from the Tasman Formation, New Zealand. Both the Minaret Formation and the Tasman Formation are folded and lie below a stratigraphy several kilometers in thickness (see Webers, 1972; Cooper, 1975). In contrast, phosphatic brachiopods from the Georgina Province, western Queensland, Australia, have experienced shallow stratigraphic burial and little tectonic disturbance (see Shergold and Druce, 1980) and are white, straw colored, or translucent. It may be that inarticulate brachiopods will eventually provide a color index of depth of burial similar to that which has been developed for conodonts (Epstein and others, 1977).

ACKNOWLEDGMENTS

Reviews of the manuscript were kindly provided by N. E. Kurtz and A. J. Rowell, who also made available recent Russian literature. The study was made possible by G. F. Webers, who furnished the collections.

REFERENCES CITED

Bell, W. C., 1938, *Prototreta,* a new genus of brachiopods from the Middle Cambrian of Montana: Papers of the Michigan Academy of Science, Arts, and Letters, v. 23, p. 403–408.

Bell, W. C., and Ellinwood, H. L., 1962, Upper Franconian and Lower Trempealeauan Cambrian trilobites and brachipods, Wilberns Formation, central Texas: Journal of Paleontology, v. 36, p. 385–423.

Cooper, R. A., 1975, New Zealand and southeast Australia in the Early Paleozoic: New Zealand Journal of Geology and Geophysics, v. 18, p. 1–20.

Epstein, A. G., Epstein, J. B., and Harris, L. D., 1977, Conodont color alteration; An index to organic metamorphism: U.S. Geological Survey Professional Paper 995, p. 1–27.

Grant, R. E., 1960, Faunas and stratigraphy of the Snowy Range Formation (Upper Cambrian) in southwestern Montana and northwestern Wyoming: Geological Society of America Memoir 96, 171 p.

Henderson, R. A., 1976, Stratigraphy of the Georgina Limestone and a revised zonation for the Upper Cambrian Idamean State: Journal of the Geological Society of Australia, v. 23, p. 423–433.

Henderson, R. A., and MacKinnon, D. I., 1981, New Cambrian inarticulate Brachiopoda from Australasia and the age of the Tasman Formation: Alcheringa, v. 5, p. 289–309.

Koneva, S. P., 1986, A new family of inarticulate Cambrian brachiopods: Paleontologicheskij Zhurnal, no. 1, p. 49–55.

Krause, F. F., and Rowell, A. J., 1975, Distribution and systematics of the inarticulate brachiopods of the Ordovician carbonate mud mound of Meiklejohn Peak, Nevada: Lawrence, University of Kansas Paleontological Contributions, Articles, no. 61, 74 p.

Öpik, A. A., 1979, Middle Cambrian agnostids; Systematics and biogeography: Australian Bureau of Mineral Resources, Geology, and Geophysics Bulletin, v. 172, 188 p.

Palmer, A. R., 1955, The faunas of the Riley Formation in central Texas: Journal of Paleontology, v. 38, p. 310–366.

Rowell, A. J., 1966, Revision of some Cambrian and Ordovician inarticulate brachiopods: Lawrence, University of Kansas Paleontological Contributions, Papers, v. 7, p. 1–36.

Rowell, A. J., and Henderson, R. A., 1978, New genera of acrotretids from the Cambrian of Australia and the United States: Lawrence, University of Kansas Paleontological Contributions, Papers, v. 83, p. 1–12.

Shergold, J. H., 1975, Late Cambrian and Early Ordovician trilobites from the Burke River structural belt, western Queensland, Australia: Australian Bureau of Mineral Resources, Geology, and Geophysics Bulletin, v. 153, 251 p.

——, 1980, Late Cambrian trilobites from the Chatsworth Limestone, western Queensland: Australian Bureau of Mineral Resources, Geology, and Geophysics Bulletin, v. 186, 109 p.

Shergold, J. H., and Druce, E. C., 1980, Upper Proterozoic and Lower Palaeozoic rocks of the Georgina Basin, *in* Henderson, R. A., and Stephenson, P. J., eds., The geology and geophysics of northeastern Australia: Brisbane, Queensland Division of the Geological Society of Australia, p. 149–174.

Shergold, J. H., Cooper, R. A., MacKinnon, D. I., and Yochelson, E. L., 1976, Late Cambrian Brachiopoda, Mollusca, and Trilobita from northern Victoria Land, Antarctica: Palaeontology, v. 19, p. 247–291.

Solov'ev, I. A., Popov, L. E., and Samsonov, V. V., 1984, New data on the Upper Cambrian fauna of the Ellsworth and Pensacola Mountains (western Antarctica): Antarktika Doklady Komissii, v. 23, p. 46–49, 66–69 (in Russian).

Webers, G. F., 1970, Paleontological investigations in the Ellsworth Mountains, West Antarctica: Antarctic Journal of the United States, v. 5, p. 162–163.

——, 1972, Unusual Upper Cambrian fauna from west Antarctica, *in* Adie, R. J., ed., Antarctic geology and geophysics: Oslo, Universitetsforlaget, p. 235–237.

Williams, A., and Rowell, A. J., 1965, Morphological terms applied to brachiopods, *in* Moore, R. C., ed., Treatise on invertebrate paleontology: Part H, Brachiopoda: Geological Society of America, p. 139–155.

Manuscript Accepted by the Society October 3, 1989

BILLINGSELLACEAN BRACHIOPODA

A. J. Rowell

Billingsella has previously been recorded from the Transantarctic Mountains of northern Victoria Land (Shergold and others, 1976). There it occurs in the upper beds of the Mariner Group (Cooper and others, 1983) of late Dresbachian (*Erixanium sentum* Zone) age. This region is presently some 2,500 km from the outcrops of the Minaret Formation of the Ellsworth Mountains, the only other area of the continent from which the genus has been noted (Webers, 1972). As is discussed below, the Minaret material resembles *Billingsella borukaevi* Nikitin more closely than any other described species, but it is debatable that the two are conspecific. *Billingsella borukaevi* is known only from the equivalent of the Dresbachian (the *Aphelaspis-Kujundaspis* fauna) of Kazakhstan (Nikitin, 1956). These relations suggest that the Minaret Formation may be of the same general age but do not preclude debate on the precise age of the Minaret fauna with the Late Cambrian (Pojeta and others, 1981).

Genus BILLINGSELLA Hall and Clarke, 1892
Billingsella cf. *borukaevi* Nikitin, 1956
(Plate 4, Figs. 1–5)

cf. 1956. *Billingsella borukaevi* sp. nov. Nikitin, 1956, pp. 30–32, Plate II, Figs. 1–23

Material. Some 56 specimens, mostly broken ventral and dorsal valves showing external ornament. The majority of complete valves are figured. The remaining specimens were recovered as byproducts of preparation for mollusks, and their valves suffered from being calcined.

Description. The shells are ventribiconvex, the height of the ventral valve being about twice that of the dorsal one. In commissural outline the valves are rounded, transversely subquadrate and slightly auriculate. The ventral valve length is about 80 percent of its width, and the ventral beak is posterior of the hinge line. The maximum width is either at or near the midlength of the valve or it occurs at the hinge line. Its position depends largely on the degree of development of the auricles. If they are well developed, then maximum width is at the hinge line. The ventral valve is orthocline or slightly apsacline; the interarea is gently curved in lateral profile with the beak slightly overhanging. The delthyrium, whose sides subtend an angle of about 50°, is closed for approximately two-thirds of its length by an externally convex pseudodeltidium. Details of the pedicle foramen are not well preserved in any specimen. In both posterior and lateral profile, the ventral valve is gently convex. The maximum height is about one-third of the valve length and occurs in the posterior third of the valve. The internal features of the ventral valve are poorly known; the principal trunks of the vascula media diverge anteriorly at about 40° to each other, but the remainder of the mantle canal pattern is not preserved.

The dorsal valve has a very low anacline or vertical interarea, and its beak barely overhangs the hinge line. The notothyrium is seemingly open. A low, narrowly diverging sulcus is present posteriorly but dies out toward the front of the valve. Internally, the socket ridges are subparallel with the hinge line and extend laterally for about one-third of its width. The notothyrial platform is well developed and extends forward as a broad conspicuous ridge to near the midlength of the valve. The platform bears a simple, ridgelike cardinal process posteriorly. The vascula genitalia diverge widely from in front of the sockets, subtending an angle of about 30° with the hinge line. Other features of the musculature and mantle canal system are not clearly discernible.

Both valves are ornamented by strong costae that increase rather irregularly by intercalation. Typically seven or nine costae extend to the ventral beak. Costae increase in height anteriorly and are strongly rounded in cross section. Prominent growth lines become almost lamellose near the anterior and lateral margins of the valves. Dorsal valve has a similar ornament of costae and strongly developed peripheral growth lines.

Dimensions of measured specimens

Specimen	Length mm	Width mm
406065	12.5	16.0
406066	6.0	8.1
406067	——	12.4
406068	7.7	11.5

Discussion. At least 46 described species are potentially referable to *Billingsella.* Unfortunately, the majority of them are not well known. The few that have been studied in some detail (Bell, 1941; Bell and Ellinwood, 1962; Cooper, 1952) exhibit considerable variation, and according to Bell and Ellinwood (1962), there is intergradation between some taxa.

The majority of the species of the genus have a finely costellate ornament. The Dresbachian material from the Mariner Group is of this type (MacKinnon in Shergold and others, 1976, pl. 38, figs. 11–19). The Minaret specimens, however, have a much coarser ornament, and in this feature and in other characteristics resemble *Billingsella borukaevi* Nikitin from Kazakhstan. The significance of the differences between the Kazakhstan and Minaret material is difficult to assess in the absence of large collections. The interspaces between the radial ornament, for example, typically are narrower in *B. borukaevi* than in the Antarctic material. The notothyrial platform of the Antarctic specimens is extended forward as a marked ridge. This development also occurs in other species of *Billingsella,* for example, *B. texana* (Bell and Ellinwood, 1962, pl. 62, fig. 21), but seemingly is much less conspicuously developed in *B. borukaevi.*

With available information it would appear preferable to draw attention to the similarities between the specimens from the Minaret Formation of the Ellsworth Mountains and those from Kazakhstan, rather than refer them to the same taxon. With our limited knowledge, it is perhaps premature to attach too much biogeographic emphasis to this relation. One notes, however, that what would appear to be the *Billingsella* sp. of northern Victoria Land has recently been reported from the Idamean of New South Wales (Percival in Powell and others, 1982).

ACKNOWLEDGMENTS

I am indebted to John Pojeta, Jr., for making the material from the Ellsworth Mountains available and to Gerry Webers for encouraging me to examine it.

REFERENCES CITED

Bell, W. C., 1941, Cambrian Brachiopoda from Montana: Journal of Paleontology, v. 15, p. 193–255.

Bell, W. C., and Ellinwood, H. L., 1962, Upper Franconian and Lower Trempealeauan Cambrian trilobites and brachiopods, Wilberns Formation, central Texas: Journal of Paleontology, v. 36, p. 385–423.

Cooper, R. A., 1952, New and unusual species of brachiopods from the Arbuckle Group in Oklahoma: Smithsonian Miscellaneous Collections, v. 117, no. 14, p. 1–35.

Cooper, R. A., Jago, J. B., Rowell, A. J., and Braddock, P., 1983, Age and correlation of the Cambrian-Ordovician Bowers Supergroup, Northern Victoria Land, *in* Oliver, R. L., James, P. R., and Jago, J. B., eds., Antarctic Earth Science: Canberra, Australian Academy of Science, p. 128–131.

Nikitin, I. F., 1956, Brakhiopody Kembriya i nizhnego Ordovika severo-vostoka tsentral'nogo Kazakhstana: Alma Ata, Isdatel'stvo Akademii Nauk Kazakhskoj SSR, 143 p. (in Russian).

Pojeta, J., Jr., Webers, G. F., and Yochelson, E. L., 1981, Upper Cambrian mollusks from the Ellsworth Mountains, West Antarctica, *in* Taylor, M. E., ed., Short Papers for the 2nd International Symposium on the Cambrian System 1981: U.S. Geological Survey Open-File Report 81-743, p. 167–168.

Powell, C. M., Neff, G., Crane, D., Jell, P. A., and Percival, I. G., 1982, Significance of Late Cambrian (Idamean) fossils in the Cupala Creek Formation, northwestern New South Wales: Proceedings of the Linnean Society of New South Wales, v. 106, p. 127–150.

Shergold, J. H., Cooper, R. A., MacKinnon, D. I., and Yochelson, E. L., 1976, Late Cambrian Brachiopoda, Mollusca, and Trilobita from northern Victoria Land, Antarctica: Palaeontology, v. 19, p. 247–291.

Webers, G. F., 1972, Unusual Upper Cambrian fauna from West Antarctica, *in* Adie, R. J., ed., Antarctic geology and geophysics: Oslo, Universitetsforlaget, p. 235–237.

Manuscript Accepted by the Society October 3, 1989

ARCHAEOCYATHA

Francoise Debrenne

ABSTRACT

A number of specimens of cylindrical fossils that have two "walls" connected by an irregular wavy skeletal structure were reported as "archaeocyathids" twenty-five years ago. Only recently, after examination of important material collected within the last few years, the conclusion of the Archaeocyathan nature was confirmed. A new genus *Antarcticocyathus* was proposed for these cylindrical solitary or colonial fossils. Probably because of their high plasticity and primitive development, these younger representatives of the phylum, until now reported only in the Lower Cambrian, have been able to survive in restricted niches up to the Upper Cambrian.

INTRODUCTION AND DISCUSSION

The presence of Archaeocyatha in the Minaret Formation of the Heritage Range, Ellsworth Mountains, Antarctica, is an unexpected and interesting fact that modifies the conception of a group until now considered as strictly restricted to the Lower Cambrian (Rosanov and Debrenne, 1974). These cylindrical bodies were reported by Craddock and Webers (1964) as "archaeocyathid-like organisms" but have only been positively identified by Debrenne and Rosanov during the Second Cambrian Symposium where photographs of the Antarctic fossils were displayed.

Investigations followed at the Institute of Paleontology, Paris, by Debrenne and Rosanov and were then carried on by F. and M. Debrenne, who prepared the outer and inner walls and the external shape of some specimens. Results of these investigations led these scientists to establish a new genus and a new species tentatively referred to the family Archaeocyathidae. Two publications have already been prepared (Debrenne, Rosanov, and Webers, 1984; Debrenne, F., Debrenne, M., and Webers, 1983).

About 60 cups were collected for study. Despite some variation in the external shape they appear to belong to one genus and species. No real data will be presented here. We will recall the main characteristics, according to the description given in Debrenne, Rozanov, and Webers (1984), and will emphasize the variations observed in shape from smooth cylindrical to colonial forms:

The fossils are cylindrical with transverse irregular annulations with an unknown broken base. In thin sections and polished sections, small isometric crystals limit the outer and inner zone of each cup, suggesting the presence of walls. Preparation of samples confirms this opinion. The cavities which are recognized on the outer surface, and with more difficulty in the inner surface, are interpreted as pores. These pores do not belong to a wall independent of the constructions in the zone between the two "walls." Structures between the two "walls" are made of discontinuous skeletal elements that are ribbon-like or twisted, and that delimit alveolar tubules or less commonly, radial partitions. The orientation of the tubes might change from the inner to the outer surface.

Investigation by ultra-thin section and SEM shows that the skeleton has a fine-grained structure (diameter 2.5 microns). The inside of the tube is filled with middle size isometric crystals different from those of the central cavity of the surrounding rock.

The conclusions of the above observations are that we are confronted with an irregular structure between "walls' and an empty central cavity (Debrenne, Rozanov, and Webers, 1984).

The bulging shape strongly suggests some Sphinctozoa, but despite some preferential orientation for the synapticulae, simulating tabulae, it is not possible to find true successive internal chambers here (plate 5, figs. 1–2). Most frequently the cups are cylindrical with smooth undulations (plate 5, figs. 3–4) and no tabulaelike structures. Twin cups are sometimes observed. The preparation of such samples shows that they are colonial forms. The evolution of the colony was displayed at a poster session of the 4th International Symposium on Antarctic Earth Sciences (Adelaide, 1982). The habit is composite colonial evolving from a single cup that becomes elongated and gives rise to two rapidly independent branches. Later the branches make contact and coalesce into a catenulate colony (plate 5, figs. 5–8). The nondifferentiation of walls and the exceptional plasticity of the Upper Cambrian forms allow the construction of this composite habit, a type not known in their Lower Cambrian ancestors.

Nevertheless, the Upper Cambrian Antarctic forms might be compared with representatives of Archaeocyathidae s.l., but they have peculiar features, such as the lack of individualized sheaths, the possibility of a hemispherical base, and the absence of vesicular tissue and exostructures. These last two features may be due to the insufficient number of specimens and the restricted biota where they are found. Compared to the high diversity reached by porosity of walls in Lower Cambrian Archaeocyatha, the Upper Cambrian ones look regressive, with direct opening of the intervallar mesh structures to restrict the aperture or direct the currents.

They apparently have the same grade of organization as sponges and are less sophisticated than the last known Lower Cambrian archaeocyathan. It is probably because of their generalized or rustic development that they survived beyond the Middle Cambrian forms to the Upper Cambrian (Debrenne, Rosanov, and Webers, 1984).

SYSTEMATIC DESCRIPTION

Phylum ARCHAEOCYATHA
?Family ARCHAEOCYATHIDAE

Genus *Antarcticocyathus* Debrenne and Rosanov (in Debrenne, Rosanov, and Webers, 1984).

Type species A. webersi Debrenne and Rosanov (in Debrenne, Rosanov, and Webers, 1984)
Holotype. USNM 333901
Diagnosis: Cylindrical to conical cups sometimes displaying horizontal corruga-

tions. Outer and inner wall with coarse pores in irregular horizontal (or less frequently vertical) undulating rows, underlined by prominent skeletal ridges continuous or limited to several pores.

Pores are the openings of the intervallar mesh made of ribbonlike taeniae delimiting irregular incomplete alveolar tubules, or more radial loculi.

Discussion. Among Lower Cambrian forms, *Antarcticocyathus* is close to the Archaeocyathinae, but because of the lack of real walls other than direct opening of the intervallum, it might not be included in any described genus. The variation of the intervallar mesh—alveoles or radial partitions linked with synapticulae—is comparable with the variation within a genus of Irregulares. The absence of vesicular tissue, not a necessary feature but a common one among Irregulares, is the main difference between *Antarcticocyathus* and Archaeocyathinae. Is it due to evolution, environmental conditions, or destruction of the structures because the fossils were highly affected by low-grade metamorphism? It is not possible to decide whether or not the absence of vesicular tissue was a genuine character or due to the process of fossilization, although I would prefer the first hypothesis that also brings *Antarcticocyathus* close to sponges.

REFERENCES CITED

Craddock, C., and Webers, G. F., 1964, Fossils from the Ellsworth Mountains, Antarctica: Nature, v. 201, no. 4915, p. 174–176.

Debrenne, F., Debrenne, M., and Webers, G. F., 1983, Upper Cambrian Archaeocyathans; New Morphotype, *in* Oliver, R. L., James, P. R., and Jago, J. B., eds., Antarctic Earth Science: Canberra, Australian Academy of Science, p. 280.

Debrenne, F., Rozanov, A. Yu., and Webers, G. F., 1984, Upper Cambrian Archaeocyatha from Antarctica: Geological Magazine, v. 121, no. 4, p. 291–299.

Rosanov, A. Yu., and Debrenne, F., 1974, Age of Archaeocyathid assemblages: American Journal of Science, v. 274, p. 833–848.

MANUSCRIPT ACCEPTED BY THE SOCIETY OCTOBER 3, 1989

PLATE 1

Figures 1–2. *Angulotreta ellsworthensis* sp. nov. External and internal view of dorsal valve, paratype USNM 333908, ×30.

Figure 3. *Angulotreta ellsworthensis* sp. nov. Internal view of ventral valve, paratype USNM 333910, ×30.

Figure 4–5. *Angulotreta ellsworthensis* sp. nov. External and posterior profile views of ventral valve, paratype USNM 333911, ×40.

Figure 6. *Treptotreta* sp. Internal view of ventral valve, USNM 333915, ×30.

Figures 7–8. *Treptotreta* sp. Lateral profile and internal view of pedicle valve, USNM 333914, ×30.

Figures 9–10. *Angulotreta ellsworthensis* sp. nov. Internal and external views of dorsal valve, paratype USNM 333909, ×30.

Figures 11–12. *Angulotreta ellsworthensis* sp. nov. Internal and lateral profile views of ventral valve, holotype USNM 333907, ×30.

Figure 13. *Angulotreta ellsworthensis* sp. nov. Lateral profile view of ventral valve, USNM 333913, ×30.

Figure 14. *Angulotreta ellsworthensis* sp. nov. Interior view of dorsal valve, USNM 333912, ×30.

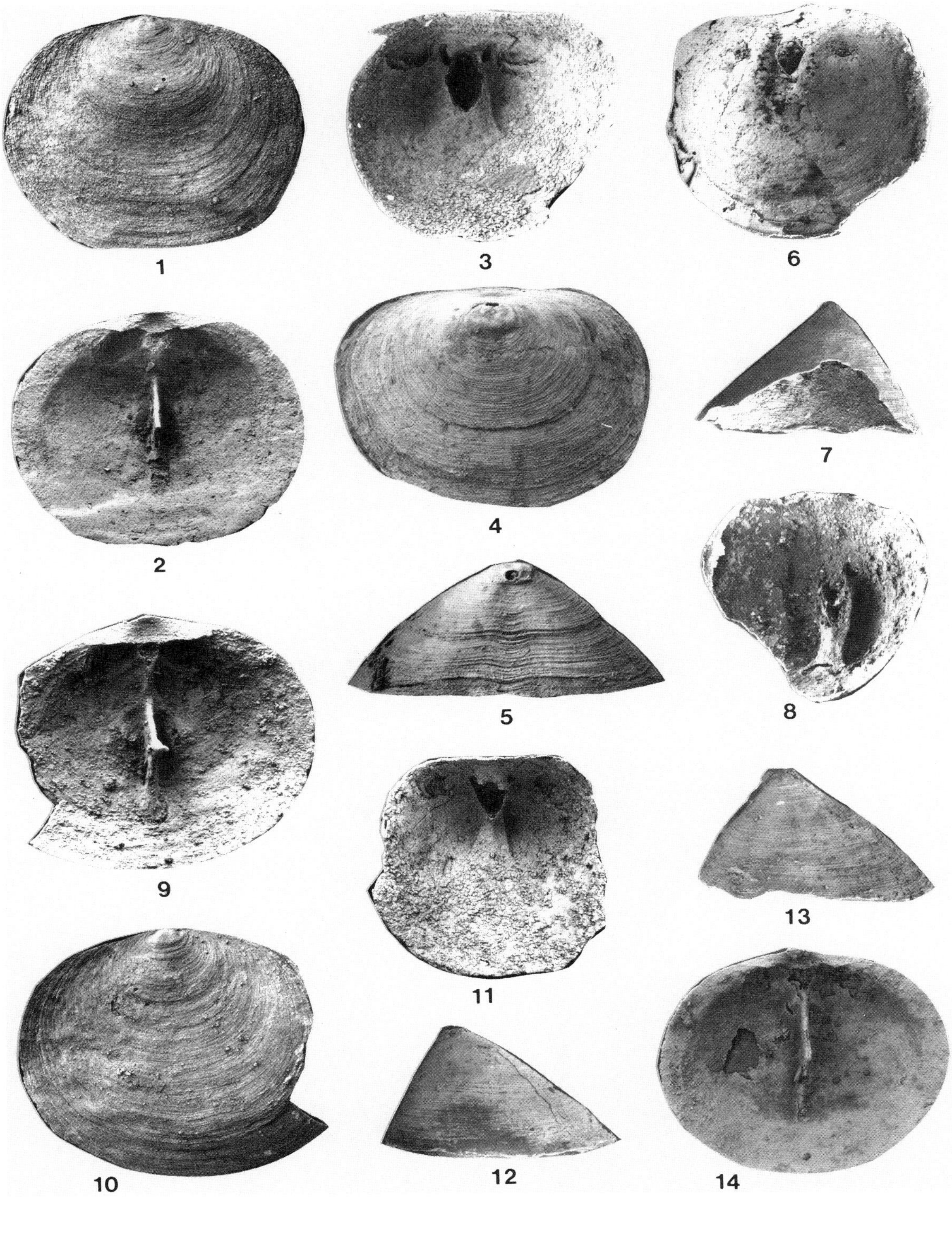

PLATE 2

Figure 1. *Quadrisonia* sp. Internal view of ventral valve, USNM 333916, ×40.

Figures 2–3. *Quadrisonia* sp. Lateral profile and internal view of ventral valve, USNM 333917, ×40.

Figure 4. *Quadrisonia* sp. Internal view of ventral valve, USNM 333918, ×40.

Figures 5–6. *Dactylotreta* sp. Lateral profile and internal view of ventral valve, USNM 333920, ×40.

Figure 7. *Dactylotreta* sp. Internal view of dorsal valve, USNM 333919, ×30.

Figure 8. *Dactylotreta* sp. Internal view of dorsal valve, USNM 333921, ×30.

Figures 9–10. *Dactylotreta* sp. External and internal view of dorsal valve, USNM 333922, ×30.

Figure 11–12. *Lingulella* sp. B. External and internal views of dorsal valve, USNM 333923, ×50.

Figures 13–14. *Schizambon* sp. External and internal views of dorsal valve, USNM 333924, ×50.

Figures 15–16. *Zhanatella* sp. External and internal views of ventral valve, USNM 333925, ×50.

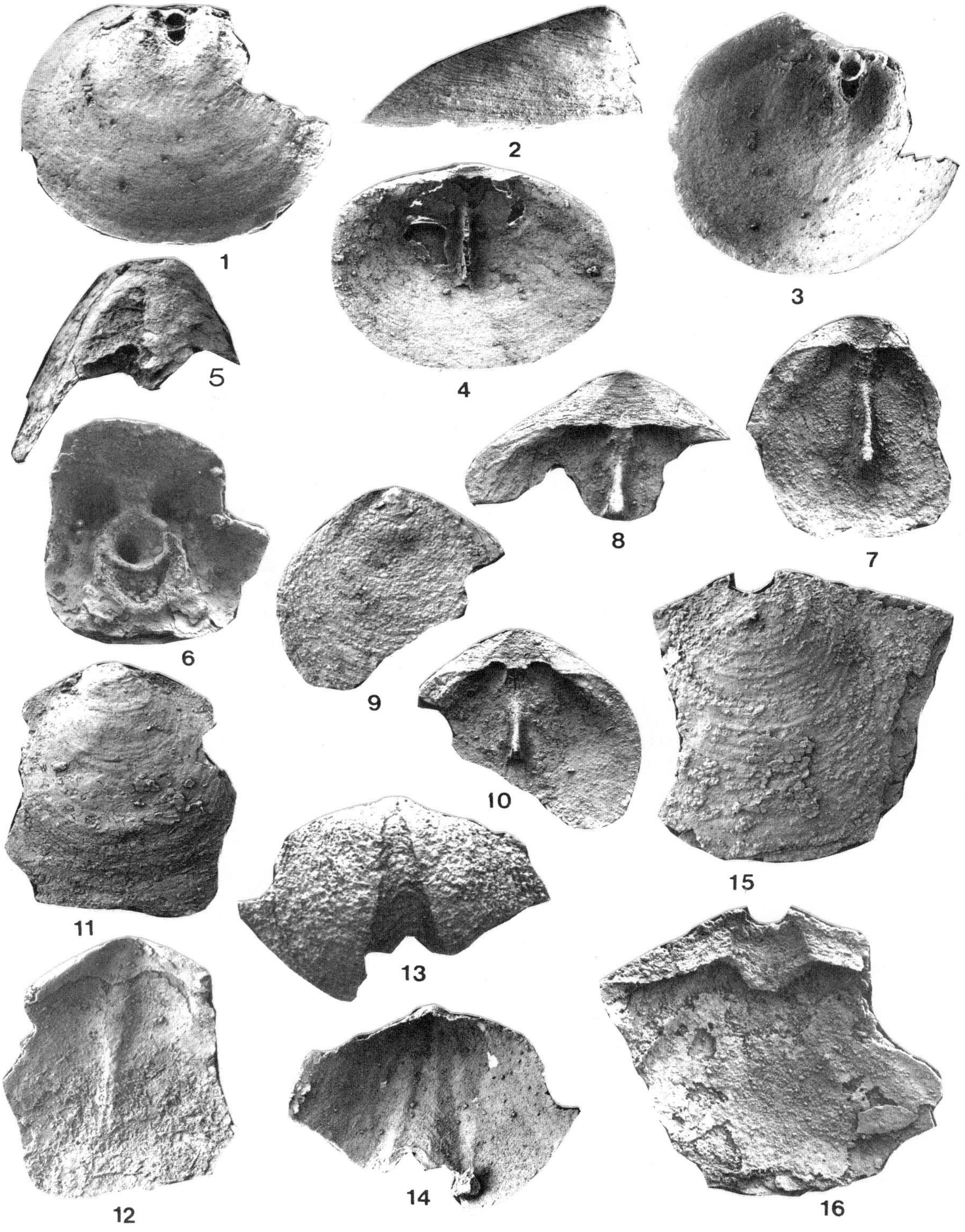
1
2
3
4
5
6
7
8
9
10
11
12
13
14
15
16

PLATE 3

Figure 1. *Micromitra* sp. External view of dorsal valve, USNM 334182, ×20.

Figures 2–3. *Micromitra* sp. External and internal views of ventral valve, USNM 333931, ×30.

Figure 4. *Micromitra* sp. External view of dorsal valve, USNM 334181, ×40.

Figures 5–6. *Micromitra* sp. External and internal views of dorsal valve, USNM 333934, ×40.

Figures 7–8. *Zhanatella* sp. Internal and external view of dorsal valve, USNM 333927, ×30.

Figure 9. *Micromitra* sp. External view of dorsal valve, USNM 333932, ×30.

Figure 10. *Zhanatella* sp. Internal view of ventral valve, USNM 333926, ×40.

Figure 11. *Micromitra* sp. Internal view of dorsal valve, USNM 333933, ×30.

Figure 12. *Lingulella* sp. A. Internal view of ventral valve, USNM 333931, ×20.

Figure 13. *Lingulella* sp. A. Internal view of dorsal valve, USNM 333928, ×20.

Figures 14–15. *Lingulella* sp. A. External and internal views of ventral valve, USNM 333930, ×20.

Figures 16–17. *Lingulella* sp. A. Internal and external views of dorsal valve, USNM 333929, ×30.

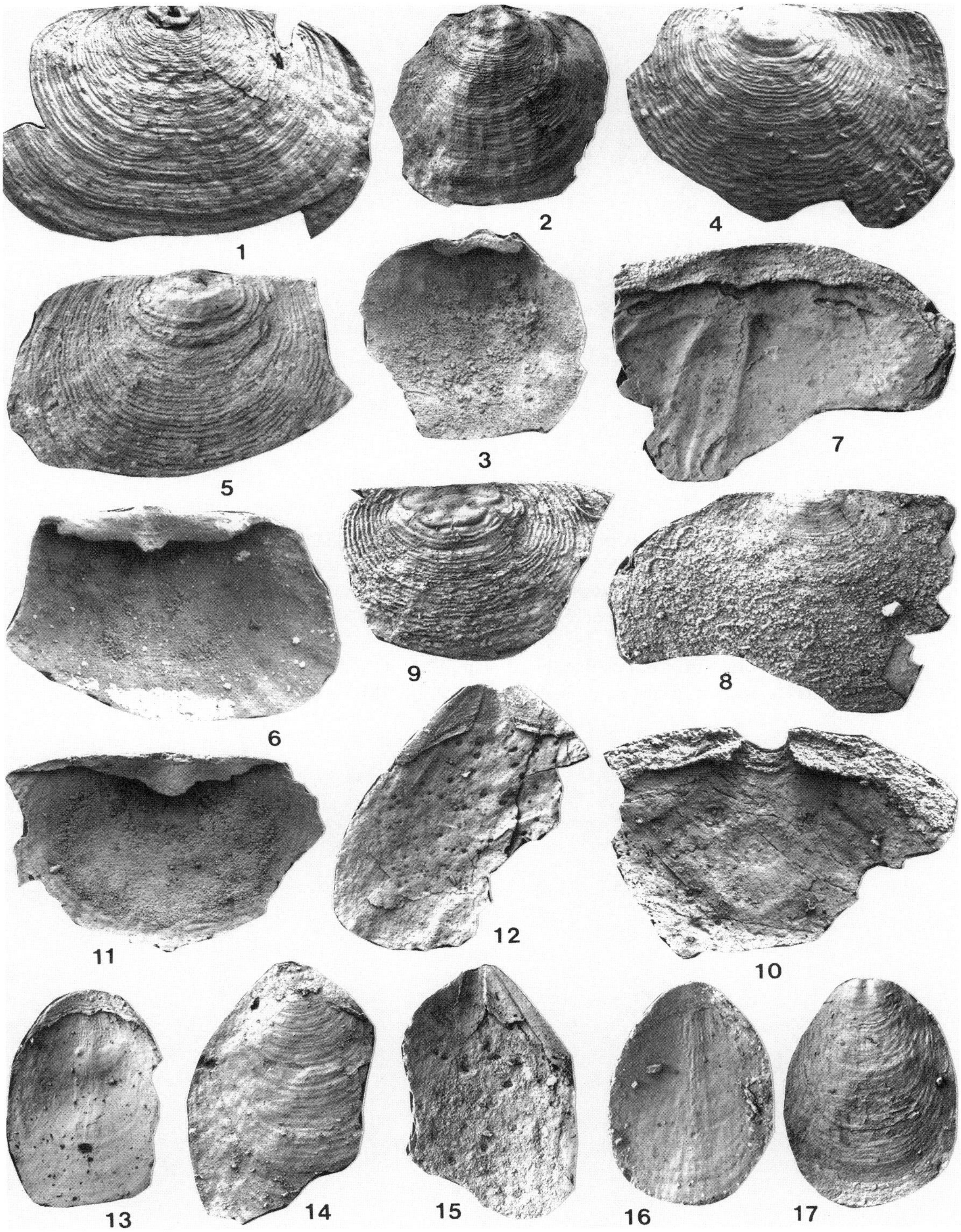
1
2
4
3
5
7
6
9
8
11
12
10
13
14
15
16
17

PLATE 4

Figures 1–5. *Billingsella* cf. *borukaevi* Nikitin. 1. Dorsal valve interior with well-developed cardinalia, USNM 406067, ×3.5. 2. Dorsal valve exterior with small posterior sulcus, USNM 406066, ×5. 3. Exfoliated dorsal valve exterior with well-developed auricles, USNM 406068, ×3.5. 4 and 5. Posterior and exterior views of ventral valve, USNM 406065, ×3.5.

Figures 6–7. Pelmatozoa. 6. Possible thecal plate, USNM 334083, ×15. 7. Scattered columnals and sections of pelma on bedding surface, USNM 334082, ×2.

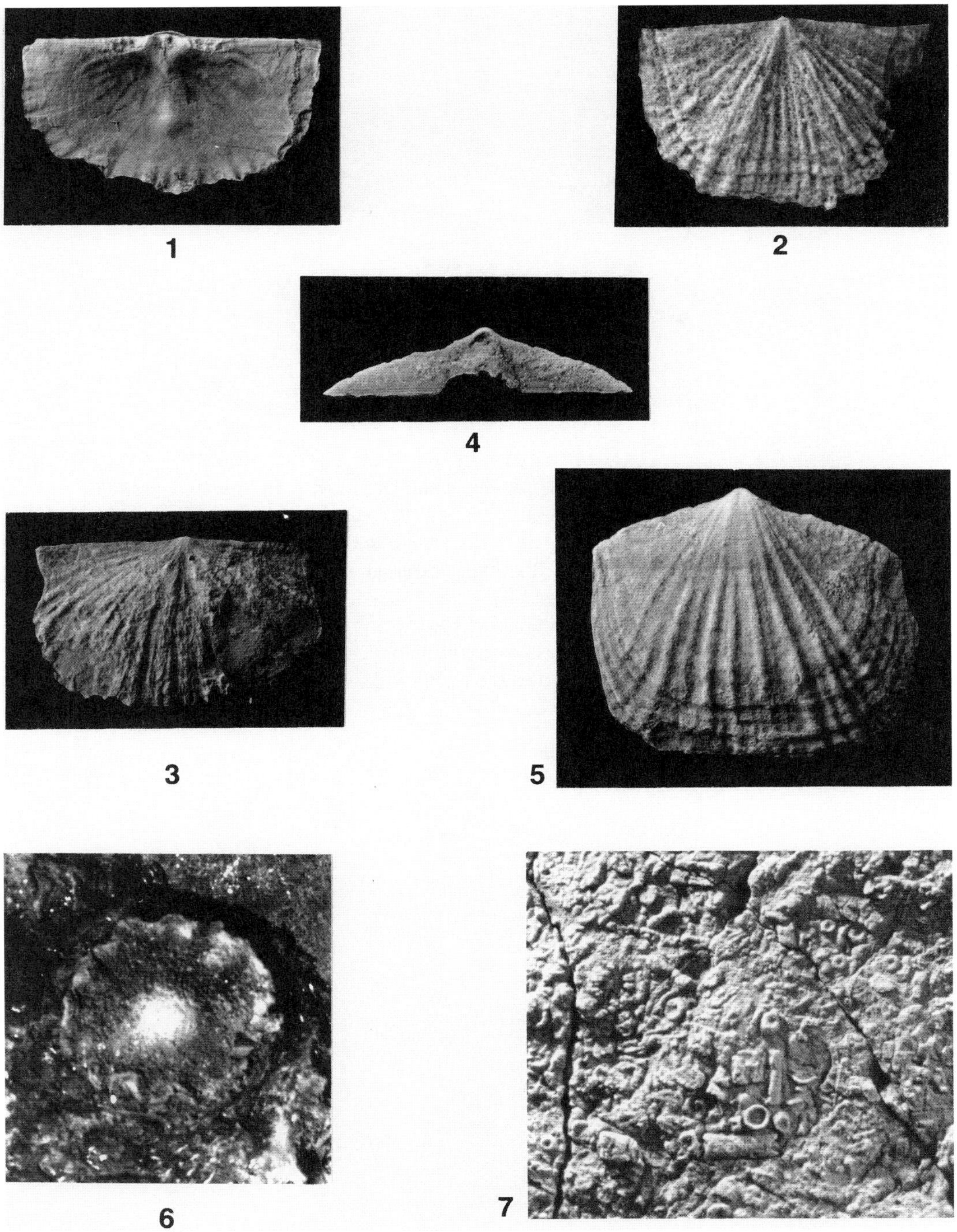
1
2
4
3
5
6
7

PLATE 5

Figures 1–8. *Antarcticocyathus webersi* Debrenne and Rosanov. 1. Holotype, polished longitudinal axial section, USNM 333901, ×2.5. 2. Paratype, polished longitudinal section through the intervallum showing the arrangement in the horizontal plane of the synapticulae, USNM 334084, ×4. 3. Paratype, external view of the outer wall, USNM 334183, ×3. 4. Paratype, polished longitudinal axial section without external corrugations, USNM 334184, ×4. 5–8. Paratype, USNM 334185. 5. External view of colonial form, ×1.5. 6. Upper transverse section of the colony, ×1.5. 7. Middle transverse section of the two separate branches of the colony, ×2. 8. Lower transverse section before the formation of the branches of the colony, ×2.

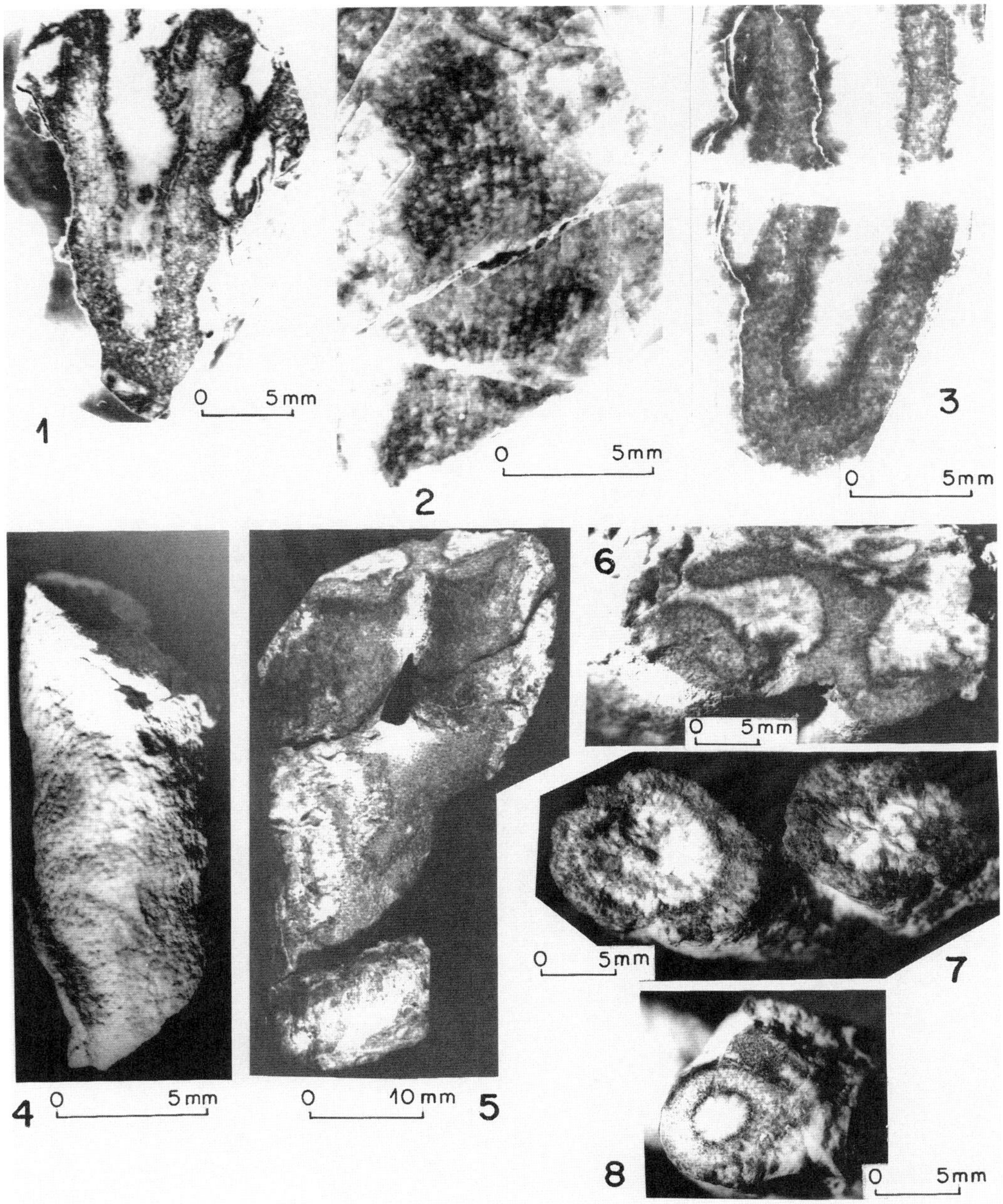
1
0 5mm
2
0 5mm
3
0 5mm
4
0 5mm
0 10mm
5
6
0 5mm
0 5mm
7
8
0 5mm

Printed in U.S.A.

Geological Society of America
Memoir 170
1992

Chapter 12

Devonian fossils from the Ellsworth Mountains, West Antarctica

Gerald F. Webers
Department of Geology, Macalester College, St. Paul Minnesota 55105
Brian Glenister
Rural Route 2, Box 148-B, North Liberty, Iowa 52317
John Pojeta, Jr.
U.S. Geological Survey, 970 National Center, Reston, Virginia 22092
Gavin Young
Bureau of Mineral Resources, Geology and Geophysics, P.O. Box 378, Canberra City, A.C.T. 2601, Australia

ABSTRACT

An Early Devonian orbiculoid brachiopod fauna was reported by Boucot and others (1967) from the Crashsite Quartzite of the northern Heritage Range of the Ellsworth Mountains. Re-collection on this site from strata now known as Mt. Wyatt Earp Formation of the Crashsite Group (Spörli, this volume) has yielded a diverse but sparse and poorly preserved fauna dominated by orbiculoid brachiopods (*Orbiculoidea* cf. *falklandensis* Rowell). Also present in the fauna are cephalopods (identifiable only to the order Orthocerida), pelecypods (*Nuculites* aff. *N. cuneiformis* Conrad; and *Grammysiodea?* sp. indt.), a rostroconch (*Hippocardia?* sp. indt.), gastropods (*Holopea?* sp. indt.), a fish spine (*Machaeracanthus* cf. *kayseri* Kegel), and single unidentifiable specimens of a conularid, a trilobite, and an articulate brachiopod.

The fauna correlates with those of the Lower Devonian Horlick Formation, Ohio Range, Horlick Mountains, Antarctica, and with those of the Lower Devonian of the Falkland Islands and represents the Malvinokaffric Faunal Province.

LITHOSTRATIGRAPHY, FAUNAL CHARACTERISTICS, AND DESCRIPTION OF BRACHIOPODS, GASTROPODS, A TRILOBITE, AND A CONULARID

Gerald F. Webers

An Early Devonian fauna was reported by Boucot and others (1967) from the Upper Dark Member of the Crashsite Quartzite in the northern Heritage Range, Ellsworth Mountains. Strata of the Upper Dark Member are now known as the Mt. Wyatt Earp Formation of the Crashsite Group (Spörli, this volume). Re-collection of this site (Fig. 1) during the 1979–1980 expedition to the Ellsworth Mountains has yielded a diverse but sparse and poorly preserved Early Devonian fauna.

The Crashsite Group (Sporli, this volume) is a 3,000-m-thick unit composed predominantly of orthoquartzite. The group was deposited from Late Cambrian to possibly early Permo-Carboniferous time and marks an interval of gentle subsidence under predominantly shallow-marine conditions. The Crashsite Group has the greatest surface exposure of any rock unit in the Ellsworth Mountains. In ascending order, the group is subdivided into the Howard Nunataks, Mount Liptak, and Mount Wyatt Earp Formations (Spörli, this volume). A Late Cambrian trilobite fauna is present in the basal transition beds of the Howard Nunataks Formation (Shergold and Webers, this volume), and a Devonian fauna is present in the Mount Wyatt Earp Formation.

The youngest formation in the Crashsite Group is the Mt. Wyatt Earp Formation, which attains a maximum thickness of 300 m in the Sentinel Range. Lithologically it consists domi-

Webers, G. F., Glenister, B., Pojeta, J., Jr., and Young, G., 1992, Devonian fossils from the Ellsworth Mountains, West Antarctica, *in* Webers, G. F., Craddock, C., and Splettstoesser, J. F., Geology and Paleontology of the Ellsworth Mountains, West Antarctica: Boulder, Colorado, Geological Society of America Memoir 170.

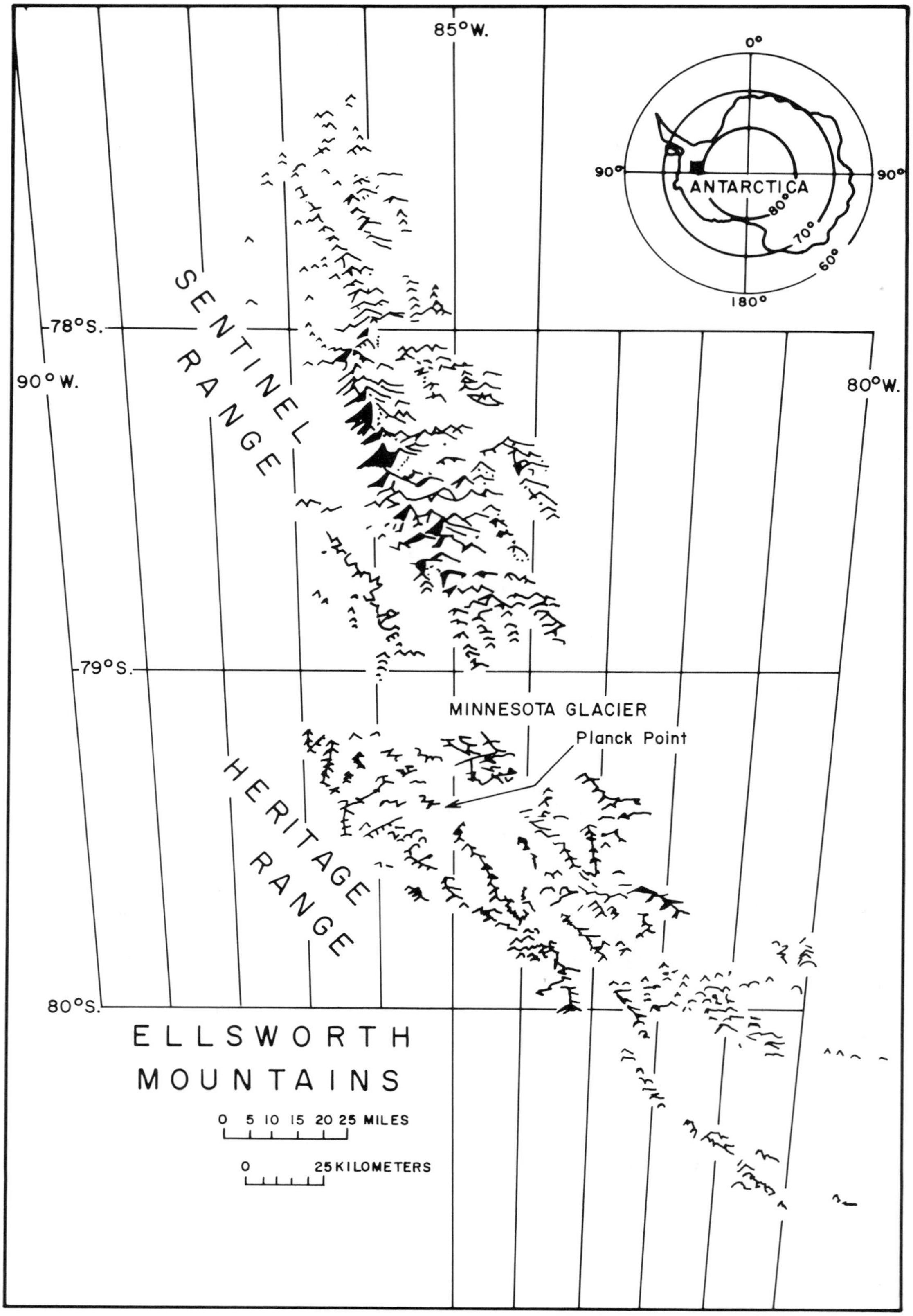

Figure 1. Location map of the Ellsworth Mountains and the Devonian locality at Planck Point.

nantly of dark gray, red, and dark brown impure quartzites with interbeds of brown argillite. The quartzites are poorly sorted and high in matrix and rock fragments. The formation is quite variable laterally. In the southeastern Heritage Range the top of the formation contains interbedded diamictite identical to that of the overlying Whiteout Conglomerate. In the northern Sentinel Range the isolated pebbles present in the uppermost beds of the formation presumably represent ice-rafted dropstones similar to those of the overlying Whiteout Conglomerate. The age of the lowermost beds of the Mt. Wyatt Earp Formation is not known with certainty but is probably Early Devonian. The age of the uppermost beds of the Mt. Wyatt Earp Formation containing interbeds of diamictites could be Devonian but is more probably early Permo-Carboniferous.

The Devonian fauna was collected from a small, unnamed nunatak (79°21′S, 85°17′W) about 3 km northwest of the highest point of Planck Point in the central Heritage Range of the Ellsworth Mountains (Fig. 1); it is believed to be from near the base of the formation. This site is hereafter referred to as the Planck Point locality. A 128-m-thick measured section exposes buff micaceous, silty, fine-grained sandstones, brown micaceous sandy siltstones, and blocky, dark gray argillites. The fossils were found in the siltstones and argillites.

The Mount Wyatt Earp Formation was deposited under shallow-marine conditions. Marked lateral variations in lithology are apparently due to variations in source directions, although no work on this subject has yet been done. The pronounced increase in argillite and the marked increase in matrix and rock fragment percentages in the quartzites compared to the underlying Mount Liptak Formation could indicate an uplift of the source area and/or the onset of glacial conditions associated with the Permo-Carboniferous glaciations. The presumed dropstones and the interbedded diamictite in the upper part of the Mount Wyatt Earp Formation support the latter conclusion.

FAUNAL CHARACTERISTICS

The fauna is diverse but sparse and poorly preserved. It is dominated by orbiculoid brachiopods. Boucot and others (1967) reported a fauna consisting of more than thirty specimens of *Orbiculoidea cf. falklandensis* and a single specimen of an articulate brachiopod possibly referable to *Chonetes.* Re-collection of the site has yielded 30 additional specimens of *Orbiculoidea* cf. *falklandensis,* two specimens of cephalopods (identifiable only to the order Orthocerida), three specimens of pelecypods (*Nuculites* aff. *N. cuneiformis* Conrad; and *Grammysiodea*? sp. indt.), one specimen of a rostroconch (*Hippocardia*? sp. indt.), six specimens of gastropods (*Holopea*? sp. indt.), a fish spine (*Machaeracanthus cf. kayseri* Kegel), and single unidentifiable specimens of a conularid and a trilobite. Flattened horizontal burrows are common in some beds. A single weathered U-shaped burrow with spreiten was observed.

Boucot and others (1967) considered the fauna to be Early Devonian in age and to represent the Malvinokaffric Province. The additional material collected is consistent with that interpretation. Large orbiculoids similar to *Orbiculoidea falklandensis* are present in the Devonian of other Southern Hemisphere continents and are considered to be one of the characteristic elements of the Malvinokaffric Province. The fauna correlates most closely with that of the Horlick Formation, Ohio Range, Antarctica, and common faunal elements include the genera *Holopea*?, *Nuculites,* and *Orbiculoidea,* as well as a placoderm spine. *Orbiculoidea falklandensis* was described from Lower Devonian strata of the Falkland Islands (Doumani and others, 1965).

INARTICULATE BRACHIOPODS, GASTROPODS, A TRILOBITE, AND A CONULARID

About 30 specimens of orbiculoid brachiopods were identified by A. J. Rowell as *Orbiculoidea cf. falklandensis* and reported by Boucot and others (1967) from the original fossil collection of the Planck Point locality. *Orbiculoidea falklandensis* was described by Rowell (Doumani and others, 1965) from Devonian strata of the Falkland Islands. An additional 30 specimens were recovered during the re-collection of the Planck Point site. The additional specimens represent almost equal numbers of pedicle and brachial valves and exhibit both internal and external surfaces. All specimens are damaged and partially exfoliated. Figured specimens are stored at the United States National Museum (USNM) Plate 1, Figure 7 (USNM 334194) shows an external view of a brachial valve; Plate 1, Figure 8 (USNM 334195) shows an external view of a pedicle valve.

Six poorly preserved specimens of gastropods (USNM 334196) were recovered. They are not figured due to poor preservation. These were examined by Ellis Yochelson of the U.S. Geological Survey and referred to *Holopea*? sp. indt. A specimen of *Holopea*? was reported from the Horlick Formation of the Ohio Range, Antarctica (Doumani and others, 1965).

Single, broken, poorly preserved specimens of both a ptychopariid trilobite pygidium (USNM 334198) and a conularid (USNM 334199) were recovered. Neither is figured due to poor preservation. The preserved portion of the conularid measures 46 mm from the base to the broken edge. Width at the broken end is estimated to be 12 mm; crests of transverse ridges at this end are separated by about 0.5 mm.

REFERENCES CITED

Boucot, A. J., Doumani, G. A., Johnson, J. G., and Webers, G. F., 1967, Devonian of Antarctica, *in* Oswald, D. H., ed., International Symposium on the Devonian System, Calgary, Alberta, 1967, Papers Volume 1: Calgary, Alberta Society of Petroleum Geologists, p. 639–648.

Doumani, G. A., and 8 others, 1965, Lower Devonian fauna of the Horlick Formation, Ohio Range, Antarctica, *in* Hadley, J. B., ed., Geology and paleontology of the Ellsworth Mountains: American Geophysical Union Antarctic Research Series, no. 1299, v. 6, p. 241–281.

DIASOME MOLLUSKS

John Pojeta, Jr.

INTRODUCTION

Three poorly preserved pelecypods and one incomplete rostroconch were found in about 44 person hours of collecting. They represent three taxa, all of which are known from marine rocks, and they do not disagree with the Early Devonian age of the Mt. Wyatt Earp Formation near Planck Point, determined from other fossils.

The major work on Antarctic Lower Devonian diasomes is McAlester (1965) whose material was from the Horlick Formation, Ohio Range, Horlick Mountains, Antarctica. McAlester's specimens, supplemented by new collections, are now under study by M. A. Bradshaw, Canterbury Museum, Christchurch, New Zealand. Among the diasomes, only the genus *Nuculites* is, at present, common to both the Ellsworth Mountains and the Ohio Range. In addition to *Nuculites,* McAlester listed eight other genera of pelecypods; he did not describe any rostroconchs.

SYSTEMATIC PALEONTOLOGY

Phylum MOLLUSCA
Subphylum DIASOMA
Class ROSTROCONCHIA
Order CONOCARDIOIDEA
Superfamily CONOCARDIACEA
Family HIPPOCARIIDAE?

Genus *Hippocardia*

***Hippocardia?* sp. indt.**
Plate 1, Figs. 1, 2

Discussion. This taxon is represented by one incomplete specimen (USNM 383603) collected 30 m below the top of the measured section. The specimen is preserved as part and counterpart internal and external molds of the snout and body of a conocardioid (Pojeta and Runnegar, 1976, Plate 49, Fig. 15); it is compressed in a right-left direction. The internal ribs on the internal mold can be seen clearly (Plate 1, Fig. 1) and poorly preserved growth increments are discernible. At the posterior end of the specimen, on what in the figure is the left side, there is a lateral expansion that lacks ribs and probably represents the anterior surface of the hood (Plate 1, Fig. 2) (Pojeta and Runnegar, 1976, Plate 49, Figs. 10, 14). It is on the basis of the probable presence of a hood that USNM 383603 is placed in *Hippocardia*?. Too little of the specimen is preserved to make the generic assignment unequivocal or to assign it to a species. *Hippocardia*? sp. indt. is preserved in a thin-bedded micaceous siltstone. The known stratigraphic range of *Hippocardia* is Middle Ordovician (Llanvirnian)–Mississippian (Tournaisian).

Class PELECYPODA
Subclass PALAEOTAXODONTA
Order NUCULOIDEA
Superfamily NUCLANACEA
Family MALLETIIDAE

Genus *Nuculites* Conrad

***Nuculites* aff. *N. cuneiformis* Conrad**
Plate 1, Figs. 3, 4

Discussion. This taxon is represented by one specimen (USNM 383604) from a float block found 10 m below the top of the section. The specimen is preserved as part and counterpart internal and external molds and has been compressed parallel to the bedding. USNM 383604 has the characteristic anterior buttress found in *Nuculites* (Plate 1, Fig. 3) and also preserves a posteroumbonal carina (Plate 1, Fig. 3). The posteroventral and anterodorsal parts of the specimen are missing. *Nuculites* aff. *N. cuneiformis* occurs in a calcareous micaceous sandstone. It differs from the Horlick Formation specimens of *Nuculites* in having a much narrower anterior buttress and in lacking a posterior buttress.

Pojeta, Zhang, and Yang (1989) have divided the Devonian species of *Nuculites* into four informal types, each of which is characterized by a typical species. Type 3 species have elongated cuneiform shells with a prominent posteroumbonal carina as in *Nuculites cuneiformis* Conrad. Bailey (1983, p. 270) pointed out the difficulty of classifying species of *Nuculites* because of the intergradation of shell form in the genus. He suggested that the specimens of *N. cuneiformis* figured by Hall (1885, Plate 47, Figs. 13–16) might be distorted specimens of *N. triqueter* Conrad. This suggestion has yet to be tested. It is known that USNM 383604 is compressed medially, yet it still has a shape much like the specimens of *N. cuneiformis* figured by Hall. Hall's specimens would have to be compressed dorsoventrally according to Bailey's suggestion. In New York, *N. cuneiformis* occurs in Givetian (Hamiltonian) age rocks; the Antarctic specimen described here is treated as having affinities to the New York species.

Subclass ANOMALODESMATA
Order PHOLADOMYOIDA
Superfamily PHOLADOMYACEA
Family GRAMMYSIIDAE

Genus *Grammysioidea*? Williams and Breger

***Grammysiodea*? sp. indt.**
Plate 1, Figs. 5, 6

Discussion. This taxon is represented by two poorly preserved composite molds (USNM 383605, 383606) found 10 m below the top of the section. The specimens are preserved in an argillaceous micaceous siltstone. The shape of the Antarctic material is much like that of *Grammysioidea arcuata* (Conrad) (Hall, 1885, Plate 61, Figs. 1, 2, 5) from the Givetian (Hamiltonian) age rocks of New York. In addition to shape, the Antarctic specimens show relict rugose comarginal ornament (Plate 1, Fig. 6). Because USNM 383605 and 383606 are so poorly preserved, and because rugose comarginal ornament is a widespread feature in various anomalodesmatans and some isofilibranchs, they are here assigned to *Grammysioidea* with question.

REFERENCES CITED

Bailey, J. B., 1983, Middle Devonian Bivalvia from the Solsville Member (Marcellus Formation), central New York State: American Museum of Natural History Bulletin, v. 174, p. 193–326.

Hall, J. B., 1885, Dimyaria of the Upper Helderburg, Hamilton, Portage, and Chemung Groups: Geological Survey of New York, Palaeontology, v. 5, part 1, Lamellibranchiata II, p. 269–561.

McAlester, A. L., 1965, Bivalves, *in* Hadley, J. B., ed., Geology and paleontology of the Antarctic: American Geophysical Union Antarctic Research Series, v. 6, p. 261–266.

Pojeta, J., Jr., and Runnegar, B., 1976, The paleontology of rostroconch mollusks and the early history of the Phylum Mollusca: U.S. Geological Survey Professional Paper 968, 88 p.

Pojeta, J., Jr., Zhang, R., and Yang, A., 1989, The Devonian rocks and Lower and Middle Devonian pelecypods of Guangxi, China and Michigan: U.S. Geological Survey Professional Paper (in press).

CEPHALOPODS

Brian Glenister

Class CEPHALOPODA
Subclass ORTHOCERATOIDEA
Order ORTHOCERIDA
cf. Orthoceratidae, Gen. Indt.

Description. Two poorly preserved internal molds of nautiloid phragmocones (USNM 334197) can be treated as a single species. They are from the same collection, and although they do not fit together, could be portions of one specimen. Both are orthocones and have circular sections and a central siphuncle. No trace of ornament or of cameral or siphuncular deposits is visible on either specimen. The apical fragment has a length of 5 cm, reaches a maximum diameter of 15 mm, and comprises 8 camerae. Its siphuncle is 2 mm in diameter and appears to be cylindrical or slightly constricted at about the mid-height of each segment. Although obscure, the septal necks appear orthochoanitic, with a length approximately one-quarter the cameral interval. The second specimen is 3 cm long, achieves a maximum diameter of 22 mm, and comprises 4 camerae. Sutures appear to be straight and directly transverse. The specimens are not figured due to poor preservation.
Comparisons. Inadequate preservation precludes confident assignment at any taxonomic level less than the order Orthocerida. However, if better material confirms the lack of ornament and the orthochoanitic septal necks, then reference to either of the orthoceroid families Orthoceratidae or Geisonoceratidae is possible. The distinction between these two groups rests largely on development of siphuncular deposits in the latter. In our present state of knowledge, the most appropriate reference is to the "waste basket" family Orthoceratidae.
Occurrence. The described specimens were recovered as float about 15 m below the top of the section near Planck Point exposing the Mt. Wyatt Earp Formation.

DESCRIPTION OF THE FISH SPINE

Gavin Young

The only vertebrate so far recorded from the Mount Wyatt Earp Formation is a portion of an acanthodian fin-spine. The acanthodian fishes were the first group of gnathostome vertebrates to appear in the fossil record (earliest confirmed report, Early Silurian), and they persisted until the Early Permian.

The single acanthodian spine (USNM 334200) described here was collected in two pieces of hard sandstone. The main portion of the spine was exposed only as cross sections on opposite surfaces of the larger sample (piece A), but an indication of overall shape was obtained from x-radiography, and two additional transverse sections were cut through the spine. The smaller sample (piece B), broken from the other piece, contains another short length of spine with part of its anterior surface exposed. The hard tissue of the spine is dark gray in color and is clearly distinguished from the lighter rock matrix. It is generally excellently preserved and shows some details of histological structure in the various transverse sections. The classification of acanthodians used below follows that of Denison (1979).

Subclass ACANTHODII Owen
Order ISCHNACANTHIDA Berg

Genus *Machaeracanthus* Newberry 1857

***Machaeracanthus* cf. *kayseri* Kegel 1913**
Figs. 2, 3; Plate 1, Fig. 9
1983 fish(?) vertebrae. Webers & Spörli

Locality. Planck Point (79°21′S, 85°11′W), Heritage Range, Ellsworth Mountains, West Antarctica (locality number YP-79-18S).
Horizon. Mount Wyatt Earp Formation (lower part).
Age. Early Devonian.
Description. The shape of this spine in transverse section shows clearly the characteristic morphology of *Machaeracanthus,* in which the central body of the spine projects anteriorly as a thin unornamented "keel" and posteriorly as a wider unornamented "wing" (e.g., Zidek, 1981). The larger section exposed on one surface of piece A has a prominent central cavity (*cc,* Fig. 2B) and clearly represents a proximal part of the spine. The much smaller exposed section on the opposite surface represents a level near the tip of the spine. About 41 mm of spine length is contained within the sample. X-radiographs of the sample only show up the thicker central part of the spine, so two additional sections were cut through the sample to determine the extent of the keel and wing. An approximation of spine shape based on this information is shown in Fig. 2. Another 7.5-mm length of the more proximal part of the spine is preserved on piece B. There is no clear indication of an inserted part of the spine, and its proximal end is probably incomplete. Judging by other completely preserved *Machaeracanthus* spines (e.g., Newberry, 1889, Plate 29, Fig. 4), it would seem that at least two-thirds and probably three-quarters of spine length is preserved (estimated total length 65 to 75 mm).

Among *Machaeracanthus* species the shape of the spine in transverse section can be distinctive (Zidek, 1981). In the new specimen, however, there is some variation in this feature over the preserved length (Fig. 2). The most proximal section as preserved on the surface of piece A shows a good general resemblance to the spines from Overath referred to *M. kayseri* by Gross (1933, Fig. 12A). At this level the spine is 10 mm wide and 5.2 mm high, also giving similar proportions to the spine figured by Gross (Fig. 3). Following Gross, the main ridged portion of the central body of the spine is assumed to be dorsal, although this is not clearly demonstrated in any of the examples of *Machaeracanthus* where paired spines from one individual are preserved (e.g., Barrande, 1872,

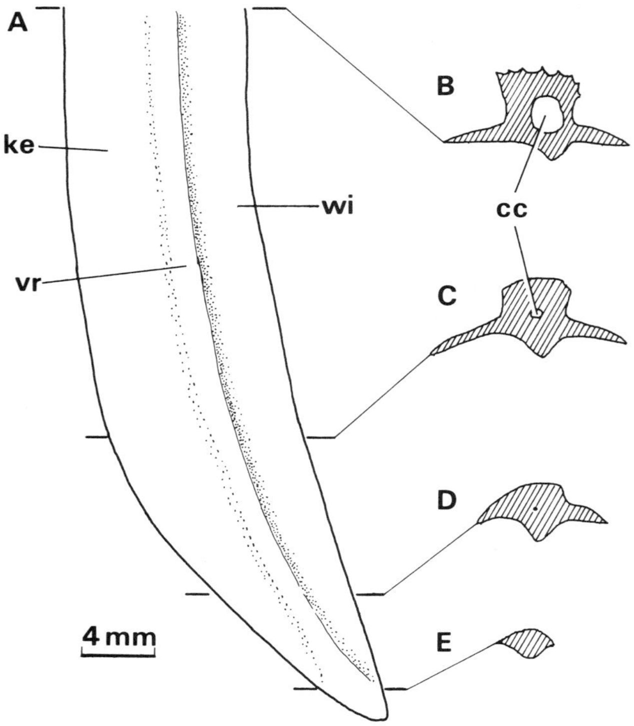

Figure 2. *Machaeracanthus* cf. *kayseri* Kegel. A. Approximation of spine shape in ventral view, based on the spine portion preserved in piece B. B–E. Four transverse sections through the spine at the levels indicated. cc = central cavity; ke = anterior keel; vr = approximate position of ventral ridge; wi = posterior wing.

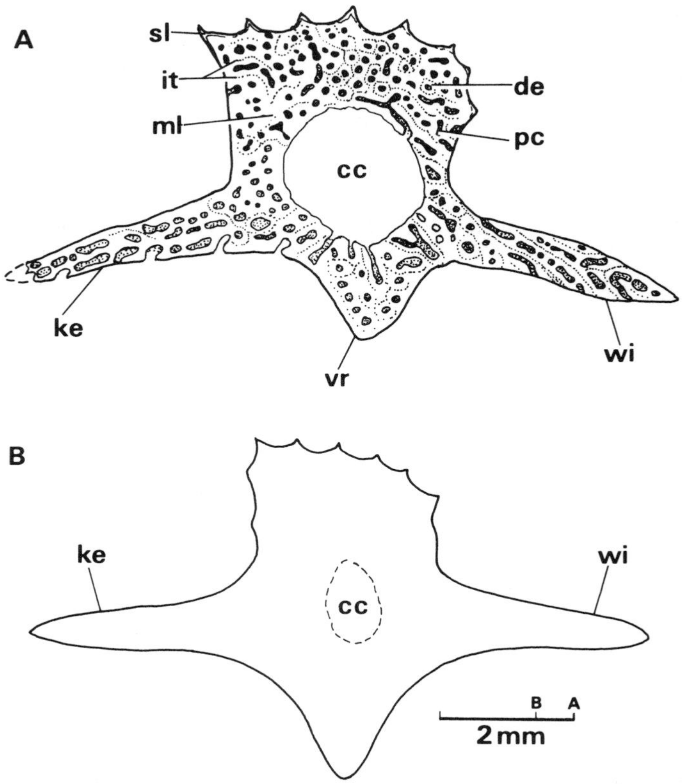

Figure 3. *Machaeracanthus* cf. *kayseri* Kegel. Detail of the proximal transverse section exposed on the surface of piece A (cf. Fig. 2B). B. *Machaeracanthus kayseri* Kegel, from the Early Devonian of Overath, Germany. Outline of spine in transverse section, after Gross (1933, Fig. 12A; central cavity restored after his Fig. 12B). de = denteon; it = interstitial tissue; ml = middle layer; pc = pulp canal; sl = superficial layer; other abbreviations as in Figure 2.

Plate 34, Fig. 29; Newberry, 1873, p. 303; Eastman, 1907; see Hussakof and Bryant, 1918, p. 167). The main difference between this section and that from Overath is the slightly wider and ventrally deflected keel and wing in the Antarctic specimen (*ke, wi,* Fig. 3). Otherwise the ventral ridge (*vr*) has a similar triangular section, and the keel and wing are of similar width. The dorsal part of the central body in both examples is relatively high, with a squarish shape and a ridged dorsal surface.

In the Antarctic specimen eight clearly defined dorsal ridges presumably formed a subparallel longitudinal arrangement down the dorsal surface, as previously figured (Gross, 1933, Plate 5, Fig. 7). In this specimen, however, a more distal section (Fig. 2C) shows that the ridges did not extend along the length of the spine. The two most posterior ridges lie on the posterior face of the dorsal portion (Fig. 3A). The anterior face is, however, smooth and slightly concave, as is also seen on the exposed part of this surface on piece B. As in the Overath specimen (Fig. 3B), the anterior edge of the dorsal surface is clearly defined by the most anterior longitudinal ridge, whereas the posterior edge is somewhat lower and more rounded in section. The only obvious difference between these specimens would appear to be that in the Overath examples there are low ridges on both anterior and posterior surfaces of the central body above the keel and wing; in the Antarctic specimen these are confined to the posterior surface.

The more distal sections (Fig. 2C–E) show that the central cavity becomes smaller and disappears toward the tip of the spine. The dorsal part of the central body loses its ridged surface and then becomes lower and more rounded at the distal end where its anterior face merges with the keel (Fig. 2D). The keel apparently increases in width (Fig. 2C) before reducing toward the tip of the spine, suggesting a somewhat similar shape to the spines figured by Lehman (1976, Plate 1B). The ventral ridge retains its triangular section until very near the tip of the spine.

Some general histological details are visible in the section exposed on the surface of piece A (Fig. 3A). The outer surface of the spine is clearly defined by a thin dark line, and on the upper surface a narrow zone can be distinguished beneath the surface ridges (*sl*), which is darker than the deeper tissue. This is the only indication of a differentiated superficial layer, and no identifiable basal layer surrounds the central cavity, as is the case in a number of other acanthodian genera (Denison, 1979). The middle layer, making up most of the spine (*ml*), is presumably osteodentine ("trabecular dentine" of Denison, 1979) as in other acanthodian fine spines, although no dentine tubules are visible in the examined material. Conspicuous are numerous pulp canals (*pc*), some in open communication with the central cavity or, in ventral parts of the spine, with the exterior. There is some suggestion of a concentric arrangement of pulp canals just above the central cavity and of a radial arrangement beneath the cavity and into the keel and wing. Otherwise, in the main parts of the central body of the spine and the ventral ridge, the canals are predominantly longitudinal. Many individual denteons (*de*) are identifiable by thin zones of interstitial tissue (*it*) that are much lighter in color, suggesting a higher degree of mineralization. Gross (1933) also commented on the distinctness of circumpulpal dentine between adjacent denteons in *M. kayseri* from Overath. According to

Ørvig (1951, Fig. 2A, B), this interstitial tissue in some species of *Machaeracanthus* was acellular but in other species consisted of true bone.

The more distal sections (Fig. 2C–E) provide a few extra details. As noted above, the central cavity becomes smaller distally, but there is no sign of a second enlarged or subcostal canal. This suggests that the large cavity incompletely preserved in the spine from Overath (see Gross, 1933, p. 67) does represent the central cavity. In the same specimen, Gross noted enlarged pulp canals in a zone around the central cavity, a feature clearly absent in the section just described (Fig. 3A). In the next more distal section, however, there are a number of enlarged pulp canals in the central region. The central cavity at this level (Fig. 2C) is apparently smaller than in the Overath specimen, and it can reasonably be suggested that the section illustrated by Gross (1933, Fig. 12B) came from a level between those represented here in Figures 2B and C, where dorsal ridges were still present, but the central cavity was reduced and surrounded with enlarged pulp canals. The central cavity in the new specimen is still present but much reduced in the next section (Fig. 2D), and in the most distal section the central region is occupied only by three or four advanced denteons.

Remarks. Zidek (1981) lists ten species of *Machaeracanthus* that he regards as valid. Each has a distinctive spine shape in transverse section, although information on variation along the spine is not always available. It would seem that sections near the tip of the spine are less reliable to differentiate species. Apart from *M. kayseri,* three of the listed species bear longitudinal ridges, but each is readily distinguished from the new form. In *M. retusus* the distal part of the spine is somewhat similar in section to that described above, but in the middle of the spine the section is much lower and broader with a shorter wing and keel (Wells, 1940). *M. sulcatus,* regarded by Gross (1933) as closely related to *M kayseri,* has the keel and wing only slightly pronounced, and the ridged central body is rounded in section (Newberry, 1889, Plate 29, Fig. 5; Lehman, 1977, p. 108). In *M. sarlei* the ridges converge toward the center of the spine, and it is again much more rounded in section (Zidek, 1981, Fig. 1). Of the remaining species, *M. polonicus* has a squarish outline in section, but the keel and wing are not as broad, and the ventral ridge is also square rather than triangular in shape. The section of *M. major* figured by Hussakof and Bryant (1918, Fig. 56B) shows some resemblance to the new specimen (Fig. 2C), but the ridge and keel are again less extensive, the central body is much lower and broader, and this species is a much larger form that does not bear ridges (e.g., Newberry, 1873). None of the other species shows any specific resemblances worthy of comment.

The similarities between the new form and *M. kayseri* as described by Gross (1933) have been discussed above but may be summarized as follows: section in the middle part of the spine showing a high dorsal portion of squarish shape, with about seven or eight sharp dorsal ridges separated by wider grooves, and the anterior ridge forming a prominent angular edge to the dorsal surface; keel and wing both fairly broad, with the keel sometimes broader; ventral ridge always triangular in section, and pulp canals around the central cavity sometimes enlarged. Other assumed resemblances (e.g., ridges absent from the distal third) are not yet established for the Overath material, and there are slight differences noted above that may prove to be specific characters, although a close relation between the two forms is strongly indicated. On this basis the Antarctic material is referred in open nomenclature to *M.* cf. *kayseri* Kegel. Gross (1933) noted differences between his Overath specimens and the type material from the Taunus Quartzite, which is lower in cross section, with a narrower keel and wing. Possibly more significant is the rounded rather than triangular section of the ventral ridge, since the Antarctic specimen and other species (e.g., *M. retusus,* Wells, 1940) show this to be a consistent feature, at least in the distal part of the spine. The Taunus Quartzite material requires restudy, and the assignment of the Antarctic specimen to *M.* cf. *kayseri* assumes that the Overath material is correctly referred to the species.

DISCUSSION

Biostratigraphy. The genus *Machaeracanthus* is known only from isolated fin spines and scales, and in only one species (*M. bohemicus*) have both spines and scales been described (Gross, 1973). *Machaeracanthus* scales are present in the Stonehouse Formation of Nova Scotia (Legault, 1968, Plate 3, Fig. 13; see Denison, 1979; Zidek, 1981), where they occur with *eosteinhornensis* zone conodonts, indicating a Late Silurian (Pridolian) age (e.g., Klapper, 1977). *Machaeracanthus* spines or scales are also known from the Early Devonian of North America (Zidek, 1975), Canada (Gardiner, 1966; Pageau, 1969), Bohemia (Chlupac, 1976), France (Goujet, 1976), Germany (Gross, 1933, 1965), Poland (Krassowska and Kulczycki, 1963; Lobanowski and Przybylowicz, 1979), Algeria (Lehman, 1963), and Morocco (Lehman, 1976, 1977). Unconfirmed reports also exist of occurrences from the Lower Devonian of Brazil (Mendes, 1971; see Janvier, 1976) and South Africa (duToit, 1954; see Chaloner and others, 1980). In the Middle Devonian there are reports of occurrences from various Eifelian and Givetian strata in the United States and Canada (e.g., Newberry, 1857; Eastman, 1907; Hussakof and Bryant, 1918) and an occurrence in the Eifelian of the Rheinland, Germany (see below). Finally, Hussakof and Bryant (1918) and Wells (1940) report *Machaeracanthus* spines from the Genesee conodont bed and Genesee shale of New York State, for which an early Frasnian age has been proposed (e.g., Rickard, 1975; Klapper and Ziegler, 1978). Thus the genus has a known stratigraphic range from latest Silurian (Pridolian) to earliest late Devonian (Frasnian).

The species *M. kayseri,* with which this new form is closely related or conspecific, was first described by Kegel (1913) from the Taunus Quartzite and then by Gross (1933) from the Wahnbach beds and the Gemund Conglomerate in the German Rheinland. The age of these beds is Siegenian (e.g., Grabert, 1967; Erben and Zagora, 1967). Another report of *M. kayseri* from the Eifelian Brandenberg Beds (Langenstrassen and others, 1979, Table 1) is uncertain, since Zidek (1981) regards this occurrence as a separate species (*M. westfalicus* Pfeiffer). In summary, the evidence suggests an Early Devonian (?Siegenian) age for the Ellsworth Mountains *Machaeracanthus* spine.

Biogeography. The association of this acanthodian spine with Malvinokaffric Province brachiopods is of some interest, since Devonian fish remains from areas considered to belong to this province are very poorly known. Placoderms occur with another Early Devonian Antarctic brachiopod fauna from the Horlick Formation of the Ohio Range (Boucot, Doumani, and others, 1967), but so far only one specimen has been described (Miles, in Doumani and others, 1965). This is an arthrodire plate tentatively compared to Early Devonian forms from the western United States (Denison, 1958). It could now be suggested, however, that the absence of *Machaeracanthus* from the western United States makes close affinity with any arthrodires from these faunas less likely. Another Antarctic Devonian fish fauna from the Aztec Siltstone of south Victoria Land (e.g., White, 1968;

Young, 1982) is somewhat younger than the faunas under consideration here (?late Middle to Late Devonian). Chaloner and others (1980), however, recently described a new fish fauna, including placoderms and acanthodians, from the Bokkeveld Group, Cape Province, South Africa. An Early/Middle Devonian Malvinokaffric invertebrate fauna occurs in lower strata of the Bokkeveld Group (Cooper, 1982), and Boucot and others (1983) reported brachiopods from the base of the overlying Witteberg Group that may be Middle Devonian. Thus the age of the Bokkeveld fish fauna is uncertain but may be slightly younger (?early Middle Devonian) than the Antarctic Early Devonian occurrences. As noted above there is also an unconfirmed report of a *Machaeracanthus* spine from the Bokkeveld Group and another from South America, in the Parnaiba Basin of northern Brazil (see Janvier, 1976). That *Machaeracanthus* should be reported from localities on three different continents within the Early Devonian Malvinokaffric Province, which has a very sparse and/or poorly known vertebrate fauna, would seem to be more than coincidental. With reference to a palaeogeographic map of the Early Devonian of these areas (Johnson, 1979, Fig. 6), the proximity of the Ellsworth Mountains and South African occurrences is noteworthy (but cf. Cooper, 1982, p. 165). The various European and north African occurrences of *Machaeracanthus* all fall within the Old World Province based on brachiopods and, as noted above, are mainly Early Devonian in age. On the other hand, North American occurrences are mainly Middle Devonian and occur within the Appalachian Province. Exceptions are the *Machaeracanthus* occurrences in the Bois d'Arc limestone (Gedinnian) of Oklahoma (Zidek, 1975), the Stonehouse Formation of Nova Scotia (Denison, 1979), and the Gaspé area, Quebec (Pageau, 1969). The latter two areas, at least in the late Lower Devonian, were also part of the Old World Province (e.g., Boucot, Johnson, and Talent, 1967, p. 1251). A somewhat similar pattern of earlier occurrence in Europe is reported for pelecypods (Bailey, 1978) and some brachiopods (Johnson, 1979, p. 293). One of the brachiopods, *Tropidoleptus,* is also widespread in the Malvinokaffric Province (see Boucot and others, 1983, p. 56). If confirmed, such similarities of pattern between different groups of organisms may be of some importance to future biogeographic investigations.

ACKNOWLEDGMENTS

I thank D. Goujet and M. Smith for advice on the material, A. Haupt for translating some texts from the German, and K. Heighway for taking some x-radiographs. G. C. Young publishes with the permission of the Director, Bureau of Mineral Resources, Canberra.

PLATE 1

Figures 1, 2. *Hippocardia*? sp. indt. USNM 383603. Fig. 1. Oblique right-lateral view of internal mold of snout and body of shell showing internal ribs and poorly preserved growth increments, ×3. Figure 2. Anterior view of the specimen seen in Figure 1 above, showing lateral expansion (arrow), which lacks ribs and probably represents the anterior surface of the hood, ×5.

Figures 3, 4. *Nuculites* aff. *N. cuneiformis* Conrad. USNM 383604. Part and counterpart of a right valve, ×2. In Figure 3, the downward-facing arrow points to the posteroumbonal carina, and the right-facing arrow points to the impression of the buttress.

Figures 5, 6. *Grammysiodea*? sp. indt. USNM 383605 and 383606. Poorly preserved right and left valve internal molds. The specimen seen in Figure 5 suggests the presence of relict comarginal rugae, ×2.

Figure 7. *Orbiculoidea* cf. *falklandensis.* External surface brachial valve, partially exfoliated. ×1.5 USNM 334194.

Figure 8. *Orbiculoidea* cf. *falklandensis.* External surface pedicle valve, partially exfoliated. ×1.5 USNM 334195.

Figure 9. *Machaeracanthus* cf. *kayseri.* USNM 343200 Proximal transverse section. ×7.

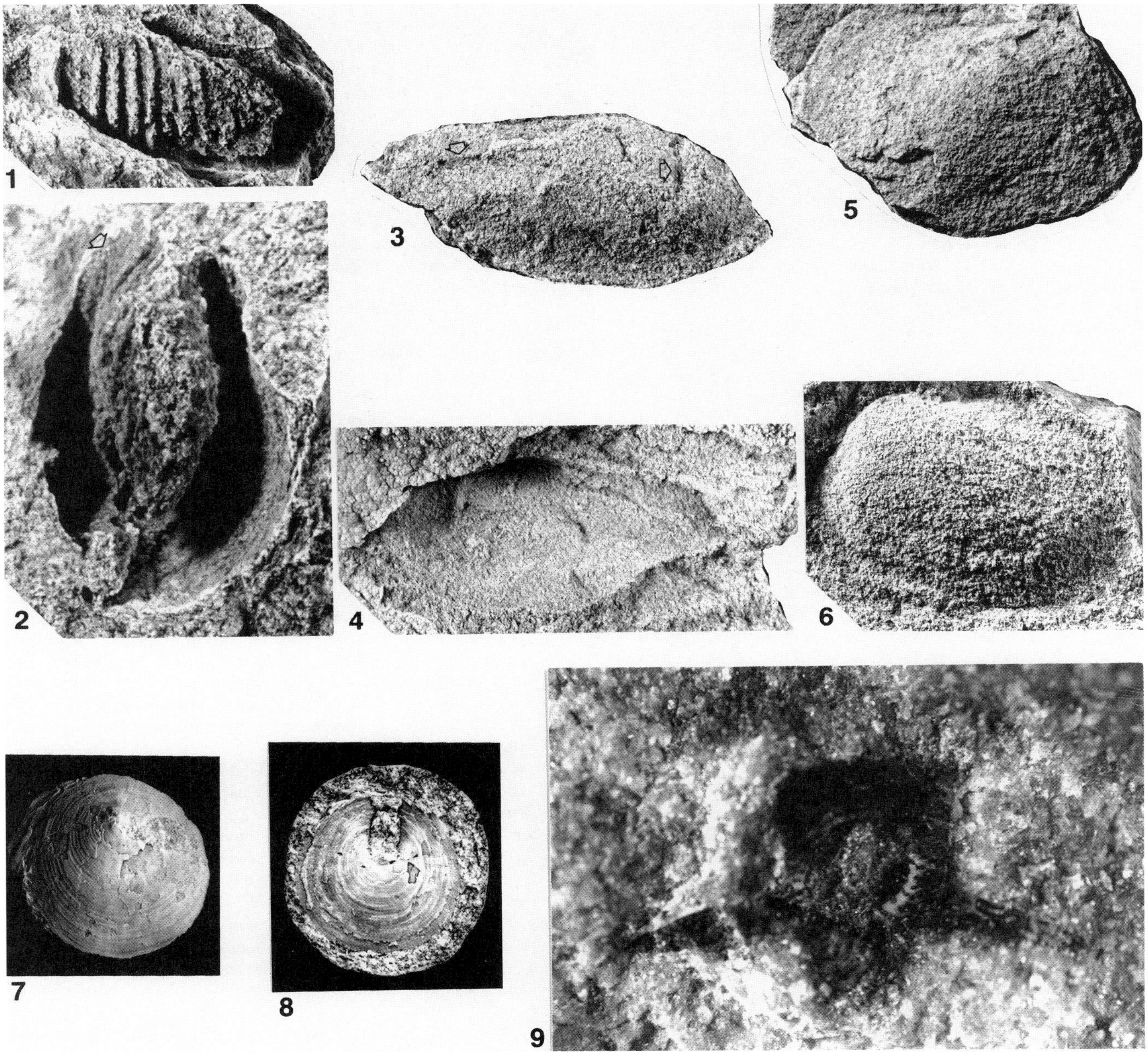
1
2
3
4
5
6
7
8
9

REFERENCES CITED

Bailey, J. B., 1978, Provincialism and migration in Lower and Middle Devonian pelecypods: Palaeogeography, Palaeoclimatology, Palaeoecology, v. 23, p. 119–130.

Barrande, J., 1872, Systeme silurien du centre de la Boheme, supplement au Vol. 1, trilobites, crustaces divers et poissons: Prague and Paris, 647 p.

Boucot, A. J., Doumani, G. A., Johnson, J. G., and Webers, G. F., 1967, Devonian of Antarctica, *in* Oswald, D. H., ed., International Symposium on the Devonian System: Calgary, Alberta Society of Petroleum Geologists, v. 1, p. 639–648.

Boucot, A. J., Johnson, J. G., and Talent, J. A., 1967, Lower and Middle Devonian faunal provinces based on Brachhiopoda, *in* Oswald, D. H., ed., International Symposium on the Devonian System: Calgary, Alberta Society of Petroleum Geologists, v. 2, p. 1239–1254.

Boucot, A. J., Brunton, C.H.C., and Theron, J. N., 1983, Implications for the age of South African Devonian rocks in which *Tropidoleptus* (Brachiopoda) has been found: Geological Magazine, v. 120, p. 51–58.

Chaloner, W. G., Forey, P. L., Gardiner, B. G., Hill, A. J., and Young, V. T., 1980, Devonian fish and plants from the Bokkeveld Series of South Africa: Annals of the South African Museum, v. 81, p. 127–157.

Chlupac, I., 1976, The Bohemian Lower Devonian stages and remarks on the lower-Middle Devonian boundary: Newsletters on Stratigraphy, v. 5, p. 168–189.

Cooper, M. R., 1982, A revision of the Devonian (Emsian-Eifelian) Trilobita from the Bokkeveld Group of South Africa: Annals of the South African Museum, v. 89, 174 p.

Denison, R. H., 1958, Early Devonian fishes from Utah; 3, Arthrodira: Fieldiana Geology, v. 11, p. 461–551.

—— , 1979, *Acanthodii, in* Schultze, H.-P., ed., Handbook of paleoichthyology, v. 5: Stuttgart, Gustav Fischer Verlag, p. 1–62.

Doumani, G. A., and 7 others, 1965, Lower Devonian fauna of the Horlick Formation, Ohio Range, Antarctica: American Geophysical Union Antarctic Research Series, v. 6, p. 241–281.

DuToit, A. L., 1954, The geology of South Africa, 3rd ed.: Edinburgh and London, Oliver and Boyd, 611 p.

Eastman, C. R., 1907, Devonic fishes of the New York formations: New York State Museum Memoir 10, 235 p.

Erben, H. K., and Zagora, K., 1967, Devonian of Germany, *in* Oswald, D. H., ed., International Symposium on the Devonian System: Calgary, Alberta Society of Petroleum Geologists, v. 1, p. 53–68.

Gardiner, B. G., 1966, Catalogue of Canadian fossil fishes: Royal Ontario Museum Life Sciences Contribution 66, p. 5–154.

Goujet, D., 1976, Les poissons, *in* Lardeux, H., ed., Les schistes et calcaires Eodevoniens de Saint-Cenere (Massif Armoricain, France): Memoires de la Societe geologique et mineralogique de Bretagne, v. 19, p. 313–323.

Grabert, H., 1967, Devonian strata in the central part of the Rhenish Massif (Rheinisches Schiefergebirge, North Rhine-Westphalia), Germany, *in* Oswald, D. H., ed., International Symposium on the Devonian System: Calgary, Alberta Society of Petroleum Geologists, v. 2, p. 1–10.

Gross, W., 1933, Die unterdevonischen fische und Gigantostraken von Overath: Abh. Preuss. Geol. Landesanst, v. 145, p. 41–77.

—— , 1965, Uber einer neuen schadelrest von *Stensioella heintzi* un schuppen von *Machaeracanthus* sp. indet. aus dem Hunsruckschiefer: Weisbaden, Notizbl. hess. Landesamt. Bodenforsch, v. 93, p. 7–18.

—— , 1973, Kleinschuppen, flossenstacheln und zahne von fischen aus europaischen und Nordamerikanischen bonebeds des devons: Palaeontographica, series A, v. 142, p. 51–155.

Hussakoff, L., and Bryant, W. L., 1918, Catalog of the fossil fishes in the Museum of the Buffalo Society of Natural Sciences: Buffalo Society of Natural Sciences Bulletin, v. 12, p. 5–198.

Janvier, P., 1976, Description des restes d'Elasmobranches (Pisces) du Devonien moyen de Bolivae: Palaeovertebrata, v. 7, p. 126–132.

Johnson, J. G., 1979, Devonian brachiopod biostratigraphy, *in* House, M. R., Scrutton, C. T., and Bassett, M. B., eds., The Devonian System: Special Papers in Palaeontology, v. 23, p. 291–306.

Kegel, W., 1913, Der Taunusquartzit von Katzenelnbogen: Abhandlunged der Preussischen geologischen Landesanstalt, v. 76, p. 1–163.

Klapper, G., 1977, Conodonts, *in* The Silurian-Devonian boundary: Stuttgart, International Union of Geological Sciences, series A, no. 5, p. 318–319.

Klapper, G., and Ziegler, W., 1978, Devonian conodont biostratigraphy, *in* House, M. R., Scrutton, C. T., and Bassett, M. G., eds., The Devonian System: Special Papers in Palaeontology, v. 23, p. 199–224.

Krassowska, A., and Kulczycki, J., 1963, Devonian in the vicinity of Ciepielow: Przeglad Geol., v. 8, p. 394–395 (in Polish).

Langenstrassen, F., Becker, G., and Groos-Uffenorde, H., 1979, Zur facies und fauna der Brandenberg-Schichten bei Lasbeck: Neues Jahrbuch fur Geologie und Palaontologie Abhandlungen, v. 158, p. 64–99.

Legault, J. A., 1968, Conodonts and fish remains from the Stonehouse Formation, Arisaig, Nova Scotia: Geological Survey of Canada Bulletin 165, p. 1–23.

Lehman, J. P., 1963, A propos de quelques arthrodires et ichthyodorulites sahariens: Institut Francasi d'Afrique Noire Memoires 68, p. 193–200.

—— , 1976, Nouveaux poissons fossiles du Devonien du Maroc: Annales de Paleontologie Vertebres, v. 62, p. 1–23.

—— , 1977, Nouveaux arthrodires du Tafilalet et de ses environs: Annales de Paleontologie Vertebres, v. 63, p. 105–132.

Lobanowski, H., and Przybylowicz, T., 1979, Tidal flat and floodplain deposits in the Lower Devonian of the western Lublin Uplands (after the boreholes Pionki 1 and Pionki 4): Acta Geologica Polonica, v. 29, p. 383–407.

Mendes, J. C., 1971, Geologia do Brasil: Rio de Janeiro, Institut Nacional do Libro Ministerio de Educasao e Cultura, 207 p.

Newberry, J. S., 1857, Fossil fishes from the Devonian of Ohio: Washington, D.C., Proceedings of the National Institute for the Promotion of Science, n.s., v. 1, p. 119–126.

—— , 1873, Descriptions of fossil fishes: Report of the Geological Survey of Ohio, v. 1, no. 2, p. 246–355.

—— , 1889, The Paleozoic fishes of North America: U.S. Geological Survey Monograph 16, 340 p.

Ørvig, T., 1951, Histological studies of placoderms and fossil elasmobranches; 1, The endoskeleton, with remarks on the hard tissues of lower vertebrates in general: Stockholm, Arkiv for Zoologi, v. 2, no. 2, p. 321–454.

Pageau, Y., 1969, Nouvelle faune ichthyologique du Devonien moyen dans les gres de Gaspe (Quebec); 2, Morphologie et systematique premiere section; A, Eurypterides; B, Ostracodermes; C, Acanthodiens et Selaciens: Naturaliste Canada, v. 96, p. 399–478.

Rickard, L. V., 1975, Correlation of the Silurian and Devonian rocks of New York State: New York State Museum and Science Service Map and Chart Series 24, p. 1–16.

Webers, G. F., and Spörli, K. B., 1983, Palaeontological and stratigraphic investigations in the Ellsworth Mountains, West Antarctica, *in* Oliver, R. L., ed., Antarctic earth science; 4th international symposium: Cambridge University, p. 261–264.

Wells, J. W., 1940, A new ichthyodorulite from the Geneseo Shale (Devonian of New York): The American Midland Naturalist, v. 24, p. 411–413.

White, E. I., 1968, Devonian fishes of the Mawson-Mulock area, Victoria Land, Antarctica; Trans-Antarctic Expedition 1955–1958: Scientific Reports, v. 16, p. 1–26.

Young, G. C., 1982, Devonian sharks from southeastern Australia and Antarctica: Palaeontology, v. 25, p. 817–843.

Zidek, J., 1975, Oklahoma paleoichthyology; Part 4, Acanthodii: Oklahoma Geology Notes, v. 35, p. 135–146.

—— , 1981, *Machaeracanthus* Newberry (Acanthodii: Ischnacanthiformes); Morphology and systematic position: Neues Jahrbuch fur Geologie und Palaontologie, Monatshefte, p. 742–748.

Manuscript Accepted by the Society October 3, 1989

Printed in U.S.A.

Geological Society of America
Memoir 170
1992

Chapter 13

The archaeocyathan fauna from the Whiteout Conglomerate, Ellsworth Mountains, West Antarctica

Francoise Debrenne
ER154 CNRS Institut de Paleontologie, 8 Rue Buffon, Paris, France 75005

ABSTRACT

A small collection of limestone clasts from the Whiteout Conglomerate (Permo-Carboniferous), Ellsworth Mountains, was examined for archaeocyathids and other fossils. Thirty or more poorly preserved archaeocyathid specimens are here referred to *Ajacicyathus* sp., *Gordonicyathus* sp., *Erismacoscinus* cf. *endutus*, ?*Graphoscyphia* sp., and *Paranacyathus* cf. *parvus*, ?*Dictyocyathus quadruplex*, and *Archaeocyathus* sp.

Other fossils observed but not described include fragmentary specimens of sponge spicules, *Chancelloria* rosettes, trilobites, and brachiopods.

The limestone clasts were derived from a source area with exposed Lower Cambrian strata.

INTRODUCTION

The material studied comes from a small collection of some thirty or more specimens of rocks that contain some archaeocyathan cups. These rocks are clasts from the Whiteout Conglomerate (a Permo-Carboniferous diamictite) and were collected from both the Sentinel Range (Mt. Lymburner) and the Heritage Range (Meyer Hills) of the Ellsworth Mountains. The Mt. Lymburner material was collected by C. Craddock in December 1962 and the Meyer Hills material by G. F. Webers during the 1979–80 season. The preservation of fossils is poor to bad, the skeleton being partly dissolved and replaced by coarse calcite. The size of the elements observed is consequently larger than the initial skeleton (Plate 1, Fig. 7). The outer walls are scarcely preserved, so the determinations are either imprecise or speculative. Nevertheless, some taxa have been recognized and are listed in Table 1 for 25 of the rock samples collected. *Archaeocyathus* (sensu lato) is the most abundant form and is present in most of the rock specimens. The others are rare, represented by only one or two specimens or fragments. Sponge spicules and *Chancelloria* rosettes are rare, together with trilobite fragments and brachiopods; no algae were observed.

SYSTEMATIC DESCRIPTION

Phylum ARCHAEOCYATHA
Class REGULARES Vologdin 1937
Family AJACICYATHUS Bedford and Bedford 1939

Genus *Ajacicyathus* Bedford and Bedford 1939
Type species: *Archaeocyathus ajax* Taylor 1910

Diagnosis. Outer wall with simple pores, septa aporose to sparsely porous; the septal pore row adjacent to the inner wall forms stirrup pores with the corresponding pores of the inner wall. The inner wall contains the pore in front of each septum and an additional row of pores in between.
Geographic and stratigraphic distribution. Cosmopolitan, Botomian and equivalents.

***Ajacicyathus* sp.**
Plate 1, Fig. 1

Material. 2 specimens; WEB5, 7.
Description. Small cups as much as 5.5 mm in diameter with an intervallum of 1 mm. Twenty septa are present, 0.5 mm apart. The outer wall is partly eroded. Two to three pores per intercept are preserved in one

Debrenne, F., 1992, The archaeocyathan fauna from the Whiteout Conglomerate, Ellsworth Mountains, West Antarctica, *in* Webers, G. F., Craddock, C., and Splettstoesser, J. F., Geology and Paleontology of the Ellsworth Mountains, West Antarctica: Boulder, Colorado, Geological Society of America Memoir 170.

TABLE 1. DISTRIBUTION OF FOSSILS IN LIMESTONE-CLAST SPECIMENS FROM THE WHITEOUT CONGLOMERATE

Names / Specimens	CRAD 1	CRAD 2a	CRAD 2b	CRAD 3	CRAD 4	CRAD 5	CRAD 7	CRAD 10	CRAD 12	CRAD 13	CRAD 14	WEB 1	WEB 2	WEB 3	WEB 4	WEB 5	WEB 6	WEB 7	WEB 9	WEB 10	WEB 13	WEB 14	WEB 15	WEB 16	WEB 17
Ajacicyathus sp.																1		1							
Gordonicyathus sp.		1	1												?										
Asterocyathidae				1																					
Erismacoscinus cf. *endutus*					1	1																			
?*Graphoscyphia* sp.							1			?															
?*Dictyocyathus quadruplex* Bedf.											1														
Paranacyathus cf. *parvus* Bedf.																							1	1	
Archaeocyathus sp.	2	?	2	1				1	1			1	1	1	1	3	2	1	1	1	1			1	1
Chancelloria			1											1	?							1			
Spicules				1																		1			

sector. The inner wall is relatively thick (0.2 mm) with stirrup pores and additional rows of pores, the diameter of which is 0.2 mm and the lintels 0.3 mm. The intervallum coefficient is 0.18, parietal coefficient 3.6, and ratio of interseptal loculi 1:2.

Discussion. This form differs from *A. ajax* (Taylor) by the spacing of the septa—less numerous, with a larger loculus—and by the relative thickness of the inner wall. This last characteristic suggests to us that the considered form could also be a young stage of *Diplocyathellus* Debrenne, in which the canals are not yet developed.

Family BRONCHOCYATHIDAE Bedford and Bedford 1936

Genus Gordonicyathus Zhuravleva 1959
Type Species: *Thalmocyathus gerassimovensis* Krasnopeeva 1955

Diagnosis. Outer wall with simple pores, V-shaped annuli at the inner wall, septa porous.

***Gordonicyathus* sp.**
Plate 1, Fig. 2

Material. 2 specimens; CRAD 2a, 2b, ?WEB4.

Description. Fragments of small cylindrical cups as much as 6 mm in diameter, with an intervallum of 1:13. Septa are 0.20 mm apart, 33 in total number. The outer wall has three rows of pores per intersept, horizontally elongated (diameter 0.05 × 0.03 mm, lintels 0.01 mm, thickness 0.02 mm). It is not possible to determine the exact shape of the inner wall annuli (oblique section suggests that they are probably V-shaped); the thickness of the annulus wall is 0.03 mm; they are vertically 0.09 mm apart. Septa are porous, with pores 0.07 mm in diameter, separated by lintels of 0.07 mm width and 0.05 mm thickness. The intervallum coefficient is 0.18, the parietal coefficient 5.5. The ratio of interseptal loculi is 1:5.6.

Discussion. Due to the scarcity of specimens it is not possible to include the cups studied here in a described species. It is close to some thalamocyathids figured by Hill (1965, Pl. VII, Fig. 2d), but the septa are more porous in this form. The material is too sparse to be certain about the lack of pectinate tabulae.

Family ASTEROCYATHIDAE Vologdin 1956

Genus *Erismacoscinus* Debrenne 1959
Type species *Erismacoscinus marocanus* Debrenne 1959

Diagnosis. Outer and inner wall simply porous, independent from the tabulae. Septa porous.

The porous tabulae are flat and distributed at remote intervals; they differ in this respect from those of *Coscinocyathus* Bornemann, which has arched tabulae participating in the construction of one or two walls.

Geographic and stratigraphic distribution. Cosmopolitan, from mid-Tommotian to Botomian.

***Erismacoscinus* cf. *endutus* Gordon 1920**
Plate 1, Fig. 3

1920. *Coscinocyathus endutus* Gordon. p. 702–704; Plate 3, Fig. 33; Pl. 5, Fig. 51–59.
1937. *Coscinocyathus endutus* Gordon. Ting, p. 363; Plate 10, Fig. 3.
1965. *Coscinocyathus* ?*endutus* Gordon. Hill, p. 100–101; Plate 7, Fig. 9–26.

Material. 2 specimens; CRAD4, 5.

Description. Oblique sections of tabulate form with porous elements; cup surrounded by exothecal laminate structure. Endostructure and stereoplasma are occasionally present within the central cavity and the intervallum. Tabulae are scarce (one is seen in an oblique section) and are thinner than septa. The diameter is as much as 9 mm, with an intervallum of 1.33 mm; 26 septa are present, 0.72 mm apart. Three pores per intersept at the outer wall (diameter 0.11 mm, lintels 0.11 mm, thickness 0.11 mm) screened by the laminate outgrowths in the section observed. Two or three pores occur at the inner wall, sometimes modified from simple to tube by the secondary development of the endotheca (diameter 0.10 mm, lintels 0.20 mm, thickness 0.11 mm, without the endothecal tissue). Septa are coarse (5 pores in the intervallum diameter 0.11 mm, lintels 0.20 mm, thickness 0.20 mm). Tabula are thin (diameter 0.10 mm, lintels 0.10 mm, thickness 0.10 mm).

The intervallar coefficient is 0.15; the parietal coefficient is close to 3.

Discussion. The size, the spacing of septa and tabulae (scarce and thin), and the development of exo- and endothecal tissue is comparable to *Erismacoscinus endutus* (Gordon). It differs from the type in a smaller parietal coefficient (3 instead of 3.6) and a larger central cavity (intervallum coefficient 0.15 instead of 0.25).
Geographic and stratigraphic distribution. Antarctica: Weddell Sea area, Whichaway Nunataks; Whiteout Conglomerate, Ellsworth Mountains. Australia: Ajax Mine. Botomian equivalent.

Class IRREGULARES Vologdin 1937
Family GRAPHOSCYPHIIDAE Debrenne 1974 (=1974a)

Genus *Graphoscyphia* Debrenne 1974
Type species: *Protopharetra graphica* (Bedford and Bedford) 1934 by original designation of Debrenne 1974a.

Diagnosis. Outer wall with simple pores, inner wall with one row of simple pores per intersept; pseudo-septa straight, perforated by large pores and connected by synapticulae at each interpore node.
Geographic and stratigraphic distribution. Australia, Antarctica. Botomian equivalent.

?*Graphoscyphia* sp.
Plate 1, Fig. 4

Material. 2 specimens; CRAD7, ?CRAD13.
Description. Oblique section of Irregulares; diameter as much as 5.15 mm, with intervallum 1.03 mm wide. Outer wall is poorly preserved. Inner wall with one row of pores per intersept (diameter 0.11 mm, lintels 0.15 mm, thickness 0.11 mm). Septa are straight and largely porous. When compared with the normal type of *Graphoscyphia,* the present form has less regularly disposed synapticulae.
Discussion. The imprecision on the structure of the outer wall and the irregularly set synapticulae prevent a more precise generic and specific attribution.

Family PARANACYATHIDAE Debrenne 1970

Genus *Paranacyathus* Bedford and Bedford 1937
Type species: *Paracyathus parvus* (Bedford and Bedford) 1936

Diagnosis. Outer wall and inner wall with one to two rows of coarse pores per intersept, irregular in shape and size. Intervallum with radial wavy septa having widely separated irregular pores.
Geographic and stratigraphic distribution. Australia, Antarctica, Siberia, Mongolia. Atdabanian and Botomian equivalent.

***Paranacyathus* cf. *parvus* (Bedford and Bedford) 1936**
Plate 1, Fig. 5

1936. *Paracyathus parvus* Bedford R. and W. R., p. 17, Plate 17, Fig. 76.
1937. *Paranacyathus parvus* (Bedford R. and W. R.), Bedford R. and J., p. 34; Plate 35, Fig. 137 A–G.
1937. *Paranacyathus magnipora* Bedford R. and J. (pars), Plate 37, Fig. 142 A–B.
1974b. *Paranacyathus parvus* (Bedford R. and W. R.), Debrenne, p. 170.

Material. 2 specimens; WEB15, 16.
Description. Small cup as much as 3.9 mm in diameter. One or two rows of irregular pores at the outer and the inner wall. Intervallum 1.23 mm wide, with 14 wavy septa, nearly aporose. Intervallum coefficient is 0.31. Parietal coefficient 3.5.
Discussion. Differs from the type material by its low parietal coefficient, may be due to small size of the cup.

Family ARCHAEOCYATHIDAE Hinde 1889

Genus *Archaeocyathus* Billings 1861
Type species: *Archaeocyathus atlanticus* Billings 1861 subsequent designation by Walcott 1886.

Diagnosis. Conical cup displaying a few smooth transverse constrictions. Outer wall with many fairly regular small pores; inner wall with one pore per intertaenial space; intervallum made of pseudosepta (or taeniae) undulating and sometimes bifurcating. Few synapticulae and dissepiments. Most of the primary elements of the skeleton, and notably the inner wall and the inner part of the intervallum adjacent to the central cavity, are surrounded by secondary laminations that can be three or four times thicker than the initial skeleton. Outgrowths, although not as important as in *Metaldetes,* are present mainly as buds of limited size. (After Debrenne, in Debrenne and James, 1981).
Discussion. The development of successive layers considerably changes the aspect of the skeletal structures. Preservation, which emphasizes the enlargement of the skeleton, must also be taken into account here.

***Archaeocyathus* sp.**
Plate 1, Figs. 7–8

Material. Twenty-three specimens; CRAD1, 2a, 2b, 3, 10, 12, WEB1, 2, 3, 4, 5, 6, 7, 9, 10, 13, 16, 17.
Description. Cup with smooth transverse constriction as much as 14.3 mm in diameter with a corresponding central cavity of 8.4 mm and an intervallum of 2.8 mm. The average size of the cups in the collection is around 10 mm diameter, with an intervallum coefficient of 0.25. Small cups (diameter 3 to 4 mm) have a larger intervallum coefficient up to 0.35 showing a narrow central cavity at the beginning of development. Outer wall is not often seen. When present it consists of small pores covering the aperture of the intervallum taenia structures (diameter 0.2 mm, lintels 0.1 mm, thickness 0.1 mm). Taeniae are thin, when not artificially enlarged by the process of dissolution-recrystallization (Fig. 7), porous, normally wavy, 0.60 mm apart. Inner wall has one pore-tube per intersept; the tube is not long in this form (0.5 mm for an aperture of 0.4 mm).
Discussion. Besides the *Archaeocyathus* forms one can observe "*Bicyathus*"-like budding or "*Archaeopharetra*" stages on small forms. Affinities of the species are difficult to establish because of the preservation. The skeletal tissue is lighter than in the species described by Gordon as "*Spirocyathus atlanticus*" from the material of the Weddell Sea area, which I consider instead as an *A. kuzmini* (Vologdin, 1932). The genus is poorly represented in Australia and Antarctica, according to the literature, but is quite abundant in the collection studied here.

INCERTAE SEDIS GENERA

?*Dictyocyathus quadruplex* Bedford and Bedford 1936
Plate 1, Fig. 6

1936. ?*Dictyocyathus quadruplex* Bedford R. and W. R., p. 13; Plate 12, Fig. 59–60.
1974. ?*Dictyocyathus quadruplex* Bedford R. and W. R., Debrenne p. 198, Fig. 6.

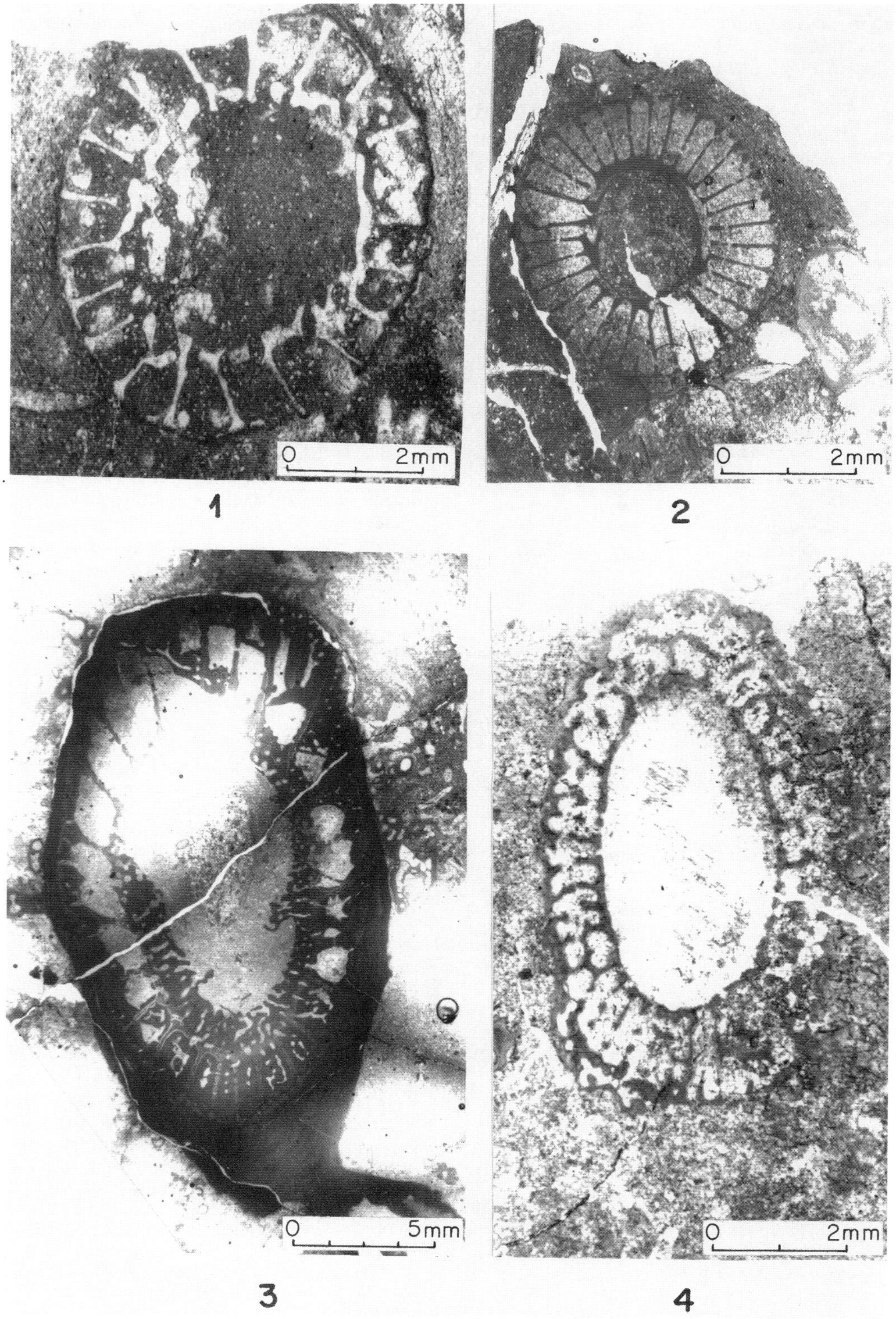

PLATE 1

Figure 1. *Ajacicyathus* sp., USNM 334186, WEB5, ×10.
Figure 2. *Gordonicyathus* sp., USNM 334187, CRAD2b, ×10.
Figure 3. *Erismacoscinus* cf. *endutus* (Gordon), USNM 334188, CRAD4, ×5.
Figure 4. ?*Graphoscyphia* sp., USNM 334189, CRAD7, ×10.

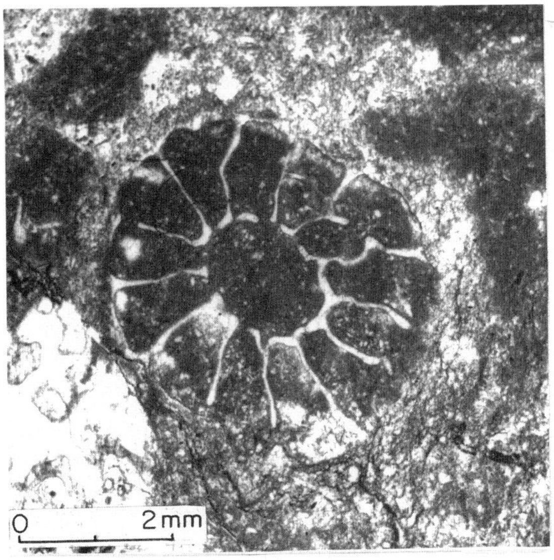

5

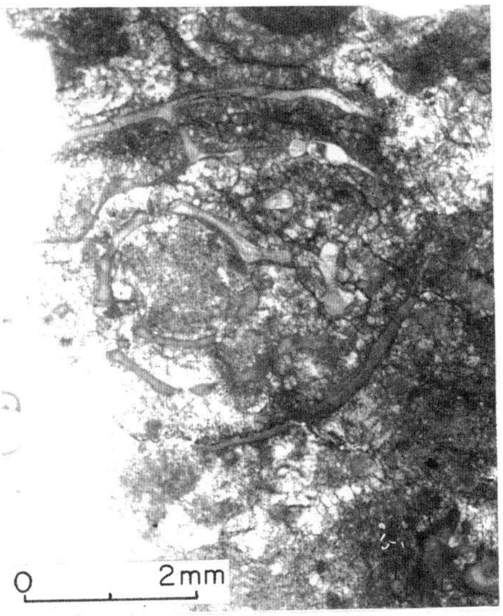

6

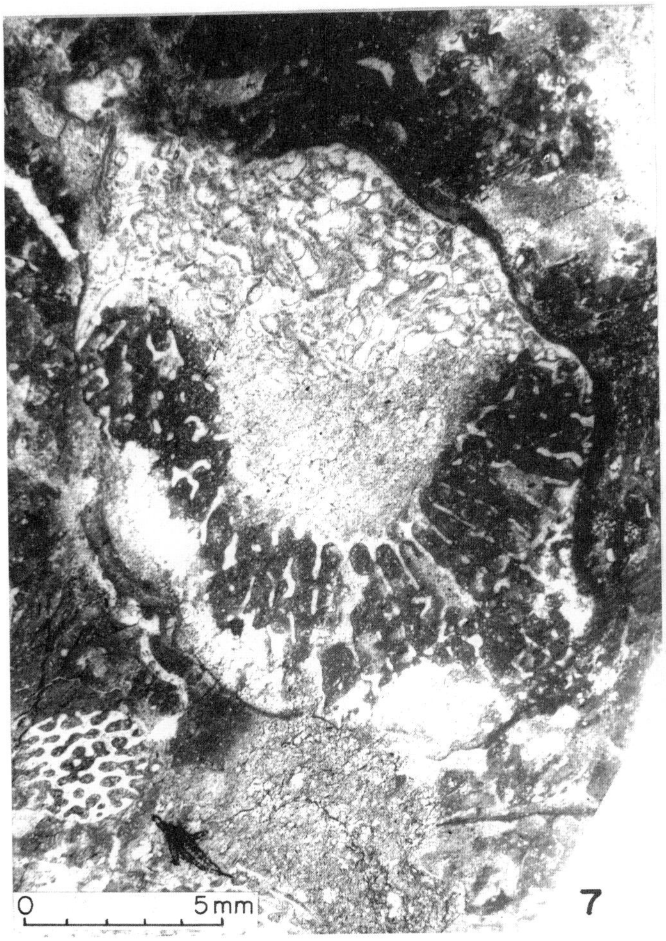

7

8

Figure 5. *Paranacyathus* cf. *parvus* (Bedford R., and W. R.), USNM 334190, WEB15, ×10.
Figure 6. ?*Dictyocyathus quadruplex* Bedford R., and W. R., USNM 334191, CRAD14, ×10.
Figure 7. *Archaeocyathus* sp., USNM 334192, WEB7, × 5.
Figure 8. *Archaeocyathus* sp., USNM 334193, WEB7, ×5.

Material. One specimen; CRAD14.
Description. Nonporous outer wall, intervallum with a loose scaffolding of diversely orientated bars, inner wall coarsely porous. Diameter 4.12 mm, intervallum 1.25 mm.
Discussion. As for all the small cups of Irregulares with few or no characteristic features, it is very difficult to decide which genus the observed form belongs to. By the presence of the rods and nonporous wall, it is close to the form called *Dictyocyathus quadruplex* Bedford R. and W. R. but is probably not a *Dictyocyathus* Bornemann 1891; possibly it is a *Chouberticyathus* Debrenne 1964 or even only a young stage of one of the genera of the order Archaeocyathida.

REFERENCES CITED

Bedford, R., and Bedford, J., 1937, Further notes on Archaeos (Pleospongia) from the Lower Cambrian of South Australia: Kyancutta Museum Memoir 4, p. 27–38.

——, 1939, Development and classification of Archaeous (Pleospongia): Kyancutta Museum Memoir 6, p. 67–82.

Bedford, R., and Bedford, W. R., 1936, Further notes on Archaeocyathi (Cyathospongia) and other organisms from the Lower Cambrian of Beltana, South Australia: Kyancutta Museum Memoir 2, p. 9–20.

Billings, E., 1861, New species of Lower Silurian fossils; On some new or little-known species of Lower Silurian fossils from the Potsdam Group (Primordial Zone): Montreal, Geological Survey of Canada, 24 p.

Debrenne, F., 1959, Un nouveau genre d'Archaeocyatha du Cambrien marocain: Soc. Géol. France, Comptes Rendus somm. Séanc., n. 1, p. 14–15.

——, 1970, A revision of Australian genera of Archaeocyatha: Transactions of the Royal Society of South Australia, v. 94, p. 21–41.

——, 1974a, Les Archeocyathes irreguliers d'Ajax Mine (Cambrien inferieur, Australie du Sud): Bulletin of the Museum of Natural History, Nature, Science de la Terre, 3rd series, v. 33, no. 195, p. 185–258.

——, 1974b, K revizii roda *Paranacyathus* Bedford R., and W. R., 1937, *in* Biostratigrafija i paleontolojija nizhnego kembrija Evropy i Severnoj: Azii. Moskva Izdat. Nauka, p. 167–178.

Debrenne, F., and James, N. P., 1981, Reef-associated archaeocyathans from the Lower Cambrian of Labrador and Newfoundland: Palaeontology, v. 24, part 2, p. 343–378.

Gordon, W. T., 1920, Cambrian organic remains from a dredging in the Weddell Sea; Scottish National Antarctic Expedition 1902–04: Transactions of the Royal Society of Edinburgh, v. 52, part 4, no. 27, p. 681–714.

Hill, D., 1965, Archaeocyatha from Antarctica and a review of the phylum: Scientific Report Transantarctic Expedition 10, Geol. 3, 151 p.

Hinde, G. T., 1889, On *Archaeocyathus* Billings and other genera allied to or associated with it, from the Cambrian strata of North America, Spain, Sardinia, and Scotland: Quarterly Journal of the Geological Society of London, v. 45, p. 125–148.

Krasnopeeva, P. S., 1955, Tip Archaeocyathi, Arkheotsiati, *in* L. L. Khalfina, ed., Atlas rukovodyashchikh form iskopaemykh fauny i flory Zapadnoy Sibiri, v. 1, p. 74–102, Gosgeoltekhizdat (Moskva).

Taylor, T. G., 1910, The Archaeocyathinae from the Cambrian of South Australia, with an account of the morphology and affinities of the whole class: Adelaide, Royal Society of South Australia Memoir 2, part 2, p. 55–188.

Ting, T. H., 1937, Revision der Archaeocyathinen: Neues Jahrb. Geologie, Mineralogie, Paläontologie, v. 78, abt. B., p. 327–379.

Vologdin, A. G., 1932, The Archaeocyathinae of Siberia; Part 2, Fossils of the Cambrian limestones of Altai Mountains: United Geological Survey, U.S.S.R., p. 1–106.

——, 1937, Archaeocyatha and the results of their study in U.S.S.R.: Moscow University Lab. Pal, Problems Paleont., v. 2–3, p. 453–500.

——, 1956, K klassifikatsii tipa Archeocyatha: Akad. Nauk SSSR, Doklady, v. 111, p. 877–880.

Walcott, C. D., 1886, Second contribution to the studies of the Cambrian faunas of North America: U.S. Geological Survey, Bulletin 30, 369 p.

Zhuravleva, I. T., 1959, Archaeocyatha of the Bazaikh horizon of the Kiya River: Doklady Akadamii Nauk S.S.S.R., tome 124, no. 2, p. 424–427.

MANUSCRIPT ACCEPTED BY THE SOCIETY MARCH 1, 1989

Geological Society of America
Memoir 170
1992

Chapter 14

Permian plants from the Ellsworth Mountains, West Antarctica

Thomas N. Taylor
Department of Botany, Ohio State University, 1735 Neil Ave., Columbus, Ohio 43210
Edith L. Taylor
Byrd Polar Research Center, Ohio State University, 125 S. Oval Mall, Columbus, Ohio 43210

ABSTRACT

Plant megafossils are described and illustrated from several localities and stratigraphic levels within the Ellsworth Mountains. The flora is dominated by various species of *Glossopteris* foliage and also includes sphenophyte stems and leaf sheaths, glossopterid reproductive organs, *Gangamopteris* leaves, and *Vertebraria*-type axes. The plants are preserved as impressions in fine-grained shale within the Polarstar Formation. Based on comparisons with other Antarctic and Gondwana floras, a Middle to Late Permian age is suggested.

INTRODUCTION

The first comprehensive paleobotanical studies on Antarctic plant fossils were completed in the early part of this century (Halle, 1913; Seward, 1914), although fragments of fossil plants were reported from some of the early exploratory expeditions to Antarctica (e.g., Scott [1901–1904; 1912]; Shackleton [1908]; Mawson [1911–1914]). Plant fossils are known from nearly every geologic period beginning in the Devonian and extending into the Tertiary. The most commonly known floras occur in Permian strata and are characterized by abundant glossopterid leaves that are preserved most often as impressions. Since the International Geophysical Year, 1957–1958, many areas of the continent have become more accessible to researchers, with the result that more detailed and systematically oriented studies of fossil plants have been possible.

Craddock and others (1965) reported the presence of possible plant fossils collected during the 1961–1962 field season from the Polarstar Formation in the Ellsworth Mountains; however, the quality of preservation was poor. In the 1962–1963 season, carbonaceous remains were discovered that conclusively demonstrated the presence of plant fossils in these sediments. Finally, after collections made in 1963–1964, these authors were able to document the occurrence of numerous *Glossopteris* leaves from a number of levels at a site near Polarstar Peak. The plant fossils and coal samples collected on these expeditions were forwarded to the late James M. Schopf for analysis (Schopf and Long, 1966). In addition to the original discovery of the fossil-bearing strata by Craddock and others (1965), fossil plants have been reported from the Polarstar Formation in the contributions of Schopf (1967), Rigby and Schopf (1969), and Rigby (1969). With the exception of the systematic work by Rigby (1969) on Antarctic sphenophytes, some of which were collected from the Polarstar Formation, the remaining contributions are largely biostratigraphic in scope.

The present report is an analysis of the original plant fossils collected by Craddock and others (1965) as well as of specimens collected during subsequent field seasons in the Ellsworth Mountains (Webers, 1981). One of the primary aims of this study is to describe and illustrate the floral components that appear at different stratigraphic levels within the section and to compare the plants from the Ellsworth Mountains with other Gondwana floras of similar age.

MATERIALS AND METHODS

Most of the specimens consist of impressions preserved in a fine-grained argillite. Although many of the leaves are abundant and often occur in dense mats, preservation is highly variable, with some lacking microscopic detail. Because of the lack of contrast between the specimens and the matrix, photography was most successful using incident light and submersion in xylene.

Taylor, T. N., and Taylor, E. L., 1992, Permian plants from the Ellsworth Mountains, West Antarctica, *in* Webers, G. F., Craddock, C., and Splettstoesser, J. F., Geology and Paleontology of the Ellsworth Mountains, West Antarctica: Boulder, Colorado, Geological Society of America Memoir 170.

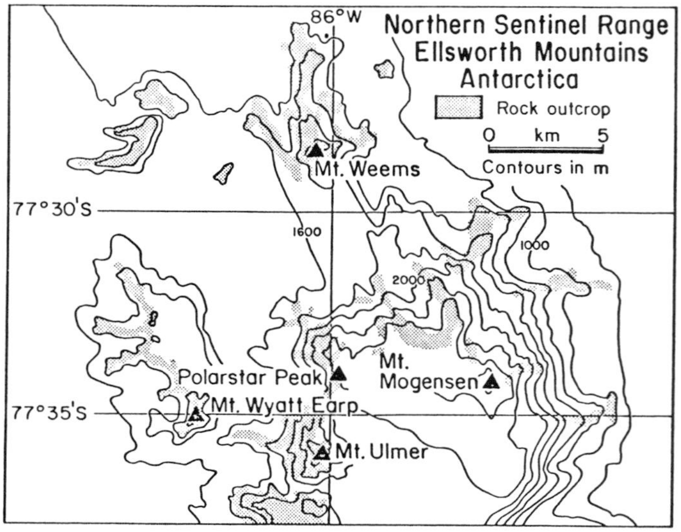

Figure 1. Locality map of the northern Sentinel Range, Ellsworth Mountains, West Antarctica. Contour interval, 200 m. (Modified from Craddock and others, 1965).

The specimens collected by Craddock and others (1965) are deposited in the Antarctic Collections of the Orton Museum of Geology at The Ohio State University under specimen numbers E-1 through E-499. Other specimens in the above collections from Polarstar Peak are designated by the prefix ANT-67-3. Additional plant specimens from the Ellsworth Mountains collected by Gerald F. Webers are housed in the Paleontological Collections, Department of Geology, Macalester College.

LOCALITIES

The plant fossils were collected from four principal localities, all within the Polarstar Formation (Fig. 1). Specimens E-1 through E-21 were discovered approximately 1.2 km north of the summit of Mt. Weems (about 77°25′S,86°0′W—about 450 m above the base of the Polarstar Formation); specimen E-22, 0.3 km west of the summit of Mt. Weems in the lower part of a 30-m-thick siltstone unit, 521 m above the base of the measured section (according to field notes compiled by James M. Schopf, the measured section is at least 160 m above the contact of the Polarstar Formation and the Whiteout Conglomerate); specimen E-23 from the summit of Mt. Weems in the uppermost 70-m-thick shale unit of the measured section (580 m above the base of the measured section). The specimens from these three localities are too poorly preserved for accurate identifications to be made. The remaining specimens were recovered from a locality on the east ridge of Polarstar Peak (77°32′S,86°10′W) that includes six stratigraphic levels designated 64-TB-68-1, -2, -3, -4, -5, -6. These levels correspond to specimens E-24 through E-88–E-499, respectively, on Figure 2. Locality No. 64-TB-68-1 (Specimen E-24) consists of poorly bedded, dark gray to black argillite, approximately 15 m thick. *Glossopteris* occurs in the upper 5 m. Locality No. 64-TB-68-2 (Specimens E-25-65) consists of an argillite unit about 18 m thick that is dark gray to black and characterized by indistinct bedding. *Paracalamites* and *Glossopteris* are associated with coal seams in the upper 3 m of the unit. Locality No. 64-TB-68-3 (Specimens E-66–72) is a dark gray, thinly bedded argillite approximately 25 m thick. Numerous *Glossopteris* leaves and coal stringers occur in the upper 8 m of the unit. Locality No. 64-TB-68-4 (Specimens E-73-75) is a 15-m-thick argillaceous unit that contains coal seams up to 15 cm thick. *Glossopteris* foliage is common in the upper 30 cm. Locality No. 64-TB-68-5 (Specimens E-76–87) is a thinly bedded dark gray to black argillite (10 m thick) containing coal seams 2.5 cm thick and *Vertebraria* axes and *Glossopteris*. Locality No. 64-TB-68-6 (Specimens E-88–499) is a zone nearly 100 m thick that consists of three argillite units separated by 15-m-thick beds of quartzite. Abundant *Vertebraria*, sphenopsids, and *Glossopteris* remains were recovered from the lowest argillite unit, which is about 30 m thick. Figure 2 is an abbreviated stratigraphic section of the Polarstar Peak locality showing the vertical distribution of the plant fossils by specimen number.

DESCRIPTION

The descriptions that follow include examples of the major floral components arranged by taxonomic categories.

Pteridospermophyta

Glossopteridales. Leaf remains of various species of *Glossopteris* constitute the most common floral element in the collections from all stratigraphic levels. Scale leaves, reproductive organs, and *Vertebraria*-type axes are also present in addition to vegetative leaves. Also present in the collection are leaves referable to the form genus *Gangamopteris*.

Glossopteris Brongniart. The genus *Glossopteris* was instituted by Brongniart in 1828 for simple, sublanceolate leaves with a prominent midrib, entire margin, and net venation. The genus is considered to be a highly artificial one, and in recent years taxonomic treatments have had little success in defining the natural limits of the species. Perhaps the primary reason for the difficulty in classifying species centers on the lack of consistent and reliable characters that can be utilized irrespective of facies and preservational differences (Banerjee, 1978). Moreover, the number of useful taxonomic characters is diminished owing to the absence of well-preserved cuticles in the majority of Permian Gondwana floras. Further compounding the problem is the difficulty in defining generic limits of the closely related taxon *Gangamopteris*, due to a morphological continuum of features between these two genera. Although these genera are known primarily from the Permo-Triassic of the Southern Hemisphere, there have been a few reports of *Glossopteris*-like leaves from sedimentary rocks of Jurassic age (e.g., Delevoryas and Person, 1975). Below is a list of

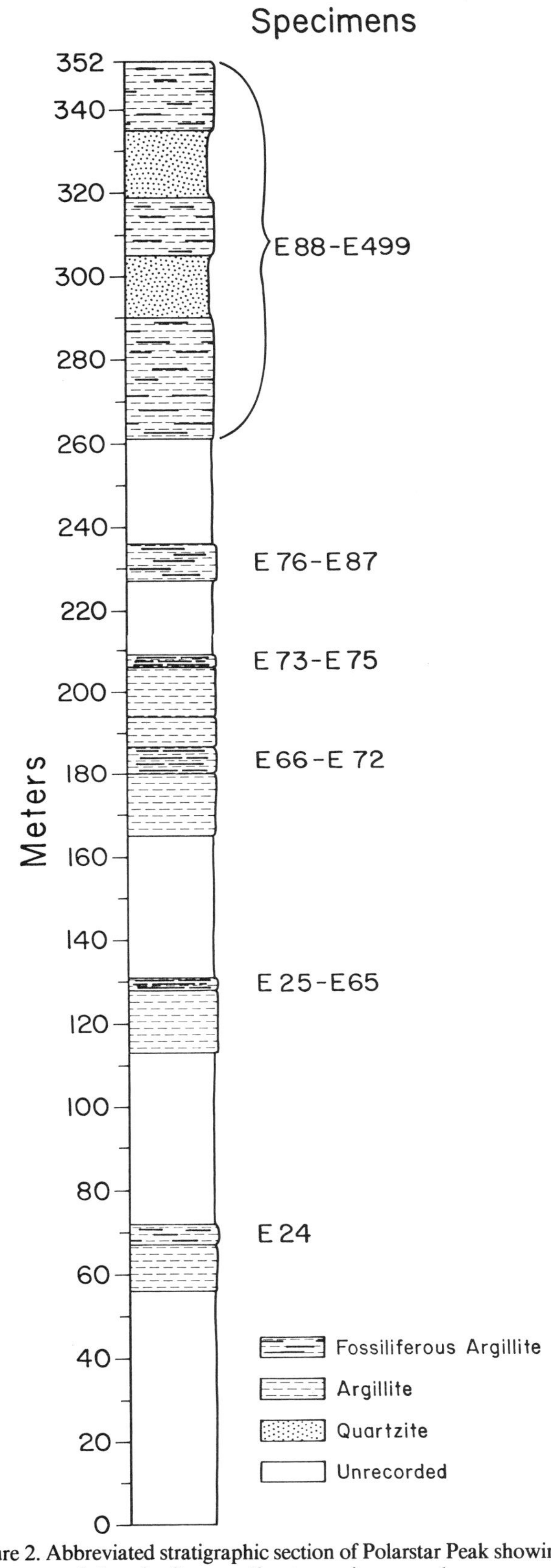

Figure 2. Abbreviated stratigraphic section of Polarstar Peak showing the occurrence of plant fossils according to specimen numbers.

the species of *Glossopteris* present in the Polarstar Formation with brief descriptions of the salient features of each taxon.

G. angustifolia Brongniart (Plate 1, Fig. 3). Perhaps the best characters for this species are the narrow, lanceolate shape of the leaf with obtuse apex and gradually tapered base. Chandra and Surange (1979) suggest that the length/width ratio of *G. angustifolia* is 8:1. However, the specimens described by Rigby and others (1980) are somewhat broader. The midrib in this species is distinct, and the lateral veins are produced at an acute angle, with the individual meshes elongate and narrow with few anastomoses.

G. browniana Brongniart (Plate 1, Fig. 4, 8; Plate 2, Fig. 7, 8). This species is known principally from Australia and may be delimited by the vein meshes that are broader and polygonal near the midrib (about five times as long as broad) and narrower near the margins. The midrib is described as being somewhat flattened but is generally distinct and strong (2.0 mm wide) and often ends just below the apex. The base of the leaf is constricted, whereas the apex varies from rounded to obtuse. The specimen illustrated in Figure 8 (Plate 1) is 3.5 cm wide.

G. communis Feistmantel (Plate 1, Fig. 5; Plate 2, Fig. 2, 5, 7). This leaf type is characterized by a narrow elliptical shape (8.5 cm long and 1.6 cm wide) with a wide midrib, acute base, and few or no cross connections between the veins, resulting in elongate-polygonal meshes. Rigby and others (1980) note that this species is often confused with *G. indica* but can be separated on the basis of cuticular structure. To date, however, no cuticles have been recovered from the Ellsworth flora.

G. damudica Feistmantel (Plate 2, Fig. 6). *G. damudica* is a large, broadly elliptical leaf type with a rounded apex that in some accounts has been described as containing a central notch (Chandra and Surange, 1979). Veins diverge from the midrib at a 90° angle, and the margin of the leaf is generally undulating.

G. indica Schimper (Plate 1, Fig. 2). This species typically includes those leaf forms that are lanceolate, up to 4.8 cm wide, and characterized by an acute apex and base. Lateral veins form meshes that are short and nearly isodiametric near the midrib but narrow and elongate near the margin.

G. linearis McCoy (Plate 2, Fig. 3). This type is characterized as narrow (1.0 cm) and linear-spathulate with parallel sides. The midrib is distinct, and the apex is rounded to slightly acute. The base is tapered.

G. cf. *cordata* Dana (Plate 1, Fig. 7). This species includes broad, elliptical leaves with a broadly rounded apex and entire margins. The base of the leaf is cordate with lobes overlapping the petiole. The midrib is wide at the base, and meshes formed by secondary veins are four times as long as wide (Rigby, 1964).

G. cf. *tortuosa* Zeiller (Plate 1, Fig. 1). This leaf type is small (4.0 cm), narrow, and elliptical with an obtuse apex. The midrib is relatively stout with lateral veins arising at an angle of approximately 45°. Meshes are broader near the midrib and narrower near the margin.

G. taeniopteroides Feistmantel (Plate 2, Fig. 5). This species is characterized by oblong spathulate leaves with a prominent striate midrib. Veins are given off at nearly right angles from the midrib with vein meshes that are elongate. Secondary veins branch infrequently and appear parallel, similar to the venation pattern of *Taeniopteris.*

Plumsteadia Rigby (Plate 1, Fig. 5). This genus was established for ovulate reproductive organs that Plumstead (1958) suggested belonged to *Glossopteris stricta* (Rigby, 1968). The organ consists of a laminar unit (0.8 cm) covered with polygonal areas, each with a circular scar in the center that represents the position of an ovule. Surange and Maheshwari (1970) describe the polygonal areas as arranged in a spiral pattern in their specimens from India.

Scale leaf (Plate 2, Fig. 1). Although scale leaves are a relatively common component of most *Glossopteris* floras, their exact affinities continue to remain obscure (Rigby, 1972). Lacey and others (1975) consider the scale leaves in their material from South Africa to represent vegetative bud-scales and sterile scales from strobili. The scale leaf illustrated from the Ellsworth Mountains is rhomboidal (2.6 cm long) in outline and possesses a wide base (0.5 cm). Scale leaves with a similar morphology have been compared with the scales of *Eretmonia* fructifications (Lacey and others, 1975; Chandra and Surange, 1979).

In addition to the taxa noted above there are several poorly preserved specimens of additional *Glossopteris* leaves, including *G. ampla, G.* cf. *decipiens, G. retifera, G.* cf. *spathulato-cordata, G. stricta,* and *G.* cf. *conspicua.* The genus *Gangamopteris* is represented by *G.* cf. *angustifolia* and *G. obovata.* Other pteridosperm taxa include *Vertebraria australis* (= *V. indica;* see Schopf, 1982), *Arberiella*-type pollen sacs, and seeds of the *Samaropsis* type.

Sphenophyta

Equisetales. *Paracalamites* cf. *australis* Rigby (Plate 1, Fig. 6). We have followed the recommendation of Rigby (1969) in which *Paracalamites* is used for articulate stems that represent both pith casts and external stems. The genus is characterized by ribs that do not alternate from node to node except to accommodate changes in the number of ribs. The internodes are extended, and delicate striations are common on the ribs and in the furrows. Although *Paracalamites* is used for isolated sphenopsid stems, Rigby (1969) suggests that the stems may be associated with any of the following Gondwana genera: *Phyllotheca, Schizoneura, Stellotheca,* and *Raniganjia.* Of these only *Raniganjia* is known currently from the Ellsworth Mountains.

Raniganjia bengalensis (Feistmantel) Rigby (Plate 2, Fig. 4). This taxon was initially described as a fern by Feistmantel (1876) under the binomial *Actinopteris bengalensis* for disc-like fossils from the Raniganj coalfields of India. Rigby (1962) proposed the genus *Raniganjia* for these fossils based on their presumed sphenopsid affinities. Pant and Nautiyal (1967) later emended the genus, including details about the cuticular and vegetative anatomy and descriptions of leaf whorls attached to slender stems.

The Ellsworth specimen of *Raniganjia* includes as many as 31 leaves fused into a flattened disc approximately 13 cm in diameter. The margin is crenulate, and the point of attachment to the axis is narrow. Based on the small diameter of the stem, Rigby (1962) suggested that the plant was small and perhaps had a scrambling type of habit.

Additional sphenophyte remains that are too poorly preserved to identify include nodal diaphragms and fragments of longitudinally ribbed stems.

DISCUSSION

The flora from the Ellsworth Mountains sites compares favorably with other Antarctic floras of similar age both in preservation and floral composition. Because the sedimentary rocks are highly metamorphosed, all of the leaves are preserved as impressions. However, in a few instances, as a result of the fine-grained nature of the rocks, it is possible to resolve epidermal features such as the outlines of cells. Despite the fact that the leaves occur in parallel layers, it is rarely possible to uncover an entire specimen due to ill-defined bedding planes. Many of the leaves appear to be intact, and their abundance might suggest that transport was limited. In addition, recent evidence concerning the histology of *Glossopteris* leaves (Gould and Delevoryas, 1977; Pigg and Taylor, 1987) indicates that they contain a large number of fibrous bands associated with the vascular bundles. This type of anatomy may also have contributed to the relatively complete condition of the leaves, irrespective of the extent of transport prior to deposition.

Rigby and Schopf (1969) previously reviewed the composition of Permian floras from 10 localities in Antarctica. They found that the Antarctic floras included a number of elements in common with those of similar age from India, Australia, and Argentina (Archangelsky, 1957; Chaloner and Lacey, 1973; Plumstead, 1973; Surange, 1975; Rigby and Shah, 1980; Chandra and Surange, 1979). These include sphenophytes such as *Paracalamites* and numerous glossopterid elements, including various species of *Glossopteris* and *Gangamopteris* and axes such as *Vertebraria australis.*

The flora from the Ellsworth Mountains appears similar to floras described from localities elsewhere in Antarctica, including those from the Ohio Range of the Horlick Mountains (Schopf, 1962; Cridland, 1963), Theron Mountains (Plumstead, 1962; Lacey and Lucas, 1981), Whichaway Nunataks (Plumstead, 1962), southern Victoria Land (Seward, 1914; Plumstead, 1962; Townrow, 1967), and the Falkland Islands (Seward and Walton, 1923). It is important to underscore that many of the floras are known from a small number of species and/or collection sites, and since the plants were originally named, a number of species designations have been changed. There appear to be no unique floral elements in the collections described to date from the Ellsworth Mountains, with the possible exception of *Raniganjia*. This taxon has not been identified to date from any other localities in Antarctica as far as we are aware but is known from India and Australia (Pant and Nautiyal, 1967; Rigby, 1969). The stratigraphic and geographic ranges of many of the Ellsworth taxa are extensive and therefore do not allow for precise dating of the flora based solely on biostratigraphy. Many of the elements are known from all of the southern continents and India. Nevertheless, the flora appears most similar to other Middle and Upper Permian floras that have been described from elsewhere in Gondwanaland.

ACKNOWLEDGMENTS

This research was completed under a grant from the Division of Polar Programs of the National Science Foundation (DPP-8213749). We wish to acknowledge John Rigby, Geological Survey of Queensland, for helpful discussion regarding certain species of *Glossopteris;* Gerald F. Webers, Geology Department, Macalester College, for providing access to his collection of plant fossils from the Ellsworth Mountains; and Stig Bergstrom, Department of Geology and Mineralogy, and David H. Elliot, Byrd Polar Research Center, The Ohio State University, for making the Antarctic specimens available to us to study. Many of the specimens used in this study were initially collected by Thomas W. Bastien and Robert H. Rutford.

PLATE 1

Figure 1. *Glossopteris* cf. *tortuosa.* [E 190] Loc. 64-TB-68-6. × 2.
Figure 2. *Glossopteris indica* [E 91] Loc. 64-TB-68-6. × 1.
Figure 3. *Glossopteris angustifolia* [E 117] Loc. 64-TB-68-6. × 2.
Figure 4. *Glossopteris browniana.* Detail of venation. [E 339] Loc. 64-TB-68-6. × 2.
Figure 5. *Glossopteris communis* and *Plumsteadia* sp. (arrow). [E 124] Loc. 64-TB-68-6. × 1.5.
Figure 6. *Paracalamites* cf. *australis.* [E 54] Loc. 64-TB-68-2. × 1.
Figure 7. *Glossopteris cordata* [E 126] Loc. 64-TB-68-6. × 2.
Figure 8. *Glossopteris browniana* [E 339] Loc. 64-TB-68-6. × 2.

1
2
3
4
5
6
7
8

PLATE 2

Figure 1. Scale leaf. Arrow indicates prominent base. [E 67] Loc. 64-TB-68-3. × 2.
Figure 2. *Glossopteris communis.* [Ant. 67-3-10]. × 1.
Figure 3. *Glossopteris linearis.* [E 111] Loc. 64-TB-68-6. × 2.
Figure 4. *Raniganjia bengalensis.* [E 279] Loc. 64-TB-68-6. × 2.
Figure 5. *Glossopteris communis* and *G. taeniopteroides* (arrow). [E 390] Loc. 64-TB-68-6. × 1.5.
Figure 6. *Glossopteris damudica.* [E 260] Loc. 64-TB-68-6. × 1.5.
Figure 7. *Glossopteris browniana* and *G. communis* (arrow) [E 318] Loc. 64-TB-68-6. × 2.
Figure 8. *Glossopteris* cf. *browniana.* [E 339] Loc. 64-TB-68-6. × 2.

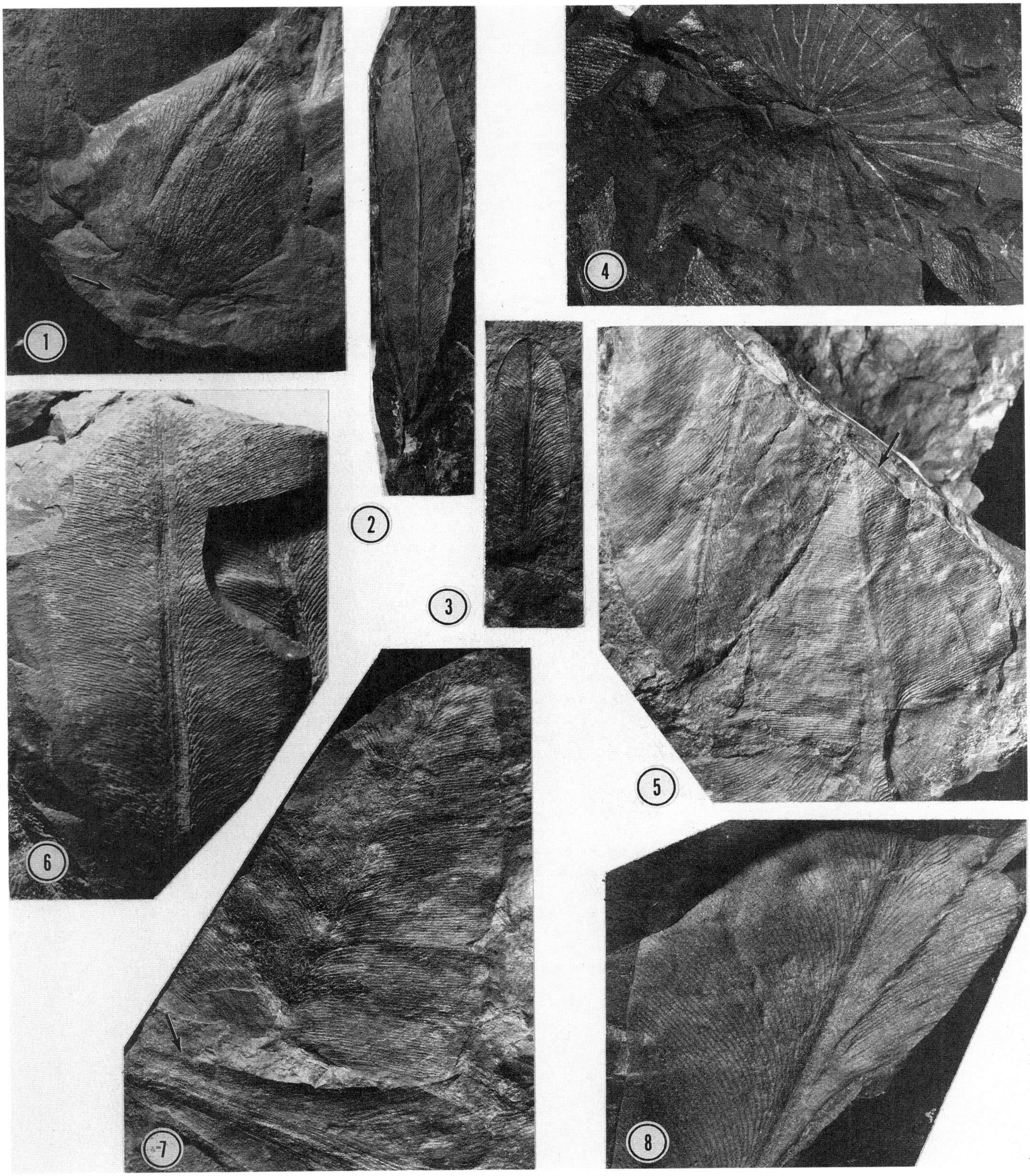
1
2
3
4
5
6
7
8

REFERENCES CITED

Archangelsky, S., 1957, Las Glossopterideas del bajo de la Leona: Revista de la Asociasión Geológica Argentina, v. 12, p. 135–164.

Banerjee, M., 1978, Genus *Glossopteris*Brongniart and its stratigraphic significance in the Palaeozoics of India; Part 1, A revisional study of some species of *Glossopteris:* Bull. Botan. soc. Bengal, v. 32, p. 81–125.

Brongniart, A., 1828, Prodrome d'une histoire des végétaux fossiles: Dictionnaire des Sciences Naturelles, v. 57, 223 p.

Chaloner, W. G., and Lacey, W. S., 1973, The distribution of Late Palaeozoic floras, *in* Hughes, N. F., ed., Organisms and continents through time: London, Palaeontological Association, Special Papers in Palaeontology 12, p. 271–290.

Chandra, S., and Surange, K. R., 1979, Revision of the Indian species of *Glossopteris:* Lucknow, Birbal Sahni Institute of Palaeobotany, Monograph 2, 291 p.

Craddock, C., Bastien, T. W., Rutford, R. H., and Anderson, J. J., 1965, *Glossopteris* discovered in West Antarctica: Science, v. 148, p. 634–637.

Cridland, A. A., 1963, A *Glossopteris* flora from the Ohio Range, Antarctica: American Journal of Botany, v. 50, p. 186–195.

Delevoryas, T., and Person, C. P., 1975, *Mexiglossa varia* gen. et sp. nov., a new genus of glossopteroid leaves from the Jurassic of Oaxaca, Mexico: Palaeontographica, v. 154B, p. 114–120.

Feistmantel, O., 1876, On some fossil plants from the Damuda Series in the Raniganj Coalfield collected by Mr. J. Wood-Mason: J. Asiat, Soc. Bengal, v. 45, p. 329–380.

Gould, R. E., and Delevoryas, T., 1977, The biology of *Glossopteris;* evidence from petrified seed-bearing and pollen-bearing organs: Alcheringa, v. 1, p. 387–399.

Halle, T. G., 1913, The Mesozoic flora of Graham Land: Wissenschaftliche Ergebnisse Schwedischer Südpolarexpeditionen, v. 3, no. 14, p. 3–124.

Lacey, W. S., and Lucas, R. C., 1981, A Lower Permian flora from the Theron Mountains, Coats Land: British Antarctic Survey Bulletin, v. 53, p. 153–156.

Lacey, W. S., van Dijk, D. E., and Gordon-Gray, K. D., 1975, Fossil plants from the Upper Permian in the Mooi River district of Natal, South Africa: Annals of the Natal Museum, v. 22, p. 349–420.

Pant, D. D., and Nautiyal, D. D., 1967, On the structure of *Raniganjia bengalensis* (Feistmantel) Rigby with a discussion of its affinities: Palaeontographica, v. 121B, p. 52–64.

Pigg, K. B., and Taylor, T. N., 1987, Anatomically preserved *Glossopteris* from Antarctica: VII Simposio Argentino Palaeobotanicy y Palinologia, Buenos Aires, Actas, p. 177–180.

Plumstead, E. P., 1958, Further fructifications of the Glossopteridae and a provisional classification based on them: Geological Society of South Africa Transactions, v. 61, p. 51–76.

——, 1962, Fossil floras of Antarctica: Trans-Antarctic Expedition 1955–1958, Scientific Reports No. 9, Geology, London, Trans-Antarctic Expedition Committee, p. 1–132.

——, 1973, The Late Palaeozoic *Glossopteris* flora, *in* Hallam, A., ed., Atlas of palaeobiogeography: Amsterdam, Elsevier, p. 187–205.

Rigby, J. F., 1962, On a collection of plants of Permian age from Baralaba, Queensland: Proceedings of the Linnean Society of New South Wales, v. 87, p. 341–351.

——, 1964, Contributions on Palaeozoic floras; 1, On the identification of *Glossopteris cordata* Dana: Proceedings of the Linnean Society of New South Wales, v. 89, p. 152–154.

——, 1968, The conservation of *Plumsteadia* Rigby, 1963, over *Cistella* Plumstead, 1958: Boletim da Sociedade Brasileira de Geologia, v. 17, p. 93.

——, 1969, Permian sphenopsids from Antarctica: U.S. Geological Survey Professional Paper 613-F, p. 1–12.

——, 1972, The flora of the Kaloola Member of the Baralaba Coal Measures, central Queensland: Geological Survey of Queensland Publication 352, p. 1–12.

Rigby, J. F., and Schopf, J. M., 1969, Stratigraphic implications of Antarctic paleobotanical studies, *in* Gondwana stratigraphy; 2, Earth sciences; International Union of Geological Sciences Symposium: Paris, United Nations Educational, Scientific, and Cultural Organization, p. 91–106.

Rigby, J. F., and Shah, S. C., 1980, The flora from the Permian nonmarine sequences of India and Australia; A comparison, *in* Cresswell, M. M., and Vella, P., eds., Gondwana five: 5th International Gondwana Symposium: Rotterdam, A. A. Balkema, p. 39–41.

Rigby, J. F., Maheshwari, H. K., and Schopf, J. M., 1980, Revision of Permian plants collected by J. D. Dana during 1839–1840 in Australia: Geological Survey of Queensland Publication 376, p. 1–25.

Schopf, J. M., 1962, A preliminary report on plant remains and coal of the sedimentary section in the central range of the Horlick Mountains, Antarctica: Columbus, Ohio State University Institute of Polar Studies Report 2, 61 p.

——, 1967, Antarctic fossil plant collecting during the 1966–1967 season: Antarctic Journal of the United States, v. 2, p. 114–116.

——, 1982, Forms and facies of *Vertebraria* in relation to Gondwana coal, *in* Turner, M. D., and Splettstoesser, J. F., eds., Geology of the central Transantarctic Mountains: American Geophysical Union Antarctic Research Series, v. 36, p. 37–62.

Schopf, J. M., and Long, W. E., 1966, Coal metamorphism and igneous associations in Antarctica, *in* Gould, R. F., ed., Coal sciences: American Chemical Society, p. 156–195.

Seward, A. C., 1914, Antarctic fossil plants; British Antarctic Terra Nova Expedition, 1910; Natural history report: British Museum of Natural History, Geology, v. 1, no. 1, p. 1–149.

Seward, A. C., and Walton, J., 1923, On a collection of fossil plants from the Falkland Islands: Quaterly Journal of the Geological Society of London, v. 79, p. 313–333.

Surange, K. R., 1975, Indian Lower Gondwana floras; A review, *in* Campbell, K.S.W., ed., Gondwana geology: Canberra, Australian National University Press, p. 135–147.

Surange, K. R., and Maheshwari, H. K., 1970, Some male and female fructifications of Glossopteridales from India: Palaeontographica, v. 192B, p. 178–192.

Townrow, J. A., 1967, Fossil plants from Allan and Carapace Nunataks, and from the Upper Mill and Shackleton Glaciers, Antarctica: New Zealand Journal of Geology and Geophysics, v. 10, p. 456–473.

Webers, G. F., 1981, Ellsworth Mountains studies, 1980–1981: Antarctic Journal of the United States, v. 16, no. 5, p. 18–19.

Manuscript Accepted by the Society March 1, 1989

Printed in U.S.A.

Geological Society of America
Memoir 170
1992

Chapter 15

Igneous petrology and geochemistry of the southern Heritage Range, Ellsworth Mountains, West Antarctica

Walter R. Vennum
Department of Geology, Sonoma State University, Rohnert Park, California 94928
Peter Gizycki
Institut für Angewandte Ökologie, Hindenburg Strasse 20, 7800 Freiburg, Federal Republic of Germany
Vladimir V. Samsonov and A. G. Markovich
Department of Antarctic Geology and Mineral Resources, Vniiokeangeologia, Moyka 120, St. Petersburg, Russian Federation 190121
Robert J. Pankhurst
British Antarctic Survey, National Environmental Research Council, c/o Institute of Geological Sciences, 64-78 Gray's Inn Road, London WC1X 8NG, England

ABSTRACT

Igneous rocks exposed in the southern Heritage Range include two diabasic sills (100 m and 300 m thick), three gabbroic stocks (outcrop areas each <0.5 km^2), a spessartite lamprophyre plug (200 m diameter), a 10-km^2 sanidine quartz phyric dacite stock, and numerous aphanitic to porphyritic dikes and lava flows. Lava flows that are largely basaltic, but range in composition from basalt to alkali rhyolite, compose about 10 to 15 percent of the 1,000-m thickness of the Liberty Hills Formation of the Heritage Group. Eighty-four samples have been analyzed for major, minor, and trace elements by x-ray fluorescence. Field relations and comparison of chemical composition of intrusive and extrusive rocks suggest that all igneous rocks in the southern Heritage Range except the dacite stock were emplaced in Cambrian time; all have undergone pumpellyite-actinolite or greenschist facies metamorphism. A four-point Rb-Sr isochron yields a mid-Devonian (369 ± 18 Ma) emplacement age for the dacite stock. The extrusive igneous rock suite is dominated by basalts but also includes rhyolites, dacites, and sparse andesites and is thus bimodal. The relict primary mineralogy, distribution of alteration-resistant trace elements (Nb, Ti, Y, Zr), and incorporation within a thick sequence of carbonate and clastic sedimentary rocks all suggest that the mafic igneous rocks are continental basalts whose original chemical compositions were largely mildly tholeiitic or transitional between tholeiitic and alkali basalt; some calc-alkaline traits are also evident. Discrimination diagrams based on alteration-resistant trace elements and distribution of large ion lithophile elements (Ba, Rb, Sr) suggest that the felsic and intermediate rocks also possess tholeiitic, calc-alkaline, and alkaline traits.

Age relations and comparison of chemical compositions indicate that the southern Heritage Range igneous rocks are not correlative with either the Jurassic Ferrar Supergroup or the Jurassic Kirwan Volcanics of the Transantarctic Mountains. The most likely possible correlatives might be diabasic sills and dikes and subordinate plugs and sills of felsic porphyry intrusive into the upper Precambrian Patuxent Formation, which crops out in the Neptune Range of the Pensacola Mountains 500 km to the southeast.

Vennum, W. R., Gizycki, P., Samsonov, V. V., Markovich, A. G., and Pankhurst, R. J., 1992, Igneous petrology and geochemistry of the southern Heritage Range, Ellsworth Mountains, West Antarctica, *in* Webers, G. F., Craddock, C., and Splettstoesser, J. F., Geology and Paleontology of the Ellsworth Mountains, West Antarctica: Boulder, Colorado, Geological Society of America Memoir 170.

Tholeiitic suites with contemporaneous calc-alkaline and alkaline traits are unusual but have been reported from the Yellowknife volcanic belt in Canada's Northwest Territories and from southeastern Maine. We conclude that the bimodal basaltic-rhyolitic suite of igneous rocks from the southern Heritage Range was emplaced in a continental area undergoing extensional tectonism and does not represent an island arc–continental margin subduction zone complex.

INTRODUCTION

Reconnaissance geological mapping of the Ellsworth Mountains (Fig. 1) was first undertaken during the 1961–1962, 1962–1963, and 1963–1964 austral summers by expeditions from the University of Minnesota (Craddock and others, 1964; Craddock, 1969). The range was next visited during the 1974–1975 austral summer by a four-man party from the Norsk Polarinstitutt (Hjelle and others, 1978, 1982) and again during the 1979–1980 austral summer by an international group of more than 40 scientists from eight countries (Splettstoesser and Webers, 1980). The Norwegian party conducted a relatively detailed stratigraphic study of the metasedimentary and metavolcanic rocks of the southern Heritage Range and briefly examined the Wilson Nunataks sill (Fig. 2). Their efforts concentrated mainly on stratigraphy, so this report and an earlier paper by Gizycki (1983) present more detailed accounts of the petrology and geochemistry of the igneous rocks. The fieldwork for this report was done during the 1979–1980 field season.

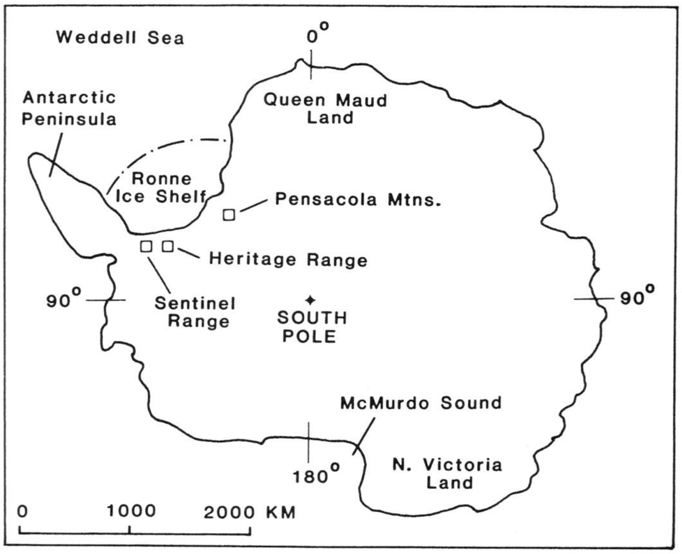

Figure 1. Map of Antarctica showing geographic features mentioned in text.

Given that the igneous rocks of the southern Heritage Range have undergone pumpellyite-actinolite and greenschist facies metamorphism (Bauer, 1983), their major-element oxide geochemistry, especially that of the mafic rocks, is most likely no longer representative of original magmatic composition. Consequently, one of the focal points of this chapter is an attempt to infer, through the use of alteration-resistant trace elements, the chemical affinities and thus the tectonic setting in which these rocks were emplaced. A comparison of the petrology, stratigraphic and structural relations, and geochemistry of these rocks is then made with the Ferrar Supergroup of the Transantarctic Mountains and the Kirwan Volcanics of Queen Maud Land.

PHYSIOGRAPHY AND GEOLOGIC SETTING

The Ellsworth Mountains are a slightly arcuate, 350-km-long, north-northwest–trending mountain chain located between the polar plateau of West Antarctica and the Ronne Ice Shelf. The range is as much as 90 km wide and constitutes one of the largest areas of exposed bedrock in West Antarctica. The Minnesota Glacier, which flows eastward from the polar plateau into the Ronne Ice Shelf, divides the Ellsworth Mountains into a northern Sentinel Range and a southern Heritage Range (Fig. 1). The 2,500-m-high western escarpment of the Sentinel Range rises abruptly from the polar plateau to elevations in excess of 5,000 m at the summit of the Vinson Massif (4,897 m), the highest peak in Antarctica. By contrast the Heritage Range is much more subdued, and toward the south its elevation gradually decreases as the range passes into a series of widely dispersed hills and nunataks.

The Ellsworth Mountains are underlain by a section of strongly folded, low metamorphic grade, largely marine metasedimentary and metavolcanic rocks that range in age from Cambrian to Permian and are at least 13,000 m thick. The entire range is folded into a major anticlinorium that plunges gently northwestward parallel to the long axis of the mountains; thus only the lower part of this section is exposed in the southern Heritage Range. The rock units that crop out in the southern Heritage Range, in decreasing order of age, are the Heritage Group, a 7,200-m-thick sequence of phyllite, argillite, black shale, quartzite, conglomerate, and minor marble of Middle and Late Cambrian age; the Minaret Formation, a 0- to 600-m-thick sequence of light gray locally oolitic Upper Cambrian carbonate rocks; and the Upper Cambrian to Devonian Crashsite Group, which is at least 3,200 m thick (Webers and Spörli, 1983).

The Heritage Group contains abundant interbedded basaltic to rhyolitic lava flows and pyroclastics and is locally intruded by gabbroic dikes, stocks, and sills as much as 300 m thick. A single felsic intrusive body, a porphyritic dacitic stock, intrudes the Heritage Group in the central Soholt Peaks (Fig. 2). Bauer (1983) has demonstrated an increase in metamorphic grade from pumpellyite-actinolite facies to greenschist facies downward within the Heritage Group and attributes this to a burial metamorphic event that may have affected the entire Ellsworth Mountains range. Mafic igneous rocks from the Wilson Nunataks possess greenschist facies metamorphic mineral assemblages. All other mafic igneous rocks described in this chapter contain pumpellyite-actinolite assemblages. R. Bauer (personal communication, 1982) suggests that the transition between these two facies requires burial of 8,000 m and temperatures of 350 to 400°C. The only igneous rocks known to occur in the Sentinel Range are a small outcrop of gabbro exposed at O'Neill Nunataks along the northern edge of the Minnesota Glacier (they are not discussed in this report). Craddock and others (1964) and Craddock (1969) have published more detailed accounts of the geology of the Ellsworth Mountains.

METHODS

Analyses for most reported major, minor, and trace elements were performed by energy dispersive x-ray fluorescence at the University of California, Davis; the Institute of the Geology of the Arctic, Leningrad, USSR (sample numbers prefixed with the letter S); and the Institut für Allgemeine und Angewandte Geologie der Universitat München, West Germany (sample numbers prefixed with the letter H). A variety of U.S. Geological Survey (AGV1, BCR1, G2, GH, GSP1, and W1) and Canadian Department of Energy, Mines, and Resources (MRG) standards were utilized in the California laboratory. Quantitative spectroscopy was used to determine Ba, Co, Cr, Cu, Ni, and V at Leningrad. Phosphorous was determined by colorimetric methods when interference from silicon prevented the use of x-ray fluorescence. Total iron was determined by x-ray fluorescence, ferrous iron by titration, and ferric iron by the difference between the other two quantities.

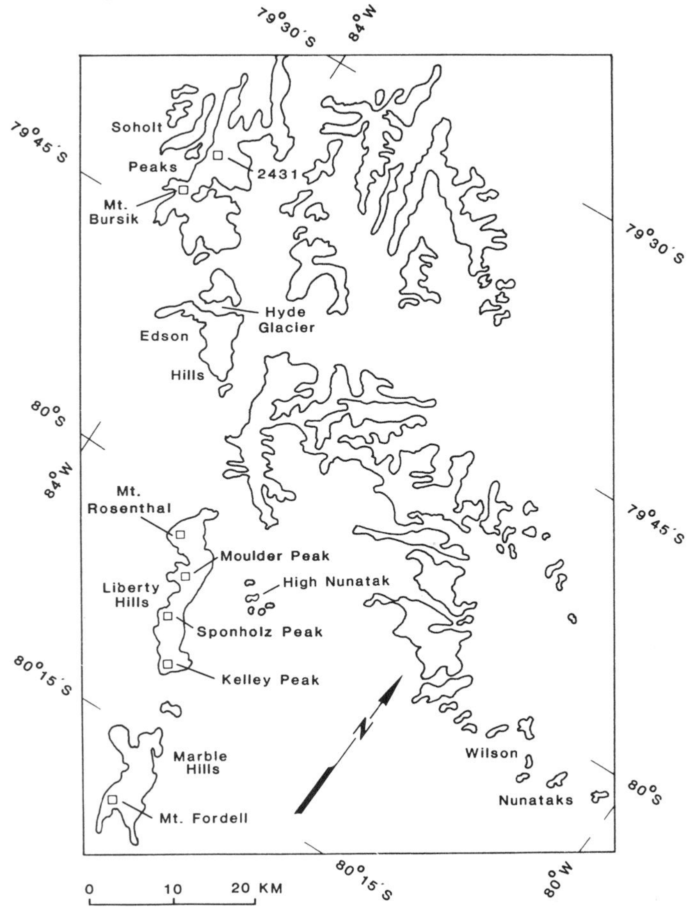

Figure 2. Map of the southern Heritage Range, Ellsworth Mountains, Antarctica, showing geographic features mentioned in text. Modified from Hjelle and others (1978). Elevation of unnamed peak given in meters. Outlines show mountains and nunataks that are surrounded by snow and ice.

FIELD RELATIONS

Mafic intrusive rocks occur in the form of diabasic sills, gabbroic stocks, a spessartite lamprophyre plug, and glacial erratics and as two mineralogically distinct types of dikes. Diabasic sills crop out in the northwestern part of the Wilson Nunataks (80°01'S, 80°44'W) and along the ridge crest immediately south of the Hyde Glacier icefall in the central Edson Hills (79°49'S, 83°57'W). Pace and compass traverses indicate that the Wilson Nunataks sill is approximately 300 m thick and that the Hyde Glacier sill is approximately 100 m thick. Detailed independent geological mapping by the senior author and by Masaru Yoshida (personal communication, 1980) indicate that the Hyde Glacier sill and its host rocks lie on the eastern limb of a west-dipping overturned anticline. The sill is intrusive into an interbedded sequence of quartzite, black slate, and siliceous lava flows, all of which are strongly cleaved. The upper (topographic) half and lower quarter of the sill are massive, but the remainder of this body is strongly jointed. The joints are oriented perpendicular to the margins of the sill, and their average spacing is 10 to 15 cm. Kink folds with amplitudes of 20 cm and wavelengths of 15 cm deform about one-third of the joint planes. The Wilson Nunataks sill and its metasedimentary host rocks are also strongly cleaved.

Gabbroic stocks crop out at High Nunatak and in the northwestern (80°01'S, 80°47'W) and central (survey control station 787, 80°03'S, 80°34'W) Wilson Nunataks. None of the gabbroic stocks is exposed over areas greater than 0.5 km^2. A spessartite lamprophyre plug 200 m in diameter crops out on the northeastern corner of High Nunatak where it is cross cut by the abovementioned gabbroic stock.

A 1-km-long glacier-carved surface along the northeastern flank of Mt. Fordell in the Marble Hills is covered with a large number of glacial erratics that range from 15 cm to 2 m in diameter. All erratics there are fine- to medium-grained gabbroic rocks. Orientation of glacial striations on this surface, the lack of outcrops of similar rock in the Marble and Liberty Hills, and the absence of erratics of other rock types that are exposed in the Marble and Liberty Hills strongly suggest that the source of the erratics is a body of gabbroic rock concealed beneath ice to the northwest of the Marble Hills.

Mafic dikes (1 to 2.5 m thick) cut every outcrop where the

sills and stocks are exposed and are also widespread throughout the rest of the southern Heritage Range. Although some of these dikes cannot be traced laterally to a parent intrusive body, many are obviously satellitic to the sills and stocks; these relations are well exposed at High Nunatak (Fig. 3). Andesitic dikes (0.5 to 1.5 m thick) crop out locally in the Wilson Nunataks and a 75-cm-wide, 7-m-long pinkish-gray quartz latite dike extensively veined with epidote cuts metasedimentary rock on the eastern flank of High Nunatak. Most dikes are strongly cleaved and often are moderately well foliated parallel to their margins.

A pyrite-bearing sanidine quartz phyric dacite stock, the only body of felsic intrusive rock yet discovered in the Ellsworth Mountains, underlies the central Soholt Peaks. Only the southern margin of this body was examined, but field relations and study of aerial photographs suggest that the stock crops out over a roughly circular area of approximately 10 km^2 centered midway between Mt. Bursik and Peak 2,431 m (Fig. 2). The rock weathers into a rugged craggy topography and is often stained a dark rusty red by the weathering of finely disseminated pyrite. A 10-m-wide chill zone surrounds the southern margin of the stock. The rock is strongly cleaved and extensively veined with quartz stringers.

Lava flows from tens of centimeters to 50 m thick, which are largely (approximately 75 percent) basaltic but which range in composition from basalt to alkali rhyolite, are abundant along the eastern flank of the Liberty Hills, where they are well exposed along the northeastern and southeastern ridges of Mt. Rosenthal, Moulder Peak, Sponholz Peak, and Kelley Peak. Outcrops of volcanic rocks along the eastern flank of the Liberty Hills all lie along the north-south strike of Heritage Group sedimentary rocks and most likely represent eruption at separate, but approximately contemporaneous, volcanic centers. The Liberty Hills are underlain by the Liberty Hills Formation of the Heritage Group, a unit composed of polymictic conglomerate, quartzite, argillite, and interbedded lava flows and volcanic breccia and considered to be at least 1,000 m thick (Webers and Spörli, 1983). Lava flows compose about 10 to 15 percent of the total thickness of this formation. Volcanic rocks of similar composition also crop out in the Edson Hills and at several unnamed nunataks adjacent to High Nunatak. All geographic locations mentioned in the text are depicted in Figure 2.

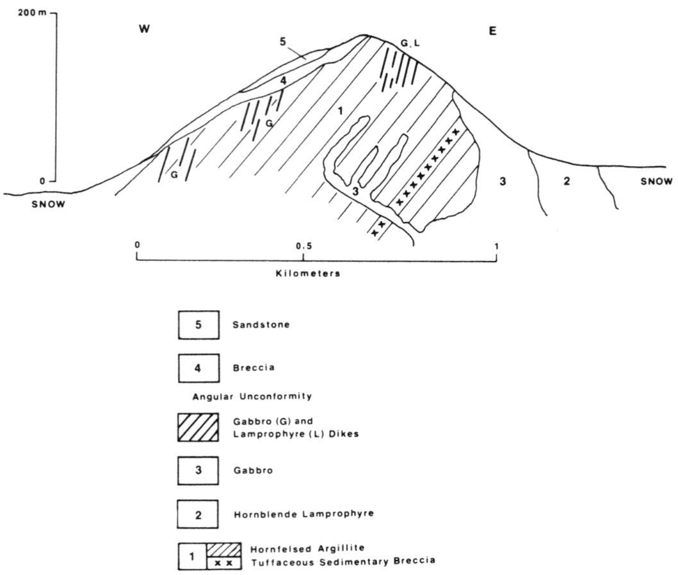

Figure 3. An east-west geological cross section of High Nunatak, southern Heritage Range.

PETROGRAPHY

Hyde Glacier sill

Although this sill has been metamorphosed, relict textures are preserved. Hand samples from the margins of the sill are fine- to medium-grained (<1 mm to 1 to 5 mm) greenish-black rocks with a subophitic to equigranular texture and a few widely scattered clinopyroxene phenocrysts (2 mm). These textures grade inward to rocks that are subophitic but that also contain poikilitic clinopyroxene crystals (5 mm). Grain size progressively increases and rock composition becomes more feldspathic toward the interior of the sill. The coarsest grain size and most felsic rocks occur slightly below the center of the sill.

Plagioclase in all thin sections examined has been converted to a mixture of albite and clinozoisite with subordinate sericite and minor carbonate. Clinopyroxene crystals, however, are relatively fresh (RIβ = 1.68, 2Vγ = 60°, ZΛc = 38°, δ = 0.030, nonpleochroic but very pale brown in plain polarized light). The groundmass is a mixture of actinolite, albite, chlorite (anomalous brown colors), clinozoisite, leucoxene, pumpellyite, sphene and minor calcite, hematite, pyrite, and apatite. Toward the interior of the sill the amounts of groundmass clinozoisite and actinolite increase, and both these minerals begin to take on a spray-like habit; apatite and clinopyroxene crystals are more prismatic, clinopyroxene grains are fringed with actinolite, and small amounts of interstitial quartz, zircon and skeletal titanomagnetite are present.

Wilson Nuntaks sill

This rock body bears many similarities to the Hyde Glacier sill. Rocks from the sill margins have an aphanitic to fine-grained greenish-black groundmass and a few widely scattered blocky greenish-white plagioclase crystals (1.5 to 2 mm). Toward the interior of the sill the grain size increases, the rocks become more feldspathic, and a diabasic texture becomes evident. Pegmatitic dikes(?) and segregations occur in the upper quarter of the sill, but most outcrop in this area is blanketed with a thick accumulation of frost-shattered talus blocks, and the relation of these coarse-grained rocks to the rest of the sill is not obvious. Clots of randomly oriented plagioclase crystals (to 5 cm) also occur in the upper quarter of the sill, and groups of these clots are usually aligned parallel to the sill margins.

Sills and stocks compared

The gabbroic stocks and erratics are petrographically identical to the two sills described above. Only slight differences in mineralogy exist between the Wilson Nunataks and Hyde Glacier sills. Pale brown clinopyroxene occurs only locally in the Wilson Nuntaks sill as relict cores of grains (1 to 3 mm) that have been almost totally replaced by an actinolitic amphibole (RIβ = 1.63, 2Vα = 70°, ZΛc = 18°, δ = 0.021, Z = moderate green, Y = pale green, X = very pale yellow, Z>Y>X). Toward the interior of the Wilson Nunataks sill the pleochroic colors of the actinolite intensify, and the extinction angle changes to ZΛc = 25°, changes that suggest a higher iron content in the actinolite crystals. Grains of reddish-brown amphibolite (RIβ = 1.66, 2Vα = 80°, ZΛc = 18°, δ = 0.020, Z = dark red brown, Y = red brown, X = pale red brown, Z>Y>X) occur only in the coarser grained parts of the Wilson Nunataks sill. They occur as rims on relic clinopyroxene grains and also apparently as primary crystals, as some grains of this mineral exhibit the euhedral diamond-shaped amphibole cross section. Electron microprobe analyses of the reddish-brown amphiboles indicate that they are magnesian hastingsite (Leake, 1978). Some of the primary magnesian hastingsite is rimmed with actinolite, and in one instance a clinopyroxene core is rimmed with magnesian hastingsite, which is in turn rimmed with actinolite. Pumpellyite is not present in rocks from the Wilson Nunataks sill except for a single anomalous occurrence in a quartz-rich sample described by Bauer (1983). The pegmatitic segregations were originally essentially bimineralic (plagioclase and magnesian hastingsite) and equigranular. Their mineralogy is now similar to the interior of the sill.

Spessartite lamprophyre plug

The ground mass of this rock body is an aphanitic to very fine grained mottled green to greenish-black mixture of albite, actinolite, epidote, calcite, chlorite, pumpellyite, pyrite, and sphene. Adjacent to the contact with the younger gabbro the plug contains approximately 10 percent black hornblende phenocrysts (3 to 4 mm). Toward the core of the plug the number of hornblende phenocrysts decreases, but their size reaches a maximum of 5 cm. The majority of the hornblende phenocrysts have been entirely replaced by chlorite and lesser actinolite. A few blocky clinopyroxene phenocrysts (5 mm) are also present; these have been partially uralitized. Amygdules (<1 percent) are filled with actinolite, quartz, and chlorite and rimmed with opaque minerals.

Mafic and andesitic dikes

Mafic dikes are largely plagioclase phyric, subophitic to poikilitic textured meta-diabases or gabbros (type I), but several of the dikes (type II) exposed at High Nunatak and in the Wilson Nunataks also contain magnesian hastingsite similar to that which occurs in the Wilson Nunataks sill. Hand samples of type I dikes have an aphanitic to very fine grained green to greenish-black groundmass in which are set <1 percent minute green to black amygdules and plagioclase laths, all of which average 0.5 to 1 mm in diameter. Clinopyroxene phenocrysts are uncommon. Petrographically the dikes are composed largely of randomly oriented or locally trachytic-textured plagioclase laths now almost totally replaced by albite and epidote ± sericite ± carbonate. The groundmass is composed largely of chlorite (pennine; RIβ = 1.60, δ = 0.0004, optically positive, α = β = pale green, γ = pale yellow) and lesser amounts of epidote, actinolite, sericite, leucoxene, hematite, quartz, magnetite, and pyrite. Celadonite was found in a single dike from the Wilson Nunataks. Clinopyroxenes have been almost totally replaced by chlorite. Amygdules are filled with epidote and/or chlorite ± quartz ± carbonate. The petrographic characteristics of these dikes very closely match the features of quartz keratophyres described by Williams and others (1982) and Nockolds and others (1978).

Type II dikes crop out only at High Nunatak and in the Wilson Nunataks. They were not found in contact with type I dikes. Hand samples have an aphanitic to very fine grained green to greenish-black groundmass in which are set scattered plagioclase and hornblende phenocrysts (3 mm). Plagioclase phenocrysts are now mixtures of epidote and albite ± sericite and carbonate, but the hornblende phenocrysts are fresh (RIβ = 1.67, 2Vα = 75°, ZΛc = 11°, δ = 0.026, Z = moderate red brown, Y = yellow brown, X = straw yellow, Z>Y>X). The groundmass is largely albite, hornblende occasionally rimmed with actinolite, and pyrite along with lesser amounts of sphene, actinolite, apatite, hematite, and quartz. Trachyitic texture is locally developed, and a very few partially chloritized clinopyroxene crystals (1 mm) occur in the groundmass of several of the dikes. Andesitic dikes previously mentioned are simply slightly more siliceous variants of the type II dikes.

Quartz latite dike

In thin section the fine-grained quartz latite dike displays a typical hypidiomorphic intergrowth of plagioclase (albite and epidote), quartz, mafic minerals, and pyrite. Most of the mafic minerals are now chloritized, but their shapes suggest they were originally biotite. A trace amount of relict hornblende (RIβ = 1.65, δ = 0.020, ZΛc = 14°, Z = dark green, X = moderate yellow green) rimmed with actinolite is present. The groundmass is a mixture of chlorite, albite, actinolite, pyrite, epidote, sphene, apatite, zircon, hematite, and calcite.

Soholt Peaks stock

Fresh hand samples are dark gray to greenish-black and consist of 10 to 20 percent phenocrysts (3 to 4 mm) of feldspar (mostly sanidine) and high-temperature (β) quartz set in an aphanitic groundmass. Many of the β quartz phenocrysts are marginally resorbed or embayed, glomeroporphyritic, and occasionally crushed and recrystallized. Feldspar phenocrysts are almost entirely unaltered. The groundmass is largely microcrystalline (0.05 to

0.10 mm) granoblastic quartz ± feldspar, minor calcite, hematite, and rare epidote. Texture and composition suggest the groundmass represents devitrified glass (Lofgren, 1971). The groundmass also contains small (0.1 to 1.0 mm) anhedral porphyroblasts of nearly isotropic, optically negative chlorite (weakly pleochroic in shades of green and pale yellow green), and abundant ragged flakes of biotite (pleochroic in shades of dark olive to pale yellow). The biotite, and locally the chlorite, define a moderately well developed foliation. Accessory minerals in approximate order of abundance are pyrite, magnetite, sphene, apatite, and zircon. Vennum and Nishi (this volume) have described an unusual assemblage of secondary materials that have formed from the oxidation of pyrite concentrated in the margins of this stock.

Lava flows

Mafic flows are generally purple to green, are locally pillowed and amygdaloidal, and are usually plagioclase phyric. Clinopyroxene phenocrysts are uncommon, and olivine phenocrysts are rare. Plagioclase phenocrysts have been converted to a mixture of albite, epidote, and sericite ± calcite, whereas olivine has been completely pseudomorphed by calcite, chlorite, opaque minerals, and epidote. Clinopyroxene phenocrysts are generally relatively fresh but are commonly rimmed by actinolite and/or opaque minerals or partially replaced by chlorite ± calcite. Amygdules are filled with epidote, chlorite, calcite, and/or pumpellyite, and the groundmass often displays a pilotaxitic to trachytic texture where plagioclase laths are enclosed in a matrix of actinolite, albite, epidote, calcite, chlorite, opaque minerals, pumpellyite, sericite, and sphene. Felsic flows tend to be yellowish brown, tan, or light gray and are often plagioclase and/or quartz phyric. Sanidine and hornblende phenocrysts are uncommon. Except for the development of smaller amounts of actinolite and the lack of pumpellyite the groundmass mineralogy of these rocks is similar to that of the mafic flows. Patches of mosaic-textured quartz and feldspar suggest that some of these rocks were originally largely glass. Although metamorphism has obliterated many of the primary features, some of the felsic rocks in the Liberty Hills appear to have been emplaced as ash flows.

GEOCHEMISTRY

Mafic rocks

Sixty-four new analyses (37 from stocks and sills, 14 dikes, and 13 lava flows) of major, minor, and trace elements are reported in tables 1A–D and 2A. CIPW norms for all 64 analyses are listed in the same tables and are plotted on Figure 4; the normative basalt classification is from Yoder and Tilley (1962). Norms were calculated after the analyses were converted to a 100 percent anhydrous total. As the oxidation state of iron has a significant effect on the amount of normative quartz, these computations were completed using an Fe_2O_3:FeO ratio of 0.15 (Brooks, 1976). When calculated in this fashion, 9 analyses are silica-saturated tholeiites (quartz normative), 28 are olivine tholeiites (normative hypersthene and olivine), and 27 are alkali basalts (normative nepheline). Analyses from High Nunatak plot mainly in the olivine tholeiite field, but those from other centers

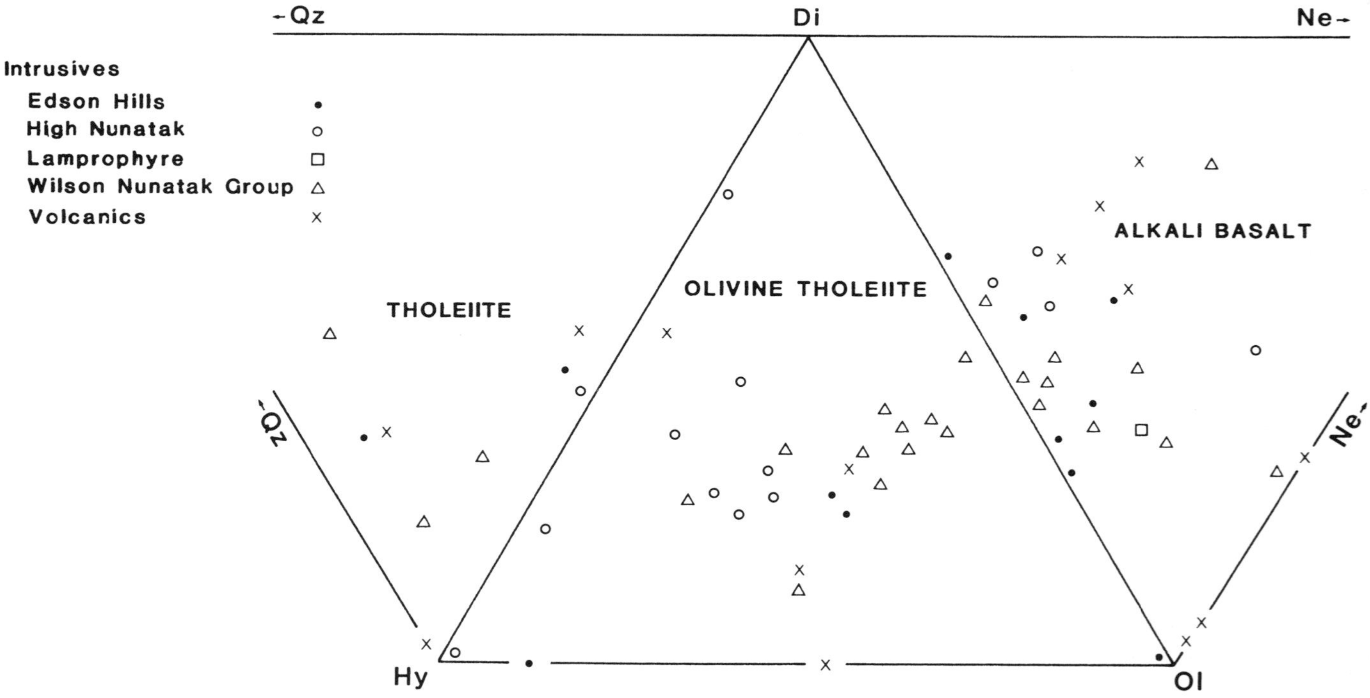

Figure 4. Plot of CIPW norms calculated from chemically analyzed samples of mafic igneous rocks from the southern Heritage Range. Classification is from Yoder and Tilley (1962).

of igneous activity are not confined to any particular part of the diagram. On an alkali-silica diagram (Fig. 5), all but 8 analyses fall inside the alkali basalt field of Macdonald and Katsura (1964). On an AFM diagram (Fig. 6) the analyses form a large cluster of points near the center of the triangle and are approximately equally divided between the tholeiitic and calc-alkaline fields of Irvine and Barager (1971). Similar results are obtained from a plot (Fig. 7) of weight percent Al_2O_3 versus normative plagioclase composition [100An/(An+Ab+5/3Ne)]. The dikes are generally higher in SiO_2 than the rest of the analyses, and although there are exceptions, they also have a tendency to be lower in Al_2O_3, MgO, CaO, and total iron oxides and higher in Na_2O than the other rock types.

It is, however, now well known that ocean-floor basalt is prone to extensive chemical modification when subjected to either hydrothermal alteration (Bass and others, 1973) or greenschist facies metamorphism (Pearce, 1975). The oxides that are most mobile during metamorphism appear to be CaO and Al_2O_3 and, to a lesser extent, Na_2O, K_2O, SiO_2, and FeO. Although Bass and others (1973) and Pearce (1975) dealt with ocean-floor basalt, it seems safe to extend Pearce's conclusions regarding element mobility during metamorphism to nonoceanic basalt. Due to the apparent mobility of silica and the alkali elements during metamorphism, the FeO*/MgO ratio (where FeO* is total iron expressed as FeO + 0.9 Fe_2O_3) is apparently a much better indicator of fractionation than the more commonly used Harker diagrams or solidification index (Saunders and others, 1979). FeO* is plotted against FeO*/MgO in Figure 8 and against SiO_2 in Figure 9. Although some calc-alkaline characteristics are still evident, these two diagrams suggest that the suite of mafic igneous rocks from the southern Heritage Range is much more strongly tholeiitic than indicated by Figures 6 and 7, both of which are partially based on abundance of alkali elements. Figure 8 also shows that as the FeO*/MgO ratio increases, there is an accompanying increase in the amount of total iron, a feature considered to be indicative of a tholeiitic differentiation trend (Miyashiro, 1974). Irvine and Barager (1971) have previously commented that in tholeiitic series, andesites and/or more felsic rocks are greatly subordinate to basalts, but that in calc-alkaline suites, this situation is reversed. As discussed in a later section of this chapter, igneous rocks of the southern Heritage Range are dominantly basaltic in composition. Cr and Ni and, to a lesser

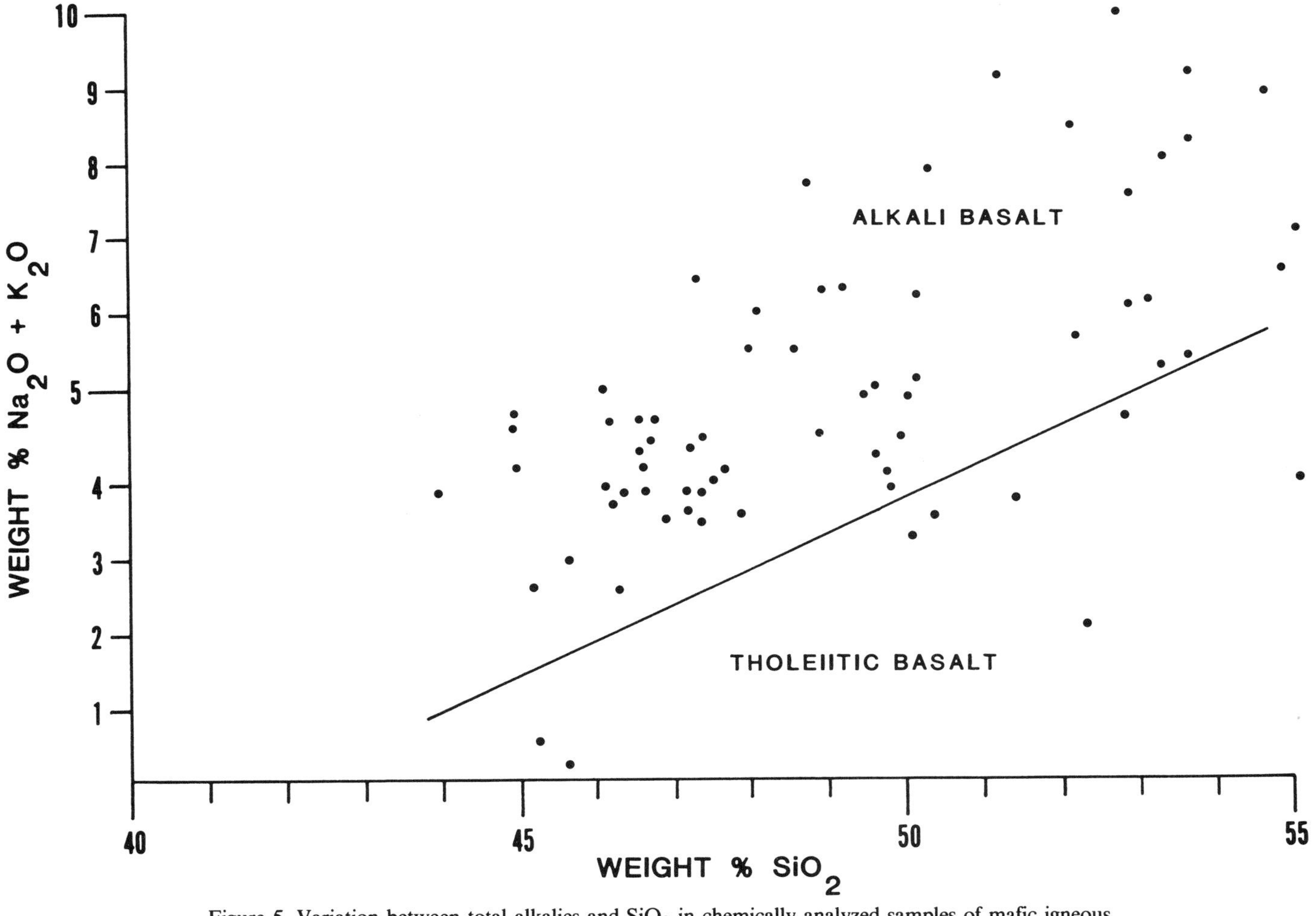

Figure 5. Variation between total alkalies and SiO_2 in chemically analyzed samples of mafic igneous rocks from the southern Heritage Range. Classification is from Macdonald and Katsura (1964).

TABLE 1A. CHEMICAL AND NORMATIVE DATA FOR INTRUSIVE ROCKS FROM THE SOUTHERN HERITAGE RANGE

	14g[1]	141[1]	H33[2]	H95[1]	8C[2]	S11[2]	S14[2]	7f[2]	S15[2]	S10[2]	7b[2]	S12[2]	S13[2]	S20[2]	S9[2]
Rock Type	OT	AB	AB	AB	OT	OT	OT	AB	OT	OT	AB	OT	AB	OT	OT
X-ray fluorescence analysis (in weight percent)															
SiO_2	43.96	44.95	45.00	45.10	45.60	46.18	46.21	46.27	46.30	46.55	46.62	46.72	46.72	46.80	46.84
TiO_2	1.11	1.14	2.50	1.79	2.00	2.51	2.87	1.99	2.51	2.45	2.03	2.58	3.35	2.45	2.51
Al_2O_3	17.10	14.63	16.70	16.20	14.45	15.72	13.92	13.75	13.96	14.71	13.34	16.22	14.65	14.83	15.72
Fe_2O_3	2.57	2.69	16.50	10.60	3.44	3.68	2.37	3.04	3.61	3.26	3.11	4.05	2.42	4.17	3.68
FeO	8.91	8.97			13.76	10.21	12.31	13.28	11.01	11.04	13.40	8.99	11.78	10.32	10.21
MnO	0.15	0.18	0.28	0.12	0.26	0.17	0.22	0.25	0.18	0.16	0.20	0.23	0.14	0.16	0.17
MgO	9.84	10.83	6.99	9.83	7.43	4.80	6.19	6.13	6.12	6.03	6.12	5.08	3.52	5.81	4.80
CaO	6.02	8.67	7.53	10.10	7.00	8.05	7.71	7.94	8.36	8.58	8.26	7.89	8.65	8.42	8.05
Na_2O	2.52	2.76	4.33	3.28	2.35	3.38	3.77	3.92	2.97	2.95	2.90	2.71	4.25	2.92	3.38
K_2O	1.36	1.40	0.40	1.50	0.64	0.76	0.19	0.50	0.95	0.98	1.23	1.86	0.49	0.76	0.76
P_2O_5	0.04	0.14	0.47	0.30	0.34	0.42	0.39	0.18	0.42	0.42	0.16	0.46	0.68	0.47	0.42
H_2O	5.81	3.67			3.16	3.25	4.23	1.96	3.24	2.90	1.57	3.03	2.89	3.21	3.25
Total	99.39	100.03	100.70	98.82	100.43	99.79	100.38	99.21	99.63	100.3	98.94	99.82	99.54	100.32	99.79
Trace element analyses (in parts per million)															
Ba	185	338	68	1031	1180	380	270		480	470	712	880	330	400	440
Cl	144	164			51			168			224				
Co			57	33		51	49		49	50		36	36	46	49
Cr	152	204	193	63	68	58	50	89	40	62	102	130		52	50
Cu	46	43	45	23	41	91	180	50	74	150	56	82	200	62	96
Ga	20	18			18			19			20				
Nb	16	15			14			13			16				
Ni	126	124	114	77	78	87	87	83	78	82	92	72	32	55	41
Pb		2	5	17	21	2	13	9	2	2		5	3	2	2
Rb	30	24	16	94	15	35	3	8	21	34	24	32	11	20	24
S	330	492			448			150			544				
Sc			67	33											
Sr	414	291	302	186	267	328	223	244	292	329	258	339	343	392	365
Th															
V			451	219		300	270		300	330		210	250	270	260
Y	31	22	40	33	19			22			21				
Zn	88	92	73	55	169			152			134				
Zr	132	118	153	146	155	156	180	147	166	161	146	228	332	173	198

TABLE 1A. CHEMICAL AND NORMATIVE DATA FOR INTRUSIVE ROCKS FROM THE SOUTHERN HERITAGE RANGE (continued)

	14g[1]	141[1]	H33[2]	H95[1]	8C[2]	S11[2]	S14[2]	7f[2]	S15[2]	S10[2]	7b[2]	S12[2]	S13[2]	S20[2]	S9[2]
Rock Type	OT	AB	AB	AB	OT	OT	OT	AB	OT	OT	AB	OT	AB	OT	OT
CIPW Norms (in weight percent)															
Ab	22.55	16.13	24.47	9.81	19.93	23.34	33.15	27.21	26.07	25.70	24.32	23.71	35.61	25.49	29.63
An	31.72	25.32	24.76	25.32	27.27	26.46	21.34	18.92	22.76	24.71	20.03	27.49	20.11	25.85	26.37
Or	8.35	6.12	2.34	8.96	3.90	6.90	1.17	2.78	5.85	5.96	7.24	11.36	3.01	4.62	4.68
Q															
C	0.51														
Ne		4.33	6.46	9.91				3.44			0.18		0.88		
Di		15.10	7.74	18.98	4.89	11.92	13.11	17.07	14.51	13.56	17.62	8.23	16.68	11.68	10.20
Hy	0.22				18.64	6.88	4.74		7.35	6.53		7.57		12.85	8.17
Ol	31.84	27.76	25.34	20.85	16.48	15.99	17.00	22.43	14.70	14.98	22.65	12.94	12.72	10.77	12.29
Il	2.28	2.28	4.70	3.44	3.79	4.76	5.66	3.79	4.93	4.78	3.95	5.07	6.59	4.78	4.93
Mt	2.55	2.55	3.10	2.04	3.47	2.80	2.89	3.47	2.87	2.78	3.47	2.55	2.78	2.82	2.71
Ap	0.09	0.33	1.11	0.71	0.80	0.94	0.97	0.43	1.01	1.01	0.38	1.11	1.65	1.14	1.01

Note: Symbols used are AB, alkali basalt (normative nepheline); OT, olivine tholeiite (normative hypersthene and olivine). 1) Edson Hills, 2) Wilson Nunataks. Blank spaces indicate oxide or element not reported.

TABLE 1B. CHEMICAL AND NORMATIVE DATA FOR INTRUSIVE ROCKS FROM THE SOUTHERN HERITAGE RANGE

Specimen	H32[2]	H30[2]	H36[2]	H97[1]	H27[2]	H37[2]	14j[1]	5n[3]	H100[1]	S21[3]	S22[3]	H42[2]	51[3]	H23[3]	S16[1]
Rock Type	AB	AB	AB	AB	OT	OT	AB	AB	OT	AB	AB	AB	AB	OT	T
X-ray fluorescence analysis (in weight percent)															
SiO_2	46.90	47.20	47.20	47.20	47.30	47.30	47.48	47.67	47.80	47.87	48.09	48.50	49.19	49.80	49.93
TiO_2	2.27	2.23	2.63	1.73	2.42	2.13	1.23	0.69	2.10	1.32	1.52	3.21	1.44	0.88	2.40
Al_2O_3	17.10	17.00	15.40	15.30	15.70	15.10	14.84	13.34	13.70	15.36	15.87	14.40	17.16	17.10	12.72
Fe_2O_3	13.90	13.50	15.00	12.40	15.00	15.20	2.98	2.25	12.30	6.28	6.03	15.50	2.05	9.91	1.84
FeO							7.17	7.21		5.19	4.25		8.71		9.89
MnO	0.21	0.21	0.24	0.19	0.22	0.24	0.18	0.13	0.22	0.16	0.13	0.29	0.18	0.18	0.20
MgO	6.22	6.60	5.94	9.35	5.87	5.80	10.37	10.06	7.20	5.97	5.45	5.68	6.25	8.25	6.80
CaO	8.49	7.93	8.61	8.02	8.73	8.77	8.07	10.76	11.40	8.99	9.65	7.51	6.23	8.80	9.12
Na_2O	3.44	4.16	3.09	3.54	2.49	3.22	3.10	1.72	2.75	3.45	2.95	4.00	3.60	1.71	2.72
K_2O	1.12	1.36	1.31	0.33	0.75	0.68	0.88	2.40	0.73	2.15	2.94	1.43	2.98	2.37	0.45
P_2O_5	0.34	0.36	0.38	0.25	0.43	0.45	0.05	0.06	0.28	0.74	0.59	0.58	0.09	0.27	0.51
H_2O							3.41	3.05		2.86	2.95		1.03		3.39
Total	99.99	100.55	99.80	98.31	98.91	98.89	99.76	99.34	98.48	100.34	100.42	101.40	98.91	99.20	99.97
Trace element analyses (in parts per million)															
Ba	423	346	466	215		366			35	890	850	626	833	512	
Cl							141	320					76		
Co	43	46	50	38	42	48			39	40	35	58		23	
Cr	52	112	149	128	96	134	232	523	267	62	82	152	42	187	
Cu	11	6	44	31	41	30	33	36	43	180	72	17	10	24	
Ga							19	21					19		
Nb							16						21		
Ni	94	110	79	43	41	65	81	138	110	27	31	64		86	
Pb	20	6	7	4	6	12	2	23	6	10	12	6		30	2
Rb	18	32	32	5	20	15	12	95		76	91	47	66	143	4
S							752	183					123		
Sc	54	55	55	39	55	57			39			56		32	
Sr	224	315	237	380	256	331	298	268	553	647	775	539	462	431	305
Th								11							
V	439	360	351	275	435	394			192	240	220	407		115	
Y	35	33	37	28	33	34	22	23	33			44	21	42	
Zn	60	45	67	33	78	63	63	103	35			54	123	21	
Zr	158	143	143	160	158	154	127	124	150	181	164	169	200	143	186

TABLE 1B. CHEMICAL AND NORMATIVE DATA FOR INTRUSIVE ROCKS FROM THE SOUTHERN HERITAGE RANGE (continued)

Specimen	H32[2]	H30[2]	H36[2]	H97[1]	H27[2]	H37[2]	14j[1]	5n[3]	H100[1]	S21[3]	S22[3]	H42[2]	51[3]	H23[3]	S16[1]
Rock Type	AB	AB	AB	AB	OT	OT	AB	AB	OT	AB	AB	AB	AB	OT	T
CIPW Norms (in weight percent)															
Ab	24.91	23.05	26.01	29.84	21.35	27.59	25.57	10.82	23.60	22.34	16.78	30.55	22.62	14.52	23.86
An	27.91	23.57	24.32	25.32	29.71	25.01	24.76	22.26	23.23	20.59	21.90	16.91	21.98	32.41	21.89
Or	6.62	7.96	7.74	1.95	4.51	4.06	5.57	14.47	4.40	13.08	17.87	8.35	17.81	14.14	2.78
Q															2.34
C															
Ne	2.27	6.49	0.11	0.34			0.64	2.38		4.12	4.80	1.62	4.30		
Di	9.96	10.95	13.36	11.06	9.43	13.41	13.33	26.69	26.60	16.72	19.19	13.54	7.60	8.04	17.66
Hy					19.29	5.21								16.73	23.25
Ol	20.61	20.38	19.72	25.20	7.21	16.68	25.55	20.06	15.09	16.62	13.09	18.79	20.40	10.06	
Il	4.31	4.22	4.99	3.34	4.66	4.08	2.43	1.37	4.05	2.56	2.96	6.02	2.73	1.68	4.70
Mt	2.62	2.52	2.85	2.38	2.87	2.89	2.08	2.08	2.36	2.22	1.99	2.89	2.32	1.88	2.29
Ap	0.81	0.84	0.91	0.61	1.01	1.07	0.12	0.14	0.67	1.78	1.45	1.34	0.21	0.64	1.24

Note: Symbols used are AB, alkali basalt (normative nepheline); OT, olivine tholeiite (normative hypersthene and olivine); T, tholeiitic basalt (normative quartz). 1) Edson Hills, 2) Wilson Nunataks, 3) High Nunatak. Blank spaces indicate oxide or element not reported.

TABLE 1C. CHEMICAL AND NORMATIVE DATA FOR INTRUSIVE ROCKS FROM THE SOUTHERN HERITAGE RANGE

Specimen	5j[3]	S17[3]	H18[3]	H14[3]	H102[1]	S18[3]	S19[3]	S24[1]	S23[2]	S25[2]	S26[2]	S27[1]	S28[3]	16a[1]	S29[3]
Rock Type	OT	OT	OT	OT	T	T	T	OT*	T*	T*	OT*	OT*	OT*	OT*	OT*
X-ray fluorescence analyses (in weight percent)															
SiO_2	50.04	50.05	50.30	51.40	52.30	53.58	55.08	45.22	45.22	45.78	46.23	49.54	49.78	49.94	50.35
TiO_2	0.89	1.11	0.96	0.81	3.53	0.44	2.27	1.00	2.14	1.68	3.67	0.65	0.98	1.82	0.41
Al_2O_3	16.56	17.63	15.10	16.40	12.80	14.93	11.56	15.68	14.41	12.85	12.13	16.00	13.77	14.71	14.22
Fe_2O_3	2.11	3.70	9.74	9.38	15.40	2.70	3.95	5.46	5.47	5.09	4.47	2.85	2.15	2.99	2.49
FeO	8.16	5.34				5.95	8.84	8.14	7.09	5.43	13.19	5.79	6.81	9.41	6.68
MnO	0.16	0.13	0.21	0.18	0.17	0.18	0.20	0.23	0.17	0.17	0.27	0.14	0.18	0.19	0.17
MgO	6.16	5.50	7.57	7.94	5.54	6.37	2.61	7.16	8.23	11.03	4.58	5.88	10.70	7.13	9.94
CaO	9.20	8.01	10.20	8.37	6.79	8.51	5.72	8.89	10.81	8.83	7.54	6.46	7.67	3.92	7.26
Na_2O	2.15	2.27	2.20	2.23	1.88	2.58	2.50	2.52	0.48	0.12	3.70	4.66	2.16	3.55	2.20
K_2O	1.75	2.85	1.21	1.60	0.20	2.40	1.38	0.16	0.05	0.05	1.01	0.25	2.00	1.59	2.36
P_2O_5	0.70	0.21	0.21	0.21	0.48	0.24	0.97	0.27	0.51	0.47	0.62	0.34	0.19	0.70	0.21
H_2O	2.64	3.63				2.60	4.46	5.09	5.76	8.06	2.56	7.60	3.54	3.46	3.70
Total	99.89	100.43	97.70	98.50	99.09	100.48	99.54	99.82	100.34	99.56	99.97	100.16	99.93	99.41	99.99
Trace element analyses (in parts per million)															
Ba		540	420	583		610	750	200	200	210	700	240	420		530
Cl	266													113	
Co		34	32	26	63	34	22	49	37	59	48	52	50		49
Cr	124	100	168	112	74	160		120	320	810	40	730	540	37	510
Cu	51	63	3	10		96	69	100	96	54	160	180	51	208	49
Ga	23													19	
Nb	14													21	
Ni	67	39	105	61	20	62	35	90	180	70	30	440	280		160
Pb	11	3	13	15	20			25	2	20	2	32	2	59	11
Rb	42	90	44	65	3			14	2	36	27	12	119	46	104
S	96													412	
Sc			37	31	58										
Sr	402	391	526	335	132			867	480	497	337	551	269	168	301
Th	6													22	
V		270	162	126	470	200	120	360	190	280	370	180	170		230
Y	22		25	29	58									28	
Zn	81		25	44	81									189	
Zr	125	71	133	139	162			101	130	174	177	162	91	454	101

TABLE 1C. CHEMICAL AND NORMATIVE DATA FOR INTRUSIVE ROCKS FROM THE SOUTHERN HERITAGE RANGE (continued)

Specimen	5j[3]	S17[3]	H18[3]	H14[3]	H102[1]	S18[3]	S19[3]	S24[1]	S23[2]	S25[2]	S26[2]	S27[1]	S28[3]	16a[1]	S29[3]
Rock Type	OT	OT	OT	OT	T	T	T	OT*	T*	T*	OT*	OT*	OT*	OT*	OT*
CIPW norms (in weight percent)															
Ab	18.36	19.88	19.04	19.19	16.10	22.24	22.24	22.50	4.30	1.10	32.15	42.59	18.93	30.94	19.30
An	30.23	30.47	28.38	30.47	26.12	22.59	17.11	32.72	39.15	37.59	13.86	23.79	22.81	15.30	22.84
Or	10.58	17.37	7.35	9.57	1.17	14.47	8.57	1.00	0.28	0.28	6.12	1.61	12.25	9.46	14.47
Q					12.31	0.73	16.79		4.31	6.83					
C														1.84	
Ne															
Di	11.90	7.52	18.05	8.44	3.92	15.50	5.16	9.90	11.49	5.62	17.19	6.93	12.40		10.65
Hy	18.26	13.72	15.37	28.84	29.54	21.35	20.69	12.98	32.35	41.71	1.80	9.35	17.11	30.07	16.67
Ol	5.05	6.58	7.57	1.63				15.51			16.81	11.78	12.35	4.52	12.83
Il	1.67	2.19	1.87	1.56	6.75	0.85	4.54	1.99	4.29	3.49	7.16	1.34	1.94	3.64	0.82
Mt	2.08	1.76	1.87	1.81	2.94	1.67	2.52	2.71	2.50	2.18	3.43	1.76	1.76	2.55	1.81
Ap	1.65	0.50	0.50	0.50	1.14	0.61	2.42	0.67	1.28	1.21	1.51	0.87	0.47	1.68	0.50

Note: Symbols used are OT, olivine tholeiite (normative hypersthene and olivine); T, tholeiitic basalt (normative quartz). 1) Edson Hills, 2) Wilson Nunataks, 3) High Nunatak. Blank spaces indicate element or oxide not reported. * = dike.

TABLE 1D. CHEMICAL AND NORMATIVE DATA FOR INTRUSIVE ROCKS FROM THE SOUTHERN HERITAGE RANGE

Specimen	S30[3]	H39[2]	5o[3]	4c[3]	S31[1]	H45[2]	S32[2]	H31[2]	S33[3]	5g[3]	17c[4]	17b[4]
Rock Type	OT*	AB*	OT*	AB*	AB*	AB*	AN*	AN*	AN*	QL*	D	D
X-ray fluorescence analysies (in weight percent)												
SiO_2	52.18	52.60	53.67	53.69	54.14	54.90	59.71	62.00	62.42	63.57	70.28	75.81
TiO_2	0.51	2.22	1.25	0.93	0.41	1.17	0.84	0.76	0.73	0.57	0.63	0.32
Al_2O_3	17.67	18.40	12.14	18.05	17.26	18.60	17.64	18.50	16.76	14.72	11.70	9.73
Fe_2O_3	3.53	8.95	1.90	2.04	4.01	9.19	2.49	5.39	2.57	0.96	1.33	0.94
FeO	6.08		5.71	6.13	1.80		2.62		3.38	3.73	6.17	4.03
MnO	0.17	0.16	0.08	0.18	0.15	0.16	0.14	0.13	0.11	0.05	0.11	0.08
MgO	4.93	2.71	5.27	5.32	2.40	3.41	0.17	0.65	0.30	1.63	0.74	0.53
CaO	5.97	6.02	8.81	3.17	5.44	3.12	3.23	2.55	3.30	1.56	1.92	0.40
Na_2O	2.60	7.49	4.34	7.76	5.48	7.45	6.63	8.97	4.35	5.16	2.44	5.91
K_2O	3.00	2.32	1.15	1.54	2.62	1.68	3.73	0.67	3.76	4.45	3.53	0.67
P_2O_5	0.25	0.72	0.06	0.08	0.39	0.43	0.30	0.23	0.33	0.07	0.13	0.12
H_2O	3.28		6.23	1.42	5.89		2.06		1.59	2.31	1.58	1.07
Total	100.17	101.59	100.61	100.31	99.99	100.11	99.56	99.86	99.60	98.78	100.56	99.61
Trace element analyses (in parts per million)												
Ba	590	335	92		650	1298		281		1110	743	
Cl			382	198						29	36	
Co	31	31			15	30		20				
Cr	30	61	201	31	58	149		85		28	2	14
Cu	36	12	23	371	6	21				36	17	12
Ga			18	24						19	19	18
Nb			19	16						17	23	23
Ni	16	22	84		32	26		38		3		
Pb	4	15	50	16	16	33	107	16	12	21	21	26
Rb	179	82	28	89	79	30	48		126	98	87	27
S				257						245	119	5355
Sc		34				27		16				
Sr	345	352	653	854	457	895	795	729	603	294	129	182
Th				11						5	6	
V	230	304			87	188		139				
Y		31	36	25		29		29		31	47	76
Zn		56	80	100		115		93		32	137	99
Zr	133	211	307	346	177	207	393	458	244	333	375	444

TABLE 1D. CHEMICAL AND NORMATIVE DATA FOR INTRUSIVE ROCKS FROM THE SOUTHERN HERITAGE RANGE

Specimen	S30[3]	H39[2]	5o[3]	4c[3]	S31[1]	H45[2]	S32[2]	H31[2]	S33[3]	5g[3]	17c[4]	17b[4]
Rock Type	OT*	AB*	OT*	AB*	AB*	AB*	AN*	AN*	AN*	QL*	D	D
CIPW Norms (in weight percent)												
Ab	22.66	33.78	38.81	43.73	47.10	51.16	59.90	75.42	40.10	45.11	46.68	46.68
An	28.60	9.60	10.57	10.02	15.72	12.35	7.70	8.29	14.60	4.17		1.39
Or	18.31	13.47	7.24	8.91	16.42	9.91	22.20	3.99	22.60	27.28	3.90	14.47
Q							1.30		14.30	9.61	35.87	25.42
C									0.70			4.18
Ne		15.48		12.11	1.16	6.38						
Di	0.20	12.71	28.98	4.82	8.61	0.25	3.60	2.54		3.25	1.69	
Hy	26.55		9.64					1.99	3.40	8.01	8.60	6.62
Ol	0.16	7.48	0.50	16.91	8.02	15.14		4.77				
Il	1.00	4.16	2.58	1.82	0.82	2.21	1.10	1.44	2.80	1.06	0.61	0.46
Mt	1.87	1.67	1.62	1.62	1.16	1.74	2.70	1.02	0.50	1.39		0.69
Ap	0.61	1.68	0.14	0.19	0.97	1.01	0.60	0.54	0.78	0.17	0.31	0.28
Ac											1.85	
Ns											0.37	

Note: Symbols used are AB, alkali basalt (normative nepheline); OT, olivine tholeiite (normative hypersthene and olivine); An, andesite; D, dacite; QL, quartz latite. 1) Edson Hills, 2) Wilson Nunataks, 3) High Nunatak, 4) Soholt Peaks. Blank spaces indicate oxide or element not reported. * = dike.

TABLE 2A. CHEMICAL AND NORMATIVE DATA FOR MAFIC VOLCANIC ROCKS FROM THE SOUTHERN HERITAGE RANGE

Specimen	H127[6]	S3[5]	S1[3]	S2[3]	H80[3]	S4[5]	H58[3]	H88[3]	S5[5]	H103[6]	H4[3]	H123[6]	H78[3]
Rock Type	OT	AB	AB	AB	AB	AB	OT	T	AB	T	AB	OT	OT
X-ray fluorescence analyses (in weight percent)													
SiO_2	46.30	48.83	49.22	50.26	51.30	52.22	53.30	53.30	53.39	53.50	53.80	55.30	55.70
TiO_2	1.72	0.60	0.69	1.44	1.38	0.69	1.38	1.33	0.65	1.33	1.52	1.46	1.22
Al_2O_3	16.20	16.08	15.35	15.25	18.40	18.44	16.80	15.90	17.40	19.00	13.60	18.60	15.60
Fe_2O_3	11.30	7.88	8.25	6.47	10.10	4.04	9.61	8.91	7.00	9.29	10.90	8.85	8.38
FeO		1.37	2.84	4.09		3.03			2.10				
MnO	0.21	0.13	0.19	0.11	0.14	0.13	0.14	0.13	0.11	0.20	0.26	0.12	0.13
MgO	8.87	3.47	5.70	5.52	9.01	6.93	6.19	4.65	6.21	4.55	5.55	4.79	3.04
CaO	9.81	7.28	9.12	6.27	2.74	1.94	5.86	9.24	2.36	5.61	7.68	3.68	7.32
Na_2O	2.50	7.10	3.72	4.67	6.04	5.00	4.39	3.57	6.46	3.54	4.20	5.78	5.14
K_2O	0.05	0.62	2.50	3.25	3.29	3.58	1.75	0.93	1.48	2.60	3.34	0.78	2.00
P_2O_5	0.29	0.42	0.54	0.56	0.50	0.51	0.52	0.41	0.46	0.50	0.56	0.43	1.15
H_2O		6.55	1.76	2.38		3.28			2.56				
Total	97.25	100.33	99.88	100.27	102.90	99.79	100.00	98.37	100.18	100.12	101.41	99.79	99.70
Trace element analyses (in parts per million)													
Ba	919	290	650	670	434	3400	545	184	1100	663		115	932
Co	43	30	36	32	30	38	31	28	35	30	36	36	27
Cr	53	160	110	54	380	210	254	59	170	54	310	179	66
Cu		16	30	58		41	108	19	120		50	18	
Ni	72	97	47	29	199	14	103	30	150	20	33	82	71
Pb	11	17	54	3	134	19	35	145	17	142	4	4	305
Rb	3	17	82	106	4	25	56	31	23	85	125		13
Sc	49				31		26	26		34	40	42	25
Sr	278	464	659	587	1982	555	375	285	918	424	781	694	700
V	282	170	240	210	192	220	137	231	150	250	207	284	169
Y	24				26			32		27	37	26	56
Zn	28				34		87	127		76	47	33	100
Zr	128	171	173	150	280	160	172	186	202	155	145	174	195

TABLE 2A. CHEMICAL AND NORMATIVE DATA FOR MAFIC VOLCANIC ROCKS FROM THE SOUTHERN HERITAGE RANGE (Continued)

Specimen	H127[6]	S3[5]	S1[3]	S2[3]	H80[3]	S4[5]	H58[3]	H88[3]	S5[5]	H103[6]	H4[3]	H123[6]	H78[3]
Rock Type	OT	AB	AB	AB	AB	AB	OT	T	AB	T	AB	OT	OT
CIPW Norms (in weight percent)													
Ab	21.77	35.40	20.35	25.42	27.22	42.32	37.13	30.73	53.44	29.84	13.32	54.13	43.58
An	33.77	10.88	18.14	11.29	10.04	6.54	20.98	24.98	8.90	24.54	15.63	15.50	13.66
Or	0.28	3.90	15.09	19.59	18.93	21.93	10.35	5.62	8.96	15.36	19.43	4.62	11.86
Q								2.27		0.14			
C					1.08	4.17			2.03	1.37		1.59	
Ne		15.51	6.33	8.04	12.16	0.82			1.39		11.76		
Di	11.69	20.68	20.21	13.90			3.85	15.57			21.27		12.76
Hy	11.34						11.47	15.58		23.29		9.22	10.39
Ol	14.90	9.56	15.10	15.45	25.06	20.23	10.50		21.11		12.59	10.10	1.15
Il	3.35	1.21	1.33	2.79	2.55	1.35	2.63	2.56	1.27	2.52	2.85	2.78	2.32
Mt	2.20	1.87	2.13	2.04	1.85	1.39	1.81	1.71	1.76	1.76	2.04	1.69	1.60
Ap	0.71	1.07	1.31	1.34	1.18	1.24	1.24	0.97	1.11	1.18	1.31	1.01	2.72

Note: Symbols used are AB, alkali basalt (normative nepheline); OT, olivine tholeiite, normative hypersthene and olivine); T, tholeiitic basalt (normative quartz). 3) High Nunatak, 5) Liberty Hills, 6) Edson-Liberty Hills. Blank spaces indicate oxide or element not reported.

TABLE 2B. CHEMICAL AND NORMATIVE DATA FOR FELSIC AND INTERMEDIATE VOLCANIC ROCKS FROM THE SOUTHERN HERITAGE RANGE

Specimen	S6[5]	H151[6]	S7[5]	10j[5]	S8[5]	H56[3]	H94[6]	H134[6]	H25[3]	H135[6]	H128[6]	H11[3]	H130[6]	H73[3]
Rock Type	AN	R	D	AN	D	R	R	R	R	AR	R	RD	R	RD
X-ray fluorescence analyses (in weight percent)														
SiO_2	56.94	58.50	65.58	67.51	69.41	69.90	71.00	71.50	71.70	72.60	73.20	74.40	74.70	75.50
TiO_2	0.44	1.06	0.93	0.19	0.18	0.22	0.34	0.61	0.43	0.56	0.55	0.49	0.49	0.40
Al_2O_3	15.77	17.50	12.83	15.64	16.18	15.90	15.10	10.70	11.40	10.90	10.90	8.92	11.00	8.88
Fe_2O_3	5.61	8.92	5.64	1.11	2.15	3.65	3.37	6.51	3.09	6.21	5.59	3.48	5.59	4.45
FeO	1.33		1.04	2.11	1.94									
MnO	0.13	0.16	0.00	0.01	0.00	0.01	0.09	0.17	0.10	0.12	0.13	0.11	0.14	0.16
MgO	3.13	3.85	3.31	0.62	0.36	0.51	1.15	0.98	1.47	0.84	0.65	2.66	0.80	2.35
CaO	4.69	6.57	1.49	0.25	0.44	0.39	2.30	2.02	5.90	1.10	0.86	5.99	1.21	5.91
Na_2O	3.64	1.41	4.50	5.27	6.26	5.49	3.22	3.32	2.63	3.98	4.48	1.43	2.70	
K_2O	3.20	2.48	1.56	2.35	1.78	1.94	2.15	2.40	2.59	1.68	1.73	2.97	0.98	2.28
P_2O_5	0.34	0.44	0.54	0.16	0.06	0.10	0.14	0.15	0.31	0.12	0.10	0.28	0.12	0.32
H_2O	4.90		2.44	5.20	1.19									
Total	100.12	100.90	99.86	100.42	99.95	98.09	98.79	98.36	99.63	98.11	98.19	100.74	97.74	99.89
Trace element analyses (in parts per million)														
Ba	890		410		670	192	408	563	285		395	284	321	115
Cl				112										
Co	26	16	24			8	7	18	6	20	15	7	18	11
Cr	40	185	710				27	5	6			16	7	18
Cu	140	10	110	11	100									
Ga				22										
Nb				24										
Ni	32	51	210	4	11	18	33	23	27	10	12	38	22	9
Pb	7	14		1	5	8	17	6	405	7	19	64	7	25
Rb	94	64		90	53	61	63	82	57	45	49	45	29	91
Sc		26				14	16							
Sr	272	881		162	117	87	558	20	16	19	17	17	18	16
Th				20				144	448	151	172	546	202	181
V	110	161	220		10	73	66	105	80	103	93	94	103	67
Y		29		21		8	24	117	28	104	110	23	138	31
Zn		66		81		23		66		81	76		108	63
Zr	197	173		323	295	315	226	390	194	402	592	338	518	163

TABLE 2B. CHEMICAL AND NORMATIVE DATA FOR FELSIC AND INTERMEDIATE VOLCANIC ROCKS FROM THE SOUTHERN HERITAGE RANGE Continued)

Specimen	S6[5]	H151[6]	S7[5]	10j[5]	S8[5]	H56[3]	H94[6]	H134[6]	H25[3]	H135[6]	H128[6]	H11[3]	H130[6]	H73[3]
Rock Type	AN	R	D	AN	D	R	R	R	R	AR	R	RD	R	RD
CIPW Norms (in weight percent)														
Ab	32.31	11.03	42.10	51.40	56.50	47.53	27.66	27.75	22.41	15.47	37.83	12.05	23.38	
An	18.11	28.96	4.30	1.11	1.70	1.31	10.65	7.37	11.73	21.48	3.70	9.11	5.34	17.45
Or	19.87	14.66	9.50	26.72	10.60	11.73	12.90	14.52	15.41	0.77	10.47	17.48	5.93	13.44
Q	6.40	16.17	25.50	3.97	22.60	26.58	35.34	34.00	33.46	54.81	34.85	39.60	47.38	49.93
C		1.60	2.40		3.70	4.40	3.68			0.84	0.34		3.67	
Ne														
Di	3.36			6.16				1.62	13.09			15.32		8.11
Hy	16.88	22.88	9.40	8.35	2.20	7.11	8.15	11.96	1.77	3.91	10.44	4.20	11.98	9.57
Ol														
Il	0.86	1.99	1.40	1.06	0.30	0.41	0.65	1.18	0.82	1.08	1.06	0.93	0.95	0.76
Mt	1.37	1.67	0.30	0.93	2.20	0.69	0.64	1.25	0.58	1.20	1.07	0.65	1.08	0.84
Ap	0.84	1.04	1.20	0.38	0.20	0.24	0.33	0.35	0.73	0.28	0.24	0.66	0.29	0.76

Note: Symbols used are AN, andesite; AR, alkali rhyolite; D, dacite; R, rhyolite; RD, rhyodacite. 3) High Nunatak, 5) Liberty Hills, 6) Edson-Liberty Hills. Blank spaces indicate oxide or element not reported

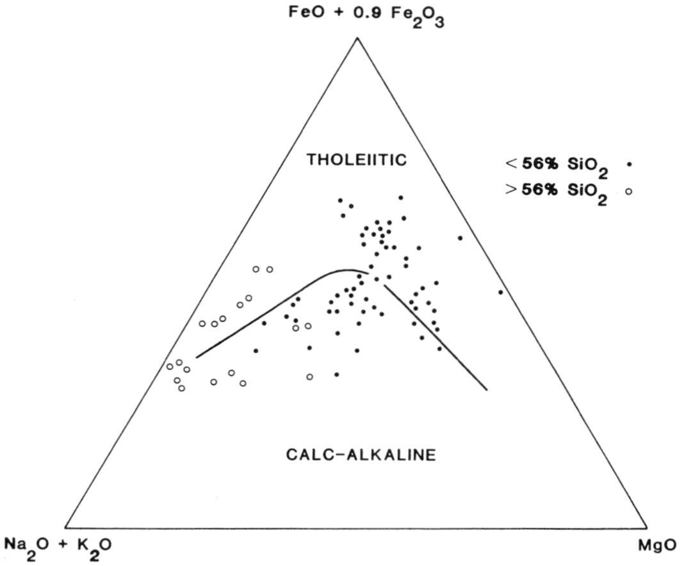

Figure 6. Ternary plot showing variation of total alkalies, magnesium, and iron expressed as oxides in chemically analyzed samples of igneous rocks from the southern Heritage Range. Classification is from Irvine and Barager (1971).

extent, Co, Cu, and V are the only trace elements with behaviors that correlate (negatively) with the FeO*/MgO ratio. Amounts of alkalis, alkali earths, and various alkali/alkali-earth ratios fluctuate randomly; this is believed due to mobility of these elements during metamorphism.

In 1970, Cann was able to show that amounts of Nb, Ti, Y, and Zr in ocean-floor basalt fall within a fairly restricted range and that abundances of these elements are strongly resistant to modification by secondary processes. A number of other petrologists (Jakeš and Gill, 1970; Pearce and Cann, 1971; Floyd and Winchester, 1975; Pearce and others, 1975) have since devised a variety of geochemical methods utilizing these four elements as well as Cr, P, Sr, and the rare earths to determine the original chemistry and tectonic setting of altered basaltic rocks. Pearce and Cann (1973) used the Y/Nb ratio to indicate whether a specific rock is alkalic or tholeiitic; $Y/Nb \leqslant 1$ is indicative of alkaline chemistry and $Y/Nb \geqslant 2$ is indicative of tholeiitic chemistry in their scheme. Examination of a Ti-Y/Nb plot (Fig. 10) indicates that all analyzed Ellsworth Mountains mafic rocks have Y/Nb ratios that fall in the transitional basalt chemistry range ($Y/Nb \geqslant 1$, but $\leqslant 2$). When these analyses are plotted in a Ti-Zr-Y discrimination diagram (Fig. 11), 22 of 26 points plot in or along the edge of the "within-plate basalt" field, a category that includes both ocean island and continental basalt (Pearce and Cann, 1973). On the previously mentioned Ti-Y/Nb plot, all analyzed samples fall in the continental tholeiite field but in areas overlapped by the fields of either oceanic tholeiites and continental alkali basalt. Twenty-five of 27 points plotted on a Zr/Y-Zr diagram (Fig. 12) fall in or just outside the "within-plate field" (Pearce and Norry, 1979).

It should be pointed out that the validity of the above-mentioned discrimination diagrams, especially the Ti-Zr-Y diagram, has been recently questioned (Holm, 1982; Prestvik, 1982). Some of these misgivings apparently result, however, because the precautions mentioned by Pearce and Cann (1973) were not followed (Brooks and Cole, 1980). In order to circumvent this problem, only analyses with CaO + MgO between 12 and 20 percent and total alkalis $\leqslant 20$ percent after calculations are made for presentation in an AFM diagram (Fig. 6) are plotted in Figures 11 and 12.

The scarcity of olivine and the relatively large amount of plagioclase relative to pyroxene are suggestive of tholeiitic chemistry. Although titanaugite is apparently not present, optical properties of relict clinopyroxenes suggest a calcic augite composition. This feature, the apparent lack of orthopyroxene, and the presence of magnesian hastingsite, are all characteristics of alkali basalts. Thus, study of alteration-resistant trace-element abundances and relict primary mineralogy suggests that the mafic igneous rocks of the southern Heritage Range are continental basalts, the original chemical compositions of which were mildly tholeiitic or transitional between tholeiitic and alkali basalt. Their intrusion into a thick sequence of carbonate and clastic sedimentary rocks is not compatible with an oceanic island setting.

Felsic and intermediate rocks (>56 percent SiO_2)

Twenty analyses (14 lava flows, 4 dikes, and 2 from the Soholt Peaks stock) of major, minor, and trace elements are reported in Tables 1D and 2B. CIPW norms are listed in the same tables and are plotted on Figure 13, the quartz-feldspar normative volcanic rock classification scheme of Streckeisen (1979). As some of the analyses reported total iron as Fe_2O_3, norms were calculated on the same basis as those for the basaltic rocks (on a 100 percent anhydrous base with an Fe_2O_3:FeO ratio of 0.15). Analzyed samples span the complete range of intermediate to felsic compositions from andesite to alkali rhyolite. It should be emphasized that these rocks are present in the field in far smaller amounts, relative to the basalts, than is indicated by the number of analyses presented here. When these analyses are plotted on an AFM diagram (Fig. 6) they appear to define a calc-alkaline differentiation trend. Plots of vanadium versus chromium (Fig. 14), vanadium versus FeO*/MgO, and weight percent SiO_2 versus chromium (the latter two not shown) all indicate that the felsic and intermediate rocks display both tholeiitic and calc-alkaline traits. These are apparently misleading observations, as several petrologists (Noble, 1968; Lipman and others, 1969; Kochhar, 1977) have shown that glassy felsic rocks are prone to modification of their chemical composition, especially of the alkali elements, by a variety of pre- and post-consolidation mechanisms. Indeed the possibility of alkali exchange during hydration while these rocks were still glassy (Lipman and others, 1969) is supported by two lines of evidence.

When these analyses are plotted in a Na_2O-K_2O-CaO triangle (not shown), they outline a random distribution pattern scattered over most of the diagram but concentrated toward the Na_2O corner (50 to 85 percent relative Na_2O). In addition, rocks with high Na_2O contents generally have low K_2O contents.

Discrimination diagrams such as those used with basaltic rocks have not been used extensively with altered igneous rocks of other compositions. Winchester and Floyd (1977) and Floyd and Winchester (1978) have, however, utilized several immobile elements (Ce, Ga, Nb, Sc, Y, Ti, Zr) to construct a series of diagrams relating altered intermediate and felsic volcanic rocks to their original chemical composition but not to tectonic setting. Their plot of SiO_2 percent versus the ratio Zr/TiO_2, reproduced as Figure 15, suggests that the intermediate and felsic volcanic rocks of the southern Heritage Range are in part calc-alkaline and in part alkalic, i.e., that the high alkali contents present in the whole rock analyses are not entirely due to alkali element mobility but are in part a primary magmatic characteristic of these rocks. Granted none of these rocks apparently contain the modal indicators of high alkalinity (feldspathoids, sodic amphibole and/or pyroxene, aenigmatite), but the components of these minerals could easily have been originally incorporated in glassy groundmasses. Contents of large ion lithophile elements (Ba, Rb, Sr) fall between those reported for calc-alkaline and shoshonitic associations (Jakeš and White, 1972) and thus support the above conclusion, but as these elements are also prone to mobility this criterion must be used with caution.

GEOCHRONOLOGY

Lack of either unaltered rocks or minerals of suitable composition has restricted radiometric geochronology in the Ellsworth Mountains. Four whole-rock K-Ar dates from intrusive rocks reported earlier by Yoshida (1982) and six new K-Ar dates performed on three intrusive rocks and three lava flows in the German Democratic Republic (DDR) are listed in Table 3. Two of Yoshida's dates and all six of the DDR dates are clearly anomalous, most likely due to Ar loss. The three DDR dates from the Hyde Glacier sill range from 161 to 245 Ma and are much lower than Yoshida's 396-Ma date from the same body. The three dated alkali basalt flows from High Nunatak (242 to 303 Ma) are all interbedded with Cambrian sedimentary rocks of

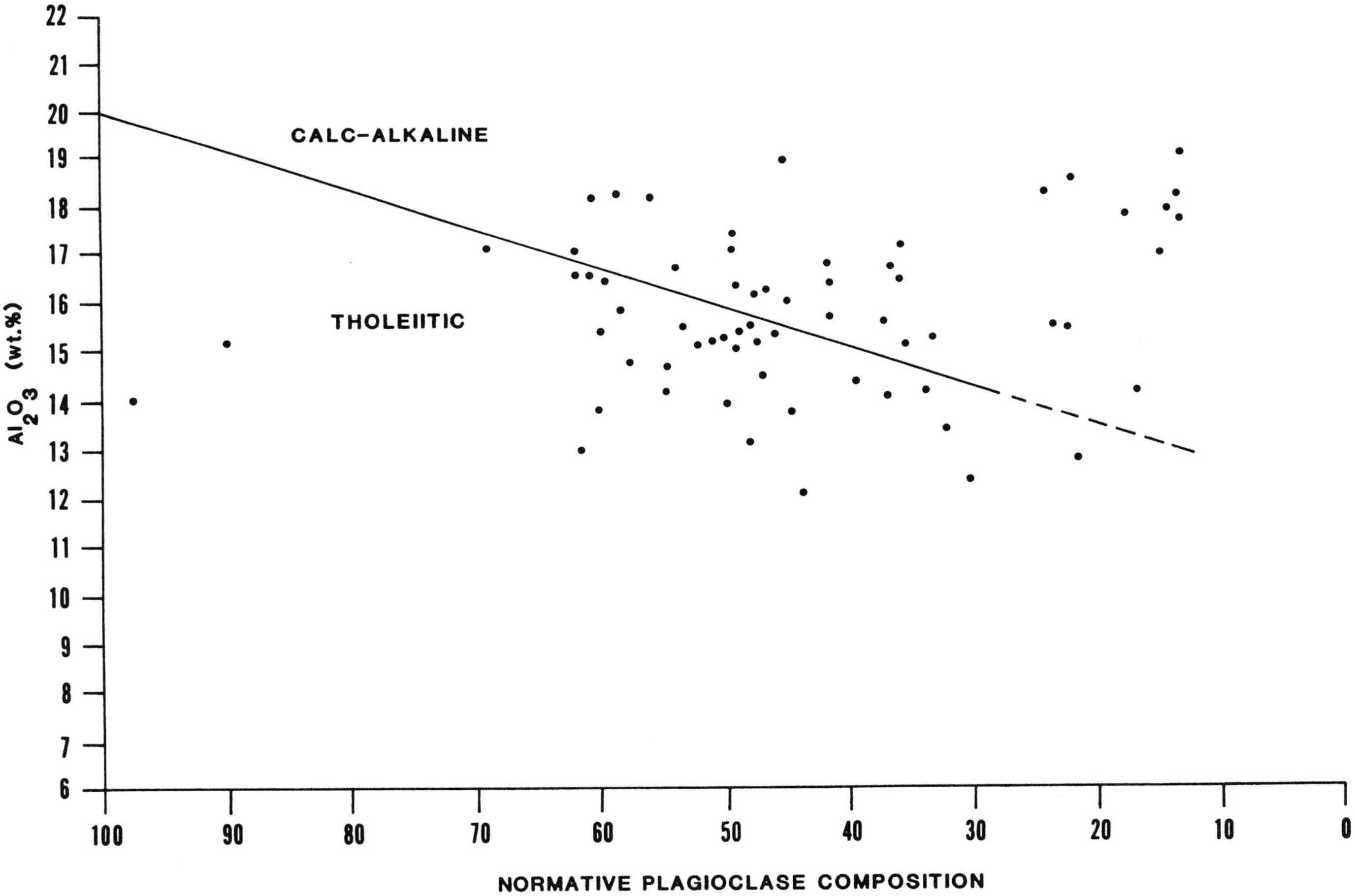

Figure 7. Variation between weight percent Al_2O_3 and normative plagioclase composition in chemically analyzed samples of mafic igneous rocks from the southern Heritage Range. Classification is from Irvine and Barager (1971).

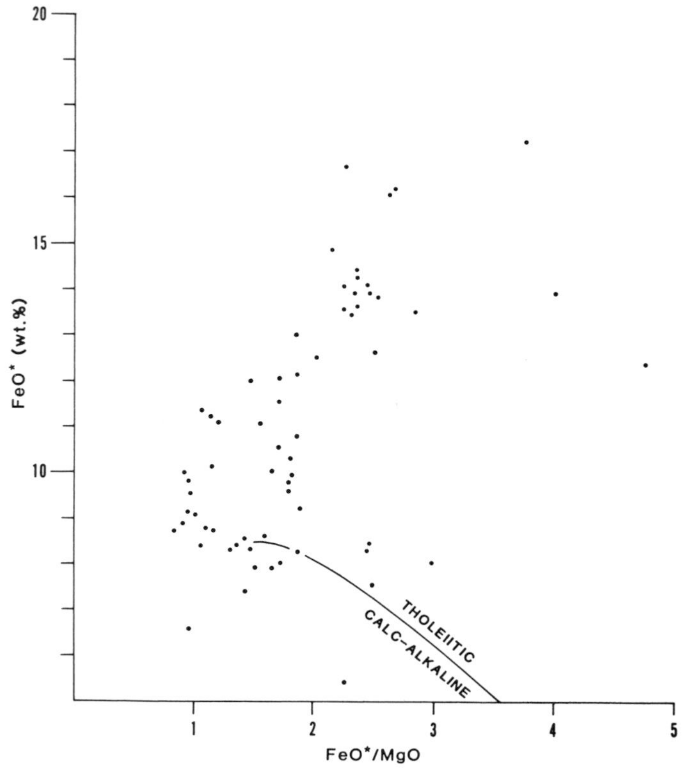

Figure 8. Plot of FeO* versus FeO*/MgO for chemically analyzed samples of mafic igneous rocks from the southern Heritage Range. Classification is from Miyashiro (1974).

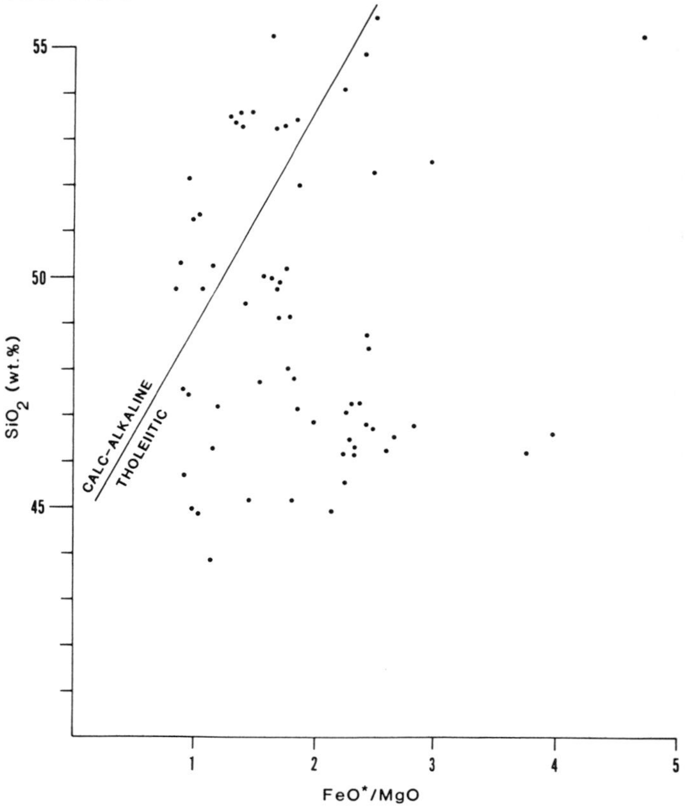

Figure 9. Plot of FeO*/MgO versus SiO_2 for chemically analyzed samples of mafic igneous rocks from the southern Heritage Range. Classification is from Miyashiro (1974).

the Heritage Group. The two dikes dated by Yoshida give ages of 237 and 254 Ma. These two dikes intrude sedimentary rocks of the lower Heritage Group but are in turn truncated by a conglomeratic unit that is also part of the Heritage Group. The significance of Yoshida's dates from the Wilson Nunatak sill (381 ± 19 Ma) and the Hyde Glacier sill (396 ± 20 Ma) is discussed below.

Although igneous rocks of the Heritage Range are also generally unsuitable for Rb-Sr whole rock dating, a four-point whole-rock isochron was obtained from the Soholt Peaks stock. The available hand samples were small (about 500 g each) and exhibited a moderate amount of groundmass alteration but contained reasonable fresh feldspar phenocrysts. Rb/Sr ratios were determined on whole-rock powders by x-ray fluorescence (Pankhurst and O'Nions, 1973), and Sr-isotope compositions were determined with a fully automated VG Micromass 30 mass spectrometer at the Institute of Geological Sciences in London. Sufficient spread in Rb/Sr (Table 4) was found to define an age fairly precisely, but some scatter of the data points outside analytical error is indicated by the MSWD figure of 7.3. If this scatter is used to define the uncertainty of a best fit line, an age of 369 ± 18 Ma and an initial $^{87}Sr/^{86}Sr$ ratio of 0.7148 ± 0.0003 are obtained (Fig. 16). These data clearly represent a discrete geological event, and the age is close to the two K-Ar ages reported by Yoshida (1982) from the Hyde Glacier and Wilson Nunataks sills. Interpretation of the isochron is, however, problematic. If the samples were fresh and unmetamorphosed, the preferred explanation would be that 369 ± 18 Ma represents the time of crystallization of the stock, in which case the high initial $^{87}Sr/^{86}Sr$ ratio would be ascribed to magma genesis by anatexis or assimilation of older crustal material (either the exposed Cambrian strata of the Heritage Range or a hypothetical Precambrian basement). Yoshida (1982) has also reported a K-Ar date of 935 ± 47 Ma from a fine-grained meta-pelite from the Wilson Nunataks, the only radiometric evidence of Precambrian rocks in the Ellsworth Mountains. In view of the altered nature and small size of the samples, however, it is equally possible that the data reflect isotopic homogenization on a hand-specimen scale during low-grade metamorphism, a phenomenon frequently claimed to occur in fine-grained felsic volcanic rocks. This would imply a mid-Devonian phase of metamorphism, an event earlier postulated by Yoshida (1982).

Most geologists who have worked in the Ellsworth Mountains feel that the range was deformed and metamorphosed by a single orogenic event (Ellsworth Orogeny) that is presently no more closely dated than Late Permian to early Mesozoic (Craddock, 1972; Ford, 1972). It is thought by Craddock (1983) to be probably Triassic in age. Hjelle and others (1982) and Yoshida (1982) have postulated other earlier periods of folding and/or low-grade metamorphism.

In choosing between the two alternative interpretations of the Rb-Sr isochron, we are impressed by the fact that the Rb-Sr data are apparently unaffected by the Ellsworth Orogeny, whereas this event apparently had a profound effect on the K-Ar whole-rock ages. Because the earlier phases of metamorphism postulated by Yoshida (1982) are of low grade, it would seem illogical that almost total resetting should have occurred in an earlier phase but none at all during the later Ellsworth Orogeny. We, therefore, consider it more likely that the Rb-Sr age represents a mid-Devonian intrusive emplacement of the Soholt Peaks stock and that low-grade metamorphism accompanying the Ellsworth Orogeny slightly disturbed the Rb-Sr systematics but had a much greater effect on the K-Ar whole-rock systems.

In relation to the three anomalously low DDR K-Ar dates from the Hyde Glacier sill (161 to 245 Ma), Yoshida's (1982) K-Ar dates of 381 ± 19 Ma from the Wilson Nunataks sill and 396 ± 20 Ma from the Hyde Glacier sill must be considered minimum ages of emplacement and also to have been partially reset. The apparent close correspondence between Yoshida's two dates and the age of emplacement of the Soholt Peaks stock would imply the lack of a genetic link between these three intrusive bodies and the volcanic rocks of the Heritage Group. All volcanic rocks discussed in this report are interbedded with sedimentary rocks containing Cambrian fossils (Webers, 1972; Hjelle and others, 1978, 1982; Webers and Spörli, 1983). All plutonic and hypabyssal rocks discussed in this study are intrusive only into Heritage Group sedimentary rocks. The lack of intrusive rocks, especially dikes, in the overlying Minaret Formation and Crashsite Quartzite and the close correspondence in chemical composition between the intrusive and extrusive rocks strongly suggests that the intrusive igneous rocks of the southern Heritage Range, except for the Soholt Peaks stock, are cogenetic with the volcanic rocks and were emplaced no later than Late Cambrian time. We regard the K-Ar dates obtained from the Hyde Glacier and Wilson Nunataks sills by Yoshida (1982) to be reset minimum ages of emplacement, and their close correspondence to the emplacement age of the Soholt Peaks stock to be coincidental.

COMPARISON WITH TRANSANTARCTIC MOUNTAINS IGNEOUS ROCKS

Mesozoic basalt and dolerite, generally tholeiitic in composition and Jurassic in age, are widely distributed on Southern Hem-

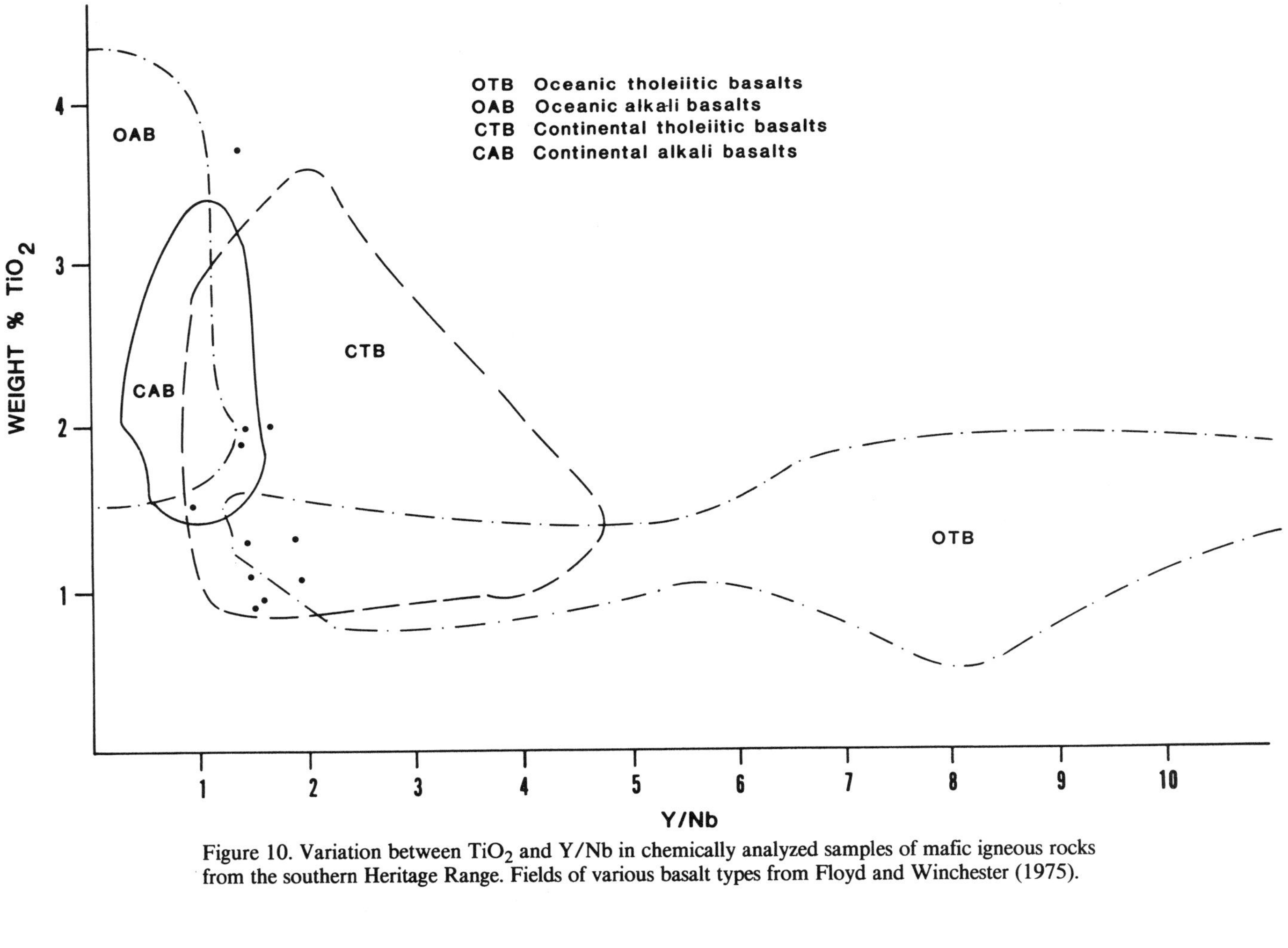

Figure 10. Variation between TiO_2 and Y/Nb in chemically analyzed samples of mafic igneous rocks from the southern Heritage Range. Fields of various basalt types from Floyd and Winchester (1975).

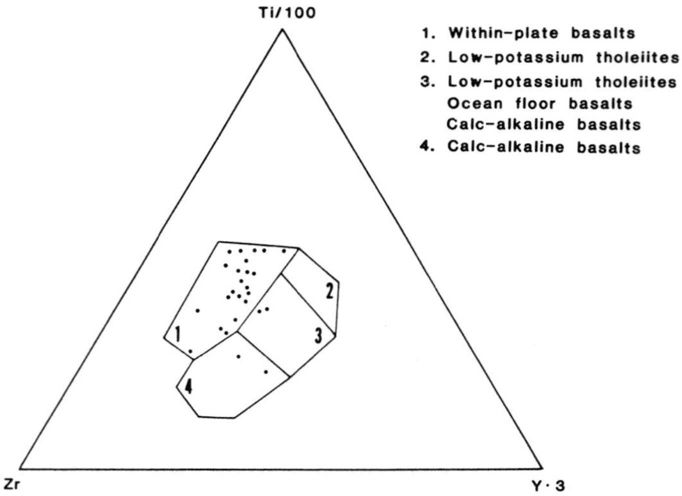

Figure 11. Ti-Zr-Y discrimination diagram from chemically analyzed samples of mafic igneous rocks from the southern Heritage Range. Fields of various basalt types from Pearce and Cann (1973).

isphere continents. In Antarctica, Ferrar Supergroup rocks crop out for 3,000 km along the Transantarctic Mountains from northern Victoria Land to the Pensacola Mountains (Fig. 1) and include lava flows (Kirkpatrick Basalt Group), dolerite sills and dikes (Ferrar Dolerite Group), and the Forrestal Gabbro Group, which is composed entirely of the Dufek intrusion, a very large (minimum area of 50,000 km^2) differentiated stratiform gabbroic intrusion (Kyle and others, 1981). The Ferrar Supergroup is chemically and isotopically similar to Tasmanian dolerite sills, and these two groups of rocks are now regarded as part of the same petrologic province (Heier and others, 1965, Compston and others, 1968; Faure and others, 1972, 1974, 1982; Kyle, 1980; Kyle and others, 1983). Basalt and dolerite (Kirwan Volcanics) that crop out in Queen Maud Land (Fig. 1) are chemically and isotopically distinct from the Ferrar Supergroup but are similar in these characteristics to the Karoo basalts of South Africa and the Serra Geral dolerites of Brazil (Faure and others, 1972, 1979). Ford and Kistler (1980) have recently shown that in Antarctica the boundary between these two Mesozoic basaltic petrologic provinces lies to the north of the Pensacola Mountains along the margin of the Weddell Sea.

In 1966, Gunn subdivided the Ferrar Dolerite in the McMurdo Sound area into three magma types: hypersthene tholeiite, pigeonite tholeiite, and much less common olivine tholeiite. Analyses of chilled hypersthene and pigeonite tholeiite

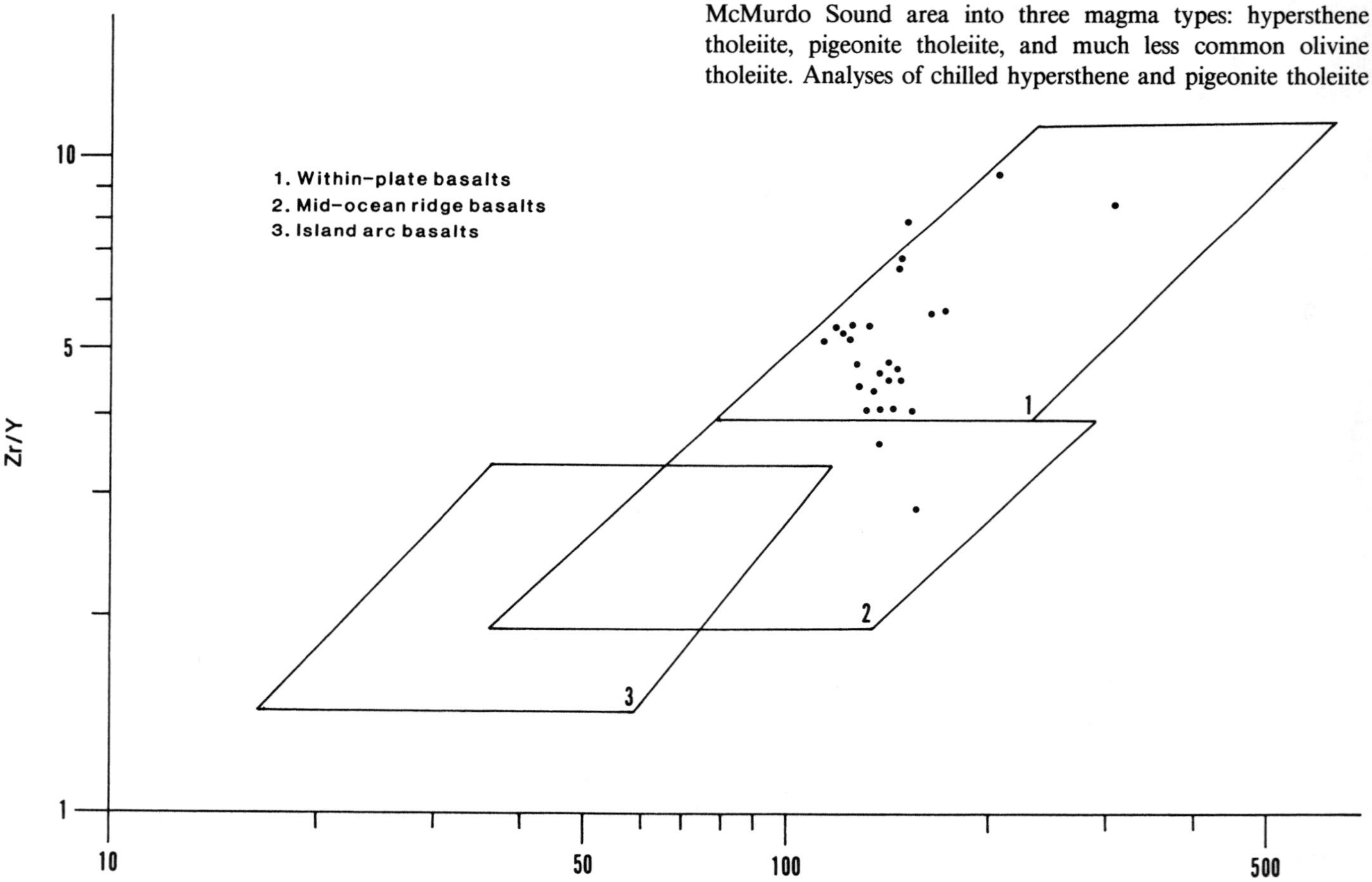

Figure 12. Zr/Y-Zr discrimination diagram from chemically analyzed samples of mafic igneous rocks from the southern Heritage Range. Fields of various basalt types from Pearce and Norry (1979).

are similar to some of the Kirkpatrick Basalt (Kyle, 1980). Chilled margins of sills from the Pecora Escarpment in the Pensacola Mountains are intermediate in composition between Gunn's (1966) hypersthene and pigeonite tholeiite, and average composition of mafic dikes from the Cordiner Peaks (also in the Pensacola Mountains) corresponds closely to hypersthene tholeiite (Ford and Kistler, 1980). Analyses of these rock types, as well as "average" Kirwan Volcanics (Faure and others, 1979), are given in Table 5 for comparison purposes. The basalts from Queen Maud Land have noticeably lower SiO_2, K_2O, Rb, and Rb/Sr and higher TiO_2, MgO, CaO, Sr, and K/Rb than Kirkpatrick Basalt.

Metamorphism of the Ellsworth Mountains basaltic rocks hinders comparison of their chemistry with nonmetamorphosed Ferrar Supergroup and Kirwan Volcanics. It should be pointed out, though, that some Kirkpatrick Basalt has been zeolitized and thus has also probably undergone some geochemical mobility (Elliot, 1970, 1972; Kyle, 1980). Consequently, only chilled margins of Heritage Range sills (analyses 7f, S10, S14, 14g, and 14l in Table 1A) are used for comparison, as these rocks should be the least differentiated and should thus more closely approximate original magma composition. These five analyses are also listed in Table 5 but are not averaged because there are significant chemical differences between margins of the Hyde Glacier and Wilson Nunataks sills. Compared to basaltic rocks of the Ferrar Supergroup, the Ellsworth Mountains sills have much lower SiO_2 (43.96 to 46.55 percent versus 50.40 to 56.60 percent), Cu, and Rb/Sr ratios and higher total iron oxides, Sr, and K/Rb ratios. The Hyde Glacier sill also has a much higher MgO content than all Ferrar Supergroup rocks except the rare olivine tholeiite magma types. Na_2O is also generally higher in the Ellsworth Mountains rocks than in the Ferrar Supergroup, although this element's distribution is prone to mobility. There are thus significant chemical differences between the Ellsworth Mountains mafic rocks and those of the Ferrar Supergroup. The same conclusion was reached earlier by Hjelle and others (1978, 1982).

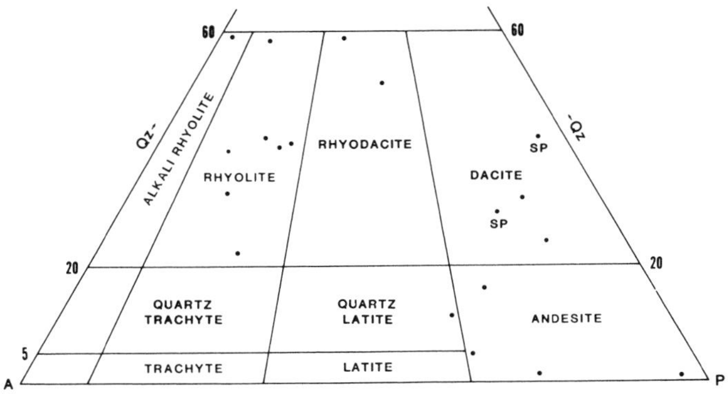

Figure 13. Ternary plot of normative quartz (Qz), orthoclase (A), and albite plus anorthite (P) from chemically analyzed samples of felsic and intermediate (>56 percent SiO_2) igneous rocks from the southern Heritage Range. Classification is from Streckeisen (1979).

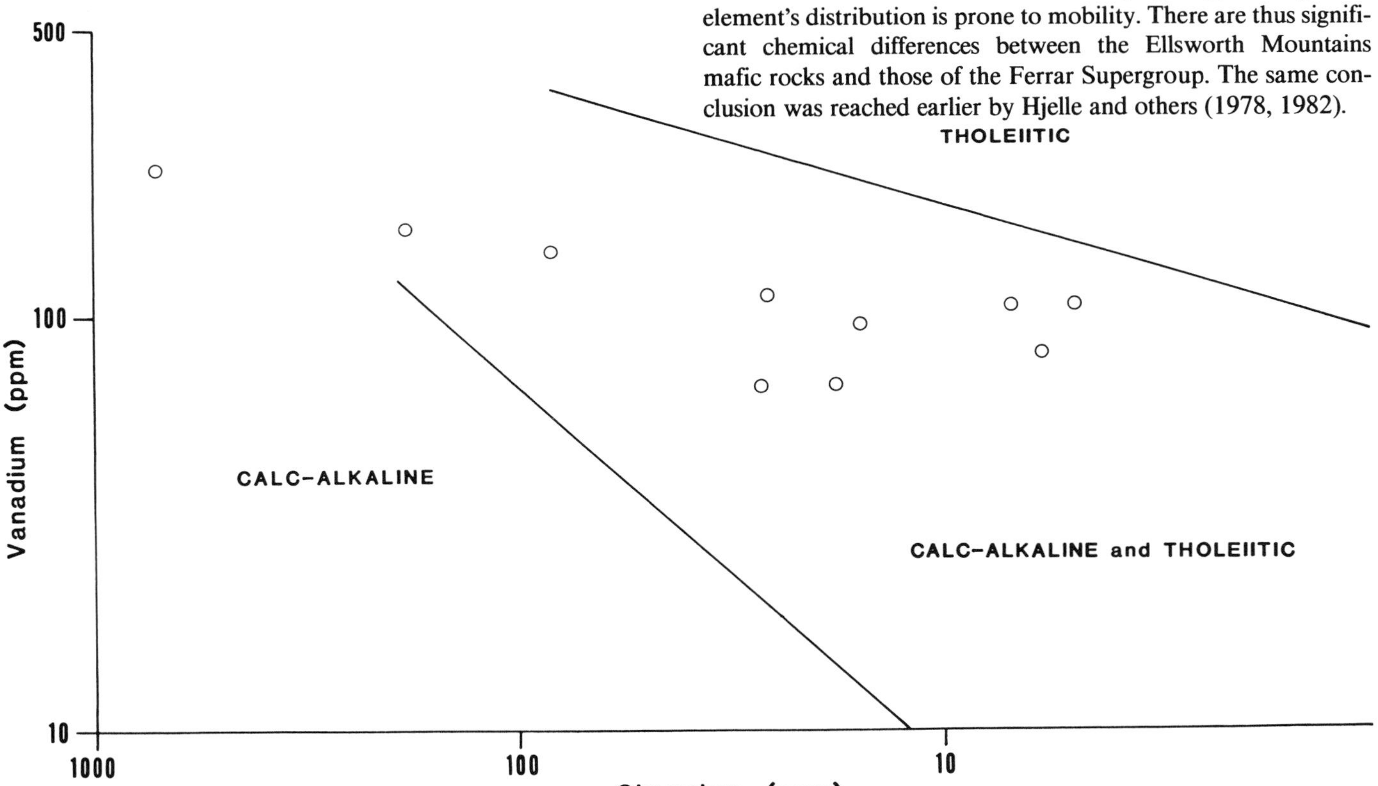

Figure 14. Plot of vanadium versus chromium for chemically analyzed samples of felsic and intermediate (>56 percent SiO_2) igneous rocks from the southern Heritage Range. Classification is from Miyashiro and Shido (1975).

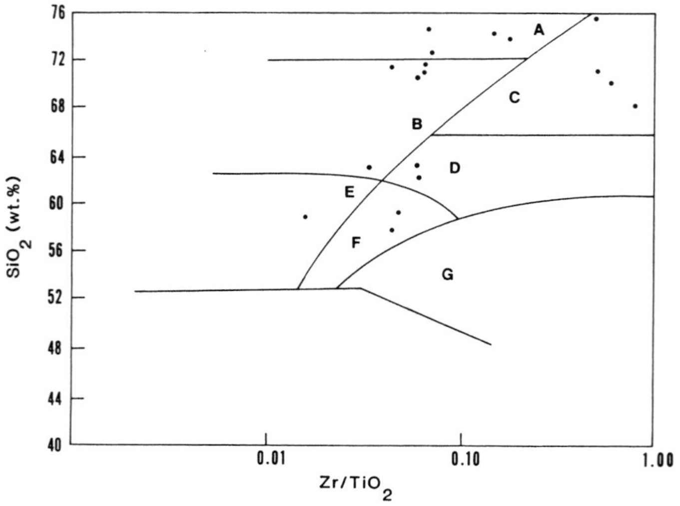

Figure 15. SiO_2-Zr/TiO_2 discrimination diagram for chemically analyzed samples of felsic and intermediate (>56 percent SiO_2) igneous rocks from the southern Heritage Range. Symbols used are A, rhyolite; B, dacite and rhyodacite; C, comendite and pantellerite; D, trachyte; E, andesite; F, trachyandesite; G, phonolite. Fields of various rock types from Winchester and Floyd (1977).

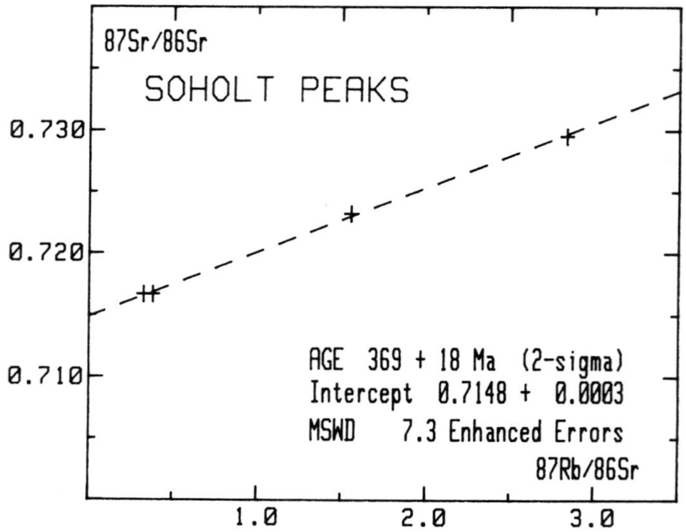

Figure 16. Rubidium-strontium whole-rock isochron diagram for the Soholt Peaks dacite stock.

When compared to average Kirwan basalt (Faure and others, 1979) the Ellsworth Mountains rocks are lower in SiO_2, TiO_2, and CaO but higher in Al_2O_3 and Na_2O. The Hyde Glacier sill has a much higher MgO and K_2O content, and the Wilson Nunataks sill has a much higher total iron oxide content. There is, however, considerable overlap in the values of Rb, Sr, K/Rb, and Rb/Sr. Thus these two groups of rocks do not seem to be chemical equivalents either, but the disparity is less and it should be pointed out that only a few samples are being compared. The transitional to mildly tholeiitic original chemistry of the Ellsworth Mountains mafic rocks predicted earlier by the use of alteration resistant trace elements is not a feature of either the Ferrar Supergroup or the Kirwan Volcanics. Felsic and intermediate rocks are relatively rare in both Ferrar Supergroup and Kirwan volcanic terranes but have been reported locally (Marshak and others, 1981).

CORRELATION

Only a few tentative correlations can be attempted between igneous rocks of the Ellsworth and Transantarctic mountains, the closest possible comparisons being in the Pensacola Mountains. Schmidt and others (1978) and Schmidt and Ford (1969) have mapped an extensive suite of basaltic and diabasic sills and dikes and subordinate plugs and sills of felsic porphyry of uncertain, but probable, late Precambrian age that are intrusive into the late Precambrian Patuxent Formation in the Neptune Range of the Pensacola Mountains. Unpublished chemical analyses of these mafic rocks are similar to those from the southern Heritage Range (A. B. Ford, personal communication, 1982). Dolerite dikes that intrude Cambrian-Ordovician sandstone in the Shackleton Range (northeast of the Pensacola Mountains) yield K-Ar dates between 297 and 457 Ma (Clarkson, 1981) but do not appear to be close chemical equivalents of the Ellsworth Mountains mafic rocks. Porphyritic metafelsic rocks also crop out in the Queen Maud Range of the central Transantarctic Mountains. These include the late Precambrian Wyatt Formation, which contains both extrusive and hypabyssal rocks as well as porphyritic metarhyolites and minor metabasalts that occur in the lower part of the Cambrian Taylor Formation along the Shackleton Glacier (Stump, 1982).

TECTONIC SETTING

Earlier in this paper, discrimination diagrams utilizing immobile trace elements were employed in an attempt to determine as much as possible about the original chemistry and tectonic setting of the metamorphosed igneous rocks of the southern Heritage Range. We concluded on their basis and on the basis of field setting (an associated thick sequence of carbonate and clastic sedimentary rocks) and scarcity of andesite that the mafic rocks were continental in setting and largely tholeiitic in composition but that they also displayed some calc-alkaline tendencies even when plotted on diagrams (FeO* versus FeO*/MgO) designed to totally negate the effect of alkali mobility. The TiO_2-Y/Nb diagram, however, suggests that some of the mafic rocks are transitional in composition between tholeiitic and alkali basalt. Intermediate to felsic igneous rocks are much less abundant than mafic igneous rocks in the southern Heritage Range, and the intermediate rocks are far less abundant than the felsic ones. The

TABLE 3. WHOLE ROCK K-AR DATES FROM IGNEOUS ROCKS OF THE SOUTHERN HERITAGE RANGE

Rock Type and Location	K (%)	$Ar^{40}Rad$ (%)	$Ar^{40}Rad$ X 10^{-9} Mole/g	K^{40} X 10^{-7} Mole/g	Isotopic Age (Ma)
Diabase-upper part Hyde Glacier Sill*	1.39	88.6	0.569	0.415	220.0 ± 9.4
Diabase-central part Hyde Glacier Sill (S16)*	0.84	52.2	0.246	0.251	161.6 ± 11.2
Diabase-lower part Hyde Glacier Sill*	1.24	88.7	0.565	0.370	245.2 ± 11.2
Alkali basalt flow High Nunatak*	2.55	92.8	1.149	0.761	242.8 ± 10.9
Alkali basalt flow High Nunatak (S2)*	2.42	91.4	1.386	0.722	303.4 ± 13.9
Alkali basalt flow High Nunatak*	2.76	88.5	1.260	0.824	245.7 ± 10.4
Diabase- Hyde Glacier Sill†	0.43	86.3	0.740§		396 ± 20
Diabase- Wilson Nunataks Sill†	0.785	90.8	1.295§		381 ± 19
Andesite Dike- Edson Hills†	4.88	97.6	4.805§		237 ± 12
Basaltic Dike- Edson Hills†	2.35	96.4	2.495§		254 ± 13

Note: *Samples analyzed by I. Pilot, G. Kaiser, and V. Klemm, German Democratic Republic. Constants: $\lambda\beta = 4.72 \times 10^{-10} a^{-1}$, $\lambda_\varepsilon = 0.584 \times 10^{-10} a^{-1}$.

†Samples from Yoshida (1982), analyzed by Teledyne Isotopes Co., New Jersey. Constants: $\lambda\beta = 4.962 \times 10^{-10} a^{-1}$, $\lambda\varepsilon = 0.581 \times 10^{-10} a^{-1}$, $K^{40} = 1.167 \times 10^{-4}$ atom/atom of natural potassium.

§scc $Ar^{40}Rad$/g X 10^{-5}.

TABLE 4. RUBIDIUM-STRONTIUM DATA FROM THE SOHOLT PEAKS DACITE STOCK

Sample Number	Rb (ppm)	Sr (ppm)	$^{87}Rb/^{86}Sr$	$^{87}Sr/^{86}Sr$
V79a	14.4	110	0.380	0.71665
V79b	13.4	119	0.325	0.71665
V79c	83.7	85.9	2.834	0.72955
V79d	71.4	133	1.555	0.72319

Error: Rb, Sr ± 2%; Rb/Sr ± 0.5%; $^{87}Sr/^{86}Sr$ ± 0.01%

TABLE 5. COMPOSITION OF FERRAR SUPERGROUP AND KIRWAN VOLCANICS COMPARED TO ELLSWORTH MOUNTAINS SILL MARGINS

	A	B	C	D	E	F	G	H	I	J
SiO_2	50.40	53.75	55.65	53.3	54.6	56.50	48.07	55.86	46.34	44.45
TiO_2	0.44	0.70	1.03	0.65	0.75	1.28	2.06	1.35	2.44	1.13
Al_2O_3	15.51	14.33	13.95	15.5	14.5	12.92	13.03	13.55	13.99	15.87
Fe_2O_3				1.5	1.8	3.2	3.64		2.89	2.63
FeO	8.72*	9.65*	10.30*	8.0	8.2	8.5	9.33	11.51*	12.21	8.94
MnO	0.17	0.18	0.17	0.13	0.18	0.22	0.19	0.18	0.21	0.17
MgO	10.60	6.64	4.50	6.6	5.6	3.44	8.59	3.50	6.12	10.33
CaO	10.87	10.60	8.51	10.3	9.5	7.91	9.70	8.12	8.08	7.35
Na_2O	1.42	1.83	2.50	1.9	1.8	2.29	2.44	2.24	3.55	2.64
K_2O	0.37	0.81	1.45	0.72	0.82	1.29	0.51	1.43	0.56	1.38
P_2O_5	0.08	0.18	0.23	0.15	0.12	0.16	0.22	0.15	0.33	0.09
Ba	157	232	376						370	262
Cr	352	142	59	168					71	178
Cu	69	118	123	112					129	45
Ni	249	85	53	89					87	125
Rb	12	30	50	21	28	61	10	55	20	27
Sr	100	126	138	172	113	129	213	128	270	353
Zr	53	83	157						162	125
K/Rb	308	270	290	267	182	215	500	260	280	510
Rb/Sr	0.12	0.24	0.36	0.12	0.25	0.47	0.47	0.43	0.07	0.08

Note: A = chilled margin olivine tholeiite sill, Gunn (1966); B = average 5 analyses chilled margins hypersthene tholeiite sills, Gunn (1966); C = average 4 analyses chilled margins pigeonite tholeiite sills, Gunn (1966); D = average 12 analyses Cordiner Peaks dikes, Ford and Kistler (1980); E = average 5 analyses chilled margins Pecora Escarpment sills, Ford and Kistler (1980); F = average 12 analyses Kirkpatrick Basalt, Storm Peak, Faure and others (1974); G = average 17 analyses Kirwan Volcanics, Faure and others (1979); H = average 14 analyses Kirkpatrick Basalt, Mt. Falla, Faure and others (1982); I = average 3 analyses chilled margins Wilson Nunataks sill; J = average 2 analyses chilled margins Hyde Glacier sill.
*Total iron as FeO.

igneous rocks thus constitute a bimodal suite characterized by large amounts of mafic rocks and very few of intermediate composition. Discrimination diagrams, although not as extensively developed for use with more siliceous rocks, and content of incompatible trace elements suggest that the felsic and intermediate rocks also show both tholeiitic and calc-alkaline tendencies along with some alkaline traits.

It is well known that igneous rock suites, especially volcanic ones, often grade from tholeiitic through calc-alkaline to shoshonitic through time and/or space across island arc–continental margin situations (Jakeš and White, 1972). This is not the case here, however. Igneous rocks in the southern Heritage Range are interlayered with, or intrusive into, the same portion of the sedimentary section and are thus regarded as being contemporaneous in time and space. If the conclusions based on the discrimination diagrams are valid, then the igneous rock association discussed in this paper is somewhat unusual. Similar sequences have, however, been reported from the Yellowknife volcanic belt in Canada's Northwest Territories (Irvine and Barager, 1971) and from southeastern Maine (Gates and Moench, 1981). In addition, Arculus and Johnson (1978) have shown that tholeiitic, calc-alkaline, and alkaline volcanism can occur simultaneously in different parts of the same terrane.

Basaltic or strongly bimodal basaltic-rhyolitic suites with scarce andesite are characteristic of continental areas subjected to plate separation and extension. In direct contrast, igneous rock suites developed along converging plate boundaries usually are unimodal and display a wide range of compositions often dominated by andesite (Christiansen and Lipman, 1972; Lipman and others, 1972; Miyashiro, 1974). Bimodal convergent suites have been noted (individual Cascade volcanoes and individual islands of the Lesser Antilles), but in these cases, andesite is an abundant rock type (McBirney, 1968; Brown and others, 1977). We therefore conclude that the igneous rocks of the southern Heritage Range were emplaced in a continental area undergoing extensional tectonism and do not represent an island arc–continental margin subduction zone complex.

ACKNOWLEDGMENTS

This project was funded by National Science Foundation grant DPP78-21720 to Gerald Webers of Macalester College.

Logistical support was provided by the Antarctic Research Program of the National Science Foundation and U.S. Navy squadron VXE-6. We thank Philip Fenn of the University of California, Davis, for his help with the chemical analyses and Ronald Leu of Sonoma State University for preparing the thin sections. Discussions in the field with Bernard Spörli and over many glasses of beer with Arthur Ford were especially helpful. All figures were drafted by Marc Druckman of Sonoma State University. A previous version of this manuscript benefited from reviews by Paul Weiblen of the University of Minnesota and Glen Himmelberg of the University of Missouri.

REFERENCES CITED

Arculus, R. J., and Johnson, R. W., 1978, Criticism of generalized models for the magmatic evolution of arc-trench systems: Earth and Planetary Science Letters, v. 39, p. 118–126.

Bass, M. N., Moberly, R., Rhodes, J. M., Shih, C., and Church, S. E., 1973, Volcanic rocks cored in the central Pacific, leg 17, deep sea drilling project, *in* Winterer, E. S., and Ewing, J. L., eds., Initial reports of the Deep Sea Drilling Project: Washington, D.C., U.S. Goverment Printing Office, v. 17, p. 429–503.

Bauer, R. L., 1983, Low-grade metamorphism in the Heritage Range of the Ellsworth Mountains, West Antarctica, *in* Oliver, R. L., James, P. R., and Jago, J. B., eds., Antarctic earth science: Canberra, Australian Academy of Science, p. 256–260.

Brooks, C. K., 1976, The Fe_2O_3/FeO ratio of basalt analyses; An appeal for a standardized procedure: Geological Society of Denmark Bulletin, v. 25, p. 117–120.

Brooks, E. R., and Cole, D. G., 1980, Use of immobile trace elements to determine original tectonic setting of eruption of metabasalts, northern Sierra Nevada, California: Geological Society of America Bulletin, v. 91, pt. 1, p. 665–671.

Brown, G. M., Holland, J. G., Sigurdsson, H., Tomblin, J. F., and Arculus, R. J., 1977, Geochemistry of the Lesser Antilles volcanic island arc: Geochimica et Cosmochimica Acta, v. 41, p. 758–801.

Cann, J. R., 1970, Rb, Sr, Y, Zr, and Nb in some ocean floor basaltic rocks: Earth and Planetary Science Letters, v. 10, p. 7–11.

Christiansen, R. L., and Lipman, P. W., 1972, Cenozoic volcanism and plate tectonic evolution of the western United States; 2, late Cenozoic: Philosophical Transactions of the Royal Society of London, ser. A, v. 271, no. 1213, p. 249–284.

Clarkson, P. D., 1981, Geology of the Shackleton Range; 4, The dolerite dikes: British Antarctic Survey Bulletin 53, p. 201–212.

Compston, W., McDougall, I., and Heier, K. S., 1968, Geochemical comparison of the Mesozoic basaltic rocks of Antarctica, South Africa, South America, and Tasmania: Geochimica et Cosmochimica Acta, v. 32, p. 129–150.

Craddock, C., 1969, Geology of the Ellsworth Mountains, *in* Bushnell, V. C., and Craddock, C., eds., Geologic maps of Antarctica: New York, America Geographical Society, Antarctic Map Folio Series, Folio 12, Plate 4, scale 1:1,000,000.

——, 1972, Antarctic tectonics, *in* Adie, R. J., ed., Antarctic geology and geophysics: Oslo, Universitetsforlaget, p. 449–455.

——, 1983, The East Antarctica–West Antarctica boundary between the ice shelves; A review, *in* Oliver, R. L., James, P. R., and Jago, R. B., eds., Antarctic earth science: Canberra, Australian Academy of Science, p. 94–97.

Craddock, C., Anderson, J. J., and Webers, G. F., 1964, Geological outline of the Ellsworth Mountains, *in* Adie, R. J., ed., Antarctic geology: Amsterdam, North-Holland, p. 155–170.

Elliot, D. H., 1970, Jurassic tholeiites of the central Transantarctic Mountains, Antarctica, *in* Gilmour, E. H., and Stradling, D., eds., Proceedings of the second Columbia River Basalt Symposium: Cheney, Eastern Washington State College Press, p. 301–325.

——, 1972, Major oxide chemistry of the Kirkpatrick Basalt, central Transantarctic Mountains, *in* Adie, R. J., ed., Antarctic geology and geophysics: Oslo, Universitetsforlaget, p. 413–418.

Faure, G., Hill, R. L., Jones, L. M., and Elliot, D. H., 1972, Isotope composition of strontium and silica content of Mesozoic basalt and dolerite from Antarctica, *in* Adie, R. J., ed., Antarctic geology and geophysics: Oslo, Universitetsforlaget, p. 617–624.

Faure, G., Bowman, J. R., Elliot, D. H., and Jones, L. M., 1974, Strontium isotope composition and petrogenesis of the Kirkpatrick Basalt, Queen Alexandra Range, Antarctica: Contributions to Mineralogy and Petrology, v. 48, p. 153–169.

Faure, G., Bowman, J. R., and Elliot, D. H., 1979, The isotopic composition of strontium of the Kirwan Volcanics, Queen Maud Land, Antarctica: Chemical Geology, v. 26, p. 77–90.

Faure, G., Pace, K. K., and Elliot, D. H., 1982, Systematic variations of $^{87}Sr/^{86}Sr$ ratios and major element concentrations in the Kirkpatrick Basalt of Mount Falla, Queen Alexandra Range, Transantarctic Mountains, *in* Craddock, C., ed., Antarctic geoscience: Madison, University of Wisconsin Press, p. 715–723.

Floyd, P. A., and Winchester, J. A., 1975, Magma type and tectonic setting discrimination using immobile elements: Earth and Planetary Science Letters, v. 27, p. 211–218.

——, 1978, Identification and discrimination of altered and metamorphosed volcanic rocks using immobile elements: Chemical Geology, v. 21, p. 291–306.

Ford, A. B., 1972, Weddell orogeny; Latest Permian to early Mesozoic deformation at the Weddell Sea margin of the Transantarctic Mountains, *in* Adie, R. J., ed., Antarctic geology and geophysics: Oslo, Universitetsforlaget, p. 419–425.

Ford, A. B., and Kistler, R. W., 1980, K-Ar age, composition, and origin of Mesozoic mafic rocks related to Ferrar Group, Pensacola Mountains, Antarctica: New Zealand Journal of Geology and Geophysics, v. 23, p. 371–390.

Gates, O., and Moench, R. H., 1981, Bimodal Silurian and Lower Devonian volcanic rock assemblages in the Machias–Eastport area, Maine: U.S. Geological Survey Professional Paper 1184, 32 p.

Gizycki, P., 1983, Die magmatischen gesteine der Ellsworth Mountains, West-Antarktis: Neues Jahrbuch für Geologie und Paläontologie, Abhandlungen, Band 167, Heft 1, p. 65–88.

Gunn, B. M., 1966, Modal and element variation in Antarctic tholeiites: Geochimica et Cosmochimica Acta, v. 30, p. 881–920.

Heier, K. S., Compston, W., and McDougall, I., 1965, Thorium and uranium concentrations and the isotopic composition of strontium in the differentiated Tasmanian dolerites: Geochimica et Cosmochimica Acta, v. 29, p. 643–659.

Hjelle, A., Ohta, Y., and Winsnes, T. S., 1978, Stratigraphy and igneous petrology of southern Heritage Range, Ellsworth Mountains, Antarctica: Norsk Polarinstitutt Skrifter, v. 169, p. 5–43.

——, 1982, Geology and petrology of the southern Heritage Range, Ellsworth Mountains, *in* Craddock, C., ed., Antarctic geoscience: Madison, University of Wisconsin Press, p. 599–608.

Holm, P. E., 1982, Non-recognition of continental tholeiite using the Ti-Y-Zr diagram: Contributions to Mineralogy and Petrology, v. 79, p. 308–310.

Irvine, T. N., and Barager, W.R.A., 1971, A guide to the chemical classification of the common volcanic rocks: Canadian Journal of Earth Sciences, v. 8, p. 523–548.

Jakeš, P., and Gill, J. B., 1970, Rare earth elements and the island arc tholeiite series: Earth and Planetary Science Letters, v. 9, p. 17–28.

Jakeš, P., and White, A.J.R., 1972, Major and trace element abundances in volcanic rocks of orogenic areas: Geological Society of America Bulletin, v. 83, p. 29–40.

Kochhar, N., 1977, Post-emplacement modifications in rapidly cooled acid volcanic rocks; American Mineralogist, v. 62, p. 333–335.

Kyle, P. R., 1980, Development of heterogeneities in the subcontinental mantle; Evidence from the Ferrar Group, Antarctica: Contributions to Mineralogy

and Petrology, v. 73, p. 89–104.
Kyle, P. R., Elliot, D. H., and Sutter, J. F., 1981, Jurassic Ferrar Supergroup tholeiites from the Transantarctic Mountains, Antarctica, and their relationship to the initial fragmentation of Gondwana, *in* Cresswell, M. M., and Vella, P., eds., Gondwana Five: Rotterdam, A. A. Balkema, p. 283–287.
Kyle, P. R., Pankhurst, R. J., and Bowman, J. R., 1983, Isotopic and chemical variations in Kirkpatrick Basalt Group rocks from southern Victoria Land, *in* Oliver, R. L., James, P. R., and Jago, J. B., eds., Antarctic earth science: Canberra, Australian Academy of Science, p. 234–237.
Leake, B. E., 1978, Nomenclature of amphiboles: American Mineralogist, v. 63, p. 1023–1053.
Lipman, P. W., Christiansen, R. L., and Van Alstine, R. E., 1969, Retention of alkalies by calc-alkalic rhyolites during crystallization and hydration: American Mineralogist, v. 54, p. 286–291.
Lipman, P. W., Prostka, H. J., and Christiansen, R. L., 1972, Cenozoic volcanism and plate tectonic evolution of the western United States; 1, Early and middle Cenozoic: Philosophical Transactions of the Royal Society of London, ser. A, v. 271, no. 1213, p. 217–248.
Lofgren, G. E., 1971, Experimentally produced devitrification textures in natural rhyolitic glass: Geological Society of America Bulletin, v. 82, p. 111–124.
Macdonald, G. A., and Katsura, T., 1964, Chemical composition of Hawaiian lavas: Journal of Petrology, v. 5, p. 82–133.
Marshak, S., Kyle, P. R., McIntosh, W. C., Samsonov, V. V., and Shellhorn, M., 1981, Butcher Ridge igneous complex, Cook Mountains, Antarctica: Antarctic Journal of the United States, v. 16, p. 54–55.
McBirney, A. R., 1968, Petrochemistry of the Cascade andesite volcanoes, *in* Dole, H. M., ed., Andesite Conference Guidebook: Oregon Department of Geology and Mineral Industries Bulletin 62, p. 101–107.
Miyashiro, A., 1974, Volcanic rock series in island arcs and active continental margins: American Journal of Science, v. 274, p. 321–355.
Miyashiro, A., and Shido, F., 1975, Tholeiitic and calc-alkalic series in relation to the behavior of titanium, vanadium, chromium, and nickel: American Journal of Science, v. 275, p. 265–277.
Noble, D. C., 1968, Systematic variation of major elements in comendite and pantelleritic glasses: Earth and Planetary Science Letters, v. 4, p. 167–172.
Nockolds, S. R., Knox, R. W., and Chinner, G. A., 1978, Petrology for students: New York, Cambridge University Press, 435 p.
Pankhurst, R. J., and O'Nions, R. K., 1973, Determination of Rb/Sr and $^{87}Sr/^{86}Sr$ ratios of some standard rocks and evaluation of x-ray fluorescence spectrometry in Rb-Sr geochemistry: Chemical Geology, v. 12, p. 127–136.
Pearce, J. A., 1975, Basalt geochemistry used to investigate past tectonic environments on Cyprus: Tectonophysics, v. 25, p. 41–67.
Pearce, J. A., and Cann, J. R., 1971, Ophiolite origin investigated by discriminant analysis using Ti, Zr, and Y: Earth and Planetary Science Letters, v. 12, p. 339–349.
——, 1973, Tectonic setting of basic volcanic rocks determined using trace element analysis: Earth and Planetary Science Letters, v. 19, p. 290–300.
Pearce, J. A., and Norry, M. J., 1979, Petrogenetic implications of Ti, Zr, Y, and Nb variations in volcanic rocks: Contributions to Mineralogy and Petrology, v. 69, p. 33–47.
Pearce, T. H., Gorman, B. E., and Birkett, T. C., 1975, The TiO_2-K_2O-P_2O_5 diagram; A method of discriminating between oceanic and non-oceanic basalts: Earth and Planetary Science Letters, v. 24, p. 419–426.
Prestvik, T., 1982, Basic volcanic rocks and tectonic setting; A discussion of the Zr-Ti-Y discrimination diagram and its suitability for classification purposes: Lithos, v. 15, p. 241–247.
Saunders, A. J., Tarney, J., Stern, C. R., and Dalziel, I.W.D., 1979, Geochemistry of Mesozoic marginal basin floor igneous rocks from southern Chile: Geological Society of America Bulletin, v. 90, pt. 1, p. 237–258.
Schmidt, D. L., and Ford, A. B., 1969, Geology of the Pensacola and Thiel Mountains, *in* Bushnell, V. C., and Craddock, C., eds., Geologic maps of Antarctica: New York, American Geographical Society, Antarctic Map Folio Series, Folio 12, Plate 5, scale 1:1,000,000.
Schmidt, D. L., Williams, P. L., and Nelson, W. H., 1978, Geologic map of the Schmidt Hills Quadrangle and part of the Gambacorta Peak Quadrangle, Pensacola Mountains, Antarctica: U.S. Geological Survey Antarctic Geologic Map A–8, scale 1:250,000.
Splettstoesser, J. F., and Webers, G. F., 1989, Geological investigations and logistics in the Ellsworth Mountains, 1979–80: Antarctic Journal of the United States, v. 15, no. 5, p. 36–39.
Streckeisen, A. L., 1979, Classification and nomenclature of volcanic rocks, lamprophyres, carbonatites, and melilitic rocks; Recommendations and suggestions of the IUGS subcommission on the systematics of igneous rocks: Geology, v. 7, p. 331–335.
Stump, E., 1982, The Ross supergroup in the Queen Maud Mountains, *in* Craddock, C., ed., Antarctic geoscience: Madison, University of Wisconsin Press, p. 565–569.
Webers, G. F., 1972, Unusual upper Cambrian fauna from West Antarctica, *in* Adie, R. J., ed., Antarctic geology and geophysics: Oslo, Universitetsforlaget, p. 235–237.
Webers, G. F., and Spörli, K. B., 1983, Palaeontological and stratigraphic investigations in the Ellsworth Mountains, West Antarctica, *in* Oliver, R. L., James, P. R., and Jago, J. B., eds., Antarctic earth science: Canberra, Australian Academy of Science, p. 261–264.
Williams, H., Turner, F. J., and Gilbert, C. M., 1982, Petrography, 2nd ed.: San Francisco, California, W. H. Freeman, 626 p.
Winchester, J. A., and Floyd, P. A., 1977, Geochemical discrimination of different magma series and their differentiation products using immobile elements: Chemical Geology, v. 20, p. 325–343.
Yoder, H. S., Jr., and Tilley, C. E., 1962, Origin of basalt magmas; An experimental study of natural and synthetic rock systems: Journal of Petrology, v. 3, p. 342–532.
Yoshida, M., 1982, Superposed deformation and its implication to the geologic history of the Ellsworth Mountains, West Antarctica: Tokyo, Memoirs of National Institute of Polar Research Special Issue 21, p. 120–171.

MANUSCRIPT ACCEPTED BY THE SOCIETY MARCH 1, 1989

Printed in U.S.A.

Geological Society of America
Memoir 170
1992

Chapter 16

Radioelement distribution in the sedimentary sequence of the Ellsworth Mountains, West Antarctica

G.A.M. Dreschhoff and E. J. Zeller
Department of Geology, University of Kansas, Lawrence, Kansas 66045
V. Thoste
Bundesanstalt für Geowissenschaften und Rohstoffe, 3000 Hannover 51, Federal Republic of Germany

ABSTRACT

Airborne gamma-ray spectrometry can be used as a remote sensing technique to measure the radioelement distribution in the sedimentary sequence that crops out in the Ellsworth Mountains. By means of a highly sensitive detector system it is possible to obtain an almost instantaneous chemical analysis for equivalent uranium, thorium, and potassium that is updated in flight each second. Data provided by this technique, together with visual information recorded during flights, can serve as a reliable basis for stratigraphic correlation and can furnish additional information on diagenetic and sedimentological factors. The radiometric signature of each of the outcropping sedimentary units was determined, and this signature was found to vary within clearly defined limits. The study revealed no significant radiometric anomalies in the Ellsworth Mountains, and this implies that the low-grade metamorphism that is widespread in the area was not sufficient to cause substantial alteration of the original radioelement distribution in the sedimentary rocks.

INTRODUCTION

Field measurements for the determination of the radioelement distribution were made using an airborne system for the detection of gamma rays (Zeller and Dreschhoff, 1980; Dreschhoff and others, 1983) in the Ellsworth Mountains during the 1979–1980 field season. Both the Heritage and Sentinel Ranges were included in the survey area, which is roughly 300 km in length by 60 km in width. The flight tracks in the survey area are indicated in Figure 1. Outcrops are limited in the region, which receives more than 50 cm of snow each year, and they are generally confined to steep slopes, ridge crests, and a few localities where high winds sweep rock surfaces free of snow. Exposures are generally discontinuous, although several small dry valleys of limited areal extent exist in the Heritage Range. Representative examples of all the geologic formations present in the Heritage Range were examined by this aerial survey. In the Sentinel Range, which rises 5,000 m above the ice, flight operations were limited to a profile on the west side of the mountain range where the best outcrops occur. Most of the rocks that were surveyed are sedimentary, ranging in age from Cambrian to Permian limited outcrops of basic igneous rocks were also examined. The airborne survey track was flown mainly over Lower Paleozoic sedimentary rocks because these outcrops were most readily accessible from the camp location. Two flights did make traverses over Upper Paleozoic sedimentary rocks, but in both cases the outcrops were near the end of helicopter range from the camp and flight time was limited.

A comprehensive review of the stratigraphy of the Ellsworth Mountains is presented by other authors in this volume and has been discussed previously in several references (Craddock and others, 1964; Webers, 1970; Hjelle and others, 1982; Webers and

Dreschhoff, G.A.M., Zeller, E. J., and Thoste, V., 1992, Radioelement distribution in the sedimentary sequence of the Ellsworth Mountains, West Antarctica, *in* Webers, G. F., Craddock, C., and Splettstoesser, J. F., Geology and Paleontology of the Ellsworth Mountains, West Antarctica: Boulder, Colorado, Geological Society of America Memoir 170.

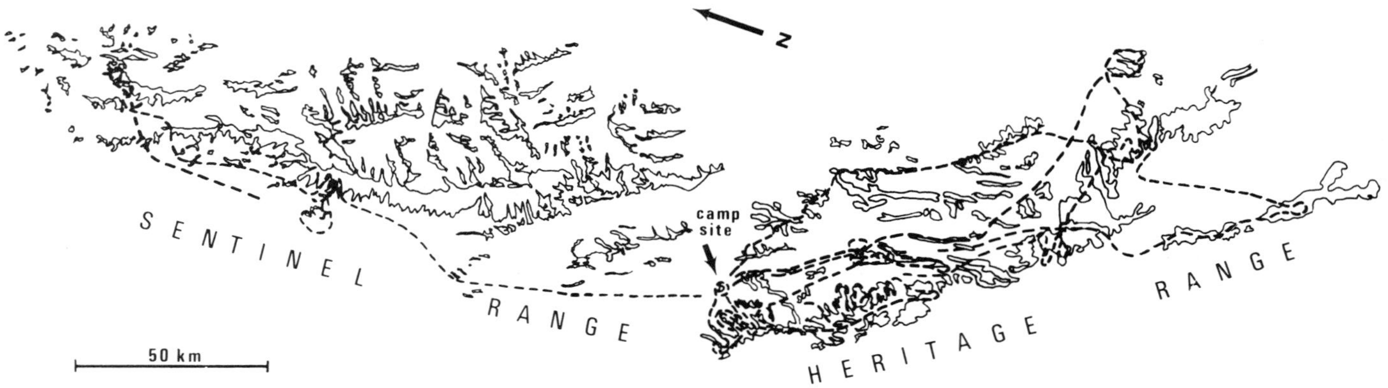

Figure 1. Map showing flight paths in the Ellsworth Mountains.

Spörli, 1983). A brief summary of the stratigraphic sequence has been included to provide a better understanding of the basis used in our regional correlations. Fortunately, many portions of the stratigraphic sequence are lithologically distinct and can be recognized with ease from the air during normal helicopter flights. When the data from the gamma-ray spectrometer are added to the visual information obtained during the flights, the geologist rapidly attains considerable skill in identifying the formations over which the airborne traverses are conducted.

GENERAL STRATIGRAPHY

The rocks of the Heritage Group consist mainly of contorted argillite, phyllite, sandstone, and black shale with gray limestone containing trilobites of Middle and Upper Cambrian age. These rocks form many of the mountain ridges of the Heritage Range. At the top of the Heritage Group, the Minaret Formation is exposed in the Heritage Range, where it forms the core of an anticline that makes up the southernmost part of the central Heritage Range. Marble is prominently developed, and massive brecciated zones occur at several localities.

The Crashsite Group overlies the Heritage Group and also shows low-grade metamorphism, with tight folds present in many areas. Quartzite is by far the most common rock type, but some localities have fairly extensive interbedding of phyllite, and a few areas in the Heritage Range show nearly unmetamorphosed sandstone and shale. The Crashsite Group has been subdivided into three formations by Spörli (this volume). The Crashsite Group forms the tops and high ridges of the Sentinel Range.

The Whiteout Conglomerate, a distinctive formation that is in conformable contact with the underlying Crashsite Group, is a poorly sorted dark diamictite with cobbles and boulders distributed throughout most of its thickness. It is usually easily recognized from the air and serves as a valuable marker in survey operations. At the northern end of the Sentinel Range, the Whiteout is overlain by the Polarstar Formation, which occurs in a broad syncline. The Polarstar contains plant fossils of Permian age and is generally characterized by dark conglomerate, siltstone, and graywacke and by thick but weakly folded black shale containing fossil plant remains. Shallow-water sedimentary structures are common in many portions of the sedimentary sequence. Some thin coal beds are also present (Collinson and others, 1980).

GEOLOGICAL USE OF GAMMA-RAY SPECTROMETER DATA

The gamma-ray spectrometer records separately the radiation from potassium (K-40), uranium (measured as Bi-214), and thorium (measured as Th-208) (Zeller and Dreschoff, 1980; Dreschhoff and others, 1983). The average abundances of these elements are known in crustal rocks, and these relations are assumed to have developed when the elements migrated upward during the early phases of crustal evolution (Gabelman, 1976). Under oxidizing conditions, uranium is much more mobile than thorium, and the ratio of these two elements can be used to determine the geochemical conditions of deposition. Because these conditions are usually maintained during the accumulation of sedimentary formations, the thorium-to-uranium ratio often provides a valuable identifier for specific lithologic units. By assuming an average thorium-to-uranium ratio of 3.5 to 1 for unaltered crustal rocks (Rogers and Adams, 1969), it is possible to separate areas of geochemical removal from those of enrichment (Darnley and others, 1977; Saunders, 1979).

Data evaluation using radiometric parameters

To evaluate the aerial gamma-ray data, the potassium, uranium, and thorium concentrations of all the individual outcrops have been divided by the mean concentrations for the entire survey area in the Ellsworth Mountains. This technique provides a basis for determining the relative amount of radioelement enrichment or depletion for individual outcrops within the area (Dreschhoff and others, 1983). The results, shown in Figure 2, are plotted in a scatter diagram of ATh (adjusted thorium) versus AU (adjusted uranium). In the Ellsworth Mountains, the mean con-

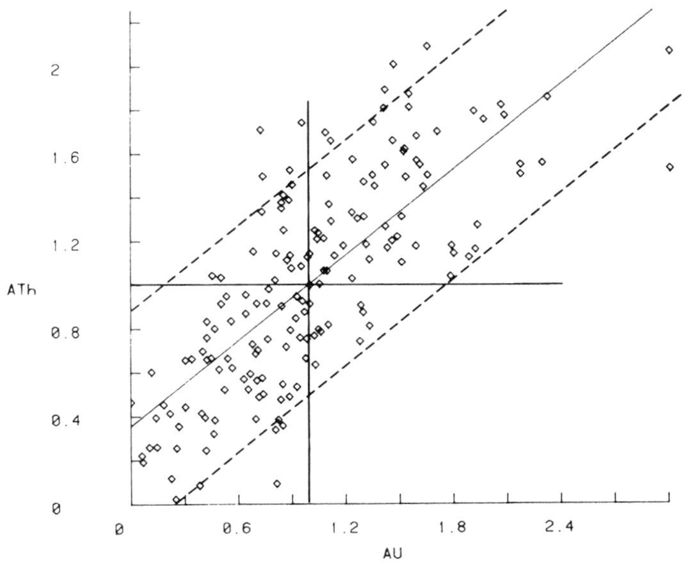

Figure 2. Scatter diagram of adjusted thorium (ATh) versus adjusted uranium (AU) of all outcrops surveyed in the Ellsworth Mountains. A linear regression line (solid) has been fitted to the data, and the broken lines indicate two standard deviations above and below the regression line. Heavy solid lines represent AU = 1 and ATh = 1, respectively.

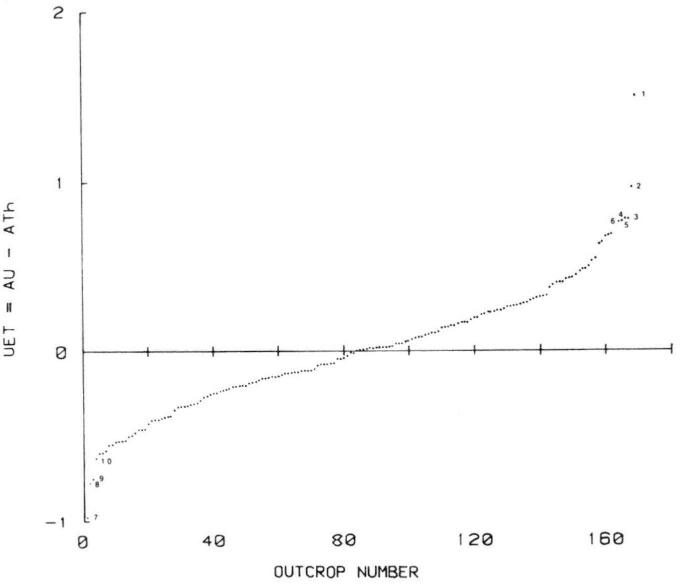

Figure 3. Plot of uranium in excess of thorium (UET = AU - ATh). Values have been plotted in ascending order for every outcrop surveyed. Each outcrop represents an average of approximately 30 individual radiation measurements. The most extreme positive and negative UET anomalies were found in the area of the Founders Peaks. Numbers assigned to the anomalous values on the graph are identified as follows: 1. Founders Escarpment near Frazier Ridge; 2. Founders Peaks near Matney Peak; 3. Webers Peaks; 4. Edson Hills; 5. Hurst Peak; 6. Webers Peaks; 7. Founders Peaks on ridge between Pipe Peak and Matney Peak; 8. Reuther Nunataks; 9. Founders Peaks near Pipe Peak; and 10. Watlack Hills near Skelly Peak.

centrations were 0.8 percent K, 1.5 ppm eU, and 5.4 ppm eTh. If the rocks in an outcrop contained exactly these mean values in thorium and uranium, it would be plotted at AU = 1 and ATh = 1. Anomalies are defined as being located above or below 2 standard deviation limit lines (broken lines) that are drawn parallel to the linear regression line (solid) for the entire data set. Two observations are immediately apparent: the number of anomalies is very small, and the size of the anomalies is insignificant. Most of the anomalies can be assigned to outcrops of the Heritage Group. However, this result is consistent with the observed low level of radioactivity in the entire region, and uranium deposits are not known to occur within districts having below-average concentrations of radioactive elements (Darnley and others, 1971).

The lack of anomalies is even more apparent in the UET (uranium in excess of thorium) gradient distribution plot shown in Figure 3. The interpretation of this plot is based on the assumption that thorium can be used as an indicator of the original regional distribution in the crust from which uranium was removed by oxidizing conditions (Darnley and others, 1977; Saunders and others, 1981; Dreschhoff and others, 1983). The graph shows that there are only a few points at the extremes representing anomalies. The uranium anomalies, indicated by positive UET values, are all the result of concentrations in the Heritage Group rocks in the Founders Peaks area. The almost total lack of anomalies in the Polarstar Formation was not expected. On the contrary, the Permian black shales, which carry plant remains, and the Middle and Upper Cambrian black shales provide the chemically reducing environment that is often conducive to the deposition of uranium.

Radiometric signatures of formations in the Ellsworth Mountains

Basically, the airborne gamma-ray spectrometer constitutes a remote sensing technique that provides an instantaneous chemical analysis for equivalent uranium, thorium, and potassium in the rocks over which it is flown. The analysis is obtained by measuring the specific gamma rays from the decay series of uranium and thorium and from the direct decay of potassium-40. The apparatus used to make this survey updates the count rate every second, and at normal forward speed in helicopter flight, this means that each counting area, usually designated as the "footprint," is roughly 35 meters in length and about 60 meters in width. In order to use radiometric survey data for stratigraphic delineation it is first necessary to determine the location of the flight path over the ground and then to relate the radiometric data to each specific outcrop. Fortunately, excellent topographic maps exist for the Ellsworth Mountains, and these were used to provide a base for the geological investigations.

If we assume equilibrium in the radioactive decay series,

individual stratigraphic zones can be expected to show characteristic radiometric signatures related to the proportions of potassium, uranium, and thorium that they contain (Potts, 1976). With the exception of arkosic sediments, potassium is usually contained in the clay mineral component. The thorium content in most sediments tends to be related to the amount of heavy minerals present in silt-size or coarser component of clastic sediments. On the other hand, in both clastics and carbonates, uranium is often selectively removed by oxidation and solution processes. Because of uranium's chemical reactivity, it frequently shows lateral variability in outcropping sedimentary formations. However, in Antarctica the lack of intense chemical weathering on exposed surfaces reduces uranium mobility, and the data from airborne surveys should provide a good indication of the average uranium content of the stratigraphic units. As shown in Figures 2 and 3, there is an almost total absence of significant anomalies above the average radioelement concentrations in the Ellsworth Mountains. This implies that the low-grade metamorphic processes that characterize much of the area were not sufficiently intense to cause substantial alteration in the original radioelement distribution of the sedimentary sequence.

The radiometric signatures of the stratigraphic units that crop out in the Heritage Range and Sentinel Range are best illustrated in the scatter diagrams that have been prepared for both areas (Figs. 4 and 5 and Figs. 6 and 7, respectively). In one of these diagrams, thorium is plotted against uranium; in the other, against potassium. The bars provide an estimate of the extent of overlap that can be anticipated when comparing the radioactivity signatures of the formations. For example, it is clear that the radiation signature of the Minaret Formation (€m) differs so sharply from that of all the other stratigraphic units that it can be unambiguously identified throughout the Heritage Range. The extreme thickness and lithologic variability of the Heritage Group rocks (€h) are directly reflected in the length of the bars. The Crashsite Group (Dc) and the Whiteout Conglomerate (Cw) show much less variability.

Furthermore, a comparison of the uranium-to-thorium ratios of the Crashsite and the Whiteout shows that the thorium

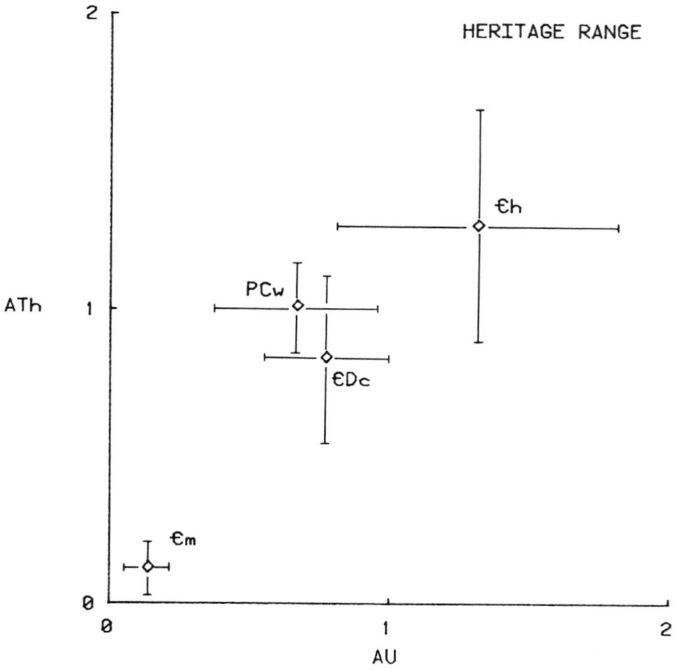

Figure 4. Scatter diagram of adjusted thorium (ATh) versus adjusted uranium (AU) for some of the major stratigraphic units in the Heritage Range: the Heritage Group (€h), consisting of the Frazier Ridge and Springer Peak formations and the Minaret Formation (€m); the Crashsite Group (CDc); and the Whiteout Conglomerate (PCw). The points characterize specific stratigraphic units and are derived by averaging each radioelement concentration from all outcrops of the unit that could be identified from survey data. Bars with a length of one standard deviation have been added to indicate the expected range of variation in the radioelement concentration. It is important to note that within the Heritage Group, the Minaret Formation is characterized by the lowest radioactivity and the Frazier Ridge and Springer Peak formations show the highest radioactivity.

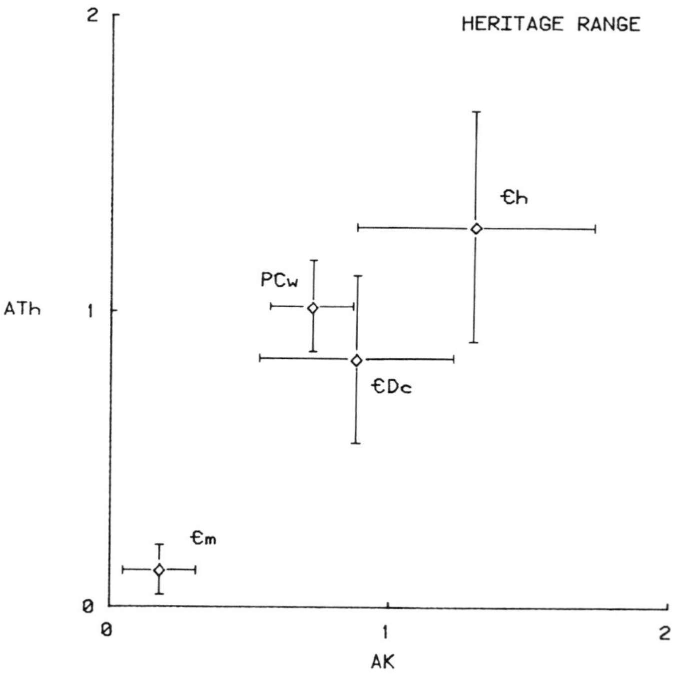

Figure 5. Scatter diagram of adjusted thorium (ATh) versus adjusted potassium (AK) for some of the major stratigraphic units in the Heritage Range: the Heritage Group (€h), consisting of the Frazier Ridge and Springer Peak formations and the Minaret Formation (€m); the Crashsite Group (CDc); and the Whiteout Conglomerate (PCw). The points characterize specific stratigraphic units and are derived by averaging each radioelement concentration from all outcrops of the unit that could be identified from survey data. Bars with a length of one standard deviation have been added to indicate the expected range of variation in the radioelement concentration. The extremely low radioactivity of the Minaret Formation is undoubtedly related to the fact that it is very low in clay minerals in the area we examined. The high levels of radioactivity in the Frazier Ridge and Springer Peak formations probably reflect the presence of a higher content of heavy minerals and clay minerals.

level in the Whiteout is substantially higher than it is in the Crashsite. The thorium-to-potassium ratio shows the same trend to a somewhat greater degree. Although a completely unambiguous separation of the two units is not possible on the basis of a few measurements, it should be possible if an adequate survey of a number of outcrops were to be made. Similarly, the Heritage Group generally shows a higher level of radioactivity than any of the other stratigraphic units, although it is overlapped by the one standard deviation bars in thorium and potassium of the Crashsite. There is no overlap by the Whiteout. Multiple measurements from a number of outcrops would permit positive identification of the Heritage Group on the basis of its radiometric signature.

In the Sentinel Range, the Mount Liptak Formation of the Crashsite Group (€Dl) is the only stratigraphic unit that can be distinguished unambiguously from all of the other units by using thorium-to-potassium and thorium-to-uranium ratio plots. A uniform, low radioactivity is typical of all the outcrops we examined, but it is especially characterized by an extremely low potassium value that is probably the result of a very low clay mineral content. In the thorium-to-uranium ratio plot the Howard Nunataks Formation (€Dh) of the Crashsite is overlapped by the high level of uranium and thorium variability in the Whiteout Conglomerate (Cw). Nevertheless, the much lower level of potassium of the Howard Nunataks Formation permits distinction of the two units. The Howard Nunataks Formation of the Crashsite is uniformly low in uranium and only slightly higher in thorium than the Mount Liptak Formation. As might be expected, it does show an increase in potassium, most probably as a result of a higher clay mineral content.

In general, the Whiteout Conglomerate of the Sentinel Range shows less total radioactivity than does the same formation in the Heritage Range. The reason for this variation is not entirely clear, but it seems to be mainly related to the higher level of thorium present in the Heritage Range. Furthermore, the Whiteout is characterized by a rather large variability in uranium content in the Sentinel Range, resulting in maximum uranium values being about the same in both areas. It is possible that the thorium present in the Whiteout is contained mainly in the cob-

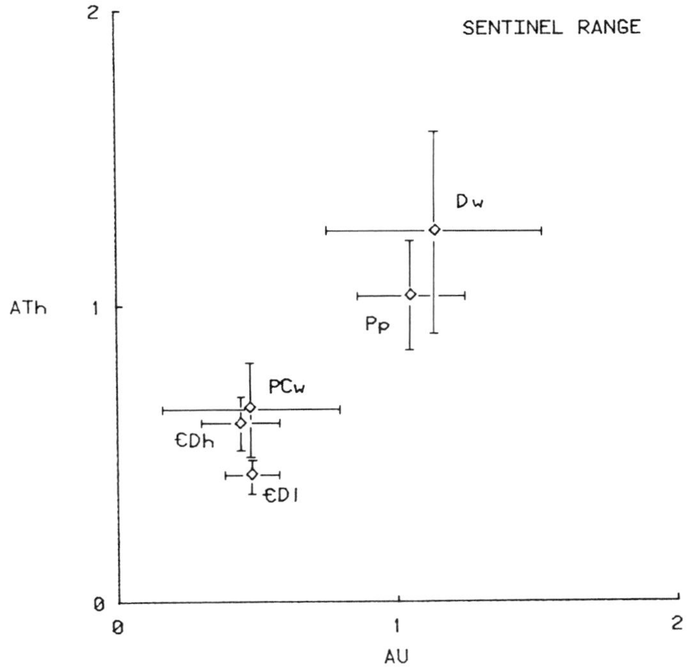

Figure 6. Scatter diagram of adjusted thorium (ATh) versus adjusted uranium (AU) for the stratigraphic units in the Sentinel Range: the Howard Nunataks Formation (€Dh), the Mount Liptak Formation (€Dl), and the Mount Wyatt Earp Formation (Dw) of the Crashsite Group; the Whiteout Conglomerate (PCw); and the Polarstar Formation (Pp). The points characterize specific stratigraphic units and are derived by averaging each radioelement concentration from all outcrops of the unit that could be identified from survey data. Bars with a length of one standard deviation have been added to indicate the expected range of variation in the radioelement concentration. Note that the light colored quartzite of the Mount Liptak Formation shows a significantly lower radioactivity than the Howard Nunataks Formation upon which it rests. The Mount Wyatt Earp Formation is substantially higher in total radioactivity than any other formation in the stratigraphic sequence in the Sentinel Range.

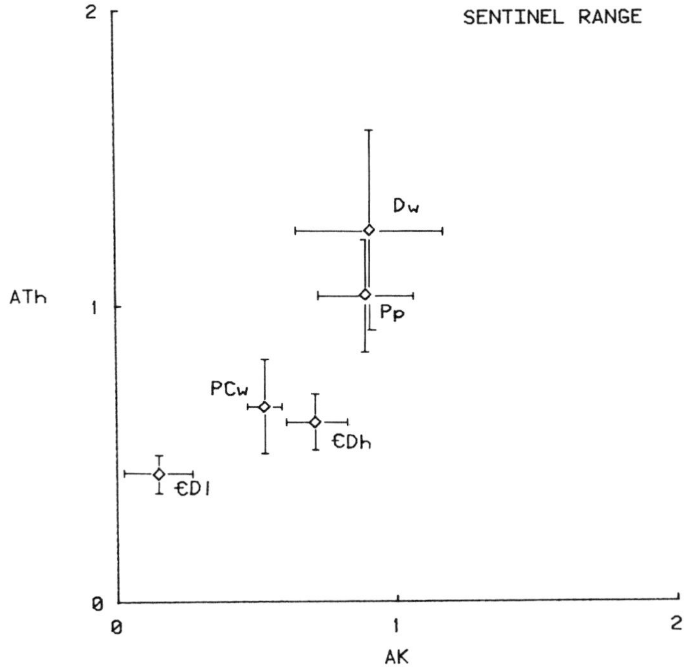

Figure 7. Scatter diagram of adjusted thorium (ATh) versus adjusted potassium (AK) for the stratigraphic units in the Sentinel Range: the Howard Nunataks Formation (€Dh), the Mount Liptak Formation (€D1), and the Mount Wyatt Earp Formation (Dw) of the Crashsite Group; the Whiteout Conglomerate (PCw); and the Polarstar Formation (Pp). The points characterize specific stratigraphic units and are derived by averaging each radioelement concentration from all outcrops of the unit that could be identified from survey data. Bars with a length of one standard deviation have been added to indicate the expected range of variation in the radioelement concentration. The low total radioactivity and especially the low AK value for the Mount Liptak Formation undoubtedly reflect its low clay mineral content. The higher radioactivity levels in the Mount Wyatt Earp Formation are most probably related to the presence of higher concentrations of heavy minerals and clay minerals.

bles and boulders that make up the conglomerate clasts. If this is true, the proportional representation of the thorium-bearing lithologies may differ between the two ranges, and this difference may reflect differing source areas for the clasts.

The Polarstar Formation (Pp) and the Mount Wyatt Earp Formation (Dw) of the Crashsite Group have similar radiometric signature and are not easily separated on this basis alone. Generally, the Mount Wyatt Earp Formation shows a higher level of total radioactivity, but differences are small. The most prominent difference is in the higher level of thorium present in the Mount Wyatt Earp Formation, but a substantial number of outcrops would have to be measured before these stratigraphic units could be separated on this basis alone.

DISCUSSION AND CONCLUSIONS

The primary objective of the airborne radiometric survey is to assess the uranium resource potential of Antarctica. Prior to the beginning of field operations, a review of existing literature on the Ellsworth Mountains area had revealed that only two sedimentary formations could be considered as primary targets with a possibility of showing accumulations of uranium. These were the black shales of the Middle and Upper Cambrian Heritage Group and the plant-fossil–carrying black shales of the Permian Polarstar Formation. Flight operations, however, detected small anomalous concentrations of uranium only in Heritage Group rocks in the Founders Peaks area, but landings failed to show any significant concentrations of radioactive elements in the exposures that were examined on the ground. The low level of radioactivity detected aerially over all the formations in the Ellsworth Mountains is consistent with the apparent absence of substantial anomalies.

In addition to their usefulness in evaluation of an area for potential uranium resources, the gamma-ray spectral data are of substantial value in providing stratigraphic information. Obviously, airborne gamma-ray surveys cannot serve as a substitute for careful ground work on the outcrops. On the other hand, once the formation type sections have been established by direct examination of the outcrops, which is the case in the Ellsworth Mountains, radiometric signatures can be determined by flying over the type sections. A radiometric survey can furnish a rapid scan of a large area, and it provides not only the visual information that is accumulated during the flights but also the highly detailed radiometric record that constitutes the "hard data" of the survey.

This is best demonstrated by the fact that we were able not only to identify the major stratigraphic units with ease but also, on the basis of our radiometric survey, to distinguish three formations within the Heritage Group cropping out in three separate areas. Founders Peaks and Webers Peaks show the highest concentration of all three radioelements. Dunbar Ridge and Edson Hills have lower concentrations, particularly in potassium and thorium. At Liberty Hills in the southern Heritage Range the Heritage Group sediments show the lowest thorium content. In addition, the Crashsite Group in the Heritage Range seems to be divisible into two units. The low potassium values characteristic of the Howard Nunataks Formation in the Sentinels are also found in the Heritage Range along the flight path in the Enterprise Hills at the Henderson Glacier. This occurrence has also been reported by Hjelle and others (1982).

The technique of associating radioactivity patterns with specific stratigraphic sequences finds daily use in the petroleum industry in well-logging operations throughout the world. In temperate and tropical regions, however, the presence of soils complicates the use of airborne gamma-ray data for correlation, and the high cost of radiometric surveys makes the technique too expensive for routine surface stratigraphy studies. In Antarctica in general, and especially in the Ellsworth Mountains, the fact that most outcrops are solid rock surrounded by snow or ice makes aerial gamma-ray spectrometry much more applicable and reliable.

In terrains like those in the Ellsworth Mountains, this type of radiometric survey may be one of the most economical methods for the rapid accumulation of information about the distribution of particular stratigraphic units. In many places in Antarctica, outcrops are widely separated and commonly inaccessible. Even where thin snow cover is present, many formations can be identified by their characteristic radiometric signatures. The data accumulated from the radiometric survey of the Ellsworth Mountains have proven to be particularly suited to demonstrating the usefulness of gamma-ray remote sensing techniques for stratigraphic correlation of outcropping sedimentary rocks in Antarctica.

ACKNOWLEDGMENTS

This research was funded in part by the National Science Foundation, the University of Kansas, and the Bundesandstalt für Geowissenschaften und Rohstoffe in Hannover, Federal Republic of Germany.

REFERENCES CITED

Collinson, J. W., Vavra, C. L., and Zawiskie, J. M., 1980, Sedimentology of the Polarstar Formation (Permian), Ellsworth Mountains: Antarctic Journal of the United States, v. 15, p. 30–32.

Craddock, C., Anderson, J. J., and Webers, G. F., 1964, Geologic outline of the Ellsworth Mountains, *in* Adie, R. J., ed., Antarctic geology: Amsterdam, North-Holland Publishing Co., p. 155–170.

Darnley, A. G., Grasty, R. L., and Charboneau, B. E., 1971, A radiometric profile across part of the Canadian Shield: Geological Survey of Canada Paper 70-46, 38 p.

Darnley, A. G., Charboneau, B. E., and Richardson, K. A., 1977, Distribution of uranium in the rocks as a guide to the recognition of uraniferous regions, *in* Recognition and evaluation of uraniferous areas: Vienna, International Atomic Energy Agency Panel Proceedings 450, p. 55–86.

Dreschhoff, G.A.M., Zeller, E. J., and Kropp, W. R., 1983, Radiometric survey in northern Victoria Land, *in* Oliver, R. L., James, P. R., and Jago, J. B., eds., Antarctic earth science: Canberra, Australian Academy of Science, v. 429–432.

Gabelman, J. W., 1976, Expectations from uranium exploration: American Association of Petroleum Geologists Bulletin, v. 60, no. 11, p. 1993–2004.

Hjelle, A., Ohta, Y., and Winsnes, T. S., 1982, Geology and petrology of the southern Heritage Range, Ellsworth Mountains, *in* Craddock, C., ed., Antarctic geoscience: Madison, University of Wisconsin Press, p. 599–608.

Potts, M. J., 1976, Computer methods for geological analysis of radiometric data, *in* Exploration for uranium ore deposits: Vienna, International Atomic Energy Agency Panel Proceedings 434, p. 55–69.

Rogers, J.J.W., and Adams, J.A.S., 1969, Uranium, *in* Wedepohl, K. H., ed., Handbook of geochemistry, v. 2/1, Berlin, Springer Verlag, Chapter 92.

Saunders, D. F., 1979, Characterization of uraniferous geochemical provinces by aerial gamma-ray spectrometry: Mining Engineering, v. 31, no. 12, p. 1715–1722.

Saunders, D. F., Jordt, D. H., and Galbraith, J. K., 1981, Interpretation of NURE aerial radiometric and hydrogeochemical and stream sediment reconnaissance data; Final report: Bendix Field Engineering Corp. Subcontract 79-300-L, document GJBX-32(81).

Webers, G. F., 1970, Paleontological investigations in the Ellsworth Mountains, West Antarctica: Antarctic Journal of the United States, v. 5, no. 5, p. 163–164.

Webers, G. F., and Spörli, K. B., 1983, Paleontological and stratigraphic investigations in the Ellsworth Mountains, West Antarctica, *in* Oliver, R. L., James, P. R., and Jago, J. B., eds., Antarctic earth science: Canberra, Australian Academy of Science, v. 261–264.

Zeller, E. J., and Dreschoff, G.A.M., 1980, Evaluation of uranium resources in Antarctica, *in* Uranium evaluation and mining techniques: Vienna, International Atomic Energy Agency, p. 381–390.

MANUSCRIPT ACCEPTED BY THE SOCIETY MARCH 1, 1989

Printed in U.S.A.

Geological Society of America
Memoir 170
1992

Chapter 17

Pretectonic burial metamorphism in the Heritage Group, southern Ellsworth Mountains, West Antarctica

Robert L. Bauer
Department of Geology, University of Missouri at Columbia, Columbia, Missouri 65211

ABSTRACT

The Middle to Upper Cambrian Heritage Group, exposed in the Heritage Range of the Ellsworth Mountains, contains interlayered basaltic lava flows and intrusive gabbroic stocks, dikes, and sills that preserve evidence of a pretectonic burial metamorphism. Three zones ranging from the upper prehnite-pumpellyite facies (zone I) through the pumpellyite-actinolite facies (zone II) to the greenschist facies (zone III) are correlated with stratigraphic depth of the mafic igneous rocks in the Heritage Group. A temperature of approximately 370°C is obtained for the boundary between the pumpellyite-actinolite facies and the greenschist facies on the basis of Fe/(Fe+Al) in epidote coexisting with pumpellyite + chlorite + actinolite + quartz. A geothermal gradient of approximately 23 to 24°C/km is estimated for the stratigraphic interval containing pumpellyite-actinolite facies assemblages.

Syntectonic recrystallization associated with the development of penetrative foliation in the metasedimentary country rocks is only sparsely developed in the prehnite- and/or pumpellyite-bearing mafic igneous rocks, whereas the more penetratively deformed igneous rocks lack evidence of subgreenschist facies conditions. This differential recrystallization may account for the variation in K-Ar whole-rock ages reported previously for the mafic igneous and metasedimentary rocks of the Heritage Group.

INTRODUCTION

The Ellsworth Mountains (Fig. 1) are an elongate north-northwest–trending mountain chain, approximately 350 km long and as much as 90 km wide, located between the polar plateau of West Antarctica and the Ronne Ice Shelf. The mountains are divided into the Sentinel Range in the north and the Heritage Range in the south by the Minnesota Glacier, which flows eastward from the polar plateau into the Ronne Ice Shelf. The ranges display a 13,000-m-thick succession of mostly marine metasedimentary and metavolcanic rocks ranging from Cambrian to Permian in age. The structure of the mountain belt is dominated by a major northwest-plunging anticlinorium that exposes the lower portions of the stratigraphic section in the Heritage Range. The lowest stratigraphic unit, the Heritage Group (Cambrian), is well exposed there, and its metamorphism is the focus of this study.

The first geologic investigations in the Ellsworth Mountains, initiated in the early 1960s (Craddock, 1962; Anderson and others, 1962; Anderson, 1965), emphasized the stratigraphy, paleontology, and structural geology, but no detailed study of the metamorphic history of the rocks was undertaken. Preliminary studies by Craddock and others (1964) tentatively assigned the recrystallization in the metasedimentary rock sequence to the greenschist facies. However, subsequent work by Castle and Craddock (1975) indicated that the uppermost unit in the sequence, the Polarstar Formation (Permian), was subjected to low-grade burial metamorphism perhaps as high as laumontite grade of the zeolite facies. Craddock (1969, 1972, 1975) attri-

Bauer, R. L., 1992, Pretectonic burial metamorphism in the Heritage Group, southern Ellsworth Mountains, West Antarctica, *in* Webers, G. F., Craddock, C., and Splettstoesser, J. F., Geology and Paleontology of the Ellsworth Mountains, West Antarctica: Boulder, Colorado, Geological Society of America Memoir 170.

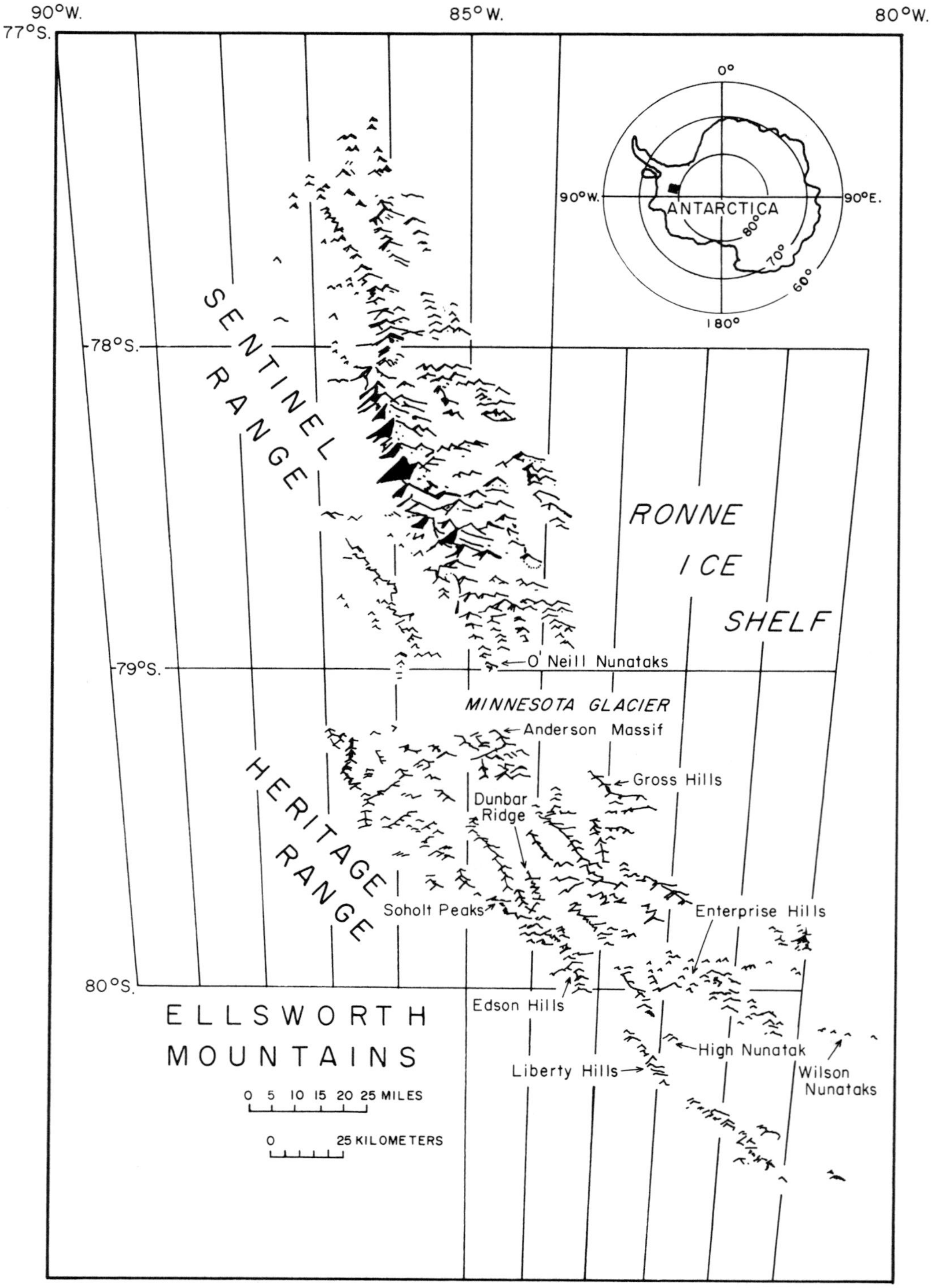

Figure 1. Location map of the Ellsworth Mountains showing the topography and geographic features in the Heritage Range that are referred to in the text.

buted both regional folding of the range and the low-grade metamorphism to orogeny during the early Mesozoic.

Recent investigations by Hjelle and others (1978, 1982) and Yoshida (1982, 1983) describe the metamorphism in the southern Heritage Range and continue to emphasize the occurrence of greenschist facies mineral assemblages. Yoshida (1982) also describes older metamorphic imprints, which he attributes to the Paleozoic Borchgrevink Orogeny and Ross Orogeny. As the work of Hjelle and others and Yoshida did not concentrate on metamorphism, the present investigation was initiated to examine in detail the metamorphic textural and mineralogical changes in rocks from the Heritage Group and to compare these findings with those of earlier studies.

The Heritage Group consists primarily of metamorphosed sedimentary and pyroclastic rocks that are locally interlayered with basaltic lava flows and intruded by small gabbroic stocks and diabasic dikes and sills. Although the age of the mafic intrusive rocks is not well defined, they have not intruded any of the rock units overlying the Heritage Group and are therefore generally considered to have been intruded in Late Cambrian time (cf. Vennum and others, this volume). Preliminary optical examination of rocks from the Heritage Group led to the identification of pumpellyite + actinolite + chlorite assemblages, indicative of the pumpellyite-actinolite facies, in the mafic igneous rocks (Bauer, 1983). No mineral assemblages indicative of subgreenschist facies conditions were observed in the schists and phyllites of the more deformed metasedimentary country rock. Therefore, this chapter has focused on the mafic igneous rocks of the Heritage Group in an effort to determine the conditions and distribution of the subgreenschist facies metamorphism and its significance in the tectonothermal evolution of the Heritage Range.

The rocks studied include 65 thin sections of basalt, gabbro, and diabase from samples collected during the 1962–1963 and 1963–1964 U.S. expeditions, and 32 thin sections of basalt from samples collected by W. Vennum in the Liberty Hills and High Nunatak (Fig. 1) during the 1979–1980 U.S. expedition. Thin sections of mafic to intermediate tuffaceous rocks and metasedimentary rocks collected in 1979 by G. Webers from the Edson Hills, and various volcanic and metasedimentary rocks collected from the Liberty Hills by J. M. Anderson were examined during the investigation of clastic units from the Heritage Group (Webers and others, this volume). Samples studied from the two early 1960s expeditions were collected from High Nunatak, Liberty Hills, Edson Hills, Soholt Peaks, Dunbar Ridge, Gross Hills, Anderson Massif, and the western Enterprise Hills in the Heritage Range and the O'Neill Nunataks of the very southern Sentinel Range (Fig. 1).

Optical and electron microprobe studies were conducted to determine the identity of coexisting minerals and the composition of selected mineral phases with the hope of defining metamorphic isograds in the Heritage Range. The analysis has been limited by the nature of the sampling, which was not undertaken for a detailed metamorphic investigation, and by the irregular distribution of rocks of appropriate mafic composition. However, variations in mineral assemblages and mineral chemistry, when compared with the Heritage Group stratigraphy proposed by Webers and others (this volume), indicate a pattern of burial metamorphism evidenced by a progressive increase in metamorphic grade from the upper prehnite-pumpellyite facies through the pumpellyite-actinolite facies to the greenschist facies with depth in the section.

This chapter presents petrographic and phase chemical evidence for (1) the change in metamorphic grade in the mafic igneous rocks from the prehnite-pumpellyite facies through the pumpellyite-actinolite facies with depth in the Heritage Group, (2) the temperature conditions prevailing during subgreenschist facies burial metamorphism, and (3) the relative timing of deformation and burial metamorphism.

STRATIGRAPHY

The Middle to Upper Cambrian Heritage Group, the lowest stratigraphic unit in the Ellsworth Mountains, is exposed primarily within the Heritage Range (Fig. 1). Webers and others (this volume) divide the 7,800-m thickness of the Heritage Group into four lower formations: the Union Glacier, Hyde Glacier, Drake Icefall, and Conglomerate Ridge Formations. These are overlain by three stratigraphically equivalent formations—the Liberty Hills, Springer Peak, and Frazier Ridge formations—and the uppermost unit, the Minaret Formation (cf., Webers and others, this volume, Figs. 2 and 3). During this study, metamorphic mineral assemblages were examined from mafic to intermediate tuffs, basalt flows, gabbros, and diabases from the Union Glacier, Drake Icefall, Liberty Hills, and Springer Peak Formations. The correlation in this chapter of sample locations with stratigraphic position in the Heritage Group is based on oral communication with G. Webers. Only brief descriptions of the formations containing mafic igneous rocks or clasts pertinent to the burial metamorphic history of the terrane are given here; Webers and others (this volume) give more detailed descriptions of all of the formations of the Heritage Group.

The basal formation in the Heritage Group, the Union Glacier Formation, has an exposed thickness of approximately 3,000 m, consisting primarily of metamorphosed tuffaceous rocks with minor amounts of calcareous sandstone and carbonate conglomerate. The tuffaceous rocks include crystal tuff, lithic tuff, tuff breccia, and lapilli tuff. The latter three rock types are heterolithic, containing lithoclasts of crystal tuff, basalt, shaley fragments, and probable scoriaceous material, all in a crystal tuff matrix.

The overlying Hyde Glacier Formation is a clastic unit, approximately 1,750 m thick, containing tuffaceous argillite, graywacke, polymictic conglomerate, and quartzite. Clasts include granite, basalt, and quartzite and range in size from 0.5 to 60 cm in the conglomerate. Portions of the graywackes contain clasts of foliated mica- and chlorite-rich metamorphic rock fragments.

The Drake Icefall Formation is a unit of black shales and interbedded limestones estimated by Webers and Spörli (1983) to be approximately 800 m thick. Samples of a 100-m-thick diabase sill, the Hyde Glacier sill of Vennum and others (this volume), were examined from this unit. The sill crops out just south of the Hyde Glacier Icefall in the central Edson Hills.

The Springer Peak Formation, estimated to be approximately 1,000 m thick, contains a basal unit about 60 m thick of limestone with interbedded black shale. The rest of the unit consists of brown argillite with locally interbedded black shale and buff quartzite. In the Anderson Massif the argillites and black shales at the top of the formation are interbedded with basalt flows estimated to be at least 150 m thick (Webers and Spörli, 1983). Similar mafic rocks in a similar stratigraphic position occur just north of the Anderson Massif across the Minnesota Glacier in the O'Neill Nunataks. In addition to these basalts, gabbroic stocks and diabasic sills and dikes intrusive into the Springer Peak Formation were examined from the Gross Hills, Dunbar Ridge, O'Neill Nunataks, Anderson Massif, western Enterprise Hills, and Soholt Peaks.

The Liberty Hills Formation, laterally equivalent to the Springer Peak Formation, is composed of basaltic flows, volcanic breccia, polymictic conglomerate, quartzite, and green argillite. The unit is at least 1,000 m thick, but stratigraphic relationships are complicated by rapid facies changes (cf. Webers and others, this volume). Basaltic flows and volcanic breccias in the Liberty Hills and High Nunatak have been studied from this formation.

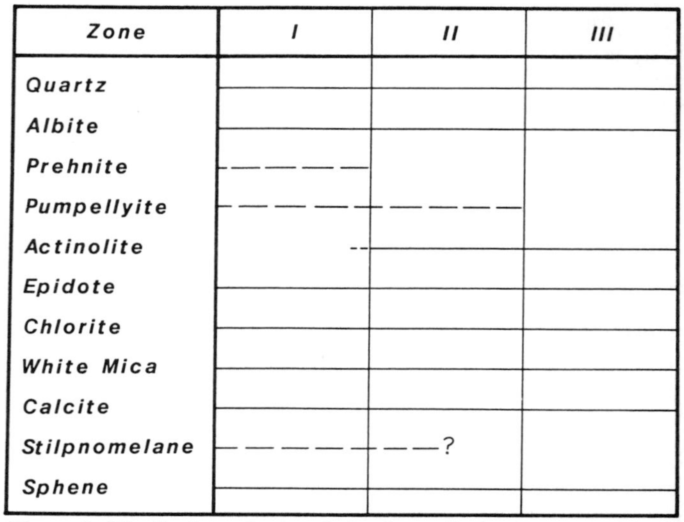

Zone
I
II
III
Quartz
Albite
Prehnite
Pumpellyite
Actinolite
Epidote
Chlorite
White Mica
Calcite
Stilpnomelane
?
Sphene

Figure 2. Distribution of minerals in the mafic igneous rocks of the Heritage Group with respect to metamorphic zones. Dashed lines indicate minerals that occur intermittantly throughout the zone.

METAMORPHIC ZONATION

Hashimoto (1966) proposed the pumpellyite-actinolite facies as an independent facies occupying a pressure-temperature field intermediate between the lower grade prehnite-pumpellyite or lawsonite-albite facies and the higher grade greenschist facies. As originally defined, rocks of this facies contain assemblages including pumpellyite, actinolite, and stilpnomelane exclusive of prehnite and lawsonite. Subsequent definitions (Coombs and others, 1976; Nakajima and others, 1977), as adopted in this chapter, are based on the assemblage pumpellyite + actinolite + chlorite; the upper limit of the facies is defined by the disappearance of pumpellyite.

A gross metamorphic zonation from the uppermost prehnite-pumpellyite facies through the pumpellyite-actinolite facies into the greenschist facies can be recognized within the mafic igneous rocks of the Heritage Group. Three metamorphic zones, coincident with these facies boundaries, are defined in the rocks based on the distribution of prehnite, pumpellyite, and their association with actinolite. The mineral occurrences within each zone are summarized in Figure 2; mineral assemblages in zones I and II, which are the principal facies of this study, are listed in Table 1.

The selection of equilibrium assemblages in rocks from zones I and II is complicated by the incomplete recrystallization of primary phases and by uncertainty regarding the size of equilibrium domains within the rocks. Watanabe (1974), for instance, considered only phases in direct contact to be in stable equilibrium domains as large as 1 mm in diameter. Obvious disequilibrium features such as compositional zoning, relict primary phases, and local intergrain inhomogeneities were observed in many of the samples examined from the Heritage Group. The assemblages listed in Table 1, although not necessarily in direct contact with one another, occur within areas of a thin section less than 1 mm in diameter and are commonly included within or adjacent to a single chlorite mass or a single amygdule. However, the representative mineral analyses listed in Tables 2 through 5 (to be discussed below) are not invariably from phases within a given sample that meet these criteria.

Zone I is characterized by the presence of assemblages containing prehnite with pumpellyite and belongs to the prehnite-pumpellyite facies. Other common metamorphic minerals include chlorite, epidote, quartz, albite, sphene, and calcite. Actinolite was observed in one of the samples but is not abundant. Stilpnomelane was observed in only one sample; no zeolites were observed. The local occurrences of actinolite, coupled with the corroded appearnce of the prehnite in these rocks and their close stratigraphic proximity to prehnite-free assemblages of zone II, are taken as an indication that the rocks examined from zone I are from the uppermost prehnite-pumpellyite facies. Samples from the volcanic flows of the upper Springer Peak Formation contain assemblages defining this zone.

Zone II is characterized by the absence of prehnite and the association of pumpellyite with actinolite. Rocks recrystallized in this zone belong to the pumpellyite-actinolite facies and can be divided into two subzones—IIa and IIb—on the basis of pumpellyite composition (cf. Bishop, 1972; Kawachi, 1975). Both subzones commonly contain chlorite, quartz, epidote, albite, white mica, calcite, and sphene as additional phases. Subzone IIa primarily contains green pumpellyite in association with epidote,

TABLE 1. MINERAL ASSEMBLAGES IN ZONES I AND II

Zone	I				II							
Rock Type*	B	B	B	B	GD	GD	GD	B	B	BS	BA	GD
Formation†	SP	SP	SP	SP	SP	SP	SP	LH	LH	LH	LH	DI
Quartz	X	X	X	X	X	X	X	X	X	X	X	
Albite	X		X	X	X	X	X	X	X	X	X	
Prehnite	X	X		X								
Pumpellyite	X	X		X	X		X	X§			X	X
Actinolite				(rare)	X	X	X	X	X		X	X
Epidote**	X	X		X	X	X	X	X	X		X	X
Chlorite	X	X	X	X	X	X	X	X	X	X		X
White Mica			X	X	X	X	X	X			X	X
Calcite		X	X	X	X	X		(rare)		X	X	
Stilpnomelane			X									
Sphene	X	X	X	X	X	X	X	X	X	X		X
Hematite								X	X	X		

*B = basalt, GD = gabbro-diabase, BS = spilitic basalt minimum assemblage, BA = basalt amygdule maximum assemblage.
†SP = Springer Peak, LH = Liberty Hills, DI = Drake Icefall.
§Pumpellyite occurs only in plagioclase with epidote and white mica; the nature of equilibration with other phases is uncertain.
**Epidote includes clinozoisite.

whereas zone IIb contains clear pumpellyite in association with clinozoisite. Rocks from zone II constitute the bulk of the samples considered during this investigation. Gabbros and diabases from the Springer Peak Formation and basalts from the Liberty Hills Formation belong to zone IIa; gabbros from the Hyde Glacier sill in the Drake Icefall Formation belong to zone IIb.

Zone III contains mineral assemblages characteristic of the greenschist facies. Both prehnite and pumpellyite are absent, and actinolite-epidote-chlorite assemblages with sphene, albite, quartz, white mica, and ± calcite occur. Mafic to intermediate tuffaceous rocks from the Union Glacier Formation contain mineral assemblages of this zone.

Although stilpnomelane was observed only in samples from zone I, numerous other studies have indicated the occurrence of stilpnomelane in all of the zones of this facies series (cf. summary in Hashimoto and Kanehira, 1975). The relative lack of stilpnomelane in rocks of the present study may be attributed to the low FeO/MgO ratios of the samples (cf. Coombs and others, 1976) or possibly to the low-P_{CO_2} conditions during metamorphism (cf. Hashimoto and Kanehira, 1975).

The use of pumpellyite composition to distinguish between zones IIa and IIb is complicated by the numerous factors that have been cited as influencing Ca-Al silicate compositions in such rocks. In addition to the variables pressure, temperature, oxygen fugacity (fO_2), and bulk rock and metamorphic fluid composition, several independent factors such as P_{fluid}/P_{total}, permeability, reaction rate, and intensity of primary silicate and/or Fe-Ti oxide alteration have been suggested as important variables in some instances (e.g., Kuniyoshi and Liou, 1976; Offler and others, 1981; Levi and others, 1982; Evarts and Schiffman, 1983; Nyström, 1983; Cortesogno and others, 1984). Furthermore, as Liou and others (1985) have emphasized, only mineral compositions from buffered assemblages can be used to constrain the intensive properties of metamorphism. In an effort to limit or eliminate the significance of as many of these variables as possible, the zone IIa/IIb distinction is based on pumpellyite composition changes only in magnetite-ilmenite–bearing metagabbros, considered to be of the same intrusive event, that display considerable though incomplete alteration of pyroxene and Fe-Ti oxide phases. Quantitative temperature estimates (discussed below) over the stratigraphic range of zone II are based on Ca-Al silicate compositions from buffered assemblages.

PETROGRAPHY AND MINERALOGY

The petrography and selected mineral chemistry of rocks from metamorphic zones I and II are described in this section. The chemistry of representative grains of prehnite, pumpellyite, chlorite, epidote-clinozoisite, and actinolite was determined for comparison with chemical variations in these minerals from similar metamorphic sequences. No attempt has been made to characterize the chemical variations of these minerals within a given zone. Chemical compositions of the minerals were obtained using the 3-channel ARL EMX-SM electron microprobe at the University of Missouri. Natural silicates and oxides were used as standards. Matrix corrections followed the method of Bence and Albee (1968) using the alpha factors of Albee and Ray (1970). Standard operating parameters were 15 kv accelerating potential, 0.03 microamp specimen current on brass, and 400,000 counts beam termination, yielding a counting time of approximately 8 seconds. Five or more points per grain were analyzed, and the results were averaged. To reduce the problem of matching the analyzed points on the three runs required for analysis, photomicrographs were used for spot locating. Structural formulas for pumpellyite were calculated assuming 16 cations per formula unit, as suggested by Coombs and others (1976). This distribution of cations in the various sites was made according to the procedure suggested by Passaglia and Gottardi (1973) as modified by Coombs and others (1976). Amphibole structural formulas and Fe^{2+} and Fe^{3+} values were calculated according to the procedure of Robinson and others (1982), normalizing total cations to 13 exclusive of K, Na, and Ca. Representative analyses of selected phases from zones I and II are tabulated in Tables 2 through 5.

Zone I—Basalts of the upper Springer Peak Formation

Mineral assemblages characteristic of this zone were observed in only four samples of basalt flows from the upper Springer Peak Formation that are interbedded with argillites and black shales in the Anderson Massif and the O'Neill Nunataks (Fig. 1). All the basalts examined probably contained pyroxene phenocrysts in an intergranular matrix dominated by plagioclase laths; however, primary clinopyroxene, as a brown to reddish-brown titanaugite, was observed in only one sample (64TB36) from this zone. Spot microprobe analyses for titanium indicated as much as 4.0 percent TiO_2 in the pyroxene. The pyroxene in the other samples examined from this zone has been completely replaced by pale green pycnochlorite (Table 2, analyses 1 and 2) with anomalous blue or brown interference colors. The chlorite is pseudomorphic after euhedral pyroxene phenocrysts in some instances or rarely may preserve a subophitic outline (Fig. 3), but most of the chlorite pods after pyroxene have been flattened during subsequent deformation and display a weak, spaced foliation approximately parallel to their long dimension (Fig. 3). Yellow-brown oxychlorite, similar to that described by Chatterjee (1966) and Brown (1967) as a product of weathering, occurs locally along the foliation planes and fractures in the chlorite pods (Fig. 3). Pale green pycnochlorite, similar to that replacing pyroxene, occurs in the interstices between plagioclase laths in all the basalts from this zone.

Plagioclase laths are partly or completely altered to combinations of albite, prehnite, pumpellyite, and ± quartz. In the one sample containing remnant pyroxene, prehnite is the principal secondary phase and in most instances is completely pseudomorphic (± quartz, ± pumpellyite) after plagioclase. Prehnite in the other samples from this zone is locally abundant and occurs as

TABLE 2. MICROPROBE ANALYSES OF REPRESENTATIVE CHLORITE FROM ZONES I AND II

Analysis	1	2	3	4	5	6	7	8	9
Zone	I	I	IIa	IIa	IIa	IIa	IIa	IIa	IIb
Specimen	64R27	64TB36	64MH12	64MH54	79Ve13	79Ve101	79Ve12b	79Ve3a	W63226
Formation	SP	SP	SP	SP	LH	LH	LH	LH	DI
SiO_2	27.34	27.52	27.80	26.32	28.97	29.88	30.80	29.11	28.70
TiO_2	0.01	0.00	0.05	0.03	0.00	0.00	0.00	0.06	0.00
Al_2O_3	19.91	18.90	19.09	18.88	16.88	17.46	18.23	16.22	20.00
FeO*	23.87	24.29	23.19	31.47	12.90	9.80	9.31	23.14	16.84
MnO	0.24	0.19	0.29	0.18	0.18	0.12	0.28	0.11	0.21
MgO	15.92	16.77	17.01	10.40	25.18	26.58	28.17	19.22	21.18
CaO	0.26	0.00	0.20	0.18	0.16	0.33	0.14	0.21	0.32
Na_2O	0.04	0.00	0.04	0.02	0.00	0.00	0.05	0.00	0.03
K_2O	0.05	0.00	0.05	0.09	0.00	0.00	0.04	0.14	0.03
Total	87.64	87.67	87.72	87.57	84.27	84.17	87.02	88.21	87.31
Structural formulae on the basis of 28 oxygens									
Si_{IV}	5.71	5.75	5.78	5.74	5.95	6.02	5.99	6.02	5.78
Al_{VI}	2.29	2.25	2.22	2.26	2.05	1.98	2.01	1.98	2.22
Al	2.61	2.41	2.48	2.59	2.03	2.17	2.17	1.97	2.53
Ti	0.00	0.00	0.01	0.01	0.00	0.00	0.00	0.01	0.00
Fe	4.17	4.25	4.03	5.74	2.21	1.65	1.51	4.00	2.84
Mn	0.04	0.03	0.05	0.06	0.03	0.02	0.05	0.02	0.04
Mg	4.95	5.23	5.27	3.38	7.70	7.99	8.16	5.93	6.36
Ca	0.06	0.00	0.05	0.04	0.04	0.07	0.03	0.05	0.07
Na	0.02	0.00	0.02	0.01	0.00	0.00	0.02	0.00	0.01
K	0.01	0.00	0.01	0.03	0.00	0.00	0.01	0.04	0.01
ΣCations	19.86	19.92	19.92	19.86	20.01	19.90	19.95	20.02	19.86
$\frac{Fe^*}{Fe^* + Mg}$	0.46	0.45	0.43	0.63	0.20	0.20	0.16	0.40	0.31

corroded patches of single, optically uniform grains partially replacing plagioclase (Fig. 4). Prehnite in both forms contains relatively low Fe and high Al (Table 3, analyses 1 and 2, respectively). Pumpellyite is abundant in some samples as an alteration product of plagioclase, or prehnite, or in chlorite pods after clinopyroxene. Microprobe analyses of pumpellyite indicate a low Fe content with relatively high Al and Mg values (Table 3, analyses 3 and 4).

Actinolite was observed in one sample from this zone as small parallel clusters of acicular grains in the basalt groundmass. Clinozoisite occurs sparsely distributed in some of the prehnite- and pumpellyite-bearing samples (Table 4, analysis 1) and is most common within or along the margins of chlorite pods.

Pyrrhotite and chalcopyrite are the only opaque phases observed in the rocks from this zone. Pyrrhotite has been largely altered to goethite in many samples, probably during the weathering that produced oxychlorite along the fractures and foliation planes in chlorite. Calcite partially replaces chlorite after pyroxene in some samples or, rarely, replaces prehnite. In the one stilpnomelane-bearing sample, calcite occurs along with white mica, green iron-rich chlorite, and albite replacing plagioclase.

Zone IIa—The lower pumpellyite-actinolite facies

Springer Peak Formation. Gabbro and diabase of the Springer Peak Formation from the Gross Hills, Dunbar Ridge, Anderson Massif, O'Neill Nunataks, western Enterprise Hills, and Soholt Peaks invariably display subophitic textures and contain primary clinopyroxene, labradorite, titanomagnetite, ilmenite, ± pyrrhotite, chalcopyrite. The clinopyroxene is a pale brown to brown augite or rarely a violet-brown titanaugite. Red-brown to green-brown hornblende of late-stage magmatic origin (cf., Otten, 1984) occurs in some of the gabbros. The hornblende partially replaces clinopyroxene grains and is especially common

TABLE 3. MICROPROBE ANALYSES OF REPRESENTATIVE PREHNITE FROM ZONE I AND PUMPELLYITE FROM ZONES I AND II

	Prehnite		Pumpellyite						
Analysis	1	2	3	4	5	6	7	8	9
Zone	I	I	I	I	IIa	IIa	IIa	IIa	IIb
Specimen	64TB36	64R27	64TB36	64R27	64MH12	79Ve12a	79Ve10l	79Ve3a	W63226
Formation	SP	SP	SP	SP	SP	LH	LH	LH	DI
SiO_2	42.80	43.52	37.56	37.50	37.74	37.47	36.72	36.96	37.79
TiO_2	0.06	0.01	0.11	0.07	0.06	0.00	0.00	0.03	0.35
Al_2O_3	23.70	24.57	25.51	25.84	24.68	24.16	20.97	21.66	25.59
Fe_2O_3*	1.27	0.45							
FeO*			3.06	2.94	5.87	3.97	9.98	9.79	2.08
MgO	0.45	0.08	3.14	3.38	2.34	3.71	2.97	2.41	3.70
CaO	26.36	26.71	23.39	22.73	22.47	22.69	22.45	22.93	23.03
MnO	0.03	0.02	0.23	0.19	0.13	0.04	0.00	0.00	0.08
Na_2O	0.00	0.04	0.00	0.05	0.06	0.00	0.00	0.02	0.09
K_2O	0.05	0.05	0.05	0.04	0.05	0.00	0.00	0.03	0.00
Total	94.73	95.45	93.05	92.74	93.39	92.04	93.09	93.83	92.71
Structural formulae on the basis of 22 oxygens (Pr) or 16 cations (Pu)									
Si_Z	5.96	5.99	5.99	5.99	6.06	6.05	5.98	5.98	6.02
Al_Y	0.04	0.01	0.01	0.01	....	....	0.02	0.02	0.00
Al_X			4.00	4.00	4.00	4.00	4.00	4.00	4.00
Al	3.89	3.98	0.79	0.85	0.67	0.59	0.00	0.11	0.81
Ti	0.01	0.00	0.01	0.01	0.01	0.00	0.00	0.00	0.04
Fe	0.13	0.05	0.41	0.39	0.79	0.54	1.36	1.32	0.28
Mg	0.09	0.02	0.75	0.80	0.56	0.89	0.72	0.58	0.89
ΣX(xy)	(4.12)	(4.05)	1.96	2.05	2.03	2.02	2.08	2.01	2.02
Mn	0.00	0.00	0.03	0.03	0.02	0.01	0.00	0.00	0.01
Ca	3.93	3.94	4.00	3.89	3.87	3.92	3.92	3.97	3.93
Na	0.00	0.01	0.00	0.01	0.02	0.00	0.00	0.01	0.03
K	0.01	0.01	0.01	0.01	0.01	0.00	0.00	0.01	0.00
ΣW	3.94	3.96	4.04	3.94	3.92	3.93	3.92	3.99	3.97
ΣCations	14.02	14.01`	(16.00)	(15.99)	(16.01)	(16.00)	(16.00)	(16.00)	(16.01)
$\frac{Fe^*}{Al + Fe^*}$	0.03	0.01	0.09	0.08	0.14	0.11	0.25	0.24	0.06

in the one sample examined from Soholt Peaks that contains interstitial granophyre. Clinopyroxene and brown hornblende grains are altered to clear to pale green actinolite (Table 5, analysis 1) along grain boundaries, and portions of pyroxene grains are pseudomorphically replaced by pale green pycnochlorite (Table 2, analyses 3 and 4). The chlorite pods after clinopyroxene (Fig. 5) may contain numerous other secondary phases, including euhedral grains of clinozoisitic epidote, radiating acicular clusters of epidote rimmed by prismatic clinozoisitic epidote (Table 4, analyses 2 and 3), spindle-shaped crystals of green pumpellyite, thin laths or radiating acicular clusters of actinolite (Table 5, analyses 2 and 3), and fine granular trains or clusters of sphene.

Plagioclase laths are generally partially altered to various combinations of white mica, epidote-clinozoisite, albite, and pumpellyite but may also contain chlorite, actinolite, or sphene. Some pumpellyite grains are pleochroic from pale green to white and display gradational internal self boundaries under crossed nicols. However, the pumpellyite after plagioclase in subophitic textures with pyroxene (Fig. 6) generally occurs as fibrous mats displaying darker green to pale brown pleochroism (Table 3, analysis 5). Titanomagnetite grains display both trellis and composite (granular) intergrowths of ilmenite (Buddington and Lindsley, 1964; Haggerty, 1976). Ilmenite and especially magnetite are largely replaced by sphene and rutile. Pyrrhotite invariably displays "bird's-eye" texture indicative of alteration to marcasite or some intermediate product (cf., Ramdohr, 1980, p. 603–607).

Liberty Hills Formation. Basaltic volcanic rocks from the Liberty Hills and High Nunatak have retained their primary igneous textures despite metamorphic recrystallization that has largely transformed them into spilites (in a textural and mineralogical sense; cf., Cann, 1969). Most of the flows examined are plagioclase phyric with minor phenocrysts of clinopyroxene and/or completely altered olivine pseudomorphically replaced by low grade minerals. Less commonly, the flows display glomeroporphyritic textures of zoned augite phenocrysts, and some contain amygdaloidal textures. The groundmass surrounding the pheno-

TABLE 4. MICROPROBE ANALYSES OF VARIOUS FORMS AND OCCURRENCES OF EPIDOTE AND CLINOZIOSITE FROM ZONES I AND II

Analysis	1	2	3	4	5	6	7
Zone	I	IIa	IIa	IIa	IIa	IIa	IIb
Specimen	64R27	64MH54 core	64MH54 rim	79Ve10I	79Ve12b core	79Ve12b rim	W63226
Formation	SP	SP	SP	LH	LH	LH	DI
SiO_2	39.06	37.49	38.22	37.68	38.35	38.04	39.10
TiO_2	0.04	0.40	0.56	0.03	0.00	0.00	0.10
Al_2O_3	29.30	23.62	26.83	21.29	22.60	22.06	29.04
Fe_2O_3	5.71	13.24	9.66	15.59	13.87	15.01	5.90
MgO	0.05	0.29	0.17	0.31	0.22	0.00	0.10
CaO	23.70	22.43	23.43	22.53	23.34	23.56	23.85
MnO	0.04	0.07	0.04	0.02	0.06	0.03	0.11
Na_2O	0.04	0.10	0.07	0.01	0.02	0.00	0.00
K_2O	0.05	0.00	0.00	0.01	0.00	0.02	0.04
Total	97.99	97.64	98.98	97.19	98.46	98.81	98.25
Structural formulae on the basis of 12.5 oxygens							
Si_Z	3.01	2.99	2.97	3.04	3.04	3.02	3.01
Al_Y		0.01	0.03				
Al	2.66	2.20	2.42	2.03	2.11	2.07	2.64
Ti	0.00	0.02	0.03	0.00	0.00	0.00	0.01
Fe	0.33	0.79	0.56	0.95	0.83	0.90	0.34
Y	2.99	3.02	3.01	2.98	2.94	2.97	2.99
Mn	0.00	0.01	0.00	0.00	0.00	0.00	0.01
Mg	0.01	0.03	0.02	0.00	0.03	0.00	0.01
Ca	1.96	1.92	1.95	1.95	1.98	2.01	1.97
Na	0.01	0.02	0.01	0.00	0.00	0.00	0.00
K	0.00	0.00	0.00	0.00	0.00	0.00	0.00
ΣW	1.98	1.98	1.98	1.95	2.01	2.01	1.99
ΣCations	7.98	7.99	7.99	7.97	7.99	8.00	7.99
$\frac{Fe}{Fe + Al}$	0.11	0.26	0.19	0.32	0.27	0.30	0.11

crysts generally retains an intergranular or pilotaxitic textural appearance despite alteration.

The flows have undergone varying degrees of metamorphic recrystallization and replacement of the primary phases, the most altered rocks taking on either a purple-red or green color caused by pervasive fine-grained hematite, epidote, or chlorite disseminated throughout the sample matrix. Not only the degree of alteration but also the assemblage of alteration products replacing plagioclase, pyroxene, and olivine varies locally within the flows, presumably as a function of local variations in activities of the various ions in the fluids present during recrystallization. Plagioclase laths, both as phenocrysts and in the matrix, may be altered to (1) fine-grained epidote, ± pumpellyite (Table 3, analysis 6), ± white mica in albite; (2) fine-grained white mica ± epidote ± chlorite in albite; (3) calcite, albite, ± white mica; or (4) coarse epidote crystals in a mosaic completely pseudomorphically replacing the plagioclase. This latter mode of replacement was observed only in rare samples from epidote metadomains (cf. Smith, 1969; Jolly and Smith, 1972) in which all matrix phases and both plagioclase and pyroxene phenocrysts are replaced by epidote.

Augite and especially olivine have undergone extensive alteration. No remnant olivine was observed in any of the samples; however, its presence is inferred from complete pseudomorphic replacement of euhedral olivine phenocrysts (Fig. 7). Dull orange iddingsite within a pseudomorphic iron oxide rim occurs in some of the basalts, but more commonly rims of hematite enclose various combinations of chlorite (Table 2, analyses 5 and 6), epidote, sphene, quartz, calcite, actinolite, albite, or more hematite. Augite is also replaced by combinations of the above phases, although it is more commonly altered to actinolite + chlorite ± epidote. Remnants of augite within partially formed hematite rims similar to those after olivine were observed in some samples.

Amygdules in the Liberty Hills flows contain various com-

TABLE 5. MICROPROBE ANALYSES OF REPRESENTATIVE ACTINOLITE FROM ZONES IIa AND IIb

Analysis	1	2	3	4	5	6	7
Zone	IIa	IIa	IIa	IIa	IIa	IIa	IIb
Specimen	62PGS35e	64-BS-88	64-BS-77	79Ve3a	79Ve3a	64-MH-54	W63226
Formation	SP	SP	SP	LH	LH	SP	DI
SiO_2	52.90	53.76	52.56	54.52	51.46	53.30	55.12
TiO_2	0.00	0.00	0.26	0.00	0.34	0.03	0.02
Al_2O_3	2.08	1.39	2.66	1.63	2.78	1.55	0.40
FeO	13.56	19.97	10.47	7.42	13.64	20.67	12.23
Fe_2O_3	3.29	0.00	5.08	2.62	4.63	0.02	0.37
MnO	0.17	0.16	0.10	0.23	0.89	0.29	0.19
MgO	13.48	10.50	14.60	18.02	12.21	9.88	16.23
CaO	12.18	11.98	11.65	12.88	11.88	11.96	12.85
Na_2O	0.38	0.21	0.76	0.23	0.43	0.55	0.20
K_2O	0.10	0.10	0.19	0.07	0.18	0.19	0.17
Total	98.14	98.07	98.33	97.62	98.44	98.44	97.78
Structural formulae on the basis of 13 cations exclusive of Ca, Na, K							
Si_{IV}	7.69	7.95	7.56	7.73	7.53	7.92	7.92
Al	0.31	0.05	0.44	0.27	0.47	0.08	0.07
tet	8.00	8.00	8.00	8.00	8.00	8.00	7.99
Al	0.05	0.19	0.01	0.00	0.01	0.19	0.00
Ti	0.00	0.00	0.03	0.00	0.04	0.00	0.00
Fe^{3+}	0.36	0.00	0.55	0.28	0.51	0.00	0.04
Fe^{2+}	1.65	2.47	1.26	0.88	1.67	2.57	1.47
Mn	0.02	0.02	0.01	0.03	0.11	0.00	0.02
Mg	2.92	2.31	3.13	3.81	2.66	2.19	3.48
ΣM_{1-3}	5.00	4.99	4.99	5.00	5.00	4.95	5.01
ΣXM_{1-3}	0.00	0.00	0.00	0.00	0.00	0.00	0.01
Ca	1.90	1.90	1.80	1.96	1.86	1.91	1.98
Na	1.10	0.06	0.20	0.04	0.12	0.09	0.01
ΣM_4	2.00	1.96	2.00	2.00	1.98	2.00	2.00
Na	0.01	0.00	0.00	0.02	0.00	0.07	0.05
K	0.02	0.02	0.04	0.01	0.03	0.04	0.03
ΣA	0.03	0.02	0.04	0.03	0.03	0.11	0.08

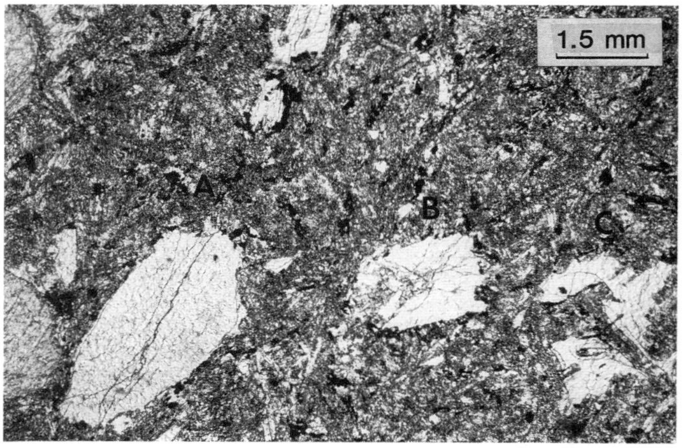

Figure 3. Basalt of the upper Springer Peak Formation containing chlorite pods after clinopyroxene. Two of the pods display euhedral (B) and subophitic (C) pyroxene outlines: the third pod (A) has been tectonically flattened and contains a weakly developed foliation normal to the direction of flattening. The dark lines in the pods are oxychlorite (Specimen 64MH7, plane light).

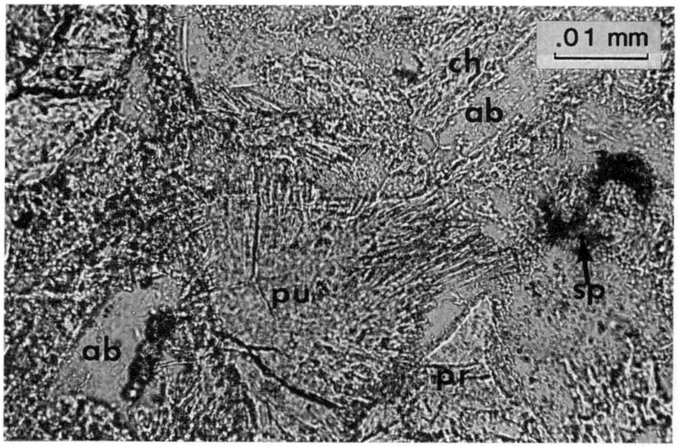

Figure 4. Prehnite (pr) and radiating crystals of pumpellyite (pu) after plagioclase in association with clinozoisite (cz), albite (ab), chlorite (ch), and sphene (sp) in basalt of the Springer Peak Formation (Specimen 64R27, plane light).

binations of epidote, calcite, chlorite, albite, pumpellyite, white mica, and quartz. Fine-grained brownish-yellow epidote (Table 4, analyses 4 and 5) or coarser grained blocky to prismatic, bright yellow epidote (Table 4, analysis 6) are most common as the outer rim phases surrounding chlorite (Table 2, analysis 7) or calcite cores and/or various combinations of the other minerals distributed irregularly within the rim. Pumpellyite in the amygdules occurs as radiating acicular masses in quartz, albite, or calcite (Fig. 8) or as acicular bundles attached to an epidote substrate (Table 3, analysis 7).

A volcanic breccia from just southeast of High Nunatak contains a basalt matrix composed of densely packed phenocrysts of plagioclase and augite. Interstitial sites between the phenocrysts contain quartz, albite, apatite, actinolite, pumpellyite, chlorite, and epidote. Opaque phases include pyrite, ilmenite, and titanomagnetite. The magnetite commonly contains trellis inclusions of ilmenite, and both the ilmenite and magnetite are largely altered to sphene ± rutile. Plagioclase phenocrysts may contain cores altered largely to white mica but are more commonly replaced by fine-grained dusky aggregates of white mica, epidote, and probable pumpellyite within clear albite rims. Radiating prisms of tourmaline displaying dark greenish blue to pale brown pleochroism were observed in one thin section of the breccia matrix. Some of the chlorite grains (Table 2, analysis 8) contain fine granular masses of sphene and remnant areas of pale brown pleochroism, indicating their formation after biotite. Green pumpellyite (Table 3, analysis 8) occurs in some of these chlorite grains as lenses aligned parallel to the chlorite cleavage planes. The clinopyroxene is locally altered to a red-brown to green-brown hornblende but is more commonly variably altered to both large crystals of a distinct green to rarely blue-green actinolite (Table 5, analysis 5) and/or mosaics of a randomly oriented patchwork of colorless to pale green actinolite (Table 5, analysis 4) ± chlorite. The brown hornblende is similar to that interpreted to be of late-stage magmatic origin in samples of gabbro from the Springer Peak Formation. The green actinolite and biotite altered to chlorite are believed to be of deuteric origin, whereas the colorless actinolite in association with chlorite and pumpellyite is considered to be a product of the zone IIa metamorphism.

Zone IIb—Gabbros of the Hyde Glacier sill. Gabbros of the Drake Icefall Formation exposed in the Hyde Glacier sill contain primary clinopyroxene, plagioclase, ilmenite, and titanomagnetite and display primary textures similar to those in the gabbros and diabases of the Springer Peak Formation. Metamorphic assemblages are also similar except for the occurrence of clinozoisite exclusive of epidote (Table 4, analysis 7) and clear, nonpleochroic pumpellyite with a rather low Fe and high Al and

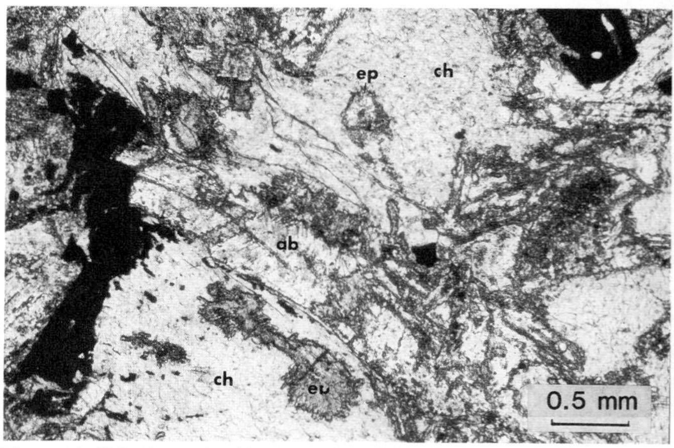

Figure 5. Chlorite pod (ch) after subophitic clinopyroxene in gabbro of the Springer Peak Formation. The pod contains radiating clusters of brown epidote (ep); ab—albite (Specimen 64MH54, plane light).

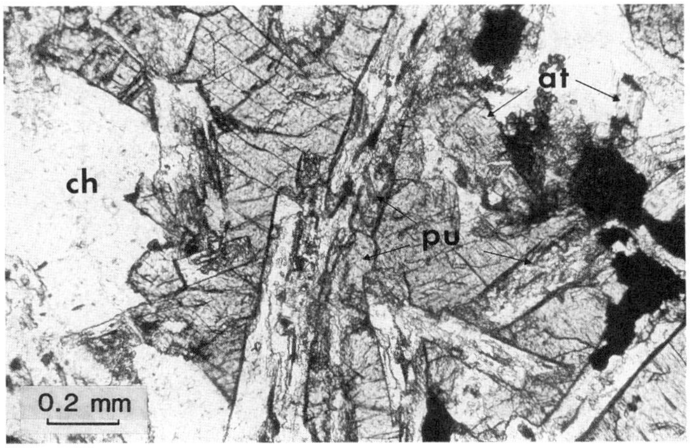

Figure 6. Pumpellyite (pu) after plagioclase in subophitic texture with clinopyroxene altering to actinolite (at) and chlorite (ch) in gabbro of the upper Springer Peak Formation (Specimen 64MH12, plane light).

Figure 8. Radiating, acicular spray of pumpellyite (pu) in an amygdule with quartz (qz) and blocky yellow epidote (ep) in a spilitic basalt of the Liberty Hills Formation (Specimen 79Vel01, plane light).

Mg content (Table 3, analysis 9). Clinozoisite and pumpellyite occur as blocky, subhedral to euhedral prisms ± actinolite (Table 5, analysis 7) in chlorite pods after clinopyroxene (Fig. 9). The chlorite is a pale green clinochlore (Table 2, analysis 9). Titanomagnetite and ilmenite are largely replaced by sphene ± rutile.

The association of aluminous pumpellyite with clinozoisite, actinolite, and chlorite has been suggested, based on field, theoretical, and experimental criteria, to mark the uppermost boundary of the pumpellyite-actinolite facies with the greenschist facies (Kawachi, 1975; Coombs and others, 1976; Nakajima and others, 1977; Nakajima, 1982; Schiffman and Liou, 1983). The gabbros of the Drake Icefall Formation mark the lowest stratigraphic occurrence of pumpellyite in the Heritage Group, in agreement with this interpretation.

Figure 7. Amygdaloidal spilitic basalt of the Liberty Hills Formation containing chlorite (ch) and epidote (ep), or calcite (cc), within a pseudomorphic opaque rim of hematite + sphene after olivine. The amygdule contains an outer rim of blocky yellow epidote and a core of calcite, quartz, white mica, epidote, and pumpellyite which are not distinguishable from one another in plane light (Specimen 79Vel01, plane light).

CONDITIONS OF METAMORPHISM

Temperature and pressure

Rocks recrystallized in the pumpellyite-actinolite facies have been described from the Sanbagawa terrane of Japan (e.g., Watanabe, 1974; Nakajima and others, 1977), the South Island of New Zealand (Bishop, 1972; Kawachi, 1974), the Alpine mountain chain of Europe (Coombs and others, 1976; Katagas and Panagos, 1979; Cortesogno and others, 1984), and the Appalachian metamorphic belt of North America (Coombs and others, 1970; Zen, 1974)—all regions characterized by convergent–plate-margin tectonics (e.g., Ernst, 1977). However, the facies series from prehnite-pumpellyite to pumpellyite-actinolite to

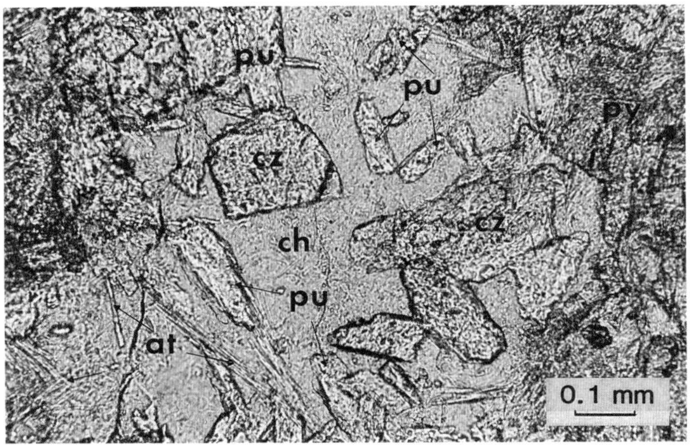

Figure 9. Coexisting pumpellyite (pu), clinozoisite (cz), and actinolite (at) in chlorite (ch) replacing clinopyroxene (py) in a gabbro from the Drake Icefall Formation (Specimen W63226-19000a, plane light).

greenschist facies is not developed in areas of especially low geothermal gradients, such as the Franciscan terrane of California (e.g., Ernst and others, 1970), and is generally characterized as a medium-pressure type of metamorphism (Seki, 1969; Coombs and others, 1970). Low-grade facies series from areas of lower pressure and higher thermal gradients, such as the Tanzawa Mountains of Japan (Seki and others, 1969) and the Karmutsen volcanics of Vancouver Island (Kuniyoshi and Liou, 1976), show a transition from the prehnite-pumpellyite facies through the prehnite-actinolite facies (cf., Liou and others, 1985) to actinolite-epidote-chlorite assemblages of the greenschist facies.

Such field observations have led many workers (e.g., Hashimoto, 1966; Seki, 1969) to conclude that the pumpellyite-actinolite facies has a minimum pressure limit and a wedge-shaped P-T region of stability with an increasing temperature range at increasing pressure (cf., Fig. 10, discussed below). This conclusion was further supported by the experimental work of Nitsch (1971) on natural Mg end-member phases in the reactions: (1) prehnite + chlorite + H_2O = pumpellyite + actinolite + quartz, and (2) pumpellyite + chlorite + quartz = epidote + actinolite + H_2O. Reaction (1) has been suggested as one of the probable reactions marking the lower temperature limit of the pumpellyite-actinolite facies (e.g., Seki, 1969; Coombs and others, 1970; Kawachi, 1975). Reaction (2) is a continuous reaction in natural systems, resulting in continuous compositional variations among the reacting phases with increasing temperature in the pumpellyite-actinolite facies. Ultimately, the loss of pumpellyite as as reactant marks the transition to the greenschist facies (Seki, 1969; Hashimoto, 1972; Kawachi, 1975; Nakajima and others, 1977).

Nitsch's (1971) experiments indicate a downward convergence of the P-T curves for these two reactions (Fig. 10) with an intersection at a pressure of 2.5 ± 1 kb marking the lower pressure limit of the pumpellyite-actinolite facies assemblage pumpellyite + actinolite + chlorite + quartz. However, Schiffman and Liou (1980) questioned the significance of this lower pressure limit because their experiments, based on an ACM system with colinear compositions of prehnite, pumpellyite, and chlorite, suggested that the prehnite + chlorite assemblage should occur at higher temperatures than the pumpellyite + actinolite assemblage. Liou and others (1985) noted that this discrepancy is due to the rather Al-rich chlorites used by Schiffman and Liou (1980) and that the use of a more Mg- and Si-rich chlorite in the reaction (3) prehnite + chlorite + quartz = pumpellyite + tremolite + H_2O yields a curve similar to but at slightly higher temperatures than that deduced by Nitsch for reaction (1) (see Liou and others, 1985, Fig. 3, for preliminary position of this curve). Nevertheless, as Liou and others (1985) emphasize, the P-T slope of this reaction is very sensitive to chlorite composition. Therefore, the intersection of reaction curves (1) and (2) or (3) and (2), marking the lower pressure limit of the pumpellyite-actinolite facies, is not well constrained.

Using the reaction curves (1) and (2) of Nitsch (1971), a temperature range of approximately 290 to 360°C is predicted

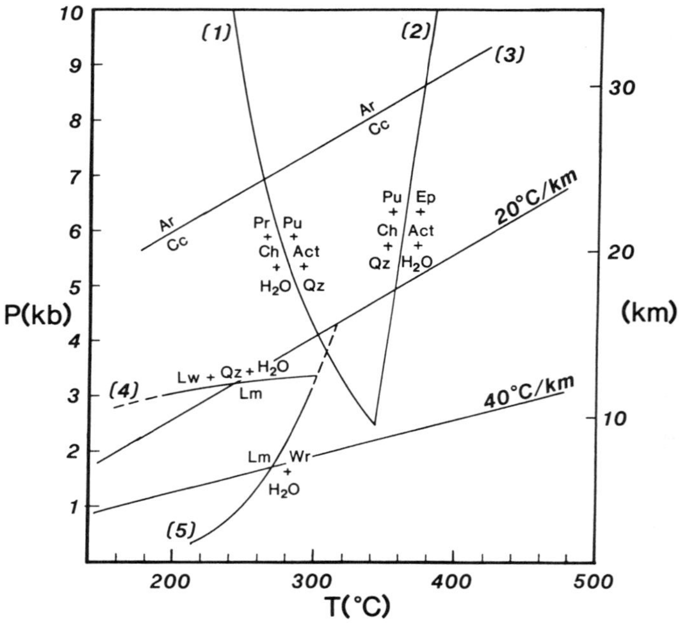

Figure 10. Pressure-temperature diagram showing reaction curves (1) and (2) determined by Nitsch (1971) and discussed in the text. Reaction curve (3) for the calcite-aragonite transition (Johannes and Puhan, 1971) and curves (4) and (5) for the breakdown of laumontite (Liou, 1971) are shown for comparison.

for the pumpellyite-actinolite facies at a pressure of 5 kb. However, reaction temperatures in this system would be expected to vary as a function of fO_2 (cf. Liou and others, 1983), which was not controlled in Nitsch's experiments, and somewhat lower temperature limits may be expected in natural iron-bearing systems (cf. Liou and others, 1985).

As noted above, reaction (2) is not univariant in natural systems due to compositional variations among the phases. Nakajima and others (1977) have considered these compositional variations and calculated a theoretical T-X_{Fe} (X_{Fe} = Fe/[Fe + Al]) relation (Fig. 11) for coexisting epidote and pumpellyite in the reaction (2) assemblage for a fixed Fe/(Fe + Mg) in chlorite of 0.45 to 0.50, arbitrarily setting a temperature of 300°C for a value of X_{Fe}^{Ep} (min.) = 0.31. Nakajima (1982) has further shown that variations in Fe/(Fe + Mg) in chlorite down to 0.40 have little effect on this relation. Brown and O'Neil (1982) obtained temperatures using the T-X_{Fe}^{Ep} curve in Figure 11 that are consistent with their oxygen isotope thermometry in rocks from the Shuksan blueschist terrane containing the reaction (2) assemblage. They concluded from this work that the T-X_{Fe}^{Ep} relationship of Nakajima and others (1977), and thus reaction (2), are relatively insensitive to higher pressures in the system. Brown and O'Neil (1982) also showed that the epidote composition used by Nitsch (1971) yields temperatures on the T-X_{Fe}^{Ep} curve that are consistent with his experimental temperatures.

Epidote compositions in the reaction (2) assemblage from

zone IIa and IIb of the present study are plotted in Figure 11. Clinozoisite from zone IIb contains X_{Fe} as low as 0.11 (Table 4, analysis 7), the lowest values yet reported in this assemblage, yielding a temperature estimate of approximately 370°C for the upper boundary of the pumpellyite-actinolite facies in the Heritage Group. This temperature is slightly higher than would be expected from the P-T curve for reaction (2) in Figure 11. The discrepancy may be a result of the lower Fe/(Fe + Mg) ratio (=0.31) in the chlorite in this reaction (2) assemblage (cf. Table 2, analysis 9) or of uncertainties in the calibration of the T-X_{Fe} curves. Despite the consistency demonstrated by Brown and O'Neil (1982) between their oxygen isotope temperatures and the T-X_{Fe}^{Ep} curve of Nakajima and others (1977), the temperatures assigned to the original curve were arbitrary. Although reaction (2) is independent of fO_2, the Fe^{3+} composition of the reacting system will depend on the fO_2 imposed on the system by the coexisting Fe-oxide + Fe-sulfide phases. This imposed fO_2 may result in variations in the equilibrium X_{Fe}^{Ep} for a given temperature. For instance, the magnetite-bearing assemblages considered by Brown and O'Neil (1982) certainly recrystallized under values of fO_2 lower than those of the hematite-bearing samples described by Nakajima and others (1977). The isotopic temperature comparison of Brown and O'Neil (1982) with the T-X_{Fe}^{Ep} curve may actually be a rough calibration of the curve for samples with an fO_2 range consistent with magnetite-bearing assemblages. The magnetite-ilmenite–bearing assemblages in the rocks from zone IIb probably recrystallized under fO_2 conditions more similar to those of the Shuksan rocks, although fO_2 was not buffered in either assemblage. A temperature of approximately 370°C is therefore considered to be a rough upper temperature limit for the pumpellyite-actinolite facies under the fO_2 and pressure conditions prevailing during the burial metamorphism.

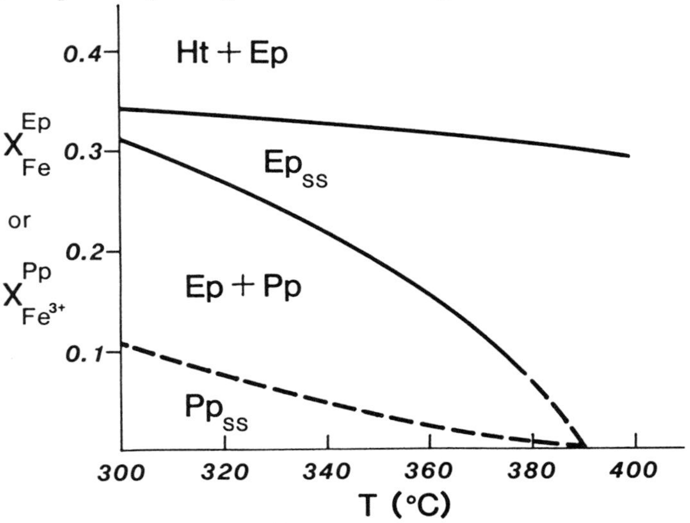

Figure 11. Part of the Al-Fe^{3+} pseudobinary phase diagram for epidote (Ep)-pumpellyite (Pu) equilibria in an assemblage with chlorite, actinolite, quartz, and albite as additional phases (after Nakajima and others, 1977). Epidote compositions in the reaction (2) assemblage of zone II are plotted from the upper Springer Peak Formation (zone IIa) and the lower Drake Icefall Formation (zone IIb).

An estimate of the prevailing pressures during metamorphism was attempted using the method of Brown (1977a) based on the Na content of the M4 sites in Ca-amphiboles. Brown (1974, 1977a) has shown that the continuous reaction crossite + epidote + H_2O = Ca-amphibole +o albite + chlorite + Fe-oxide buffers the NaM_4 composition of amphiboles in the transition from the blueschist to the greenschist facies. Therefore, in rocks of the greenschist facies containing the assemblage Ca-amphibole + chlorite + albite + Fe-oxide, the NaM_4 content of actinolite varies systematically with the pressure of metamorphism. Brown (1977a) suggested a tentative quantitative pressure relationship as a function of actinolite NaM_4 content as plotted agains Al^{IV}, a rough temperature-sensitive variable (Brown, 1977a, Fig. 10). Brown's calibration is based primarily on data from greenschist and blueschist facies rocks; however, he suggests that the technique may also be applicable in the pumpellyite-actinolite facies. One actinolite analysis published by Katagas and Panagos (1979) from pumpellyite-actinolite facies rocks in Greece is consistent with this application. Also one actinolite in the buffering assemblage analyzed by Kawachi (1975) contains NaM_4 consistent with the medium-pressure (ca. 5 kb) Western Otago greenschist rocks analysed by Brown (1977a). The other analyzed actinolites of Kawachi (1975) contain significantly lower NaM_4 but occur in unspecified mineral assemblages.

Yardley (1982) reported actinolite analyses from pumpellyite-actinolite facies rocks in the Caples Terrane of New Zealand. His plot of NaM_4 versus Al^{IV} includes all Caples actinolite analyses, irrespective of coexistence with the buffering assemblage, and is therefore difficult to interpret in terms of consistency with Brown's (1977a) calibration. Among the complete actinolite analyses published by Yardley from the Caples Terrane, only one sample (BY 1936) contains the buffering assemblage, and in this sample, both high and low NaM_4 actinolite analyses are reported.

Hynes (1982), in a comparison of previously published medium- and low-pressure metamorphic amphibole compositions, found NaM_4 values to be consistently low or at least commonly indistinguishable between medium- and low-pressure greenschist facies rocks. He suggested that, in part, pressure distinctions based on amphibole NaM_4 contents may be blurred as a result of variations in oxidation state during metamorphism.

The actinolites analyzed from pumpellyite-actinolite facies assemblages in the Heritage Group (Table 5), despite their occurrence in the buffered assemblage, have NaM_4 concentrations indicative of pressures around 2 kb or less in the calibration of Brown (1977a). However, these same analyses plot in the range of medium-pressure amphiboles on the formula proportion diagrams of Laird and Albee (1981), and they have low Ti contents, consistent with the medium-pressure amphibole analyses evaluated by Hynes (1982). Pressure estimates for the burial metamorphism of the Heritage Group are therefore inconclusive.

X_{CO_2} and fO_2 in the fluid phase

Numerous investigators have recognized the effects of fluid composition, especially X_{CO_2} and fO_2, on the phase relations among metamorphic minerals during low-grade metamorphism of basic rocks (e.g., Zen, 1961; Coombs and others, 1970; Glassley, 1974; Coombs and others, 1976). In rocks metamorphosed under high X_{CO_2} conditions, the pumpellyite-actinolite join becomes unstable relative to the calcite-epidote join (Brown, 1977b), and prehnite may break down to calcite + epidote + quartz (Glassley, 1974). Therefore, although the presence of prehnite and/or pumpellyite is a useful indicator of subgreenschist facies conditions, the absence of these phases in many of the rocks of appropriate composition in the Heritage Group may be a result of high X_{CO_2} rather than inconsistent P-T conditions. Such high X_{CO_2} conditions are indicated by the abundance of carbonate minerals throughout much of the metasedimentary portion of the Heritage Group (Webers and others, this volume) and locally by the association stilpnomelane + calcite in zone I basalt (Hashimoto and Kanehira, 1975). However, in most instances the more massive volcanic flows and diabase dikes and sills of the Springer Peak and Drake Icefall formations have recrystallized under significantly lower X_{CO_2}, as indicated by their sphene-, prehnite-, and/or pumpellyite-bearing assemblages. The occurrence of sphene in these rocks is controlled by the reaction sphene + CO_2 = rutile + calcite + quartz. Experimental evaluation of this reaction by Hunt and Kerrick (1977) indicates that for P_{total} of 3,000 bars, X_{CO_2} in the fluid phase is restricted to values of less than about 0.08 in sphene-bearing assemblages at temperatures below 400°C.

Unlike the mafic igneous rocks of the Springer Peak Formation, the locally spilitized basaltic flows from the Liberty Hills Formation have been metamorphosed in contact with fluids of quite variable compositions. Such variations are indicated by the wide variety of phases that replace plagioclase, olivine, and pyroxene in different samples. The most reconstituted volcanic rocks contain the assemblages calcite + chlorite + albite + white mica + hematite + quartz, indicative of relatively high X_{CO_2} conditions (Coombs and others, 1970; Pluysnima and Ivanov, 1981). The pervasive hematite in these flows is also indicative of relatively high fO_2 in the associated fluid phase, especially as compared to fO_2 during recrystallization of the magnetite-ilmenite–bearing assemblages in zone II gabros and the pyrrhotite-bearing assemblages in zone I basalts of the Springer Peak Formation. The rather low fO_2 required for the persistence of pyrrhotite without magnetite in the zone I basalts may account for the low iron concentrations in the pumpellyite and clinozoisite in these prehnite-bearing samples.

RELATION OF DEFORMATION AND METAMORPHISM

Yoshida (1982, 1983) has proposed a four-stage polymetamorphic-polyorogenic sequence for the geologic development of the Ellsworth Mountains, with events ranging in age from the Precambrian to the early Mesozoic Ellsworth Orogeny of Craddock (1969). It is not possible based on the scope of the present study to fully evaluate the sequence of metamorphism and cleavage formation proposed by Yoshida. However, certain petrographic observations made during the course of this study have significant implications for the relative timing of deformation, the low-grade metamorphism described in this chapter, and other metamorphic events as suggested by Yoshida (1982, 1983).

Pretectonic burial metamorphism

Numerous textural observations in rocks throughout the Heritage Group indicate that the low-grade metamorphism described here was initiated prior to the deformation affecting the rocks. For instance, some of the gabbros, diabases, and basalts of the Springer Peak Formation contain an irregular spaced cleavage that is concentrated in the chlorite pods after clinopyroxene. Some of the pods are flattened normal to the cleavage plane (Fig. 3) and may contain a more closely spaced foliation that is deflected by epidote, pumpellyite, or sphene crystals within the pods. Actinolite in the chlorite pods is partially aligned in the foliation plane, and chlorite adjacent to the cleavage planes is partially altered to oxychlorite. Cleavage planes cutting plagioclase crystals contain aligned crystals of chlorite and actinolite that truncate all of the low-grade phases after plagioclase, including pumpellyite, epidote-clinozoisite, and white mica. Chlorite- and epidote-filled amygdules in some of the basalts of the Liberty Hills Formation have been flattened and contain a second generation of chlorite or white mica ± calcite that is aligned parallel to the axis of amygdule elongation and is deflected around epidote grains (Fig. 12). These observations and the correlation of increasing metamorphic grade with depth in the stratigraphic section of the Heritage Group suggest that the sequence of metamorphic assemblages in the mafic rocks was initiated under burial metamorphic conditions prior to deformation.

An estimate of the geothermal gradient during burial metamorphism can be made from estimates of the temperatures marking the upper and lower boundaries of the pumpellyite-actinolite facies assemblages and the stratigraphic thickness of this interval. A temperature range of approximately 40°C is indicated by the range in epidote compositions plotted in Figure 11 from the reaction (2) assemblage. The lowest temperature data were obtained from samples in the upper Springer Peak Formation; the highest temperature data are from the Hyde Glacier sill in the Drake Icefall Formation. G. Webers (personal communication, 1983) estimates a stratigraphic thickness of 2,050 m over this interval, yielding a calculated geothermal gradient over the pumpellyite-actinolite facies of approximately 23 to 24°C/km. This gradient compares favorably with the range of 20 to 27°C/km estimated by Turner (1981, p. 430) for the prehnite-pumpellyite to pumpellyite-actinolite to greenschist facies series in the Haast Schist block of southern New Zealand. Lines marking constant geothermal gradients of 20° and 40°C/km are plotted on Figure 10 for comparison.

Hjelle and others (1978) calculated a geothermal gradient of 200°C/kb for the 7,000-m-thick stratigraphic sequence from the Whiteout Conglomerate to their Edson Hills Formation (upper

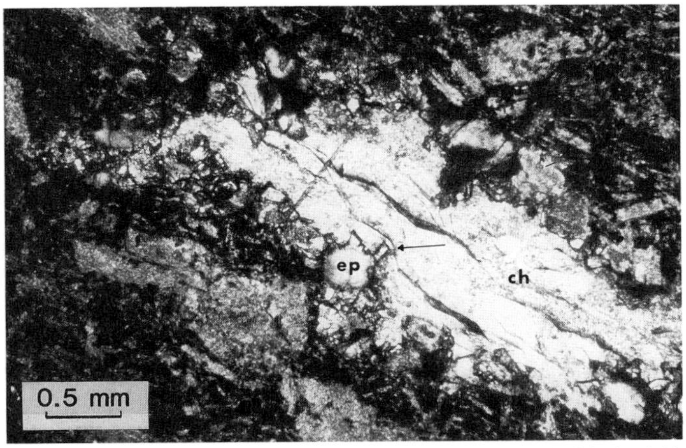

Figure 12. Flattened amygdule containing epidote (ep) and chlorite (ch). The chlorite contains a localized foliation, approximately normal to the direction of flattening, that is deflected by an adjacent epidote cluster (at arrow) (Specimen 79Ve1sb, plane light).

Heritage Group of this study; cf. Webers and others, this volume, Fig. 2). This calculation relies heavily on the identification of dark brown metamorphic biotite in the gabbros of their Middle Horseshoe Formation (Upper Heritage Group). Brown biotite altering to chlorite and to fine granules of sphene was observed during the present study in some of the gabbros; however, such grains were only observed in granophyre-bearing samples and are interpreted to be of late-stage magmatic origin.

Syntectonic recrystallization

Buggisch and Webers (1982; this volume) observed an increase in deformation from north to south in the Minaret Formation overlying the Hertiage Group and suggested a parallel north to south increase in metamorphic grade on the basis of illite crystallinity data and the Sr content of calcite. Such a trend was not observed in the mafic rocks considered in the present study. For instance, the highest grade mineral assemblages of the pumpellyite-actinolite facies (zone IIb) in rocks of the Drake Icefall Formation occur in the Edson Hills to the north of lower grade assemblages (zone IIa) of the Liberty Hills Formation in the western Enterprise Hills, the Liberty Hills, and High Nunatak (Fig. 1). However, it is possible that low-grade recrystallization, continuing or reinitiated during the Ellsworth Orogeny, was aided kinetically by the mechanical deformation so that the foliated rocks attained a different distribution of metamorphic grade and possibly a higher grade while the more massive mafic igneous rocks retained their burial metamorphic imprint. Although high P_{CO_2} in the metasedimentary rocks may be called on to explain their lack of subgreenschist facies assemblages, it is also possible that these rocks underwent syntectonic recrystallization in the greenschist facies. The generation of a second chlorite or white mica and calcite along foliation planes parallel to the long axes of flattened amygdules (Fig. 12) and chlorite + actinolite along cleavage planes cutting plagioclase were the only obvious manifestations of this syntectonic recrystallization observed in the pumpellyite-bearing mafic rocks examined during this study. Neither prehnite nor pumpellyite was observed in mafic igneous rocks displaying a more penetrative deformation fabric.

A result of this type of differential recrystallization would be the determination of younger metamorphic ages on the more deformed country rocks and dikes than on the less deformed mafic stocks, sills, and lava flows. This, in fact, is the type of relative age distribution shown by the whole-rock K-Ar ages obtained by Yoshida (1982, Table 1).

SUMMARY AND CONCLUSIONS

The gabbroic stocks, diabasic sills, and basaltic lava flows of the Heritage Group show a variation in metamorphic grade from the upper prehnite-pumpellyite facies throught the pumpellyite-actinolite facies into the greenschist facies. The spatial variation in metamorphic grade of the rocks is correlated with their stratigraphic position in the Heritage Group and is interpreted to be a result of pretectonic burial metamorphism developed under a geothermal gradient of approximately 23 to 24°C/km. The metamorphic temperature at the boundary between the upper pumpellyite-actinolite facies and the lower greenschist facies is estimated to have been approximately 370°C using the T-X^{Ep}_{Fe} curve of Nakajima and others (1977). Attempts to estimate pressures of metamorphism using the NaM_4 concentrations of actinolite in the buffering assemblages of Brown (1977a) generally yield pressures of 2 kb or less. This value is unrealistically low when compared with experimental data and pressure estimates from other regions displaying the same facies series.

During the early Mesozoic Ellsworth Orogeny, deformation and associated regional metamorphism recrystallized thin mafic dikes and the sedimentary rocks of the Heritage Group to a grade no higher than chlorite grade of the lower greenschist facies. The relatively undeformed portions of mafic stocks, sills, and basaltic lava flows retained their burial metamorphic imprint, although partial reequilbration of actinolite during the orogeny may have destroyed its utility as an indicator of pressure during the burial event. The Late Devonian K-Ar whole-rock ages reported by Yoshida (1982) for these rocks may be another result of such a partial reequilibration during the orogeny.

ACKNOWLEDGMENTS

I thank G. F. Webers for assistance in obtained the samples considered in the study and for numerous discussions of the geology and stratigraphy of the Ellsworth Mountains. I also thank E. H. Brown and K. B. Spörli for helpful reviews of the manuscript. The study was supported by National Science Foundation grant DPP 78-21720.

REFERENCES CITED

Albee, A. L., and Ray, L., 1970, Correction factors for electron probe microanalysis of silicates, oxides, carbonates, phosphates, and sulfates: Analytical Chemistry, v. 42, p. 1408–1414.

Anderson, J. J., 1965, Bedrock geology of Antarctica; A summary of exploration, 1831–1962, *in* Hadley, J. B., ed., Geology and paleontology of the Antarctic: American Geophysical Union Antarctic Research Series, v. 6, p. 1–70.

Anderson, J. J., Bastien, T. W., Schmidt, P. G., Splettstoesser, J. F., and Craddock, C., 1962, Antarctica; Geology of the Ellsworth Mountains: Science, v. 138, p. 824–825.

Bauer, R. L., 1983, Low-grade metamorphism in the Heritage Range of the Ellsworth Mountains, West Antarctica, *in* Oliver, R. L., and others, eds., Antarctic geoscience: Canberra, Australian Academy of Science, p. 256–260.

Bence, A. E., and Albee, A. L., 1968, Empirical correction factors for the electron microanalysis of silicates and oxides: Journal of Geology, v. 76, p. 382–403.

Bishop, D. G., 1972, Progressive metamorphism from prehnite-pumpellyite to greenschist facies in Dansey Pass area, Otago, New Zealand: Geological Society of America Bulletin, v. 83, p. 3177–3198.

Brown, E. H., 1967, The greenschist facies in part of eastern Otago, New Zealand: Contributions to Mineralogy and Petrology, v. 14, p. 259–292.

—— , 1974, Comparison of the mineralogy and phase relations of blueschists from the north Cascades, Washington, and greenschists from Otago, New Zealand: Geological Society of America Bulletin, v. 85, p. 333–344.

—— , 1977a, The crossite content of Ca-amphibole as a guide to pressure of metamorphism: Journal of Petrology, v. 18, p. 53–72.

—— , 1977b, Phase equilibria among pumpellyite, lawsonite, epidote, and associated minerals in low grade metamorphic rocks: Contributions to Mineralogy and Petrology, v. 64, p. 123–136.

Brown, E. H., and O'Neil, J. R., 1982, Oxygen isotope geothermometry and stability of lawsonite and pumpellyite in the Shuksan suite, north Cascades, Washington: Contributions to Mineralogy and Petrology, v. 80, p. 240–244.

Buddington, A. F., and Lindsley, D. H., 1964, Iron-titanium oxide minerals and synthetic equivalents: Journal of Petrology, v. 5, p. 310–357.

Buggisch, W., and Webers, G. F., 1982, Zur Facies der Karbonatgesteine in den Ellsworth Mountains (Paläozoikum, Westantarktis): Facies (Erlangen), v. 7, p. 199–228.

Cann, J. R., 1969, Spilites from the Carlsberg Ridge, Indian Ocean: Journal of Petrology, v. 10, p. 1–19.

Castle, J. W., and Craddock, C., 1975, Deposition and metamorphism of the Polarstar Formation (Permian), Ellsworth Mountains: Antarctic Journal of the United States, v. 10, p. 239–241.

Chatterjee, N. D., 1966, On the widespread occurrence of oxidized chlorites in the Pennine Zone of the western Italian Alps: Beitrage zur Mineralogie und Petrographie, v. 12, p. 325–339.

Coombs, D. S., Horodyski, R. J., and Naylor, R. S., 1970, Occurrence of prehnite-pumpellyite facies metamorphism in northern Maine: American Journal of Science, v. 268, p. 142–156.

Coombs, D. S., Nakamura, Y., and Vuagnat, M., 1976, Pumpellyite-actinolite facies schists of the Taveyanne Formation near Loeche, Valais, Switzerland: Journal of Petrology, v. 17, p. 440–471.

Cortesogno, L., Lucchetti, G., and Spadea, P., 1984, Pumpellyite in low-grade metamorphic rocks from Ligurian and Lucanian Apennines, Maritime Alps, and Calabria (Italy): Contributions to Mineralogy and Petrology, v. 85, p. 14–24.

Craddock, C., 1962, Geological reconnaissance of the Ellsworth Mountains: U.S. Antarctic Projects Officer Bulletin, v. 3, p. 34–35.

—— , 1969, Geology of the Ellsworth Mountains, *in* Bushnell, V. C., and Craddock, C., eds., Geologic maps of Antarctica: New York, American Geographical Society Antarctic Map Folio Series, Folio 12, Plate 4, scale 1:1,000,000.

—— , 1972, Antarctic tectonics, *in* Adie, R. J., ed., Antarctic geology and geophysics: Oslo, Universitetsforlaget, p. 449–455.

—— , 1975, Tectonic evolution of the Pacific margin of Gondwanaland, *in* Campbell, K.S.W., ed., Gondwana geology: Canberra, Australia National University Press, p. 609–618.

Craddock, C., Anderson, J. J., and Webers, G. F., 1964, Geologic outline of the Ellsworth Mountains, *in* Adie, R. J., ed., Antarctic geology: Amsterdam, North-Holland, p. 155–170.

Ernst, W. G., 1977, Mineral parageneses and plate tectonic settings of relatively high pressure metamorphic belts: Fortschrite der Mineralogie, v. 54, p. 192–222.

Ernst, W. G., Seki, Y., Onuki, H., and Gilbert, M. C., 1970, Comparative study of low-grade metamorphism in the California Coast Ranges and the outer metamorphic belt of Japan: Geological Society of America Memoir 124, 276 p.

Evarts, R. C., and Schiffman, P., 1983, Submarine hydrothermal metamorphism of the Del Puerto ophiolite, California: American Journal of Science, v. 283, p. 289–340.

Glassley, W., 1974, A model for phase equilibria in the prehnite-pumpellyite facies: Contributions to Mineralogy and Petrology, v. 43, p. 317–332.

Haggerty, S. E., 1976, Oxidation of opaque mineral oxides in basalts, *in* Rumble, D., III, ed., Oxide minerals: Mineralogical Society of America Short Course Notes, v. 3, p. Hg1–Hg100.

Hashimoto, M., 1966, On the prehnite-pumpellyite metagraywacke facies: Journal of the Geological Society of Japan, v. 72, p. 253–265.

—— , 1972, Reactions producing actinolite in basic metamorphic rocks: Lithos, v. 5, p. 19–31.

Hashimoto, M., and Kanehira, K., 1975, Some petrological aspects on stilpnomelane in glaucophanitic metamorphic rocks: Journal of the Japanese Association of Mineralogists, Petrologists, and Economic Geologists, v. 76, p. 377–389.

Hjelle, A., Ohta, Y., and Winsnes, T. S., 1978, Stratigraphy and igneous petrology of southern Heritage Range, Ellsworth Mountains, Antarctica: Norsk Polarinstitutt Skrifter, no. 169, p. 5–43.

—— , 1982, Geology and petrology of the southern Heritage Range, Ellsworth Mountains, *in* Craddock, C., ed., Antarctic geoscience: Madison, University of Wisconsin Press, p. 599–608.

Hunt, J. A., and Kerrick, D. M., 1977, The stability of sphene; Experimental redetermination and geologic implications: Geochimica et Cosmochimica Acta, v. 41, p. 279–288.

Hynes, A., 1982, A comparison of amphiboles from medium- and low-pressure metabasites: Contributions to Mineralogy and Petrology, v. 81, p. 119–125.

Johannes, W., and Puhan, D., 1971, The calcite-aragonite transition, reinvestigated: Contributions to Mineralogy and Petrology, v. 31, p. 28–38.

Jolly, W. T., and Smith, R. E., 1972, Degradation and metamorphic differentiation of the Keweenawan tholeiitic lavas of northern Michigan, U.S.A.: Journal of Petrology, v. 13, p. 273–309.

Katagas, C., and Panagos, A. G., 1979, Pumpellyite-actinolite and greenschist facies metamorphism in Lesvos Island (Greece): Tschermaks Minerlogische und Petrographische Mitteilungen, v. 26, p. 235–254.

Kawachi, Y., 1974, Geology and petrochemistry of weakly metamorphosed rocks in the upper Wakatipu district, southern New Zealand: New Zealand Journal of Geology and Geophysics, v. 17, p. 169–208.

—— , 1975, Pumpellyite-actinolite and contiguous facies metamorphism in part of upper Wakatipu district, South Island, New Zealand; New Zealand Journal of Geology and Geophysics, v. 18, p. 401–441.

Kuniyoshi, S., and Liou, J. G., 1976, Burial metamorphism of the Karmutsen volcanic rocks, northeastern Vancouver Island, British Columbia: American Journal of Science, v. 276, p. 1096–1119.

Laird, J., and Albee, A. L., 1981, Pressure, temperature, and time indicators in mafic schists; Their application to reconstructing the polymetamorphic history of Vermont: American Journal of Science, v. 281, p. 127–175.

Levi, B., Aguirre, L., and Nyström, J. O., 1982, Metamorphic gradients in burial metamorphosed vesicular lavas; Comparison of basalt and spilite in Cretaceous basic flows from Central Chile: Contributions to Mineralogy and

Petrology, v. 80, p. 49–58.
Liou, J. G., 1971, P-T stability fields of laumontite, wairakite, lawsonite, and related minerals in the system $CaAl_2Si_2O_8$-SiO_2-H_2O: Journal of Petrology, v. 12, p. 379–411.
Liou, J. G., Kim, H. S., and Maruyama, S., 1983, Prehnite-epidote equilibria and their petrologic applications: Journal of Petrology, v. 24, p. 321–342.
Liou, J. G., Maruyama, S., and Cho, M., 1985, Phase equilibria and mineral parageneses of metabasites in low-grade metamorphism: Mineralogical Magazine, v. 49, p. 321–333.
Nakajima, T., 1982, Phase relations of pumpellyite-actinolite facies metabasites in the Sanbagawa metamorphic belt in central Shikoku, Japan: Lithos, v. 15, p. 267–280.
Nakajima, T., Banno, S., and Suzuki, T., 1977, Reactions leading to the disappearance of pumpellyite in low-grade metamorphic rocks of the Sanbagawa metamorphic belt of central Shikoku, Japan: Journal of Petrology, v. 18, p. 263–284.
Nitsch, K. H., 1971, Stabilitätsbeziehungen von prehnit- und pumpellyit-haltigen paragenesen: Contributions to Mineralogy and Petrology, v. 30, p. 240–260.
Nyström, J. O., 1983, Pumpellyite-bearing rocks in central Sweden and extent of host rock alteration as a control of pumpellyite composition: Contributions to Mineralogy and Petrology, v. 83, p. 159–168.
Offler, R., Baker, C. K., and Gamble, J., 1981, Pumpellyites in two low grade metamorphic terranes north of Newcastle, NSW, Australia: Contributions to Mineralogy and Petrology, v. 76, p. 171–176.
Otten, M. T., 1984, The origin of brown hornblende in the Artfjallet gabbro and dolerites: Contributions to Mineralogy and Petrology, v. 86, p. 189–199.
Passaglia, E., and Gottardi, G., 1973, Crystal chemistry and nomenclature of pumpellyite and julgoldite: Canadian Mineralogist, v. 12, p. 219–223.
Pluysnima, L. P., and Ivanov, I. P., 1981, Thermodynamic regime of greenstone metamorphism of basic volcanic rocks after experimental data: Canadian Journal of Earth Sciences, v. 18, p. 1303–1309.
Ramdohr, P., 1980, The ore minerals and their intergrowths, 2nd ed.: Oxford, Pergamon, 1205 p.
Robinson, P., Spear, F. S., Schumacher, J. C., Laird, J., Klein, D., Evans, B. W., and Doolan, B. L., 1982, Phase relations of metamorphic amphiboles; Natural occurrence and theory, *in* Veblen, D. R., and Robbe, P. H., eds., Amphiboles; Petrology and experimental phase relations: Mineralogical Society of America Reviews in Mineralogy, v. 9B, p. 1–227.
Schiffman, P., and Liou, J. G., 1980, Synthesis and stability relations of Mg-Al pumpellyite, $Ca_4Al_5MgSi_6O_{21}(OH)_7$): Journal of Petrology, v. 21, p. 441–474.
——, 1983, Synthesis of Fe-pumpellyite and its stability relations with epidote: Journal of Metamorphic Geology, v. 1, p. 91–101.
Smith, R. E., 1969, Zones of progressive regional burial metamorphism in part of the Tasman Geosyncline, eastern Australia: Journal of Petrology, v. 10, p. 144–163.
Turner, F. J., 1981, Metamorphic petrology; Mineralogical, field, and tectonic aspects, 2nd ed.: New York, McGraw-Hill, 524 p.
Watanabe, T., 1974, Metamorphic zoning of the Sambagawa and Chichibu belts in the Koshibu-Gawa River area, Oshika district, central Japan, with special reference to pumpellyite-actinolite schist facies mineral assemblage: Journal of the Geological Society of Japan, v. 80, p. 525–538.
Webers, G. F., and Spörli, K. B., 1983, Paleontological and stratigraphic investigations in the Ellsworth Mountains, West Antarctica, *in* Oliver, R. L., and others, eds., Antarctic earth science: Canberra, Australian Academy of Science, p. 261–264.
Yardley, B.W.D., 1982, The early metamorphic history of the Haast Schists and related rocks of New Zealand: Contributions to Mineralogy and Petrology, v. 8, p. 317–327.
Yoshida, M., 1982, Superposed deformation and its implications to the geologic history of the Ellsworth Mountains, West Antarctica: Tokyo, Memoirs of the National Institute of Polar Research Special Issue 21, p. 120–171.
——, 1983, Structural and metamorphic history of the Ellsworth Mountains, West Antarctica, *in* Oliver, R. L., and others, eds., Antarctic earth science: Canberra, Australian Academy of Science, p. 266–269.
Zen, E., 1961, The zeolite facies; An interpretation: American Journal of Science, v. 259, p. 401–409.
——, 1974, Prehnite- and pumpellyite-bearing mineral assemblages, west side of Appalachian metamorphic belt, Pennsylvania to Newfoundland: Journal of Petrology, v. 15, p. 197–242.

MANUSCRIPT ACCEPTED BY THE SOCIETY MARCH 1, 1989

Printed in U.S.A.

Geological Society of America
Memoir 170
1992

Chapter 18

Stratigraphy and structure of the Marble, Independence, and Patriot hills, Heritage Range, Ellsworth Mountains, West Antarctica

K. B. Spörli
Department of Geology, University of Auckland, Auckland, New Zealand
Campbell Craddock
Department of Geology and Geophysics, University of Wisconsin, Madison, Wisconsin 53706

ABSTRACT

The Marble, Independence, and Patriot hills at the southern end of the Ellsworth Mountains consist dominantly of Cambrian limestones that occur in two major facies: (1) gray, well-bedded limestones, and (2) white, massive, marblelike limestone. The latter rock type may in part be of tectonic origin. Conglomerates in the Patriot Hills underlie the limestones, which in turn are correlated with limestones at the top of the Heritage Group at Webers Peaks to the north in the Heritage Range. Folds mainly verge northeastward and trend northwestward, and the major structures are anticlines along the western and eastern margins of the area with an intervening synclinorium. The westernmost anticline is cut by a thrust fault that also displaces some postfolding breccia bodies. Late-phase conjugate strike-slip faults, en echelon extension gashes, and calcite fiber striations indicate a changed orientation of the principal horizontal stress, probably from a northeast-southwest to a northwest-southeast orientation. The main phase of deformation is assigned to the Ellsworth or Gondwanide Orogeny, probably early Mesozoic in age.

INTRODUCTION

The Marble, Independence, and Patriot hills are the southernmost groups of peaks in the Heritage Range, Ellsworth Mountains (Figs. 1, 2). They are noteworthy in that they consist predominantly of limestone and recrystallized carbonate rocks that may be termed marble. The rocks were originally classified as the Minaret Group (Craddock, 1969), but they are now considered to be a formation in the Cambrian Heritage Group (Webers and Spörli, 1983). They have been deformed in the Ellsworth or Gondwanide Orogeny, probably in early Mesozoic time (Craddock, 1972; Dalziel and Elliot, 1982). On the basis of a revised stratigraphic-tectonic position (to be illustrated in this chapter) and of fossils discovered in the 1979–1980 field season, Buggisch and Webers (1982) consider the carbonate rocks of the Marble Hills to be part of a continuous, southward-thickening, 150-km-long band of limestone that terminates in the north at the much-studied fossil site at Springer Peak (Webers, 1972). Microscopic textures, illite crystallinity, and Sr contents indicate increasing intensity of deformation and/or metamorphism from north to south within this band (Buggisch and Webers, 1982).

Limestones of the Marble and Independence hills contain a great number of breccia bodies (Spörli and others, Chapter 19, this volume), which were formed after the main deformation but have themselves been locally affected by faults and cleavage.

Tectonically these ranges lie immediately east of or in the axial zone of the main anticline of the Heritage Range (Spörli and others, this volume).

Spörli, K. B., and Craddock, C., 1992, Stratigraphy and structure of the Marble, Independence, and Patriot hills, Heritage Range, Ellsworth Mountains, West Antarctica, *in* Webers, G. F., Craddock, C., and Splettstoesser, J. F., eds., Geology and Paleontology of the Ellsworth Mountains, West Antarctica: Boulder, Colorado, Geological Society of America Memoir 170.

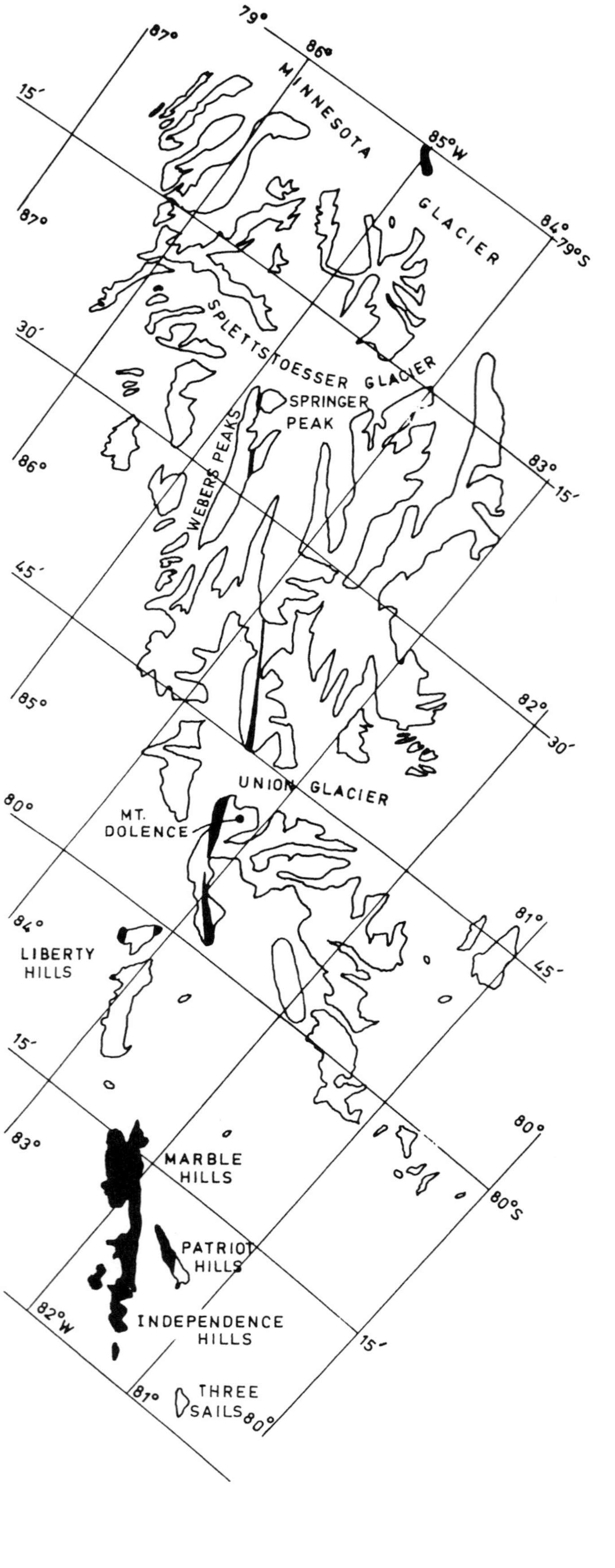

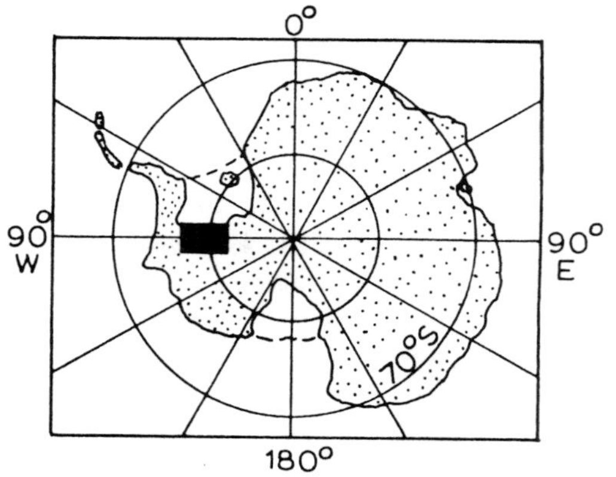

Figure 1. Location map for the Heritage Range, Ellsworth Mountains. Antarctica is in the conventional position in the inset, but in the large map meridians trend from upper right to lower left. Scale of Heritage Range map approximately 1:1,180,000.

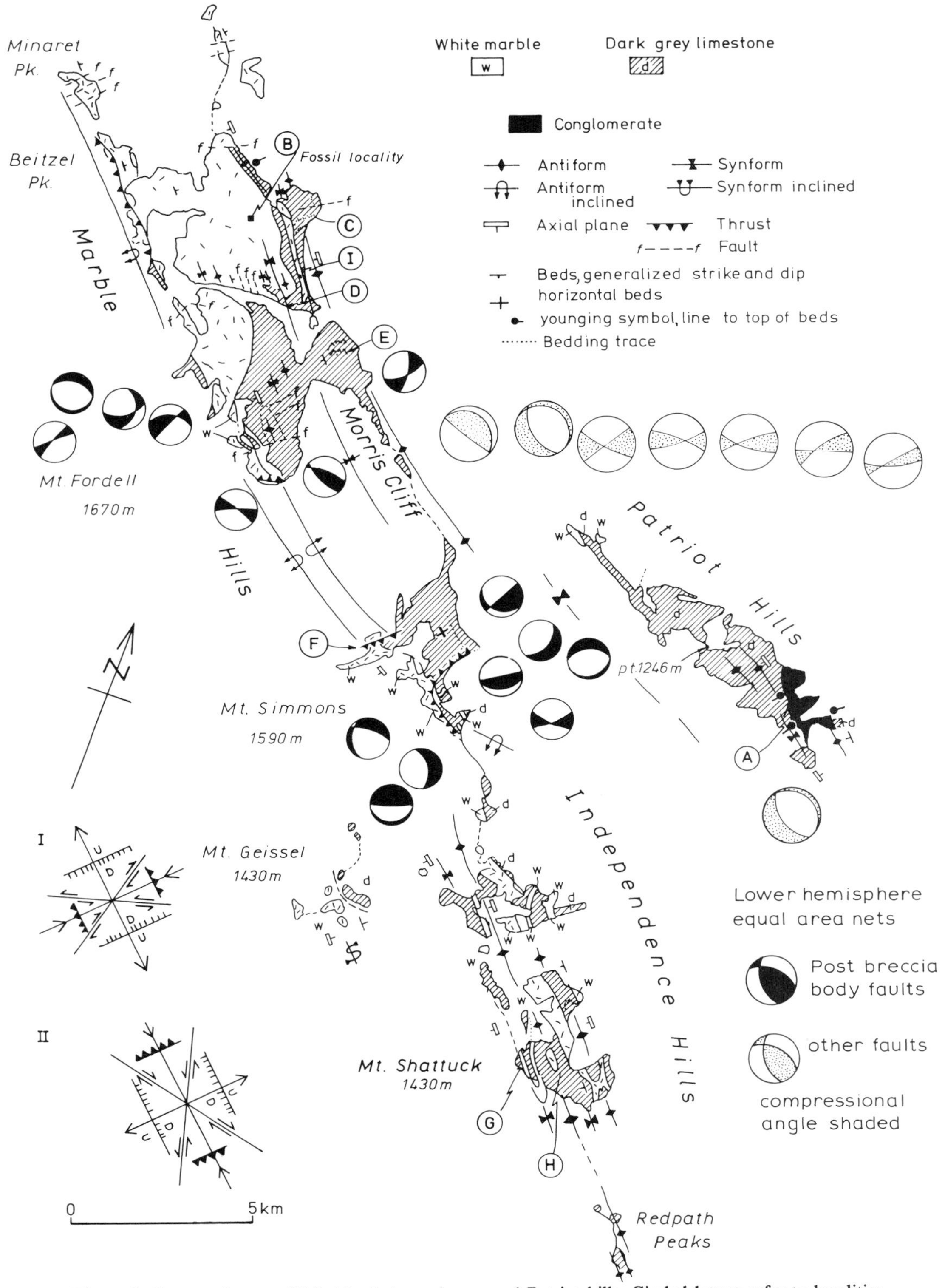

Figure 2. Structural map of Marble, Independence, and Patriot hills. Circled letters refer to localities mentioned in text. Circles with shaded angles: lower hemisphere equal-area diagrams of post–breccia body faults (compressional angle shaded). Diagrams I and II: idealized arrangement of strike-slip, thrust, and normal faults for compression across (I) and along (II) the structural trends. (Thrust symbols have teeth on hanging-wall side.) Stippled equal-area nets above Patriot Hills are all from the Mt. Fordell–Morris Cliff area.

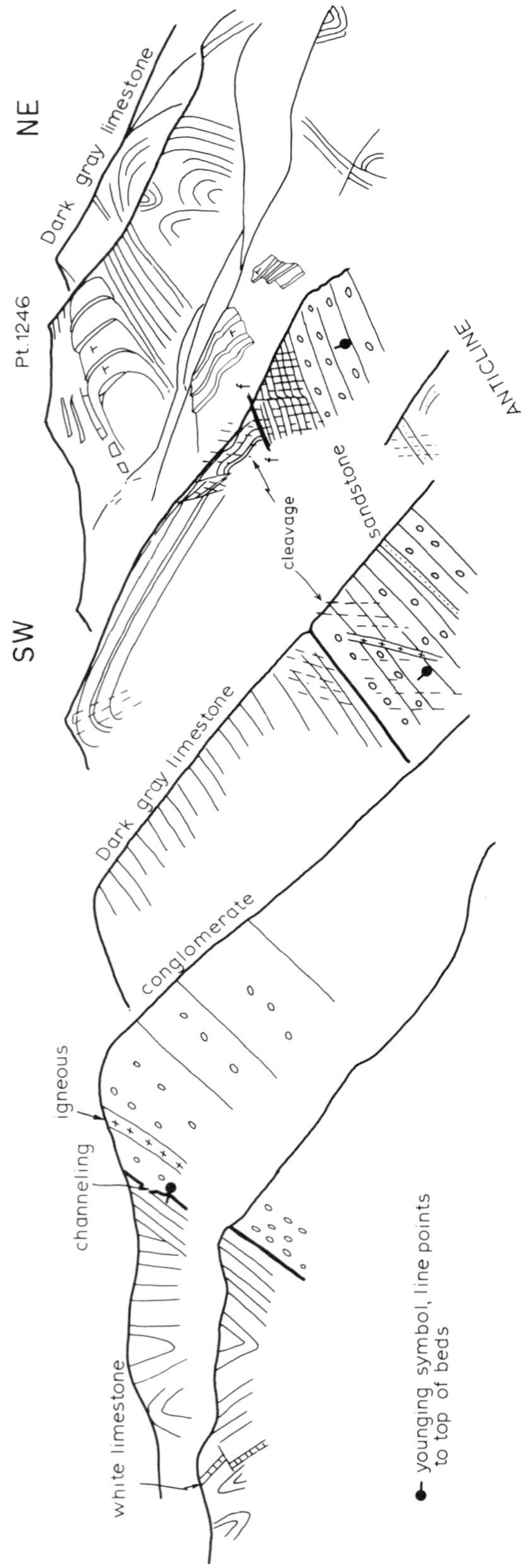

Figure 3. Field sketch of slopes of Patriot Hills, showing limestone overlying clastics on west flank of major anticline. Younging symbols mark actual locations of field determinations. f-f: fault.

STRATIGRAPHY

Conglomerates

A 100+-m section of conglomerate and brown to green, coarse-grained sandstones, associated with some purple and green basic volcanic rocks, represents the oldest exposed stratigraphic unit in the area; these rocks are located on the eastern side of the Patriot Hills (Figs. 2, 3). Younging directions in the conglomerates and cleavage/bedding relations confirm that the conglomerates underlie the limestones of the Patriot Hills (Fig. 3) and occupy the core of an anticline. A measured section of a steeply overturned succession at the southeastern end of the Patriot Hills on the eastern limb of the anticline (Fig. 2) is described briefly in Table 1.

The outcrop of the conglomerate does not extend along the entire northeast side of the Patriot Hills, because it is successively displaced eastward in the north by a set of cross faults (Fig. 2). The conglomerates are also exposed in the Three Sails Nunataks still farther southeast, and they can be correlated with conglomerates in the Liberty Hills to the northwest (Webers and Spörli, 1983).

Limestones

The limestones, which are about 750 m thick, can be subdivided into two major rock types: (1) dark gray, well-bedded limestone, and (2) white, more massive limestone. In places the two limestone varieties can be mapped as large-scale units (Figs. 4, and 5), but elsewhere they are intricately interbedded. Generally, the white massive limestone occurs to the west of and higher in the section than the dark gray, well-bedded limestones (Fig. 4). The limestones are mainly of an oolitic, oncolitic facies (Buggisch and Webers, 1982), and they contain scattered trilobite fragments.

Dark gray, well-bedded limestones. These beds lie conformably, but with a channeled contact, on the conglomerates of the Patriot Hills (Loc. A, Fig. 2; Fig. 3). Thicknesses are difficult to assess because of the strong folding and the ice-covered gap between the Patriot Hills and the Marble Hills/Independence Hills area. The stratigraphic position of a massive white limestone underlying dark gray limestones at the northern end of the Patriot Hills (Fig. 2) is not yet clear because tectonic structures may be responsible for the juxtaposition.

Some of the most thinly bedded units occur in the Patriot Hills, where marly limestone beds 30 to 80 cm thick are interbedded with fissile shales of similar thickness. Elsewhere limestone beds can attain thicknesses of several meters, and in some sections they are the dominant component.

Layers with quartz sand have only been encountered near the basal contact in the Patriot Hills, and elsewhere the rocks appear to consist dominantly of carbonate and clay minerals.

TABLE 1. MEASURED SECTION AT SOUTHEASTERN END OF PATRIOT HILLS, ON NORTHEAST LIMB OF ANTICLINE

TOP (east)	
8.	Dark gray, thin-bedded limestones, nodular, with yellow matrix due to dolomitization. Sandy layers 50 cm thick at 5-m intervals.
7.	~6 m light gray to resistant oolitic limestone.
6.	~20 m dark gray, thin-bedded limestones.
5.	~5 m yellowish gray to green and red sandstone, in part strongly sheared.
4.	~100 m massive conglomerate with green sandy matrix. Components from pebble to cobble size (sandstone, quartzite, igneous, and limestone conglomerates identical to unit 2).
3.	10+ m limestone conglomerate, with irregular, deep pockets on the east side and a smoother, sharper contact in the west. Some limestones have regular textures that may be organic (algae?). Some spherulitic textures.
2.	10 m red-weathering conglomerate without limestone components.
1.	30 m reddish-brown, silty argillite with altered, reddish-brown-weathering, porphyritic mafic volcanic rocks.
BOTTOM (west)	

Although micritic, homogeneous limestones—some of them containing shale clasts—are common, parts of the unit are strongly pisolitic and oolitic. Pisolites, as much as 2 cm in longest diameter, are embedded in a clay matrix, which in some cases is yellow or reddish in color. Pisolites are commonly dolomitized. A few isolated dolomitic beds as much as 30 cm in thickness also occur in this facies. Some limestones display pelletoidal texture, and oolites have been found at several localities. Burrowing structures are common.

Sedimentary structures are rare, but they include fine parallel lamination, cross-bedding, ripple scours and ripple lamination, indistinct graded bedding in some very thinly bedded sequences, and possible water expulsion structures associated with complex, irregular, intraformational folds.

White massive limestone. Bedding is less discernible in these rocks than the gray limestones. Dark shale interbeds commonly form only discontinuous stringers, and the rocks are more massive than the dark gray limestones. The grain size of the crystals is greater, and the white color is due to the clearness of the calcite crystals. Sedimentary structures that can still be recognized are few but include indistinct possible cross-beds. In the northern Marble Hills, there are a few thin horizons of sandy limestone pebble conglomerate. Irregular nodules and streaks (pimples on surface) in some of the white limestones may be strongly altered and sheared burrows. Contortions in bedding, which predate cleavage, may indicate early slumping.

The boundaries of white against gray limestone do not everywhere follow bedding planes but rather form irregular appendages to bedding or—in extreme cases—complex networks, apparently along orthogonal fractures. This raises the possibility that the change from gray to white limestone is due not entirely to primary sedimentation but also to secondary (hydrothermal) alteration. Small nodules of sucrose calcite in the white limestone are probably also of secondary (diagenetic) origin. In addition, many parts and bands of the white limestone have been especially strongly sheared and thinned during deformation. In some areas it appears that narrow bands of white marble have been formed by shearing of gray limestone (Fig. 6).

Environment of deposition. There is no decisive evidence that indicates whether the Liberty Hills conglomerates are marine or nonmarine or whether they are shallow- or deep-water deposits.

The existence of burrows in the limestones indicates the presence of an infauna that either had no hard parts or that was destroyed by diagenetic or metamorphic processes. Oolites and pisolites indicate subtidal conditions, perhaps a carbonate bank or shoal. Evidence for emergence, such as evaporite beds, birdseye fabrics, or stromatolites, is lacking. Buggisch and Webers (1982; this volume) regard the limestones as having been deposited in open, shallow water on a platform with uniform subsidence.

Fossils. A trilobite-mollusk fauna (Webers and others, this volume; Shergold and Webers, this volume) was found in the 1979–80 season by John Anderson, Victoria University, Wellington, New Zealand, at locality B (Fig. 2) in the northern Marble Hills. It indicates a Late Cambrian age, supporting the stratigraphic correlation of the limestones with the uppermost unit in the Heritage Group.

STRUCTURE

Overall structure

The structural pattern is dominated by a large, eastward-verging, in places almost recumbent anticline along the western margin of the Independence and Marble hills (Figs. 2, 5). A broad, irregular syncline with many minor folds and generally low limb dips can be inferred between these two ranges and the Patriot Hills, where older rocks (e.g., conglomerates) are brought to the surface by a major east-verging anticline. Axial planes, with a few exceptions in the Patriot Hills and in the northern Marble Hills, dip to the southwest; fold hinges are mainly subhorizontal and trend northwest. Some post–breccia body (Spörli and others, Chapter 19, this volume) tilting occurred if the concave-upward layering in the breccia body at Mt. Shattuck (Fig. 7c) can be interpreted as an indicator of the horizontal during its formation.

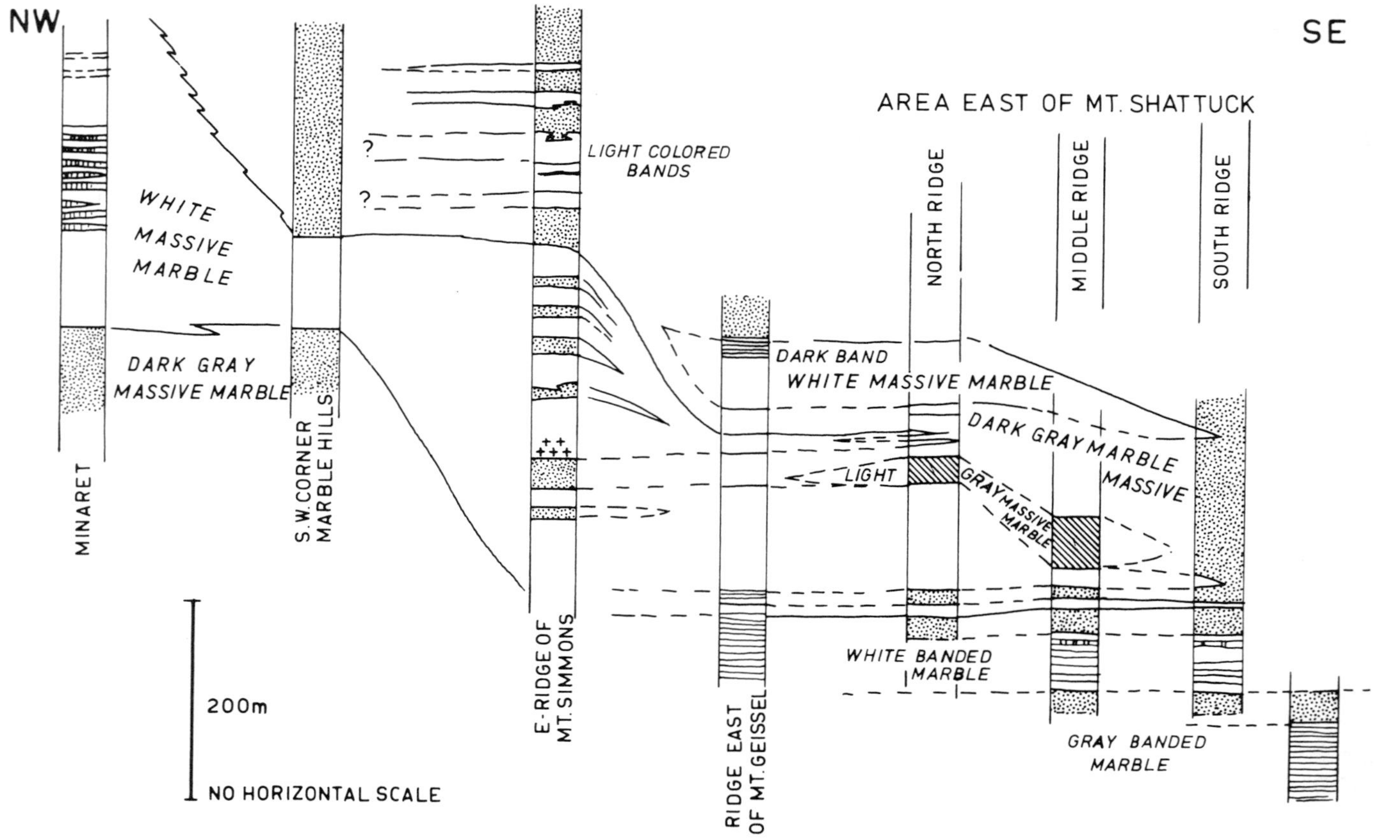

Figure 4. Correlation diagram of limestones along the length of the Marble Hills and the Independence Hills. For locations, see Figures 2 and 5.

Folds

The structure in the limestones appears to be strongly controlled by rock type and unit thickness. Massive white limestone and thick-bedded gray limestones form large-scale, simple, open structures, whereas folds in the thin-bedded gray limestones are tighter and have many complex minor folds (Figs. 7, 8).

It is not clear whether some complex, irregular, precleavage folds are due to an earlier tectonic deformation, disharmonic folding, inhomogeneous strain, or soft-sediment slumping/sliding of the limestones. In the Marble Hills some of these folds have steeply plunging axes, are highly sheared out and tightened, and cause some of the irregularities and discontinuities in the dark bands within the white limestones.

The more regular, syncleavage folds are chevron style in the thin-bedded gray limestones (Fig. 7), in places with sharp hinges, in other places with rounded hinges. The open folds, such as the major anticline/syncline couple at Mt. Simmons (Fig. 2) or its continuation to the north (Fig. 9), approach the geometry of parallel folds (class 1B of Ramsay, 1967).

Although the major fold hinges are horizontal or subhorizontal, mesoscopic hinges commonly deviate from this attitude, and several reclined folds have been found (e.g., the reclined fold at loc. C, Fig. 2, also described by Yoshida, 1982). We do not know whether the reclined geometry is due to primary or early strong curvature of the hinges or whether less important irregularities have been accentuated by stretching of the rocks during progressive deformation (Ramsay, 1979). We found no evidence that the plunging folds belong to an earlier phase of deformation.

A prominent down-dip stretching lineation in white marble at Morris Bluff (Loc. E, Fig. 2) is folded around tight, northwest-trending folds that are coaxial with the major structure. This suggests that the stretching lineation precedes or is a part of the main phase of folding.

Tectonic refolding is definitely present, however, in one locality in the Marble Hills (Fig. 2, Loc. D) and has been tentatively inferred for another. At the Marble Hills locality, downward-verging later-phase folds affect the western limb of an anticline, the cleavage of which they fold. Extension gashes have opened on the cleavage during the deformation.

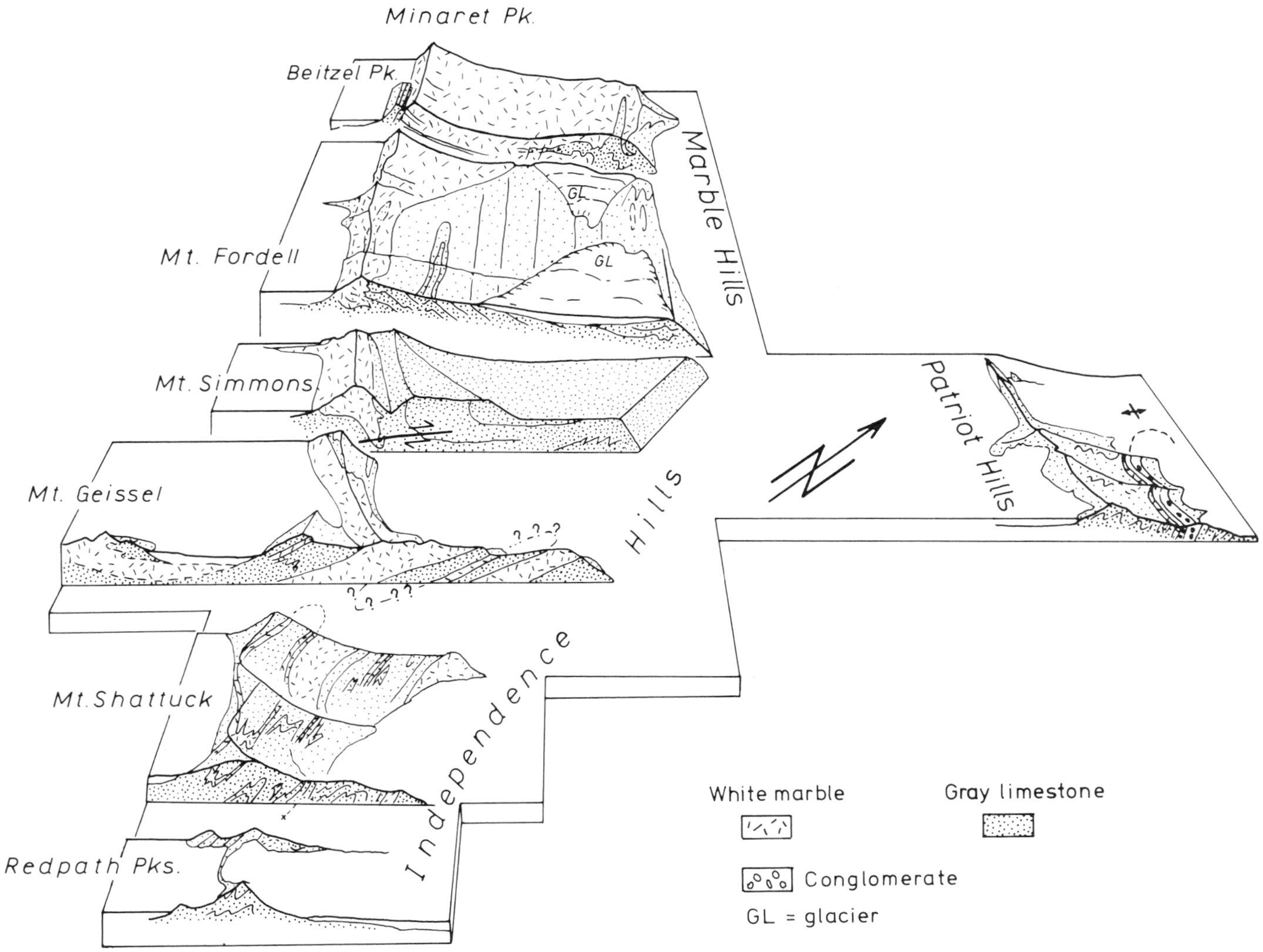

Figure 5. Idealized block diagram of the three ranges: Marble, Independence, and Patriot hills. View to the northwest.

At Mt. Geissel (Figs. 2, 5) on the west side of Independence Hills, beds dip moderately to the east and include some minor folds from which it is possible to infer a recumbent, eastward-closing major fold with an axial plane dipping gently to the east. If this major fold exists, the generally west-dipping axial planes of folds farther east may have been refolded to produce this anomalous attitude.

Kink folds affecting strongly laminated and attenuated white limestones in the northern Marble Hills (see also Yoshida, 1982, Fig. 17) postdate the main deformation and may be local deformations near major faults, rather than structures sympathetic to large folds.

Poles to bedding form a broad girdle consistent with the dominant northwest-southeast tectonic grain (Fig. 10a); the spread is due to slight local deviations and to some northeast-trending folds. The plot of fold hinges (Fig. 10b) shows an indistinct conical pattern, indicating concentric refolding of north-northwest–trending axes around subhorizontal folds trending 110° to 120°. This fold pattern may be due to the late thrust regime (see below).

Cleavage

Trends of cleavage cluster around 140°, and dips vary mostly from 35°W through vertical to 55°E (Fig. 10c). As would be expected with the generally northeastward vergence, south-

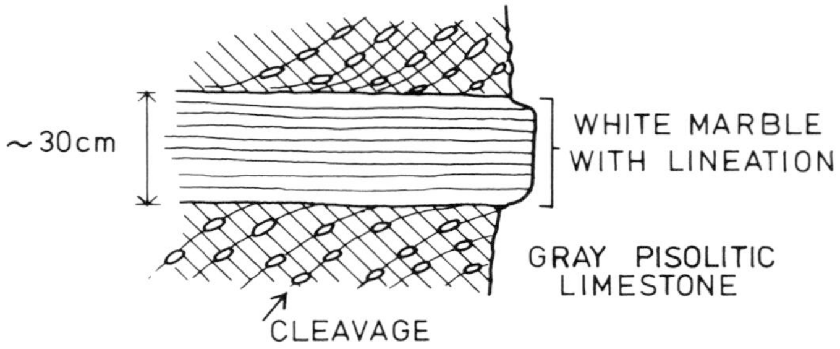

Figure 6. Sketch showing possible origin of white marble by ductile shearing of gray pisolitic limestones.

west-dipping cleavage planes are more common. The presence of some east-dipping cleavage indicates that cleavage fans have been developed to a certain extent, although in many mesoscopic folds the cleavage planes appear to be about parallel to the axial plane throughout the fold. The variations in trend of the cleavage are matched by variations in the trends of fold hinges (Fig. 10b) and may be related in part to the regional curvature of the trends of fold axes (Fig. 2). Macroscopic folds with northeastward-inclined cleavage occur in the northern Marble Hills, the south end of the Patriot Hills, and at Mt. Geissel (Fig. 2).

In the pisolitic gray limestones, the pisolites are strained into

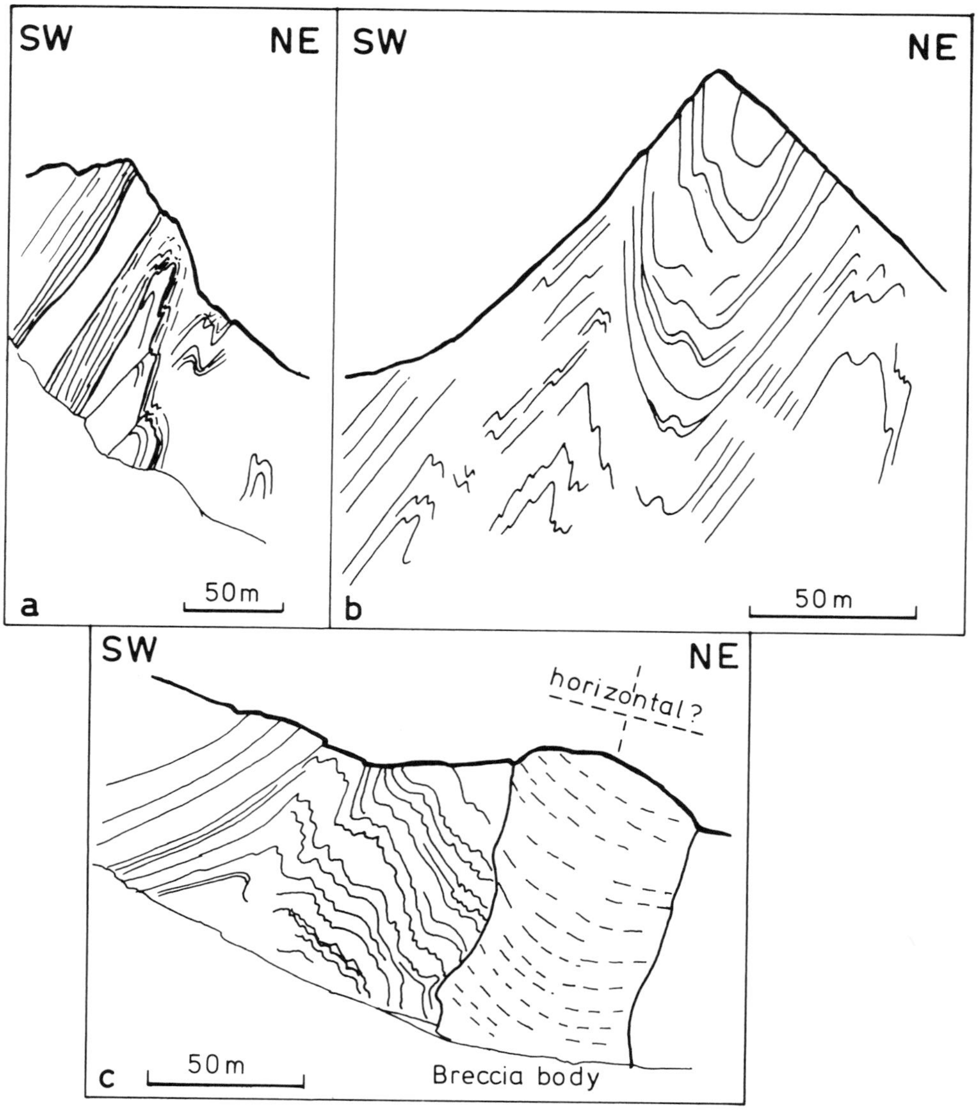

Figure 7. Chevron fold styles at southern end of the Independence Hills: a, middle ridge, Mt. Shattuck (Loc. G, Fig. 2), west limb of synform; b, south end of Redpath Peaks (see Fig. 2); and c, middle ridge, Mt. Shattuck (Loc. H, Fig. 2). Aligned clasts in breccia body may indicate tilt of formerly horizontal lines.

Figure 8. North side of middle valley of Marble Hills (west of Loc. D, Fig. 2). Note interfingering of white marble and dark gray limestone in the northeast-verging folds. View northward.

ellipsoids with the longest, and in places the intermediate, axes lying in the cleavage plane. Axial ratios of still recognizable pisolites range to about 3:1:1. The orientation of the long axes forms a mostly down-dip lineation. In the more fine grained gray limestones, cleavage commonly consists of closely spaced microstylolites. Pellets and other subspherical objects in the intervening microlithons, however, have been strained in a ductile manner and now are highly ellipsoidal. It appears that this more penetrative strain (immediately?) predates formation of the stylolites. It is interesting to note that Buggisch and Webers (1982) observed an earlier cleavage in the matrix of oolites farther north along the upper Heritage Group limestone band within the Heritage Range; there it is oblique to, and predates, the stretching lineation formed by the ooids.

Strain associated with cleavage is much higher in some of the white limestones. Cleavage, which forms angles of 20° to 40° to bedding in the surrounding gray limestones, is deflected into an orientation parallel to bedding in the white limestones; it has been converted into an extremely closely spaced foliation (Fig. 6) with a well-developed downdip lineation. It is not clear whether this local deflection is simply due to a volume loss within this zone or whether the white limestones are loci of very strong simple shear (i.e., ductile shear zones in the sense of Ramsay, 1980a). Sigmoidal structures of cleavage next to the contacts between gray and white limestones may favor the simple-shear model. The down-dip lineation in places grades into calcite fiber striations on more gently dipping planes, indicating low-angle thrusting to the northeast. This points to the close relation between folding and the subsequent first phase of faulting.

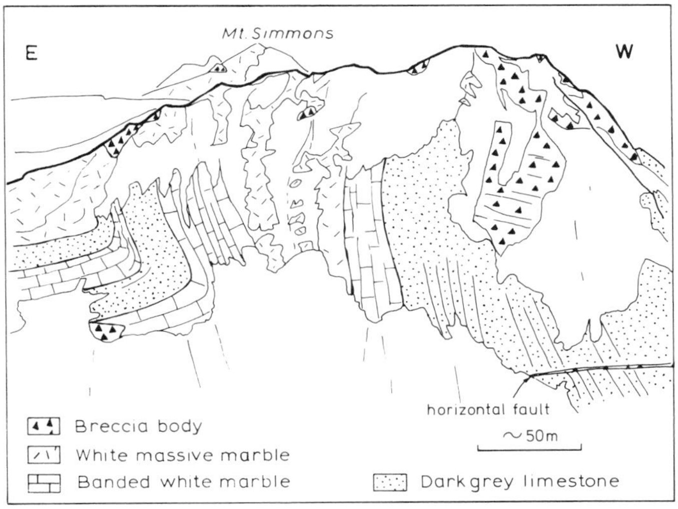

Figure 9. Example of open fold in more massive carbonate rocks at north end of the Independence Hills (Loc. F, Fig. 2). This outcrop displays the major east-verging antiform/synform couple along the west side of the Marble Hills and northern Independence Hills (see also Fig. 16 in Spörli and others, Chapter 19, this volume).

Microscopic observation indicates that the cleavage in the white limestones is formed by a fabric of clear, reoriented calcite crystals. Short, discontinuous microstylolites are distributed irregularly within this fabric. Concentric ooids have been transformed into radial aggregates or spots of pseudosparite (Buggisch

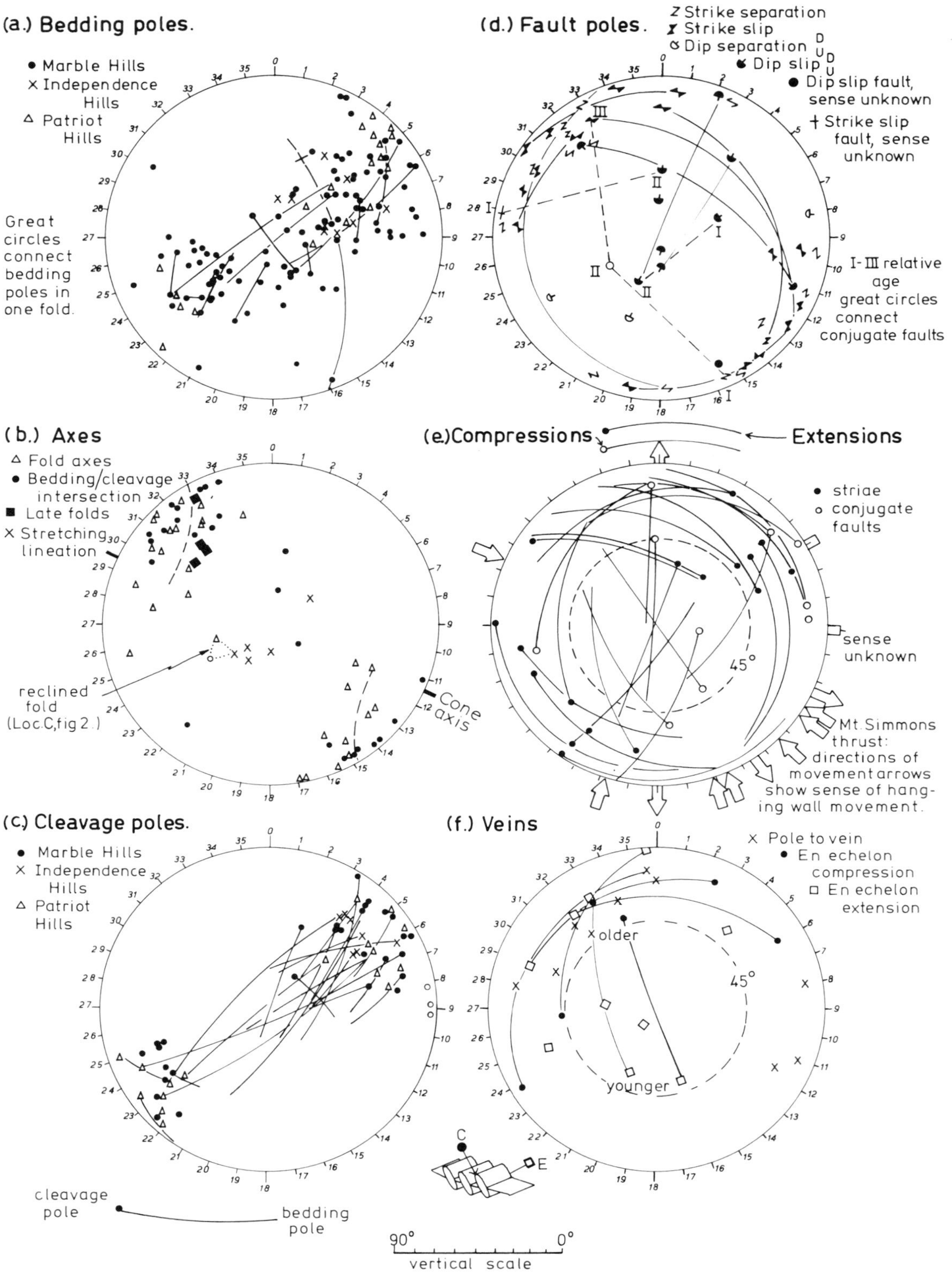

Figure 10. Structural data for the area, on lower hemisphere equal-area nets. In d, note that only NE-trending normal faults are compatible with the stress regime of the strike-slip faults shown. Dashed lines connect measurements from the same locality (I and II show only local sequence of deformation and cannot necessarily be correlated from one locality to the other). The 45° cone plotted in e and f is to help distinguish dip-slip arrays (intersect cone) from strike-slip arrays (lie outside cone).

and Webers, 1982). Dolomitic pisolites have resisted this alteration and yielded by brittle fracture; they are elongated parallel to the down-dip lineation. Spaces between the separated fragments are occupied by the fabric of reoriented and recrystallized calcite described above. Echinoderm fragments also have resisted deformation. Lateral propagation of cleavage is indicated in a vein in the Marble Hills (Fig. 11) where pisolites have been stretched parallel to the cleavage, probably by a crack-seal mechanism (Ramsay, 1980b). Stylolitic seams have also been formed along the borders of breccia body material, and they are in turn cut by en echelon extension gashes (Fig. 12).

Faulting

Reverse faulting and thrusting. A steep, westward-dipping thrust fault follows the axial zone of the main western anticline (Figs. 2, 5), in part cutting out its short eastern limb. At the southern end of the Marble Hills, this fault turns into a low-angle fault that can be traced south into the Mount Simmons cirque (Fig. 14 of Spörli and others, Chapter 19, this volume) where it gently dips to the south and displaces a breccia body by several tens of meters.

At Mt. Simmons the fault splits into several branches and on a smaller scale exhibits ramp structures. Movement directions as deduced from striations, en echelon structures, and conjugate faults are variable and range from northeast-southwest to northwest-southeast (Fig. 10e). In many places the fault is stained red by fine-grained fill material of the breccia bodies.

Small-scale thrusts of similar orientation occur throughout the Marble and Independence hills. At the locality with second-phase folding (Fig. 2, Loc. D), a conjugate set of reverse faults postdates a mylonitic calcite shear. Reverse movement is also indicated by many gently dipping surfaces with calcite fiber striations displaying accretion steps. Although most directions of transport on these surfaces are northeast-southwest, perpendicular to the structural trends, locally there are also north-south and northwest-southeast directions (Fig. 2), as on the Mt. Simmons thrust.

East-west– to northwest-southeast–trending and northeast-dipping reverse faults are present near the northern limit of the conglomerate outcrop in the Patriot Hills. A change to northeasterly dips of the axial planes of folds in the area may be due to drag on this fault system, and it may in part be responsible for bringing older rocks to relatively high levels in this area.

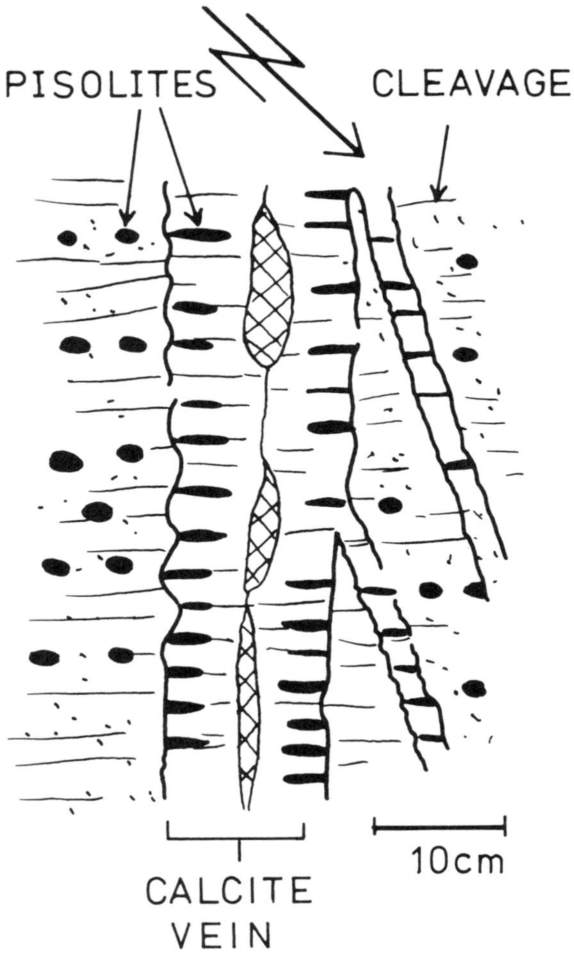

Figure 11. Pisolites (oncoids) stretched by propagation of cleavage, probably involving the crack-seal mechanism. Exposure dips gently toward observer. North-central part of Mt. Fordell massif, Marble Hills.

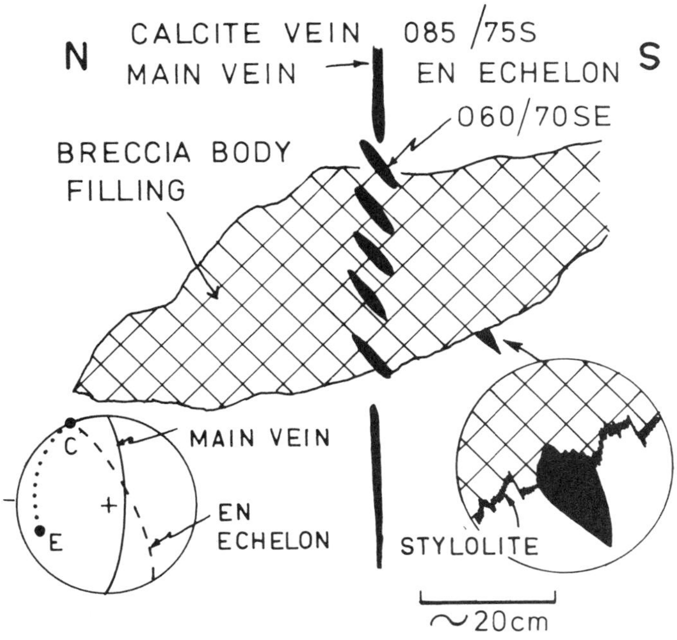

Figure 12. En echelon calcite vein postdates breccia body filling. Central part of Mt. Fordell massif, Marble Hills. Lower hemisphere equal-area net shows three-dimensional geometry of en echelon array and compression (C) and extension (E) axes.

Other faults. Normal faults and vertical strike-slip faults with striations indicating subhorizontal movement are relatively rare. Most of these normal faults strike northwest and slightly oblique to the range. They are incompatible with simple compression normal to the major northwest-trending folds, but they could be explained by bending and stretching during late coaxial arching or by compression along the ranges (Fig. 2, II).

The main system of cross faults with strike-slip movements has trends of north-northeast and east-west and a subhorizontal σ_1 oriented northeast-southwest, normal to structural grain and compatible with thrusting in the northeast-southwest direction. At one locality in the Marble Hills (Fig. 13) the cross faults offset a strike fault with reverse fault separation.

In the Marble Hills, faults with dextral strike-slip are more prominent and have larger displacements than those with sinistral movement. These faults are loci for the formation of some breccia bodies, but they also displace a few breccia bodies, indicating a close interaction of the two processes.

The cross faults in turn are displaced by strike faults with a sinistral sense of movement. Offset relations and drag structures along a north-south–trending fault in the northern Marble Hills (Fig. 14) indicate a first phase of dextral and a second phase of sinistral movement. This sequence ties in well with time relationships inferred elsewhere in the Heritage Range (Spörli et al., this volume, Chapter 20), and this suggests that σ_1 changed from an original orientation perpendicular to, to one parallel with, the structural grain of the range. Although we know no relative ages for the striations with varying orientations, it is possible that those oriented northeast on low-angle thrusts predate those oriented northwest. However, as can be seen from Figure 10a, the various faults can be fitted into a relatively uniform kinematic system based on compression across the northwest-trending fold axes. The rotation of σ_1 to an orientation parallel with the tectonic grain may be related to a rotation of the Ellsworth Mountains, but it could also reflect a change in the boundary conditions of the rocks undergoing deformation.

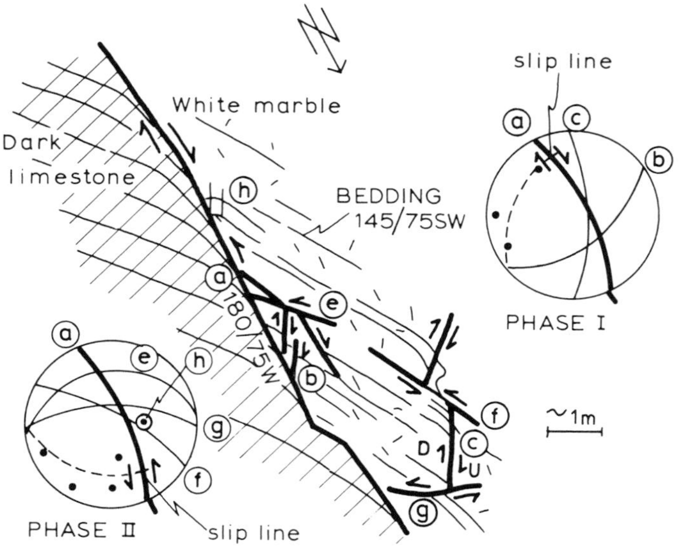

Figure 14. Sketch map of evidence for phase I, dextral strike-slip (NE compression), followed by phase II, sinistral strike-slip (NW compression); e.g., faults f and g postdate fault c. Point h is hinge of sinistral kink fold. Lower hemisphere equal-area nets show construction of slip lines, using best-fit great circle (dashed) fitted to poles of minor faults (is perpendicular to statistical intersection line of major and minor faults).

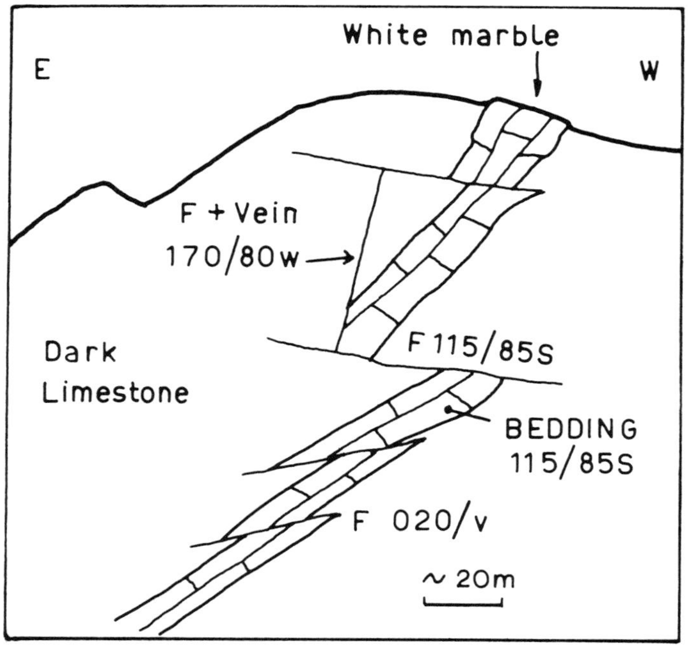

Figure 13. Sketch of strike fault (reverse?) that probably predates the cross faults. South-central Mt. Fordell massif.

Veins

Calcite veins both pre- and postdate cleavage. The precleavage veins are strongly crenulated, and in the white marbles they become indistinguishable from the surrounding rocks. The vein calcite can be white or—especially near breccia bodies—tan to red colored as a result of ferruginous impurities. Many veins follow fault planes; commonly it appears that the extensional opening took place after the fault-parallel sliding that produced the major offset on the faults. Veins also pre- and postdate the breccia bodies (Fig. 12 and Spörli and others, Chapter 19, this volume).

Major and minor principal stress axes deduced from en echelon extension gash arrays indicate two clusters (Fig. 10f), one with σ_1 subhorizontal and northeast-trending (populated by strike-slip arrays) and one with σ_1 northwest and represented by thrust arrays. In one location in the Patriot Hills, northeast-trending veins predate the en echelon gashes associated with the thrust regime. This may be an indication that compression first acted perpendicular to the tectonic grain and later changed to a northwest orientation. Movement on the thrusts offsetting breccia bodies in the Mt. Simmons cirque (Figs. 2, 5, and Fig. 2 of Spörli and others, Chapter 19, this volume) is compatible with this later regime.

DISCUSSION

The stratigraphy in the area is difficult to unravel because the origin of the white "marble" is unclear. It is possible that some of these rocks may be tectonic rather than stratigraphic markers. What has been definitely shown is that the limestones overlie the conglomerates of the Patriot Hills and are to be correlated with the other limestones at the top of the Heritage Group. Although scattered evidence suggestive of sequential deformation has been found, these observations have not coalesced into recognition of clear, temporally distinct phases of deformation. Rather they can all be attributed to incremental steps preserved from one phase of deformation, possibly under a change in regime from deeper to higher crustal levels (Table 2). Formation of the sheared white marbles could be an exception, although structures do not display the necessary consistent crosscutting relations. If they represent a distinct (older?) phase, structures of this phase must be very discontinuous because the gray limestones show no analogues to the inferred shear in the white marble.

There is no direct structural evidence for the exact environment of deformation. Buggisch and Webers (1982) estimate from illite crystallinity correlations that the rocks in the Marble Hills area attained very low to low-grade metamorphic grade. In the mafic igneous rocks of the Heritage Group, Bauer (this volume) recognizes an earlier burial metamorphism tied to stratigraphic level and ranging from upper prehnite-pumpellyite facies to greenschist facies, which during the Ellsworth (Gondwanide) Orogeny was overprinted by regional metamorphism no higher than lower greenschist facies. The carbonate rocks of the Marble Hills area almost certainly were subjected to this same regime.

TABLE 2. POSSIBLE SEQUENCE OF DEFORMATIONS IN THE MARBLE HILLS AREA

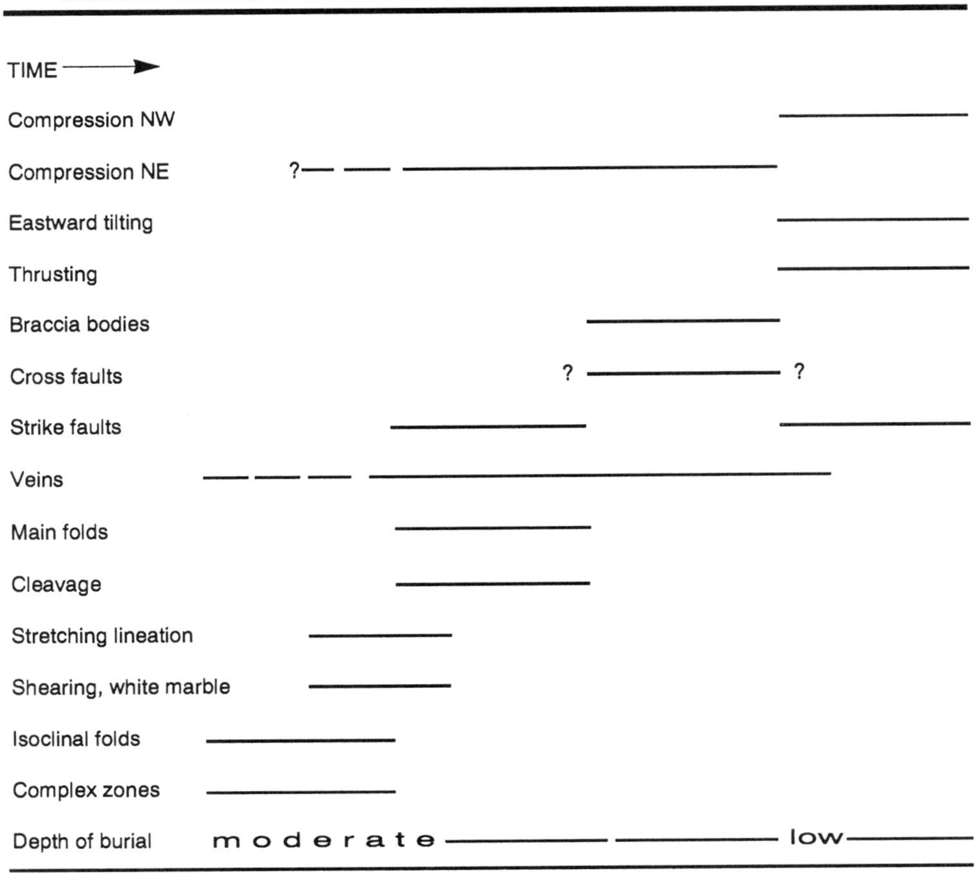

REFERENCES CITED

Buggisch, W., and Webers, G. F., 1982, Zur Fazies der Karbonatgesteine in den Ellsworth Mountains (Paläozoikum, Westantarktis): Facies, v. 7, p. 199–228.

Craddock, C., 1969, Geology of the Ellsworth Mountains, *in* Bushnell, V. C., and Craddock, C., eds., Geologic maps of Antarctica: New York, American Geographical Society, Antarctic Map Folio Series, Folio 12, Plate 4.

—— , 1972, Antarctic tectonics, *in* Adie, R. J., ed., Antarctic geology and geophysics: Oslo, Universitetsforlaget, p. 449–455.

Dalziel, I.W.D., and Elliot, D. H., 1982, West Antarctica: Problem child of Gondwanaland, Tectonics, v. 1, p. 3–19.

Ramsay, D. M., 1979, Analysis of folds during progressive deformation: Geological Society of America Bulletin, v. 90, p. 732–738.

Ramsay, J. G., 1967. Folding and fracturing of rocks: New York, McGraw Hill, 568 p.

—— , 1980a, Shear zone geometry: A review: Journal of Structural Geology, v. 2, p. 83–99.

—— , 1980b, The "crack seal" mechanism of rock deformation: Nature, v. 284, p. 135–139.

Webers, G. F., 1972, Unusual Upper Cambrian fauna from West Antarctica, *in* Adie, R. J., ed., Antarctic geology and geophysics: Oslo, Universitetsforlaget, p. 236–238.

Webers, G., and Spörli, K. B., 1983, Palaeontological and stratigraphic investigations in the Ellsworth Mountains, West Antarctica, *in* Oliver, R. L., James, P. R., and Jago, J. B., eds., Antarctic earth science: Canberra, Australian Academy of Science, p. 261–264.

Yoshida, M., 1982, Superposed deformation and its implication to the geologic history of the Ellsworth Mountains, West Antarctica: Memoirs National Institute of Polar Research, Special Issue 21, p. 120–171.

Manuscript Accepted by the Society January 28, 1992

Printed in U.S.A.

Geological Society of America
Memoir 170
1992

Chapter 19

Breccia bodies in deformed Cambrian limestones, Heritage Range, Ellsworth Mountains, West Antarctica

K. B. Spörli
Department of Geology, University of Auckland, Auckland, New Zealand
Campbell Craddock
Department of Geology and Geophysics, University of Wisconsin, Madison, Wisconsin 53706
R. H. Rutford
Department of Geosciences, University of Texas, Dallas, Texas 75083
John P. Craddock
Department of Geology, Macalester College, St. Paul, Minnesota 55105

ABSTRACT

Breccia bodies in the carbonate rocks of the Minaret Formation (Cambrian) of the Ellsworth Mountains were formed at depth by a combination of cavelike processes and contemporaneous low-temperature hydrothermal activity during the latest stages (early Mesozoic?) of compressive deformation in this fold belt. The breccia bodies consist of clasts of Minaret Formation limestone/marble with cleavage, embedded in a matrix of crystalline and detrital calcite; both clast shape alignments and the matrix locally show layering. Clast angularity and edge dissolution vary greatly as does the distance of clast transport (mainly downward). The boundaries between a breccia body and the adjacent country rock may be diffuse or sharp. Breccia body shapes, sizes, and orientations are varied; bodies are as high as approximately 250 m and as wide as 50 m. Many breccia bodies crosscut each other, showing that breccia body formation continued through a period of time. Fluid inclusions in calcite crystals from a vuggy vein yield a crystallization temperature of 160+ ± 5°C. That temperature and the inferred thickness of the overlying stratigraphic succession suggest that the breccia bodies may have formed at a depth as great as 5,000 m.

INTRODUCTION

The Ellsworth Mountains (79°S, 85°W) in West Antarctica consist of a northern Sentinel Range and southern Heritage Range. Spectacular breccia bodies occur in the southern Heritage Range within deformed Minaret Formation limestones of the Heritage Group (Cambrian) in the Marble, Independence, Patriot, and nearby hills (Fig. 1) (Craddock and others, 1964; Craddock and Webers, 1981; Craddock and others, 1982). The Heritage Group consists of sandstones, argillites, conglomerates, limestones, and mafic igneous rocks, and it is the oldest and lithologically most diverse stratigraphic unit of the ~13,000-m Ellsworth Mountains stratigraphic succession (Webers and Spörli, 1983; Webers and others, this volume). The tan, red, and green quartzites and argillites of the Crashsite Group that overlies the Heritage Group have yielded invertebrate fossils of Late Cambrian and Devonian age. The Crashsite Group is conformably overlain by the glaciomarine strata of the Whiteout Conglomerate, which is in turn overlain by the *Glossopteris*-bearing Polarstar Formation of Permian age. This typical, but unusually thick, Gondwana succession was deformed and metamorphosed to lower greenschist facies in Permian or later time, and it now forms a major north-plunging fold belt (the Ellsworth orogen). Nearly all outcrop-scale folds are cylindrical or chevron in style,

Spörli, K. B., Craddock, C., Rutford, R. H., and Craddock, J. P., 1992, Breccia bodies in deformed Cambrian limestones, Heritage Range, Ellsworth Mountains, West Antarctica, *in* Webers, G. F., Craddock, C., and Splettstoesser, J. F., Geology and Paleontology of the Ellsworth Mountains, West Antarctica: Boulder, Colorado, Geological Society of America Memoir 170.

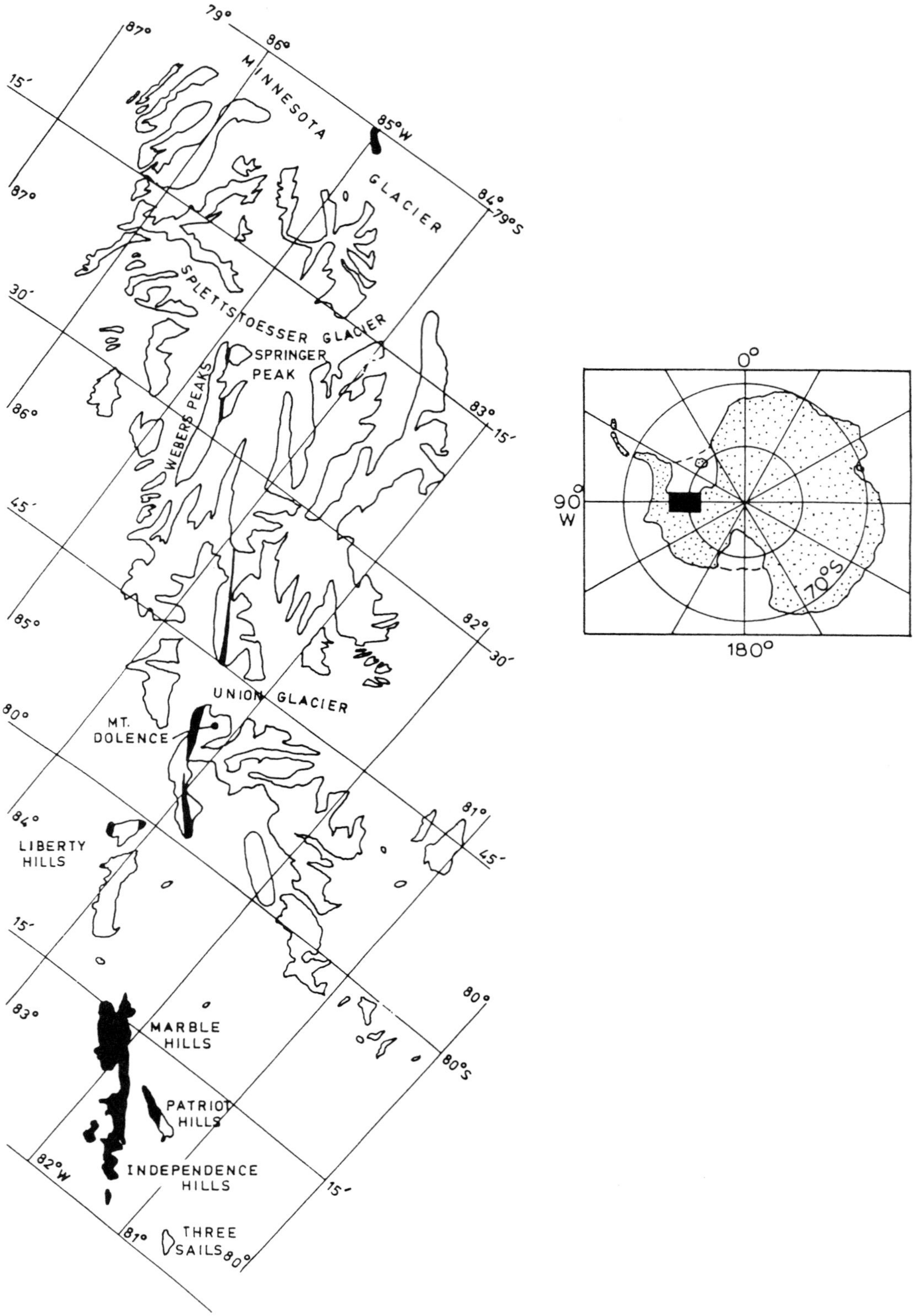

Figure 1. Index map of the Heritage Range with major rock exposures outlined. Black: limestone and marble of the Minaret Formation. White: Heritage Group, Crashsite Group, and Whiteout Conglomerate. Inset shows location of the Ellsworth Mountains within Antarctica.

and axial planes commonly dip steeply southwestward. Fold axes trend northwest-southeast in the southern Heritage Range, and locally they plunge at low to moderate angles in either direction.

In the Marble, Independence, and Patriot hills (Fig. 2) there are three major rock units: (1) conglomerate with limestone boulders, (2) well-bedded dark gray limestones with minor shale, and (3) mainly white, locally attenuated limestone/marble beds (Fig. 3) with a layer-parallel extension lineation. The contacts between these units trend approximately parallel to the local ranges (Fig. 2).

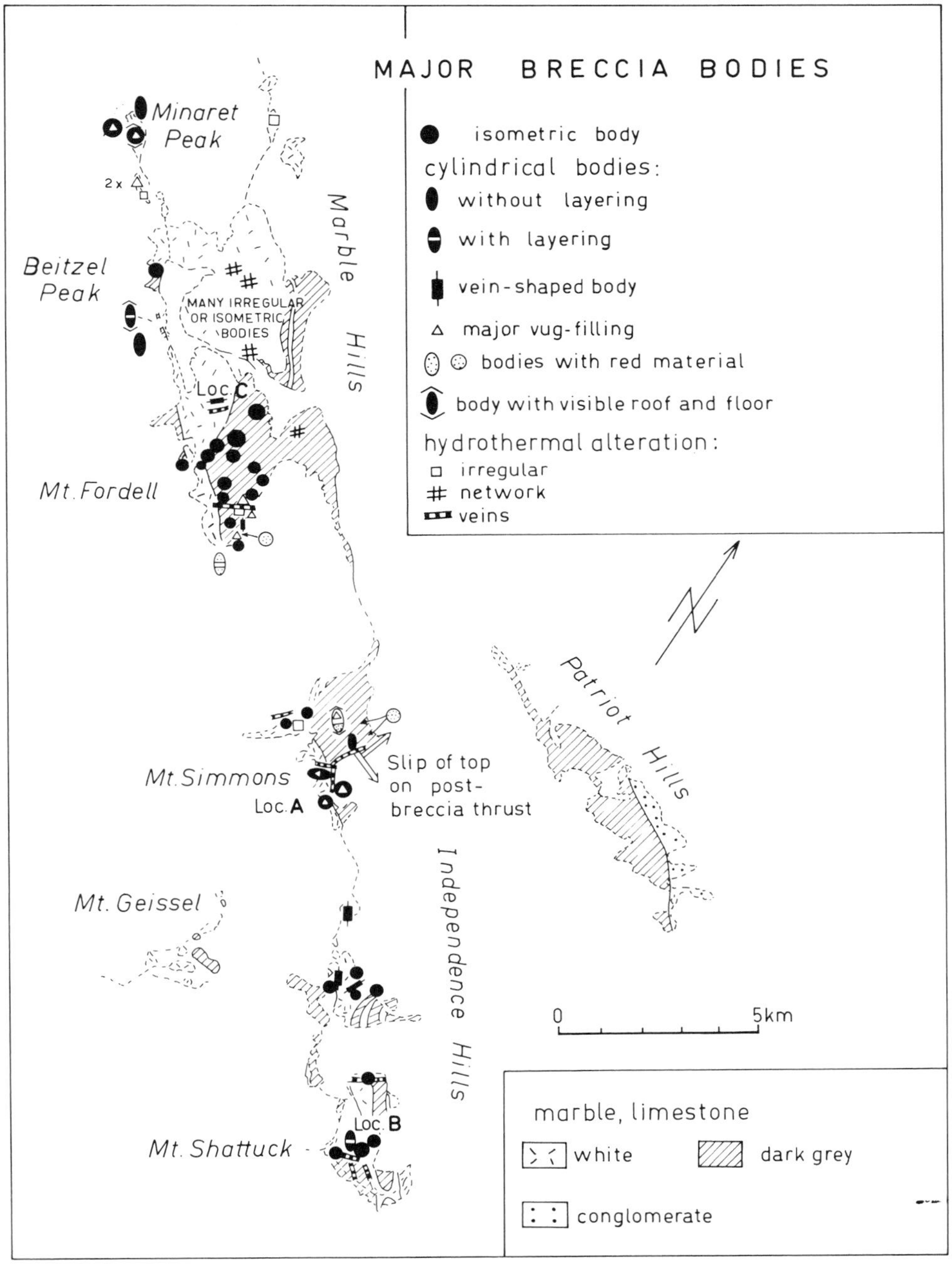

Figure 2. Map of southern Heritage Range showing distribution of breccia bodies and associated features within the Minaret Limestone (and marble) in the Marble Hills and Independence Hills.

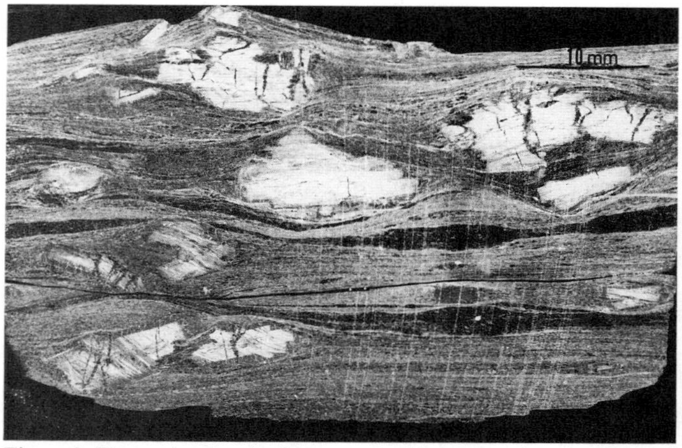

Figure 3. Photomicrograph of protomylonitic textures in a specimen of Minaret limestone from the lower (eastern) slope of Mount Fordell.

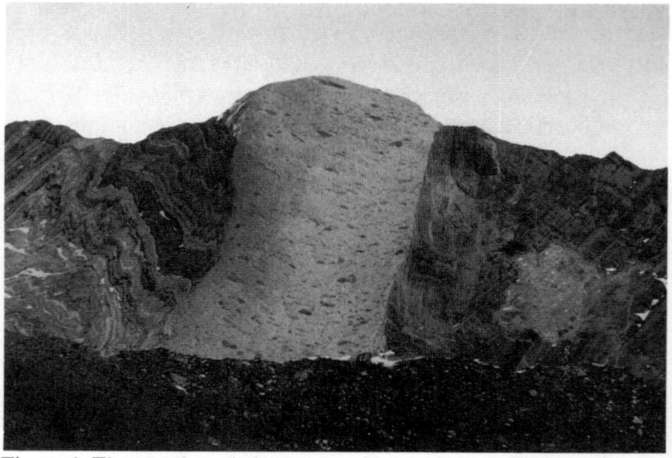

Figure 4. The type breccia body on a ridge northeast of Mt. Shattuck in the Independence Hills. Note the concave-upward layering of Minaret clasts in the nearly vertical breccia body and the fold hinges adjacent to both the large and the small breccia bodies. An axial planar cleavage is also present in the fold, and it is locally occupied by matrix next to the breccia bodies. The large breccia body is approximately 50 m high. View to northwest.

DESCRIPTION OF BRECCIA BODIES

Distribution

The largest concentration of breccia bodies is present in the Marble Hills and the Independence Hills, but some breccia bodies also occur in other carbonate rock exposures, particularly in the Mt. Dolence area and in the northwestern end of the Liberty Hills (Fig. 1). No breccia bodies have been found in the shaly limestones of the Patriot Hills (Fig. 2); however, thin veins of limonitic, clayey material there may represent incipient breccia formation. Most of the breccia bodies are located preferentially in the white and dark gray limestones and marbles of the Marble Hills and Independence Hills. Dozens of mappable bodies have been found, and there are many other smaller bodies.

Breccia body shapes

The shapes and orientations of the breccia bodies are highly variable, and they show no consistent relationship to either the cleavage or bedding. The main breccia body shapes are: (1) cylindrical (Fig. 4), with the long axis either subvertical or subhorizontal; (2) veinlike, either parallel to or oblique to bedding or cleavage; or (3) wedge shaped and highly irregular, especially in the case of the largest bodies, which have horizontal and/or vertical dimensions of several hundred meters. At Location A (Fig. 2, 5) one breccia body crosscuts one or two others, indicating different ages for these breccia bodies (see also Fig. 6).

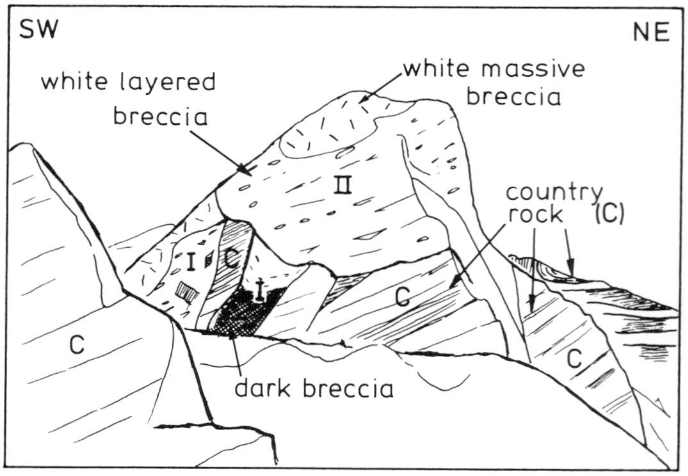

Figure 5. Line drawing of a peak southeast of Mt. Simmons showing two crosscutting breccia bodies (I and II) in southwest-dipping limestone country rock (C). Note the large angular clast in the westernmost exposure of I. The height of the visible part of the peak is about 100 m. Line drawing by K.B.S. as traced from a photograph taken by R. H. Rutford in 1980.

Breccia body clasts

Carbonate rock clasts in the breccia bodies range in size from a few centimeters to tens of meters in long dimension. In places, small limestone fragments appear as wedges within layered, veinlike breccia bodies (Fig. 7). Just south of Mt. Simmons (Loc. A, Fig. 2), on the other hand, a 100-m rectangular mass of dark gray limestone is surrounded on four sides by breccia body material. The most common clast shapes are rectangular and polygonal because of the tendency for blocks to sepa-

Figure 6. Crosscutting breccia bodies on the northern slope of Mount Fordell. The older body contains Minaret clasts, and the younger breccia body contains only concentrically layered calcite. View to the south.

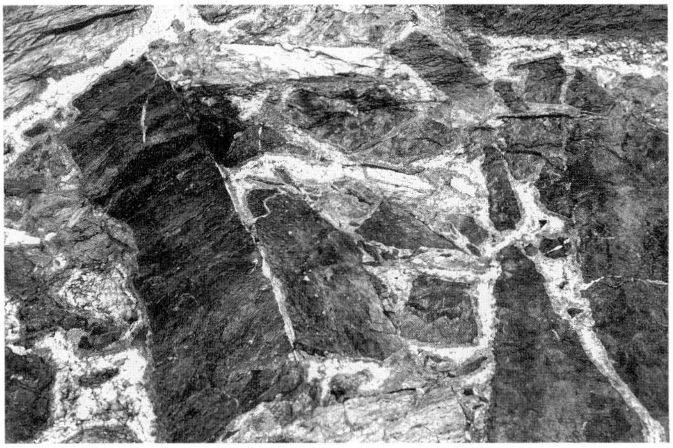

Figure 8. Randomly oriented light and dark Minaret Limestone clasts (note bedding and cleavage in clasts) within a calcite matrix. Northeast slope of Beitzel Peak. Field of view is 3 m × 5 m.

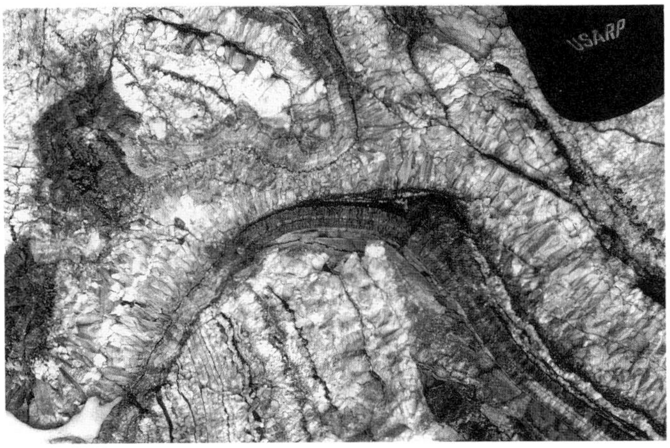

Figure 7. The convergence of multiple phases of calcite layering with included Minaret clasts (gray). One layer of calcite appears folded and forms a surface along which a younger package of calcite layers is terminated. Northern slope of Mount Fordell; view is downward. Hat for scale.

Figure 9. Layered calcite matrix (light) and interbedded detrital calcite sands (dark) surrounding a clast of Minaret limestone with fluted edges. Northeast slope of Beitzel Peak; view to northwest. Ice axe for scale.

rate from the country rock along bedding and cleavage surfaces (Fig. 8). Some breccia clasts are slightly rounded; some clasts modified by dissolution display fluted-to-wavy rims within a complex calcite-filled matrix (Fig. 9).

The breccia fragments represent the full range of largely unsheared rock types present in the surrounding Minaret Formation; one limestone block observed contains a kink band. The clast orientations within the breccia bodies can be seemingly random (Fig. 10), or the clasts can display a distinctive layering that is subhorizontal and concave upward (Fig. 4). The slight overall inclination of the concave layering in that body may be the result of subsequent tilting. Some limestone blocks can be traced to their place of origin in the adjacent wall rock of the breccia body (Fig. 8). Other clasts, however, differ greatly from the enclosing wall rocks, suggesting transport distances of tens or hundreds of meters. We have observed both apparently downward and subhorizontal clast displacements but no upward displacement of clasts from the source to the present position. The random fabric of the brecciated clasts in some ways resembles that of talus deposits or cave breccias, and the clasts commonly form a self-supported framework. Many breccia bodies give the observer the impression that little net change in the bulk volume of the rock has occurred, that is, the volume of clasts, matrix, and small voids is about equal to the original volume of affected

Figure 10. Multiple stages of breccia body formation within the Minaret Formation in the Independence Hills to the northwest of Mount Simmons (see Fig. 16 for location). The upper breccia body can be divided into three parts: layered crystalline calcite matrix (upper and youngest), Minaret clasts and detrital calcite (middle), and Minaret clasts in a calcite matrix (lower and oldest). View to west; cliff face is approximately 40 m high.

Figure 11. Vertical edge of a breccia body showing the intersection of the dipping Minaret limestones (right), horizontal solution channels filled with layers of detrital calcite, and younger vertical layers of calcite crystals. Northeast slope of Mount Fordell; view to northwest. Ice axe for scale.

country rock. This suggests that the breccia bodies formed in a closed system.

Breccia matrix

The breccia body matrix consists of (1) detrital grains, (2) crystalline calcite, and (3) calcite veins. The granular sediments are rusty red to brown to whitish yellow and are composed of limonite, hematite, calcite, and minor albite. The sediments are commonly layered, and in places they are discordant to the walls of the breccia body, indicating gravity-controlled sedimentation (Fig. 11). In other places the matrix layering is conformable to the walls of the void that has been filled (Fig. 12). Some of the sediments preserve cut-and-fill structures, and some disharmonic (slump?) folds of the layered calcite matrix have been observed. Larger fragments of coarsely crystalline calcite within the layered detrital deposits indicate that matrix crystals have been reworked and redeposited; locally changes in texture or cross-bedding are present (Fig. 13). Limonite in the fine-grained matrix in places has formed extremely complicated and delicate networks (Fig. 9). The most spectacular limonitic deposit is exposed at Mt. Simmons (Figs. 2, 10), where it appears to form a 1- to 10-m-thick sedimetary layer above coarse carbonate rock breccia and below void-filling, concentric layers of calcite crystals; this locality suggests deposition in a cave and indicates a three-stage process for the development of this particular breccia body.

The coarsely crystalline calcite in the matrix appears to have formed in voids between, along, or above the breccia clasts, as in the example above. Crystals of prismatic habit attain lengths of as much as 1.0 m. Flat rhomboidal shapes are common for crystals

Figure 12. Breccia body boundary with country-rock bedding (horizontal, bottom) and cleavage (dipping to the left). Note the horizontal layers of detrital calcite near the base of the breccia body. Northeast slope of Mount Fordell; view to northwest. Person for scale.

Figure 13. Calcite detritus (coarse on the right, fine on the left) along a solution seam within the Minaret Formation. East slope of Mount Fordell; view to southwest. Area shown in photo is 1 m across.

that developed in formerly open vugs. Many accumulations of crystals are zoned, with varying concentrations of iron minerals in adjacent layers. Layers of coarse crystals are commonly intercalated with layers of fine-grained detritus, suggesting repeated intervals of standing water and moving water. In some breccias, the crystalline calcite surrounds each clast so that the clasts appear to float in a matrix of layers of calcite crystals (Figs. 4, 7, 8, 9). In some cases the layers of calcite lie parallel to the original planar edges of the clasts, but in other blocks the layers of crystals were deposited in varied embayments formed by dissolution of the edge of the limestone block. Small void spaces are common in some of the breccia bodies.

Calcite veins identical to the layered crystalline calcite matrix in the breccia bodies extend out into the surrounding country rock, mainly along bedding, cleavage, joint, and fault planes. Some of these calcite veins are marked by rows of ellipsoidal vugs, a feature that may indicate a locally increased solution rate due either to mixing or fluids with differing chemical composition (Bogli, 1964) or to elevated fluid pressures. Diffuse alteration zones, within which darker limestones are bleached, also extend out into the surrounding country rocks along preexisting planes or zones, in places forming intricate networks.

Breccia body contacts

Most of the contacts between breccia bodies and the adjacent country rock are distinct and planar (Figs. 4, 12). Only in complexly shaped breccia bodies is there a gradation between the breccia body and unbrecciated country rock (Figs. 8, 10). In some cases calcite veins within a breccia body extend into the srrounding rock, as if to isolate blocks of carbonate rock for incorporation into the breccia body. Some breccia bodies are bounded by faults that either have acted as pathways for migrating fluids or have truncated the breccia body (Fig. 14). Two adjacent cylindrical breccia bodies crosscut folded limestones, but one axis is subvertical, the other subhorizontal (Fig. 4). The vertical breccia body is layered and has distinct boundaries subparallel to cleavage, whereas the breccia body with a horizontal axis has diffuse borders and random clast orientations.

Age of the breccia bodies

The breccia bodies contain carbonate rock clasts that preserve an older, regional axial planar cleavage, and these blocks have been rotated into various orientations within the breccia bodies. The breccia bodies clearly crosscut major F_1 folds (Fig. 4). However, in the central part of the Marble Hills (Loc. C, Fig. 2), where a second phase of deformation is apparent, a veinlike breccia body has developed a crosscutting cleavage (Fig. 15) that is concordant with the cleavage in the adjacent limestone unit. Breccia bodies near Mt. Simmons are displaced by a thrust fault (Figs. 14, 16); apparent slip directions along this thrust surface are quite varied, with striation trends ranging from north-northwest–south-southeast to northeast-southwest. En echelon veins and stylolitic surfaces that crosscut breccia bodies provide additional evidence for post–breccia body deformation, as do striated surfaces with calcite accretion steps on fine-grained breccia body fillings. Some of the en echelon calcite veins and calcite-filled faults displace breccia bodies and are themselves replaced along strike by breccia body matrix material, indicating a close temporal relationship between brittle fracturing and breccia body formation. This kinematic scenario is further supported by the observation that some breccia clasts have been formed from country-rock septa between en echelon extension gash arrays.

The faulting regime most closely related to breccia body formation consists of steep, conjugate strike-slip faults of minor

Figure 14. Breccia body offset slightly by a thrust fault. Breccia body is about 8 m high (see Figure 16 for location).

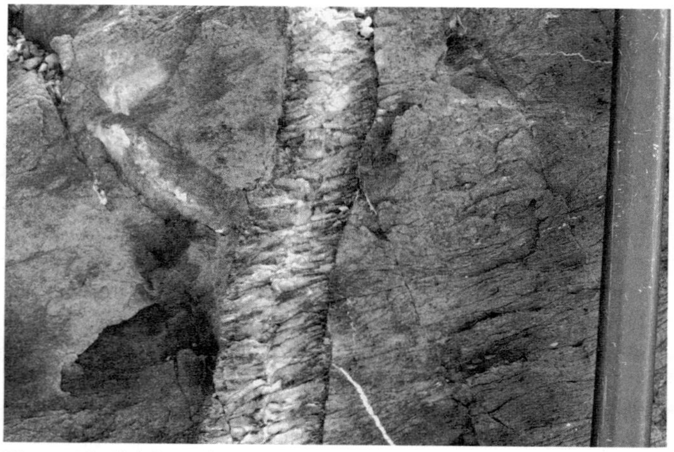

Figure 15. Calcite vein that has been affected by the development of cleavage, which is dipping gently to the left, parallel to that in the surrounding carbonate rock. Other observed calcite veins lack cleavage. East slope of Mount Fordell; view to west. Ice axe handle for scale.

displacement; they yield a subhorizontal greatest principal stress axis that trends northeast-southwest, perpendicular to the local structural grain.

Microstructures

Twinning, solution cleavage, elongate ooids (aspect ratio 3:1:1; Spörli and Craddock, this volume), and protomylonitic fabrics (Fig. 3) occur locally in the Minaret country rock. Mechanical twins are present in the calcite of the breccia body matrix, and each of the specimens observed in thin section revealed two twin lamellae sets, indicating a somewhat complex post–breccia body deformation history. Oriented specimens were not collected, so a twin analysis on the breccia material or the country rock was not possible. However, the twinning in the calcite matrix does indicate post–breccia body differential stress of at least 330 bars (Jamison and Spang, 1976).

Fluid inclusions

Fluid inclusions were observed in some crystals of calcite from a vuggy vein near a breccia body below Mt. Fordell. Preliminary heating studies of these inclusions give a crystallization temperature of 160° ± 5°C. (oral communication, P.R.L. Browne, Geothermal Institute, Auckland, New Zealand, to K.B.S., 1986).

ORIGIN OF BRECCIA BODIES

Since their discovery in 1963, we have considered many possible models for the late synorogenic formation of the breccia bodies in the Minaret Formation carbonate rock. The evidence in hand indicates to us a formation by dissolution from hydrothermal solutions in association with cavelike processes such as roof collapse, deposition of detritus in layers, and growth of calcite crystals in voids. The breccia bodies may have formed at a depth as great as 5,000 m.

Cavelike processes must be considered as a partial explanation because of the shapes and some internal structures of the breccia bodies. Basal and intercalated detrial sedimentary layers (Figs. 8, 9, 11, 12, 13) are suggestive of sediments deposited in a cave system. Karst features in Cambrian limestones in the Transantarctic Mountains of Antarctica have been described, but these features are associated with a regional unconformity (Lindsay, 1970); in the Ellsworth Mountains no unconformity has been demonstrated in the succession above the Minaret Limestone. The Ellsworth breccia bodies, moreover, do not contain any typical speleothem features such as stalactites, nor are there any mineralogic impurities in the breccia bodies, other than minor amounts of albite, despite the stratigraphic proximity of diverse detrital metasedimentary rocks. These features indicate that near-surface cave formation is not the correct explanation. But a subterranean cavern-forming process, probably with no conduits to the surface or to the enclosing strata, is a possible explanation.

The volume of calcite matrix material in the breccia bodies, the numerous calcite veins, and the zones of bleached limestone in and around the breccia bodies are suggestive of hydrothermal fluid migration at the time of breccia body formation. Preliminary studies of fluid inclusions from crystals of vein calcite give a crystallization temperature of 160° ± 5°C. If all the younger beds in the Ellsworth Mountains succession overlay the Minaret Limestone when the breccia bodies were formed, they may have been 5,000 m below the surface; the 160°C. temperature is compatible with such a depth. Significant erosion of overlying beds and metamorphic elevation of temperature would make the inferred depth less.

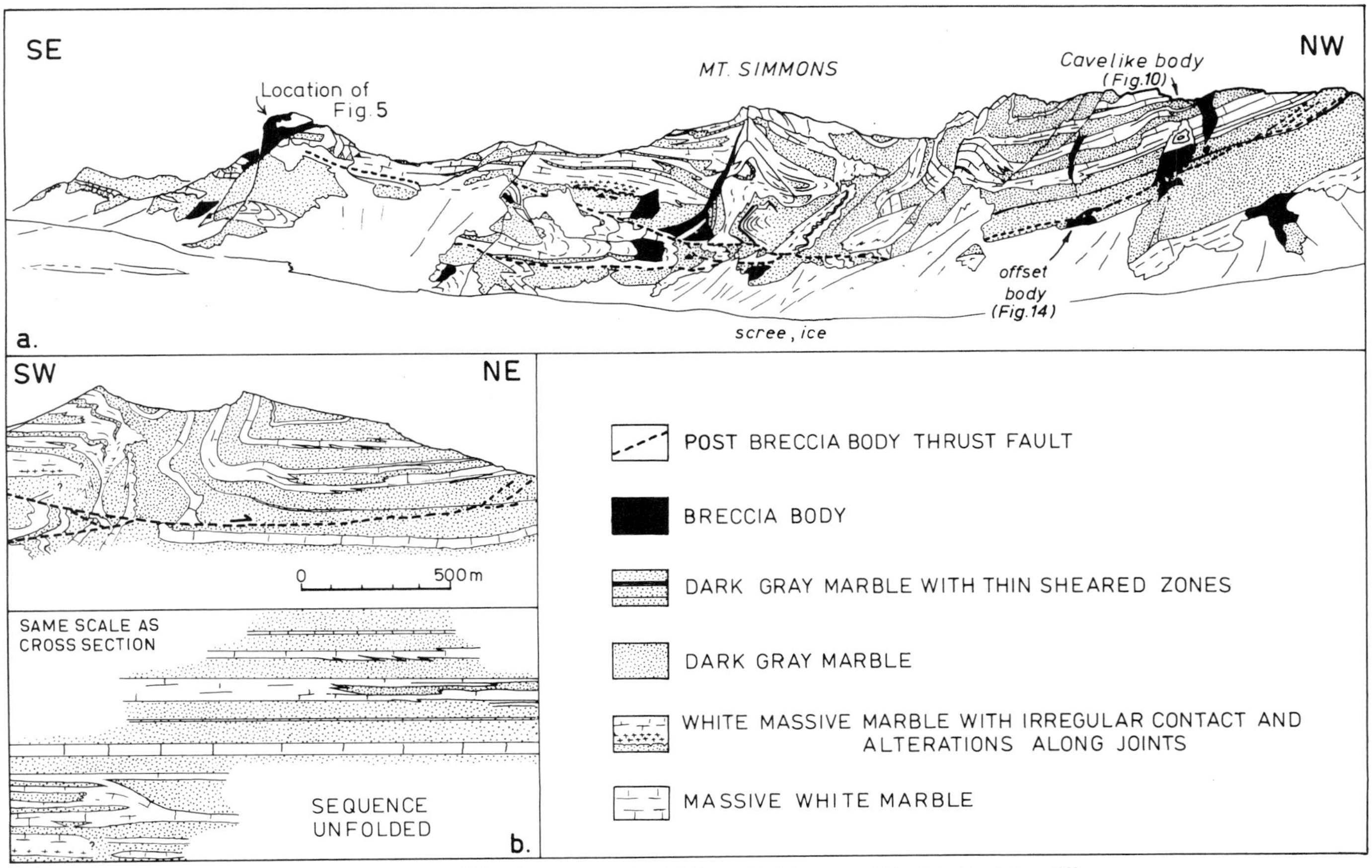

Figure 16. a, Panorama of the cirque east of and below Mt. Simmons. The height of the cliff exposure below Mt. Simmons is about 600 m. Note references to Figures 5, 10, and 14. b, Cross section through Mt. Simmons and the ridge to the northwest and that section unfolded and restored.

The interaction of brecciation, faulting, stylolite formation, and calcite veining indicates that the breccias were formed in an active tectonic environment dominated by minor faulting, limited cleavage formation in one or more veins, and active fluid movement. The breccia bodies probably formed over a period of time, with the older bodies containing the majority of the layers of detrital sedimentary calcite (Figs. 5, 6, 10). One early breccia body is offset slightly along a thrust fault (Fig. 14), and the internal clast layering in one breccia body is tilted (Fig. 4). The later-phase breccia bodies crosscut the early bodies, commonly are devoid of Minaret clasts, and usually are lacking in sedimentary detritus or layering (Fig. 6). These younger breccia bodies are less numerous, and displacements of these bodies by faults were not observed.

White (1988, p. 297) acknowledges that dissolution of carbonate rocks may occur at depths of thousands of meters, and Ford (1988) provides a good discussion of hydrothermal caves. Examples of likely hydrothermal caves are given by Bogacz and others (1973) (in Poland), Bakalowicz and others (1987) (in South Dakota), and Bridges and McCarthy (1990) (in Colorado).

CONCLUSIONS

The Heritage Range breccia bodies were formed at depth primarily by a combination of cavelike processes and synorogenic hydrothermal fluid interaction with the Minaret Formation limestones. Brecciation was in part due to the brittle response of these rocks toward the end of the Ellsworth orogeny; some breccia fragments were formed by brittle disaggregation of the country rock by *en* echelon extension gashes, and some breccia bodies are offset by faults. The breccia body clasts moved mainly downward, and in some bodies they form a rough layering; in most bodies, however, the orientations of clasts are varied and without pattern. Brecciation and subsequent formation of the calcite matrix were augmented by high-pressure, latest synorogenic fluids that caused dissolution of the carbonate rock and reprecipitation of calcite in voids and along fractures.

ACKNOWLEDGMENTS

We thank the National Science Foundation for grants to support our four expeditions and the U.S. Army and the U.S. Navy support force for transportation and other logistic support. We are grateful to many other expedition members for discussion and companionship in the field. We thank Calvin Alexander and Rich Lively for helpful discussion of the problem. The fluid inclusion data of P.R.L. Browne were important in the formulation of our model. Rich Lively and an anonymous reviewer provided valuable reviews.

REFERENCES CITED

Bakalowicz, M. J., Ford, D. C., Miller, T. E., Palmer, A. N., and Palmer, M. V., 1987, Thermal genesis of dissolution caves in the Black Hills, South Dakota: Geological Society of America Bulletin, v. 99, p. 729–738.

Bogacz, K., Dzulynski, S., and Haranczyk, C., 1973, Caves filled with clastic dolomite and galena mineralization in disaggregated dolomites: Polskie Towarzystwo Geologiczne Rocznik, v. 43, p. 59–72.

Bogli, A., 1964, La corrosion par melange des eaux: International Journal of Speleology, v. 1, p. 61–70.

Bridges, L. W., and McCarthy, K., 1990, Tertiary subsurface solution in the Mississippian Leadville Limestone geothermal aquifer of Colorado: The Mountain Geologist, v. 27, p. 57–67.

Craddock, C., Anderson, J. J., and Webers, G. F., 1964, Geologic outline of the Ellsworth Mountains, *in* Adie, R. J., ed., Antarctic geology: Amsterdam, North Holland Publishing Co., p. 155–170.

Craddock, C. Spörli, K. B., and Rutford, R. H., 1982, Breccia bodies in Cambrian limestones of the Ellsworth Mountains, West Antarctica: Geological Society of America Abstracts with Programs, v. 14, p. 468.

Craddock, J. P., and Webers, G. F., 1981, Probable cave deposits in the Ellsworth Mountains of West Antarctica, *in* Beck, B. F., ed., Proceedings, Eighth International Congress of Speleology, Bowling Green, Kentucky: Huntsville, Alabama, National Speleological Society, p. 395–397.

Ford, D., 1988, Characteristics of dissolutional cave systems in carbonate rocks, *in* James, N. P., and Choquette, P. W., eds., Paleokarst: New York, Springer-Verlag, p. 25–57.

Jamison, J., and Spang, J. H., 1976, Use of calcite twin lamellae to infer differential stress: Geological Society of America Bulletin, v. 87, p. 868–872.

Lindsay, J. F., 1970, Paleozoic cave deposits in the Central Transantarctic Mountains: New Zealand Journal of Geology and Geophysics, v. 13, p. 1018–1023.

Webers, G. F., and Spörli, K. B., 1983, Paleontological and stratigraphic investigations in the Ellsworth Mountains, West Antarctica, *in* Oliver, R. L., James, P. R., and Jago, J. B., eds., Antarctic earth science: Canberra, Australian Academy of Science, p. 261–264.

White, W. B., 1988, Geomorphology and hydrology of karst terrains: New York, Oxford University Press, 464 p.

Manuscript Accepted by the Society January 23, 1992

Geological Society of America
Memoir 170
1992

Chapter 20

Structure of the Heritage Range, Ellsworth Mountains, West Antarctica

K. B. Spörli
Department of Geology, University of Auckland, Auckland, New Zealand
Campbell Craddock
Department of Geology and Geophysics, University of Wisconsin, Madison, Wisconsin 53706

ABSTRACT

The main structure of the Heritage Range is an anticlinorium that trends slightly oblique to the range and is flanked on the east by a major synclinorium. This structural pattern is due to probably simultaneous interference folding, with fold axes trending both northwest-southeast and north-south. Second-phase folding occurs only locally, and it is coaxial with the main folds; the folding and cleavage development were followed by the formation of thrust, strike-slip, and normal faults. During the initial faulting the axis of greatest principal stress was perpendicular to the range, but this axis subsequently rotated (counterclockwise) until it was oriented parallel to the tectonic grain. Breccia bodies in Cambrian limestones at the southern end of the range postdate most of the deformation, but they have been affected by late fault movements. All these structures are thought to have formed during the Ellsworth (Gondwanide) Orogeny, probably in early Mesozoic time.

INTRODUCTION

The Ellsworth Mountains stand in isolation within West Antarctica and lie perpendicular to the tectonic trends of the Transantarctic Mountains to the south. Their enigmatic tectonic situation has been interpreted as (1) a curved primary configuration on the margin of Gondwanaland (Ford, 1972), (2) the result of secondary, counterclockwise rotation from a source in the eastern Weddell Sea (Schopf, 1969; Clarkson and Brook, 1977; Dalziel and Elliot, 1982), or (3) the result of clockwise movement from a position adjacent the the Transantarctic Mountains (Schmidt and Rowley, 1986). The Heritage Range makes up the southern half of the Ellsworth Mountains and trends northwest-southeast, as do the major structures (Fig. 1). In contrast to the formidable Sentinel Range to the north, the Heritage Range is less rugged, is easier of access, and offers excellent exposures in several dry valleys (Craddock and others, 1964).

The major stratigraphic units exposed in the range are: (1) the 7,200-m Heritage Group—sandstones, argillites, conglomerates, limestones with large and complex breccia bodies, and mafic igneous rocks of Cambrian age; (2) the 3,200-m Crashsite Group—buff, red, and green quartzites and argillites, the youngest unit of which has yielded Devonian fossils; and (3) the 1,000-m Whiteout Conglomerate—a probably Carboniferous glaciogenic deposit. The youngest unit of the Ellsworth Mountains, the *Glossopteris*-bearing Polarstar Formation of Permian age, is not exposed in the Heritage Range. This succession was deformed and regionally metamorphosed during the Ellsworth (Gondwanide) Orogeny, probably in early Mesozoic time (Fig. 2).

One of the major points of debate about the structural history of the Heritage Range is whether deformation of the strata was singular or multiphase (Yoshida, 1982). We first describe the structures and then return to that question.

MAJOR STRUCTURAL FEATURES

Fold hinges trend mainly northwest-southeast, and most display gentle plunges (Fig. 3). However, some steeply plunging folds are present and fold-hinge orientations are more variable in limestones of the Marble and Independence Hills (Spörli and Craddock, Chapter 18, this volume), along the Mount Dolence fault, and in sandstones and argillites of the Heritage Group in the Edson Hills.

Spörli, K. B., and Craddock, C., 1992, Structure of the Heritage Range, Ellsworth Mountains, West Antarctica, *in* Webers, G. F., Craddock, C., and Splettstoesser, J. F., eds., Geology and Paleontology of the Ellsworth Mountains, West Antarctica: Boulder, Colorado, Geological Society of America Memoir 170.

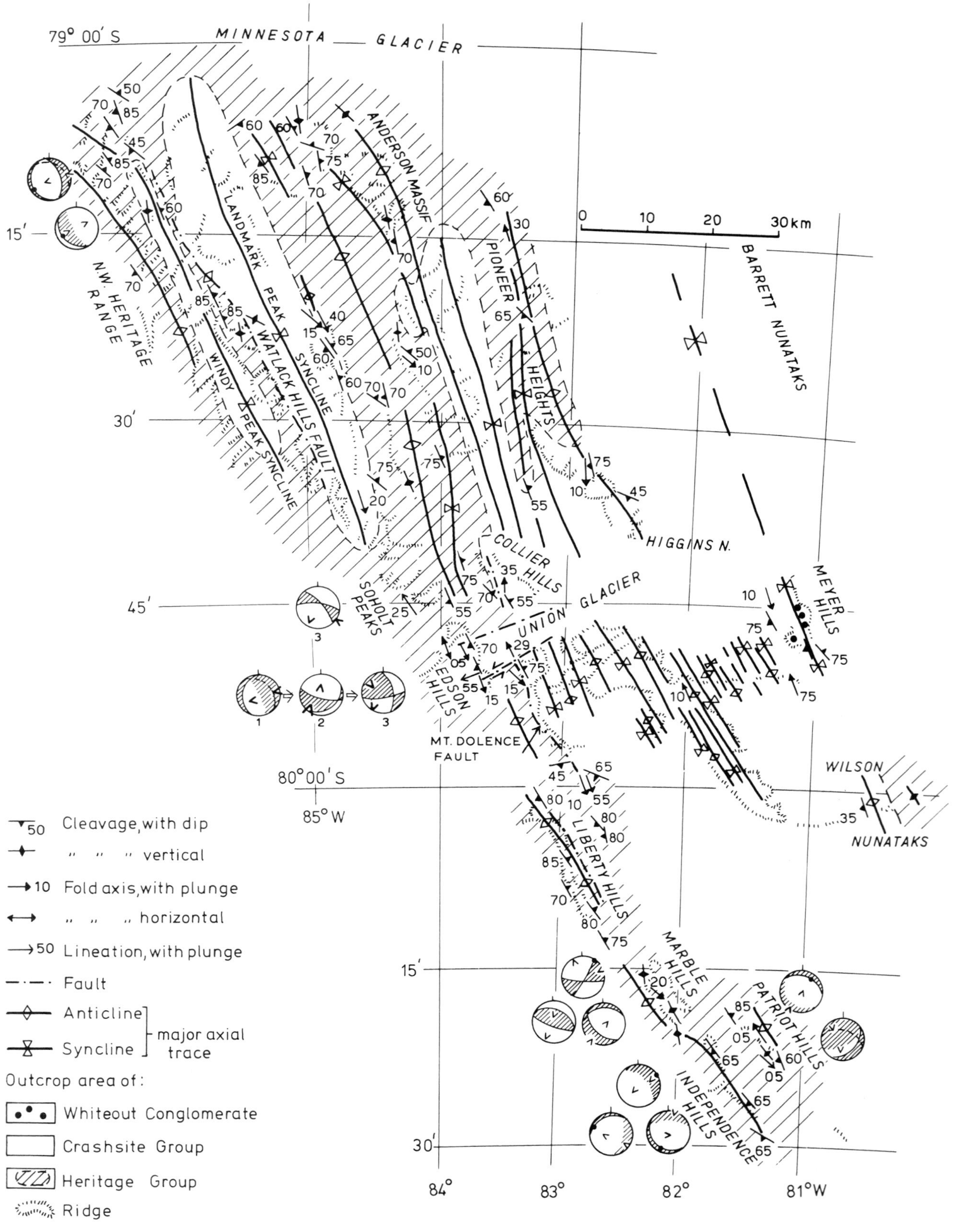
79° 00′ S
MINNESOTA GLACIER
15′
30′
45′
80°00′S
85° W
15′
30′
84°
83°
82°
81°W
0 10 20 30km
N.W. HERITAGE RANGE
WINDY PEAK SYNCLINE
WATLACK HILLS FAULT
LANDMARK PEAK SYNCLINE
ANDERSON MASSIF
PIONEER HEIGHTS
BARRETT NUNATAKS
HIGGINS N.
COLLIER HILLS
SOHOLT PEAKS
UNION GLACIER
EDSON HILLS
MT. DOLENCE FAULT
MEYER HILLS
WILSON NUNATAKS
LIBERTY HILLS
MARBLE HILLS
PATRIOT HILLS
INDEPENDENCE HILLS
Cleavage, with dip
" " " vertical
Fold axis, with plunge
" " " horizontal
Lineation, with plunge
Fault
Anticline
Syncline
major axial trace
Outcrop area of:
Whiteout Conglomerate
Crashsite Group
Heritage Group
Ridge

Axial planes generally dip steeply to the southwest, suggesting vergence to the northeast. Exceptions are found, however, in the northwest Heritage Range, along the southern side of the Union Glacier, and in the Liberty Hills, where axial planes dip steeply to the northeast (Fig. 2). The reason for the occurrence of these northeast-dipping axial planes is not clear. They appear to be located in areas where the more common chevron-style folds yield to more open, straight-limbed folds with subvertical main axial surfaces and a stronger fanlike divergence of minor axial planes across the fold profile.

The overall structural pattern in the Heritage Range indicates a low-angle interference pattern in plan of two fold trends (northwest and north–to–north-northwest). A major anticlinorium exposes the oldest rocks in the range and transects the range somewhat obliquely, lying near the western edge of the Heritage Range at the southern end and at the eastern edge near the northern end. The youngest rocks are exposed on the eastern side of the Heritage Range, in the Meyer Hills; these beds define a major synclinorium, the east flank of which is exposed in the Barrett Nunataks (Fig. 1).

The regional metamorphic grade throughout the Heritage Range seems rather constant; the rocks examined lie in the greenschist facies (Bauer, this volume).

FOLD STYLES

Chevron and kink-type folds (Fig. 4) are most common and reflect the generally thin- to medium-bedded nature of the stratigraphic succession. Where more competent thick-bedded or massive units (e.g., the Whiteout Conglomerate) are involved, folds are larger, more open, and rounded. This is the case in the northern part of the Pioneer Heights and in the northwest Heritage Range (Figs. 2, 5a, b). The most tightly appressed, isoclinal folds occur in limestones within the Heritage Group, especially in the Independence and Marble hills (Spörli and Craddock, Chapter 18, this volume). Near the northern end of the Marble Hills folds become more open and have fewer parasitic folds. This change seems to be a result of greater bed thickness and competence.

In some repetitive, very thin bedded units the folds are sharp hinged, and there is little change in thickness of the beds across the fold profile. The steeply dipping axial surfaces can commonly be traced through several hundred meters of section. In many places it is possible to see a change in style going up- or down-section along the axial plane (Fig. 5d) from concentric shapes, compatible with flexural slip, into chevron folds that are pinched at the hinges in the manner described by Ramsay (1974). In some cases an anticline will change into a syncline up or down along the axial plane, giving rise to peculiar appendages in the fold (Fig. 5c). Such structures may indicate that a décollement surface, separating structural members of flexural-slip folds, passed along that particular horizon at some time during the folding. Some quartzite layers that are isolated within an argillitic unit have taken up layer-parallel shortening by developing a train of buckle folds.

In other localities, however, individual sandstone or argillite beds as well as structural members consisting of a succession of strata change thickness with position on the fold. Both thickening and thinning in the hinge can be observed. In some examples adjacent folds in the same horizon will show opposite thickness variations (Fig. 5). Some of these irregularities may be the result of uneven original thicknesses of the beds in the folds.

Minor structures incongruent with the predominant upward-facing folds with subhorizontal axes include (1) steeply plunging dextral and subhorizontal up-dip-verging folds in the northwest Heritage Range, and (2) down-dip-verging folds in the Collier Hills, in the nunataks to the east (south end of Pioneer Heights, Fig. 4), in the Dolence Peaks area, and in the Marble Hills (Spörli and Craddock, Chapter 18, this volume). In the Marble Hills example it is possible to determine the sequence of events: The incongruent folds at this locality fold the major cleavage and therefore postdate the major folds. Down-dip-verging folds in the Collier Hills and to the east are probably older than the main folding and may record (gravitational) cascading off rising anticlines.

CLEAVAGE

Cleavage is best developed in argillites, and in places it is difficult to detect macroscopically in more competent rocks such as sandstones and quartzites. Some massive igneous rocks in the Heritage Group display no cleavage at all, some contain rhombic joints and shears instead of a cleavage, and others display cleavage throughout. Cleavage at an angle to bedding is well developed in the gray limestones in the Marble Hills. In the white marbles it appears to have been sheared out and rotated parallel to the layering. Fine-grained units in the Whiteout Conglomerate commonly display a closely spaced cleavage that almost obliterates bedding; however, this cleavage becomes more widely spaced and irregular in the coarser diamictite units.

In most folds, cleavage planes form an approximately parallel set across the fold or diverge only 5 to 10° to form a cleavage fan. Exceptions are found in some localities in the northwest Heritage Range where bedding anisotropy is stronger and intermixed divergent and convergent cleavage fans have been developed on a small scale.

Microscopically, the cleavage in argillites consists of closely spaced trains of similarly oriented phyllosilicates (mostly chlorite

Figure 1. Map showing main structures of the Heritage Range. Small lower hemisphere equal-area nets show the geometry of deformation deduced from veins, slickenside striations, and faults. The compressional angles of directly observed (two bounding great circles) or constructed (one bounding line only) conjugate couples are shaded. V pointing outward = extension axis; v pointing inward = compression axis; solid dot = striation. Note temporal change from NE-SW to NW-SE compression in Edson Hills.

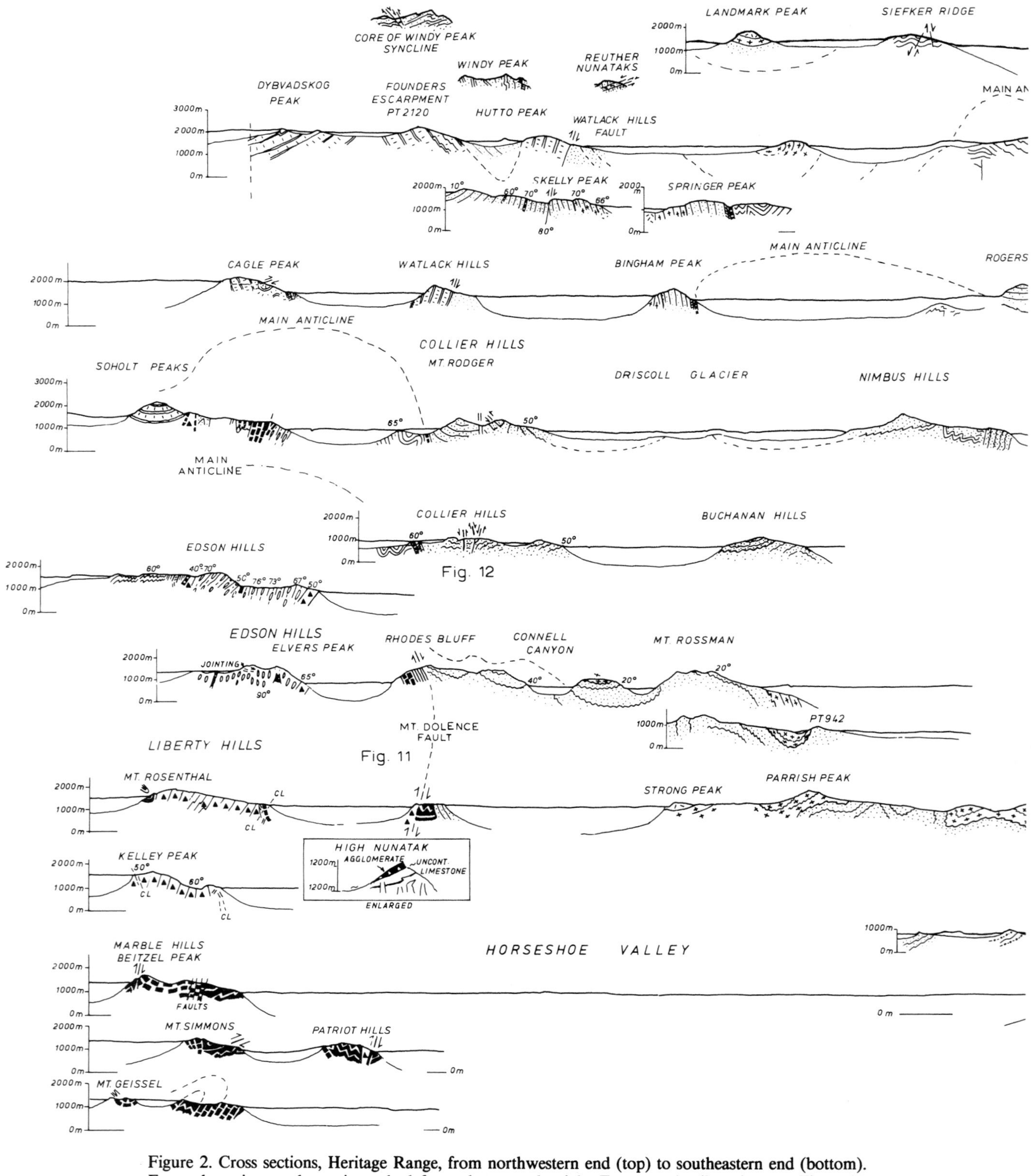

Figure 2. Cross sections, Heritage Range, from northwestern end (top) to southeastern end (bottom). For each section southwest is on the left, northeast on the right. For more detail sections see Figure 11 (Mount Dolence fault), Figure 12 (Collier Hills faults), and Spörli and Craddock (Chapter 18, this volume) (Marble, Independence, and Patriot hills).

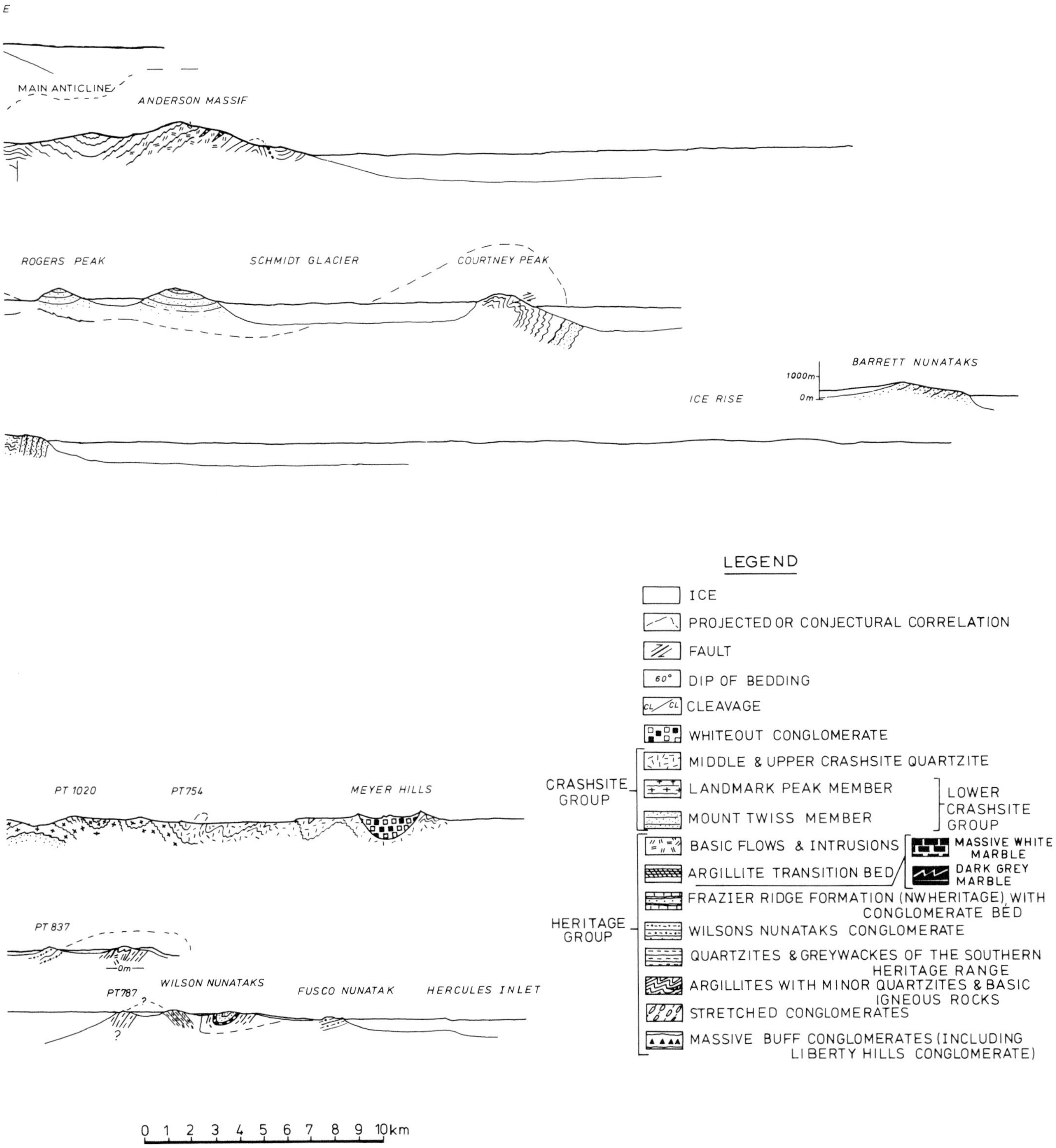
E
MAIN ANTICLINE
ANDERSON MASSIF
ROGERS PEAK
SCHMIDT GLACIER
COURTNEY PEAK
BARRETT NUNATAKS
1000m
0m
ICE RISE
PT 1020
PT 754
MEYER HILLS
PT 837
0m
PT787
WILSON NUNATAKS
FUSCO NUNATAK
HERCULES INLET
0 1 2 3 4 5 6 7 8 9 10km
LEGEND
ICE
PROJECTED OR CONJECTURAL CORRELATION
FAULT
60°
DIP OF BEDDING
CL CL
CLEAVAGE
WHITEOUT CONGLOMERATE
CRASHSITE GROUP
MIDDLE & UPPER CRASHSITE QUARTZITE
LANDMARK PEAK MEMBER
MOUNT TWISS MEMBER
LOWER CRASHSITE GROUP
HERITAGE GROUP
BASIC FLOWS & INTRUSIONS
MASSIVE WHITE MARBLE
ARGILLITE TRANSITION BED
DARK GREY MARBLE
FRAZIER RIDGE FORMATION (NW HERITAGE), WITH CONGLOMERATE BED
WILSONS NUNATAKS CONGLOMERATE
QUARTZITES & GREYWACKES OF THE SOUTHERN HERITAGE RANGE
ARGILLITES WITH MINOR QUARTZITES & BASIC IGNEOUS ROCKS
STRETCHED CONGLOMERATES
MASSIVE BUFF CONGLOMERATES (INCLUDING LIBERTY HILLS CONGLOMERATE)

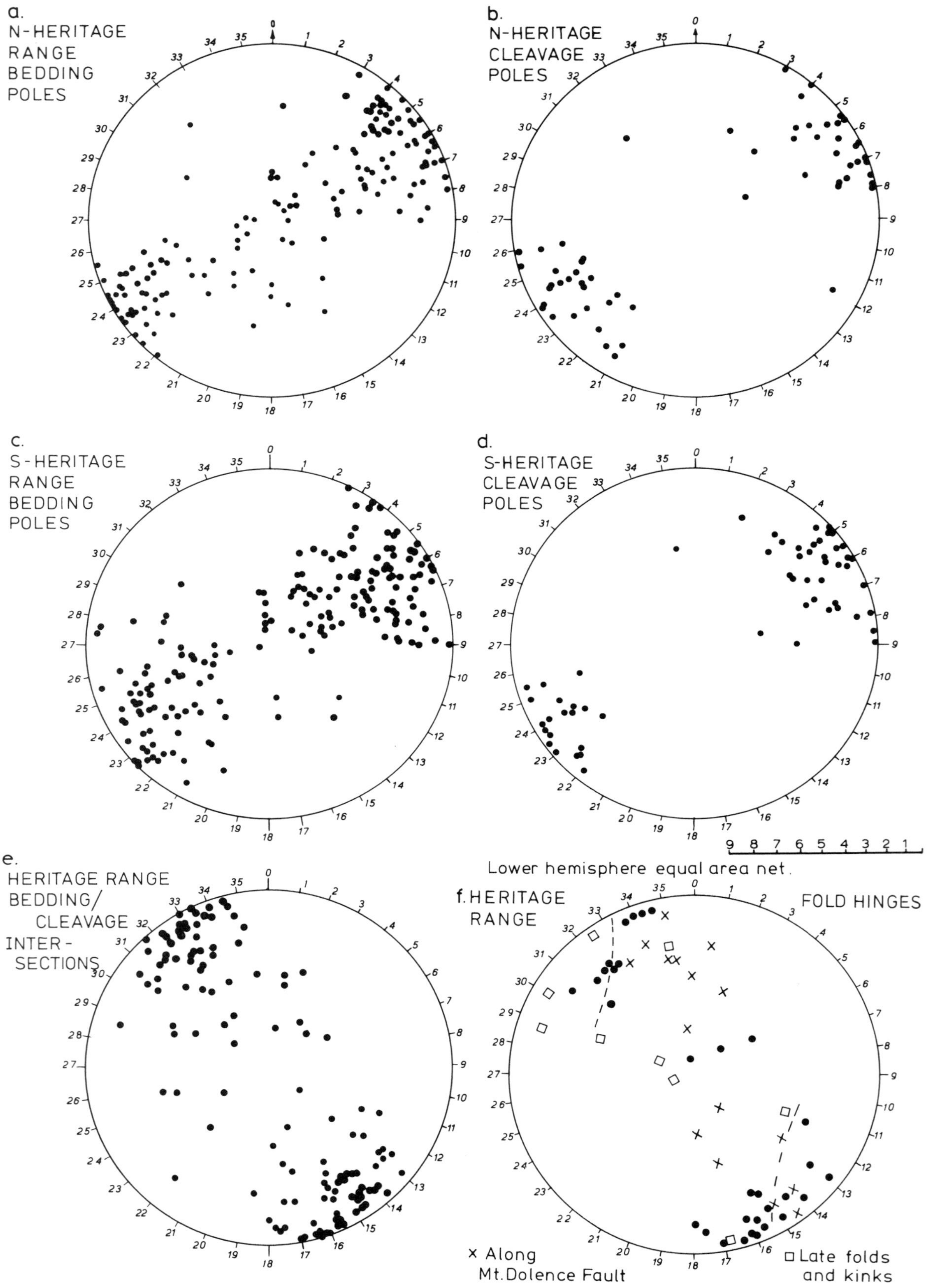
a.
N-HERITAGE
RANGE
BEDDING
POLES
b.
N-HERITAGE
CLEAVAGE
POLES
c.
S-HERITAGE
RANGE
BEDDING
POLES
d.
S-HERITAGE
CLEAVAGE
POLES
9 8 7 6 5 4 3 2 1
Lower hemisphere equal area net.
e.
HERITAGE RANGE
BEDDING/
CLEAVAGE
INTER-
SECTIONS
f. HERITAGE
RANGE
FOLD HINGES
× Along
Mt. Dolence Fault
□ Late folds
and kinks

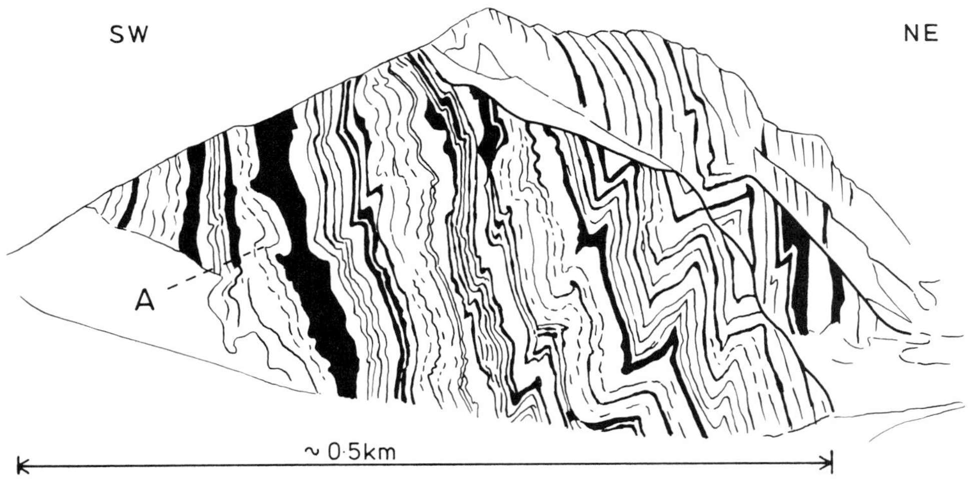

Figure 4. Asymmetric minor chevron folds in quartzite and argillite beds of the Mt. Twiss Member, Howard Nunataks Formations, Crashsite Group, on the east limb of major anticline, Higgins Nunatak, southeastern end of Pioneer Heights. Traced from photographs. Note downward-verging, rounded-hinge fold near left edge of exposure at A.

and sericite). Competent grains, such as quartz, feldspar, or heavy minerals, have "mica beards" of these minerals, elongated in the pressure shadows. Unstrained calcite commonly replaces or grows over these beards. In a few very finely laminated argillites a crenulation cleavage has been developed.

Beards and zones of phyllosilicates are also present in quartzites with sufficient matrix. In those with little matrix, the quartz grains have been sutured by solution surfaces and may or may not form a somewhat irregular cleavage. All quartz grains are strained, showing gradations from undulatory extinction to those that show development of subgrains and deformation bands that are oriented parallel to each other across a large number of grains.

Cleavage in carbonate rocks ranges from isolated to closely spaced stylolitic surfaces, some of which cut through pisolites (oncolites) that have already been strained. In the white marble the cleavage (or schistosity) is due to reorientation of the calcite grains into foliation planes. Competent grains such as dolomitic pisolites have yielded by brittle fracture parallel to the direction of shortening. These fractures are filled with clear calcite, thus defining the cleavage fabric.

The matrix of some strongly heterogeneous conglomerates displays very closely spaced sutures and zones of oriented phyllosilicates that end at the borders of the individual conglomerate clasts. Deformation is taken up by more widely spaced microfaults within these clasts and in quartz-chlorite fiber veins along their border. The fibers are oriented parallel to the cleavage.

Figure 3. Structural measurements of folds and cleavage, Heritage Range. Note dominance of NW-trending, subhorizontal fold axes. Dashed lines in f indicate possible conical refolding pattern of earlier fold hinges around later axis trending 110°.

THE DIKE PROBLEM

Some mafic dikes in the Heritage Group of the Edson Hills and the Liberty Hills pose a problem for the determination of the structural sequence. Most of these dikes are parallel to the cleavage and could therefore be interpreted as postfolding in age. Yoshida (1982) proposed that the Edson Hills dikes were intruded during or after cleavage formation. To explain the parallelism of cleaved dikes to cleavage he postulated an extensional pause during cleavage formation. His Figure 16, however, can as easily be explained by deformation of an existing dike as by two phases of syn-dike cleavage formation.

Nevertheless, there are also many dikes that trend oblique to the cleavage. We have also observed dikes that are affected by the cleavage throughout their width (Fig. 6b); exceptions occur in very massive dikes that show only a rhombic fracture pattern in the center (Fig. 6a). However, if a less competent dike rim is present, this part of the dike will have a cleavage. In one case a dike crosscutting the cleavage does not show much evidence of cleavage as long as it traverses relatively competent rock but displays strong cleavage in a more highly strained argillitic layer (Fig. 6c).

We conclude that all the dikes are precleavage in age but have reacted inhomogeneously to deformation; as with other rocks in the Heritage Range, the presence or absence of a macroscopic cleavage depends on the competence of the rock. Those dikes that now are parallel to cleavage were originally perpendicular to bedding, and their strike was close to that of the axial plane of the folds. During folding they maintained a position parallel to the cleavage. The alternative interpretation, that intrusion of the dikes occurred in the waning stages of folding and outlasted the deformation, is not plausible because all the units in the Ellsworth Mountains are affected by the same deformation;

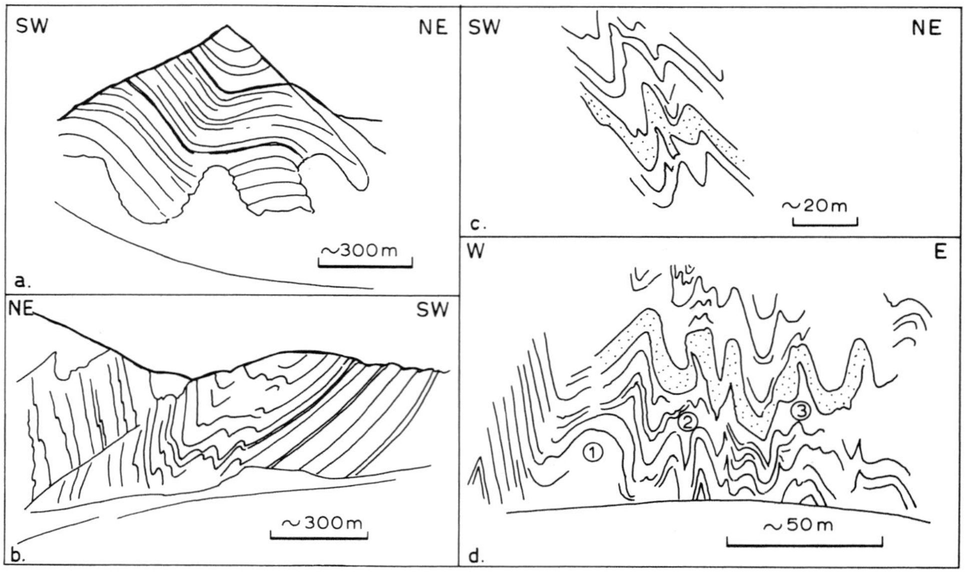

Figure 5. Styles of folding. All sketches traced from photographs. a, kink fold in strata of the Crashsite Group (impure green and gray quartzites and argillites); ridge west of Matney Peak, Founders Peaks, Heritage Range. b, southwest-verging syncline, Mt. Twiss Member, Crashsite Group, Collier Hills. c, Linder Peak Member, Crashsite Group, west ridge of Mount Dolence, with synclinal appendage in anticline. d, Mt. Twiss Member, Crashsite Group, northern Enterprise Hills, showing (1) disharmonic folding and hinge collapse, (2) both synclines and anticlines along axial surface, and (3) irregular thickening of bed on limb of anticline.

this deformation postdates the Permian Polarstar Formation, yet the dikes are only present in the Heritage Group and are in some cases truncated by channels. Dikelike bodies reported in the Crashsite Group of the eastern Heritage Range have all turned out to be clastic dikes. These observations indicate that mafic dike intrusion predated deposition of the Crashsite Group (Devonian) and must be considerably older than the early Mesozoic (?) deformation.

In support of our interpretation, it is interesting to note that Bauer (this volume) has found evidence for an earlier burial metamorphism preserved only in the massive igneous rocks of the Heritage Group. These intrusions were shielded from the inhomogeneous deformation accompanying the Ellsworth (Gondwanide) regional metamorphism.

The only locality where we have observed two macroscopically visible phases of cleavage is in the northern Marble Hills, where the main cleavage predates the formation of breccia bodies. Some breccia bodies are, however, affected by a coarse second cleavage, which has the same orientation as the one formed during the first phase (Spörli and Craddock, Chapter 18, this volume). Two cleavages also occur locally in the limestones along the Mt. Dolence fault; one cleavage here may result from displacement along the fault. Yoshida (1982) has illustrated microscopic evidence for two cleavages in the Cambrian rocks of the Wilson Nunataks.

STRETCHING LINEATION

The orientations of bedding/cleavage intersection lines are mainly subhorizontal and similar to those of fold hinges (Fig. 3e, f). In addition, a down-dip stretching lineation has been developed in some areas. White limestone (marble) in the Marble and Independence hills is highly laminated and preferentially lineated; a lineation is less well developed in the dark gray limestones of the area. Nevertheless, even in these rocks it can be recognized by the alignment of strained ellipsoidal pisolites. Other limestones in the Heritage Group, such as in the Mt. Dolence and the Soholt Peak areas, also display down-dip lineations. Such lineations also occur locally in the more argillaceous units of the Heritage and the Crashsite groups.

Cleavage is more strongly developed in the Edson Hills than elsewhere, giving the rocks a phyllitic appearance. Stretched pebbles constitute the down-dip lineation in a volcaniclastic-conglomeratic unit. Epidote veins are also affected by this deformation. Bedding is so strongly transposed here that it can only rarely be recognized. In the southern part of the Edson Hills an intrafolial fold hinge of quartzites (50 m long and 30 m wide) is embedded in the stretched conglomerates (Fig. 7).

Axial dimensions measured in the stretched conglomerates on three mutually perpendicular surfaces, one of which is the cleavage or schistosity (Fig. 8), indicate that besides the variation

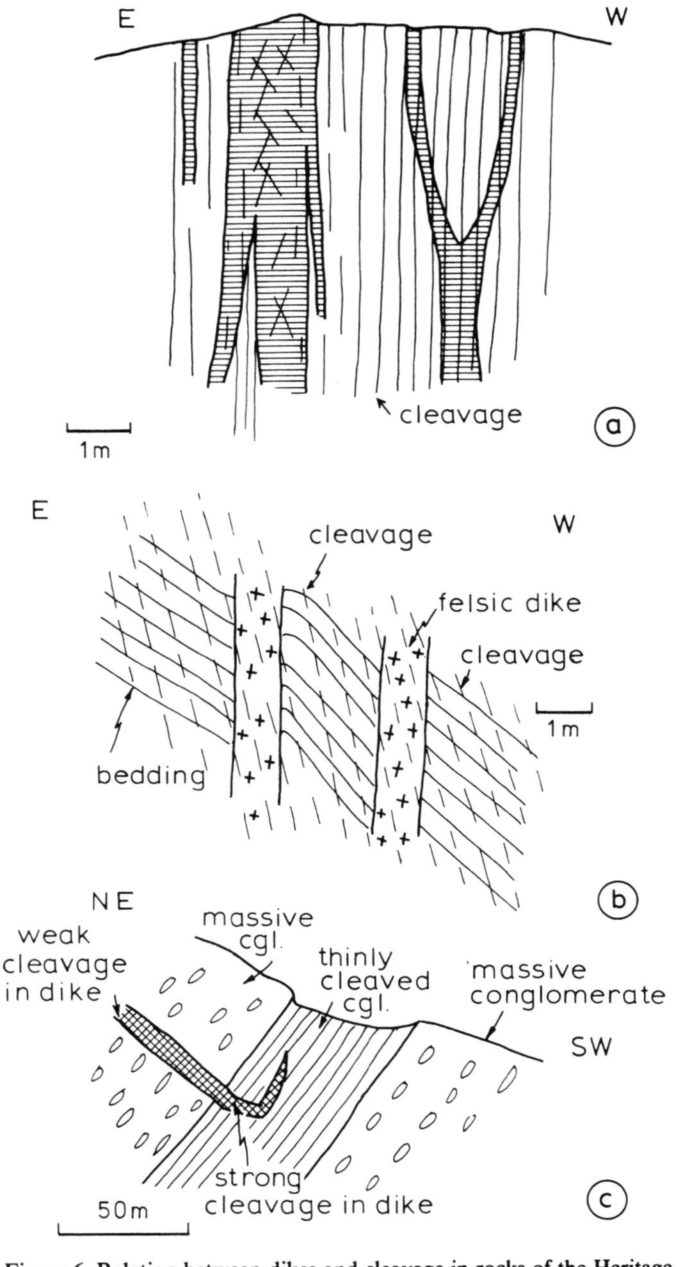

Figure 6. Relation between dikes and cleavage in rocks of the Heritage Group, south side of Hyde Glacier, Edson Hills. All sketches are of moderately steep slopes dipping toward the observer. a, note suppression of cleavage and development of rhombic set of fractures in competent center of thicker dike. b and c, drawings illustrate control of intensity of cleavage development in dike by competence of host rock.

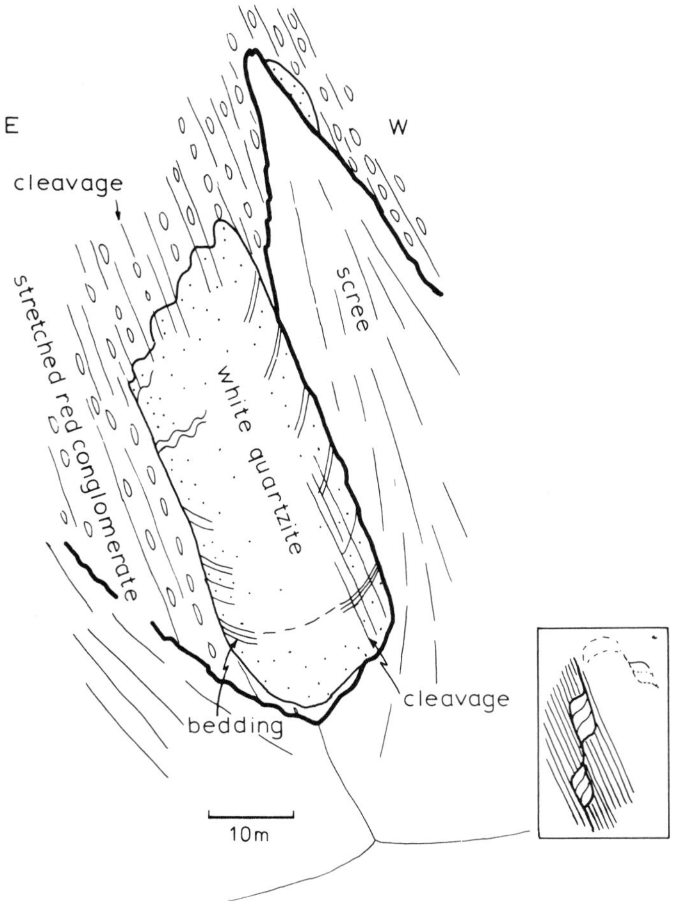

Figure 7. Intrafolial fold hinge of quartzite within stretched conglomerates of the Heritage Group, Edson Hills, in second valley south of Hyde Glacier. Strained zone is in core of main anticlinorium of Heritage Range. Inset: schematic transposition model for this exposure, indicating position on east flank of asymmetric antiform.

of axial ratios due to strain of ellipsoidal fragments of random orientation there are also systematic variations probably due to different mechanical behavior and to sedimentary sorting of the pebbles. These are expressed in different rocks by both increases and decreases in axial ratio with increasing size (Fig. 8f). In addition, competent fragments, mainly olive green, epidotized, porphyritic igneous rocks, have typical mean axial ratios of 1:2:3 and 1:2:5 (Fig. 8), whereas the incompetent, mainly argillitic and tuffaceous (pumiceous?) fragments display mean axial ratios of 1:5:15 and 1:6:14 (Fig. 8).

In another stratigraphic unit, some massive conglomerates, with pebbles of quartzite in a calcareous sandy matrix, have no stretching lineation. Thus, the amount of stretching depends on the difference in competence of matrix and clasts.

In some places the down-dip stretching lineation appears to grade laterally into calcite or quartz-chlorite fiber striations also perpendicular to the fold hinges, indicating a transition from pure strain into a flexural slip regime.

BOUDINAGE, EN ECHELON EXTENSION GASHES

Boudinage structure can be observed in the limbs of folds in the Crashsite Group, but it is even better developed in the

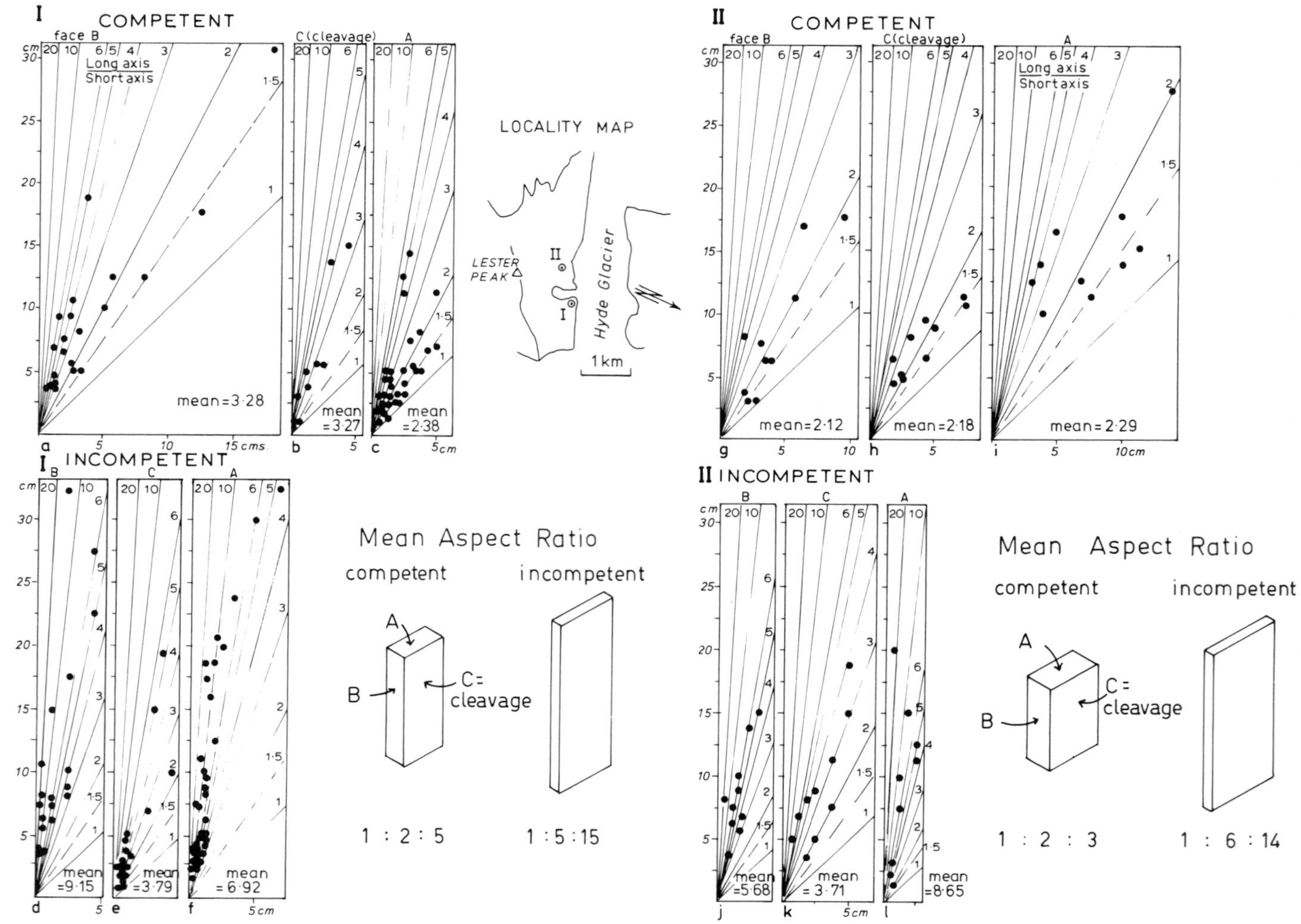

Figure 8. Axial ratios of fragments (clasts) in Edson Hills stretched conglomerates measured on three mutually perpendicular faces at two localities (I and II). Graph h shows decrease, graph l increase in axial ratio with fragment size. Graph f shows constant, then increasing, then decreasing axial ratio with increasing fragment size. Aspect ratios show difference in shape between competent and incompetent fragments. Mean shapes were obtained by taking the arithmetic mean axial ratio on each face. Three-dimensional axial ratios were determined by pairwise combination of all three faces and setting the smallest dimension equal to 1.

stretched conglomerates of the Edson Hills, especially in some mafic dikes that take up the stretching strain in this manner (Fig. 9). Boudin necks are nearly horizontal, indicating the same direction of stretching as that given by the down-dip lineation (Fig. 10), and they are commonly marked by arrays of quartz-chlorite veins, locally containing traces of chalcopyrite. No clear "chocolate tablet" structures have been recognized. A local crenulation cleavage is developed in some of the boudin necks (Fig. 9d). The orientation of en echelon vein arrays, where present, is consistent with the boudinage deformation (Fig. 10). There are, however, some en echelon arrays that do not conform to this regime of deformation. They may be associated with the later faults described below.

Quartz-chlorite veins associated with the boudinage postdate green epidote veins and the cleavage with the stretching lineation. Epidote veins are crenulated by the cleavage, but they appear not to have undergone the same amount of strain as neighboring stretched pebbles. Hence the epidote veins may have formed at a late stage of cleavage formation.

FAULTS

Strike faults

Major strike faults in the Heritage Range are the Watlack Hills fault in the northern part of the range, the Mount Dolence

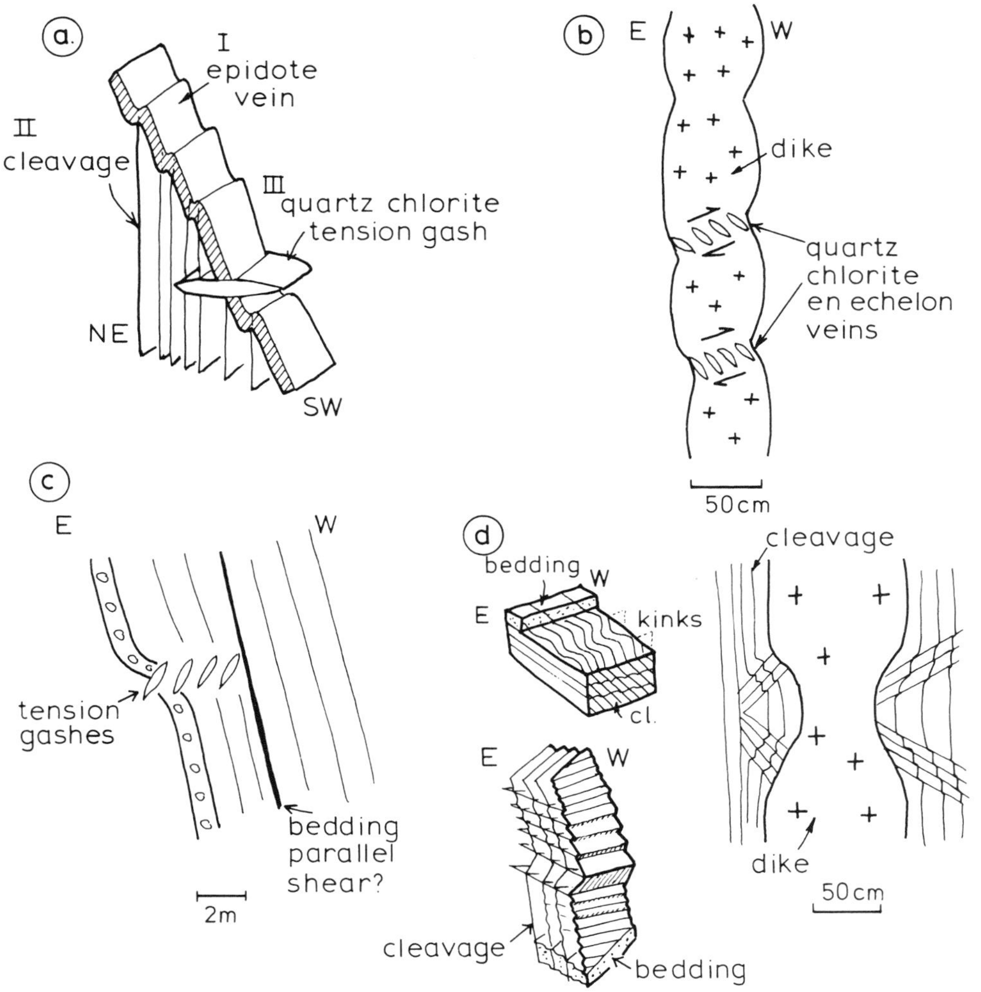

Figure 9. Relationships between some planar and linear structural elements. a, b, c, veins and boudinage; d, development of local second-phase crenulation, as observed in Heritage Group beds, eastern part of Edson Hills, mainly along the south side of Hyde Glacier. c, possibly a boudin structure displaced by bedding-plane slip. d, crenulation cleavages are developed in necks of dikes (also at much larger scale than shown), near rusty fault zones (upper left), and near later conjugate kinks (lower left).

fault in the central part, and a fault along the eastern side of the Liberty Hills in the south (Fig. 1).

The Watlack Hills fault is located in the core of an anticline, has a reverse displacement with an unknown throw, and places Frazier Ridge Formation (Heritage Group) against the lower Crashsite Group (Fig. 2). The slip vector orientation is unknown.

The Mount Dolence fault lies on the east flank of the main anticlinorium of the Heritage Range, and its western side is upthrown. At the north end (Fig. 11) it lies within the Crashsite Group, but to the south its throw probably increases, and Minaret Formation carbonate rocks (Heritage Group) are faulted against the lower Crashsite Group. At the southern end of the fault trace, the Minaret Formation is in stratigraphic contact with the Crashsite Group on the east side of the fault and is faulted against lower parts of the Heritage Group. The dip of the fault changes along the strike; it is to the west at the north and south ends (a reverse fault) and is vertical or eastward in the middle section (a normal fault), indicating that the fault plane has been folded. This is most obvious at the southern end, where the fault appears to be involved in a minor fold (Fig. 11G).

A complex history is also displayed in the Collier Hills at the northern end of the Mount Dolence fault trace (Fig. 12), where two strands of the fault are visible. The older strand in the east brings down upper Crashsite Group (pale green quartzite) into discordant contact with lower Crashsite Group strata that contain downward-verging folds, a situation incongruent with the major structures (Fig. 12A). Minor structures along this fault suggest normal dip-slip movement. This strand is warped and cut by a

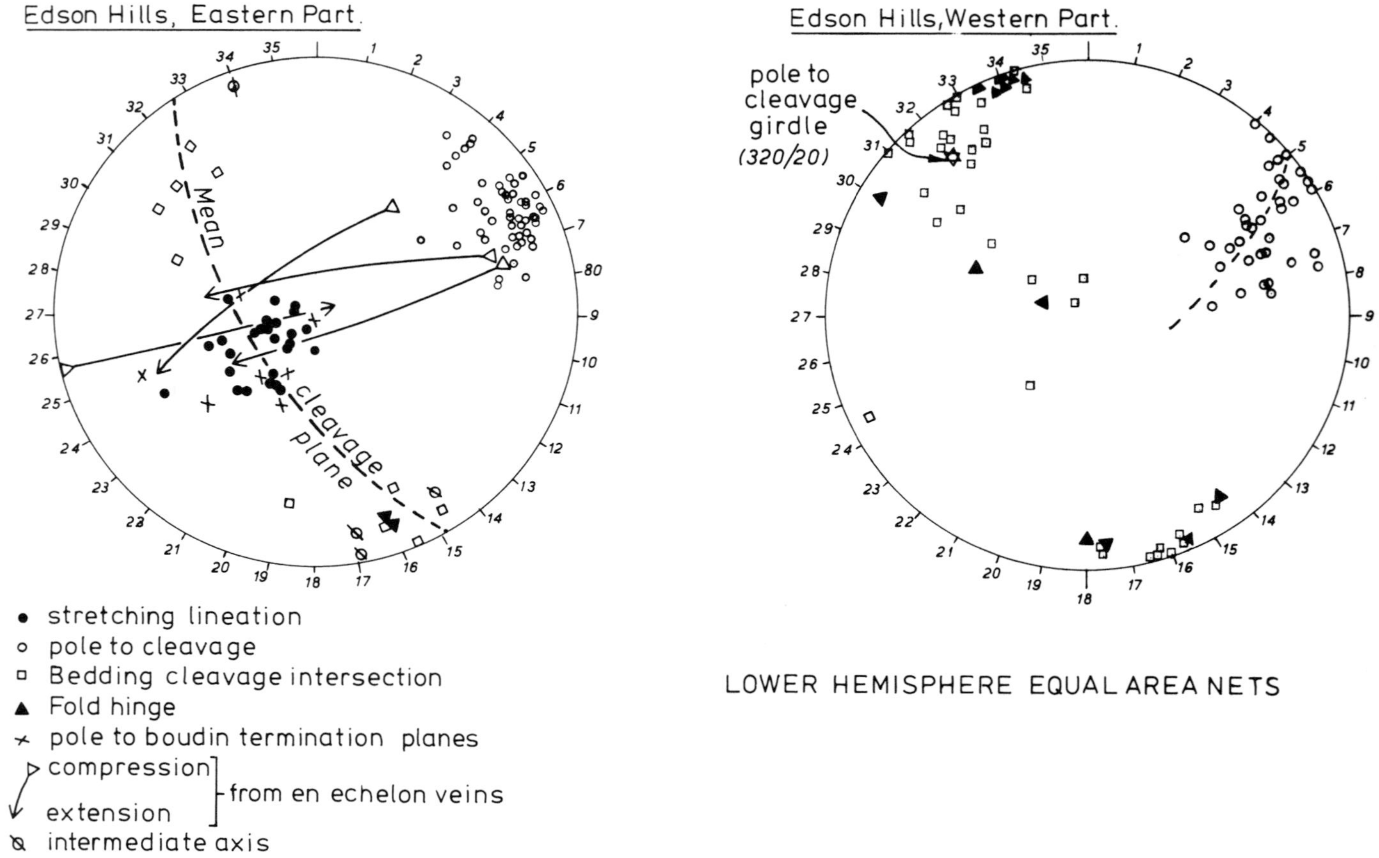

Figure 10. Structural data from Edson Hills. Note correspondence of poles to boudin termination poles and extension axes from en echelon veins with stretching lineations in the higher-strain eastern part (left). Cleavage planes are subparallel to each other in eastern part and tend to form a fan in the lower-strain western part (right).

reverse fault on the west. Fiber striations with accretion steps near the younger fault confirm reverse dip-slip movement. This fault and axial planes of folds adjacent to it are also folded by a structure of which the west side verges upward. These observations lead to the following inferred sequence of events:

1. "Cascade" folding (with downward vergence of folds) develops in the beds to the east of the faults.
2. Fold hinges form, trending north-south to north-northeast, oblique to the major northwest-southeast fold trend. Normal faulting occurs, down on the west.
3. Reverse faulting develops (probably during major regional folding); this fault is upthrown on the west.
4. Shearing continues with the same sense as in step 3. All previous structures are folded.

Immediately south of the Union Glacier, the Mount Dolence fault is displaced sinistrally by northeast-southwest–trending cross faults (Fig. 11). Limestones west of the Mount Dolence fault have steeply plunging folds with steep, northwest-trending axial planes. In the vicinity of the cross fault these axial planes have been rotated into a northeast strike and a steep dip parallel to the fault plane, indicating a strong strike-slip component of fault movement.

The Mount Dolence fault has a zone of sheared and completely transposed Crashsite Group beds along its eastern margin, and strongly laminated limestones with kink folds and striations lie to the west. Slip lines deduced from conjugate kinks pitch 45° to the north, but striations pitch 70°S in the fault plane. From structural analysis elsewhere in the Heritage Range, the kinks,

→

Figure 11. Mount Dolence fault, central Heritage Range, on the east limb of the main anticlinorium. Note that the fault changes position from east of the Heritage Group limestones in the north to west of the limestones in the south. Lower hemisphere equal-area nets give structural data at north end; black dots are fold hinges in limestone; open circles are fold hinges in Crashsite Group. On the net with conjugate folds: Arrowhead pointing inward = compression; pointing outward = extension axis. For more detailed Collier Hills section, see Figure 12.

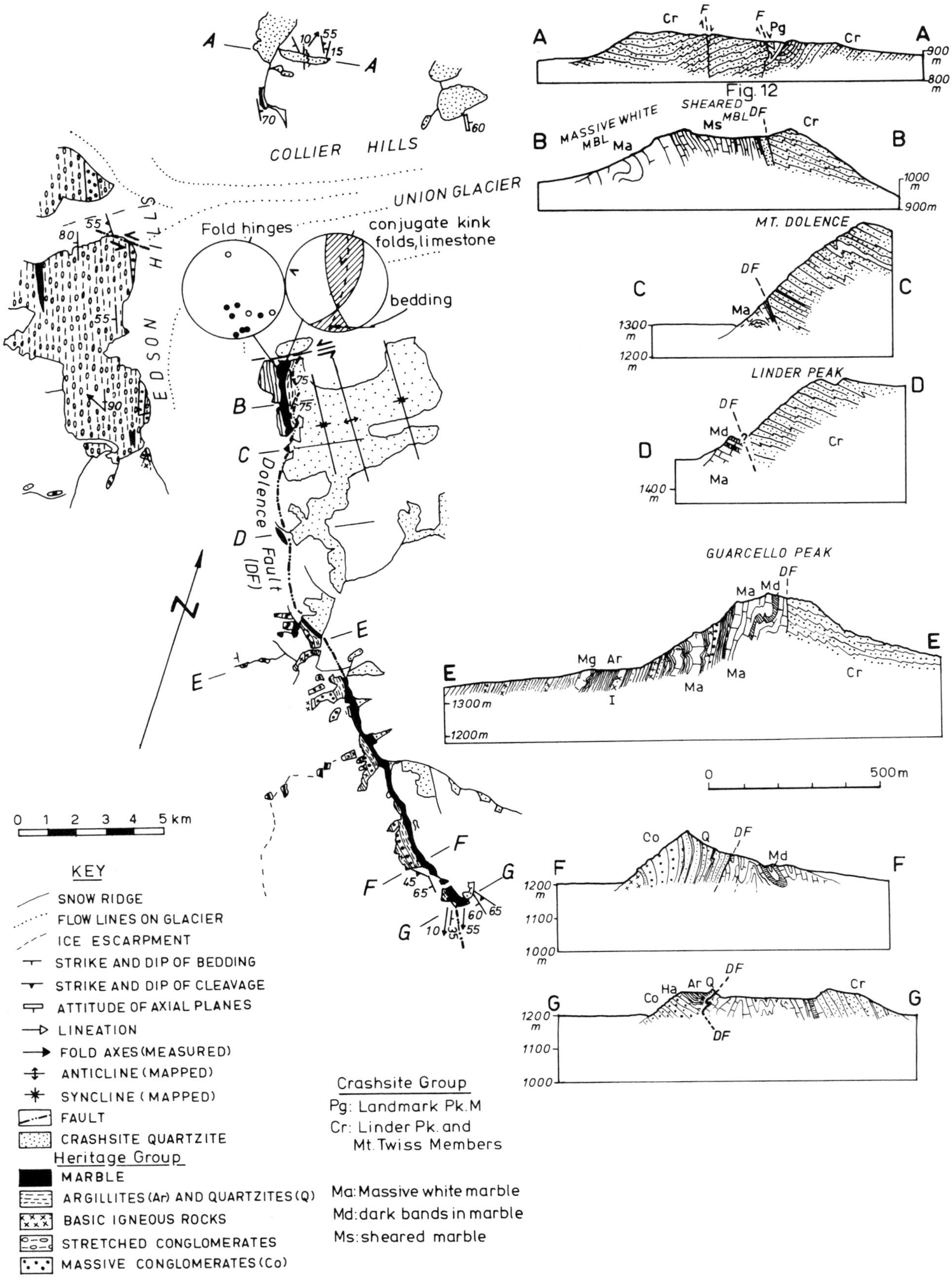
Fig. 12
COLLIER HILLS
UNION GLACIER
EDSON HILLS
Fold hinges
conjugate kink folds, limestone
bedding
Dolence Fault (DF)
N
0 1 2 3 4 5 km
KEY
SNOW RIDGE
FLOW LINES ON GLACIER
ICE ESCARPMENT
STRIKE AND DIP OF BEDDING
STRIKE AND DIP OF CLEAVAGE
ATTITUDE OF AXIAL PLANES
LINEATION
FOLD AXES (MEASURED)
ANTICLINE (MAPPED)
SYNCLINE (MAPPED)
FAULT
CRASHSITE QUARTZITE
Heritage Group
MARBLE
ARGILLITES (Ar) AND QUARTZITES (Q)
BASIC IGNEOUS ROCKS
STRETCHED CONGLOMERATES
MASSIVE CONGLOMERATES (Co)
Crashsite Group
Pg: Landmark Pk. M
Cr: Linder Pk. and Mt. Twiss Members
Ma: Massive white marble
Md: dark bands in marble
Ms: sheared marble
MASSIVE WHITE MBL
SHEARED MBL
MT. DOLENCE
LINDER PEAK
GUARCELLO PEAK
0 500m

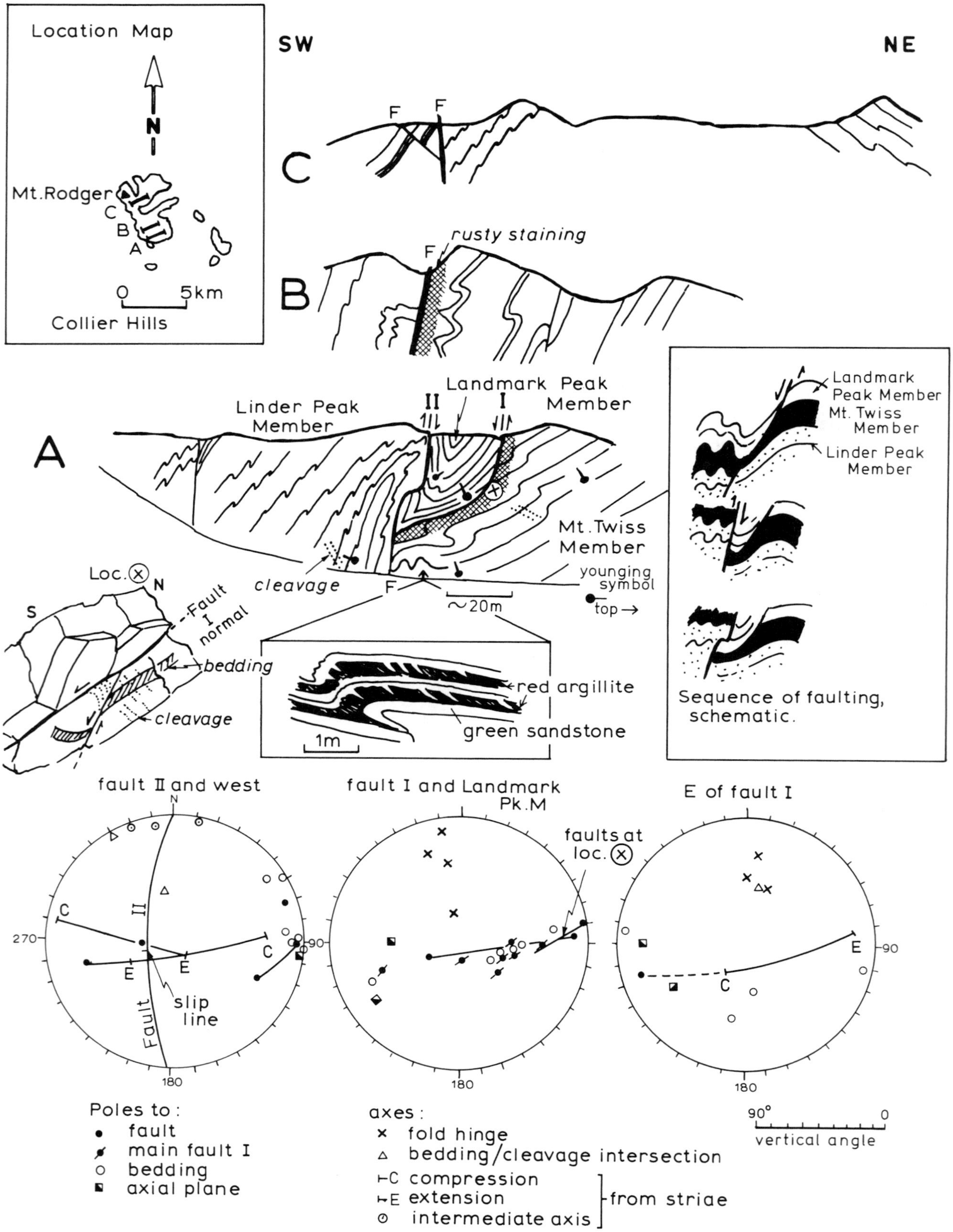

Location Map
N
Mt.Rodger
C
B
A
0 5km
Collier Hills
SW
NE
F
F
C
rusty staining
F
B
Landmark Peak Member
Linder Peak Member
II
I
A
Mt.Twiss Member
cleavage
F
~20m
younging symbol
top →
Loc. ⊗
N
S
Fault I normal
bedding
cleavage
red argillite
green sandstone
1m
Landmark Peak Member
Mt. Twiss Member
Linder Peak Member
Sequence of faulting, schematic.
fault II and west
fault I and Landmark Pk.M
E of fault I
faults at loc. ⊗
270
90
180
slip line
Fault
Poles to:
fault
main fault I
bedding
axial plane
axes:
fold hinge
bedding/cleavage intersection
compression
extension
intermediate axis
from striae
90° 0
vertical angle

which indicate a dextral component of movement on the Mount Dolence fault, may be the youngest deformational phase.

Minor folds incongruent to a major anticline are visible at several localities near the Mount Dolence fault (e.g., Section E, Fig. 11), and they are compatible with upward movement of the west side of the fault. They are accompanied by a steeply eastward dipping crenulation cleavage that displaces the major regional cleavage. Two phases of cleavage are also present at the southern end of the Mount Dolence fault outcrop belt. They both dip to the west, and the younger one has a shallower dip.

Like the other two major strike faults, the fault in the Liberty Hills is located in an anticline. It appears that the locus of these faults is closely related to the folding process.

Mesoscopic conjugate strike faults in the northwest Heritage Range are associated with quartz-chlorite en echelon extension gashes and record a subhorizontal shortening at right angles to the fold hinges, compatible with thrusting on the conjugate planes, that is also indicated by quartz chlorite fiber striations (Fig. 13). These mesoscopic faults may be part of a system of large conjugate kinks, the dominant planes of which dip steeply to the east, that are part of the fold pattern in the northwest Heritage Range. One example of incongruent folds suggests that these kinks are younger than the main folding phase.

A low-angle thrust fault at Mt. Simmons, Independence Hills, slightly displaces a breccia body. Striations on the fault plane indicate a complex history, with movement directions ranging from southwest-northeast to northwest-southeast (Spörli and others, this volume).

Later faults

The strike faults in the Heritage Range, closely associated with the folding process, are associated with or immediately postdated by a system of conjugate strike-slip faults that have their greatest shortening axis at right angles to the structural northwest-southeast trend; hence the east-west- to east-southeast-trending faults are sinistral, and the northeast-trending faults are dextral. Such faults are important in the Marble and Independence hills (Spörli and Craddock, Chapter 18, this volume). Breccia bodies have been formed along some of these faults, but others of these faults displace breccia body material, indicating a close temporal relation between faulting and breccia body formation. Some of the calcite fiber striations in the area belong to the same deformational system.

In the Edson Hills area, quartz-chlorite en echelon extension gashes indicate the same down-dip stretching directions as those given by the long axes of the stretched conglomerates (Fig. 10). They immediately postdate the stretching and are associated with the boudinage process; the major principal shortening axis that can be deduced from corresponding en echelon arrays trends about 60 to 240°, perpendicular to the structural grain. They indicate dip-slip movement, but there are also some en echelon arrays consistent with strike-slip movement.

The Edson Hills system of northeast-southwest compression is postdated by a set of faults and shears marked by zones of rusty alteration; their significance is not yet clear. Calcite fiber striations now indicate major principal stress axes oriented north-northeast–south-southwest, mostly associated with dip-slip thrusting but also with some strike-slip movements. In this system of deformation, dextral strike-slip planes should be oriented northwest, parallel to the structural grain, and sinistral planes northeast to east-northeast. The latter correspond well with the sinistral cross faults at the northern end of the Mount Dolence fault (Fig. 11) and with smaller faults of this type in the Edson Hills–Soholt Peaks area. In the Marble Hills area this deformational system is represented by calcite fiber striations and by a first-phase dextral movement on a north-northwest–trending fault (Spörli and Craddock, Chapter 18, this volume). Some of the striations on the low-angle thrust at Mt. Simmons in the Marble Hills, those oriented north-south to north-northeast, are consistent with this system of deformation.

In the Edson Hills and Soholt Peaks areas still later faults consist of kink planes; the most prominent strike northwest, parallel to the structural grain, and indicate sinistral strike-slip. Dextral strike-slip planes are minor and are oriented approximately east-west. The second phase of a two-phase fault in the Marble Hills (Spörli and Craddock, Chapter 18, this volume) shows sinistral movement. Similar late movement is indicated along northwest-trending bedding planes and faults elsewhere in the Marble Hills. The low-angle fault at Mt. Simmons displays some northwest-southeast slip directions, consistent with this later deformational regime.

SUMMARY

The sequence of events shown in Table 1 can be recognized in the Heritage Range. The deformational structures cannot be divided into distinct groups that indicate separate orogenies or pulses, but rather they seem to represent a continuous succession of events. It has been impossible to establish different ages for the two directions of fold axes recognized (northwest and north-northwest). Therefore, simultaneous cross-folding appears likely.

Many second-phase crenulation cleavages are local effects, formed in necks of boudins after the main cleavage but still under the same strain regime. More work is necessary to determine

Figure 12. Faults in the Collier Hills. Structural data from cross section A shown in the three lower hemisphere equal-area nets at the bottom of the figure. Fault poles connected by partial great circle are part of the same fault zone. Detailed sketch at location x provides evidence for normal movement on fault I. Diagram to the right of it shows fingers (clastic dikes?) of green sandstones in cleaved red argillite. The downward-facing folds may predate the redox processes that caused the red and green colors of the beds.

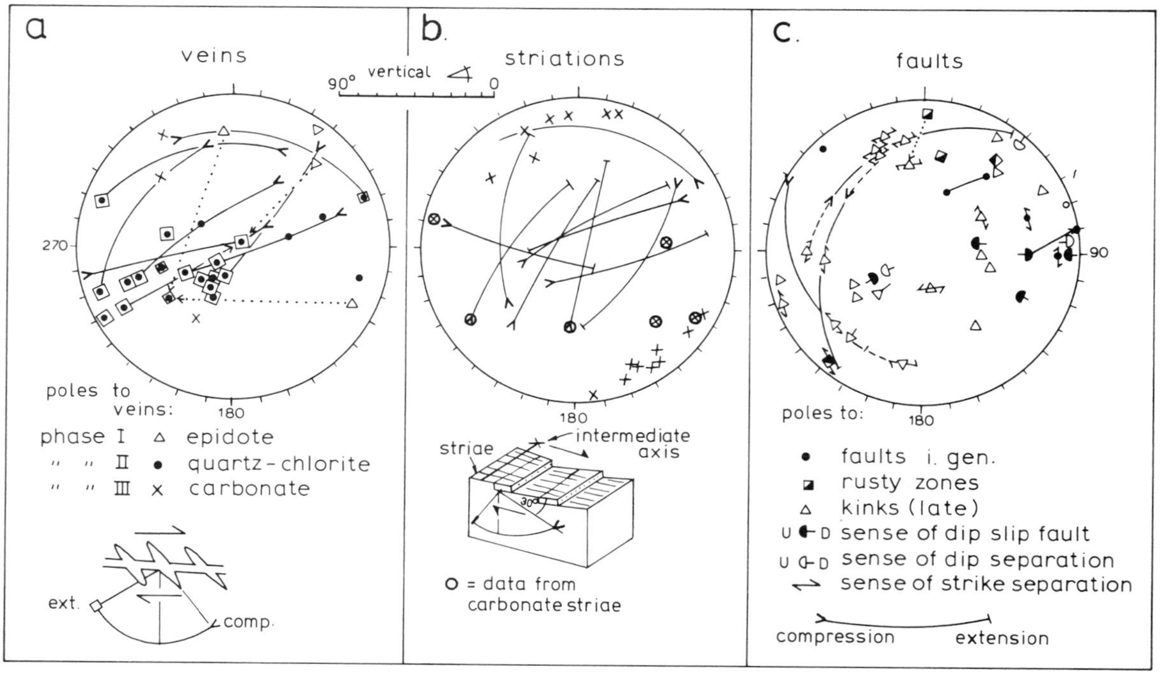

Figure 13. Structural data from veins (a), slickenside striations (b), and faults (c), generally representing deformation in the Heritage Range late in the folding or postdating folding. Does not include data from Marble, Independence, and Patriot hills (see Spörli and Craddock, Chapter 18, this volume). Structural data are shown on lower hemisphere equal-area nets. Straight dotted lines in a indicate age relations of structures. Partial great circles connect compression and extension axes of one vein system (see sketch below net). Epidote veins intersect on a steeply southwest plunging axis; quartz-chlorite veins on a subhorizontal, northwest-trending axis (parallel to fold axes); and carbonate veins on a moderately northwest plunging axis. Partial great circles are used similarly in b and c, and symbols in b are explained in the block diagram of a striated surface below the net. Those data not especially labeled are from quartz chlorite and epidote fibers. Compression and extension axes in c are from conjugate fault couples. Note northeast trending strike-slip compression on late kinks. Smallest great circle segments also connect faults in the same fault zone.

whether the two-phase cleavages reported by Yoshida (1982) from the Wilson Nunataks are due to one deformation or whether they indeed represent two quite different phases.

Fault, en echelon veins, fiber striations, and kink surfaces suggest an anticlockwise rotation of the axes of maximum horizontal principal stress from a position perpendicular to the range, representing the waning stages of the folding process, to one parallel to the tectonic grain and with a lesser horizontal principal stress axis normal to the range (Fig. 14a). Such an exchange of principal stress axes might be explained as a simple consequence of the folding process in the Ellsworth Mountains. Strain rotations of this type have been recognized in the Helvetic nappes of the European Alps (Durney and Ramsay, 1973).

On the other hand, the question arises as to how this rotation of horizontal principal stress axes may relate to a proposed counterclockwise rotation of the Ellsworth Mountains, as postulated by Schopf (1969) and others. We do not yet know the timing of such an event, but a post–Early Jurassic age is most likely.

If a simple model of a rigid, passive block in a sinistral shear system is considered for the postulated rotation of the Ellsworth Mountains, then the principal horizontal stress axis within the block should appear to be rotating in the opposite sense (clockwise) in relation to the Ellsworth Mountains (Fig. 14G). The same result would follow if the rotation was considered to be due to backarc spreading around an apex at the southern end of the Weddell Sea–Ronne Ice Shelf embayment.

If the Ellsworth Mountains have been detached from East Antarctica in the manner suggested by Schopf (1969) and others, the initial movement must have involved rifting away from that

TABLE 1. RELATIVE TIMING OF TECTONIC EVENTS

Down-dip vergent (cascade) folding	----
Main folding (early Mesozoic?)	-------------
Local second-phase folding	--------
Early strike faulting	-------
Down-dip extension boudinage	-------
Stretching lineation formed	--------------
ESE sinistral, NE dextral faulting (NE compression)	--------
N-S compression, corresponding movement on faults	---------
NNW sinistral, E-W dextral faulting (NW compression)	----------
Formation of breccia bodies	--------------------
Formation of epidote veins	----
Formation of quartz-chlorite veins	----------
Iron-staining along faults	---------
Time ------→	

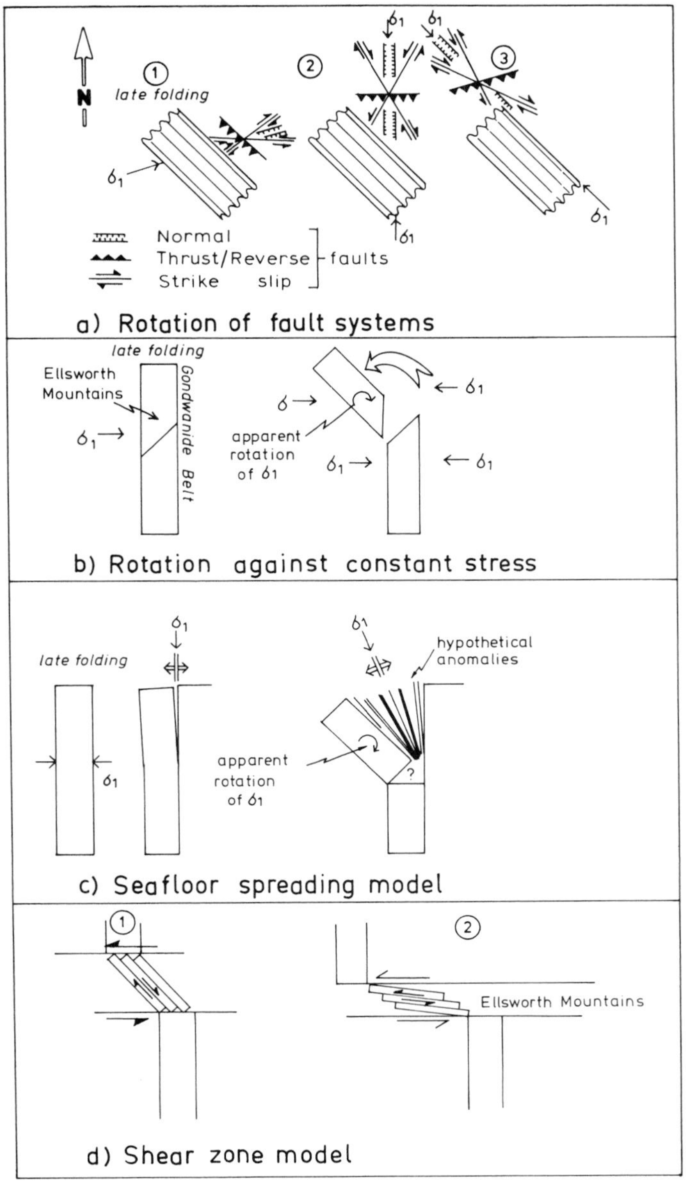

Figure 14. a, rotation of fault systems relative to Heritage Range, as inferred from observations. b, c, are models of how rotation of the Ellsworth Mountains could cause reorientation of stresses responsible for faulting. (Note, however, that if the sequence in a is correct, models b and c produce the wrong sense of rotation.)

continent, as is well documented for many continental fragmentations (Spörli, 1980; Scrutton, 1982). Under such a regime, a change from a σ_1 orientation perpendicular to the fold hinges, representing the waning stages of the folding process, to one with σ_3 perpendicular to the tectonic trend (either from normal or strike-slip faults), representing the rifting phase, could be expected (Fig. 14c). It is unlikely that the pre–reverse fault phase of normal movement recognized on the Mount Dolence Fault in the Collier Hills (Fig. 12) represents this stress change because this normal fault is thought to precede the main folding. However, limited normal fault movement that postdates the faults associated with late stages of folding and that shows northeast-southwest orientations of σ_3 could perhaps be due to this rifting. Furthermore, the late conjugate strike-slip faults with a northwest-southeast maximum horizontal principal stress have their σ_3 oriented in the same direction and could therefore be correlated with this regime of extensional faulting. If this is the case (available field observations do not yet warrant a final conclusion), these faults document the tectonic history of the Ellsworth Mountains up to its separation from East Antarctica but prior to its complete rotation of about 90°.

However, with internal deformation of the block in a simple shear regime (Fig. 14d), it is conceivable that in the early stages of the rotation dextral faults with an orientation consistent with a north-south σ_1 (in terms of modern range orientation) would be active (Fig. 14). On further rotation of the Ellsworth Mountains the sinistral fault set of the conjugate couple would come to lie parallel to the range. However, the angle of rotation required for sinistral slip on faults subparallel to the range would be nearly 90° (Fig. 14).

REFERENCES CITED

Clarkson, P. D., and Brook, M., 1977, Age and position of the Ellsworth Mountains crustal fragment, Antarctica: Nature, v. 265 (5595), p. 615–616.

Craddock, C., Anderson, J. J., and Webers, G. F., 1964, Geologic outline of the Ellsworth Mountains, *in* Adie, R. J., ed., Antarctic geology: Amsterdam, North Holland Publishing Co., p. 155–170.

Dalziel, I.W.D., and Elliot, D. H., 1982, West Antarctica: Problem child of Gondwanaland: Tectonics, v. 1, p. 3–19.

Durney, D. W., and Ramsay, J. G., 1973, Incremental strains measured by syntectonic crystal growths, *in* De Jong, K. A., and Scholten, R., eds., Gravity and tectonics: New York, Wiley, p. 67–96.

Ford, A. B., 1972, Weddell Orogeny—latest Permian to early Mesozoic deformation of the Weddell Sea margin of the Transantarctic Mountains, *in* Adie, R. J., ed., Antarctic geology and geophysics: Oslo, Universitetsforlaget, p. 419–425.

Ramsay, J. G., 1974, Development of chevron folds: Geological Society of America Bulletin, v. 85, p. 1741–1754.

Schmidt, D. L., and Rowley, P. D., 1986, Continental rifting and transform faulting along the Jurassic Transantarctic rift: Tectonics, v. 5, p. 279–291.

Schopft, J. M., 1969, Ellsworth Mountains: Position in West Antarctica due to sea-floor spreading: Science, v. 164, p. 63–66.

Scrutton, R. A., 1982, Passive continental margins: A review of observations and mechanisms, *in* Scrutton, R. A., ed., Dynamics of passive margins: American Geophysical Union Geodynamic Series, Vol. 6, p. 5–11.

Spörli, K. B., 1980, New Zealand and oblique-slip margins: Tectonic development up to and during the Cenozoic, *in* Ballance, P. F., and Reading, H. G., eds., Sedimentation in oblique-slip mobile zones, International Association of Sedimentologists Special Publication 4, p. 147–170.

Yoshida, M., 1982, Superposed deformation and its implication to the geologic history of the Ellsworth Mountains, West Antarctica: Tokyo, Memoirs National Institute of Polar Research, Special Issue 21, p. 120–171.

Manuscript Accepted by the Society January 31, 1992

Printed in U.S.A.

Geological Society of America
Memoir 170
1992

Chapter 21

Structure of the Sentinel Range, Ellsworth Mountains, West Antarctica

Campbell Craddock
Department of Geology and Geophysics, University of Wisconsin, Madison, Wisconsin 53706
K. B. Spörli
Department of Geology, University of Auckland, Auckland, New Zealand
John J. Anderson
Department of Geology, Kent State University, Kent, Ohio 44242

ABSTRACT

The 5,500+-m Paleozoic succession of the Sentinel Range includes the Heritage Group (Cambrian), the Crashsite Group (Cambrian and younger, including Devonian), the Whiteout Conglomerate (Carboniferous?), and the Polarstar Formation (Permian). All these predominantly clastic rocks were strongly folded during the early Mesozoic (?) Ellsworth (Gondwanide) Orogeny, and the folds formed mainly by a flexural-slip mechanism. Axial planar cleavage is abundant throughout the range; it may be a solution cleavage in the massive diamictite beds of the Whiteout Conglomerate. One major, eastward-dipping reverse (?) fault cuts Crashsite quartzite beds along the west side of range. The Sentinel Range succession was subjected to mild burial metamorphism, and later dynamothermal metamorphism reached lower greenschist facies. The folds in the range plunge gently northward, and the entire Sentinel Range block may be slightly tilted eastward. Stratigraphic similarities with rocks of the Transantarctic Mountains make it likely that the Ellsworth Mountains have moved and rotated away from the margin of East Antarctica, but the sense of such a rotation cannot be determined at present. In addition, similarities exist between the Sentinel Range younger rocks and those in the Cape Fold Belt of southern Africa, as foretold by Du Toit (1937), but the Cambrian System of the Sentinel Range lacks a definite counterpart in Africa.

INTRODUCTION

The Ellsworth Mountains lie in the interior of West Antarctica, centered at 79°S, 85°W (Fig. 1). They were discovered in November 1935 by Lincoln Ellsworth, who observed their northern part from a distance during his pioneering flight across West Antarctica (Ellsworth, 1936). These mountains are of special interest because (1) they display a discordant north-south trend compared to the Transantarctic Mountains, and (2) they are the largest area of rock exposure between the dissimilar tectonic provinces of the Transantarctic Mountains and the ranges of coastal West Antarctica. They show stratigraphic similarities to the former and some structural affinities with the latter province.

The Ellsworth Mountains are divided by the east-flowing Minnesota Glacier into the Sentinel Range to the north and the Heritage Range to the south (Fig. 2). The Sentinel Range extends for almost 210 km in a north-northwest direction and has a maximum width of about 82 km. The most prominent landscape feature in the range is a massive, asymmetric mountain block (Vinson Massif) with a steep escarpment on the west and a network of crevassed glaciers, generally narrow ridges, and steadily lower peaks on the east. Crowning the western escarpment are some peaks with elevations greater than 5,000 m above sea level, the highest in Antarctica.

The purpose of this chapter is to report what is known about the geologic structure of the Sentinel Range and to speculate

Craddock, C., Spörli, K. B., and Anderson, J. J., 1992, Structure of the Sentinel Range, Ellsworth Mountains, West Antarctica, *in* Webers, G. F., Craddock, C., and Splettstoesser, J. F., eds., Geology and Paleontology of the Ellsworth Mountains, West Antarctica: Boulder, Colorado, Geological Society of America Memoir 170.

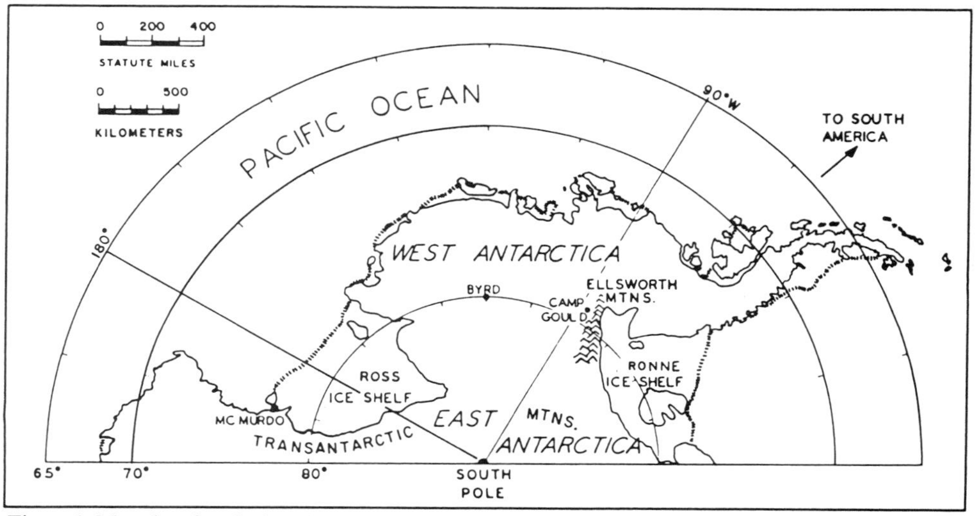

Figure 1. Map showing the location of the Ellsworth Mountains in West Antarctica. The Sentinel Range comprises the northern half of the Ellsworth Mountains. Reprinted by permission from Nature, v. 201, p. 174–175. Copyright © 1964 Macmillan Magazines Ltd.

about its tectonic significance. The data were acquired by expeditions from the University of Minnesota in 1961–1962, 1962–1963, and 1963–1964. The first two expeditions used snowmobiles for field travel; the third had both snowmobiles and helicopter support. The distribution of field observations is irregular; most are from the northern and southern ends, the lower west side, and the southern part of the eastern side of the range. This situation results from the steepness of many of the high peaks, the difficult terrain of the remote eastern side of the range, height and distance limits of the helicopters, unstable weather conditions, and crevasse fields. However, the eastern side of the range is known to a degree from photographs, contact tracing by helicopter, and strategic landings. Some information about the high peaks was gathered during the 1966–1967 U.S. mountaineering expedition (Evans, 1973). The 1979–1980 U.S. scientific expedition to the Ellsworth mountains had helicopter support, but geologic work was carried out almost solely in the Heritage Range.

STRATIGRAPHY

The rocks exposed in the Sentinel Range are nearly all clastic sedimentary rocks that have been folded and slightly metamorphosed. The total succession is more than 5,500 m in thickness, and no definite unconformity has been discovered. Fossils are rare in the Sentinel Range, but they show that the oldest unit is Cambrian and the youngest Permian. The Paleozoic formations of the Gondwana sequence in the Transantarctic Mountains to the south resemble the upper formations in the Sentinel Range, but the latter are thicker. Only a brief summary of Sentinel Range stratigraphy is included here. Detailed descriptions of these formations are given in other chapters in this volume, and the areal distributions of rock units can be seen on the Geologic Map of the Ellsworth Mountains (Plate I, in pocket).

Heritage Group

The oldest rocks known in the Sentinel Range are assigned to the Heritage Group. They crop out in the Sentinel Range only in the vicinity of the Nimitz Glacier in the south, although they are widely exposed southward in the Heritage Range. They consist of a few hundred meters of diverse sedimentary and igneous rocks. These beds may include the only exposures of carbonate rocks north of the Minnesota Glacier, which separates the two ranges.

Crashsite Group

The type area for this group is the Howard Nunataks, site of a plane crash in November 1961, during the arrival of our first expedition. This group is more than 3,000 m thick, and it is widely distributed throughout the western part of the Sentinel Range. It occurs in the highest peaks, where the beds are usually tightly folded (Fig. 3) or locally homoclinal in structure (Fig. 4). Most of the layers are resistant quartzite in beds as thick as 6 m; some beds of original siltstone and shale are also present. The Crashsite Group has been divided into three formations and three members that can be recognized on the basis of the weathered color.

Whiteout Conglomerate

The type area for this formation is at and near Whiteout Nunatak south of Mt. Wyatt Earp in the northern end of the range. This unit crops out in the eastern and northern parts of the range, and its thickness is about 1,000 m. Most of the unit consists of massive beds of medium gray-weathering diamictite with abundant clasts of diverse sizes, shapes, orientations, and compositions.

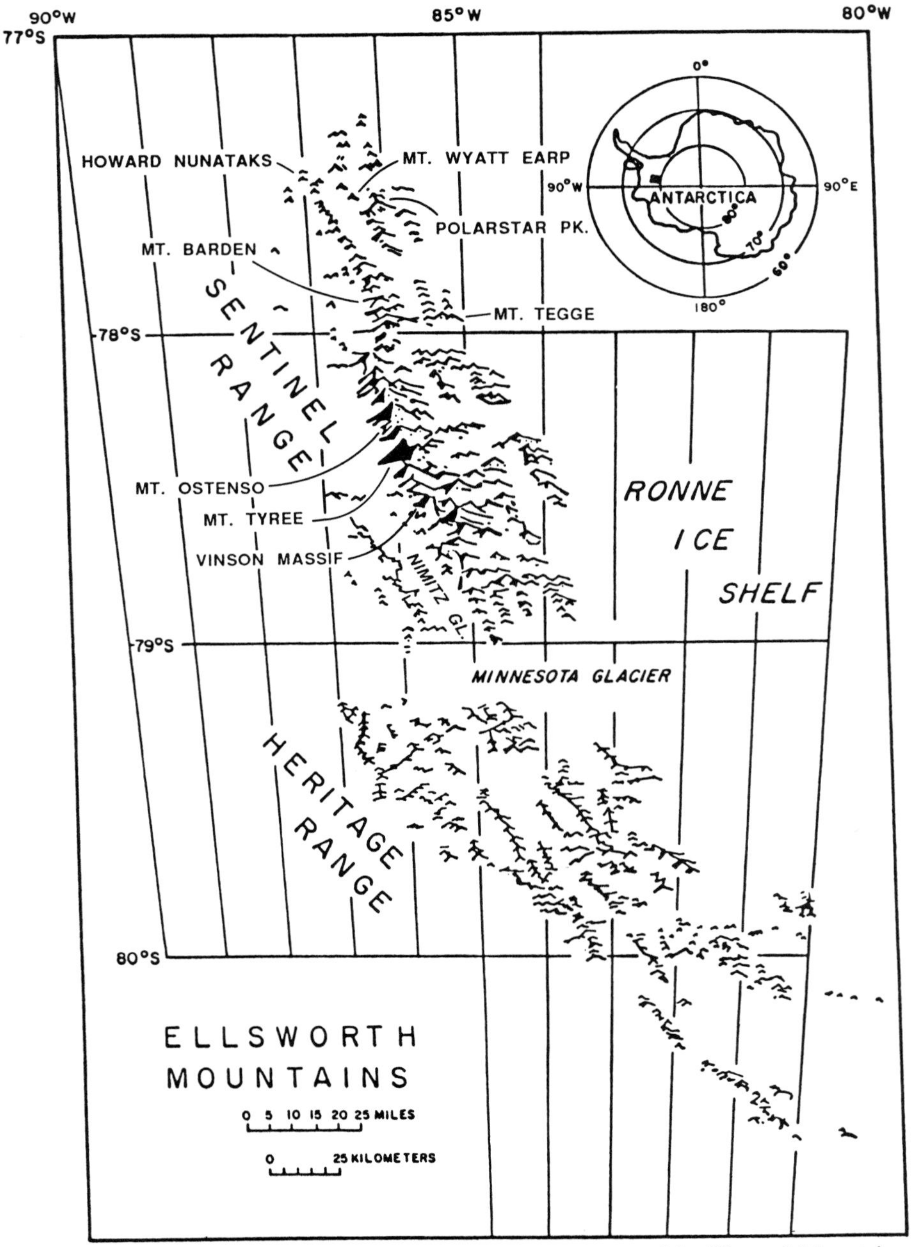

Figure 2. Map showing the geography and significant place-names of the Ellsworth Mountains.

Because of these very thick diamictites, the formation tended to behave as a single mechanical unit during deformation. Bedding in this formation is defined by a few slate beds as thick as 12 m.

Polarstar Formation

This formation includes the youngest beds that are exposed in the Sentinel Range; it occurs at the northern end and in the northern part of the eastern side of the range. The type area is in the vicinity of Polarstar Peak, and on this summit there occur *Glossopteris* leaves of Permian age. This formation consists mainly of graywacke in beds as thick as 6 m, sandstone, and slate, all typically medium to dark gray as seen from a distance. The upper part of the formation contains beds of coal 150 cm or less in thickness. No upper contact has been found, but the exposed succession has been measured—and estimated across snow cover—to be more than 1,300 m in thickness.

Figure 3. Photograph of Mount Tyree from Mount Gardner. The chevron folds occur in rocks of the Crashsite Group. View to the southeast. Photograph by John P. Evans.

Figure 4. Photograph of Mount Ostenso showing quartzite beds of the Crashsite Group. Figures on the peak show the scale. View to the northwest. Photograph by John P. Evans.

STRUCTURAL GEOLOGY

The thick succession of layered rocks in the Sentinel Range has been deformed by lateral compression into a variety of folds, with widespread cleavage lying mainly parallel to the axial planes. Only a few faults have been found in the Sentinel Range despite the pervasiveness of folds. Descriptions of these and other structural elements follow. Available structural data have been plotted on a summary structural map (Plate III, Fig. 5, in pocket).

Folds

Folds are the most prominent structural feature, and they are present in nearly every sizable outcrop on one scale or another. The range as a whole consists of an anticlinorium on the west and a synclinorium on the east (see Plate I, sections A-A′, B-B′); both plunge gently northward. Fabric diagrams of poles to bedding (Fig. 6) show that bedding and fold axis trends change gradually from near-northwesterly in the south to north-northwesterly in the north. The bedding-cleavage intersection lineations in Figure 6 confirm this range of fold axis orientations

Most of the folds show parallel fold geometry, and they have formed by the flexural-slip mechanism. In terms of appression of the limbs, folds range from open to tight. Fold styles range from rounded to chevron, and symmetric, asymmetric, and overturned fold geometries have been observed. Numerous cross-beds and ripple marks help to establish the younging direction of beds.

The style and size of folds vary from one stratigraphic unit to another. In the Crashsite Group some large folds with wavelengths of 3 km or more display a rounded, asymmetric profile (Fig. 7). Smaller folds in Crashsite strata are closed to tight, tend toward chevron style, and commonly have wavelengths in the 15- to 200-m range (Figs. 8, 9). The Whiteout Conglomerate behaves as a single massive layer in gentle or open folds (Fig. 10),

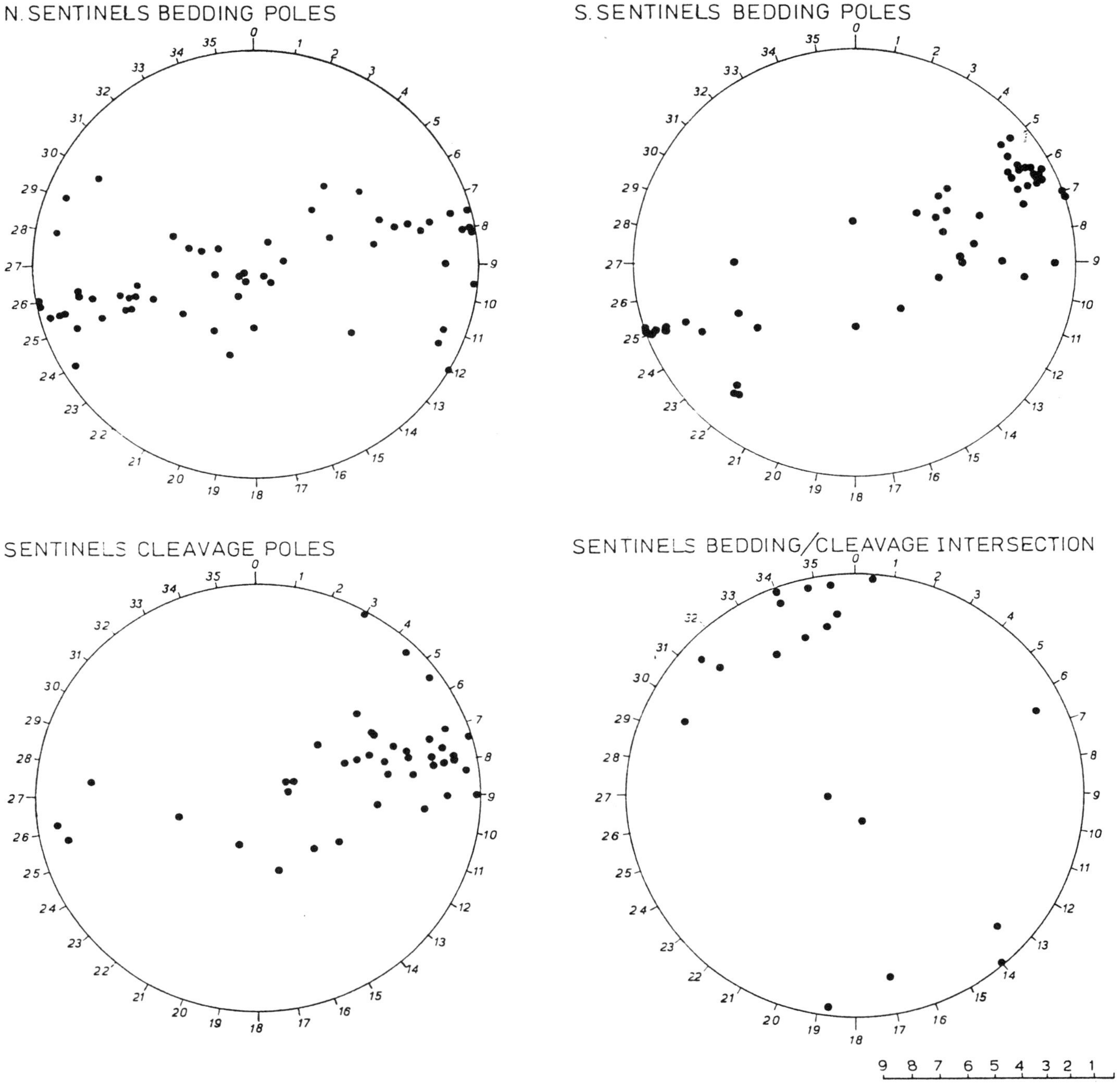

Figure 6. Lower hemisphere equal-area fabric diagrams of structural measurements from the Sentinel Range.

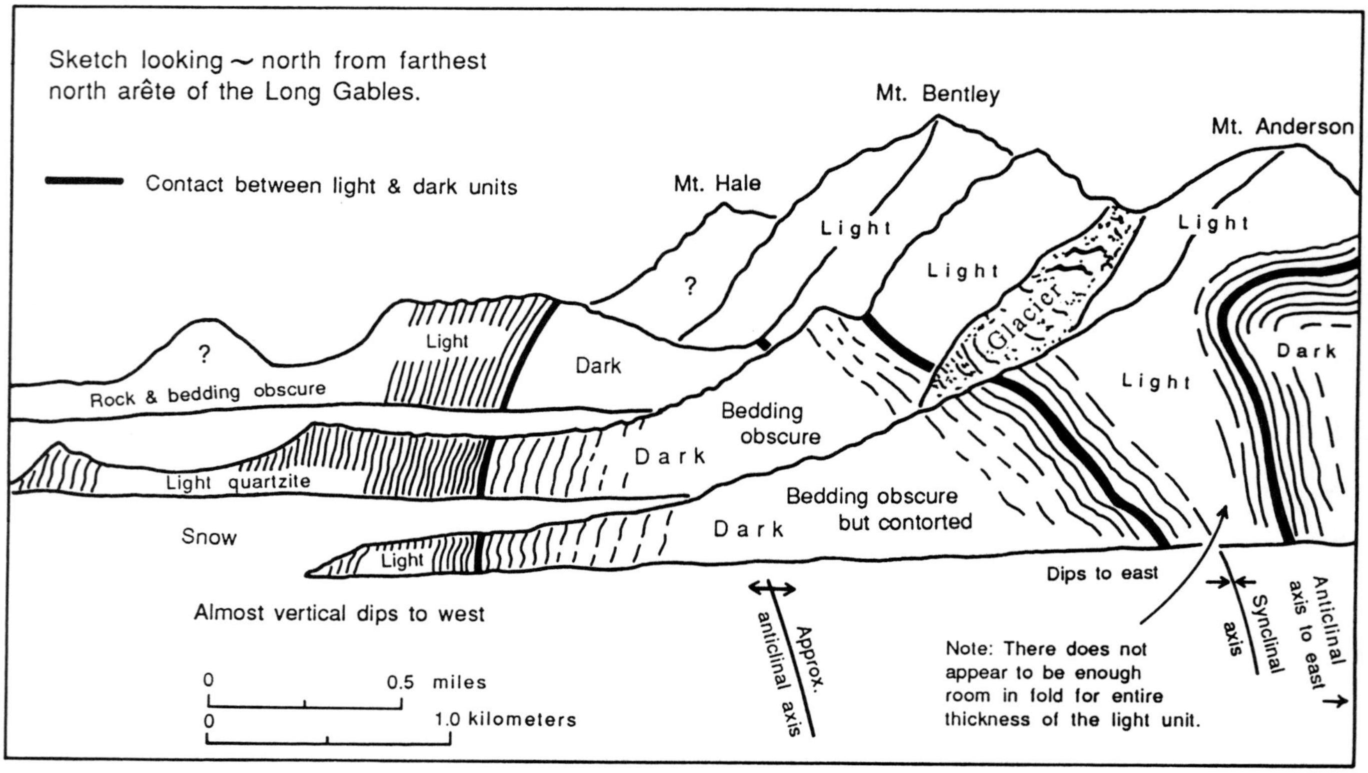

Figure 7. Notebook sketch of large folds in Crashsite Group beds along the west side of the Sentinel Range. Sketch by John J. Anderson.

Figure 8. Photograph from helicopter of folds in Crashsite Group quartzites, southeast ridge, Vinson Massif. View to north-northwest. Photograph by T. W. Bastien.

producing a fold with a wavelength of 5 km or more. Locally it has also been sharply bent into an overturned anticline (Fig. 11). Individual beds in the Polarstar Formation are 6 m or less in thickness, but observed folds have wavelengths of 800 to 1,600 m. Spectacular overturned folds occur on the northwest ridge of Polarstar Peak (Fig. 12).

The axial planes of most folds have westward dips, most at angles of 65° or more. There are, however, some folds with axial planes that dip in an eastward direction (Fig. 7).

Bedding plane striations

Many outcrops of quartzite show well-developed bedding plane striations. In localities where quartz-chlorite veins lie parallel to bedding these striations are especially prominent, and grooves with a relief of more than 1 mm may be present. These striations trend perpendicular to the local fold axis and testify to the flexural-slip mechanism of folding.

Rock cleavage

Layered rocks throughout the range commonly display secondary cleavage that lies parallel to the axial planes of individual folds. In quartzite and graywacke beds cleavage is generally

Figure 9. Photograph of chevron folds in Crashsite Group quartzite beds. West face of Mount Tyree. View to south; person for scale. Photograph by John P. Evans.

Figure 10. Photograph of anticline in Whiteout Conglomerate, here about 800 m in exposed thickness. View to SSE from Mt. Wyatt Earp.

Figure 11. Photograph of anticline with Whiteout Conglomerate (dark) and Crashsite Group beds (light). View to SE across Embree Glacier from Mount Tegge. Distance from lowest exposed rocks at base of photograph to highest exposed rock near center of image about 250 m.

Figure 12. Photograph of overturned folds in Polarstar Formation, northwest ridge of Polarstar Peak. View to NNW.

poorly developed or absent, but original shales have well-developed slaty cleavage that weathers into chips and pencils. Argillaceous beds in the Crashsite Group are slates or phyllites, and cleavage plane surfaces are commonly dark greenish-gray with a waxy luster. Laminated shales within the Whiteout Conglomerate have slaty cleavage, and even the massive diamictite units display pronounced cleavage (Fig. 13). In hand specimens the spacing of cleavage surfaces is 1 mm or less, and the non-planar character of some surfaces suggests that the foliation may be a solution cleavage.

Fractures

Many outcrops display abundant joints, commonly with spacings between 30 cm and 3 m. Measured joints strike N 60–90°E and dip 80–90°S. On a small peak near the northern end of the range conjugate shear fractures and joints are both present, all dipping about 90°; the joints strike about east-west, perpendicular to the strike of the steeply eastward-dipping bedding.

Faults

Despite the intensive folding of the strata in the Sentinel Range, only a few faults have been found; most of these have small displacements of 1 m or less. The only fault with a large displacement presently known is in an unnamed nunatak about 12 km west of Mount Barden (Fig. 14). This fault has a breccia zone about 3 m in width, and the fault is considered to be a reverse fault on the basis of the stratigraphic offset.

The Sentinel Range block appears to be slightly tilted toward the east. This tilt could be related to some unseen major fault along the western side of the range.

Deformed clasts

At Whiteout Nunatak and a nearby locality cleavage surfaces in the Whiteout Conglomerate show clasts that are some-

Figure 13. Photograph of Whiteout Nunatak showing contact between Whiteout Conglomerate (below) and Polarstar Formation (above) and well-developed cleavage in both formations. Figure on left skyline gives scale.

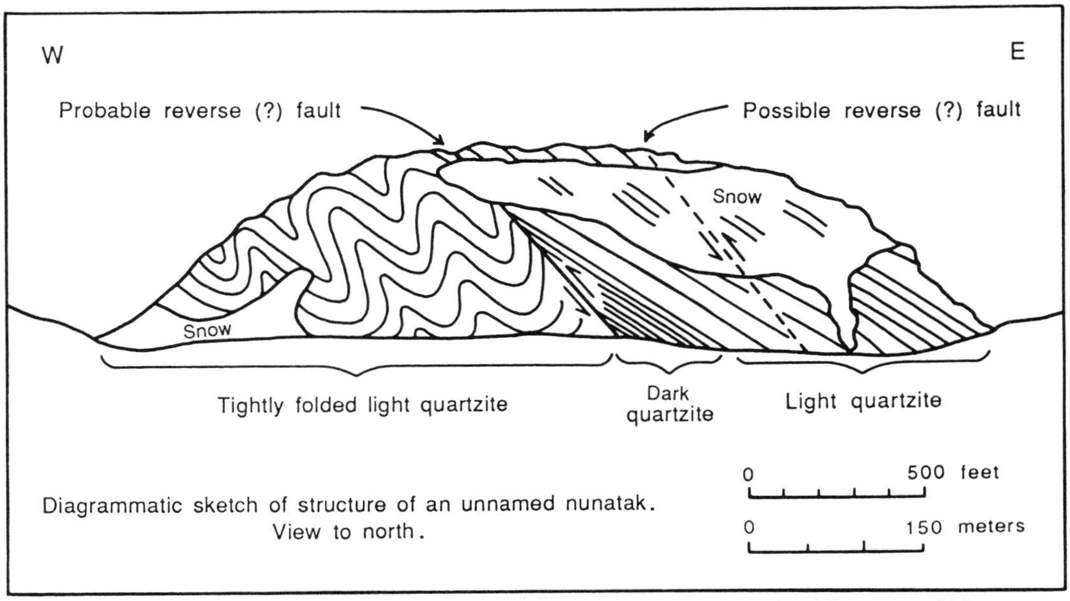

Figure 14. Notebook sketch showing reverse (?) fault in unnamed nunatak 12 km west of Mount Barden. Bedding thickness not to scale. Relief of nunatak about 200 m. Sketch by John J. Anderson.

what elongated in the cleavage dip direction, or about vertically. No measurements were attempted. One clast shows several extension fractures perpendicular to the long axis. Similar subvertical clast elongations have been observed in older rocks in the Heritage Range (Spörli and Craddock, this volume).

METAMORPHISM

Bauer (this volume) describes burial metamorphism that is recorded in mafic igneous rocks in the Heritage Range, and Castle (1974) concluded that laumontite-facies metamorphism of the Polarstar Formation also represents an early burial metamorphism. In addition, all the rocks of the Sentinel Range commonly contain veins of quartz as well as chlorite and/or epidote; many of these veins follow bedding or fractures, but some definitely postdate the formation of cleavage. Coals in the Polarstar Formation are of medium- to high-rank bituminous grade. In the slaty beds of the Crashsite Group flakes of chlorite have grown on the cleavage planes. Although no regional study of metamorphism in the Sentinel Range has been carried out, it seems likely that regional metamorphism associated with the early Mesozoic (?) deformational event reached lower greenschist facies, at least in the lower stratigraphic units.

TECTONIC INTERPRETATION

The stratigraphic succession in the Sentinel Range is thicker than Paleozoic successions in the Transantarctic Mountains, and the Sentinel Range strata are mainly or entirely of marine origin. Based on regional geology it is likely, though not definite, that the Sentinel Range beds were deposited along the margin of East Antarctica on a subsiding continental shelf. The Sentinel succession was compressed and deformed in early Mesozoic (?) time; the Permian Polarstar Formation is deformed, and granitic intrusions with ages of about 175 Ma (middle Jurassic) in the ranges south of the Ellsworth Mountains appear to be posttectonic. The vergence direction during the Sentinel deformation was probably eastward, but some cleavage orientations and fold shapes vary from the norm (Figs. 6, 7).

Little information has been available about the time of uplift of the Sentinel Range, but a recent paper by Fitzgerald and Stump (1991) provides welcome insight on the problem. They collected apatite-bearing quartzite specimens from the Vinson Massif area through a vertical range of about 4.2 km and then carried out fission-track analysis on apatite grains. They conclude that uplift of at least 4 km occurred during some 20 m.y. in Early Cretaceous time and that at least 1.8 km of relief has persisted since then.

Because of compositional similarity between the younger Sentinel Range stratigraphic units and correlative units in the Transantarctic Mountains, it is reasonable to postulate that the Ellsworth Mountains moved to their present position from an original location closer to East Antarctica. Such movement would also require some rotation because the present trend of the Ellsworth Mountains, at least in the north, is about perpendicular to the trend of the Transantarctic Mountains and the structures there. The arcuate shape of the Ellsworth Mountains may be a primary feature, but it may imply some bending during rotation. Schopf (1969) and others since have argued for a model of counterclockwise rotation through the area of the present Weddell Sea and shelf. Schmidt and Rowley (1986), however, suggest a model involving clockwise rotation of the Ellsworth Mountains away from a position to the southwest of their present position. At present there is no compelling evidence with which to answer the question of rotational direction, if such rotation has occurred.

The Ellsworth Mountains may represent a continuation of the Cape Fold Belt of southern Africa, as predicted by Du Toit (1937) with no knowledge of their geology. There are many similarities between the upper Paleozoic strata of the two areas, and the times and styles of deformation are comparable; however, the thick Cambrian System in the Ellsworth Mountains seems to lack a certain counterpart in the Cape Fold Belt.

The Ellsworth Mountains offer many interesting problems for future workers, including the following:

1. Did these mountains move from their original location? If so,what are the history and direction of that movement?

2. Is the arcuate map shape of the Ellsworth Mountains primary, or was there tectonic bending?

3. What is the complete uplift history of these mountains? Was any uplift related to rotation and/or bending of the mountain chain?

4. How are the Ellsworth Mountains related to coastal West Antarctica and the Antarctic Peninsula? The recent discoveries of *Glossopteris* in Ellsworth Land to the north (Laudon and others, 1987) and fossiliferous Cambrian limestone boulders on King George Island off the Antarctic Peninsula (Wrona, 1989) are exciting clues to this question.

ACKNOWLEDGMENTS

We thank the Division of Polar Programs, National Science Foundation, for grants in support of our work, and the U.S. Army and the U.S. Navy for the necessary logistic support. Many other members of the expeditions provided field data and analysis, and we are especially grateful for their discussions of problems. We thank the National Geographic Society and John P. Evans for the photographs from the 1966–1967 American Antarctic Mountaineering Expedition and T. W. Bastien for compiling the structural map of the Sentinel Range. T. S. Laudon and P. R. Farnham provided useful reviews.

REFERENCES CITED

Castle, J. W., 1974, The deposition and metamorphism of the Polarstar Formation (Permian), Ellsworth Mountains, Antarctica [M.S. thesis]: University of Wisconsin–Madison, 102 p.

Du Toit, A. L., 1937, Our wandering continents; an hypothesis of continental drifting: Edinburgh, Oliver and Boyd, 366 p.

Ellsworth, L., 1936, My flight across Antarctica: National Geographic Magazine, vol. 70, p. 1–35.

Evans, J. P., 1973, Scientific aspects of the American Antarctic Mountaineering Expedition: National Geographic Society Research Reports, 1966 Projects, p. 113–115.

Fitzgerald, P. G., and Stump, E., 1991, Early Cretaceous uplift in the Ellsworth Mountains of West Antarctica: Science, vol. 254, p. 92–94.

Laudon, T. S., Lidke, D. J., Delevoryas, T., and Gee, C. T., 1987, Sedimentary rocks of the English Coast, eastern Ellsworth Land, Antarctica, *in* McKenzie, G. D., ed., Gondwana six: Structure, tectonics, and geophysics: American Geophysical Union Monograph 40, p. 183–189.

Schmidt, D. L., and Rowley, P. D., 1986, Continental rifting and transform faulting along the Jurassic Transantarctic Rift, Antarctica: Tectonics, vol. 5, p. 279–291.

Schopf, J. M., 1969, Ellsworth Mountains: Position in west Antarctica due to sea-floor spreading, Science, vol. 164, p. 63–66.

Wrona, R., 1989, Cambrian limestone erratics in the Tertiary glacio-marine sediments of King George Island, West Antarctica: Polish Polar Research, vol. 10, p. 533–553.

MANUSCRIPT ACCEPTED BY THE SOCIETY JANUARY 30, 1992

Printed in U.S.A.

Geological Society of America
Memoir 170
1992

Chapter 22

Glacial history of the Ellsworth Mountains, West Antarctica

George H. Denton
Department of Geological Sciences and Institute for Quaternary Studies, University of Maine, Orono, Maine 04469
James G. Bockheim
Department of Soil Science, 1525 Observatory Drive, University of Wisconsin, Madison, Wisconsin 53706
Robert H. Rutford
Department of Geosciences, University of Texas, Dallas, Texas 75083
Björn G. Andersen
Department of Geology, University of Oslo, Oslo, Norway

ABSTRACT

The West Antarctic Ice Sheet flows seaward around and through the Ellsworth Mountains to feed the Ronne-Filchner Ice Shelf. A high ice-sheet surface, featuring a major inland divide, abuts the western mountain flank. The present-day grounding line of the Ronne-Filchner Ice Shelf is near the eastern mountain flank.

Two major erosion glacial features characterize the Ellsworth Mountains. First, the exposed mountains show classic features of alpine glacier erosion. These are best developed in the Sentinel Range of the northern Ellsworth Mountains where cirque erosion has left horns, arêtes, and sharp spurs. Only the summit plateau of Vinson Massif (4,897 m) remains as undulating, pre-alpine topography. Well-developed alpine landforms also mark the Heritage Range of the southern Ellsworth Mountains. The West Antarctic Ice Sheet today engulfs many of the lower ridges of the Ellsworth Mountains. Second, a glacial trimline is etched into alpine ridges and spurs throughout the Ellsworth Mountains. Elevations of this trimline show a remarkably consistent pattern. In the Sentinel Range, trimline elevations generally diminish along the range north and south of where the high inland ice divide today abuts the mountains; they also decline from west to east across the range. Trimline elevations in the Heritage Range likewise show a consistent pattern, with higher values near the central, inland flank of the range and lower values to the east and south. Bedrock ridges above the trimline are serrated, whereas those below the trimline lack serrations, and some show glacial polish and striations. Drift patches and erratics occur below the trimline.

We infer that extensive alpine glacial erosion of the Ellsworth Mountains antedated etching of the trimline. The trimline and striations together show major thickening of the West Antarctic Ice Sheet subsequent to this alpine erosion. Thicker West Antarctic ice flowed seaward around and through the Ellsworth Mountains. The major divide on the inland ice sheet maintained its present position and was 400 to 650 m higher than at present (ignoring isostatic compensation). Ice-surface elevations increased 1,300 to 1,900 m along the eastern mountain flank near the ice shelf (ignoring isostatic compensa-

Denton, G. H., Bockheim, J. G., Rutford, R. H., and Andersen, B. G., 1992, Glacial history of the Ellsworth Mountains, West Antarctica, *in* Webers, G. F., Craddock, C., and Splettstoesser, J. F., Geology and Paleontology of the Ellsworth Mountains, West Antarctica: Boulder, Colorado, Geological Society of America Memoir 170.

tion). We infer that substantial grounding of West Antarctic ice occurred in the Weddell Sea embayment when the West Antarctic Ice Sheet stood at the trimline.

We propose two widely opposing age models for ice-sheet expansion to the trimline. In one model the expansion is late Wisconsin/Holocene in age. In the other model the expansion is pre–late Quaternary, and perhaps Tertiary, in age.

INTRODUCTION

General statement

The late Quaternary history of the marine ice sheet in West Antarctica has been debated vigorously for several decades. Denton and Armstrong (1968), Mercer (1968), Denton and others (1971), Stuiver and others (1981), Kellogg and others (1979), and Hughes and others (1985) argued from geologic evidence that the West Antarctic Ice Sheet has repeatedly grounded over the peripheral continental shelf; the greatest expansion occurred in the Ross Sea and Weddell Sea embayments (Fig. 1). They followed Hollin (1962) in attributing these fluctuations to eustatic sea-level variations driven by waxing and waning of Northern Hemisphere ice sheets. Stuiver and others (1981, p. 376) presented the CLIMAP three-dimensional reconstruction of an expanded West Antarctic Ice Sheet during the last (late Wisconsin) glacial maximum, based on geologic data available in 1978. Hughes and others (1985) updated this reconstruction on the basis of additional geologic control points and an improved glaciological model. In sharp contrast, Mayewski (1975) and Fillon (1974, 1975) argued from geologic evidence in the central Transantarctic Mountains and the Ross Sea, respectively, that the West Antarctic Ice Sheet did not expand significantly into the Ross Sea embayment during Quaternary time. Whillans (1976) suggested from internal layering that the thickening of interior West Antarctic ice has not changed significantly for the past 30,000 years. While accepting earlier Quaternary expansions of grounded ice into the Ross Sea embayment, Drewry (1979) inferred that late Wisconsin sea-level depression was not sufficient in magnitude or duration to cause extensive grounding in the Ross Sea, and he interpreted available field data accordingly.

The pre–late Quaternary history of the Antarctic Ice Sheet is, likewise, a matter of considerable debate and is undergoing fundamental revisions (Webb and others, 1984; Denton and others, 1984, 1991). Drastic ice-volume fluctuations, ranging from mountain overriding to extensive collapse, have been proposed for the East Antarctic Ice Sheet. Whether the West Antarctic Ice Sheet was included in such events is not yet known with any certainty.

Ellsworth Mountains topography and present-day glaciers

The Ellsworth Mountains occur near the present-day grounding line between the West Antarctic Ice Sheet and the Ronne-Filchner Ice Shelf (Figs. 1 and 2) and are thus uniquely situated to record past surface levels of expanded West Antarctic ice flowing into the Weddell Sea embayment. Hence, geologic field data from the Ellsworth Mountains potentially bear on both late Quaternary and late Tertiary history of the Antarctic Ice Sheet. To gather these field data we mapped glacial erosional and depositional features as part of the Ellsworth Mountains Project of 1979–1980. One objective was to relate local glacial history to overall fluctuations of the West Antarctic Ice Sheet and hence to test the varying scenarios of ice-sheet history; another was to study local soil development to derive relative ages for drift.

The Ellsworth Mountains extend for 350 km with a NNW-SSE trend and are bounded by longitudes 78° and 87°W and latitudes 80°30′ and 77°15′S (Plate II, Fig. 3, in pocket). The eastward flowing Minnesota Glacier at latitude 79°04′S separates the mountains into the Sentinel Range in the north and the Heritage Range in the south.

The Sentinel Range is 220 km long and about 55 km wide. Extensive NNW-SSE longitudinal ridges, and short ENE-WSW transverse ridges dominate the topography. A prominent longitudinal ridge, which forms the high western margin of the range, is a 50-km-long continuous barrier to seaward flow of inland ice. This high ridge, with a steep western scarp, causes the range to have asymmetric topography. Arrayed along the narrow and continuous ridge crest are numerous peaks exceeding 3,500 m; these include Mount Bentley, Mount Giovinetto, Mount Shinn, and Vinson Massif, which is the highest peak in Antarctica at 4,897 m (Knutzen and others, 1981). Somewhat lower peaks occur along eastern ridges of the Sentinel Range.

The ice cover in the Sentinel Range lies almost entirely within accumulation areas. There are few blue-ice ablation areas, apparently because katabatic winds from the inland ice sheet do not sweep through this high range (Rutford, 1969). An extensive system of cirque-headed, egress, and compound mountain glaciers drains into the Rutford Ice Stream from the eastern side of the Sentinel Range (Rutford, 1969). Extensive compound mountain glaciers, such as the Ellen Glacier, flow eastward and northeastward for 40 km toward the Rutford Ice Stream from cirques on the western ridge. Numerous local cirque-headed and short egress glaciers feed these compound mountain glaciers. The entire local glacier complex in the eastern side of the range receives sufficient snowfall to afford a nearly complete snow and/or ice blanket over all but the steepest ridges and faces, which therefore form the only ice-free areas. The west-facing cirques on the dominant western ridge feed short glaciers that flow directly inland, where they are deflected by inland ice either southeastward into the Nimitz Glacier, which flows around the southern end of the range, or northward around the range as eventual inflow into the Rutford Ice Stream. Near the central portion of this western

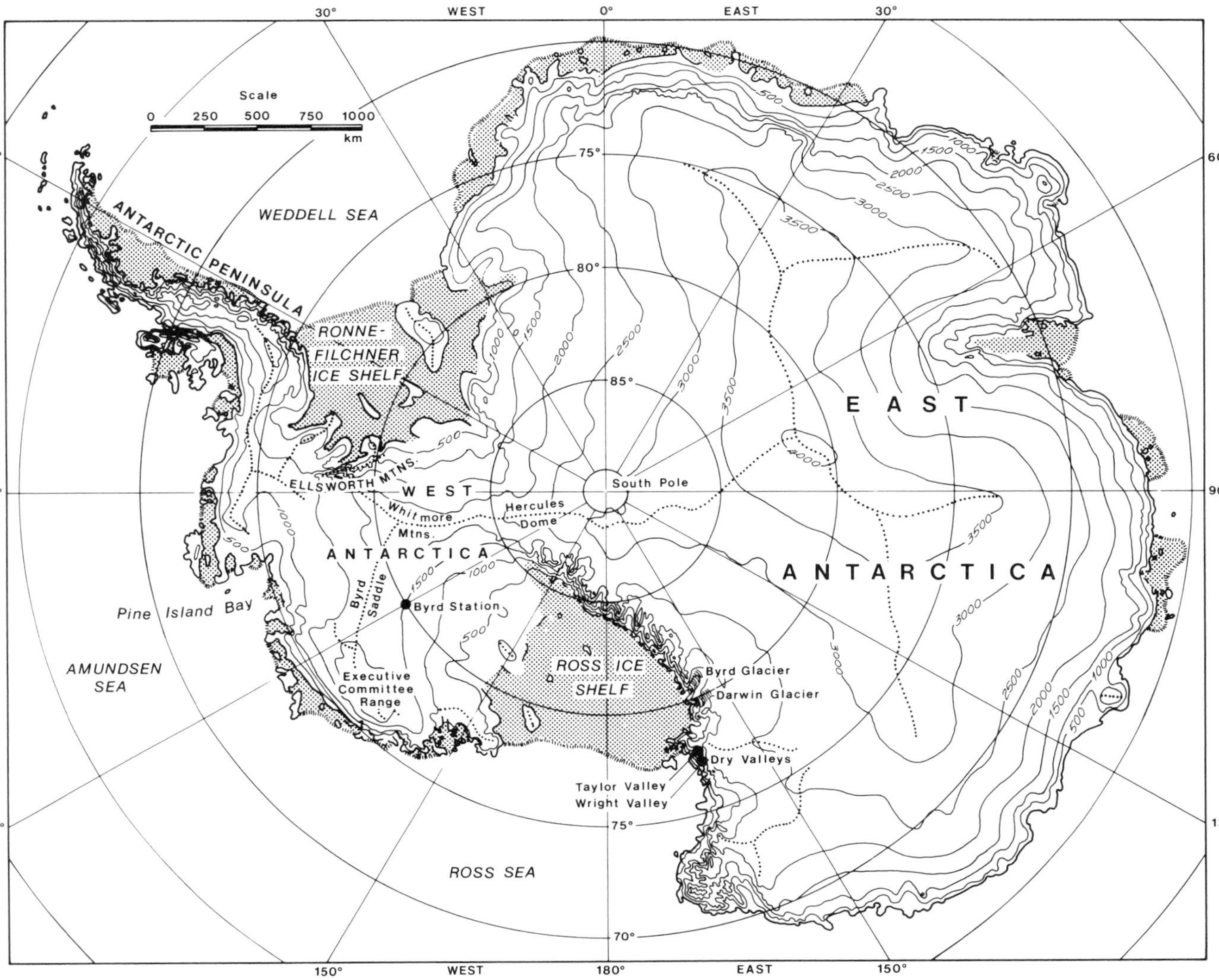

Figure 1. Index map of Antarctica (adapted from Drewry, 1983). Ice shelves are shaded.

escarpment, cirque-headed glaciers are relatively small, and ridges and cols are largely ice free.

The Heritage Range is 190 km long and about 90 km wide. The NNW-SSE–trending longitudinal ridges that dominate the topography exhibit classic glacial alpine morphology (Plate II, Fig. 3, in pocket). The Heritage Range is considerably lower than the Sentinel Range. The highest peaks are 2,000 to 2,500 m in elevation; most are less than 2,000 m high. Inland ice nearly inundates the western ridge of the Heritage Range. Katabatic winds from the ice sheet sweep across the Heritage Range (Rutford, 1969), enhancing snow accumulation in cirques and along ridge slopes that face westward toward the prevailing wind. In sharp contrast, many east- and northeast-facing cirques, as well as mountain ridges and slopes, are ice free because of ablation by katabatic winds flowing around and over these topographic obstacles. Blue-ice ablation areas also occur on glacier surfaces downwind from topographic obstacles. Because of these katabatic winds, the Heritage Range has more ice-free areas than the Sentinel Range.

Aside from ice-free areas caused by katabatic winds, the Heritage Range is heavily glacierized. Internal ice drainage is channeled largely through three extensive glacier systems. Two of these systems, the Splettstoesser and Union glaciers in the northern and central Heritage Range, are spillover glaciers (Rutford, 1969) that head on the western ridge of the Heritage Range. They drain inland ice and local snow domes banked against the ridge by katabatic winds. Both glaciers flow ENE through the range, fed by numerous tributaries from internal cirques and accumulation basins between mountain ridges. A third large valley glacier system, situated in Horseshoe Valley in the southern Heritage Range, drains internal accumulation basins as well as some inland ice that spills over cols in the western ridge.

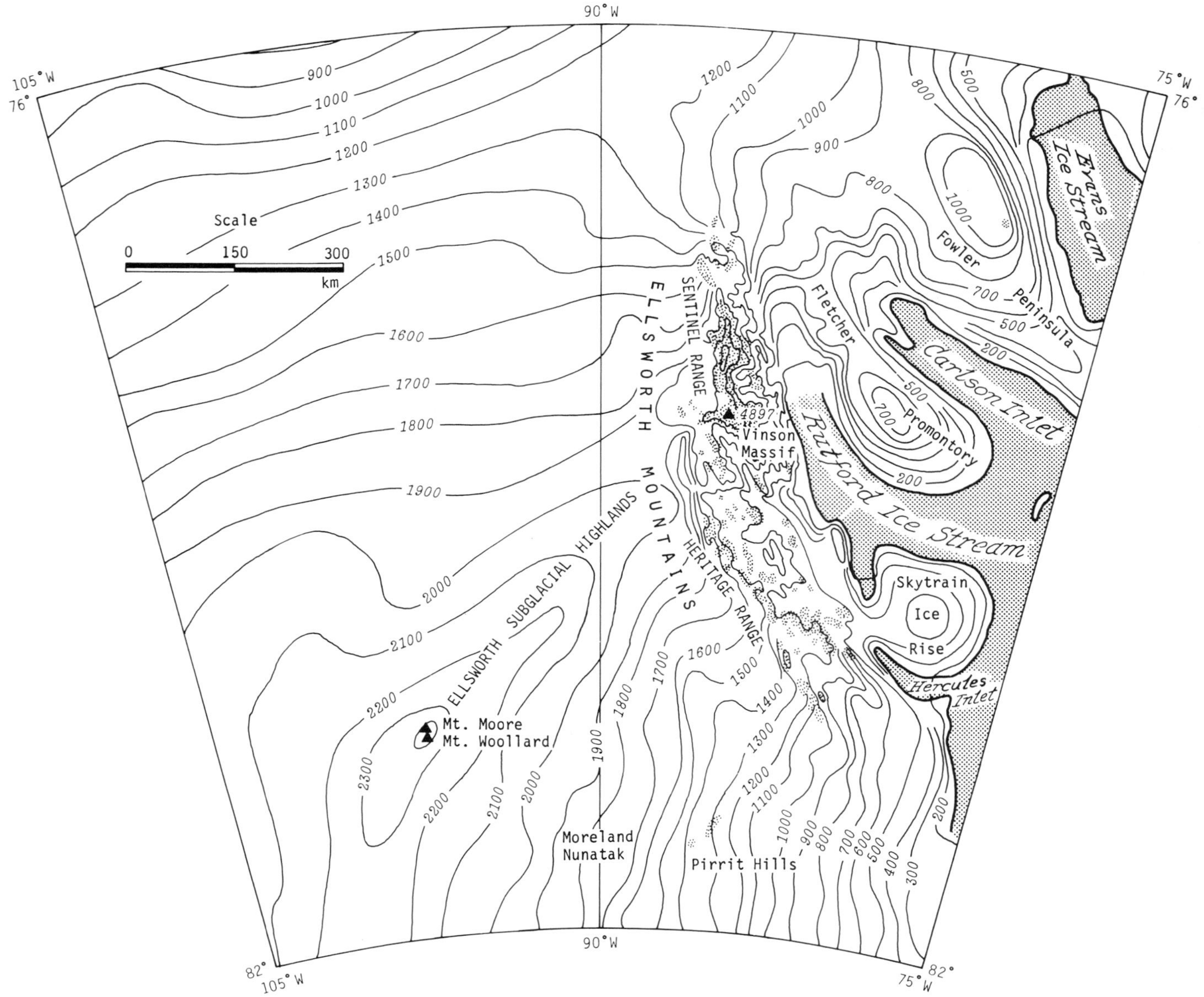

Figure 2. Surface topography in meters of West Antarctic Ice Sheet near Ellsworth Mountains (from Drewry, 1983).

The West Antarctic Ice Sheet now surrounds and partly inundates the Ellsworth Mountains (Figs. 2 and 3). The high, inland ice-sheet surface abuts the western margin of the exposed mountains. A major inland ice divide extends in an arc from near the northwestern Heritage Range, where it has a surface elevation of about 2,000 m, southwestward over the Ellsworth Subglacial Highlands and through the Whitmore Mountains (GEBCO, 1980; Crabtree and Doake, 1982; Drewry, 1983). Although some southeastward-flowing ice from this divide feeds the Minnesota Glacier, most drains around the southern Heritage Range into the Ronne-Filchner Ice Shelf. Inland-flowing ice from this divide near the Ellsworth Mountains drains northwestward into Pine Island Bay in the Amundsen Sea. A saddle separates this long ice ridge from a small, high dome adjacent to the western side of the Sentinel Range. This small dome is partly fed by local alpine glaciers. In turn, some ice from this dome feeds the Minnesota and Nimitz glaciers, and some flows northward around the mountains into the Rutford Ice Stream.

The present grounding line between the West Antarctic Ice Sheet and the Ronne-Filchner Ice Shelf is near the eastern margin of the Ellsworth Mountains (GEBCO, 1980; Drewry, 1983). Several extensive ice rises separated by embayments adjoin the eastern Ellsworth Mountains (Figs. 2 and 3). Surface elevations of these ice rises are 400 to 700 m, in contrast to inland ice-sheet surfaces of 1,400 to 2,000 m on the western margin of the mountains. The huge Rutford Ice Stream that flows along the eastern mountain flank drains a low inland ice divide between the Ronne-Filchner Ice Shelf and Pine Island Bay, in addition to local Ellsworth ice, as it flows seaward into the Ronne-Filchner Ice Shelf. This ice stream is constrained on the west by the Sentinel Range and the Skytrain Ice Rise and on the east by the Fletcher Ice Rise. The grounding line of the Rutford Ice Stream is complex and occurs opposite the Flowers Hills in the Sentinel Range (Stephenson and Doake, 1982).

Published ice-depth soundings near the Ellsworth Mountains reveal a bedrock high, the Ellsworth Subglacial Highlands, extending southwestward from the mountains beneath the major inland ice divide (GEBCO, 1980; Drewry, 1983). Depths of bedrock range to –1,600 m beneath the Rutford Ice Stream near the grounding line (Stephenson and Doake, 1982) and to –1,000 m inland of the Sentinel Range and near the southwestern tip of the Heritage Range (Drewry, 1983).

Bedrock geology and previous glacial geology

The bedrock geology of the Ellsworth Mountains, described by Craddock and others (1964), Craddock (1969), Hjelle and others (1978, 1982), Splettstoesser and Webers (1980), and Webers and Spörli (1983), is important to this discussion because much of the glacial history is inferred from erosional features cut in bedrock.

The stratigraphic column presented by Craddock (1969) resulted from fieldwork carried out during the 1960s. Hjelle and others (1978, 1982) proposed revisions to the base of the section as a result of fieldwork in 1974–75. A group of geologists examined selected sites within the Ellsworth Mountains to complete detailed studies during the 1979–1980 field season. The results of that work, reported in detail in other papers in this volume, suggest a major revision in the base of the stratigraphic section. The revised section proposed by Webers and Spörli (1983) will be followed here. Maps elsewhere in this volume show the areal distribution of bedrock units.

Four major stratigraphic units are now recognized in the Ellsworth Mountains. The areal distribution of rock units shown by Craddock (1969) remains basically unchanged. However, marble units of the Minaret Group are now demoted to formation status and included within the Heritage Group. With the exception of the Minaret Formation, the rocks of the Heritage Group are mainly clastic rocks, but both mafic and felsic igneous rocks are known, as well as some carbonate rocks within units other than the Minaret Formation.

Overlying the Heritage Group is the Crashsite Group, which is composed of quartzite and is the most widespread unit in the Ellsworth Mountains. This quartzite forms the high peaks of the Sentinel Range (Craddock, 1969). Three subdivisions are recognized: the Howard Nunataks Formation, a lower dark unit of impure quartzite; the Mount Liptak Formation, a middle light unit of light-colored, thin-bedded, clean quartzite; and the Mount Wyatt Earp Formation, an upper dark unit of impure quartzite with volcanic rock fragments. The total thickness of the Crashsite Group quartzite is about 3,200 m. The upper contact with the Whiteout Conglomerate is conformable.

The Whiteout Conglomerate, a dark gray unit of massive diamictite with minor sediments and boulder pavements, extends from the Meyer Hills in the south to the northern end of the Sentinel Range. Megaclasts, which compose up to 10 percent of the diamictites and range up to 5 m (rare), consist of quartzite, quartz, granite, gneiss, schist, phyllite, argillite, chert, sandstone, conglomerate, and limestone. The upper contact with the Polarstar Formation is sharp but conformable (Matsch and Ojakangas, this volume).

The Polarstar Formation crops out along the east side of the Sentinel Range north of the Flowers Hills. The rocks are predominantly dark gray to black carbonaceous argillite and arenite. Bedding is well developed, and coal seams as much as 30 cm thick are present near the top of the formation. A Permian *Glossopteris* flora occurs in the upper part of the formation (Craddock and others, 1965; Taylor and Taylor, this volume).

The igneous rocks of the Ellsworth Mountains are mainly basic intrusions or flows, although felsic rocks are reported from near Soholt Peaks (Vennum and others, this volume). All igneous rocks appear to be older than the Crashsite Group except for a felsic intrusion of Devonian age (Vennum and others, this volume) and in most cases were deformed with associated rocks.

The thick sequence of rocks in the Ellsworth Mountains has undergone one or two periods of deformation, is folded, has been uplifted a minimum of 6,100 m, and has a tectonic grain parallel to the long axis of the mountains. The size and type of folds varies. The major style is large open folds. Minor drag folds commonly occur on flanks of larger folds, which in turn may be part of still larger folds. The north-trending linear ridges are parallel to the trace of folds, especially in outcrops of the Crashsite Group.

Minor structural elements occur in all rock units. They are important to physical weathering and the formation of pinnacles on serrated ridge crests above the widespread glacial trimline discussed later. Two dominant joint sets occur within all quartzite units. One is generally parallel to traces of fold axes; the other is normal to the first; both are perpendicular to bedding. The result is an orthogonal system of three intersecting structural features, one a bedding plane and the other two joint surfaces.

The entire sequence of rocks exposed in the Ellsworth Mountains shows effects of low-grade regional metamorphism that increases with stratigraphic depth (see Bauer, this volume). Rock cleavage is well developed, and carbonate rocks are partially recrystallized to marble.

Previous workers described two major phases of glacial history of the Ellsworth Mountains (Craddock and others, 1964; Rutford, 1969, 1972). The earliest phase involved mountain glaciation, perhaps beginning in late Mesozoic or early Tertiary time when the Ellsworth Mountains were an island or peninsula (Rutford, 1972). Valley glaciers modified a preglacial, structurally controlled fluvial drainage pattern (Rutford, 1972). The second phase involved continental glaciation that inundated most of the Heritage Range while leaving high ridges of the Sentinel Range as nunataks mantled with mountain glaciers (Rutford, 1972). Craddock and others (1964) estimated a former ice-sheet surface 300 to 500 m above present-day ice surfaces. A former high ice-sheet surface is also shown by erratic boulders and cobbles as much as 500 m above the present-day ice surfaces. Many of these erratics were derived locally, but some are of bedrock types not yet discovered in the Ellsworth Mountains (Craddock and others, 1964).

Rutford (1972) and Craddock and others (1964) suggested that lowering of the ice-sheet surface from its high level has occurred recently. Craddock and others (1964, p. 159) stated:

> However, the freshness of some glacial features well above the surface of present glaciers suggests that deglaciation has been both recent and rapid. Throughout the ranges many examples of glacially striated bedrock have been observed hundreds of metres above the surfaces of adjacent glaciers. For example, excellent striations are preserved on a resistant conglomerate 6 ft (1.8 m) below the summit of Mount Lymburner, a peak in the northern Sentinel Range with a minimum relief above the present glacial surface of 300 m. Most of these striae trend in directions consistent with the flow pattern of the modern adjacent ice streams. The freshness of these scratches even where preserved on slate or marble suggests these rocks have not been long exposed to weathering.

Several other features suggested that deglaciation was recent. Craddock and others (1964, p. 159) stated that many erratics were large, fresh blocks. Moreover, ice-surface lowering has caused abnormal westward flow into several valleys in the Heritage Range, in addition to leaving tongues of stagnant ice and moraines in some of these valleys (Rutford, 1972; Craddock and others, 1964).

Rutford (1969, 1972) emphasized that the alpine erosional features that emerged from the thinning continental ice cover were preserved in detail. From this he inferred that many alpine features predated a protective, higher-than-present continental ice cover.

Present study

Our glacial geology and soil studies in the Ellsworth Mountains were necessarily a reconnaissance. Our headquarters was the main Ellsworth Project Camp, adjacent to the Minnesota Glacier in the northern Heritage Range at latitude 79°05′S and longitude 85°58′W. With helicopter support from this camp, we placed parties at a total of 182 ground localities. We examined soil development and surface boulder weathering at 22 sites on drift patches and moraines mapped in the Heritage and Sentinel Ranges below the trimline (Plate II, Fig. 3, in pocket). Bockheim (1979) previously described field procedures for soil and weathering studies. Using methods described in American Public Health Association and others (1975), we analyzed 41 soil samples for pH, electrical conductivity (EC), and water-soluble salts, including Na^+, Ca^{2+}, Mg^{2+}, K^+, Cl^-, NO_3^-, SO_4^{2-}, and HCO_3^-.

We mapped a conspicuous glacial trimline on aerial photographs by ground examination and by slow helicopter traverses between ground localities. We plotted striations at the 94 sites below the trimline (Plate II, Fig. 3, in pocket).

We determined elevations of ground localities and the trimline in four ways. First, map elevations were used for crude estimates. Second, these map elevations were supplemented by absolute values for a network of surveyed sites established by the U.S. Geological Survey (Knutzen and others, 1981). Third, Paulin altimeter measurements were made by using the surveyed sites as control. Fourth, trimline elevations were determined from helicopter altimeters that were repeatedly checked for accuracy by landing at control surveyed sites, by flying past peaks of known elevations, and by repeatedly flying critical areas. We estimate that most elevations are accurate to within ±50 m.

FIELD AND LABORATORY RESULTS

Glacial alpine landforms

Landforms of alpine glacial erosion dominate the Sentinel and Heritage Ranges. They are best exposed in the Sentinel Range, which towers above the ice sheet. Alpine landforms also occur in the lower Heritage Range, but they are less well exposed because the range is now partially engulfed by inland ice.

Numerous cirques at heads of alpine mountain glaciers cut the linear quartzite ridges of the Sentinel Range. These cirques are best developed on the high western ridge, where they have particularly steep headwalls; sharp bedrock spurs act as divides (Figs. 4 and 5). Opposing cirques have commonly eroded the divide into a thin wall that forms sharp arêtes in the high cols separating cirques. Only the summit of Vinson Massif reveals undulating, prealpine topography. Many peaks at ridge-and-spur junctions are steep, and some are alpine horns. In plan view, the western ridge is sinuous (Plate II, Fig. 3, in pocket). Alternating concave sectors mark headwalls of opposing cirques. The lower, eastern ridges of the Sentinel Range also have cirques, arêtes, spurs, and steep peaks. Here, however, greater snow and ice cover partly obscures these features. The Bastien Hills, located on the western margin of the Nimitz Glacier, likewise exhibit alpine erosional features that are partly submerged by inland ice.

Well-developed alpine landforms also mark the Heritage Range, although they are more subdued than in the Sentinel Range. These alpine landforms include long series of opposing cirques separated by sinuous ridges. Steep headwalls and intervening spurs characterize most cirques. Arêtes and steep peaks, some forming horns, are common along high ridges. Only Anderson Massif has not been carved into sharp alpine forms.

Four factors strongly suggest that most alpine erosion occurred before the West Antarctic Ice Sheet engulfed lower ridges of the Ellsworth Mountains. (1) On satellite photographs (Plate II, Fig. 3, in pocket) the image of glacial alpine topography is refracted through the thin ice sheet cover on the Ellsworth Subglacial Highlands that extend southwest from the Ellsworth

Figure 4. Aerial photograph of cirques, arêtes, and horns in Crashsite Group quartzite of western ridge of Sentinel Range. View is eastward (U.S. Navy photograph).

Mountains beneath a major inland ice divide. (2) The inland ice sheet nearly inundates the southwestern ridge of the Heritage Range, with some inland ice feeding through-flowing spillover glaciers. This southwestern ridge has numerous cirques that face toward and are nearly submerged by the ice sheet (Plate II, Fig. 3, in pocket). (3) Katabatic winds that sweep across the Heritage Range from the inland ice sheet have piled snow onto cirques and slopes that face toward the prevailing wind. These katabatic winds have also created blue-ice and ice-free areas downwind from topographic obstacles. Several cirques and valleys facing downwind are ice free except where reversal of ice flow has infilled their distal portions with tongues of blue ice. (4) The trimline discussed below is etched into the glacial alpine landforms. This trimline represents a higher surface level of the inland ice sheet. Therefore, the extensive alpine topography into which the trimline is etched must antedate the higher inland ice surface.

Figure 5. Cirques in Crashsite Group of western ridge of Sentinel Range. View is northeastward.

The trimline

Throughout the Ellsworth Mountains a well-defined erosional trimline is etched into alpine ridges and spurs (Fig. 6). This trimline persists with consistent elevations across varying bedrock lithology and structure. It separates low alpine ridges and crests that are smoothed and striated from high ridges and crests that are serrated and lack striations. At the trimline itself, there is commonly a transition zone between lower, blunt ridge crests and higher, serrated ridge crests (Figs. 6, 7, 8, and 9). This transition zone generally extends for about 50 m vertically and shows a combination of coarse striations and stubs of former spires. Depositional features do not occur at the trimline. However, drift patches, erratics, and sporadic moraines near blue-ice areas occur below the trimline, particularly in the Heritage Range.

In the Sentinel Range the trimline occurs on nearly all steep ridges of the Crashsite Group. The trimline is most conspicuous on relatively ice-free, steep western ridges (Fig. 6). However, it also occurs on heavily glacierized eastern ridges. Trimline elevations on all ridges in the Sentinel Range show a remarkably consistent pattern (Plate II, Fig. 3, in pocket). The steep western flank of the Sentinel Range illustrates this pattern. Here, the trimline is highest (3,000 m) on the long spur that projects westward from near Mount Bentley toward Mount Hubley (Plate II, Fig. 3, in pocket). From here southward to the head of Nimitz Glacier, there are twenty major westward-projecting spurs with the following trimline elevations: 2,960 m, 3,000 m (Mount Anderson); 3,010 m, 3,000 m (Long Gables); 2,960 m, 2,960 m, 2,930 m (Mount Viets); 2,930 m (Mount Giovinetto); 2,900 m, 2,900 m (Mount Ostenso); 2,930 m, 2,900 m, 2,900 m (Mount Shear); 2,870 m (Mount Gardner); 2,840 m, 2,870 m (Mount Epperly); and 2,840 m, 2,840 m, and 2,840 m (Mount Shinn). From the head of Nimitz Glacier to its foot near the southern tip of the Sentinel Range, 10 additional spurs show the following

Figure 6. Spur at the foot of Mount Epperly, west-central Sentinel Range. The trimline at 2,870 m elevation is clearly shown and separates a lower smooth ridge crest from a higher serrated crest. Reproduced with permission from Craddock and others (1964). View to southeast.

Figure 7. Trimline (2,870 m elevation) on spur in foreground near Mount Epperly, west-central Sentinel Range. Trimline separates lower blunt ridge crest from higher serrated crest and occurs one-third of the way up the ridge. Ridge composed of Crashsite Group quartzite. View is northward.

trimline elevations: 2,590 m (Vinson Massif); 2,500 m (Vinson Massif); 2,470 m (Vinson Massif); 2,320 m, 2,330 m, 2,380 m (Mount Strybing); 2,130 m, 2,140 m (Mount Southwick); 2,040 m, 1,920 m. Northward from Mount Bentley, west-facing spurs and ridge crests show the following trimline elevations: 2,930 m (Mount Davis); 2,870 m (Mount Hale); 2,820 m, 2,770 m, 2,770 m, 2,770 m, 2,770 m, 2,770 m (Mount Dalrymple); 2,770 m (Mount Alf); 2,680 m (Mount Sharp); 2,680 m (Mount Barden); and 2,680 m (Mount Dawson).

Ridges on the east side of the Sentinel Range also show consistent trimline elevations (Plate II, Fig. 3, in pocket). Trimline elevations range between 2,680 m and 2,440 m along short spurs projecting eastward from the central portion of the western ridge. Trimline elevations generally diminish north and south of this high, central area. Also, the trimline is lower on eastern ridges near the Rutford Ice Stream. For example, trimline elevations on ridges and spurs adjacent to the lower Rutford Ice Stream are 1,950 m, whereas those opposite higher reaches of the Rutford Ice Stream are 2,470 m (Plate II, Fig. 3, in pocket).

The trimline in the Heritage Range is discontinuous, occur-

Figure 8. Trimline on spur of dominant west ridge of central Sentinel Range. Serrated crest is above trimline; nonserrated, blunt crest is below trimline. View to northeast.

ring only on high peaks and ridges. Trimline elevations (Plate II, Fig. 3, in pocket) are: Soholt Peaks (2,200 to 2,400 m), Mount Sporli and Robinson Peak (2,000 m), Donald Ridge and Mount Capley (1,700 m), the Enterprise Hills (1,900 m on the inland side to 1,750 m on the seaward side), Parrish Peak (1,700 m), Mount Rosenthal (1,820 m), and Moulder Peak (1,800 m). Elsewhere, all peaks fall below the trimline. This means that the trimline exceeds 2,200 m along the western ridge north of Soholt Peaks.

The trimline in the Heritage Range shows a consistent elevation pattern, with higher values near the central, inland flank of the range and lower values to the east and south. The trimline thus slopes downward from west to east across the range; for example, from Soholt Peaks to Mount Capley and from the western to the eastern Enterprise Hills. The trimline also slopes downward from northwest to southeast along the western boundary ridge of the Heritage Range.

In both the Sentinel and Heritage ranges the trimline is cut largely into quartzite of the Crashsite Group, which forms the core of both ranges. It is not confined to any of the formations of the Crashsite Group, to contacts between formations, or to particular structural elements within formations. Rather, the trimline exhibits consistent elevations, as discussed above and shown in Figure 3 (Plate II, in pocket) and passes indiscriminately across quartzite units and structural elements.

Although the trimline most commonly occurs in the Crashsite Group of the mountain core, it is also well preserved in basic igneous rocks of the Soholt Peaks in the Heritage Range and in pre-Crashsite rocks in the Heritage Range. Outcrops of both Whiteout Conglomerate and the Polarstar Formation lie below trimline elevations within our map area in the eastern Sentinel and Heritage ranges.

The trimline marks changes in the morphology of sharp ridges and spurs. Above the trimline, ridges and spurs are highly serrated with closely spaced spires. Individual spires are as much as 30 m high. Many have weathered into thin, delicate pinnacles; some have holes weathered into their bases; several precariously support each other. Spires characterize ridges with differing attitudes of bedding, jointing, and cleavage. Spires are well developed on quartzite ridge segments composed of flat-lying or steeply tilted bedding or with bedding attitude transitional between these extremes. They occur in all formations of the Crashsite Group and show no change in character across contacts between formations. As a result, serrations are not confined to selected ridge segments but rather occur consistently on nearly all steep ridges and spurs above the trimline, thus allowing accurate mapping of the trimline. Figures 10, 11, and 12 show serrations that occur above the trimline on ridge crests composed of differing quartzite units with various bedding attitudes. In all cases the quartzite ridges show joint systems described previously. Figure 13 shows spires above the trimline in basic igneous rocks at Soholt Peaks.

In sharp contrast, bedrock crests and ridges below the trimline are never serrated. Rather, they are blunt and relatively smooth, with lower portions commonly planed (Figs. 14–16). This change from serrated to smooth ridge crests is not associated with changes in bedrock lithology or structure. Many blunt ridge crests below the trimline exhibit glacial striations. Well-preserved striations are scattered on bedrock outcrops at all elevations between present ice surfaces and the trimline. In the Heritage Range, erratics are common below the trimline, and extensive fields of erratics litter some ice-free areas. Such erratics are rare, however, in the Sentinel Range.

At several localities in the northern Sentinel Range (for example, east of Mount Reimer and Mount Dalrymple in Fig. 3) and in the Heritage Range (for example, between Hall Peak and Robinson Peak and near site 90 in Fig. 3), serrated spurs occur on the east side of major north-south trending ridges, which are not serrated but show scattered striations.

Figure 9. Trimline on unnamed peak composed of Crashsite Group quartzite in the southern Sentinel Range. Trimline is just below summit. Only the summit, which is marked by pinnacles in flat-lying and steeply inclined beds of the Crashsite Group, lies above trimline. All lower ridges, which lack pinnacles in similar flat and inclined quartzite, lie below trimline. View to south.

Figure 10. Spires on serrated spur above trimline near Mount Epperly, west-central Sentinel Range. Spires formed in tightly folded Crashsite Group quartzite at 3,300 m elevation. View to northeast.

Figure 11. Serrated spurs of Crashsite Group quartzite between cirques in eastern Sentinel Range near Mount Craddock. The spires continue nearly unbroken across numerous changes in bedding attitude and across dark and light quartzite units. The proximity of serrated spurs to present accumulation surfaces indicates that upper portions of east-flowing glaciers in the central Sentinel Range did not thicken significantly during the last glaciation. View to southwest.

Glacial features below the trimline

Figure 3 (Plate II, in pocket) shows 94 localities where we found striations below the trimline. At some localities, only the trend of former ice flows is shown. At many localities, however, former ice-flow direction is evident from the distribution of striations on bedrock faces of differing orientations or from small stoss-and-lee forms. Table 1 gives pertinent striation data. We concentrated our search near the trimline and on high peaks and ridges; therefore, high-elevation sites are disproportionately represented.

Several general statements can be made about striations in the Ellsworth Mountains. The widespread Crashsite Group holds striations well, apparently because the quartzite does not weather easily by exfoliation or granular disintegration. Likewise, many fine-grained slate, argillite, and phyllite units of the Heritage Group show well-preserved striations. However, carbonate rocks common to the Minaret Group in the southern Heritage Range do not hold striations because of surface granular disintegration. Likewise, the Whiteout Conglomerate and Polarstar Formation in the eastern fringes of the field area do not commonly preserve striations because of rapid frost shattering along cleavage planes.

Figure 12. Spires on serrated spur above trimline on Mount Strybing in the southern Sentinel Range. Spires formed on steeply inclined Crashsite Group quartzite. View to north.

On competent bedrock below the trimline, striations occur on steep ridge crests and summits of alpine horns as well as where they would be more naturally expected in lower cols and on gently smoothed bedrock hills. Even on bedrock lithologies that typically preserve striations, only a few sites show polish and closely spaced, fine striations over extensive areas. Such rare sites are either in cols (Figs. 17 and 18) where former ice flow was constricted, or at low elevations where former ice thickness was greatest. More typically, striations occur as coarse, broad scratches and gouges scattered on oxidized bedding planes, joint surfaces, or beveled edges of upturned beds. Commonly, these scratches occur only on upper surfaces of bedrock projections and are absent elsewhere. The corners of many bedrock projections facing toward former ice flow are smoothed, whereas those facing away from former ice flow are chipped.

Bedrock ridges with striations commonly show overall smoothing on a large scale (Figs. 14–16). Striations are widely scattered on oxidized bedding or joint planes that mark these blunted ridges. Most such bedrock surfaces have oxidation rinds 2 to 10 mm thick. The extensive oxidized bedrock surfaces between striations show no signs of abrasion by overriding ice, even

Figure 13. Spires cut in serrated ridge of basic igneous rocks in Soholt Peaks. This ridge lies above trimline. View to north.

Figure 14. Spurs that separate cirques along the western escarpment of the central Sentinel Range. The two spurs in the foreground lie below the trimline, are blunted, and lack serrations. On the third spur, the trimline separates serrated from nonserrated spur crest. View to north.

Figure 15. Unnamed alpine peak composed of Crashsite Group quartzite in the southern Sentinel Range. Nimitz Glacier and Bastien Hills in background. Peak lies below trimline, lacks serrated ridges, and has survived overriding by inland ice sheet. Aside from blunted ridges, the original alpine morphology has survived the overriding intact. View to south.

Figure 16. Unnamed alpine peak composed of Crashsite Group quartzite in the Enterprise Hills, Heritage Range. Peak lies below trimline, lacks serrated ridges, and has survived overriding by inland ice. View to south.

though they appear to have experienced only minor postglacial weathering. In many cases, the age relation of striations and oxidation rinds is not apparent. Specific situations occur near Mount Virginia (Crashsite Group) and Lishness Peak (Crashsite Group) (Fig. 19) as well as in the Flowers Hills (Whiteout Conglomerate). In all three cases, striations are widespread and well preserved, but only at Mount Virginia did a few striation troughs penetrate the oxidation rind into unweathered bedrock. At both Mount Virginia and Lishness Peak (Fig. 19), unoxidized and angular quartzite blocks have been glacially transported onto the oxidized striated surfaces. This relation may demonstrate a second glacial overriding of the oxidized bedrock surface and strongly suggests that at least some striations postdate the deeply oxidized surface rind even though they do not penetrate this rind. However, many striations at these localities may predate substantial bedrock oxidation and thus may be older than at least one glacial overriding.

In some cases where striations have differing depths, they appear to postdate most or all of the surface oxidation. Shallow striation troughs are in the upper deeply oxidized portion of the rind, whereas deeper troughs penetrate the less-oxidized portion of the rind, and some break through the rind entirely into underlying fresh bedrock. Progressively deeper troughs in quartzite commonly show progressively lighter colors because they pene-

TABLE 1. STRIATION DATA COLLECTED IN THE ELLSWORTH MOUNTAINS DURING THE 1979–1980 AUSTRAL SUMMER FIELD SEASON*

Field Locality No.	Rock Type	Approximate Elevation (m)	Trend (T) or Direction (D)† of Striations
1	conglomerate	2,000	N80°W (T)
2	argillite	2,280	N10°E–N20°E (T)§
3	quartzite	1,900	(a) N40°E (D); N10°E (D); and N30°W (D)** in saddle to west of Shockey Peak
	quartzite	2,000	(b) N40°E–N55° (D) E and N70°E (D) on Shockey Peak
4	conglomerate	1,900	N80°W (T)
5	quartzite	1,800	N40°E (D); N80°E (D); and S70°E (D)
6	quartzite	1,800	(a) N40°E–N55°E (D) in col
	quartzite	2,050	(b) N45°E (D) on peak to north of col
	quartzite	2,050	(c) N60°E (D) on peak to south of col
7	conglomerate	2,050	(a) N40°E–N55°E (T), and N90°E (T) on one ridge sector
	argillite	2,050	(b) N80°E (T) on nearby ridge sector
8	quartzite	2,118	N85°E (T)
9	quartzite	2,790	N20°E–N25°E (T)
10	quartzite	2,400	(a) N20°E (D) on lower part of ridge
	quartzite		(b) S20°W (D) and N90°E (D) on upper part of ridge
11	quartzite	2,400	N90°E (T) and S60°E (T)
12	quartzite	2,385-2,957	(a) N85°W–N60°W (T) on ridge below 2,660 m
	quartzite		(b) N75°W–N60°W (T) on same ridge above 2,660 ms
13	quartzite	2,400	N80°W (T) and N70°E (T)
14	quartzite	2,700-2,826	(a) N80°E-N70°E (T) below 2,660 m on ridge
	quartzite		(b) N70°W–N40°W (T) above 2,660 m on same ridge
15	quartzite	2,350	N80°E (T) and N°35E (T)
16	quartzite	2,500-2850	N85°E–N70°W (T)
17	quartzite	2,300	S20°W (T)
18	quartzite	2,200	N90°W (T)
19	quartzite	1,900	N90°E (D) and S45°E (D)
20	quartzite	2,000	N80°E (D)
21	quartzite	2,100	N75°E (D)
22	quartzite	1,600	N45°W (T) and N0°E (T)
23	quartzite	2,200	S80°E–S10°E (D)
24	quartzite	1,900	S70°E (T)
25	quartzite	2,350	S80°E–S70°E (D)
26	quartzite	1,800	N50°E (D)
27	quartzite	2,200	S80°E (D) and S85°E (D)
28	quartzite	1,600	N70°E (D)
29	quartzite	1,450	N70°E (D)
30	quartzite	1,300	N80°E (D)
31	quartzite	1,240	N65°E–N75°E (D)
32	conglomerate	1,503	N90°E (D), N75°E (D), and N80°E (D)
33	quartzite	1,000	N80°E (D)
34	conglomerate	1,200	N60°E (D)
35	quartzite	1,450	S15°E (D) and S30°E (D)
36	conglomerate	700	N45°E (D) and N40°E (D)
37	quartzite	1,930	S40°W (T) (oldest and dominant, parallel to stoss-and-lee)‡ N75°E (T) (intermediate, faint) N40°E (T) (youngest, few scattered grooves)
38	quartzite	2,000	N0°E (D) and N45°E (D)
39	quartzite	1,750	N70°E (D)
40	quartzite	1,650	N10°E (T)
41	quartzite	400	N65°E–N70°E (D)
42	marble	1,100	N90°E (D)
43	quartzite	1,476	Two sites on the summit of Welcome Nunatak show striations. One (a) is on the north side and the other (b) is on the south side of the summit. (a) S40°E (T) (oldest), N90°E (T) (intermediate), and N65°E (T) (oldest). (b) N90°E (T)
44	quartzite	1,600	S85°E (T) and N45°E (T) in saddle
		1,700	N90°E (T) on summit
45	quartzite	1,700	N20°E (T)
46	quartzite	1,840	N35°E (T) and N70°E (T)
47	argillite and dirty quartzite	2,120	N90°E–S70°E (T)
48	quartzite	2,150	S45°E (T)
49	quartzite, argillite, and phyllite	2,150	N55°E (T)

TABLE 1. STRIATION DATA COLLECTED IN ELLSWORTH MOUNTAINS DURING THE 1979–1980 AUSTRAL SUMMER FIELD SEASON* (continued)

Field Locality No.	Rock Type	Approximate Elevation (m)	Trend (T) or Direction (D)† of Striations
50	quartzite	2,080	N35°E–N45°E (D)
51	argillite and silty quartzite	2,000	N40°E (D)
52	dirty quartzite and slate	1,910	N90°E (D) and N65°E (D)
53	dirty quartzite	1,810	N50°E (T)
54	argillite and quartzite	1,060	N65°E (T)
55	quartzite	1,150	N70°E (D) (oldest) and N15°E (D) (youngest)
56	slate	2,170	N90°E (D)
57	quartzite	1,360	N80°E (D)
	quartzite		N45°E (D)
58	argillite and phyllite	1,120	N45°E–N50°E (T)
59	argillite and siltstone	2,100	N50°E (T) (dominant, plotted) and N10°E (T) (weak, not plotted)
60	argillite	1,220	N80°E
61	dirty quartzite	1,850	N50°E (T) and N10°E (T)
62	quartzite	1,600	N40°E (D)
63	gabbrodiorite	1,340	N90°E (D)
64	dirty quartzite	2,350	N75°W (T)
65	quartzite	1,900	N30°E (D)
66	quartzite	2,000	N70°E (D)
67	quartzite	1,920	S85°E (D)
68	quartzite	1,850	N55°E (D)
69	quartzite	1,500	N25°E–N30°E (D)
70	quartzite	2,070	N60°E (D) and N90°E (D)
71	quartzite	1,500	N40°E (D) and N20°E (D)
72	quartzite	1,790	N0°E (T)
73	quartzite and conglomerate	1,600	N65°W (T)
74	quartzite	1,980	S45°E (D)
75	conglomerate	1,520	N90°E (D)
76	quartzite	1,110	N50°W (T) (oldest); N80°E (T) (youngest)
77	quartzite	880	N45°E (D)
78	quartzite	1,400	N50°E (T)
79	intrusion	2,200	N60°E (D)
80	quartzite	990	N90°E (T)
81	conglomerate	800	N45°W–N90°E (T)
82	quartzite	1,800	N60°E–N70°E (D)
83	quartzite	1,400	N60°E (D)
84	phyllite	1,550	N65°E (T)
85	quartzite	1,410	N50°E (D) and N45°E (D)
86	quartzite and argillite	1,600	N35°E (D) and N85°E (D)
87	quartzite	1,700	N0°E–N20°E (T)
88	quartzite and argillite	1,600	N25°E (T) and N55°E (T)
89	quartzite	2,000	N75°E (D)
90	marble	1,800	N50°E (D)
91	quartzite	1,700	N80°E (T)
92	quartzite	1,650	N70°E (T)
93	marble	1,000	N43°E (T)
94	marble and breccia	1,220	N50°E (D)

*Field locality numbers shown in Figure 3.

†Direction of striation determined where possible from distribution of striations on bedrock faces of differing orientations or from small stoss-and-lee forms in the relatively few sites where they are present.

§N10°E–N20°E means that striation trends or directions showed a range between these values. The value shown on Figure 3 is intermediate between the two extremes.

**N40°E (D); N80°E (D); and S70°E (D) means that there were three distinct directions. The relative age relations cannot be determined because the striations do not cross each other.

‡Age relations can be determined because striations cross. The relative ages are numbered from oldest to youngest in Figure 3.

Figure 17. Striations and polish imparted by wet-based ice in col of Crashsite Group quartzite on eastern side of Sentinel Range at site 37, in Figure 3, in southern Sentinel Range. The col is close to the trimline. Serrated ridge in background is above trimline. View to northwest.

Figure 18. Striations in Crashsite Group quartzite in col in southern Sentinel Range. Striations cut by wet-based ice.

Figure 19. Cross-cutting striations etched by wet-based ice into Crashsite Group quartzite in col near Lishness Peak in southern Sentinel Range at site 26 in Figure 3. Here, the striation troughs are all within the oxidation rind.

trate into the less-oxidized portion of the rind. Figures 20 and 21 show examples of fresh-appearing striations cut into oxidized bedrock surfaces. In each of these cases, the striations postdate most or all of the oxidation of the bedrock surface.

We found few striations on bedrock in the southern Heritage Range. Here carbonate units of the Minaret Group show surface granular disintegration that may have rapidly removed striations. However, basic igneous erratics litter the bossed carbonate outcrops. These erratics, as discussed below, are derived from the Soholt Peaks of the Heritage Range. Many are unweathered and show surface polish and striations.

Alpine topography is well preserved below the trimline. Horns, ridges, and cirques (Figs. 14–16) are nearly unmodified, particularly in the Crashsite Group. The major difference in ridge-crest morphology across the trimline is that (1) ridges below the trimline are blunt and lack spires present on sharp ridges above the trimline and (2) cols below the trimline have been rounded, whereas those above the trimline remain sharp. Extensive bossing and stoss-and-lee molding of alpine peaks and ridges are largely absent. The two exceptions to this general observation are that low-level quartzite and conglomerate ridges beside major glaciers show planing in the Heritage Range and eastern Sentinel Range (Fig. 22) and that carbonate bedrock of the Minaret Group in the southern Heritage Range shows bossing.

Drift and moraines are relatively rare in the Ellsworth Mountains. Most striated bedrock surfaces lack covering drift and erratics. Moraines, drift, and erratics are absent at the trimline. The rare drift patches are concentrated in ice-free areas ablated by katabatic winds flowing through the Heritage Range, although a few patches occur in the Sentinel Range. Drift patches are partic-

Figure 20. Striation gouge with unoxidized trough that penetrates oxidation rind on Crashsite Group quartzite in the northern Sentinel Range at site 3 in Figure 3.

ularly common in the southern Soholt Peaks, Edson Hills, and Enterprise Hills. Many small drift patches occur on the lee slopes of the Watlack Hills and Gifford Peaks. Most drift patches are thin and discontinuous and lack constructional morphology; they therefore cannot be shown at the scale of Figure 3 (Plate II, in pocket). Aside from ice-cored medial and lateral moraines in or beside blue-ice areas of active glaciers, we found moraine ridges only in the Edson Hills.

Soil development below the trimline

Soil development and surface boulder weathering were examined on three types of drift surfaces. The first type was young ice-cored drift of moraines adjacent to blue-ice areas near Mount Dolence and Mount Twiss in the Heritage Range. These ice-cored moraines were typical of those that occur discontinuously beside blue-ice areas elsewhere in the Heritage Range. The second type was intermediate-age drift patches of ground moraine and sparse lateral moraine segments between the trimline and present ice level in the Heritage Range as well as in the low, eastern foothills of the Sentinel Range. The third type was old drift that projects as "windows" through the more extensive, intermediate-age drift in several restricted sites.

Soils on young, ice-cored moraines (profiles 79-24 and 79-32 in the Heritage Range in Fig. 3) feature a loose, unoxidized layer 3 to 17 cm thick (Cn) resting directly on glacial ice (Tables 2 and 3). The soils contain no visible salts or ghosts. Surface boulder weathering features were measured on two plots (79-24 and 79-32 in Fig. 3). None of the quartzite boulders is striated, 50 percent are ventifacted, 17 percent are spalled, and only 2 percent are pitted (Table 4). Chemistry of 1:5 soil:water extracts from profile 79-32 shows that these soils contain very low concentrations of salts (Table 5). Electrical conductivity is 0.05 dSm^{-1}. Calcium and sulfate are the dominant cation and anion, respectively. In the classification scheme of Campbell and Claridge (1975), these soils are classified as weathering stage 1 (no salts or soil horizonation; fresh surface boulders).

Soils on intermediate-age drift are unoxidized or have an 8- to 23-cm thick weakly oxidized layer (Cox) over unoxidized drift (Table 2). Permafrost with associated interstitial ice occurs at depths ranging between 18 cm and >110 cm, depending on aspect and on proximity to glaciers or snowpatches. Salt encrustations occur beneath coarse fragments to depths exceeding 100 cm. All but two of these soils reacted vigorously to 10 percent HCl, indicating the presence of carbonates, to depths from 25 to >100 cm. Argillite and marble ghosts occur to depths of >70 and 38 cm in profiles 79-17 and 79-23, respectively. Average surface boulder frequency is 530 boulders per 314 m^2 (Table 4). An average of 6 percent of the quartzite boulders are striated, 69 percent are ventifacted, 24 percent are spalled, and 11 percent are pitted. Electrical conductivity of the salt-enriched layer in these soils of the Heritage Range ranges between 0.46 and 3.2 dS m^{-1} (average = 1.7 dS m^{-1}) (Table 5). Calcium and sulfate are the dominant cation and anion, respectively. Because they have a few salt flecks and weak horizonation, and because their surface boulders have light staining and show some disintegration, these soils are classified as weathering stage 2 (Campbell and Claridge, 1975).

"Windows" of old drift occur on the Anderson Massif (profile 79-18), in the Edson Hills (79-20), on a peak in the Sentinel Range (79-30), and probably elsewhere in the Ellsworth Mountains. Profile 79-20 is representative of soils on this old drift; it is deeply oxidized and features a salt-indurated horizon (Table 2). Profile 79-18 has ghosts derived from igneous rocks to a depth of 20 cm. Surface boulder frequency is slightly lower (310) than on intermediate-age drift (Table 4). No striated boulders were found on the surface of the old drift. Although the percentage of spalled boulders is similar on the two drifts, there are more pitted and ventifacted boulders on the old drift. However, these differences

Figure 21. Striation gouges with unoxidized trough that penetrates oxidation rind on Crashsite Group quartzite on summit of Landmark Peak (1,840 m) in the northern Heritage Range at site 46 in Figure 3. Landmark Peak is a steep glacial horn that lies below the trimline.

TABLE 2. MORPHOLOGY OF SOILS ON DRIFT AND MORAINES IN THE ELLSWORTH MOUNTAINS*

Profile Number	Location and Approximate Elevation of Soil Pit (m)	Thickness of Solum (cm)	Depth of Coherence (xm)	Depth of Ghosts (cm)	Depth to ice or Ice Cement (cm)	Salt Stage†
			Weathering stage 1			
79-24	Mt. Dolence (1,200)	0	17	0	17	0
79-32	Mt. Twiss (1,200)	0	3	0	3	0
			Weathering stage 2			
79-15	Edson Hills (1,000)	0	57	0	>60	0
79-16	Edson Hills (1,100)	0	80 +	0	>80	I
79-17	Edson Hills (900)	0	15	70 +	>70	0
79-19	Carnell Peak (1,300)	0	42	0	>42	I
79-21	Welcome Nanatak (1,476)	0	24	0	>24	I
79-22	Edson Hills (1,200)	0	17	0	>17	I
79-23	Edson Hills (1,100)	0	50	38	50	I
79-25	Mt. Dolence (1,100)	0	75 +	0	>80	I
79-26	Mt. Twiss (1,200)	0	75 +	0	>80	I
79-27	Dickey Peak (1,390)	0	18	0	18	III
79-28	Flowers Hills (1,200)	0	20	0	>20	I
79-29	Flowers Hills (600)	0	20	0	20	I
79-31	Mt. Twiss (1,400)	0	100 +	0	>100	I
79-33	Meyer Hills (600)	0	14	0	33	I
79-34	Mt. Fordell (1,300)	0	100 +	0	>100	I
79-35	Parrish Peak (950)	0	75 +	0	>85	I
79-36	Cagle Peaks (1,300)	0	33	5	>110	I
			Weathering stage 3			
79-18	Anderson Massif (1,800)	12	13	20	>25	I
79-20	Edson Hills (1,200	0	59	0	>100	IV
79-30	Peak 1600 (1,600)	0	15	0	15	IV

*Locations of soil profiles shown in Figure 3.
†0 = none visible; I = salt encrustations beneath clasts; II = few salt flecks (0.5 to 2 mm diameter); III = abundant salt flecks; IV = weakly cemented salt pan.

are not statistically significant because of high standard deviations. Soils on the old drift have EC values in excess of 1.6 dS m^{-1} throughout the profiles, indicating some movement of soluble salts (Table 5). As with the younger soils, these old soils contain dominantly calcium and sulfate. Because they contain abundant salts and have distinct horizonation and because their surface boulders have been pervasively ventifacted, these old soils represent weathering stage 3 (Campbell and Claridge, 1975).

Data pertaining to the distribution and origin of salts were obtained by sampling soils on drifts composed predominantly of argillite (79-17), igneous rocks (79-18), marble (79-34), conglomerate (79-27 to 79-29), quartzite-phyllite (79-15, 79-16), and quartzite (79-19 to 79-26, 79-30 to 79-33, 79-35, 79-36). In all of these soils, the dominant cation in water extracts is calcium and the dominant anion is sulfate (Fig. 23). In addition, x-ray diffraction analysis of salt encrustations reveals the widespread occurrence of gypsum ($CaSO_4 \cdot 2H_2O$) in the Heritage Range. Other salt minerals include calcite ($CaCO_3$) and mirabilite ($NaSO_4 \cdot 10H_2O$). Soda niter ($NaNO_3$) occurs on Dickey Peak (78°20′S, 84°30′W) in the Sentinel Range.

Because $CaSO_4$ is dominant in these soils regardless of parent materials, we suggest that the salts probably originated from an external source such as marine aerosols. To test this hypothesis, the ratio of SO_4^{2-} to Ca^{2+} in soils of the Heritage Range was compared to that of seawater and igneous rocks. Whereas SO_4^{2-}/Ca^{2+} in the soils is 1.0 ± 0.2, SO_4^{2+} in seawater is 13.8, and in igneous rocks is 0.0089. Therefore, the ratio of SO_4^{2-} to Ca^{2+} in soils is closer to that of seawater than that of igneous rocks. Chemistry of soil and water extracts was compared to that of snow collected in the Ellsworth Mountains and analyzed by Rutford (1969). Sodium was the dominant cation in snow, followed by Ca^{2+}, Mg^{2+}, and K^+ in descending order (Table 6). Chloride was the dominant anion, followed by SO_4^{2-} and Cl^-. SO_4^{2-}/Ca^{2+} of snow ranged between 0.33 and 0.84 (average = 0.60). That sodium was not the dominant cation in soils of the Ellsworth Mountains may be attributable to leaching. In most of the soils, EC remained in excess of 1.0 dS m^{-1} to depths greater than 50 cm. Likewise, sodium concentrations increased with depth in soils of the Ellsworth Mountains. In a field experiment involving $Na_{22}Cl$, Ugolini and Anderson (1973) demonstrated that Na^+

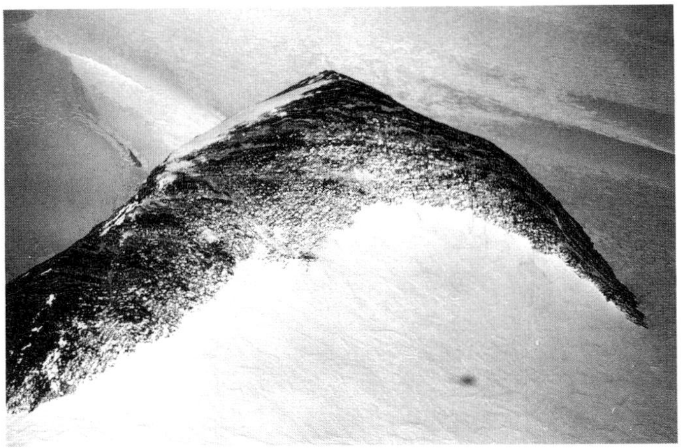

Figure 22. Higgins Nunatak (883 m), a low-level planed spur of Crashsite Group quartzite beside Union Glacier in the Heritage Range. Site 77 in Figure 3 is on Higgins Nunatak. View to east.

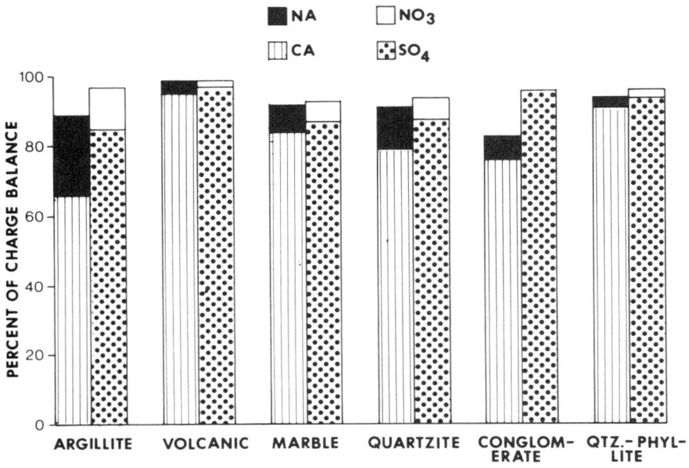

Figure 23. Charge balance of ions in water extracts from the horizon of maximum salt accumulation in soils in drift at various locations in the Ellsworth Mountains.

migrated from a shallow point source in permafrost in a soil in eastern Wright Valley.

In summary, soils of three different weathering stages were identified in drift in the Ellsworth Mountains. The youngest soils, which occur in ice-cored drift, show no horizonation or visible salts and feature fresh surface boulders. The intermediate-age soils, which are most common on drift patches in the Heritage Range between the trimline and present-day glaciers, have salt encrustations below coarse fragments to depths exceeding 50 cm, may be slightly oxidized in the upper 8 to 23 cm, and have surface boulders with light staining and some disintegration. The oldest soils, which occur on drift that projects as "windows" through intermediate-age drift, have numerous salt flecks or a salt pan, distinct horizonation, and ventifacted and pitted quartzite boulders. Gypsum ($CaSO_4 \cdot 2H_2O$) is the dominant salt in these soils and occurs in xerous and ultraxerous regions. Subxerous soils of the Ellsworth Mountains show evidence of leaching, particularly of Na^+ and Cl^-.

Our geologic field mapping, as well as soil and boulder-weathering studies, shows no surface weathering differences or widespread morphologic breaks in the intermediate-age drift that occurs as patches from close to the trimline down to young, ice-cored moraines beside present-day glaciers in the Heritage Range. Instead, intermediate-age drift shows relatively uniform surface morphology and weathering characteristics over this entire elevation range. Moreover, this intermediate-age drift in the Heritage Range is similar in morphology and weathering to the few scattered drift patches in the Sentinel Range.

In view of these data, we conclude that the intermediate-age drift, although discontinuous in distribution, represents a single drift unit. We here named this surface drift the Ellsworth Drift. There are no stratigraphic exposures of Ellsworth Drift, and therefore, soil profiles were used for type sections and parasections. The type section is chosen as soil profile 79-17 (Fig. 24) in

Figure 24. Type section of Ellsworth Drift (soil locality 79-17 in Fig. 3).

TABLE 3. DESCRIPTIONS OF SOIL PROFILES IN THE ELLSWORTH MOUNTAINS*

Profile Number	Horizon	Depth (cm)	Munsell Color (dry)	Field Texture	Structure	Consistency Dry	Consistency Wet	Reaction to Acid	Lower Boundary	Salt Morphology	Weathering Stage
79-24	D	0–1/2	var.	gs	osg	dlo	wso/wpo	e	vas	I	1
	Cln	1/2–17	2.5Y 7/4	vgs	osg	dso	wso/wpo	e	aw	I	
	C2ic	17–23	—	vgs	m-osg	dsh	wso/wpo	ved-e	—	I	
79-32	D	0–3	var.	g	osg	dlo	wso/wpo	nr	vaw	I	1
	Cic	3–8	—	—	—	—	—	—	vaw	0	
79-15	D	0–1/2	var.	vgs	osg	dlo	wso/wpo	—	vas	I	2
	Cln	1/2–39	2.5Y 7/4	vgsl	m-osg	dso	wss/wps	es	aw	0	
	$IIC2_n$	39–57	5Y 7/1	vgl	osg	dlo-dso	ws/wp	es	—	0	
79-16	D	0–18	2.5Y 7/2	vstls	osg	dlo	—	e	aw	I	2
	Cln	18–10	2.5Y 7/4	vgls	osg	dso	wss/wps	ve	cw	I	
	C2n	40–58	2.5Y 7/4	vgls	osg	dso	wss/wps	e	aw	I	
	C3n	58–80	2.5Y 7/2	vstls	osg	dso	wss/wpo	e	—	0	
79-17	D	0–1/2	var.	vgs	osg	dlo	wso/wpo	ve	aw	I	2
	Cln	1/2–15	2.5Y 7/4	vstls	m-osg	dlo-dso	wso/wpo	ve	aw	0	
	C2n	15–70	2.5Y 7/4	vstls	osg	dlo	wss/wpo	ve	—	0	
79-19	D	0–1	var.	vgs	osg	dlo	swo/wpo	ve	as	I	2
	Cln	1–25	10YR 7/4	vgs	osg	dso	wso/wpo	ve-e	aw	I	
	$IIC2_{ox}$	25–42	10YR 7/6	vgls	osg	dso	wss/wpo	nr	aw	I	
79-21	D	0–1/2	var.	vgs	osg	dlo	wso/wpo	—	as	I	2
	Cln	1/2–13	2.5Y 8/4	gs	m-=osg	dh	wso/wpo	nr	aw	I	
	CR	13–24	2.5Y 7/4	vks	m-osg	dsh	wso/wpo	nr	sw	I	
79-22	D	0–1/2	var.	vgs	osg	dlo	wso/wpo	ev	as	I	2
	Clox	1/2–14	10YR 7/4	vgs	m-osg	dsh	wso/wpo	es	vaw	I	
	IIC2n	14–17	5Y 8/1	gs	osg	dso	wso/wpo	e	aw	0	
79-23	D	0–1	5Y 6/2	vgs	osg	dlo	wso/wpo	e	vas	I	2
	Clox	1–10	2.5Y 7/4	vgs	osg	dso	wso/wpo	es	aw	I	
	C2n	10–3–1	2.5Y 7/4	vgs	osg	dlo-dso	wss/wpo	es	aw	I	
	C3n	34–46	2.5Y 7/4	vgs	osg	dso	wso/wpo	e-es	aw	I	
	C4ox	45–50	2.5Y 7/4	vgs	osg	dso	wso/wpo	es	aw	I	
	C5ic	50+	—	—	osg	—	wso/wpo	e	—	I	
79-25	D	0–1/2	var.	vgs	osg	dlo	swo/wpo	—	as	I	2
	Clox	1/2–10	2.5Y 7/2	vgs	m-osg	dso	wso/wpo	es	aw	I	
	C2ox	10–20	2.5Y 7/4	vgs	m-osg	dh	wso/wpo	es	aw	I	
	C3ox	20–30	5Y 8/2	gs	osg	dso	wss/wps	es	aw	I	
	C4n	30–50	5Y 8/1	vgls	osg	dso	wss/wps	es	aw	I	
	C5n	53–75	5Y 8/1	vstls	osg	dso	wss/wps	es	—	0	
79-26	D	0–5	—	g	osg	dlo	wso/wpo	—	aw	1	2
	B2	5–11	10YR 7/4	vgs	m-osg	dh	wso/wpo	nr	aw	I	
	Cln	11–48	10YR 7/4	gs	osg	dlo-dso	wso/wpo	nr	aw	I	
	C2n	48–81	10YR 7/3	vks	osg	dlo-dso	wso/wpo	ve	—	0	
79-27	D	0–1/2	var.	g	osg	dlo	wso/wpo	—	as	I	2
	C1OX	1/2–18	10YR 6/6	vgs	m-osg	dsh	wso/wpo	nr	ai	I	
	IIC_{ic}	18+	—	vgs	—	dsh	wso/wpo	—	—	0	

TABLE 3. DESCRIPTIONS OF SOIL PROFILES IN THE ELLSWORTH MOUNTAINS* (Continued)

Profile Number	Horizon	Depth (cm)	Munsell Color (dry)	Field Texture	Structure	Consistency Dry	Consistency Wet	Reaction to Acid	Lower Boundary	Salt Morphology	Weathering Stage
79-28	D	0–1/2	var.	vgs	osg	dlo	wso/wpo	—	as	I	2
	Cox	1/2–20	10YR 7/4	vgls	m-osg	dsh	wso/wpo	—	aw	I	
79-29	D	0–1/2	var.	g	osg	dlo	wso/wpo	—	as	I	2
	Cln	1/2–9	10YR 7/4	vgs	osg	dso	wso/wpo	—	aw	I	
	IIC2n	9–20	10YR 4/1	vgs	m-osg	dso	wso-wpo	—	—	I	
79-31	D	0–4	var.	vs	osg	dlo	swo/wpo	nr	vas	I	2
	Clox	4–2	—	vgs	m-osg	dso	wso/wpo	nr	aw	I	
	C2n	26–34	10YR 7/4	vgs	m-osg	dso	wso/wpo	nr	aw	I	
	C3n	34–65	10YR 7/4	vgs	osg	dso	wso/wpo	nr	—	I	
	C4n	65–100	10YR 7/3	vgs	osg	dso	wso/wpo	nr	—	I	
79-33	D	0-1	var.	vgcos	osg	dlo	wso/wpo	e	vas	I	2
	Clox	1–6	5Y 7/4	vgls	m-osg	dso	wss/wpo	e	aw	I	
	C2ox	6–14	5Y 6/3	vgls	m-osg	dsh	wss/wpo	e-ve	aw	I	
	C3n	14–33	5Y 6/3	vgls	osg	dlo-dso	wss/wpo	ve	aw	I	
79-34	D	0–2	var.	vgcos	osg	dlo	wso/wpo	ev	vas	I	2
	Cln	2–17	10YR 8/2	gls	osg	dso	wss/wpo	ev	aw	I	
	C2n	17–60	10YR 8/2	vgs	osg	dso	wso/wpo	ev	—	I	
	C3n	60–100	10YR 8/2	vgs	osg	dso	wso/wpo	ev	—	I	
79-35	D	0–2	var.	vgcos	osg	dlo	wso/wpo	e	vas	I	2
	Clox	2–25	10YR 7/3	vgls	osg	dso	wss/wpo	e	aw	I	
	C2n	25–37	10YR 8/3	vgls	osg	dso	wss/wpo	e	cw	I	
	C3n	37–75	2.5Y /2	vgls	osg	dso	wss/wpo	e	—	I	
79-36	D	0–3	var.	vgcos	osg	dlo	wso/wpo	—	vaw	I	2
	Cln	3–33	2.5Y 7/2	vgls	osg	dso	wss/wpo	—	aw	I	
	C2n	33–67	2.5Y 7/4	vgls	osg	dlo-dso	wss/wpo	—	aw	I	
	C3n	67–100	2.5Y 7/4	vgls	osg	dlo-dso	wss/wpo	—	—		
79-18	D	0–1/2	var.	vgs	osg	dlo	wso/wpo	es	as	I	3
	B2	1/2–13	7.5YR 7/4	vgs	osg	dso	wso/wpo	es	ai	I	
79-20	D	0–1/2	var.	vgs	osg	dlo	wso/wpo	—	as	I	3
	Cln	1/2–7	10YR 7/2	vgls	osg	dlo-dso	was/wps	ev	aw	I	
	C2sa	7–18	10YR 7/3	vgls	m-osg	dh	wss/wps	ev	vaw	IV	
	IIC3ox	18-37	10YR 7/6	vgs	m-osg	dsh	wss/wpo	es	aw	I	
	IIC4ox	37–59	10YR 7/4	vgs	osg	dso	wso/wpo	es	aw	I	
	IIC5ox	59–90	10YR 7/4	vls	osg	dlo-dss	wso/wpo	es-ev	—	I	
79-30	D	0–1/2	var.	vgs	osg	dlo	wso/wpo	—	as	0	3
	Clsa	1/2–15	—	—	m-osg	dsh	—	—	aw	IV	

Abbreviations: Munsell soil color - var. = variegated; field texture - s = sand, ls = loamy sand, sl = sandy loam, l = loam, cos = coarse sand, g = gravelly, k = cobbly, st = stony, v = very; structure - osg = structureless, single grain, m = massive; dry consistency - dlo = loose, dso = weakly coherent, dsh = slightly hard, dh = hard; wet consistency - wso = nonsticky, wss = sticky, ws = sticky, wpo = nonplastic, wps = slightly plastic, wp = plastic; reaction to HCl - ve = very slightly effervescent, e = slightly effervescent, es = strongly effervescent, ev = violently effervescent; lower boundary - va = very abrupt, a = abrupt, c - clear, s = smooth, w = wavy, i = irregular; salt morphology - 0 = no salts, I = salt encrustations, IV = weakly cemented salt pan.

TABLE 4. WEATHERING OF SURFACE BOULDERS AT SOIL-PROFILE SITES ON DRIFT AND MORAINES IN THE ELLSWORTH MOUNTAINS*

Profile Number	Surface Boulder Frequency (per 314 m^2)	Percent of Quartzite Boulders			
		Striated	Ventifacted	Spalled	Pitted
	Data on soils assigned to weathering stage 1 (see Table 2)				
79-24	420	0	69	30	1
79-32	250	0	32	4	4
Average	335	0	50	17	2
	Data on soils assigned to weathering stage 2 (see Table 2)				
79-15	525	14	90	24	2
79-16	710	0	34	4	7
79-17	525	37	92	14	28
79-19	205	3	93	25	3
79-23	405	0	93	47	24
79-25	710	1	34	20	0
79-26	575	0	55	18	7
79-35	365	0	81	60	31
79-36	750	0	45	6	0
Average	530	6	69	24	11
	Data on soils assigned to weathering stage 3 (see Table 2)				
79-20	310	0	97	18	17

*Locations of soil profiles shown in Figure 3.

the Edson Hills, and reference sections are profiles 79-22 (Fig. 25) in the Edson Hills and 79-36 at Cagle Peaks. Table 7 gives descriptions of the type section and reference sections. Descriptions of the other soils are given in Table 3. Figure 3 (Plate II, in pocket) shows locations of soil profiles.

We do not name the young ice-cored drift because we cannot determine if it is a widespread drift unit that represented a significant glacial event. Nor do we name the old drift, because it occurs only in isolated "windows," and we have no way of determining whether it represents single or multiple units.

Glacial erratics occur on bedrock surfaces in some ice-free areas in both the Sentinel and Heritage Ranges. Most commonly, the erratics are isolated and widely scattered. However, they litter low cols in the Sentinel Range and bossed carbonate bedrock of the Marble Hills in the southern Heritage Range. For example, rounded cols between opposing cirques in the southern Sentinel Range show concentrated erratics and glacially transported clasts of local bedrock resting on striated surfaces. In the Sentinel Range, the domination of bedrock outcrops of the Crashsite Group limits usefulness of glacially transported clasts for determining higher ice levels except in specific areas. The Crashsite Group has been folded and uplifted so that it is topographically higher than younger stratigraphic units. In the Heritage Range the greater diversity of bedrock types and limited outcrop area for some units allows better definition of higher ice levels from the distribution of erratics.

In the Sentinel Range, the presence of blocks of the distinctive middle light unit (Mount Liptak Formation) of the Crashsite Group resting on surfaces both stratigraphically and topographically higher than adjacent outcrops of the unit strongly suggests ice levels higher than today's. In the northeastern and eastern Sentinel Range, erratics of the Crashsite Group resting on outcrops of Whiteout Conglomerate and Polarstar Formation afford strong evidence for higher ice levels and in some cases show overriding of the entire outcrop.

In the Heritage Range the diversity of rock types allows tracing of erratics upglacier to their source. For example, erratics mantling the Marble Hills are dominated by basic igneous rocks that crop out in the Soholt Peaks to the north and west. In the Mount Dolence area, Crashsite Group blocks can be traced to their source area to the west and north.

The distribution of erratics shows that the Heritage Range has been largely inundated by inland ice. The high peaks probably were nunataks, as suggested by the lack of erratics on outcrop areas 900 to 1,000 m above present ice levels. In the Sentinel Range the evidence from erratics of higher ice levels is less compelling because of the more uniform rock types and because of the structural and topographic relations. However, erratics on Whiteout Conglomerate and Polarstar outcrops provide evidence for strong eastward ice flow from the quartzite core of the range.

Figure 25. Reference section of Ellsworth Drift in the Edson Hills (soil locality 79-22 in Fig. 3).

DISCUSSION

Former ice surface and flowlines: Ellsworth Mountains

The interpretation of the widespread trimline in the Ellsworth Mountains, as well as the Ellsworth Drift and striations that lie below the trimline, is central to our discussion, which is similar to that concerning trimlines in the mountains of northern Victoria Land (Denton and others, 1986a). The Ellsworth Mountains trimline is an erosional feature etched into alpine topography. It separates two alpine landscapes: The higher is dominated by ridges showing serrations and no evidence of ice passage; the lower by ridges that lack serrations and show signs of ice overriding. In these characteristics the trimline in the Ellsworth Mountains is similar to that previously described for northern Victoria Land (Denton and others, 1986a).

Several possible interpretations of this trimline can be made. One is that it represents a change in thermal regime beneath thick covering ice. In this scenario, warm ice below the thermal boundary had a thawed bed and eroded pinnacles, leaving drift patches and glacially striated bedrock as record of its passage. Shallower cold ice above the thermal boundary had a frozen bed that protected pinnacles and left no evidence of ice passage. If the trimline marks such an internal thermal boundary, it affords a minimum value for former ice expansion, as thick cold ice would have occurred above the thermal boundary.

A second interpretation of the trimline is that it represents a former ice limit superimposed on preexisting alpine glacial topography. Of course, the trimline cannot precisely mark a former ice surface, because erosion of pinnacles would have required ice whose thickness may have amounted to 100 m. We favor this alternative for several reasons. It explains adequately the physical characteristics of the trimline. Trimline elevations vary smoothly, are not determined by bedrock lithology or structure, and are consistent with a former ice surface. Further, the striations below the trimline are perpendicular to former ice-surface contours as determined by trimline elevations. Moreover, above the trimline, there is no evidence of ice passage, such as glacially transported clasts lodged between pinnacles or preferential removal of delicate pinnacles. We conclude by noting that very few erratics occur near the trimline. This suggests that the trimline was close to an upper ice surface that was largely in the accumulation area. A glacier generally transports few erratics in its upper zone in areas above the firn line, where ice flowlines are directed downward. Therefore, the small number of erratics is in accord with our conclusion that the trimline represents an upper ice surface.

The remainder of our discussion is based on the assumption that the trimline approximately represents a former ice surface. Should the internal thermal argument prove to be correct, the expanded ice system would have been even thicker than is outlined below.

Does the former ice surface marked by the trimline merely represent expansion of local alpine glaciers or does it reflect a thicker West Antarctic Ice Sheet that flowed seaward around and through the Ellsworth Mountains? This question is more easily answered for the Heritage Range. Here, the trimline occurs only on high peaks and exhibits a regional slope consistent with inland ice overflow across the range rather than local outward flow of alpine ice. In accord with this, the regional striation pattern similarly indicates ice-sheet flow across the range. In particular, striations show ice flow directed eastward across high ridges and cols between opposing cirques, as well as across summits of several glacial horns (Plate II, Fig. 3, in pocket). Moreover, erratics have been transported by southeastward flow from near the Soholt Peaks to the southern Heritage Range, as well as by eastward flow across the range. It is highly unlikely that expansion of alpine glaciers could have caused this combination of features. Therefore, we conclude that the trimline in the Heritage Range records a higher-than-present West Antarctic Ice Sheet that flowed through and around the range, leaving only local serrated peaks projecting as nunataks above the ice surface.

The interpretation of the trimline in the Sentinel Range in terms of inland or alpine ice is less straightforward. Here, striation trends on bedrock below the trimline generally indicate outward

TABLE 5. CHEMISTRY OF 1:5 SOIL-WATER EXTRACTS FROM SOILS IN THE ELLSWORTH MOUNTAINS*

Profile	Depth	Ph	EC	Na^+	Ca^{2+}	Mg^{2+}	K^+	Cl^-	NO_3^-	SO_4^{2-}	HCO_3^-
			$dS\ m^{-1}$				meq l^{-1}				
			Soils assigned to weathering stage 1 (see Table 2)								
79-32	3–8	7.4	0.05	0.04	0.27	0.03	0.03	0.11	0.009	0.20	0.1
	8+	7.3	0.05	0.09	0.22	0.04	0.04	0.11	0.02	0.12	0.08
			Soils assigned to weathering stage 2 (see Table 2)								
79-15	1/2–20	7.2	1.6	1.2	20	1.6	0.05	0.85	1.4	18	0.06
	20–39	7.2	2.3	1.2	32	2.0	0.06	1.4	1.1	30	0.06
	39–57	7.1	1.5	1.3	19	1.2	0.04	0.39	0.93	18	0.05
79-17	1/2–15	7.9	0.18	0.52	1.0	0.10	0.02	0.03	0.05	1.0	0.32
	15–40	7.3	0.97	2.7	7.6	1.3	0.04	0.25	1.2	8.4	0.08
	40–70	7.4	0.92	2.4	7.4	1.2	0.03	0.25	1.0	9.3	0.10
79-19	1–25	7.3	1.4	2.6	14	2.4	0.12	1.5	2.3	12	0.08
	15–42	6.8	0.96	2.7	5.9	1.7	0.08	1.4	1.9	5.6	0.02
79-21	1/2–13	7.3	0.46	0.41	4.1	0.23	0.08	0.03	0.003	4.6	0.08
	13–24	7.3	0.18	0.31	1.0	0.13	0.04	0.03	0.006	1.2	0.08
79-25	1/2–10	8.1	0.07	0.02	0.28	0.18	0.07	0.01	0.01	0.08	6.3
	10–20	8.0	0.13	0.11	0.65	0.23	0.10	0.03	0.01	0.54	0.4
	20–30	8.4	0.98	1.5	6.4	2.0	0.36	2.3	0.69	6.3	1.0
	30–52	8.0	1.6	1.6	15	2.7	0.41	2.1	1.0		0.4
	52–75	8.1	1.7	4.4	12	1.8	0.44	3.6	2.1	9.3	0.5
79-26	0–6	7.7	0.23	0.96	0.70	0.09	0.19	0.08	0.64	1.0	0.2
	6–43	7.3	2.6	4.7	31	3.0	0.46	2.2	2.1	32	0.08
	43–75	7.1	2.6	3.8	31	2.3	0.38	1.7	1.8	31	0.05
79-27	1/2–18	7.4	0.31	0.39	4.2	0.78	0.13	0.09	0.02	2.4	0.10
79-33	1–6	8.2	0.57	0.26	2.7	0.81	0.10	0.96	0.16	4.6	0.63
	6–14	7.6	0.08	1.4	0.15	0.16	0.06	0.45	0.26	0.50	0.16
	14–33	7.8	0.05	0.16	0.23	0.07	0.03	0.01	0.004	0.10	0.25
	33 +	8.4	0.07	1.9	0.32	0.08	0.02	0.08	0.03	0.09	1.0
79-34	2–17	7.4	2.7	3.2	33	2.8	0.18	2.7	2.1	31	0.10
	17–60	8.5	0.80	2.0	3.7	1.9	0.10	2.0	2.1	2.9	1.3
	60–100	8.0	0.56	1.3	2.9	1.2	0.08	1.4	2.8	2.1	0.40
79-35	2–25	8.0	3.2	7.8	34	2.8	0.44	2.9	6.4	33	0.40
	25–37	7.1	2.5	6.0	33	3.2	0.49	2.9	4.3	31	0.05
	37–75	7.3	2.5	4.7	25	2.3	0.41	2.4	4.1	22	0.08
79-36	3–33	7.3	1.9	2.7	20	1.8	0.23	1.6	1.3	14	0.08
	33–67	7.2	2.7	3.7	31	3.3	0.26	2.5	2.0	34	0.06
	67–100	7.2	2.7	3.8	32	3.1	0.26	2.4	2.5	31	0.06
			Soils assigned to weathering stage 3								
79-18	1/2–13	7.5	2.2	1.2	32	0.47	0.17	0.39	0.06	30	0.13
79-20	1/2–7	7.7	1.7	0.61	22	1.6	0.16	0.58	0.04	17	0.20
	7–18	7.5	1.6	0.40	18	1.1	0.12	0.48	0.16	19	0.13
	18–37	7.5	2.8	3.6	33	4.0	0.18	5.1	2.1	30	0.13
	37–59	7.2	2.7	4.4	24	4.9	0.18	5.9	3.0	19	0.06
	59–90	7.4	3.0	4.5	32	4.3	0.18	5.4	2.6	29	0.10
79-30	1/2–15	5.4	11	162	14	28	0.77	1.9	11	173	0

*Locations of soil profiles shown in Figure 3.

TABLE 6. CHEMISTRY OF SNOW SAMPLES COLLECTED IN THE ELLSWORTH MOUNTSINS*

Station	Na^+	K^+	Ca^{2+}	Mg^{2+} (g/ml)	$C1^-$	SO_4^{2-}	NO_3	SO_4^{2-}/Ca^{2+} (meq/1)
320	0.40	0.04	0.13	0.05	0.52	0.24	0.08	0.77
464	0.23	0.10	0.09	0.03	0.46	0.18	0.05	0.84
496	0.16	0.05	0.07	0.02	0.20	0.14	0.05	0.83
940	0.12	0.03	0.08	0.02	0.32	0.08	0.06	0.42
1008	0.15	0.07	0.09	0.03	0.34	0.07	0.05	0.33
E-1	0.04	<0.02	0.43	0.02	0.16	<0.05	0.01	....
E-2	0.04	<0.02	0.02	0.01	0.18	0.02	0.06	0.42
E-3	0.05	<0.02	0.04	0.02	0.18	<0.1	0.02	....
E-4	0.27	<0.02	0.15	0.02	0.15	<0.1	0.38	....

*Location of stations given in Rutford (1969). Analysis by D. W. Fisher, U.S. Geological Survey.

flow of thicker alpine glaciers, except in the northern and southern sectors of the range. Thicker cirque glaciers flowed outward from the western ridge toward the inland ice sheet, while thicker egress glaciers drained eastward into the Rutford Ice Stream. Several factors indicate that these local glaciers thickened in concert with a higher-than-present West Antarctic Ice Sheet and therefore that the trimline indirectly records higher inland ice. First, striations near Mount Hubley trend north-south (Plate II, Fig. 3, in pocket), suggesting that higher West Antarctic inland ice deflected cirque glaciers flowing westward from the Sentinel Range. Second, striations on high peaks in the Bastien Hills, which lie below the trimline and occur west of the Nimitz Glacier, record outflow from the local West Antarctic dome west of the Sentinel Range (Figs. 2 and 3). Third, cols between opposing cirques in the northern and southern sectors of the range have been overrun by eastward flowing ice (Plate II, Fig. 3, in pocket), a situation best explained by thicker inland rather than alpine ice. Fourth, trimline elevations in the eastern Sentinel Range indicate a significantly higher Rutford Ice Stream. This ice stream drains both local alpine ice and the east side of the low inland divide. Finally, regional trimline elevations in the Sentinel Range match without break those in the Heritage Range, where the trimline is clearly related to higher inland ice (Plate II, Fig. 3, in pocket). For these reasons we believe that the trimline in the Sentinel Range represents thickening of alpine and local intermontane ice fields in concert with a higher-than-present West Antarctic Ice Sheet flowing around the range and through low southern and northern cols. In several cases in the Heritage Range and the northern Sentinel Range, serrated spurs on lee slopes of smooth ridges show localities where inland ice flowing eastward across the mountains barely spilled over major north-south– trending ridges.

Field data given in Figure 3 (Plate II, in pocket) allow reconstruction of former ice-sheet flowlines from the regional slope of the trimline and from striation trends, particularly those at high elevations close to the trimline. The overall conclusion is that the directions of former flowlines were similar to those of present-day flowlines. Cirque-headed glaciers flowing westward from the high, west-central Sentinel Range were deflected north and south to join inland ice flowing around the range. Ice from the West Antarctic divide inland of the Heritage Range and from the dome west of the Sentinel Range flowed seaward through the thicker Minnesota and Nimitz Glaciers, merging with local alpine ice to inundate low peaks and cols in the southern Sentinel Range. Likewise, thicker eastward-flowing inland and alpine ice covered low peaks and cols in the northern Sentinel Range. Cirque-headed outlet glaciers and local intermontane ice fields in the eastern Sentinel Range formed tributaries of a higher Rutford Ice Stream.

Seaward-flowing inland ice in the Heritage Range inundated all but the highest peaks of the western ridge. Inland ice spilled over the western ridge and flowed eastward and northeastward through the range, largely following drainage systems that now feed Splettstoesser and Union Glaciers. However, some ice also spilled seaward across ridges between drainage systems. In the southern Heritage Range, flow was southeastward along Horseshoe Valley. The presence of Soholt Peaks erratics in the Marble Hills shows that, although some thicker West Antarctic ice spilled eastward over ridges of the Heritage Range, a major flow component was southeastward around the range, just as it is today.

The increase of West Antarctic ice thickness can be estimated by comparing trimline and present-day ice elevations. The present-day ice surface is highest on the inland ice divide west of the Sentinel and Heritage Ranges (Fig. 2). It slopes eastward and southeastward around and through the mountains. The trimline shows the same general trends, although with a shallower seaward slope (Plate II, Fig. 3, in pocket). As described in detail previously, the trimline attains its highest elevation of 3,000 m in the west-central Sentinel Range and slopes downward to 2,600 m in the north and 1,920 m in the south. In the east-central Sentinel Range, trimline elevations decrease eastward from 2,650 m near

TABLE 7. TYPE SECTION AND REFERENCE SECTIONS OF ELLSWORTH DRIFT*

Type section (holostratotype) of Ellsworth Drift

Profile: 79-17. Type locality of Ellsworth Drift.
Location: Heritage Range; first lateral moraine above Hyde Glacier in Edson Hills (79°48'S, 83°04'W).
Parent Material: Ellsworth Drift derived from quartzite and argillite with <5 percent phyllite.
Topography: Convex, simple slope; western aspect; elevation ~900 m.
Climate: Subxerous (Campbell and Claridge, 1969).
Date Described and Sampled: December 9, 1979.

Depth (cm)	Horizon	Description
0-1/2	D	Poorly expressed desert pavement ~85 percent cover of stained and striated quartzite and weathered argillite; variegated color; very gravelly sand; structureless, single grain; loose when dry, nonsticky and nonplastic when wet; carbonate encrustations, very slightly effervescent; abrupt, wavy lower boundary.
1/2-15	Cln	Pale yellow (2.5Y 7/4 d) very cobbly loamy sand, massive breaking down to structureless, single grain; slightly coherent to loose when dry, nonsticky and nonplastic when wet; very slightly effervescent; vesicular porosity; weathered argillite coarse fragments; clear, wavy lower boundary.
15-70	C2n	Pale yellow (2.5Y 7/4 d) very stony loamy sand; structureless, single grain; loose when dry; slightly sticky and nonplastic when wet; very slightly effervescent; seathered argillite coarse fragments.

Reference section (hypostratstype) of Ellsworth Drift

Profile: 79-36.
Location: Heritage Range; ground moraine at Cagle Peaks (79°35'S, 85°30'W).
Parent Material: Ellsworth Drift derived from quartzite and phyllite.
Topography: Simple, flat slope; elevation ~1,300 m.
Date Described and Sampled: December 27, 1979.

Depth (cm)	Horizon	Description
0-3	D	Somewhat poorly expressed desert pavement ~98 percent cover of varnished, angular quartzite; variegated color; very gravelly coarse sand; structureless, single grain; loose when dry, nonsticky and nonplastic when wet; <1 mm thick carbonate encrustations beneath coarse fragments; very abrupt, wavy boundary.
3-33	C1n	Light gray (2.5Y 7/2 d) very gravelly loamy sand; structureless, single grain; slightly coherent when dry, slightly sticky and nonplastic when wet; 1-2 mm thick carbonate encrustations; vesicular porosity in places; mineral soil coatings beneath coarse fragments; abrupt, wavy boundary.
33-67	C2n	Pale yellow (2.5Y 7/4 d) very gravelly loamy sand; structureless, single grain; loose and slightly coherent when dry, slightly sticky and nonplastic when wet; mineral coatings; clear, wavy boundary.
67-100	C3n	Pale yellow (2.5Y 7/4 d) very gravelly loamy sand; structureless, single grain; loose and slightly coherent when dry, slightly sticky and nonplastic when wet; mineral soil coatings beneath coarse fragments.

TABLE 7. TYPE SECTION AND REFERENCE SECTIONS OF ELLSWORTH DRIFT* (continued)

Reference section (hypostratotype) of Ellsworth Drift

Profile: 79-22.
Location: Heritage Range; ground moraine in Edson Hills (79°44'S, 83°55'W).
Parent Material: Ellsworth Drift derived from quartzite, phyllite, and conglomerate.
Topography: Simple, flat slope; elevation !1,200 m.
Date Described and Sampled: December 14, 1979.

Depth (cm)	Horizon	Description
0-1/5	D	Well expressed desert pavement ~90 percent cover of varnished quartzite coarse fragments with a matrix of phyllite flakes; variegated color; very gravelly sand; structureless, single grain; loose when dry, nonsticky and nonplastic when wet; carbonate encrusted beneath coarse fragments; violently effervescent; abrupt, smooth boundary.
1/5-14	Clox	Very pale brown (10YR 7/4 d) very gravelly sand; massive, breaking down to structureless, single grain; slightly hard when dry, nonsticky and nonplastic when wet; strongly effervescent carbonate encrustations beneath coarse fragments; very abrupt, wavy boundary.
14-17	IIC2n	White (5Y 8/1 d) gravelly sand; structureless, single grain; slightly coherent when dry, nonsticky and nonplastic when wet; slightly effervescent; mineral soil coatings beneath coarse fragments; abrupt, wavy boundary.
17+	IIR	Fractured phyllite and conglomerate bedrock.

*Locality numbers shown in Figure 3.

the western ridge to 2,250 m near the Rutford Ice Stream. Trimline elevations along the western margin of the Rutford Ice Stream decrease from 2,350 m in the north to 1,890 m in the south.

In the Heritage Range the trimline is generally significantly lower than in the Sentinel Range. The trimline slopes downward from west to east across the range and from northwest to southeast along the western boundary ridge of the Heritage Range. These elevations are consistent with inland ice flow southeastward around the range as shown by Soholt Peaks erratics in the Marble Hills.

The trimline thus shows a former ice-sheet surface similar to today's but with a shallower slope through and around the Ellsworth Mountains. Ice thickened 400 to 650 m along the western flank of the Ellsworth Mountains. In the Sentinel Range a value for thickening of 600 to 650 m was obtained by comparing trimline elevations on spurs between westward-facing cirques with the elevations at which these spurs intersect present ice level. In the western Heritage Range, values for ice thickening of >400 m at the head of Splettstoesser Glacier, 500 m near Soholt Peaks, 500 m near Enterprise Hills, and 400 m near Liberty Hills were obtained by comparing trimline elevations with present ice-sheet surface elevations west of the range.

The amount of ice thickening along the eastern flank of the Ellsworth Mountains cannot be accurately determined, because this flank is close to the grounding line and is marked by floating ice tongues, ice shelves, and ice rises with poorly defined grounding lines and submarine basal topography. Thus, it is uncertain where ice is floating or grounded. However, minimum estimates of ice thickening come from differences in elevations of the trimline and present-day ice surfaces. For the eastern Sentinel Range the surface of the Rutford Ice Stream was 1,900 m higher in its middle reaches opposite Mount Bentley and 1,500 m higher in its lower reaches. For the eastern Heritage Range, the lower Minnesota Glacier was 1,000 m higher opposite Anderson Massif. The ice-sheet surface near Mount Capley and Parrish Peak, which both lie close to the present-day grounding line, was 1,300 to 1,500 m higher than now.

Taken together, the trimline and striation data indicate that a higher-than-present West Antarctic Ice Sheet covered all but a few high peaks of the Heritage Range and that local glaciers simultaneously inundated lower slopes of the Sentinel Range. Before correction for isostatic compensation or any tectonic uplift, this ice-sheet surface attained elevations of 3,000 to 1,800 m on the interior flank and 2,350 to 1,700 m on the ice-shelf flank. This compares to modern ice-surface elevations of 2,300 to 1,400 m on the interior and 600 to 400 m on the ice-shelf flank. Therefore, thickening occurred on both flanks of the mountains but was greatest along the eastern flank close to the present-day grounding line between the West Antarctic Ice Sheet and the Ronne-Filchner Ice Shelf.

What was the basal thermal regime of ice that eroded the trimline? Rutford (1969, 1972) first noted that deglaciation from the trimline revealed remarkably well-preserved alpine glacial topography that had been buried beneath inland ice. We concur with this observation (Figs. 14–16). Only in low cols or near major glaciers has significant planing occurred. Major stoss-and-lee forms are not present. By itself, this would suggest that most ice passing across high-elevation ridges in both the Sentinel and Heritage ranges was cold based. On the other hand, it seems unlikely to us that cold-based ice could have eroded such a consistent trimline throughout the Ellsworth Mountains and at the same time left excellent striated bedrock surfaces near the trimline (Fig. 17). We think it more likely that ice with a thawed base accomplished the relatively minor erosion necessary to form the trimline, while at the same time leaving the overall alpine topography of the Ellsworth Mountains largely intact. If this inference is correct, it has important implications about paleoclimate. The presence of a thawed bed near the trimline requires extensive summer melting in the accumulation area, with consequent percolation and refreezing to raise the temperature of the snowpack. This process requires summer temperatures considerably higher than today's.

Former ice surface: West Antarctic Ice Sheet

Our conclusions from the Ellsworth Mountains bear on the configuration of the West Antarctic Ice Sheet when ice stood at the trimline. At this time a higher-than-present West Antarctic Ice Sheet poured seaward around and through the Ellsworth Mountains. Flowlines reconstructed from striation trends, distribution of erratics, and trimline elevations show that the major West Antarctic ice divide inland of the mountains (located over subglacial highlands and shown in Figs. 1 and 2) maintained its present position and attained a surface elevation 400 to 650 m higher than present (ignoring isostatic compensation). Trimline and striation data also indicate increased ice-surface elevations of 1,300 to 1,900 m along the eastern flank of the Ellsworth Mountains (ignoring isostatic compensation). We infer that substantial grounding of West Antarctic ice occurred in the Weddell Sea embayment when ice stood at the trimline.

Former ice surface: Multiple glaciation and age of the trimline

An important question concerns multiple glaciation. That is, were the trimline and associated erosional features cut during one expansion episode, or were several episodes involved? There are two indications that inland ice expansion may have been multiple. The first is the relation of striations to oxidized bedrock surfaces, particularly on the Crashsite Group. As noted previously, deep striations from the latest glaciation penetrate the oxidation rind and expose relatively fresh bedrock in their troughs. The oxidized surfaces that bear these scattered striations extend for considerable areas on high, blunted ridges below the trimline in both ranges, as well as on bossed and smoothed hills close to present ice levels beside major spillover glaciers in the Heritage Range. This relation suggests to us that much of the glacial erosion by overriding ice below the trimline may have occurred prior to the latest glaciation. But this conclusion must remain tentative until we understand bedrock weathering in the Ellsworth Mountains. The second is the few "windows" of old drift with well-developed soils that project through Ellsworth Drift in the Anderson Massif, Edson Hills, and eastern Sentinel Range. This argument is not as strong, because conceivably the drift "windows" could have been deposited by alpine glaciers or could have been patches of Ellsworth Drift that suffered extensive weathering.

What was the age of the West Antarctic ice-sheet expansion that carved the trimline? Current data allow at least two widely differing age models, each based on a distinct set of assumptions. One age model is that ice expansion to the trimline occurred during late Wisconsin time. This model, which was followed by Rutford and others (1980), is based on the assumption that poor soil development on drift, combined with excellent preservation of striations on isolated bedrock patches, indicates a young age for this expansion. The other age model, which makes no assumptions about soil development or preservation of striations, is that the expansion is pre–late Quaternary in age.

In our previous description, we purposely did not draw inferences about the origin of Ellsworth Drift in terms of West Antarctic or local alpine ice. However, given the previous interpretation of the trimline in terms of expanded West Antarctic ice, we can now state that Ellsworth Drift was most probably deposited during this West Antarctic expansion. We reach this conclusion about Ellsworth Drift because of its areal extent and distribution, its association with striations etched by overriding ice, and its superposition on older alpine erosional topography.

If these inferences are correct, we can estimate the age of the last West Antarctic ice expansion from soil development and surface rock weathering in Ellsworth Drift. The accuracy of such age estimates varies widely, depending on the area of Antarctica involved. In the McMurdo Sound area, two of us (JGB and GHD) are developing a master relative chronology of drift sheets based on soil development, surface rock weathering, geometric relation of drift units, and drift stratigraphy (Bockheim, 1979; Pastor and Bockheim, 1980; Stuiver and others, 1981). We are establishing an overall absolute chronology by using radiometric and fossil dates of individual drift sheets that are tied into the master relative chronology (Stuiver and others, 1981; Armstrong, 1978). We control the variables of climate, parent lithology, and topography in soil development by constructing the regional relative chronology in piecemeal fashion with independent geological and soil data. Hence, we consider our control in the McMurdo Sound area to be relatively accurate. Three of us (JGB, BGA, and GHD) have also developed a relative chronology of drift sheets for the Darwin-Byrd Glacier area (Denton, 1979; Bockheim and Wilson, 1979). Again, we consider this relative chronology to be reasonably accurate, because we could circumvent most soil-forming variables by constructing the chronology piecemeal, using interlocking regional geological and soil data as control.

However, our absolute chronology is bracketed only by radiocarbon dates at the younger end of the scale. Therefore, we can correlate only one end member of the Darwin-Byrd relative chronology accurately with the nearby McMurdo Sound master relative chronology.

In the Ellsworth Mountains we use soil and rock-weathering data in two senses: first for regional correlation, and then for estimates of absolute chronology. We could not develop a relative chronology using independent soil and geological data because drifts of several ages were not present except in "windows" or in scattered ice-cored moraines. Fortunately, most of our soil data came from the Heritage Range, where we could limit variability in climate, parent lithology, and aspect. Therefore, we are sufficiently confident in our regional use of soil development to conclude that the Ellsworth Drift represents a single glacial unit, particularly because this conclusion is consistent with the geologic evidence. We are less confident in using soil and rock-weathering data to estimate the absolute age of Ellsworth Drift. To do so, we must make long-range comparisons of soil development without any independent control of several soil-forming variables. Moreover, we have no independent radiometric or geologic control.

Within these limitations, we now make a best estimate of the age of the last West Antarctic ice expansion from soil development and surface rock weathering in Ellsworth Drift. Table 8 shows that these features in Ellsworth Drift are very similar to those in Ross Sea drift in eastern Taylor Valley (77°31′S and 163°45′E) (Bockheim, 1977; Pastor and Bockheim, 1980; Stuiver and others, 1981). In eastern Taylor Valley the outer edge of Ross Sea drift dates to 17,000 to 21,000 ^{14}C yr B.P., and recessional drift dates between 17,000 and 8,340 ^{14}C yr B.P. (Stuiver and others, 1981). Ellsworth Drift is also similar in soil development and surface rock weathering to Britannia drift in the Byrd-Darwin Glacier area (Bockheim and Wilson, 1979). On the basis of comparison of relative chronologies, the outer edge of Britannia drift is correlated with the outer edge of Ross Sea drift. Young Britannia recessional moraines are radiocarbon dated between 10,700 and >5100 ^{14}C yr B.P. In both the Dry Valleys and the Darwin-Byrd Glacier areas, the penultimate drift shows considerably more soil development than Ross Sea or Britannia drifts and, therefore, more soil development than Ellsworth Drift. On the basis of this comparison of soil development and surface rock weathering, the Ellsworth Drift is young, perhaps dating to late Wisconsin and Holocene times.

Another line of evidence involves striations that occur on bedrock outcrops between present ice surfaces and the trimline. These striations are best preserved on Crashsite Group quartzite and fine-grained slate and argillite, less well preserved on Whiteout Conglomerate, and rarely preserved on carbonate rocks of the Minaret Group that exhibit surface granular disintegration. It is a widespread but poorly based perception from the Northern Hemisphere that well-preserved molded and striated bedrock surfaces on subaerially exposed bedrock suggest a late Wisconsinan age for ice passage. This accords with the conclusions of Craddock and others (1964) and Rutford (1969, 1972) that deglaciation was recent from their observations of well-preserved striations in the Ellsworth Mountains.

The late Wisconsinan age model, however, is flawed by several deficiencies. The model is based on the unproved assumptions that development of soil in drift and preservation of molding and/or striations on bedrock are accurate age indicators in the Ellsworth Mountains. But the Ellsworth Mountains are far distant from the ice-free areas in the Transantarctic Mountains where soil development is a key in delineating drifts and developing a master relative chronology. Neither a relative chronology nor absolute age control is available in the Ellsworth Mountains. Therefore, long-range comparisons have to be made without independent control of several soil-forming variables and in the absence of radiometric and geologic control. Moreover, the use of striated pavements for age control will also remain suspect until it has been demonstrated conclusively that late Wisconsinan ice actually striated bedrock and that where such striations have been subaerially exposed, they are restricted to terrain covered by late Wisconsinan ice. That the latter condition does not hold is shown by a recent study in the Beardmore Glacier area, where pre–late Wisconsin striated bedrock surfaces occur high in the Transantarctic Mountains (Prentice and others, 1986; Denton and others, 1991). In addition, the concept that a late Wisconsinan ice surface represents the highest glacial limit recognized in the Ellsworth Mountains is not in accord with geological data from farther south in the Transantarctic Mountains near the same major inland ice divide that abuts the Ellsworth Mountains. Here, the late Wisconsin ice limit, determined from lateral moraines, was far short of maximum glaciation (Mercer, 1968, 1972; Denton and others, 1986b). In fact, the inland ice surface remained unchanged or even declined slightly during late Wisconsin time. Finally, it is most unlikely that late Wisconsin summer temperature was sufficiently high to produce ice with the thawed base that we previously argued was necessary to carve the trimline.

A second model is that the trimline predates late Quaternary time and represents enormous expansion equivalent to ice-sheet overriding of the Transantarctic Mountains (Denton and others, 1984, 1991). This model has several strengths. Pre–late Quaternary maximum glaciation better fits the glacial history in the Transantarctic Mountains farther south (Mercer, 1968, 1972; Denton and others, 1986b; Prentice and others, 1986). Further, the preservation of striations in the Transantarctic Mountains is not well understood and therefore cannot be used as a chronological argument for recent ice passage. Several examples of exposed glacially striated bedrock are preserved well above moraines thought to be late Wisconsin in age; examples occur near Byrd Glacier (J. Anderson, personal communication, 1979) and Beardmore Glacier (Prentice and others, 1986; Denton and others, 1991). Finally a pre–late Quaternary age is more consistent with the warmer-than-present climate inferred to be associated with the trimline.

The values of ice thickening given earlier may have to be corrected for any tectonic uplift if the pre–late Quaternary age model proves to be valid. At present such a correction obviously

TABLE 8. COMPARISON OF SOILS IN THE ELLSWORTH MOUNTSINS WITH THOSE IN THE McMURDO SOUND AND DARWIN–BYRD GLACIER AREAS

Location	Drift Unit	Age*	Number of Profiles	Thickness of Solum (cm)	Depth of Ghosts (cm)	Salt Stage†	Depth of Coherence (cm)	Depth to Ice Core or Cement (cm)	Weathering Stage	References
Taylor Valley	Ice-cored moraine adjacent to modern glaciers	<3,100 B.P.	2	0	0	0	22	22	1	Bockheim and Leide (1980; Stuiver and others (1977).
Wright Valley	Ice-cored moraine adjacent to modern glaciers	<3,100 B.P.	3	0	0	0	12	12	1	Leide (1980)
Heritage Range	Ice-cored moraine adjacent to modern glaciers	—	2	0	0	0	0	12	1	Bockheim and Leide (1980); this paper.
Taylor Valley	Ross Sea drift	17,000-21,000 B.P.	1	0	0	0	0	18	1	Stuiver and others (1981)
Wright Valley	Trilogy drift	17,000-21,000 B.P.	5	0	8	I	0	72	2	Bockheim and Wilson (1979); Stuiver and others (1981).
Royal Society Range	Ross Sea drift	17,000-21,000 B.P.	4	0	0	I	0	31	2	Bockheim (1977); Stuiver and others (1981)
Darwin–Byrd Glacier Area	Britannia drift	Younger portion is >10,700 B.P., outer edge is correlated with Ross Sea drift and therefore dates to 17,000-21,000 B.P.	15	4	4	I	5	80	2	Bockheim and Wilson (1979)
Heritage Range	Ellsworth Drift	—	13	0	10	I	58	>80	2	Bockheim and Leide (1980); this paper
Sentinel Range	Ellsworth Drift	—	4	0	0	I	18	18	2	Bockheim and Leide (1980); this paper
Wright Valley	Loop drift§	—	2	13	8	II	42	>115	2	Bockheim (1979)
Taylor Valley	Rhone drift**	—	33	9	4.5	I	23	47	2	Bockheim (1977)
Royal Society Range	Pre-Ross Sea drift‡	—	7	11	0	I	0	37	3	Bockheim (unpublished)
Darwin–Byrd Glacier	Danum drift§§	—	10	13	19	II	28	>90	2	Bockheim and Wilson (1979)
Heritage Range	Older drift***	—	3	6	0	I	54	>100	3	Bockheim and Leide (1980); this paper.

*Age derived directly from radiocarbon dates or indirectly from master relative chronology tied to radio carbon dates.

†0 = none visible; I = salt encrustations beneath clasts; II = few salt flecks (0.5–2 mm diameter).

§Loop drift is next-oldest to Trilogy drift in eastern Wright Valley.

**Rhone drift, which comprises moraines deposited by the Ross Sea ice sheet in Taylor Valley, is older than Ross Sea drift and younger than Chad drift. Chad drift is the same age as Loop drift and is the next-oldest drift deposited by the Ross Sea ice sheet in Taylor Valley. We show soil data for Rhone drift instead of Chad drift because Chad drift has been covered by a young lake and therefore is not suitable for soil analysis.

‡This is the next-oldest to Ross Sea drift in the Royal Society Range.

§§Danum drift is the next-oldest to Britannia drift in the Darwin-Byrd Glacier area.

***This is the old drift described in the text as "windows" projecting through Ellsworth Drift.

cannot be made. We note here, however, that the trimline shows smoothly consistent elevations across faults in the Ellsworth Mountains. Further, trimline elevations allow reconstruction of a former ice surface that would require internal flowlines consistent with striations below the trimline. To us, this suggests that the trimline postdates faulting and perhaps significant tectonic uplift. If significant tectonic uplift occurred, it would have to have been without any faulting that can be recognized from trimline data and would have fortuitously left trimline surfaces that could reflect a former ice surface.

It is clear that a unique chronology cannot yet be applied to glacial deposits and erosional features in the Ellsworth Mountains. It is even possible that glaciations of several ages affected the Ellsworth Mountains. Two techniques, however, offer promise of resolving the chronologic problem. First, combined isotopic and gas-content studies of a deep ice core from the inland divide near the Ellsworth Mountains would show if late Wisconsinan ice expanded to the trimline. Such data would supersede our qualitative age estimate from soil development. Second, measurement of cosmogenic ^{3}He concentrations shows real promise of affording exposure times for striated surfaces that would replace our qualitative estimates based on their preservation (Kurz, 1986).

We conclude by noting the implications of our field work in the Ellsworth Mountains to the CLIMAP reconstruction of the West Antarctic Ice Sheet in late Wisconsinan time (Stuiver and others, 1981). First, a late Wisconsinan age model for the trimline and lower striations would be consistent with the CLIMAP reconstruction of late Wisconsinan grounded ice in the Weddell Sea embayment (Stuiver and others, 1981, p. 376), as previously noted by Rutford and others (1980). Only minor adjustments in the CLIMAP surface ice contours would be necessary (Hughes and others, 1985). The hypothesis of late Wisconsinan grounded ice in the Weddell Sea might well fit with mapping and radiometric dating of sea-floor tills (Anderson and others, 1980; Elverhøi, 1981) and with glacial geological studies in mountains at the base of the Antarctic Peninsula adjacent to the Ronne Ice Shelf (Carrara, 1979, 1981). The Ellsworth Mountains data, however, would imply significant changes in the CLIMAP reconstruction (Stuiver and others, 1981) of the surface of the interior West Antarctic Ice Sheet. The CLIMAP reconstruction shows a high, central interior dome rather than the complex (Figs. 1 and 2) of saddles and domes that exists today. In the absence of geological data, this interior dome was simply the artifact of applying a glaciological model to an arbitrary selection of interior flowlines. The Ellsworth Mountains data show late Wisconsinan flowlines similar to today's in this portion of the ice sheet. The major ice divide inland of the Ellsworth Mountains did not migrate but simply increased 450 to 600 m in surface elevation (ignoring isostatic compensation) during late Wisconsinan time. This suggests that inland West Antarctic ice divides maintained nearly their present positions rather than merging to form a high, central dome (Stuiver and others, 1981).

Next, we consider the implications of our pre–late Quaternary age model to the CLIMAP reconstruction. If this age model is correct, we cannot yet draw any conclusions about late Wisconsinan grounding in the Weddell Sea embayment. But we can infer that the single high interior CLIMAP dome must be incorrect if the interior divide near the Ellsworth Mountains did not change position even when ice stood at the trimline.

ACKNOWLEDGMENTS

This research was supported by National Science Foundation Grants DP-78-23832 and DPP 78-21720 to the University of Maine and Macalester College, respectively. We thank the officers and enlisted men of VXE-6 for helicopter and LC-130 support in the Ellsworth Mountains. Gerald F. Webers and John F. Splettstoesser, chief scientist and scientific coordinator of the Ellsworth Mountains Project, respectively, cooperated in every way to assure the success of this study. H. B. Conway, J. E. Leide, and M. L. Prentice assisted in the field work. R. Kelly drafted the diagrams and maps. V. Sirois typed the manuscript. We thank H. E. Wright, Jr., and C. L. Matsch for very helpful comments on an early version of this paper, and T. J. Hughes for extensive discussion about the glaciologic and paleoclimatic implications of our field data.

REFERENCES CITED

American Public Health Association, American Water Works Association, and Water Pollution Control Federation, 1975, Standard methods for the examination of water and wastewater, 14th ed.: Washington, D.C., American Public Health Association, 1193 p.

Anderson, J. B., Kurtz, D. D., Domack, E. W., and Balshaw, K. M., 1980, Glacial and glacial marine sediments of the Antarctic continental shelf: Journal of Geology, v. 88, p. 399–414.

Armstrong, R. L., 1978, K-Ar dating; Late Cenozoic McMurdo Volcanic Group and Dry Valley glacial history, Victoria Land, Antarctica: New Zealand Journal of Geology and Geophysics, v. 21, p. 685–698.

Bockheim, J. G., 1977, Soil development in the Taylor Valley and McMurdo Sound area: Antarctic Journal of the United States, v. 12, p. 105–108.

—— , 1979, Relative age and origin of soils in eastern Wright Valley: Soil Science, v. 128, p. 142–152.

Bockheim, J. G., and Leide, J. E., 1980, Soil development and rock weathering in the Ellsworth Mountains, Antarctica: Antarctic Journal of the United States, v. 15, p. 33–34.

Bockheim, J. G., and Wilson, S. C., 1979, Pedology of the Darwin Glacier area, Antarctica: Antarctic Journal of the United States, v. 14, p. 58–59.

Campbell, I. B., and Claridge, G.G.C., 1969, A classification of frigic soils; The zonal soils of the Antarctic continent: Soil Science, v. 107, p. 75–85.

—— , 1975, Morphology and age relationships of Antarctic soils; Quaternary studies: New Zealand Royal Society Bulletin, v. 13, p. 83–88.

Carrara, P., 1979, Former extent of glacial ice in the Orville Coast region, Antarctic Peninsula: Antarctic Journal of the United States, v. 14, p. 45–46.

—— , 1981, Evidence for a former large ice sheet in the Orville coast–Ronne Ice Shelf area, Antarctica: Journal of Glaciology, v. 27, p. 487–491.

Crabtree, R. D., and Doake, C.S.M., 1982, Pine Island Glacier and its drainage basin; Results from radio echo-sounding: Annals of Glaciology, v. 3, p. 65–70.

Craddock, C., 1969, Geology of the Ellsworth Mountains, *in* Bushnell, V. C., and Craddock, C., eds., Geologic maps of Antarctica: New York, American Geographical Society, Antarctic Map Folio Series, Folio 12, Plate 4.

Craddock, C., Anderson, J. J., and Webers, G. F., 1964, Geologic outline of the Ellsworth Mountains, *in* Adie, R. J., ed., Antarctic geology: Amsterdam, North-Holland, p. 155–170.

Craddock, C., Bastien, T. W., Rutford, R. H., and Anderson, J. J., 1965, *Glossopteris* discovered in West Antarctica: Science, v. 148, p. 634–637.

Denton, G. H., 1979, Glacial history of the Byrd–Darwin Glacier area, Transantarctic Mountains: Antarctic Journal of the United States, v. 14, p. 57–58.

Denton, G. H., and Armstrong, R. L., 1968, Glacial geology and chronology of the McMurdo Sound region: Antarctic Journal of the United States, v. 3, p. 99–101.

Denton, G. H., Armstrong, R. L., and Stuiver, M., 1971, The late Cenozoic glacial history of Antarctica, *in* Turekian, K. K., ed., The late Cenozoic glacial ages: New Haven, Connecticut, Yale University Press, p. 267–306.

Denton, G. H., Prentice, M. L., Kellogg, D. E., and Kellogg, T. B., 1984, Late Tertiary history of the Antarctic Ice Sheet; Evidence from the Dry Valleys: Geology, v. 12, p. 263–267.

Denton, G. H., Bockheim, J. G., Wilson, S. L., and Schlüchter, C., 1986a, Late Cenozoic history of Rennick Glacier and Talos Dome, northern Victoria Land, Antarctica, *in* Stump, E., ed., Geological investigations in northern Victoria Land: American Geophysical Union Antarctic Research Series, v. 46, p. 339–375.

Denton, G. H., Andersen, B. G., and Conway, H. B., 1986b, Late Quaternary surface ice level fluctuations of the Beardmore Glacier, Antaractica: Antarctic Journal of the United States, v. 21, p. 90–92.

Denton, G. H., Prentice, M. L., and Burckle, L. H., 1991, Cainozoic glacial history of the Antarctic Ice Sheet, *in* Tingey, R., ed., The geology of Antarctica: Oxford University Press, p. 365–433.

Drewry, D. J., 1979, Late Wisconsin reconstruction for the Ross Sea region, Antarctica: Journal of Glaciology, v. 24, p. 231–244.

——, ed., 1983, Antarctica; Glaciological and geophysical folio: University of Cambridge Scott Polar Research Institute, 1:6,000,000.

Elverhøi, A., 1981, Evidence for a late Wisconsin glaciation of the Weddell Sea: Nature, v. 293, p. 292–293.

Fillon, R. H., 1974, Late Cenozoic foraminiferal paleoecology of the Ross Sea, Antarctica: Micropaleontology, v. 20, p. 129–151.

——, 1975, Late Cenozoic paleo-oceanography of the Ross Sea, Antarctica: Geological Society of America Bulletin, v. 86, p. 839–845.

GEBCO, 1980, General bathymetric chart of the oceans (GEBCO): Ottawa, Ontario, Canadian Hydrographic Service, scale 1:12,000,000.

Hjelle, A., Ohta, Y., and Winsnes, T. S., 1978, Stratigraphy and igneous petrology of the southern Heritage Range, Ellsworth Mountains: Norsk Polarinstitutt Skrifter, v. 169, p. 5–43.

——, 1982, Geology and petrology of the southern Heritage Range, Ellsworth Mountains, *in* Craddock, C., ed., Antarctic geoscience: Madison, University of Wisconsin Press, p. 599–608.

Hollin, J. T., 1962, On the glacial history of Antarctica: Journal of Glaciology, v. 4, p. 173–195.

Hughes, T. J., Denton, G. H., and Fastook, J. L., 1985, The Antarctic Ice Sheet; An analogue for Northern Hemisphere Ice Sheets? *in* Woldenberg, M., ed., Models in geomorphology: London, Allen and Unwin, p. 25–72.

Kellogg, T. B., Truesdale, R. S., and Osterman, L. E., 1979, Late Quaternary extent of the West Antarctic Ice Sheet; New evidence from Ross Sea cores: Geology, v. 7, p. 249–253.

Knutzen, D., Henderson, T., Pearsall, R., and Stoner, J. E., 1981, Ellsworth Mountains control, Antarctica, Sentinel Range: U.S. Geological Survey National Mapping Division unpublished report.

Kurz, M. D., 1986, In situ production of terrestrial cosmogenic helium and some applications to geochronology: Geochimica et Cosmochimica Acta, v. 50, p. 2855–2862.

Leide, J. E., 1980, Soils and relative age dating of alpine moraines in Wright and Taylor Valleys, Antarctica [M.S. thesis]: Madison, University of Wisconsin, 40 p.

Mayewski, P. A., 1975, Glacial geology and late Cenozoic history of the Transantarctic Mountains, Antarctica: Columbus, Ohio State University Institute of Polar Studies Report 56, 168 p.

Mercer, J. H., 1968, Glacial geology of the Reedy Glacier area, Antarctica: Geological Society of America Bulletin, v. 79, p. 471–486.

——, 1972, Some observations on the glacial geology of the Beardmore Glacier area, *in* Adie, R. J., ed., Antarctic geology and geophysics: Oslo, Universitetsforlaget, p. 427–433.

Pastor, J., and Bockheim, J. G., 1980, Soil development on moraines of Taylor Glacier, lower Taylor Valley, Antarctica: Soil Science Society of America Bulletin, v. 44, p. 341–348.

Prentice, M. L., Denton, G. H., Lowell, T. V., Conway, H. B., and Heusser, L. E., 1986, Pre–Late Quaternary glaciation of the Beardmore Glacier region, Antarctica: Antarctic Journal of the United States, v. 21, p. 95–98.

Rutford, R. H., 1969, The glacial geology and geomorphology of the Ellsworth Mountains, West Antarctica [Ph.D. thesis]: Minneapolis, University of Minnesota, 289 p.

——, 1972, Glacial geomorphology of the Ellsworth Mountains, *in* Adie, R. J., ed., Antarctic geology and geophysics: Oslo, Universitetsforlaget, p. 225–232.

Rutford, R. H., Denton, G. H., and Andersen, B. G., 1980, Glacial history of the Ellsworth Mountains: Antarctic Journal of the United States, v. 15, p. 56–57.

Splettstoesser, J. F., and Webers, G. F., 1980, Geological investigations and logistics in the Ellsworth Mountains, 1979–80: Antarctic Journal of the United States, v. 15, p. 36–39.

Stephenson, S. N., and Doake, C.S.M., 1982, Dynamic behavior of Rutford Ice Stream: Annals of Glaciology, v. 3, p. 295–299.

Stuiver, M., Denton, G. H., Kellogg, T. B., and Kellogg, D. E., 1977, Glacial geologic studies in the McMurdo Sound region: Antarctic Journal of the United States, v. 13, p. 44–45.

Stuiver, M., Denton, G. H., Hughes, T. J., and Fastook, J. L., 1981, History of the marine ice sheet in West Antarctica during the last glaciation; A working hypothesis, *in* Denton, G. H., and Hughes, T. J., eds., The last great ice sheets: New York, Wiley-Interscience, p. 319–436.

Ugolini, F. C., and Anderson, D. M., 1973, Ionic migration and weathering in frozen Antarctic soils: Soil Science, v. 115, p. 461–470.

Webb, P. N., Harwood, D. M., McKelvey, B. C., Mercer, J. H., and Stott, L. D., 1984, Cenozoic marine sedimentation and ice-volume variation on the East Antarctic craton: Geology, v. 12, p. 287–291.

Webers, G. F., and Spörli, K. B., 1983, Palaeontological and stratigraphic investigations in the Ellsworth Mountains, West Antarctica, *in* Oliver, R. L., James, P. R., and Jago, J. B., eds., Antarctic earth science: Canberra, Australian Academy of Science, p. 261–264.

Whillans, I. M., 1976, Radio-echo layers and the recent stability of the West Antarctic Ice Sheet: Nature, v. 264, p. 152–155.

MANUSCRIPT ACCEPTED BY THE SOCIETY MARCH 1, 1989

Geological Society of America
Memoir 170
1992

Chapter 23

Chemical weathering of Cu, Fe, and Pb sulfides, southern Ellsworth Mountains, West Antarctica

Walter R. Vennum
Department of Geology, Sonoma State University, Rohnert Park, California 94928
James M. Nishi
465 South Wright Street, No. 116, Lakewood, Colorado 80028

ABSTRACT

Forty-eight samples of green blue, yellow, orange, and red surficial salts and efflorescences were collected at widely scattered localities in the southern half of the Heritage Range of the Ellsworth Mountains. Green and blue salts include azurite, chalcanthite, malachite, paratacamite, and malachite-paratacamite mixtures. Yellow and orange salts include alunogen, fibroferrite, an aragonite-natrojarosite mixture, natrojarosite-gypsum mixtures (± quartz), and an anglesite-beaverite mixture. All red salts are hematite-quartz mixtures (± muscovite, ± calcite). All salts form by the oxidation of pyrite, chalcopyrite, or galena and then are preserved by the cold, arid Antarctic climate. The assemblage of copper salts is different from that recently described from the Orville Coast (atacamite, antlerite, brochantite, plancheite) 500 km to the northeast. The difference in copper salts found at these two localities, coupled with studies of marine-derived Antarctic aerosols, suggests that malachite, azurite, and chalcanthite will be the common secondary copper minerals found deeper in the Antarctic interior and that copper chloride compounds will become less abundant farther away from the coast. This is the second reported occurrence of fibroferrite and the first reported occurrence of alunogen, anglesite, beaverite, chalcanthite, and paratacamite from Antarctica.

INTRODUCTION

Studies of sulfide alteration minerals developed in oxidized caps above the large porphyry copper deposits in the Andean foothills of Chile and Peru indicate that chemical weathering in the hot dry climate of that region has produced very complex secondary mineral assemblages (Bandy, 1938; Jarrel, 1944; Cook, 1978). The predominant secondary minerals are mainly copper and iron sulfates, oxides, carbonates, and hydroxides, some of which are hydrated.

Although numerous reports of green and yellow surficial salts have appeared in the Antarctic geological literature, it is commonly assumed that these salts are malachite, chrysocolla, and/or limonite. Detailed studies of varicolored salts from the Prince Olav Coast (Hirabayashi and Ossaka, 1976; Kaneshima and others, 1973), the Orville Coast (Vennum, 1980), and the area surrounding McMurdo Station (Jones and others, 1983) suggest, however, that chemical weathering of sulfide minerals in the cold, dry climate of Antarctica also produces complex assemblages of secondary minerals comparable to those described from South America and the arid regions of the southwestern United States. Secondary minerals reported in these four Antarctic studies include atacamite [$Cu_2(OH)_3Cl$], antlerite [$Cu_3(SO_4)(OH)_4$], brochantite [$Cu_4(SO_4)(OH)_6$], plancheite [$3CuSiO_3{\cdot}H_2O$], natrojarosite [$NaFe_3(SO_4)_2(OH)_6$], hydronium jarosite [$H_3O\ Fe_3(SO_4)_2(OH)_6$], fibroferrite [$Fe(SO_4)(OH){\cdot}5H_2O$], copiapite [$(Fe, Mg)\ (Fe_4{}^{3+})(SO_4)_6{\cdot}20\ H_2O$], and alunite [$KAl_3(SO_4)_2(OH)_6$]. The above mentioned Antarctic localities are shown in Figure 1.

During the 1979–1980 austral summer, 48 samples of

Vennum, W. R., and Nishi, J. M., 1992, Chemical weathering of Cu, Fe, and Pb sulfides, southern Ellsworth Mountains, West Antarctica, *in* Webers, G. F., Craddock, C., and Splettstoesser, J. F., Geology and Paleontology of the Ellsworth Mountains, West Antarctica: Boulder, Colorado, Geological Society of America Memoir 170.

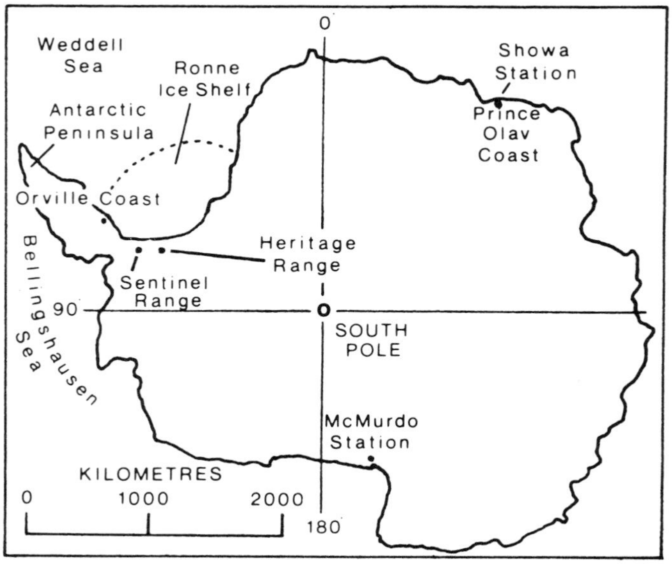

Figure 1. Index map of Antarctica showing location of Ellsworth Mountains and other geographic localities mentioned in text.

green, blue, yellow, orange, and red salts were collected at widely scattered locations (Fig. 2) in the southern half of the Heritage Range of the Ellsworth Mountains (79°30′–80°30′S, 80°–85°W). This chapter describes the identity and origin of the minerals that make up these salts. As supergene minerals rarely develop in the frigid Antarctic environment, a thorough knowledge of the products of sulfide oxidation in cold, dry climates might eventually aid economic mineral exploration in both polar regions.

PHYSIOGRAPHY AND GEOLOGIC SETTING

The Ellsworth Mountains are a slightly arcuate 350-km-long north-northwest–trending mountain chain that is bordered on the west by the polar plateau of West Antarctica and on the east by the Ronne Ice Shelf. The range is as much as 90 km wide and constitutes one of the largest areas of exposed bedrock in West Antarctica. The Minnesota Glacier, which flows eastward from the polar plateau into the Ronne Ice Shelf, divides the Ellsworth Mountains into a northern Sentinel Range and a southern Heritage Range. The 2,500-m-high western escarpment of the Sentinel Range rises abruptly from the polar plateau to elevations in excess of 5,000 m at the summit of the Vinson Massif (5,140 m), the highest peak in Antarctica. By contrast, the Heritage Range is much more subdued, and toward the south its elevation gradually decreases as the range passes into a series of widely dispersed hills and nunataks.

The Ellsworth Mountains are underlain by a 13,000-m-thick section of strongly folded, low-metamorphic-grade, largely marine metasedimentary and metavolcanic rocks that range in age from Cambrian to Permian. The entire range appears to be folded into a north-plunging anticlinorium so that only the lower part of this section is exposed in the southern Heritage Range. The rock units that crop out in the southern Heritage Range, in order of decreasing age, are the Heritage Group, a 7,000+-m-thick sequence of phyllite, argillite, black shale, quartzite, conglomerate, and marble of Middle and Late Cambrian age, the Minaret Formation, an 800 m thick sequence of light grey locally oolitic Upper Cambrian carbonate rocks; and the Upper Cambrian to Devonian Crashsite Group, which is at least 3,200 m thick (Webers and others, this volume). The Heritage Group contains abundant interbedded basaltic lava flows and pyroclastics and is locally intruded by gabbroic stocks and sills as much as 300 m thick. Craddock and others (1964), Craddock (1969), and various authors in this volume have published more detailed accounts of the geology of the Ellsworth Mountains.

CLIMATE

Climatic data from the Ellsworth Mountains are scarce. Daily weather records were maintained from late October 1979 until late January 1980 at Camp Macalester, at an elevation of approximately 1,250 m near the head of the Minnesota Glacier (79°05′S, 85°58′W). No above-freezing air temperatures were recorded during this period, and the overall daily temperature averaged about –15°C. Borehole studies on the Lassiter Coast 900 km to the northeast (immediately north of the Orville Coast) indicate that the mean annual temperature in that area is approximately –22°C (Williams, 1970). Average annual precipitation in the Ellsworth Mountains area has been estimated at 20 cm water equivalent, and average annual temperature at –30°C (Rubin and Weyant, 1965).

Although above-freezing air temperatures apparently are rare, certain features—such as clear, sheet-like masses of ice in gullies, icicles as long as 30 cm on the lips of overhanging rock slopes, and small frozen pools of glacial meltwater in shallow depressions adjacent to rock outcrops or moraines—indicate that at times melting does occur and that water is free to move over rock surfaces. Studies on the Lassiter Coast indicate that melting is caused by heat radiating from dark-colored rocks on sunny days. During the week of 28 December 1969 to 3 January 1970, a Taylor temperature recorder located on the Lassiter Coast recorded above-freezing rock temperatures on a dark gray hornfels for an average of nine hours per day. During that week, rock temperatures exceeded +15°C for an average of two hours per day and reached a maximum of +21°C. Air temperatures during the same week ranged from –3°C to –14°C (Boyer, 1975).

ANALYTICAL METHODS

Debye-Scherrer x-ray powder films (Ni-filtered $Cu_{K\alpha}$ radiation operated at 35 kV and 18 ma) were obtained for all 48 samples with a 114.6-mm-diameter powder camera. Samples that strongly fluoresced in $Cu_{K\alpha}$ radiation were x-rayed again with Zr-filtered $Mo_{K\alpha}$ radiation operated at 42 kV and 15 ma.

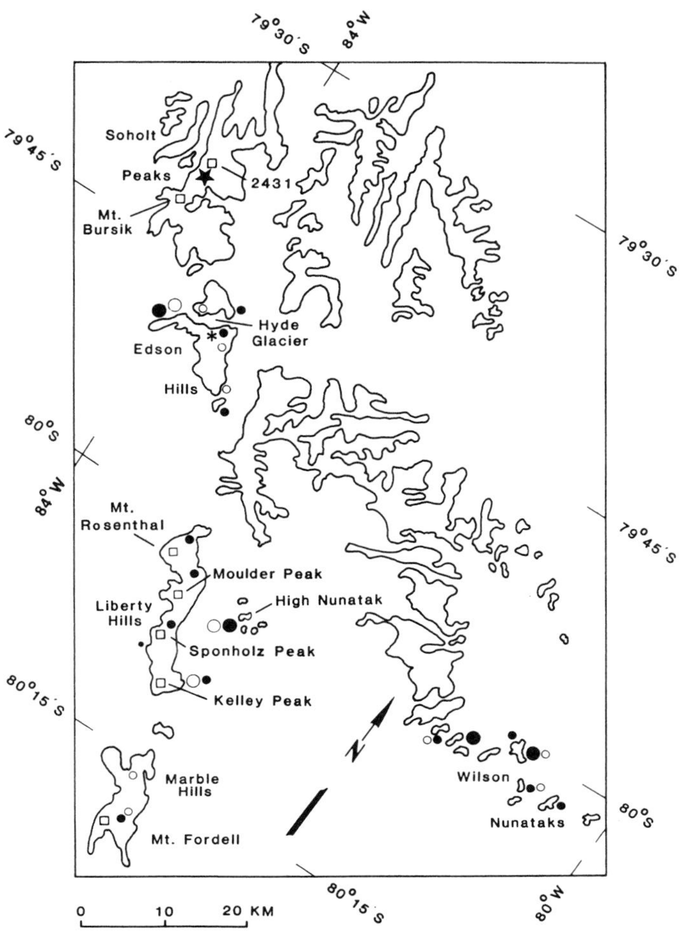

Figure 2. Map of southern Heritage Range showing locations of sample sites. Solid dots represent copper salts, open circles represent natrojarosite sites, star represents fibroferrite-alunogen site, and snowflake represents lead salt site. The large solid dots and large open circles indicate multiple sample sites. Open squares represent mountain peaks.

Because all iron sulfate salts are intergrown with gypsum and most copper salts are intergrown with quartz, these minerals provided a convenient internal standard. Attempts to use an internal standard with the lead salts were unsatisfactory, as its addition either greatly reduced or eliminated many of the x-ray reflections produced by the sample. Corrections were made for film shrinkage. Least-squares refinement of unit cell dimensions for jarosite-group compounds were obtained using a Fortran IV program (Evans and others, 1963). Starting parameters were derived from the 006 and 220 reflections, and space group R3m was assumed. Ambiguous or questionable x-ray identifications were resolved with atomic absorption spectrophotometry or the use of a CWIKSCAN 100 field emission scanning electron microscope equipped with qualitative analysis capabilities. Color codes are taken from the Geological Society of America's Munsell Rock-Color Chart.

MINERALOGY

Copper salts

Twenty-two samples of light (5B7/6) to moderate (5B5/6) blue, light blue green (5BG6/6), light (5G7/4) to moderate green (5G5/6), and very pale green (10G8/2) salts were examined. Six samples are malachite, four are chalcanthite ($CuSO_4 \cdot 5H_2O$), two are paratacamite ($Cu_2(OH)_3Cl$) (Fig. 3), two are azurite, and eight are paratacamite-malachite mixtures. Quartz is intergrown with all samples except those of chalcanthite and two of the paratacamite-malachite mixtures. One of the paratacamite-malachite mixtures contains intergrown gypsum.

Both azurite samples were collected at outcrops where paratacamite-malachite mixtures were found. Synthetic preparations of paratacamite usually contain admixed malachite if special care is not taken to exclude CO_2 (Frondel, 1950), and the occurrence of either of the copper carbonates with paratacamite should thus not be construed as unusual. Eight of the 22 samples contain finely disseminated sulfide grains. X-ray analyses of hand-picked grains indicate that all of the remaining primary sulfide is chalcopyrite.

Efflorescences and crusts of white salts are abundant in the southern Heritage Range and occur at all sites where green and blue salts were collected. X-ray diffraction patterns of 30 randomly collected white salts indicate that they are calcite, gypsum, and calcite-gypsum mixtures. In a similar, but more extensive study, Bockheim and Leide (1980) found white encrustations from the Heritage Range to be composed largely of gypsum and lesser amounts of calcite and thenardite ($NaSO_4$). Vennum (1980) has shown that similar salts found on the Orville Coast form by evaporation of salt-bearing solutions that have risen to the surface of fractured rocks by capillary action. Their presence in the southern Heritage Range indicates that $CO_3^{=}$- and $SO_4^{=}$-bearing waters are present on outcrop surfaces and can react with chalcopyrite to form malachite, azurite, and chalcanthite.

Figure 3. Scanning electron microscope photomicrograph of rhombohedral paratacamite crystals from the Wilson Nunataks in the southern Heritage Range, Ellsworth Mountains. Approximately 5000× magnification.

TABLE 1. X-RAY DIFFRACTION DATA FOR JAROSITES

Heritage Ridge Natrojarosite a = 7.325 ± 0.001, c = 16.741 ± o.004Å			Synthetic $NaFe_3(SO_4)_2(OH)_6$* a = 7.329 ± 0.002, c = 16.703 ± 0.004Å		
d meas	I est	hkl	d meas	I est	hkl
5.96	30	101	5.94	30	101
5.584	50	003	5.57	40	003
5.048	80	012	5.06	90	012
3.667	20	110	3.66	30	110
3.474	20	104	3.49	20	104
3.116	100	021	3.12	90	021
3.054	100	113	3.06	100	113
2.970	10	015	2.960	20	202,015
2.791	30	006	2.783	30	006
2.526	40	024	2.527	30	024
2.233	50	107	2.308	10	112
1.976	50	033	2.236	50	107
1.906	20	027	1.979	50	033
1.830	40	220	1.909	30	027
1.577	10	128	1.857	10	009
1.560	10	315	1.834	50	220
1.530	20	226	1.743	20	208,223
1.482	20	404	1.724	20	312
1.422	10	137	1.693	5	217
			1.657	5	119
			1.623	20	134
			1.578B	20	401,128
			1.560	20	042,315
			1.532	30	226
			1.484	10	404
			1.479B	40	02.10,01.11

Very few data exist regarding the oxidation of chalcopyrite in air at T<100°C. Steger and Desjardins (1978) have shown, however, that at 52°C and 68 percent relative humidity, chalcopyrite will react with oxygen to produce cupric sulfate by the following reaction:

$$4CuFeS_2 + 15½O_2 \rightarrow Fe_2(SO_4)_3 + Fe_2O_3 + 4CuSO_4 + S^0.$$

Temperature and humidity values used in this experiment obviously are never approached in Antarctica. They are, however, necessary in the laboratory in order for the reaction to occur in a reasonable amount of time. Steger and Desjardins thus assumed that the oxidation products of this reaction will be the same, especially under less severe conditions. Essentially the same conclusion was reached in an earlier study of the system Fe_2O_3-Cu^0-SO_3-H_2O (Tunell and Posnjak, 1931). These workers showed that cupric sulfate that had formed by the oxidation of copper sulfides aided in the oxidation of ferrous sulfate and thus tended to cause iron to precipitate in the hydrous ferric state as jarosite or limonite. A high content of ferric iron would, in turn, inhibit precipitation of copper minerals. Jarrell (1944) later substantiated these results in an excellent study of the oxidized zone overlying the well-known porphyry copper deposit at Chuquicamata, Chile.

Posnjak and Tunell (1929) concluded that paragenesis of the hydrous copper sulfates is controlled by HSO_3^- concentration and that with decreasing amounts of HSO_3^- (higher pH) the order of formation is chalcanthite, antlerite, and brochantite followed by cuprite (Cu_2O). If, according to Posnjak and Tunell (1929), only a portion of the sulfate is removed in jarosite, then the HSO_3^- content would remain relatively high and chalcanthite would form, but if pH were to rise due to lower pyrite content, then antlerite or brochantite would form. If Posnjak and Tunell were correct, then oxidizing solutions in the southern Heritage Range evidently contain enough HSO_3^- to favor the formation of chalcanthite and to prevent the formation of antlerite and brochantite. According to Garrels (in Anderson, 1955), the controlling factor is a combination of pH and cupric sulfate concentration; the paragenetic sequence with rising pH and decreasing cupric sulfate concentration is chalcanthite, antlerite, brochantite, and cuprite. In either case, this suggests that chalcanthite found in the southern Heritage Range formed approximately in situ and that oxidizing solutions have not migrated very far from their sources.

Copper (II) trihydroxychloride, $Cu_2(OH)_3Cl$, occurs in

TABLE 1. X-RAY DIFFRACTION DATA FOR JAROSITES (continued)

SYNTHETIC $KFe_3(SO_4)_2(OH)_6$*
a = 7.315 ± 0.001, c = 17.119 ± 0.004Å

d meas	I est	hkl
5.96	30	101
5.71	20	003
5.10	70	012
3.66	30	110
3.54	5	104
3.12	60	021
3.08	100	113
3.02	5	015
2.977	5	202
2.859	30	006
2.549	30	024
2.307	5	122
2.283	50	107
1.982	60	033
1.938	10	027
1.903	10	009
1.830	60	220
1.774	5	208
1.743	5	223
1.723	10	312
1.711	<5	217
1.689	<5	119
1.625	10	134
1.597	10	128
1.579	5	401
1.563	10	315
1.559	10	042
1.541	30	226
1.507	40	02.10
1.487	10	404

SYNTHETIC $H_3OFe_3(SO_4)_2(OH)_6$*
a = 7.347 ± 0.002, c = 16.994 ± 0.005Å

d meas	I est	hkl
5.96	20	101
5.67	30	003
5.10	100	012
3.67	20	110
3.54	10	104
3.13	90	021
3.08	100	113
3.00	10	015
2.840	30	006
2.548	30	024
2.317	5	122
2.270	40	107
1.987	40	033
1.932	20	027
1.890	10	009
1.838	40	220
1.769	5	208
1.750	10	223
1.728	10	312
1.631	10	134
1.594	10	128
1.583	<5	401
1.566	10	315,042
1.542	20	226
1.501	20	01.11
1.490	5	404
1.441	10	045,232
1.427	5	137
1.412	5	039

*From Dutrizac and Kaiman (1976), B - Broadline

three different polymorphic forms: botallackite (monoclinic), atacamite (orthorhombic), and paratacamite (rhombohedral). Walter-Levy and Goreaud (1969) have shown that under most conditions of mineral deposition, botallackite readily transforms into either or both of the other two polymorphs, and thus it is found only rarely in nature.

Both atacamite and paratacamite are widely reported in the oxidized zone of copper-bearing ore deposits, especially in arid climates, as corrosion products on metals and as patinas on art and archeological objects. Sharkey and Levin (1971) attribute the formation of the particular polymorph (atacamite or paratacamite) to the amount of Cl^- in solution. Their work agrees with that of Walter-Levy and Goreaud (1969) and suggests that pure paratacamite forms if the Cl^- concentration is small, and that as the concentration of Cl^- rises, mixtures of paratacamite and atacamite form. Pure atacamite forms as Cl^- concentration reaches values around 0.1F. Thus, paratacamite forms under conditions of low Cl^- concentrations, should form slowly, and should, in favorable conditions, produce macroscopic crystals. Atacamite would form only if the amount of Cl^- is large and would then precipitate rapidly as a powdery material. Paratacamite samples from the southern Heritage Range do contain microcrystals, whereas atacamite collected on the Orville Coast occurs as powdery green aggregates.

Frondel (1950) and Miller and Taylor (1966) attributed the origin of paratacamite and atacamite, respectively, to the reaction of seawater with primary sulfides. To our knowledge, atacamite has been found in Antarctica only on the Orville Coast and on the Prince Olav Coast in the vicinity of the Japanese facilities at Showa Station (Fig. 1). Open water is found annually at Showa Station, but the Orville Coast locality is at least 400 km from open water of the Bellingshausen Sea. Chlorine-bearing sea spray would have to be carried 800 km inland to reach the southern Heritage Range, and if this is the source of the chlorine in the paratacamite found there, then the occurrence of paratacamite and not atacamite can be attributed to low concentrations of Cl^-. Studies of Antarctic atmospheric halogens, presumably derived from a marine source, show a greater than seven-fold decrease in the concentration of particulate Cl when values for McMurdo Station (Fig. 1) are compared to those recorded at the South Pole (Duce and others, 1973). The known distribution of halite in Antarctic evaporite deposits follows a similar pattern. This mineral is commonly found in coastal localities such as the Dry Valleys near McMurdo Station (Nishiyama, 1979) and near Showa Station (Hirabayashi and Ossaka, 1976) but has not been reported from the interior of the continent. Ericksen (1981) described a similar concentration of marine-derived Cl in evaporite deposits along the coast of northern Chile; the amount of Cl declines rapidly inland.

The assemblage of secondary copper salts in the southern Heritage Range (malachite, azurite, chalcanthite, paratacamite) is entirely different from that found on the Orville Coast (atacamite, antlerite, brochantite, plancheite) and suggests a drier, colder climate in the former locality. Although some of the Cl may be derived in situ, the factors mentioned above suggest that malachite, azurite, and chalcanthite will be the common secondary copper minerals deeper in the Antarctic interior and that copper chloride compounds will become less abundant farther away from the coast. Stewart (1964) does not record chalcanthite or paratacamite in a list of Antarctic minerals, and a literature search has shown that these two minerals have not previously been reported from Antarctica.

Iron and aluminum salts

Fifteen samples of moderate yellow (5Y7/6), grayish yellow (5Y8/4), pale yellowish orange (10YR8/6), and dark yellowish orange (10YR6/6) salts were examined. One sample is aragonite stained with natrojarosite, and the other 14 are mixtures of natrojarosite and gypsum ± quartz. Natrojarosite is more abundant in the orange-colored samples, whereas the yellow-colored salts contain more gypsum.

Brophy and Sheridan (1965) have shown that most naturally occurring jarosite-group compounds belong to a solid solution series involving three end members: jarosite [$KFe_3(SO_4)_2(OH)_6$], natrojarosite, and hydronium jarosite (carphosiderite in older literature). In addition, Kubisz (1970) has pointed out that both natural and synthetic potassium and sodium jarosites contain an excess of H_2O and are deficient in alkali ions when compared to the quantities of these substances predicted from stoichiochemical considerations. He concluded that electrostatic neutrality was satisfied by substitution of the hydronium ion (H_3O^+) into the alkali position. Dutrizac and Kaiman (1976) extended these conclusions to the entire jarosite group and stated that all jarosites contain some H_3O^+ substituting for both alkali and metal ions and that the extent of substitution probably is dependent on the conditions of formation.

Consequently, the various jarosite group minerals have very similar x-ray diffraction patterns. There are, however, sufficient differences in d spacings and intensities in the patterns of the pure jarosites for a fairly reliable identification to be made. Although the entire pattern should be examined, the 003, 006, and 107 reflections and the spread between the 021 and 113 reflections and their relative intensities are the most critical (Warshaw, 1956). A representative x-ray diffraction pattern from the purest jarosite-group sample collected is compared with patterns for synthetic jarosite, natrojarosite, and hydronium jarosite in Table 1. Gypsum lines were first deleted from this representative pattern, and the visually estimated intensities of the jarosite-group compound were then recalculated on the basis of I = 100 for the 113 reflection. The x-ray pattern of the sample from the Ellsworth Mountains most closely resembles that of natrojarosite. Analysis by atomic absorption spectrophotometry confirms this conclusion and indicates that the Na:K ratio ranges from 7:1 to 10:1 for the 14 samples studied.

Brophy and Sheridan (1965) have shown that the c unit cell dimension systematically increases in the series natrojarosite–hydronium jarosite–jarosite due to the larger size of H_3O^+ and K^+ ions relative to Na^+. The c unit cell dimension for all 14 Ellsworth Mountains samples ranges from 16.730 to 16.762 ± 0.004 Å. These values also suggest that the jarosite-group mineral is natrojarosite and that it contains no more than a small percent of potassium and/or hydronium ions.

The jarosite compounds mentioned above are all secondary minerals formed by the oxidation of pyrite or other iron-bearing sulfides. Although not restricted to arid climates, they are commonly most extensively developed in such environments. Posnjak and Merwin (1922) and Jarrell (1944) have shown that the formation of various hydrated iron sulfates is controlled by the amount of pyrite initially present in the host rock. The iron in pyrite is taken into solution either as ferrous ion or as ferrous sulfate and is then oxidized to an insoluble oxide such as hematite. These reactions are shown in equations 1 and 2.

$$2FeS_2 + 7O_2 + 2H_2O \rightarrow 2Fe^{2+} + 4SO_4^{=} + 4H^+ \quad (1)$$

$$2Fe^{2+} + \tfrac{1}{2}O_2 + 2H_2O \rightarrow Fe_2O_3 + 4H^+. \quad (2)$$

In massive pyrite bodies or where pyrite is initially abundant, excess sulfuric acid will keep the pH low and retain the iron in the soluble ferrous state without forming much hematite or limonite. If the solutions involved in these reactions are rich in oxygen, pyrite may alter directly to ferric sulfate without going through a ferrous sulfate state (equation 3).

$$2FeS_2 + \frac{15}{2}O_2 + H_2O \rightarrow Fe_2(SO_4)_3 + SO_4^{=} + 2H^+. \quad (3)$$

Either the disseminated nature of pyrite in rocks of the Ellsworth Mountains restricts the amount of sulfuric acid that is formed or the oxidation of Fe^{2+} and sulfur proceeds simultaneously. In either case, aqueous solutions have a high pH, and pyrite tends to alter directly to ferric sulfate without passing through the ferrous sulfate sequence (melanterite → rosenite → copiapite → coquimbite) proposed by Nordstrom and Dagenhart (1978). Another factor that may help to explain the absence of ferrous sulfate compounds in the Ellsworth Mountains and on the Orville Coast is the very small amount of microbiological activity that in some cases can decrease the redox potential and lead to the reduction of ferric iron. In more humid climates iron sulfate compounds, because of their high solubilities, persist only for a very short period of time before oxidizing. They remain on the outcrop in Antarctica because of the extremely arid climate.

In nature, jarosite is generally more abundant than natrojarosite. In more hospitable climates, production of acids (largely carbonic) by microorganisms is a major mechanism whereby rock-forming minerals are decomposed and K^+ is released as a soluble ion. The presumed low content of microorganisms in rocks of the Ellsworth Mountains and the Orville Coast could thus contribute to the prevalence of the Na-rich end member in the jarosite-group compounds described from both those localities. Pure hydronium jarosite is a relatively rare mineral and forms only from alkali-deficient iron-bearing sulfate solutions. Its formation thus requires rapid oxidation of pyrite accompanied by a rapid neutralization of the resulting acidic solution (possibly by carbonate rocks) before alkali ions in the host rocks could be dissolved. The small amount of hydronium ion incorporated in the Ellsworth Mountains natrojarosites evidently results from a combination of relatively high feldspar content in the host rocks and oxidizing solutions with acidic pH values.

Eight samples of moderate red (5R4/6), moderate red brown (10R4/6), very dark red (5R2/6), and dusky red (5R3/4) salts were examined. All are mixtures of hematite and quartz ± muscovite ± calcite. Two of the more interesting samples were collected at 2,400 m elevation in the central Soholt Peaks (79°43′S, 84°23′W). The host rock for these samples is a pyrite-bearing, sanidine-quartz, phyric Devonian hypabyssal stock, the only felsic igneous body so far known in the Ellsworth Mountains. The collecting site (Fig. 2) is adjacent to the contact of this stock with metasedimentary rocks of the Heritage Group. Very pale orange (10YR8/2) powder that fills pockets as much as 2 cm in diameter in the igneous rock is composed of fibroferrite [$Fe(SO_4)(OH)\cdot5H_2O$]. A very pale orange, bitter-tasting powder that composes a 2-m-long, 10-cm-wide nearby vein is composed of alunogen [$Al_2(SO_4)_3\cdot17H_2O$] and a small amount of admixed fibroferrite. This is the first reported occurrence of alunogen (Vennum and Nishi, 1981) and the second reported occurrence of fibroferrite from Antarctica. The mode of origin of fibroferrite is similar to that of natrojarosite. Alunogen formed by reaction between sulfate-bearing solutions derived by oxidation of pyrite and either feldspars in the felsic igneous rocks or micas and clay minerals of the Heritage Group metasedimentary rocks.

TABLE 2. X-RAY DIFFRACTION DATA FOR LEAD SALTS

Antarctic Mixture		Anglesite*		Beaverite†	
d meas	I est	d meas	I/I	d meas	I est
5.887	100	5.381	3	5.88	100
5.415	10	4.260	87	5.07	40
4.267	100	3.813	57		
3.806	50	3.622	23	3.62	80
3.619	50	3.479	33	3.55	40
3.485	20	3.333	86		
3.339	70	3.220	71		
3.224	40	3.001	100	3.06	100
3.066	80	2.773	35	2.945	60
3.006	90	2.699	46	2.866	60
2.948	10	2.618	8	2.533	60
2.874	10	2.406	17		
2.762	40	2.355	<1	2.343	60
2.695	40	2.276	20	2.284	80
2.545B	10	2.235	5	2.246	40
2.405	10	2.193	7		
2.343	10	2.164	26		
2.286	30	2.133	5		
2.166	20	2.067	76	2.080	20
2.127	20	2.031	34		
2.068	60	2.028	48		
2.030	40	1.973	21	1.964	60
1.968B	20	1.905	3		
1.814	10	1.879	6	1.810	60
1.796	10	1.793	15	1.770	40
1.704	20	1.741	8	1.730	40
1.620	20	1.716	3		
		1.703	16	1.699	40
		1.656	7	1.686	40
		1.648	3	1.658	40
		1.621	19		
		1.611	10		
		1.571	6		
		1.542	2	1.544	40
		1.525	1	1.529	40
1.493	10	1.493	15	1.507	40
		1.467	7	1.469	40
1.441	10	1.441	8	1.433	40
		1.429	4		
1.370	10	1.406	3		

*ASTM card no. 5-0577

†ASTM card no. 15-70, B = Broad line

Lead salts

The sample with the most complex and interesting mineralogy was collected from a sulfide-bearing gabbroic dike that is in turn cut by a quartz vein. A crust of light blue-green slightly botryoidal chalcanthite occurs intergrown with a light-green mixture of malachite, paratacamite, and gypsum. A dark yellowish orange to pale yellowish orange powder occurs in small (2 to 5 mm) pockets and pits in the green crust. No primary sulfides were observed at this site. X-ray data (Table 2) indicate that the orange powder is largely anglesite ($PbSO_4$) but that a second mineral, beaverite [$Pb, (Cu, Fe, Al)_3 (SO_4)_2 (OH)_6$], is also present (Fig. 4).

Figure 4. Scanning electron microscope photomicrograph of prismatic anglesite and platy beaverite (upper left and upper right) crystals from the Edson Hills in the southern Heritage Range, Ellsworth Mountains. Approximately 1500× magnification.

Steger and Desjardins (1980) have shown that galena, when oxidized in air at 52°C and 68 percent relative humidity for periods of up to five weeks, produces $PbSO_4$ by the following reaction:

$$PbS + 2O_2 \rightarrow PbSO_4$$

Beaverite is a relatively rare mineral that forms in the oxidized caps of Cu-Pb ore deposits in arid regions, where it has been reported in association with anglesite and plumbojarosite [$PbFe_6(SO_4)_4(OH)_{12}$] (summary in Palache and others, 1951). It most likely forms by a reaction between galena and pyrite in a low-pH oxidizing environment (Mumme and Scott, 1966). Dutrizac and Kaiman (1976) commented that synthesis of plumbojarosite is complicated by the fact that the addition of lead to a sulfate solution results in the immediate precipitation of $PbSO_4$, and this factor may account for the lack of plumbojarosite in the sample described above. This is the first reported occurrence of both anglesite and beaverite from Antarctica.

DISCUSSION

Detailed studies on the Orville Coast and in the Ellsworth Mountains indicate that oxidation of sulfide ore minerals in the cold, dry climate of Antarctica produces a complex assemblage of secondary alteration minerals comparable to those found in hot, dry climates elsewhere in the world. Marine-derived aerosols, especially Cl, deposited on rock outcrops by snow, apparently contribute to the development of these secondary mineral assemblages, but effects of supergene enrichment are lacking.

Most exposed Antarctic bedrock has been deeply scoured by ice. During glaciation, soft, more easily eroded mineralized or hydrothermally altered rock will be quickly removed or buried beneath the extensive ice cover in topographic lows. These factors and the lack of alluvial sediments and placer deposits produced by a surface drainage system suggest that the most effective method of prospecting for ore deposits in Antarctica will be through a study of alteration minerals (Rowley and others, 1983).

Distribution of copper salts in the Ellsworth Mountains and on the Orville Coast suggests that oxidizing solutions in the Ellsworth Mountains have lower pH and migrate shorter distances than those on the Orville Coast. This suggests that alteration zones developed around copper sulfide mineralization will be less aerially extensive toward the interior of the continent. Pyrite from both the Orville Coast and the Ellsworth Mountains appears to alter directly to ferric sulfate (natrojarosite or fibroferrite) without passing through a ferrous sulfate stage, a process that might be aided by a presumed low content of microorganisms in rocks of these two areas. We have not been able to demonstrate a correlation between any of the salts described in this chapter and host rock type, elevation, or geographic location.

Most alteration mineral occurrences described in this chapter are exposed over areas of no more than a few cm^2 and are localized either within or along the contacts of igneous bodies.

We regard them as alteration "microzones" developed around widely disseminated sulfide accessory minerals and not indicative of economic mineral deposits. Although our study was necessarily of a reconnaissance nature, we agree with Rowley and others (1983) that at least the southern one-third of the Ellsworth Mountains has a low potential for metallic mineral deposits.

ACKNOWLEDGMENTS

This project was funded by National Science Foundation grant DPP 78-21720 to Gerald Webers of Macalester College. Logistical support was provided by the U.S. Antarctic Research Program of the National Science Foundation and U.S. Navy Squadron VXE-6. Atomic absorption analyses were undertaken by Craig Cheevers of the University of California at Riverside. Bob Noble, Gary Roberts, and Mary Stone of Hewlett-Packard Inc., Santa Rosa, California, kindly performed the scanning electron microscopy work. Figures 1 and 2 were drafted by Marc Druckman of Sonoma State University. A previous version of this manuscript was reviewed by Thomas Anderson and Steven Norwick of Sonoma State University, Ian Lange of the University of Montana, and two anonymous reviewers. The tedious task of reviewing the x-ray data was accomplished by Barbara Young.

REFERENCES CITED

Anderson, C. A., 1955, Oxidation of copper sulfides and secondary sulfide enrichment: Economic Geology, 50th Anniversary Volume, p. 324–340.

Bandy, M. C., 1938, Mineralogy of three sulphate deposits of northern Chile: American Mineralogist, v. 23, p. 669–760.

Bockheim, J. G., and Leide, J. E., 1980, Soil development and rock weathering in the Ellsworth Mountains, Antarctica: Antarctic Journal of the U.S., v. 15, no. 5, p. 33–34.

Boyer, S. J., 1975, Chemical weathering of rocks on the Lassiter Coast, Antarctic Peninsula, Antarctica: New Zealand Journal of Geology and Geophysics, v. 18, p. 623–628.

Brophy, G. P., and Sheridan, M. F., 1965, Sulfate studies; 4, The jarosite–natrojarosite–hydronium jarosite solid solution series: American Mineralogist, v. 50, p. 1595–1607.

Cook, R. B., 1978, Famous mineral localities; Chuquicamata, Chile: Mineralogical Record, v. 9, p. 321–333.

Craddock, C., 1969, Geology of the Ellsworth Mountains: American Geographical Society Antarctic Map Folio Series, Folio 12, Plate 4.

Craddock, C., Anderson, J. J., and Webers, G. F., 1964, Geological outline of the Ellsworth Mountains, *in* Adie, R. J., ed., Antarctic geology: Amsterdam, North-Holland, p. 155–170.

Duce, R. A., Zoller, W. H., and Moyers, J. L., 1973, Particulate and gaseous halogens in the Antarctic atmosphere: Journal of Geophysical Research, v. 78, p. 7802–7811.

Dutrizac, J. E., and Kaiman, S., 1976, Synthesis and properties of jarosite-type compounds: Canadian Mineralogist, v. 14, p. 151–158.

Ericksen, G. E., 1981, Geology and origin of the Chilean nitrate deposits: U.S. Geological Survey Professional Paper 118, 37 p.

Evans, H. T., Appleman, D. E., and Handwerker, D.J., 1963, The least squares refinement of crystal unit cells with powder diffraction data by an automatic indexing method: Annual Meeting American Crystallography Association, Cambridge, Massachusetts, 1963, Program and Abstracts, E-10, p. 42–43.

Frondel, C., 1950, On paratacamite and some related copper chlorides: Mineralogical Magazine, v. 29, p. 34–45.

Hirabayashi, J., and Ossaka, J., 1976, The x-ray diffraction patterns and their mineral components of evaporites at Prince Olav Coast, Antarctica: Japanese Antarctic Research Expedition Report 32, 58 p.

Jarrell, O. W., 1944, Oxidation at Chuquicamata, Chile: Economic Geology, v. 39, p. 251–286.

Jones, L. M., Faure, G., Taylor, K. S., and Corbató, C. E., 1983, The origin of salts on Mount Erebus and along the coast of Ross Island, Antarctica: Isotope Geoscience, v. 1, p. 57–64.

Kaneshima, K., Torii, T., and Miyahira, K., 1973, Mineralogical composition of white evaporites and yellow salts found around Showa Station, Antarctica: U.S. Army Cold Regions Research and Engineering Laboratory Translation 391, 13 p.

Kubisz, J., 1970, Studies on synthetic alkali-hydronium jarosites; I-synthesis of jarosite and natrojarosite: Mineralogia Polonica, v. 1, p. 47–59.

Miller, J. M., and Taylor, K., 1966, Uranium mineralization near Dalbeattie, Kirkcudbrightshire: Geological Survey of Great Britain Bulletin 25, p. 1–18.

Mumme, W. G., and Scott, T. R., 1966, The relationship between basic ferric sulfate and plumbojarosite: American Mineralogist, v. 51, p. 443–453.

Nishiyama, T., 1979, Distribution and origin of evaporite minerals from Dry Valleys, Victoria Land: National Institute for Polar Research (Tokyo) Memoirs, Special Issue 13, p. 136–147.

Nordstrom, D. K., and Dagenhart, T. V., 1978, Hydrated iron sulfate minerals associated with pyrite oxidation; Field relations and thermodynamic properties: Geological Society of America Abstracts with Programs, v. 10, p. 464.

Palache, C., Berman, H., and Frondel, C., 1951, Dana's system of mineralogy, 7th ed., v. 2: New York, John Wiley and Sons, 1124 p.

Posnjak, E., and Merwin, H. E., 1922, The system Fe_2O_3-SO_3-H_2O: Journal of the American Chemical Society, v. 44, p. 1965–1994.

Posnjak, E., and Tunell, G., 1929, The system CuO-SO_3-H_2O: American Journal of Science, v. 5, no. 18, p. 1–34.

Rowley, P. D., Williams, P. L., and Pride, D. E., 1983, Mineral occurrences of Antarctica, *in* Behrendt, J. C., ed., Petroleum and mineral resources of Antarctica: U.S. Geological Survey Circular 909, p. 25–50.

Rubin, M. J., and Weyant, W. S., 1965, Antarctic meteorology, *in* Hatherton, T., ed., Antarctica: New York, Frederick A. Praeger, p. 375–402.

Sharkey, J. B., and Levin, S. Z., 1971, Conditions governing the formation of atacamite and paratacamite: American Mineralogist, v. 56, p. 179–192.

Steger, H. F., and Desjardins, L. E., 1978, Oxidation of sulfide minerals; 4, Pyrite, chalcopyrite, and pyrrhotite: Chemical Geology, v. 23, p. 225–237.

——, 1980, Oxidation of sulfide minerals; 5, Galena, sphalerite, and chalcocite: Canadian Mineralogist, v. 13, p. 365–372.

Stewart, D., 1964, Antarctic mineralogy, *in* Adie, R. J., ed., Antarctic geology: Amsterdam, North-Holland, p. 395–401.

Tunell, G., and Posnjak, E., 1931, A portion of the system Fe_2O_3-CuO-SO_3-H_2O: Journal of Physical Chemistry, v. 35, p. 929–946.

Vennum, W. R., 1980, Evaporite encrustations and sulphide oxidation products from the southern Antarctic Peninsula: New Zealand Journal of Geology and Geophysics, v. 23, p. 499–505.

Vennum, W. R., and Nishi, J. M., 1981, New Antarctic mineral occurrences: Antarctic Journal of the U.S., v. 16, p. 14–15.

Walter-Levy, L., and Goreaud, M., 1969, Sur la formation des chlorures basiques cuivriques en solution aqueuse de 25 à 200°C: Société Française de Chimica Bulletin, v. 8, p. 2623–2634.

Warshaw, C., 1956, The occurrence of jarosite in underclays: American Mineralogist, v. 41, p. 288–296.

Williams, P. L., 1970, Geology of the Lassiter Coast: Antarctic Journal of the U.S., v. 5, p. 98–99.

MANUSCRIPT ACCEPTED BY THE SOCIETY MARCH 1, 1989

Printed in U.S.A.

Appendix I: Economic Resources of the Ellsworth Mountains, West Antarctica

John F. Splettstoesser

With a few minor exceptions, no economic deposits have been found in the Ellsworth Mountains. There are minor occurrences of coal in thin beds in the Polarstar Formation in the northern Sentinel Range, and several sulfide minerals and their secondary weathering products occur in the southern Heritage Range. There is no indication of any hydrocarbon potential in this part of Antarctica.

The Ellsworth Mountains lie within the early Mesozoic Ellsworth orogen (Craddock, 1972), which appears to be devoid of mineralized localities of any consequence. The probability that significant mineral deposits occur within this belt seems low, largely because thick sequences of clastic rocks, such as those in the Ellsworth Mountains, ordinarily do not host metallic ores (Wright and Williams, 1974). The presence of Mesozoic intrusive rocks in the Whitmore Mountains to the southwest suggests more favorable conditions there, but no mineralized areas have been reported (Webers and others, 1982).

Coal

Coal of Permian age was first reported in the Ellsworth Mountains by Craddock and others (1965) in a paper on the discovery of *Glossopteris* fossils in West Antarctica. The high-rank coal occurs in beds as much as 30 cm thick in the upper part of the Polarstar Formation (Permian) at Polarstar Peak. Coal beds there are interbedded with argillite and occur in several units of the 381-m measured section. Schopf and Long (1966, p. 174, 187) published analyses of three samples of the coal described by Craddock and others (1965). No estimates or measurements have been made of the extent of the coal at this location.

Metallic minerals

Sulfide mineralization occurs sparsely in various parts of the Ellsworth Mountains, and it is also found associated with weathering or oxidation products of the primary Cu, Fe, and Pb sulfide minerals. Vennum and Nishi (1981, this volume) reported on the analytical results of 49 samples of varicolored salts collected during the 1979–80 austral summer in the southern Heritage Range; they suggest that additional knowledge of the products of sulfide oxidation in cold dry climates might eventually aid economic mineral exploration in the polar regions. No economically significant amounts of sulfides or their weathering products have been found in the Ellsworth Mountains.

Other

An airborne radiometric survey was conducted in the 1979–80 austral summer, which showed no significant readings above background in any of the areas that were traversed (Dreschhoff and others, this volume).

REFERENCES CITED

Craddock, C., 1972, Antarctic tectonics, *in* Adie, R. J., ed., Antarctic geology and geophysics: Oslo, Universitetsforlaget, p. 449–455.

Craddock, C., Bastien, T. W., Rutford, R. H., and Anderson, J. J., 1965, *Glossopteris* discovered in West Antarctica: Science, v. 148, no. 3670, p. 634–637.

Schopf, J. M., and Long, W. E., 1966, Coal metamorphism and igneous associations in Antarctica, *in* Gould, R. F., ed., Coal science: Washington, D.C., American Chemical Society, p. 156–195.

Vennum, W. R., and Nishi, J. M., 1981, New Antarctic mineral occurrences: Antarctic Journal of the U.S., v. 16, no. 5, p. 14–15.

Webers, G. F., Craddock, C., Rogers, M. A., and Anderson, J. J., 1982, Geology of the Whitmore Mountains, *in* Craddock, C., ed., Antarctic Geoscience: Madison, University of Wisconsin Press, p. 841–847.

Wright, N. A., and Williams, P. L., 1974, Mineral resources of Antarctica: U.S. Geological Survey Circular 705, 29 p.

Printed in U.S.A.

Appendix II: Bibliography of the Ellsworth Mountains, West Antarctica

John F. Splettstoesser and Gerald F. Webers

This bibliography was compiled as a guide to the literature on the geology of the Ellsworth Mountains, Antarctica. Some citations on other subjects are included as they relate to the Ellsworth Mountains. The bibliography was originally intended as a basis for literature search by authors of chapters in this volume. The list includes citations of all published literature about this region known to us (see Webers and Splettstoesser, 1982). The time period covered begins with literature pertaining to Lincoln Ellsworth's discovery of the range in 1935, but all other entries are post-IGY (1957–58). Articles in this GSA Memoir are not included in this bibliography. This edition of the bibliography was completed in May 1990, with a few citations added in proof.

Anderson, J. J., 1962a, Current geologic investigations in Antarctica [abs.]: Texas Journal of Science, v. 14, no. 4, p. 409–410.

—— , 1962b, Bedrock geology of Antarctica [M.S. thesis]: Minneapolis, University of Minnesota, 234 p.

—— , 1965, Bedrock geology of Antarctica; A summary of exploration, 1831–1962, *in* Hadley, J. B., ed., Geology and Paleontology of the Antarctic: American Geophysical Union, Antarctic Research Series, v. 6, p. 1–70.

Anderson, J. J., Bastien, T. W., Schmidt, P. G., Splettstoesser, J. F., and Craddock, C., 1962, Antarctica; Geology of the Ellsworth Mountains: Science, v. 138, no. 3542, p. 824–825.

Anderson, V. H., 1960, The petrography of some rocks from Marie Byrd Land, Antarctica; Report 825-2, Part 8, USNC-IGY Antarctic Glaciologic Data, Field Work 1958 and 1959: Columbus, The Ohio State University Research Foundation, 27 p.

—— , 1961, Geographic features observed on the Marie Byrd Land traverse 1957–1958, *in* Reports of Antarctic Geological Observations 1956–1960; N.Y., American Geographical Society, IGY World Data Center A: Glaciology, IGY Glaciological Report Series, No. 4, p. 143–149.

Bauer, R. L., 1982, Pumpellyite-actinolite facies metamorphism in the Heritage Range of the Ellsworth Mountains, West Antarctica: Geological Society of America Abstracts with Programs, v. 14, no. 7, p. 440.

—— , 1983, Low-grade metamorphism in the Heritage Range of the Ellsworth Mountains, West Antarctica, *in* Oliver, R. L., James, P. R., and Jago, J. B., eds., Antarctic earth science: Canberra, Australian Academy of Science, p. 256–260.

Birkenmajer, K., and Wieser, T., 1985, Petrology and provenance of magmatic and metamorphic erratic blocks from Pliocene tillites of King George Island (South Shetland Islands, Antarctica): Studia Geologica Polonica, v. 81, no. 5, p. 53–97.

Bockheim, J. G., and Leide, J. E., 1980, Soil development and rock weathering in the Ellsworth Mountains, Antarctica: Antarctic Journal of the U.S., v. 15, no. 5, p. 33–34.

Boucot, A. J., Doumani, G. A., Johnson, J. G., and Webers, G. F., 1967, Devonian of Antarctica, *in* Oswald, D. H., ed., International Symposium on the Devonian System, Calgary, Alberta, 1967, Papers, Volume 1: Calgary, Alberta Society of Petroleum Geologists, p. 639–648.

Buggisch, W., 1981, Erste paläozoische Conodonten aus der Antarktis (Oberkambrium: Ellsworth Mountains) [abs.]: Erlangen, Paläontologische Gesellschaft; 51. Jahresversammlung-Programm und Exkursions führer, 28-30 Sept 1981, p. 23.

—— , 1982a, Conodonten aus den Ellsworth Mountains (Oberkambrium/ Westantarktis): Zeitschrift der deutschen Geologischen Gesellschaft (Hannover), v. 133, p. 493–507. (In German with English abstract.)

—— , 1982b, The depositional environments of Cambrian carbonate rocks of the Ellsworth Mountains: Geological Society of America Abstracts with Programs, v. 14, no. 7, p. 454.

—— , 1983, Paleozoic carbonate rocks of the Ellsworth Mountains, West Antarctica [abs.], *in* Oliver, R. L., James, P. R., and Jago, J. B., eds., Antarctic earth science: Canberra, Australian Academy of Science, p. 265.

Buggisch, W., and Webers, G. F., 1982, Zur Facies der Karbonatgesteine in den Ellsworth Mountains (Paläozoikum, Westantarktis): Facies (Erlangen), v. 7, p. 199–228. (In German with English summary.)

Castle, J. W., 1974, The deposition and metamorphism of the Polarstar Formation (Permian), Ellsworth Mountains, Antarctica [M.S. thesis]: Madison, University of Wisconsin, 102 p.

Castle, J. W., and Craddock, C., 1975, Deposition and metamorphism of the Polarstar Formation (Permian), Ellsworth Mountains: Antarctic Journal of the U.S., v. 10, no. 5, p. 239–241.

Clarkson, P. D., 1983, The reconstruction of Lesser Antarctica within Gondwana [abs.], *in* Oliver, R. L., James, P. R., and Jago, J. B., eds., Antarctic earth science: Canberra, Australian Academy of Science, p. 599.

Clarkson, P. D., and Brook, M., 1977, Age and position of the Ellsworth Mountains crustal fragment, Antarctica: Nature, v. 265, no. 5595, p. 615–616.

Clarkson, P. D., Cooper, R. A., Hughes, C. P., Thomson, M.R.A., and Webers, G. F., 1981, Ordovician of Antarctica—A review, *in* The Ordovician System in Australia, New Zealand and Antarctica; Correlation chart and explanatory notes, by Webby, B. D., and 11 others, p. 46–50; plus various refs. on p. 50–64, figs., plate: Paris, International Union of Geological Sciences, Publication No. 6.

Clinch, N. B., 1967, First conquest of Antarctica's highest peaks: National Geographic Magazine, v. 131, no. 6, p. 836–863.

Collinson, J. W., and Vavra, C. L., 1982, Sedimentology of the Polarstar Formation (Permian), Ellsworth Mountains, Antarctica: Geological Society of America Abstracts with Programs, v. 14, no. 7, p. 466.

Collinson, J. W., Vavra, C. L., and Zawiskie, J. M., 1980, Sedimentology of the Polarstar Formation (Permian), Ellsworth Mountains: Antarctic Journal of the U.S., v. 15, no. 5, p. 30–32.

Craddock, C., 1962, Geological reconnaissance of the Ellsworth Mountains: U.S. Antarctic Projects Officer, Bulletin, v. 3, no. 9–10, p. 34–35. (Additional summary articles on field seasons in: 1963, v. 4, no. 9, p. 77–78; 1964, v. 5, no. 10, p. 107–108; 1965, v. 6, no. 7, p. 66.)

—— , 1966, The Ellsworth Mountains fold belt—a link between East and West Antarctica [abs.]: Geological Society of America Special Paper 87, p. 37–38.

—— , 1969, Geology of the Ellsworth Mountains, *in* Bushnell, V. C., and Craddock, C., eds., Geologic maps of Antarctica: New York, American Geographical Society, Antarctic Map Folio Series, Folio 12, Plate 4.

——, 1970, Map of Gondwanaland, *in* Bushnell, V. C., and Craddock, C., eds., Geologic maps of Antarctica: New York, American Geographical Society, Antarctic Map Folio Series, Folio 12, Plate 23.

——, 1971, The structural relation of East and West Antarctica, *in* Report of the 22nd International Geological Congress, India, 1964, Part 4, p. 278–292.

——, 1972, Antarctic tectonics, *in* Adie, R. J., ed., Antarctic geology and geophysics: Oslo, Universitetsforlaget, p. 449–455.

——, 1983, The East Antarctica–West Antarctica boundary between the ice shelves; A review, *in* Oliver, R. L., James, P. R., and Jago, J. B., eds., Antarctic earth science: Canberra, Australian Academy of Science, p. 94–97.

Craddock, C., and Spörli, K. B., 1982, Structure and tectonic setting of the Ellsworth Mountains, West Antarctica: Geological Society of America Abstracts with Programs, v. 14, no. 7, p. 468.

Craddock, C., and Webers, G. F., 1964, Fossils from the Ellsworth Mountains, Antarctica: Nature, v. 201, no. 4915, p. 174–175.

——, 1977, Geology of the Ellsworth Mountains to Thiel Mountains Ridge: Antarctic Journal of the U.S., v. 12, no. 4, p. 85.

Craddock, C., Anderson, J. J., and Webers, G. F., 1964, Geologic outline of the Ellsworth Mountains, *in* Adie, R. J., ed., Antarctic Geology, Proceedings of the First International Symposium on Antarctic Geology, Cape Town, 16–21 September 1963: Amsterdam, North-Holland, p. 155–170.

Craddock, C., Bastien, T. W., Rutford, R. H., and Anderson, J. J., 1965, *Glossopteris* discovered in West Antarctica: Science, v. 148, no. 3670, p. 634–637.

Craddock, C., Spörli, K. B., and Rutford, R. H., 1982, Breccia bodies in Cambrian limestones of the Ellsworth Mountains, West Antarctica: Geological Society of America Abstracts with Programs, v. 14, no. 7, p. 468.

Craddock, C., Webers, G. F., and Anderson, J. J., 1982, Geology of the Ellsworth Mountains–Thiel Mountains Ridge [abs.], *in* Craddock, C., ed., Antarctic geoscience: Madison, University of Wisconsin Press, p. 849.

Craddock, C., Webers, G. F., Rutford, R. H., Spörli, K. B., and Anderson, J. J., 1986, Geologic map of the Ellsworth Mountains, Antarctica: Geological Society of America, Map and Chart Series MC-57, in color, scale 1:250,000.

Craddock, J. P., and Webers, G. F., 1981, Probable cave deposits in the Ellsworth Mountains of West Antarctica, *in* Beck, B. F., ed., Proceedings of the Eighth International Congress of Speleology, Bowling Green, Kentucky, July 1981, p. 395–397.

Dalziel, I.W.D., 1980, The Ellsworth Mountains as a displaced fragment of the Transantarctic Mountains; Present status of the Schopf hypothesis [abs.]: EOS American Geophysical Union Transactions, v. 61, p. 1196.

——, 1981, Tectonic studies in the Scotia Arc region and West Antarctica: Antarctic Journal of the U.S., v. 16, no. 5, p. 7–8.

Dalziel, I.W.D., and Pankhurst, R. J., 1984, West Antarctica: Its tectonics and its relationship to East Antarctica: Antarctic Journal of the U.S., v. 19, no. 5, p. 35–36.

——, 1987, Joint U.K.-U.S. West Antarctic tectonics project; An introduction, *in* McKenzie, G. D., ed., Gondwana Six: Structure, tectonics, and geophysics: Washington, D.C., American Geophysical Union, Geophysical Monograph 40, p. 107–108.

——, 1987, Tectonic development of West Antarctica and its relation to East Antarctica: Joint U.S./U.K. program 1986–1987: Antarctic Journal of the U.S., v. 22, no. 5, p. 50–51.

Dalziel, I.W.D., Garrett, S. W., Grunow, A. M., Pankhurst, R. J., Storey, B. C., and Vennum, W. R., 1987, The Ellsworth-Whitmore Mountains crustal block; Its role in the tectonic evolution of West Antarctica, *in* McKenzie, G. D., ed., Gondwana Six: Structure, tectonics, and geophysics: Washington, D.C., American Geophysical Union, Geophysical Monograph 40, p. 173–182.

Debrenne, F., Debrenne, M., and Webers, G. F., 1983, Upper Cambrian archaeocyathans: New morphotype [abs.], *in* Oliver, R. L., James, P. R., and Jago, J. B., eds., Antarctic earth science: Canberra, Australian Academy of Science, p. 280.

Debrenne, F., Rozanov, A. Yu., and Webers, G. F., 1984, Upper Cambrian Archaeocyatha from Antarctica: Geological Magazine, v. 121, no. 4, p. 291–299.

Doake, C.S.M., Crabtree, R. D., and Dalziel, I.W.D., 1981, Airborne radioecho sounding in Ellsworth Land and Ronne Ice Shelf: Antarctic Journal of the U.S., v. 16, no. 5, p. 83–84.

——, 1982, New constraints on the original location of the Ellsworth Mountains crustal block, Antarctica: Geological Society of America Abstracts with Programs, v. 14, no. 7, p. 476.

——, 1983, Subglacial morphology between Ellsworth Mountains and Antarctic Peninsula; New data and tectonic significance, *in* Oliver, R. L., James, P. R., and Jago, J. B., eds., Antarctic earth science: Canberra, Australian Academy of Science, p. 270–273.

Doake, C.S.M., Frolich, R. M., Mantripp, D. R., Smith, A. M., and Vaughan, D. G., 1987, Glaciological studies on Rutford Ice Stream, Antarctica: Journal of Geophysical Research, v. 92, no. B9, p. 8951–8960.

Dreschhoff, G.A.M., Zeller, E. J., Thoste, V., and Bulla, K., 1980, Resource and radioactivity survey in the Ellsworth Mountains: Antarctic Journal of the U.S., v. 15, no. 5, p. 32.

Ellsworth, L., 1936, My flight across Antarctica: National Geographic Magazine, v. 70, p. 1–35.

——, 1937, Beyond horizons: N.Y., Doubleday, Doran, Garden City, 403 p.

Evans, J. P., 1973, Scientific aspects of the American Antarctic Mountaineering Expedition: National Geographic Society Research Reports 1966 Projects, p. 113–115.

Fitzgerald, P. G., and Stump, E., 1991, Early Cretaceous uplift in the Ellsworth Mountains of West Antarctica: Science, v. 254, p. 92–94.

Frakes, L. A., Matthews, J. L., and Crowell, J. C., 1971, Late Paleozoic glaciation; Part III, Antarctica: Geological Society of America Bulletin, v. 82, no. 6, p. 1581–1603.

Frolich, R. M., Mantripp, D. R., Vaughan, D. G., and Doake, C.S.M., 1987, Force balance of Rutford Ice Stream, Antarctica, *in* Waddington, E. D., and Walder, J. S., eds., International Symposium [on] the Physical Basis of Ice Sheet Modeling, Vancouver, B.C., Aug. 9–22, 1987: International Association of Hydrological Sciences, Publication No. 170, p. 323–331.

Funaki, M., Yoshida, M., and Matsueda, H., 1991, Palaeomagnetic studies of Palaeozoic rocks from the Ellsworth Mountains, West Antarctica, *in* Thomson, M.R.A., Crame, J. A., and Thomson, J. W., eds., Geological Evolution of Antarctica: Cambridge, Cambridge University Press, p. 257–260.

Garrett, S. W., Herrod, L.D.B., and Mantripp, D. R., 1987, Crustal structure of the area around Haag Nunataks, West Antarctica; New aeromagnetic and bedrock elevation data, *in* McKenzie, G. D., ed., Gondwana Six; Structure, tectonics, and geophysics: Washington, D.C., American Geophysical Union, Geophysical Monograph 40, p. 109–115.

Garrett, S. W., Maslanyj, M. P., and Damaske, D., 1988, Interpretation of aeromagnetic data from the Ellsworth Mountains–Thiel Mountains Ridge, West Antarctica: Journal of the Geological Society of London, v. 145, no. 6, p. 1009–1017.

Grunow, A. M., Dalziel, I.W.D., and Kent, D. V., 1987, Ellsworth-Whitmore Mountains crustal block, western Antarctica; New paleomagnetic results and their tectonic significance, *in* McKenzie, G. D., ed., Gondwana Six; Structure, tectonics, and geophysics: Washington, D.C., American Geophysical Union, Geophysical Monograph 40, p. 161–171.

Hjelle, A., Ohta, Y., and Winsnes, T. S., 1978, Stratigraphy and igneous petrology of southern Heritage Range, Ellsworth Mountains, Antarctica: Norsk Polarinstitutt Skrifter No. 169, p. 5–43.

——, 1982, Geology and petrology of the southern Heritage Range, Ellsworth Mountains, *in* Craddock, C., ed., Antarctic geoscience: Madison, University of Wisconsin Press, p. 599–608.

Jankowski, E. J., and Drewry, D. J., 1981, The structure of West Antarctica from geophysical studies: Nature, v. 291, p. 17–21.

Jankowski, E. J., Drewry, D. J., and Behrendt, J. C., 1983, Magnetic studies of upper crustal structure in West Antarctica and the boundary with East Antarctica, *in* Oliver, R. L., James, P. R., and Jago, J. B., eds., Antarctic earth science: Canberra, Australian Academy of Science, p. 197–203.

Joerg, W.L.G., 1936a, Lincoln Ellsworth's flight across West Antarctica: Geographical Review, v. 26, no. 2, p. 329–332.
——, 1936b, The topographical results of Ellsworth's transAntarctic flight of 1935: Geographical Review, v. 26, no. 3, p. 454–462.
——, 1937, The cartographical results of Ellsworth's transAntarctic flight of 1935: Geographical Review, v. 27, no. 3, p. 430–444.
Kamenev, E. N., and Ivanov, V. L., 1983, Structure and outline of geologic history of the southern Weddell Sea basin, *in* Oliver, R. L., James, P. R., and Jago, J. B., eds., Antarctic earth science: Canberra, Australian Academy of Science, p. 194–196.
Laudon, T. S., and Craddock, C., 1988, Possible Paleozoic stratigraphic linkages of the Antarctic Peninsula and the Ellsworth Mountains from sandstone petrofacies [abs.]: EOS American Geophysical Union Transactions, v. 69, no. 44, p. 1492.
Longshaw, S. K., and Griffith, D. H., 1983, A palaeomagnetic study of Jurassic rocks from the Antarctic Peninsula and its implications: Journal of the Geological Society of London, v. 140, part 6, p. 945–954.
Lucchitta, B. K., Edwards, K., Eliason, E. M., and Bowell, J., 1985, Landsat multispectral images of Antarctica applied to mapping and glaciology: Antarctic Journal of the U.S., v. 20, no. 5, p. 256–259.
Lucchitta, B. K., Bowell, J.-A., Edwards, K. L., Eliason, E. M., and Ferguson, H. M., 1987, Multispectral Landsat images of Antarctica: U.S. Geological Survey Bulletin 1696, 21 p.
Matsch, C. L., and Ojakangas, R. W., 1982, A late Paleozoic glacial and glacial-marine sequence, Ellsworth Mountains, West Antarctica: Geological Society of America Abstracts with Programs, v. 14, no. 7, p. 558.
——, 1985, Models for glacial marine deposition: Late Paleozoic Whiteout Conglomerate, Ellsworth Mountains, West Antarctica: Geological Society of America Abstracts with Programs, v. 17, no. 5, p. 300.
Matthews, J. L., Crowell, J. C., Coates, D. A., Neder, I. R., and Frakes, L. A., 1967, Late Paleozoic glacial rocks in the Sentinel and Queen Alexandra Ranges: Antarctic Journal of the U.S., v. 2, no. 4, p. 108.
McGibbon, K. J., and Smith, A. M., 1991, New geophysical results and preliminary interpretation of crustal structure between the Antarctic Peninsula and Ellsworth Land, *in* Thomson, M.R.A., Crame, J. A., and Thomson, J. W., eds., Geological Evolution of Antarctica: Cambridge, Cambridge University Press, p. 475–479.
Millar, I. L., and Pankhurst, R. J., 1987, Rb-Sr geochronology of the region between the Antarctic Peninsula and the Transantarctic Mountains: Haag Nunataks and Mesozoic granitoids, *in* McKenzie, G. D., ed., Gondwana Six; Structure, tectonics, and geophysics: Washington, D.C., American Geophysical Union, Geophysical Monograph 40, p. 151–160.
Ojakangas, R. W., and Matsch, C. L., 1981, The Late Palaeozoic Whiteout Conglomerate: A glacial and glaciomarine sequence in the Ellsworth Mountains, West Antarctica, *in* Hambrey, M. J., and Harland, W. B., eds., Earth's pre-Pleistocene glacial record: Cambridge University Press, p. 241–244.
——, 1984, Criteria for distinguishing between continental and mountain glaciation origins of ancient glacial sequences: Geological Society of America Abstracts with Programs, v. 16, no. 6, p. 612–613.
Pankhurst, R. J., Storey, B. C., and Millar, I. L., 1991, Magmatism related to the break-up of Gondwana, *in* Thomson, M.R.A., Crame, J. A., and Thomson, J. W., eds., Geological Evolution of Antarctica: Cambridge, Cambridge University Press, p. 573-579.
Pojeta, J., Jr., Webers, G. F., and Yochelson, E. L., 1981, Upper Cambrian mollusks from the Ellsworth Mountains, West Antarctica, *in* Taylor, M. E., ed., Short Papers for the Second International Symposium on the Cambrian System, Golden, Colorado, August 1981: U.S. Geological Survey Open-File Report 81-743, p. 167–168.
Reynolds, J. M., 1981, The distribution of mean annual temperatures in the Antarctic Peninsula: British Antarctic Survey Bulletin, no. 54, p. 123–133.
Rigby, J. F., 1969, Permian sphenopsids from Antarctica: U.S. Geological Survey Professional Paper 613-F, p. F1–F13.
Rigby, J. F., and Schopf, J. M., 1969, Stratigraphic implications of Antarctic paleobotanical studies, *in* Gondwana stratigraphy; IUGS Symposium, Buenos Aires, 1-15 October 1967: Paris, UNESCO, p. 91–106.
Ritzwoller, M. H., and Bentley, C. R., 1983, Magnetic anomalies over Antarctica measured from MAGSAT, *in* Oliver, R. L., James, P. R., and Jago, J. B., eds., Antarctic earth science: Canberra, Australian Academy of Science, p. 504–507.
Rutford, R. H., 1969, The glacial geology and geomorphology of the Ellsworth Mts., West Antarctica [Ph.D. dissertation]: Minneapolis, University of Minnesota, 374 p.
—— 1970, The glacial geology and geomorphology of the Ellsworth Mountains, West Antarctica: Dissertation Abstracts International, v. 31, no. 1, p. 254-B–255-B.
——, 1972a, Drainage systems of the Ellsworth Mountains area [abs.], *in* Adie, R. J., ed., Antarctic geology and geophysics: Oslo, Universitetsforlaget, p. 233.
——, 1972b, Glacial geomorphology of the Ellsworth Mountains, *in* Adie, R. J., ed., Antarctic geology and geophysics: Oslo, Universitetsforlaget, p. 225–232.
Rutford, R. H., and Smith, P. M., 1966, The use of turbine helicopters in United States Antarctic operations, 1961–66: Polar Record, v. 13, no. 84, p. 299–303.
Rutford, R. H., Denton, G. H., and Andersen, B. G., 1980, Glacial history of the Ellsworth Mountains: Antarctic Journal of the U.S., v. 15, no. 5, p. 56–57.
Rutford, R. H., Bockheim, J. G., Andersen, B. G., and Denton, G. H., 1982, Glacial history of the Ellsworth Mountains, West Antarctica: Geological Society of America Abstracts with Programs, v. 14, no. 7, p. 605–606.
Samsonov, V. V., 1982, Petrography and geochemistry of magmatic rocks in the Heritage Range, Ellsworth Mountains: Geological Society of America Abstracts with Programs, v. 14, no. 7, p. 607.
Schmidt, P. G., 1962, Geological reconnaissance of the Ellsworth Mountains: U.S. Antarctic Projects Officer, Bulletin, v. 3, no. 8, p. 17–20.
Schopf, J. M., 1967, Antarctic fossil plant collecting during the 1966–1967 season: Antarctic Journal of the U.S., v. 2, no. 4, p. 114–116.
——, 1969, Ellsworth Mountains: position in West Antarctica due to sea-floor spreading: Science, v. 164, p. 63–66.
——, 1982, Forms and facies of *Vertebraria* in relation to Gondwana coal, *in* Turner, M. D., and Splettstoesser, J. F., eds., Geology of the Central Transantarctic Mountains: Washington, D.C., American Geophysical Union, Antarctic Research Series, v. 36, p. 37–62.
Schopf, J. M., and Long, W. E., 1966, Coal metamorphism and igneous associations in Antarctica, *in* Gould, R. F., ed., Coal science: Washington, D.C., American Chemical Society, p. 156–195.
Shergold, J. H., 1983, Post-Mindyallan Late Cambrian trilobite faunas from Antarctica [abs.], *in* Oliver, R. L., James, P. R., and Jago, J. B., eds., Antarctic earth science: Canberra, Australian Academy of Science, p. 132.
Shergold, J. H., and Webers, G. F., 1982, Cambrian trilobite faunas from the Ellsworth Mountains, West Antarctica: Geological Society of America Abstracts with Programs, v. 14, no. 7, p. 615–616.
Silversteen, S. C., 1967, The American Antarctic Mountaineering Expedition: Antarctic Journal of the U.S., v. 2, no. 2, p. 48–50.
Soholt, D. E., and Craddock, C.,1964, Motor toboggan sled trains in Antarctica: Arctic, v. 17, no. 2, p. 99–104.
Soloviev, I. A., Popov, L. E., and Samsonov, V. V., 1984, Novye fannye o verkhnekembriiskoi faune gor Elsuert i Pensakola (Zapadnaia Antarktida) [New data on Upper Cambrian fauna from the Ellsworth and Pensacola Mountains (West Antarctica)]: Antarktika: Doklady komissii, no. 23, p. 46–71. (in Russian.)
Splettstoesser, J. F., 1963, Ellsworth Mountains, Antarctica: Geotimes, v. 8, no. 4, p. 6.
——, 1981, Bird sightings in the Ellsworth Mountains and other inland areas: Antarctic Journal of the U.S., v. 16, no. 5, p. 177–179.
Splettstoesser, J. F., and Turner, M. D., 1982, Logistics of remote field camps for earth science programs in Antarctica: Geological Society of America Abstracts with Programs, v. 14, no. 7, p. 624.
Splettstoesser, J. F., and Webers, G. F., 1980, Geological investigations and logistics in the Ellsworth Mountains, 1979–80: Antarctic Journal of the U.S.,

v. 15, no. 5, p. 36–39.

Splettstoesser, J. F., Webers, G. F., and Waldrip, D. B., 1982, Logistic aspects of geological studies in the Ellsworth Mountains, Antarctica, 1979–80: Polar Record, v. 21, no. 131, p. 147–159.

Splettstoesser, J. F., Webers, G. F., and Craddock, C., 1984, Geologic studies in the Ellsworth Mountains, Antarctica—A 25-year journey from the unknown: Geological Society of America Abstracts with Programs, v. 16, no. 6, p. 665.

Spörli, B., 1966, The Ellsworth Mountains of Antarctica; An American geological expedition, *in* Mountain World, 1964/65: London, Allen & Unwin: Chicago, Rand McNally, p. 171–182.

Spörli, B., and Craddock, C., 1968, Analysis of Ellsworth Mountains and Ruppert Coast geologic data: Antarctic Journal of the U.S., v. 3, no. 5, p. 179.

Stephenson, S. N., 1984, Glacier flexure and the position of grounding lines; Measurement by tiltmeter on Rutford Ice Stream, Antarctica: Annals of Glaciology, v. 5, p. 165–169.

Stephenson, S. N., and Doake, C.S.M., 1982, Dynamic behaviour of Rutford Ice Stream: Annals of Glaciology, v. 3, p. 295–299.

Storey, B. C., and Macdonald, D.I.M., 1987, Sedimentary rocks of the Ellsworth-Thiel Mountains ridge and their regional equivalents: British Antarctic Survey, Bulletin No. 76, p. 21–49.

Storey, B. C., and Dalziel, I.W.D., 1987, Outline of the structural and tectonic history of the Ellsworth Mountains–Thiel Mountains ridge, West Antarctica, *in* McKenzie, G. D., ed., Gondwana Six; Structure, tectonics, and geophysics: Washington, D.C., American Geophysical Union, Geophysical Monograph 40, p. 117–128.

Stump, E., and Fitzgerald, P. G., 1990, Field collection for apatite fission track analysis, Ellsworth Mountains: Antarctic Journal of the U.S., v. 25, no. 5, p. 38–39.

Swithinbank, C., 1977, Glaciological research in the Antarctic Peninsula, *in* Scientific Research in Antarctica: London, Royal Society, p. 161–183. (Philosophical Transactions of the Royal Society of London B, v. 279, p. 161–183.)

——, 1988, Antarctic Airways: Antarctica's first commercial airline: Polar Record, v. 24, no. 151, p. 313–316.

Swithinbank, C., and Lucchitta, B. K., 1986, Multispectral digital image mapping of Antarctic ice features: Annals of Glaciology, v. 8, p. 159–163.

Tasch, P., 1967, Antarctic fossil conchostracans and the continental drift theory: Antarctic Journal of the U.S., v. 2, no. 4, p. 112–113.

——, 1968, Trace fossils from the Permian Polarstar Formation, Sentinel Mountains, Antarctica: Kansas Academy of Science, Transactions, v. 71, no. 2, p. 184–194.

Tasch, P., and Angino, E. E., 1968, Sulphate and carbonate salt efflorescences from the Antarctic interior: Antarctic Journal of the U.S., v. 3, no. 6, p. 239–241.

Tasch, P., and Riek, E. F., 1969, Permian insect wing from Antarctic Sentinel Mountains: Science, v. 164, p. 1529–1530.

Taylor, T. N., and Smoot, E. L., 1985, Plant fossils from the Ellsworth Mountains: Antarctic Journal of the U.S., v. 20, no. 5, p. 48–49.

Thiel, E., 1961, Antarctica, one continent or two?: Polar Record, v. 10, no. 67, p. 335–348.

U.S. Geological Survey, 1:250,000 Reconnaissance Series Maps: Liberty Hills, 1967; Newcomer Glacier, 1962; Nimitz Glacier, 1962; Union Glacier, 1967; Vinson Massif, 1962.

——, 1976, Ellsworth Mountains, Antarctica; Satellite Image Map, scale 1:500,000.

Vavra, C. L., and Collinson, J. W., 1981, Sandstone petrology of the Polarstar Formation (Permian), Ellsworth Mountains: Antarctic Journal of the U.S., v. 16, no. 5, p. 15–16.

Vennum, W. R., 1982, Intrusive igneous rocks of the southern Heritage Range, Ellsworth Mountains, Antarctica: Petrology and geochemistry: Geological Society of America Abstracts with Programs, v. 14, no. 7, p. 637.

Vennum, W. R., and Nishi, J. M., 1981, New Antarctic mineral occurrences: Antarctic Journal of the U.S., v. 16, no. 5, p. 14–15.

Vennum, W. R., and Storey, B. C., 1987, Correlation of gabbroic and diabasic rocks from the Ellsworth Mountains, Hart Hills, and Thiel Mountains, West Antarctica, *in* McKenzie, G. D., ed., Gondwana Six; Structure, tectonics, and geophysics: Washington, D.C., American Geophysical Union, Geophysical Monograph 40, p. 129–138.

——, 1987, Petrology, geochemistry, and tectonic setting of granitic rocks from the Ellsworth-Whitmore Mountains crustal block and Thiel Mountains, West Antarctica, *in* McKenzie, G. D., ed., Gondwana Six; Structure, tectonics, and geophysics: Washington, D.C., American Geophysical Union, Geophysical Monograph 40, p. 139–150.

Von Gizycki, P., 1983, Die magmatischen Gesteine der Ellsworth Mountains, West-Antarktis [The magmatic rocks of the Ellsworth Mountains, West Antarctica]: Neues Jahrbuch für Geologie, Paläontologie, Abhandlungen, Band 167, Heft 1, p. 65–88. (In German, with English abstract.)

Watts, D. R., and Bramall, A. M., 1980, Paleomagnetic investigation in the Ellsworth Mountains: Antarctic Journal of the U.S., v. 15, no. 5, p. 34–36.

——, 1981a, Palaeomagnetic evidence for a displaced terrain in Western Antarctica: Nature, v. 293, no. 5834, p. 638–641.

——, 1981b, Palaeomagnetic evidence from the Ellsworth Mountains supports microplate nature of western Antarctica [abs.]: Geophysical Journal of the Royal Astronomical Society, v. 65, no. 1, p. 271.

Webers, G. F., 1966, An Upper Cambrian archaeocyathid from Antarctica [abs.]: Geological Society of America Special Paper 87, p. 183.

——, 1970a, Invertebrate faunas of Antarctica: Geological Society of America, 1970 Annual Meeting, Abstracts with Programs, v. 2, no. 7, p. 717–718.

——, 1970b, Paleontological investigations in the Ellsworth Mountains, West Antarctica: Antarctic Journal of the U.S., v. 5, no. 5, p. 162–163.

——, 1972a, Geological research in Antarctica, *in* Schwartz, G. M., ed., A century of geology, 1872–1972, at the University of Minnesota: Minneapolis, University of Minnesota Press, p. 63–66.

——, 1972b, Unusual Upper Cambrian fauna from West Antarctica, *in* Adie, R. J., ed., Antarctic geology and geophysics: Oslo, Universitetsforlaget, p. 235–237.

——, 1977, Paleontological investigations in the Ellsworth Mountains, West Antarctica: Antarctic Journal of the U.S., v. 12, no. 4, p. 120–121.

——, 1981a, Cambrian rocks of the Ellsworth Mountains, West Antarctica, *in* Taylor, M. E., ed., Short papers for the Second International Symposium on the Cambrian System, Golden, Colorado, August 1981: U.S. Geological Survey Open File Report 81-743, p. 236–238.

——, 1981b, Ellsworth Mountains studies, 1980–1981: Antarctic Journal of the U.S., v. 16, no. 5, p. 18–19.

——, 1982, Upper Cambrian mollusks from the Ellsworth Mountains, *in* Craddock, C., ed., Antarctic geoscience: Madison, University of Wisconsin Press, p. 635–638.

Webers, G. F., and Craddock, C., 1986, Geologic history of the Ellsworth Mountains, Antarctica: Geological Society of America Abstracts with Programs, v. 18, no. 4, p. 329–330.

Webers, G. F., and Splettstoesser, J. F., 1982, Geology, paleontology, and bibliography of the Ellsworth Mountains: Antarctic Journal of the U.S., v. 17, no. 5, p. 36–38.

——, 1983a, Ellsworth Mountains studies, 1982–1983: Antarctic Journal of the U.S., v. 18, no. 5, p. 62.

——, 1983b, Marine fossils link Minnesota and Antarctica: Minnesota Volunteer, v. 46, no. 270, p. 2–10.

——, 1984, Geology of the Ellsworth Mountains—Data reduction, 1983–1984: Antarctic Journal of the U.S., v. 19, no. 5, p. 8.

——, 1985, Volume on geology of the Ellsworth Mountains—Progress in 1984–1985: Antarctic Journal of the U.S., v. 20, no. 5, p. 50.

——, 1986, Ellsworth Mountains geology in a reference volume—Progress in 1985–1986: Antarctic Journal of the U.S., v. 21, no. 5, p. 289–290.

——, 1987, Geology and paleontology of the Ellsworth Mountains: Antarctic Journal of the U.S., v. 22, no. 5, p. 298–299.

Webers, G. F., and Spörli, K. B., 1982, Stratigraphic investigations in the Heritage Range of the Ellsworth Mountains, West Antarctica: Geological Society of

America Abstracts with Programs, v. 14, no. 7, p. 643.

—— , 1983, Palaeontological and stratigraphic investigations in the Ellsworth Mountains, West Antarctica, *in* Oliver, R. L., James, P. R., and Jago, J. B., eds., Antarctic earth science: Canberra, Australian Academy of Science, p. 261–264.

Webers, G. F., and Yochelson, E. L., 1989, Late Cambrian molluscan faunas and the origin of the Cephalopoda, *in* Crame, J. A., ed., Origins and evolution of the Antarctic biota: Geological Society of London Special Publication No. 47, p. 29–42.

Webers, G. F., Pojeta, J., Jr., and Yochelson, E. L., 1982, Cambrian Mollusca from the Ellsworth Mountains, Antarctica: Geological Society of America Abstracts with Programs, v. 14, no. 7, p. 643.

Webers, G. F., Craddock, C., Rogers, M. A., and Anderson, J. J., 1982, Geology of the Whitmore Mountains, *in* Craddock, C., ed., Antarctic geoscience: Madison, University of Wisconsin Press, p. 841–847.

Webers, G. F., Craddock, C., Rogers, M. A., and Anderson, J. J., 1983, Geology of Pagano Nunatak and the Hart Hills, *in* Oliver, R. L., James, P. R., and Jago, J. B., eds., Antarctic earth science: Canberra, Australian Academy of Science, p. 251–255.

Winsnes, T. S., 1975, Med Sydpolstein i bagasjen [With South Polar stone in the baggage]: Forsknings Nytt, v. 20, no. 3, p. 18–21. (In Norwegian.)

Yochelson, E. L., Flower, R. H., and Webers, G. F., 1972, A theory of origin of the Cephalopods: Geological Society of America, 1972 Annual Meeting, Abstracts with Programs, v. 4, no. 7, p. 712.

—— , 1973, The bearing of the new Late Cambrian monoplacophoran genus *Knightoconus* upon the origin of the Cephalopoda: Lethaia, v. 6, p. 275–310.

Yoshida, M., 1981a, Participation in the U.S. Ellsworth Mountains operation of the 1979–1980 austral summer, Antarctica: Nankyoku Shiryô (Antarctic Record), v. 72, p. 101–107. (In Japanese.)

—— , 1981b, Tectonic and metamorphic studies of the Ellsworth Mountains: Antarctic Journal of the U.S., v. 16, no. 5, p. 16–18.

—— , 1982, Superposed deformation and its implication to the geologic history of the Ellsworth Mountains, West Antarctica: National Institute of Polar Studies, Tokyo, Memoirs, Special Issue No. 21, p. 120–171.

—— , 1983, Structural and metamorphic history of the Ellsworth Mountains, West Antarctica, *in* Oliver, R. L., James, P. R., and Jago, J. B., eds., Antarctic earth science: Canberra, Australian Academy of Science, p. 266–269.

Zawiskie, J. M., 1984, Permian ichnofossils from East Antarctica (Nimrod-Ohio basin) and the Ellsworth Mountains: Geological Society of America Abstracts with Programs, v. 16, no. 6, p. 704.

Printed in U.S.A.

Index

[Italic page numbers indicate major references]

Typeset by WESType Publishing Services, Inc., Boulder, Colorado
Printed in U.S.A. by Malloy Lithographing, Inc., Ann Arbor, Michigan